ASTRONOMY AND ASTROPHYSICS LIBRARY

Artist view of the Thirty Meter Telescope project (TMT) based on a three-mirror design (Courtesy of the TMT Observatory Corporation)

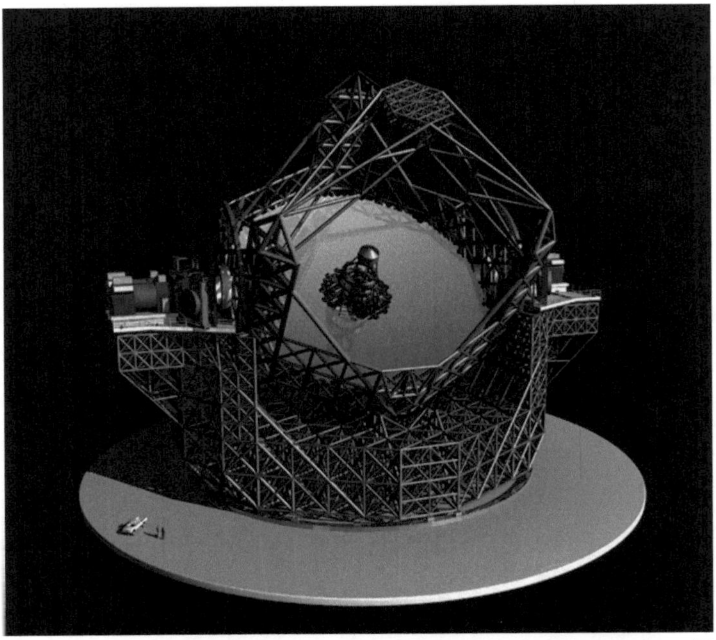

Artist view of the European Extremely Large Telescope project (E-ELT) based on a five-mirror design (Courtesy of the European Southern Observatory)

Gérard René Lemaitre

Astronomical Optics and Elasticity Theory

Active Optics Methods

Springer

Gérard René Lemaitre
Observatoire Astronomique Marseille Provence (OAMP)
Technopôle Château Gombert
38 rue Frédéric Joliot-Curie
13388 Marseille CX 13
France
gerard.lemaitre@oamp.fr

Cover image: General view of the giant reflective Schmidt telescope LAMOST, in Xinglong Station, which started operations in 2008 (courtesy National Astronomical Observatories, Chinese Academy of Sciences)

Corrected 3rd printing

ISBN 978-3-540-68904-1 ISBN 978-3-540-68905-8 (eBook)

DOI 10.1007/978-3-540-68905-8

Astronomy and Astrophysics Library ISSN 0941-7834

Library of Congress Control Number: 2008936630

Cover design: eStudio Calamar S.L.

Printed on acid-free paper

9 8 7 6 5 4 3 2 1

springer.com

Foreword

It is an honor as well as a pleasure to write this Foreword for this major work – "Astronomical Optics and Elasticity Theory" – by my friend and colleague, Prof. Gérard Lemaître. My situation is unusual in that I have not even seen, still less read, the manuscript. My Foreword must therefore be based on my general knowledge of Lemaître's work and our close contact over many years.

The initiator of "stress polishing" for astronomical purposes was the great optician Bernhard Schmidt, who proposed the technique for the manufacture of Schmidt plates for his newly invented Schmidt cameras, as reported in 1932. Surprisingly, there was no significant advance on the theoretical basis of the method until Lemaître started his work, reported in 1972. Since then, he has established himself as the acknowledged world expert on both the complete theory and its manifest practical applications. In my "Reflecting Telescope Optics II," I give a brief account of his work on pages 23–27. He has not only performed fundamental work for the classical case of spherical aberration, but also for correction of astigmatism, coma and other aberration modes. This includes applications to Cassegrain secondary mirrors. In 1989, I suggested also the application to primary mirrors for which manufacturing tolerances can be relaxed by Active Optics control in the operating phase. In this application, stress polishing can be seen as the first fundamental manufacturing step in a complete system of Active Optics.

More recently, Lemaître has extended his techniques to systems of variable focal length (optical power), which have found important application in the optical train of the 4-telescope complex of the ESO VLT. The function of the interferometric mode VLTI is dependent on this application. The correction applies not only for the axis but also in the field of the interferometric image and is achieved by a system of Variable Curvature Mirrors.

Prof. Lemaître's book, giving a full account of all these developments, is a most valuable addition to the literature in this important branch of optical manufacture. Since its major applications are in astronomical optics, it has also been my pleasure to support the publication of this work by Springer-Verlag in the same Astronomy and Astrophysics Library Series as my own books "Reflecting Telescope Optics I and II." I am confident it will become a worthy and widely recognized standard work.

Rohrbach *R. N. Wilson*

Preface

Astronomical Optics and Elasticity Theory is intended to serve both as a text and as a basic reference on *"active optics methods."* Mainly elaborated for astronomy, and following a conceptual idea originated by Bernhard Schmidt, the first developments of active optics began in the 1960s. These methods allow one to transform by a highly continuous process a spherical surface into the desired *aspherical* surface, as well as to correct tilt and decentering errors between telescope mirrors, to control the focal position by curvature variation, etc, so as to achieve diffraction-limited performance. The recent spectacular increase in telescope sizes, active image correction of telescope errors and atmospheric degradation, and the advent of detectors having nearly perfect quantum efficiencies has led to remarkable progress in observational astronomy, whose large telescopes now currently operate with active optics.

The first chapter concerns optical design and elasticity theory; I thought it useful to introduce these two topics by brief historical accounts. Most of the following chapters are dedicated to the generation of axisymmetric aspheric mirrors, as well as non-axisymmetric mirrors. Active optics methods are investigated for corrections of focus, and for aberrations of third and higher orders. Optical aberration modes that can be superposed by elastic flexure belong to a subfamily that I called Clebsch-Seidel modes. Such aberration correction modes are generated by multimode deformable mirrors. Depending on the adopted thickness class – constant or variable – various active mirror configurations are discussed using the so-called *tulip, cycloid, vase, meniscus*, and *double-vase mirrors*. Two chapters are dedicated to optical designs with the Schmidt concept; the first includes my 1985 high-order analysis of the axial wavefront reflected by a spherical mirror, the system resolving power for each option – with either a refractive, a reflective, or a diffractive corrector – and the optimal corrector shape for each design type; in the second, active optics aspherization methods of the corrector element are developed for catadioptric or all-reflective telescope types and for aspherized grating spectrographs. Another chapter on large mirror support systems treats the minimization of flexure against gravity and in situ active optics control on large telescopes. A short chapter concerns the flexure of thin lenses when bent by a uniform load; this is useful to produce stigmatic singlet lenses by active optics. Grazing incidence X-ray telescopes can also greatly benefit from the ripple-free active aspherization process for various two-mirror designs and particularly for a mirror pair strictly satisfying the sine condition; a theory of weakly conical shells is proposed in a special chapter where the aspherization of the mirrors is obtained by pure extension (or contraction).

The book provides a foundation for finding a mirror thickness geometry and an associated load configuration which can generate one or several fixed surface optical modes – this in the most practicable conditions. Computational modeling, the third branch of science which bridges analytical theory and experimentation, is the ultimate method for accurately solving the deformations of a solid for any configuration of equilibrium-force sets. In the final design stage for an active optics mirror, finite element analysis of the three-dimensional deformations allows optimizing its thickness geometry to obtain the desired mirror figure. However, geometrical optimizations with such codes must require sufficient user knowledge in elasticity theory, and a preliminary analytic solution of the problem by a first approximation theory. This preliminary approach with the theory – the aim of this book – is all the more necessary since there are generally several alternatives for generating a given surface type – as, for instance, with the various solutions presented here for *variable curvature mirrors*.

The beautiful theory of axisymmetric shallow shells, elaborated by Erik Reissner in 1946, is one of the greatest analytic achievements in elasticity theory. In the axisymmetric flexure case, this theory is here used for the aspherization of fast f-ratio mirrors. In addition, a convergent iteration vector which acts towards the required flexure is implemented for determining the thickness distribution of *meniscus-*, *vase-*, and *closed-form mirror shells*. The method has proved sufficiently accurate that no significant corrections were found necessary from finite element analysis. Active optics aspherizations of primary and secondary telescope mirrors were carried out by the *Laboratoire d'Optique de l'Observatoire de Marseille* (LOOM). The results of stress figuring or in-situ stressing of all the axisymmetric mirrors directly designed from Reissner's theory – as for instance with the modified-Rumsey anastigmatic telescope presented here – show that the axial wavefront correction errors are within conventional diffraction limited criteria.

I am grateful to M. Ferrari for his contributions in the second chapter, to J. Caplan, S. Mazzanti and K. Dohlen for fruitful discussions on several points of the book, to optician engineer P. Montiel, G. Moreaux and P. Lanzoni for their active implication in the development of the processes presently described, and also to P. Joulié for the preparation of many figures in the book.

Marseille, October 2008 *G. R. Lemaitre*

Contents

Notations

Optics Notation

$x,\ y,\ z$	rectangular coordinates
$\rho,\ \theta,\ z$	cylindrical coordinates
λ	wavelength of a monochromatic propagation of light
ν	frequency of a monochromatic propagation of light
D	clear aperture diameter
r_m	semi-radius of clear aperture
$f,\ f'$	focal length of a system in the object and image space respectively
Ω	focal ratio, or f-ratio
$n,\ n',\ N$	index of refraction of a medium
$i,\ i'$	conjugate incidence and emerging angles
R	radius of curvature of axisymmetric optical surface
c	curvature of an axisymmetric optical surface (the velocity of light in vacuum is also denoted c)
$c_x,\ c_y$	principal curvatures of an optical surface
C_P	Petzval curvature
κ	conic constant of conicoid optical surface
$u,\ u'$	conjugate aperture angles
$\eta,\ \eta'$	conjugate ray heights
$\bar{\eta}$	normalized ray height
φ	field angle
$\varphi_{\max}$	maximum field angle
$z,\ z'$	usual object and image conjugate distances
$\zeta,\ \zeta'$	Newton object and image conjugate distances
M	transverse magnification
K	optical power
H	Lagrange invariant
E	étendue invariant
T	telephoto effect
$W_{[4]}$	aberration wavefront function of third order theory
$\rho,\ \theta\ ,\ \bar{\eta}$	normalized radius, azimuth angle and image height of a wavefront aberration function

S_I to S_V	Seidel's five coefficients of third order theory
$z_{n,m}(\rho, \theta)$	cylindrical coordinate representation of a wavefront mode or an optical surface. Also simply denoted z_{nm}
Z, $Z_{\rm Opt}$	representation of a wavefront mode or an optical surface in active optics coaddition law
ω	angular frequency of a wave propagation
k	wave number
S	Strehl intensity ratio

Elasticity Notation

$x,\ y,\ z$	rectangular coordinates
$\rho,\ \theta,\ z$	cylindrical coordinates
t	thickness of a plate or a shell
$\mathcal{T}$	dimensionless thickness
q	intensity of a uniform load
F	intensity of a force at a single point
μ	weight per unit volume
g	intensity of the gravity field
I	moment of inertia of a beam about a perpendicular axis
I_p	polar moment of inertia of a beam about its axis
E	Young's elasticity modulus in uniaxial tension and compression
ν	Poisson's ratio
G	shear modulus of elasticity, sometimes called torsion modulus
K	isotropic modulus of elasticity in 3-D tension and compression
D	flexural rigidity
R	radius of curvature generated by axisymmetric flexure
$R_x,\ R_y$	principal radii of curvature generated by flexure of a surface
$\varepsilon_{xx},\ \varepsilon_{xx},\ \varepsilon_{xx}$	normal strain components in rectangular coordinates
$\varepsilon_{yz},\ \varepsilon_{zx},\ \varepsilon_{xy}$	shear strain components in rectangular coordinates
$\sigma_{xx},\ \sigma_{xx},\ \sigma_{xx}$	normal stress components in rectangular coordinates
$\sigma_{yz},\ \sigma_{zx},\ \sigma_{xy}$	shear stress components in rectangular coordinates
$u,\ v,\ w$	components of the displacement vector in rectangular coordinates (sometimes denoted $u_x,\ u_y,\ u_z$, respectively)
$M_x,\ M_y$	bending moments per unit length of sections of a plate perpendicular to x and y axes, respectively
M_{xy}	twisting moment per unit length of section of a plate perpendicular to x axis
$Q_x,\ Q_y$	shearing forces per unit length of sections of a plate perpendicular to x and y axes, respectively
$V_x,\ V_y$	net shearing forces per unit length of sections of a plate perpendicular to x and y axes, respectively
$N_x,\ N_y$	normal forces per unit length of sections of a plate in direction of x and y axes, respectively

$\varepsilon_{rr},\ \varepsilon_{tt},\ \varepsilon_{zz}$	normal strain components in cylindrical coordinates
$\varepsilon_{tz},\ \varepsilon_{zr},\ \varepsilon_{rt}$	shear strain components in cylindrical coordinates
$\sigma_{rr},\ \sigma_{tt},\ \sigma_{zz}$	normal stress components in cylindrical coordinates
$\sigma_{tz},\ \sigma_{zr},\ \sigma_{rt}$	shear stress components in cylindrical coordinates
$u,\ v,\ w$	components of the displacement vector in cylindrical coordinates (sometimes denoted $u_r,\ u_t,\ u_z$, respectively)
$M_r,\ M_t$	radial and tangential bending moments per unit length of sections of a plate perpendicular to r and t axes, respectively
M_{rt}	twisting moment per unit length of section of a plate perpendicular to r axis
$Q_r,\ Q_t$	shearing forces per unit length of sections of a plate perpendicular to r and t axes, respectively
V_r	net shearing forces per unit length of section of a plate perpendicular to r-axis
$N_r,\ N_t$	normal forces per unit length of sections of a plate in direction of r and t axes respectively, or for a shallow shell, in normal direction to curvilinear sections
$z,\ Z,\ Z_{\text{Elas}}$	equivalent representations of flexure with respect to z-axis used in active optics coaddition law
ζ	dimensionless flexure with respect to z-axis
$\mathcal{W}$	dimensionless flexure with respect to z-axis used in cylindrical shells and weakly conical shells

Chapter 1
Introduction to Optics and Elasticity

1.1 Optics and Telescopes – Historical Introduction

The idea of using a mirror that provides optical rays all converging to a single point originated in the Hellenistic period of Greece more than two millennia ago. Because of the straightforward geometrical properties of conics as meridional sections of stigmatic mirrors, catoptrics was constituted long before dioptrics. Nevertheless, and surprisingly, the first telescopes built were not reflectors but refractors.

In Greece it was long known that some problems of a geometric nature were not soluble by straightedge and compass. A legend of the Classic period of Greece states that the gods were unhappy because geometry was not sufficiently studied. The oracle of Delos, which was consulted before major decisions, said ~430 BC that recovery of the gods' clemency would require solving three problems: the angle trisection, the cube duplication, and the circle squaring. The first two problems were rapidly solved, but the third has baffled mathematicians for 2,300 years until F. Lindemann (1882) demonstrated that π is transcendental, and thus showing that the construction of the circle length by purely geometrical means is insoluble.

1.1.1 The Greek Mathematicians and Conics

• **Menaechmus** (~375–325 BC) who lived in Macedonia and Greece and was a tutor of Alexander the Great, formalized the notion of conics and found many of their geometrical properties. He solved the famous problem of *cube duplication* by determining $\sqrt[3]{2}$ from the intersection of two conics: a parabola and a hyperbola – with the Cartesian formalism $y = x^2$ and $xy = 2$, respectively (Fig. 1.1). He suspected that the solution was impossible with the classic method of straightedge and compass since a few decades earlier Hippocrates of Chios (471–410 BC, Athens) has been able to solve the not less difficult problem of the *angle trisection* by the less restrictive method of "declination" – inclinatio in Latin and $\nu\epsilon\upsilon\sigma\iota\varsigma$ (neusis) in Greek – which is now called the *marked straightedge method*.

This method consists of revolving a straight line of a given length through a fixed point until it intersects two fixed straight lines (cf. Arnaudiès & Delezoide [6]).

G.R. Lemaitre, *Astronomical Optics and Elasticity Theory,* Astronomy and Astrophysics Library, DOI 10.1007/978-3-540-68905-8_1,

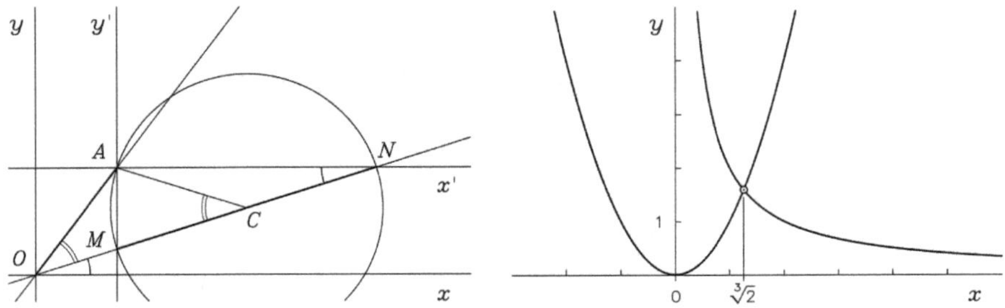

Fig. 1.1 (*Left*) *Angle trisection* obtained from the marked straightedge by Hippocrates of Chios in ~430 BC. (*Right*) *Cube duplication* obtained from the two-conic intersection method by Menaechmus in ~340 BC. The two-conic intersection method is equivalent to the *marked straightedge method*

In practice one uses a straightedge on which the given length is copied in two marks by revolving and sliding it at the fixed point. In Fig. 1.1, the angle to trisect is xOA, a length MN = 2×OA is reported on the straightedge that pivots in O, and M, N must be on the straight lines x', y' that are parallel to x, y through A. This method is equivalent to the two-conic intersection method.

• **Aristaeus the Elder** (~365–300 BC, Greece) wrote five books, now lost, entitled *Solid Loci* and concerning the conic sections. This is known from commentaries by Pappus (290–350 AD, Alexandria), one of the last Greek mathematicians.

• **Euclid** (325–265 BC, Alexandria) who is the classical reference in the founding of geometry with his 13 books, well known as *Elements*, also wrote several other works including *Optics*, *Conics*, and *Surface Loci*. *Optics* solely concerns perspective. According to Pappus, Euclid's *Conics*, now lost, was mostly a compilation whereas Aristaeus's book gave a more thorough discussion of the discoveries and properties of conics.

• **Apollonius of Perga** (262–~190 BC, born in Perga, now known as Murtana, on the south coast of Turkey) lived in Ephesus and Alexandria. It is to him that we owe the names ellipse, hyperbola, and parabola. He wrote a treatise in eight books, seven of which still exist, entitled *Conics*, which contains about four hundred propositions. He differentiated the various conic types by the angle of the intersecting plane with respect to the cone angle. Pappus states that "Apollonius, having completed Euclid's four books of conics and added four others, handed down eight books on conics". He also refers to Euclid's last work on conics that is included in the *Treasury of Analysis* under *Surface Loci* but states that Apollonius's *Conics* had become the classical reference on those questions. He gives some indications of the contents of six other works by Apollonius: *Cutting of a ratio, Cutting an area, Inclinations* (presently lost), *On determinate sections, Plane loci, On verging constructions and Tangencies*, but none of them seem to mention any optical property of a conic.

• **Diocles** (~240–~180 BC, lived in Arcadia and Karystos near Athens) published *On burning mirrors*. It is known from comments by Eutocius (480–540 AD) that he solved the problem of angle trisection by inventing a method based on cissoid curves which thus differs from Hippocrates of Chios's method. None of his writings were known in the West before 1920, but more recently (~1970) a complete Arabic translation of his work was found in the Astan Quds Library of Mashhad, Iran. In the first translations, by Toomer and Rashed [162], it appears that Diocles discovered the property that a parabola can be defined as the locus of points satisfying a constant distance ratio equal to unity from a given point (focus) and a directrix line. One of his other studies leads to the following fact.

→ *The fundamental optical property of stigmatism of a parallel beam reflected by a paraboloid was known by Diocles in ~200 BC.*

• **Anthemius of Tralles** (~474 AD–before 558), co-architect of Hagia Sophia in Constantinople, is credited with discovering the relation for ellipses MF+MF′ = *constant* and for the classical method of drawing them with a string. He found the *catoptric stigmatism between the two foci of an ellipsoid.*

1.1.2 The Persian Mathematicians and Mirrors

After the end of the Alexandrine superiority in 641, a scientific renaissance took place in Persia – mainly in Baghdad – where most of the main Greek works were translated into Arabic during the following centuries (Rouse Ball [9]). Some advances in algebra had already been assimilated by the Persians along with mathematical developments previously achieved in India, such as decimal numbering and the important invention by Bramaguptas in 629 of the "0" symbol for zero. In 820 Al-Khwarizmi published *Al-jabr wa'l Muqabala* – from which the word algebra is derived – where one of the positive roots of some second degree equations is solved (although symbolic notation did not yet exist). In around 900, Persians geometers were aware of Archimedes's failure to construct the regular heptagon and of the equally fruitless attempts by subsequent Greek geometers; they provided the first constructions of the heptagon around 970. Among their numerous contributions, we will limit our comments to the main works of Alhazen in optics.

• **Al-Haytham,** known in the West as Alhazen (965–1021), published in 1008 the *Treatise on Optics: Kitab ul Manzir*, containing seven books. In it he gives a detailed description of the human eye, explaining the function of each part. Here also Alhazen is the first to mention the *camera obscura*, some of which he built, noticing that the image is inverted. He gives the first explanation of atmospheric refraction. He investigates lenses as well as spherical and paraboloidal mirrors, and is aware of the spherical aberration and of the stigmatic property of a paraboloid already demonstrated by Diocles.

An important question which had been introduced by Ptolemy (85–165 AD, Alexandria) in his famous *Almagest* is known as *Alhazen's Problem*: given a spherical mirror of center O, a point source A and another given point B, how can one geometrically construct the intersection point R at the mirror surface where the ray

AR is reflected towards B ? Except for the trivial arrangements where the center O is on AB or on the median plane of AB, the general solution is impossible with the straightedge and compass method. Considering the AOB plane, the solution can be derived from the intersection of the circular section of the mirror with one of the homofocal ellipses of foci A and B, by selecting the ellipse which is tangent to this circle – there are generally two of them – by imposing two conditions: the equality of the ordinates and the equality of slopes. This leads to solving fourth degree equations where the solution is with a double root for the tangency. It seems that Alhazen solved this problem with the marked straightedge [6]. Huygens later gave a solution using the intersection of conics. A construction of point R by the intersection of a circle and a hyperbola is displayed in Fig. 1.2.

Al-Haytham notices that if a point source is at infinity, the image given by a spherical mirror of radius R is located at a distance equal to or a little larger than $R/2$ from the mirror: this distance is the focal length. In some examples where the source point is at finite distance, he also gives the location of the image point after reflection on a concave mirror: this is the conjugate distance.

→ *Although lacking symbolic notation, the Persians knew how to calculate the conjugate distance in ∼1000. This may be considered as the prelude to Gaussian optics.*

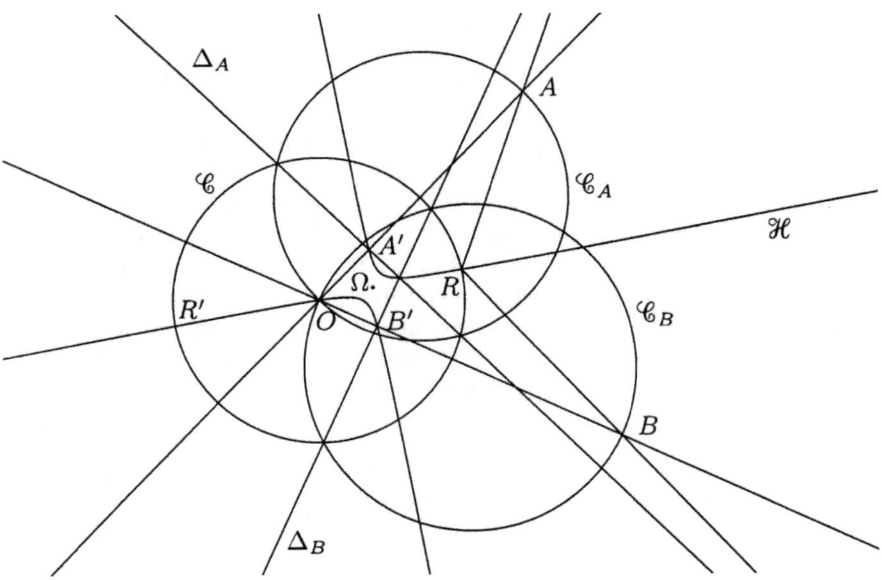

Fig. 1.2 *Al-Haytham's Problem*: Given two points A, B and a spherical mirror of center O, how can one find the point R where the ray AR is reflected through point B ? In the AOB plane where the mirror intersection is the circle $\mathcal{C}$, one constructs the circles of diameter OA, OB and lines Δ_A, Δ_B that intersect these diameters in A′, B′. A hyperbola $\mathcal{H}$, and only one, can be constructed passing by O, A′, B′ with its center Ω at the middle of A′B′ and its asymptotes in the direction of the bisections of lines OA and OB. Among the four points on C and $\mathcal{H}$, the figure shows the two solutions as points R or R′ for a convex or a concave mirror, respectively (Arnaudiès & Delozoide [6])

Al-Haytham engaged in remarkable technological developments, for instance by constructing mirrors of steel and probably of steel-silver alloy and of pure silver, but like others was unable to obtain accurate spheres. He also gave comments on the development of turning lathes.

It is mainly from Spain under Arab domination and not directly from Persia, that Persian scientific writings, rendered into Latin by Adelard, Gherard and many other translators, were introduced into Europe during several centuries after 1150, thus including the Greek heritage also. During this period most of the old European universities were created. This favored the assimilation of these heritages and gave rise to important developments which were materialized by the Renaissance.

1.1.3 End of European Renaissance and Birth of Telescopes

The emergence of small blown objects in glass originates in Phoenicia, Syria, and Egypt around ∼200 BC and much earlier for non-blown objects. The magnification effects of transparent materials was known in Antiquity. The blown glass technique had passed through the ages via the Romans and settled in Venice before the first millennium. The first lenses – *lense* was the Latin name given to a biconvex disk – used as spectacles appear in Italy before 1300 for correcting presbyopia because hand-lenses were unappropriate for writing; this was followed up around 1450 by divergent spectacles for correcting myopia. During the European Renaissance (1400–1600), the crystal- and glasswork of Murano – a small island near Venice – was flourishing. From ∼1300, the remarkable developments of Murano's furnaces and the skilfulness of glass blowers allowed the manufacturing of bottles, drinking glasses, chandeliers, polychromatic vases, and ornaments, etc. Around 1550, it had become easy in Murano to procure positive or negative lenses of small optical powers for correcting usual defects of the human eye. In this context, by empirically separating a negative lens from a positive lens, Digges in *Pantometrie* (1571) and Della Porta in *Magia Naturalis* (1589) noticed that an object at some distance is seen enlarged; this device is generally considered as the primitive ancestor of the telescope and may be called an "*enlarging monocular*" (sometimes improperly called *spyglass*).

Historics on the enlarging monocular are commented in noticeable works by Danjon and Couder [44] and King [85] where the development of early telescope is also described. The construction of enlarging monoculars between ∼1608 and 1609 in Holland, is mainly the result of technological advances in fine glass manufacturing and lens polishing in Italy.

Because of technological difficulties encountered in making accurate and efficient metal mirrors, the first telescopes were not reflectors; the refractor telescopes emerged first from transformations of the enlarging monocular by Galileo Galilei. Accounts on the development of telescopes are in Riekher [132] and Wilson [170]. The major milestones in these developments are resumed hereafter.

1.1.4 Refractive Telescopes

• **Galileo Galilei** (1564–1642) heard from France, in 1609, that Lippershey in Holland had constructed a sort of "*enlarging monocular*." This device, made of a single lens of positive power at the first end of a tube and of a negative lens on a sliding tube, was in fact a chance arrangement of eyeglass lenses available on the market which thus may only magnify distant objects by two or three times. It must be considered as a poor half part of our ancient opera glasses, and was totally useless for astronomical observations. Over a few months's time Galileo fully understood its principle and transformed it into a "*telescope*" by constructing three of them known as telescopes No. 1, 2 and 3.[1] In 1610, with telescope No. 3 he discovered Jupiter's satellites, Venus' phases, and the Sun's rotation (observations of Sunspots by the naked eye and natural *camera obscuras* were reported in China since 28 BC and later in Persia).

From the rustic enlarging monocular, Galileo discovered the basic optical features for obtaining two-lens systems with large magnifications, i.e. a *telescope* – an invention which he must be credited with –: with a plano-convex lens as his objective and one of various divergent lenses as an eyepiece, he derived afocal systems of *large* magnifications, i.e. large beam compressions. His second difficult task, and not the least, was to build accurate lenses able to provide such high magnifications (Fig. 1.3).

It is remarkable that all his objective lenses are close to a plano-convex shape which come up from currently available equiconvex lenses that, in order to obtain larger focal lengths, he probably re-figured by himself; this is relevant to his objective lens of telescope No. 3 – the only surviving piece of this telescope – which shows two concentric shapes on the same side: a flat or quasi-flat central zone defining the clear aperture surrounded by a *useless* convex surface. Galilieo obtained magnifications somewhat higher than 20 requiring deep divergent lenses down to

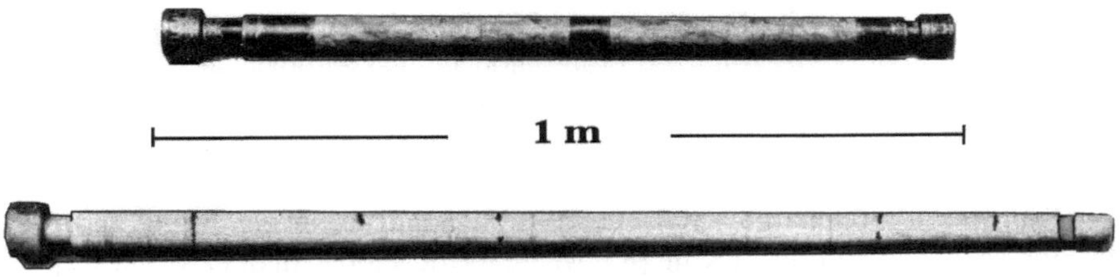

Fig. 1.3 Galileo's first refracting telescopes. (*Up*) No. 1: length 980 mm, magnification 21, clear aperture ~16 mm, f/61. (*Down*) No. 2: length 1,360 mm, magnification 14, clear aperture ~26 mm, f/51. (Institute and Museum of History of Sciences, Florence)

[1] Astronomers generally admit that Giambattista della Porta (1535-1615) built, in Murano, the first "monocular opera glasses" (or low power spyglasses) around 1580 – as he clearly mentioned in private correspondence and in *Magia Naturalis* – and that Lippershey was one among several Dutch authors who built replicas of an Italian model, about one foot long, dated 1580, and then did not obtained patent letters (cf. Danjon & Couder [1.4], p. 589 to 601). However none of them proposed or elaborated a design with sufficiently large magnification to be called a telescope.

47 mm focal length which he figured by himself because eyeglass makers did not make strong enough negative lenses for correcting such a huge myopia. Although his afocal design gave a view of objects at infinity, he naturally moved the diverging lens towards the objective by a slight amount in order to vision the sky at the eye's *punctum proximum* distance. His mother found it lucrative to sell lenses to persons who asked for them. Interferometric analyses of some Galileo telescope optics conserved at the Science Museum in Florence show that the emerging wavefronts were "diffraction limited" at a single wavelength; Galileo would not have grasped the nuance of such a compliment, however.

Galileo published his astronomical discoveries with Telescope No. 3 – with a first lens focal length $f_1 = 1{,}650$ mm, $f_1/D = \sim 50$ – in *Sidereus Nuncius* (1610), where he states having used a magnification up to 30 and recognizes that the objective lens could be replaced by a concave mirror.[2]

• **Johannes Kepler** (1571–1630) introduced the term "focus" in his work of 1604, *Ad Vitellionem Paralipomena*. In optics, he was the first to establish the conjugate distance relation for a given focal length. He noticed that the human eye works with an inverted image on the retina. In *Dioptrice* (1611), Kepler discussed the theory of telescope and enounced the rule giving the magnification as the ratio of the focal lengths of the two lenses. He described a refractor with a positive eyepiece but he never used one. The first positive eyepieces were used by C. Scheiner and later by F. Fontana in 1646; however this is generally known as *Kepler's eyepiece*.

• **Willebrord Snell** (1580–1626) discovered the sine law of refraction in 1621 from experiment. He died in 1626 without publishing his discovery. It was first published by Descartes in his *Dioptrique* (1637) without reference to Snell who communicated it privately to several people including Descartes (cf. Born and Wolf [17]).

• **René Descartes** (1595–1650) thoroughly elaborated the general *theory of stigmatic curves* based on *analytic geometry* – that he created for this purpose – and simultaneously introduced the standard symbolic writing which we are familiar with. It was known from Diocles (cf. Toomer [162]) for the paraboloid, and probably by Pappus for the ellipsoid and hyperboloid, that only *conicoid mirrors* provide a perfect reflected image of an axial source point.

In *La Géométrie* (1637), Descartes [45] introduces the complete theory of perfect axial imagery by aspherical surfaces that cancel the spherical aberration. It contains the equations of stigmatic surfaces of mirrors or lenses for a finite or infinite distance conjugate. For mirrors, the meridian sections of *stigmatic conicoids* appear as degenerated second degree curves. For lenses, the analytic geometry allowed

[2] A few years later, N. Peiresc observing the Moon with a Galileo refractor begun drawing a map of it with the help of P. Gassendi and of a distinguished engraver C. Mellan; at mid work he discovered the *Moon libration* – oscillations of $\sim 8°$ and $6°$ in longitude and latitude – and then completed his task with three maps done. He was one of the few scientists to defend Galileo against the Vatican Inquisition which condemned all published works on heliocentrism – among them *De Revolutionibus Orbium Coelestium* by N. Copernicus (1543) – and, in 1600, condemned the Copernician G. Bruno to the stake.

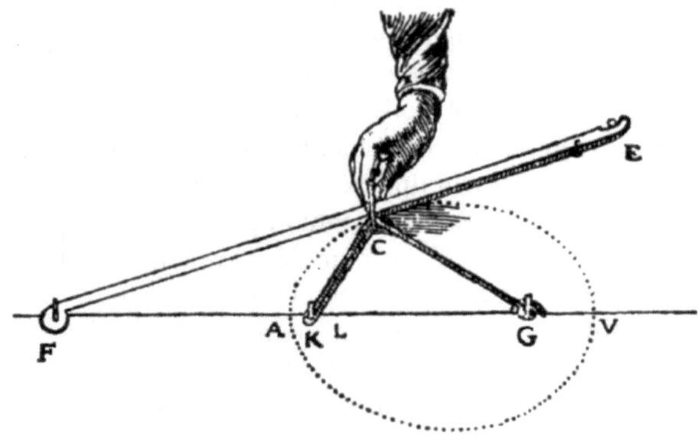

Fig. 1.4 *Descartes' ovals*: Consider a given source point F in a medium of refractive index unity, and its conjugate G in medium n. A refracting surface – diopter – of stigmatic shape satisfies $FC + n\,CG = \text{constant}$. The locus of C points is drawn, with a constant tension of the string ECKCG, along the marked straight edge FE as it pivots around F (*La Géométrie*, 1637 [45]) (cf. Chap. 9)

him to derive the *stigmatic ovoids* (cf. Chap. 9), which meridian sections, namely *Descartes's ovals*, are fourth degree curves.

Using the formalism of Greek geometrical methods, Descartes gave a famous construction of the ovals with the marked straight edge and a string (cf. Arnaudiès and Delozoide [6]) which provides, through a refractive surface, the stigmatism of axial conjugates at finite distance (Fig. 1.4). All possible shapes of stigmatic lenses, designed with one spherical surface which is centered on the object or image, are displayed in *La Dioptrique* [45], which is part of *Discours de la Méthode*.

No further advance was made in *Descartes' theory of stigmatic surfaces* until Petzval (1843) and Seidel (1856) established the complete theory with field aberrations, more than two centuries later.

- **Christiaan Huygens** (1629–1695), who recognized the importance of atmospheric *seeing*, built a 5.7 cm aperture refractor of 4 m focal length in 1655 (singlet objective lens at f/70) with which he discovered Titan. Refractors then increased further in size with Hevelius, Cassini, and others. In 1686, Constantin Huygens built several refractors so-called *"aerials"* – the tubes were open to the air – reaching 22-cm aperture for a focal length of 70 m (objective lens at f/300). Another example is the ∼f/500 singlet objective at the Marseille Observatory (Caplan [24]), apparently used around 1700.

Throughout the period 1609–1740, single lens objectives evolved towards slower f-ratios, which still did not require any asphericity correction, but suffered hugely from *chromatic aberrations* and mainly from axial chromatism.

The axial chromatism provides a first order variation of the focal length with the wavelength. Further slow down of the f-ratio was not the right way to minimize its angular size: with such huge focal lengths, the human eye could not see any image at all by lack of sensitivity or integration time.

Lead oxide glass, known in Antiquity, was reinvented in English glass factories around 1620. A standard production process was set up in 1675 by Ravenscroft. This material, so-called crystal of England or *light flint* (LF) glass, offered the white brightness of (quartz-) crystal and was easy to elaborate from closed crucibles. Its refractive index at the yellow helium line was $n_d = 1.58$ instead of 1.52 for the *crown* (K) or *borosilicate* (BK) glasses.

• **Chester Moor Hall** invented in 1728 the *achromatic objectives* – corrected from axial chromatism – by combining two lenses together: a negative flint lens and a positive crown lens. First, experimenting with flint and crown prisms, he carefully measured both their mean deviation angles and color dispersion angles. Then he determined the ratio of the prism angles of a matching prism pair that minimizes the resulting color dispersion which, thus, provided an achromatic deviation. Next, considering a lens pair, Hall stated that if at any given axial height the local prism angle ratio is the same, the chromatism correction will be achieved. Denoting K_1 and K_2 the optical power (cf. Sect. 1.4.3) of each lens in glass of respective dispersive power $\delta n_1/(n_1-1)$ and $\delta n_2/(n_2-1)$, this means that Hall discovered the *achromatism condition* $K_1\,\delta n_1/(n_1-1) + K_2\,\delta n_2/(n_2-1) = 0$. After designing a lens-pair, in 1733, Hall sub-contracted the optical figuring of the two lenses which, when assembled as a 3.5-cm aperture telescope, revealed results in accordance to his theory.

His results were well understood by Peter Dollond (renowned instrument and lens maker; his son John later succeeded in obtaining a Dollond patent for doublet lens achromats which was Hall's results ! [85]), and proved that the *dispersive power* of a glass $\delta n/(n_d-1)$ completely differs between a flint and a crown. This brings to evidence Newton's error who, by supposing that the dispersive power was linearly the same for all glasses, hastily concluded that achromatization was impossible. Essays by L. Euler in 1742 and by S. Klingenstierna some years later confirmed this error. In establishing his theory of primary chromatism correction, Hall made possible the major advance in the development of refractors.

• **Alexis Clairaut** (1713–1765) elaborated the *theory of achromatic doublet lenses* in the period 1756–1762. He more accurately repeated the refractive index measures of crown and flint by Hall and P. Dollond, and concluded that a doublet-lens *never* could be exactly matched for obtaining achromatism because chromatic residuals will remain (these residuals were later called *secondary spectrum*). In a first memoir to the Royal Academy of Sciences [32], Clairaut discussed achromats with a crown first element. In this case, considering that the chromatic aberration of the crown positive lens must be set exactly opposite to that of the flint negative lens, he showed that continuous pairings are possible (with more or less spherical aberration). The assembled lenses provide a net positive power with the same focal length at two different wavelengths.

In the second memoir he investigated various shape achromats and discovered the second solution class with a negative flint lens as the first element. Investigating the two classes, and by varying the mean curvature of the lenses – *cambrure* in French – he derived relationships for achromats with nulled spherical aberration. Denoting c_1, c_2 the surface curvatures of the first lens and c_3, c_4 those of the second lens, Clairaut introduced an equal curvature for both internal surfaces, $c_2 = c_3$. Among the infinite number of solutions, this particular solution is known as *Clairaut's equal curvature condition* of minimizing the number of surfacing tools (which later allowed cementing the lenses for a higher throughput). It falls that this particular solution $c_2 = c_3$ is not far from the other particular solution with four differing curvatures which, in addition to the spherical aberration correction, gives the correction of the (off-axis) coma.

In the third memoir of 1762 Clairaut investigated the field imagery and noticed that the focused images asymmetrically aberrated (coma and astigmatism) and did not remain in a plane (field curvature). In a figure, he displays an off-axis blur image that he derived from trigonometrical ray trace. Finally, he derived the two simultaneous algebraic equations for non-cemented achromatic objectives corrected from both spherical aberration and coma. This is the *Clairaut aplanatism conditions* in the third-order aberration theory which was usually solved algebraically by Clairaut and soon after by J. D'Alembert. An equivalent graphical solution was found much later by A.E. Conrady: considering a (c_2, c_3) Cartesian plane and a positive crown lens as the first element, Clairaut's conditions are represented by a two-branch hyperbola $(c_3 - b)^2/B^2 - (c_2 + a)^2/A^2 = 1$ for zero spherical aberration and a straight line for zero coma. From the two solutions corresponding to the two intersection points, only the solution with curvatures $c_3 = 0.987\,c_2$ both negative is useful for aplanatic objectives; the second and freakish solution is with two meniscus lenses of curvatures $c_3 = 2.520\,c_2$ both positive for a Schott glass BK7-F2

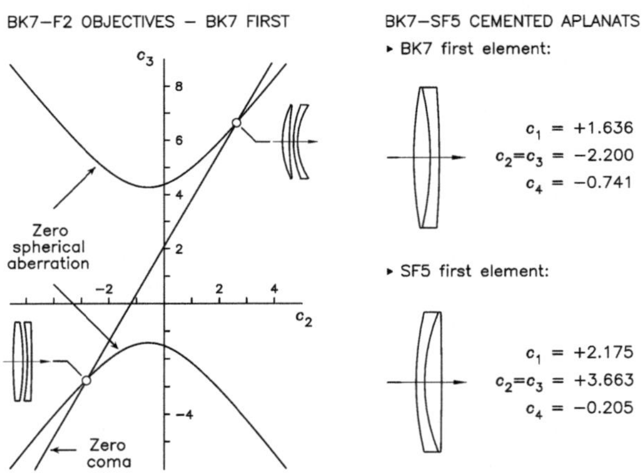

Fig. 1.5 Doublet-lens objectives achromatized for an object at infinity in the spectral range $[\lambda_C = 486; \lambda_F = 656\,\text{nm}]$, the blue and red hydrogen lines, and $\lambda_d = 587\,\text{nm}$, the yellow helium line. Optimizations with Kerber's condition of focus defined from ray height at $\sqrt{3}/2$ on the first surface, the entrance pupil. Effective focal length $f' = 1$. Focal-ratio f/16. The curvatures are exaggerated on the drawings. *Left*: *Clairaut's algebraic conditions*, like the later ones by A.E. Conrady, produce a graph of this sort which shows $Sphe\,3 = 0$ as a hyperbola and $Coma\,3 = 0$ as a straight line, for thin air-spaced doublets (cf. for instance Szulc [152]). The intersection points give two *aplanats* for crown first; with BK7-F2 from Schott, these are with $c_2 = -2.813$, $c_3 = -2.777$ and $c_2 = 2.631$, $c_3 = 6.636$. There are also two other *aplanat* solution classes for flint first. *Right*: *Clairaut-Mossotti aplanat* also called *Cemented aplanat*. Glass materials developed later, such as Schott SF5 and others, avoid half of the reflected light by cementing the two elements, $c_2 = c_3$; however this is only possible for small aperture objectives. In addition they provide a reduced secondary spectrum. With BK7-SF5 objectives and a field diameter of 1 degree, both aplanat solution classes provide the same resolution of 2 arsec; this becomes two times better with CaF2-KZFSN4

objective (Fig. 1.5-Left). A similar representation with other two solutions can be obtained with a negative flint lens as the first element.

A detailed historical account on Clairaut's optics work and the latter geometric representation of his algebraic conditions for aplanatic objectives is given by J.A. Church [31].[3]

• **Joseph Fraunhofer** (1787–1826) investigated the diffraction of light by gratings, published his theory of diffraction in 1823, and laid down the basis of spectroscopy. In 1752, the first spectral lines were observed by T. Melvill in the spectra of flames into which metals or salts have been introduced. Each chemical element is associated with a set of spectral lines. Fraunhofer studied the absorption lines of the Sun's spectra, originally observed by W. Wollaston (1802), and determined the wavelengths of the brightest lines of hydrogen, helium, oxygen, sodium, magnesium, iron, and calcium. The yellow helium d-line at 587.561 nm, and the blue H_β and red H_α hydrogen lines, i.e. the F- and C-lines at 486.132 and 656.272 nm, have been mainly used to characterize the refractive index n_d and the *reciprocal dispersive power* $\nu_d = (n_d - 1)/(n_F - n_C)$ – sometimes called *V-number* – of optical materials. Fraunhofer's works on spectral lines made possible a major advance in the achromatization accuracy of doublet lenses.

Then without apparently using the algebraic results of Clairaut which includes the exact correction of coma – at least in theory – as reformulated by d'Alembert in 1764 and 1767, Fraunhofer designed achromatic doublet lenses by successive iterations of trigonometric ray traces. This allowed him to build excellent objectives with a reduced coma although this latter aberration was not exactly nulled.[4]

• **Ottaviano Mossotti** (1791–1863), professor of geodesics at the University of Pisa, elaborated the *theory of cemented aplanatic doublet lenses* in the period 1853–1859. Let us denote $n \equiv n_d$ the refractive index of a glass, $\nu \equiv \nu_d$ its reciprocal dispersive power and K the power of a lens of this glass, so that two lenses of different glass can be characterized by (K_1, ν_1) and (K_2, ν_2). In Gaussian optics, the *Hall achromatic condition* $K_1/\nu_1 + K_2/\nu_2 = 0$, entails that the resulting focal length is exactly the

---

[3] Clairaut's optical results continue to be misinterpreted in modern literature. For example, H.C. King's *The History of the Telescope* [85] is very wide off the mark on p. 157, where we read: "Clairaut managed to reduce astigmatism and [field] curvature to within reasonable limits, but he could do nothing for coma and considered it an irremediable evil of two-lens combinations..." which is totally false since he established the algebraic conditions for thin-lens achromatic aplanats; furthermore Clairaut and D'Alembert realized correctly that astigmatism and field curvature are not too serious in the narrow fields of view of telescopes.

[4] Concerning some astronomical results obtained with Fraunhofer objectives, we must mention the Königsberg heliometer as the famous instrument with which Bessel discovered and measured the first stellar parallax, that of 61 Cygni. Referring to this instrument after publishing his aberration theory (1856), Seidel noticed that the coma was corrected [partly] and wrote that "this objective perfectly satisfies the Fraunhofer condition" but there was no such existing Fraunhofer condition since he used iterations of trigonometric ray traces.

This inappropriate claim by Seidel has often been repeated, thus introducing much confusion in the past and present literature. This condition is in fact the Seidel sum $C_{II} = 0$ (implicitly included in Claurault's algebraic formulation) and could have only been approximatively satisfied in Fraunhofer objectives.

same for the wavelengths λ_F and λ_C, whilst the effective focal length slightly differs since it is defined for the wavelength λ_d.

Considering the Seidel aberration theory (1856) and doublets that satisfy Clairaut's condition of equal internal curvatures $c_3 = c_2$, in 1857 Mossotti [110–112] derived particular glass pairs n_1, ν_1 and n_2, ν_2 which provide aplanatic cemented objectives: the *Clairaut-Mossotti aplanats* corrected from axial chromatism, spherical aberration, and coma.

This led him to iteratively solve fifth degree equations – called *Mossotti's equations* – with respect to the power ratio K_2/K_1 of the two lenses. These equations always have three real roots of which only the root $K_2/K_1 \to -1$ when $n_2 \to n_1$ is of practical interest (cf. the grid curves by Chrétien [29] in his Sect. 442). A solution can be derived for any case whether crown or flint comes first provided appropriate glass pairs which in fact were not really existing at this period. These results led H. Harting [72] to develop tables, which have been reproduced many times, giving the three computed curvatures of cemented aplanatic achromats in function of the glass pairing entries, i.e. indexes and dispersive powers. After 1900, similar theoretical results of this problem were also obtained by E. Abbe and co-workers. This boosted the elaboration of such new glass types also allowing lowered secondary spectrums, notably by O. Schott.

• **Arthur Kerber** (1842–1919, Dresden) improved the performances of doublet lenses in 1886. He showed [84] that, after the above aberration corrections, the *chromatic variations of spherical aberration* – or *sphero-chromatism* – at the extreme wavelengths λ_C and λ_F now remain dominating at the image residuals because the two least confusion images are not located at the same axial position for these aberration variations (cf. Chrétien [29] in Sect. 367). Kerber concluded that these residuals can be set to a minimal value. Since for first-order spherical aberration the axial location of the least confusion circle is defined by a zonal ray height h_K equal to $\sqrt{3}/2 = 0.866$ times that of the clear aperture h_{max}, *Kerber's condition* of achromatization states that:

→ *The axial location of the C-red and F-blue images of least confusion must be set in coincidence. This is achieved for the zonal ray height ratio $h_K/h_{max} = \sqrt{3}/2$ of the clear aperture height. Then the F and C blur images have the same size and their extremal radii are algebraically balanced for rays at half- and full-aperture heights.*

Hence, the Gaussian first-order achromatism condition $K_1/\nu_1 + K_2/\nu_2 = 0$ is now replaced by the non-Gaussian *Kerber condition* $K_1/\nu_1 + K_2/\nu_2 = K^2\delta z$, where $K = K_1 + K_2$ and the small axial shift δz is set for the coincidence of C and F blur images. This focus is defined by zonal rays of optical height h_K at the entrance pupil, i.e. at first surface of the objective. With automatized raytrace optimizations, the introduction of Kerber's condition provides an extremely robust operand for finding solutions. Another use of Kerber's condition is with the refractive correctors of Schmidt systems (cf. Sect. 4.2.1).

Around 1920, E. Turrière and H. Chrétien [29] introduced the general name *Clairaut-Mossotti doublet* to refer to the aplanatic cemented doublet lens. Today's optical glasses allow a much better correction of the secondary spectrum (Fig. 1.5). An historical introduction and analysis for various f-ratio Clairaut-Mossotti aplanats is given by Szulc [152] with a non-Gaussian condition for achromatism.[5]

These advances, mainly due to Hall, Dollond, Ramsden, Euler, Clairaut, Fraunhofer, Mossotti, Kerber and optical glass factories led by Guinand and successors Mantois and Parra, opened the way to the construction of large discs for achromatic refractors that culminates just before 1900 with 0.8–1 m aperture telescopes in Potsdam, Paris, Mount Hamilton (Lick), and – the largest – Williams Bay (Yerkes), the two later at f/19. With these large telescopes, the residual spherical aberration – partly due to the stress residuals from the casting that led to a center-edge variation of the refractive index – was annulled by slightly aspherizing one surface. However, materials of high purity and homogeneity are difficult to obtain for large lenses and, furthermore the residual secondary spectrum residual is inevitable. This severely limited the spectral domain of astronomical observations so that, even for small size, a telescope objective could not be designed for both photographic and visual work.

Added to these problems is the fact that the telescope length was very long ~20-times the clear aperture diameter. Attempts at introducing a telephoto effect, by using a faster f-ratio objective and a negative doublet-lens focal expander of the Barlow type (1834) – in a kind of *dialytic system* –, showed that, compared to the single objective case with the same focal length, the image quality cannot be improved even for smaller fields of view.

The conclusion is that the refractor concept had become stagnant soon after 1900. All these problems were next solved with reflectors which significantly progressed since 1800.

1.1.5 Reflective Telescopes

• **Nicolas Zucchi** [132, 170] made the first attempt to build a reflecting telescope in 1616, i.e. soon after Galileo developed the refractor. He states that he procured a bronze concave mirror "executed by an experienced and careful artist in the trade" and used it directly with a Galilean eyepiece. In order to avoid obstruction by the observer's head, his design introduced a significant beam deviation at the mirror, similarly to the *front-view* type later introduced by W. Herschel. Depending on the mirror f-ratio (which is not known), the coma and astigmatism due to this deviation may have partly deteriorated the image quality. The attempt failed; however it is generally assumed that in fact the mirror figuring was of poor quality.

---

[5] Unfortunately in his analysis Szulc did not take the correct Kerber's value for the zonal ray height ratio h_K/h_{max}. However, Szulc's comments leading to the non-Gaussian achromatism condition, $K_1/\nu_1 + K_2/\nu_2 = \delta z/f'^2$, are of great historical interest.

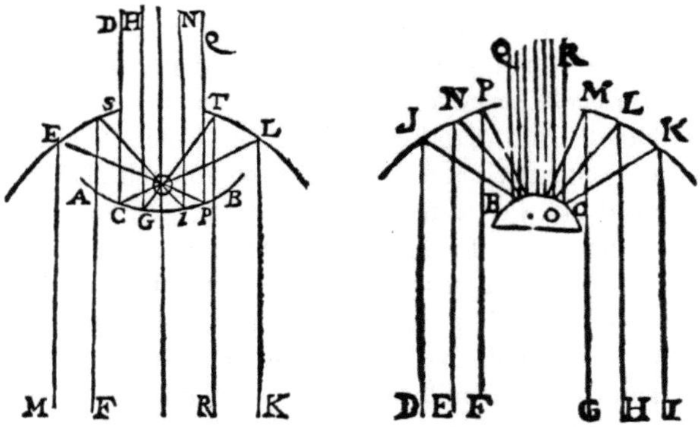

Fig. 1.6 Afocal two-mirror telescopes by Mersenne in *L'Harmonie Universelle* [108]. Forms 1 and 2 are the Cassegrain and Gregory afocal limit forms (facsimile in Danjon & Couder [44])

• **Marin Mersenne** (1588–1648) published *L'Harmonie Universelle* [108] in 1636, in which he introduced the afocal forms of two-mirror reflectors (Fig. 1.6). As noted by Wilson [170], "the full significance of novel and remarkable features presented in Mersenne's works – although often referred to – certainly could not be fully appreciated by Mersenne and his contemporaries." His confocal paraboloid concept may today be described by the following features:

(a) This is the first telescope proposed combining two mirrors.
(b) It includes forms later introduced by Gregory and Cassegrain as *afocal limit forms*.
(c) It includes two other afocal forms which are *retro-reflective* (cf. Sect. 2.3).
(d) With enough *beam compression*, and similarly to the Galileo refractive form, the Cassegrain and Gregory designs are directly usable for observing a small field of view.
(e) The Cassegrain afocal form, as well as its focal form which can be immediately derived from it, appears to provide a larger *telephoto effect* than the Gregory, i.e. the resulting focal length is larger compared to the instrument length. This is a fundamental feature in the development of large and compact reflectors. However, its importance was not emphasized by Descartes, nor by Newton.
(f) In forms 1 and 2, the paraboloid confocal pair provide aberration corrections other than just spherical aberration only: third-order coma and astigmatism are also corrected.

For these features, Mersenne must be credited for inventing the basic geometrical form of the modern telescope [170]. It was demonstrated in the mid-twentieth century – i.e. more than 300 years later – from on-axis and field aberration analysis, that Mersenne's forms 1 and 2 are *aplanats* but also *anastigmats*, i.e. "*quasi-perfect*" optical systems (cf. Sect. 2.3). These remarkable properties escaped Schwarzschild who in 1905 elaborated the two-mirror telescope theory, Chrétien who investigated two-mirror aplanatic systems, and also escaped Paul with his three-mirror theory.

Although potentially included in these theories, quite surprisingly none of these authors derived the specific properties of Mersenne afocal systems.

• **James Gregory** (1638–1675) proposed a two-mirror reflector in his *Optica Promota* (1663): a paraboloidal concave mirror provides a primary focus which is reimaged by an ellipsoidal concave mirror located after this focus. The resulting focus – the ellipsoid conjugate focus – is formed back towards the paraboloid mirror where a hole in this mirror allows the light to emerge for using an eyepiece [85]. As mentioned above, the stigmatic property of a parabola was demonstrated by the Greek geometer Diocles. However, in the case of finite conjugates, it remains unclear whether the stigmatic properties for the ellipse and hyperbola were known by his successors of the second Alexandrian school such as Pappus (290–350) or first known female mathematician Hypatia (~370–415). The first mention of the stigmatic property of the ellipse foci is credited to Anthemius of Thralles (cf. Sect. 1.1.1). Gregory sub-contracted the figuring of a low f-ratio mirror – which would not require any aspherization – but his attempt ended with useless results.

• **Isaac Newton** (1643–1727) presented his second reflector to the Royal Society in 1672 (Fig. 1.7): The beam reflected from a concave paraboloid mirror is focused

Fig. 1.7 Newton's reflecting telescope with the original mirror and eyepiece (The Royal Society, from King [85])

at the side on the tube side via a small flat mirror before the focus inclined at 45°. The concave mirror, 34 mm in diameter, was used with a 25 mm clear aperture; its 175 mm focal length – f/7 reflected beam – did not require any aspherization. A positive-lens eyepiece of 5 mm focal length provided a magnification of 35. Newton succeeded in polishing a sufficiently good spherical surface by himself. He explains his choice of *speculum metal* for the mirrors (bell-metal or CuSn25 alloy i.e bronze [115]), and proposed adding arsenic for a better polish. In 1704, he wrote that he used *pitch polishers* [85]; this appears to be the first written mention of pitch polishers for metal mirrors. Although Newton's two telescopes remained demonstration models only, his selective process leading to the speculum metal marked an unrivalled progress in mirror figuring technology.

• **Laurent Cassegrain** (1629–1693) proposed a more compact form of two-mirror reflector: A convex mirror is located before the focus of the beam reflected by a concave mirror. The resulting focus – the hyperboloid conjugate focus – is formed back towards the concave mirror through a hole which allows the light to emerge for using an eyepiece (Fig. 1.8). He confided a résumé of the principle to de Bercé who described it to the French Academy in 1672. The description of the telescope published in *Journal des Sçavans* [27] was de Bercé's, who introduced it as "plus spirituel" (more astute) than Newton's reflector. Subsequently, Newton (and also Huygens) criticized Cassegrain's proposal with a list of disadvantages compared to his own design or Gregory's; he did not see that Cassegrain's more compact design could be a fundamental advantage – providing a larger telephoto effect – in the development of large reflectors. Added to the context was the fact that Descartes favored the development of lenses. Under these circumstances Cassegrain never attempted

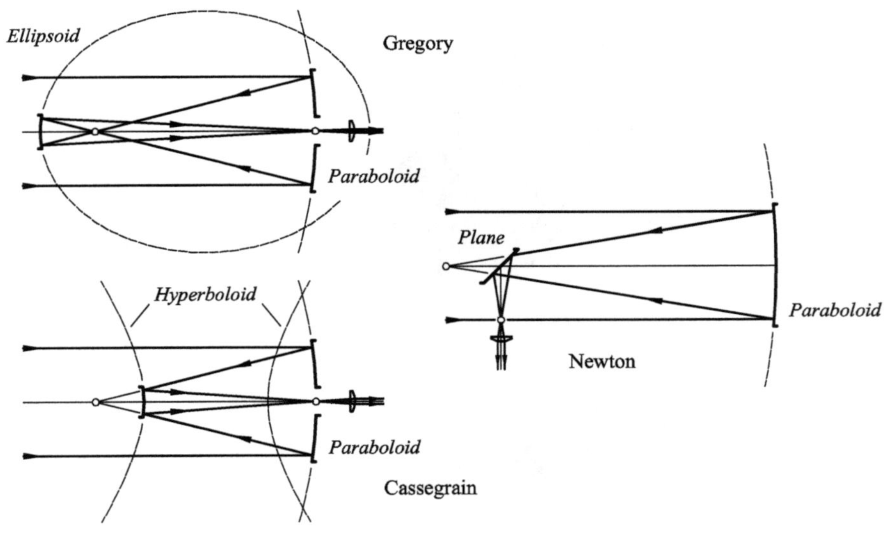

Fig. 1.8 Gregory, Newton, and Cassegrain forms of two-mirror telescopes. P: paraboloid, E: ellipsoid, H: hyperboloid

to build his reflector. The first historical account on Cassegrain has been recently given by A. Baranne and F. Launay [10].

The whole basic framework – theory as well as speculum mirror surfacing techniques – for the development of two-mirror metal reflectors was completely established by these four latter scientists. In 1674, Robert Hooke built with some success a 180 mm aperture Gregorian with an output beam at f/15. In 1721, John Hadley realized a 150 mm aperture Newtonian at f/10 and also built several small Cassegrain and Gregorian reflectors. Around 1740, James Short became a renowned reflector maker in London allowing him to progressively develop an industry which built more than a thousand reflectors. Up to 1768, he proposed in his catalog [85] 12 differing aperture diameters from 28 to 450 mm, primary mirror f-ratios from f/3 to f/8, and eyepieces magnifying from 18 to 1,000 times. Short was involved in metallurgy in order to cast fine speculum mirror blanks; he also developed accurate mechanical systems to keep the Gregorian mirrors aligned when focusing. Claude-Siméon Passement, Engineer of King Louis XV and scientific instrument maker, built microscopes and many Gregorian telescopes in the period 1730–1769.

Now, it must be noted that the early success of the Gregorian form is due to the fact that an *erect image* is viewed – instead of an *inverted image* as in the Newton or Cassegrain forms –, which better satisfied the public demand for terrestrial observing.

• **William Herschel** (1738–1822), after concluding in 1773 that available telescopes were not convenient enough for astronomical observations, invested himself in the design and construction of reflectors. He patiently and accurately ground various concave speculum mirrors of 220 mm aperture, from f/7 to f/15, using convex metal tools, starting with emery, and ending with pitch polishers. After succeeding in casting large speculum disks and building a polishing machine, Herschel completed his 1.22 m – f/10 aperture reflector (40-foot telescope) in 1789, which was last used in 1815. That same year (1789) he completed the construction of the famous 0.47 m – f/13 aperture reflector (20-foot telescope), which he observed with until 1826 and which was later used by his son John Herschel in South Africa (1834–1838). Having low f-ratios, most of Herschel's reflectors used a direct "front-view" vision at the upper end of the tube (Figs. 1.9 and 1.10): the focal surface was set near the tube wall by a convenient mirror tilt. H. Draper later commented on Herschel front-view telescopes that his mirrors showed astigmatism from figuring so that he compensated or at least minimized the astigmatism of all tilted beams by choosing the best azimuth position of the mirror. Assuming that the mirror perfectly corrected the front-view astigmatism by a toroid deformation, his 40- and 20-foot reflectors gave, at the eyepiece center, a coma blur of 5.9 and 3.1 arcsec rms, respectively.

• **Lord Rosse** i.e. **William Parsons** (1800–1867), after constructing several telescopes, undertook the development of larger size reflectors. He erected a foundry, workshops and a polishing machine that was powered by a steam engine. In 1839, he completed a reflector with a 0.91 m – f/13 aperture spherical mirror in a CuSn32 speculum alloy. The next year he undertook the construction of a 1.83 m – f/10

Fig. 1.9 (*Right*) W. Herschel 40-foot focal length telescope (from King [85]). (*Left*) Lord Rosse's 6-foot or 1.83-m aperture telescope (from Imago Mundi)

aperture Newtonian reflector which was put into service in 1847 (Fig. 1.9). Because the mirror tarnished rapidly in the air, it had to be repolished every six months; thus two mirrors were built so that one could be used while the other was being repolished. This reflector was also used by many other scientists and operated until 1878. Although the Irish sky allows only a few observable nights, Lord Rosse obtained excellent images of nebulae and spiral galaxies. He discovered an extended ring object he named the Crab Nebula, and the spiral structure of galaxies, detecting the Whirlpool Galaxy M 51 and many others he drew with fine details closely resembling photographs.

• **William Lassel** (1799–1880), in 1859, completed a 1.22 m – f/9.5 aperture Newtonian reflector with an equatorial mount which was used in Malta until 1865. In order to minimize the deformation due to gravity, he invented for it the *astatic levers* mirror support system – replacing the system by *stacked triangles* or *whiffletrees* – where each weight on a lever generates an amplified force proportional to the cosine of zenith angle. His system came into general use for all large reflectors with passive support.

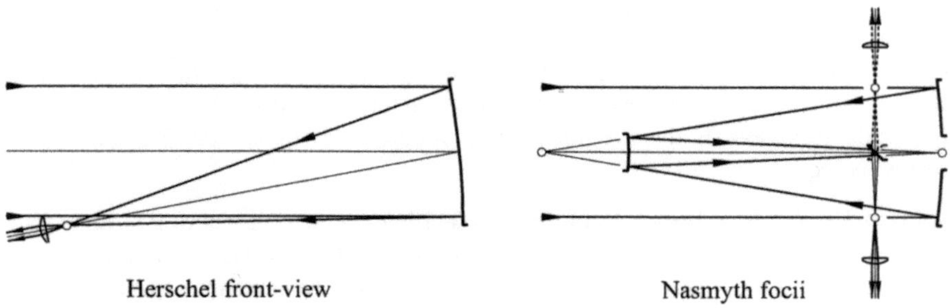

Fig. 1.10 Optical arrangement of Herschel front-view telescope. Cassegrain telescope with Nasmyth foc*ii*

• **James Nasmyth** (1808–1890), in 1845, designed and built a 0.51 m aperture f/9 – f/25 modified-Cassegrain reflector with an *alt-az mount* and with a third plane mirror giving a 90° deviation. This tertiary mirror, located at the node of the axes, provided horizontal focused beams he observed in a fixed position through the hollow altitude axis while seated on the rotating azimuth platform. This was the first large telescope of Cassegrain form. From the point of view of residual aberrations, if one assumes that his primary and secondary mirrors were both spherical, it provided paraxial images of 1.7 arcsec in diameter. His three-mirror concept – providing the now-called *Nasmyth foci* (Figs. 1.10 and 1.11) – always associates an alt-az mount and has now become of general use for all large reflectors over 5 m aperture.

From the focal ratios of the above constructed reflectors, it can be seen that the size of axial image residuals are in accordance with an atmospheric seeing limitation of 2–3 arcsec for mirrors simply requiring an accurate *spherical polishing*. Considering a conicoid mirror (Sect. 1.7), the theoretical asphericity correction of fourth degree – i.e. when unbalanced with a curvature term – can be expressed by

$$z_{\max} = A\, r_{\max}^4 = \frac{\kappa}{1024\,\Omega^3}\, D\,, \tag{1.1}$$

where $\Omega = f/D$ and κ are the f-ratio and conic constant (cf. Sect. 1.7).

For all reflectors by Short, the 1.22 m – f/10 by Herschel and the 1.83 m – f/10 by Rosse, this asphericity amplitude $z_{\max}$ is smaller than 1 – 1.5 μm. Aspherizing these mirrors would not have significantly improved the 1.5–2 arcsec image

Fig. 1.11 (*Left*) Nasmyth's 20-inch aperture telescope (from King [85]). (*Right*) Foucault's 0.8-m aperture telescope (Marseille Observatory)

quality because of the seeing limitation. In a Cassegrain form with spherical mirrors, the spherical aberration of the primary is partly compensated by the secondary. Because of the lack of accurate testing methods and the necessity to frequently repolish speculum mirrors because of tarnishing, spherical figuring was the only way to promptly overcome those difficulties. For a long time, the surfacing problem of mirrors has not been the inability of executing a correct aspherization, but the inability to execute a sphere:

→ *The period of reflectors with "spherical or quasi-spherical mirrors" ended around 1860 when Foucault obtained accurate paraboloids by introducing the powerful knife-edge optical test.*

• **Léon Foucault** (1819–1868) applied Drayton's chemical process of *cold silvering* on Saint-Gobain glass disks polished by Secretan-Eichens and concluded that the method could be easily repeated on similar glass mirrors because glass is chemically neutral. The silver is deposited from a solution of silver nitrate and ammoniac when reduced by glucose [60, 160]. Steinheil, in Munich, had previously obtained silver-on-glass mirrors using Liebig's process; however, this was by *hot silvering* i.e. requiring the risky use of a boiling solution. Compared to the tarnishing of speculum mirrors, silver-on-glass mirrors eliminated the repolishing problem; chemical removal of the tarnished silver layer conserves the original polished shape of the glass even after many repeated re-silverings. In 1858, at the suggestion of Moigno, Foucault examined the shape of a 36 cm aperture mirror reputed to be spherical. With the sensitive optical *knife-edge test* he had just invented, he found an axisymmetric center-edge defect. Instead of re-figuring the entire mirror surface as was the usual practice, Foucault proceeded by local retouches and in a few hours obtained a perfectly spherical mirror; he states that "local retouch method [is] ... an established fact" [160]. A basic quantitative *Foucault test* uses a multi-aperture screen placed in front of the mirror; the screen defines radial and concentric aperture zones with radial widths set narrower from center to edge. The mirror is illuminated by a source-slit and some reflected beams can be intersected by a mobile knife-edge when observing through an eyepiece. Assuming that the slit and knife-edge are maintained on a same line and that the aperture zones expand in a perpendicular direction to it, appropriate axial and lateral displacements of the knife-edge allow one to observe the simultaneous auto-collimation of two symmetric given zones. The correct set up of the axial position of a focused zone is obtained if the light vanishes totally in a tiny movement of the knife. The table of successive axial positions of the auto-collimated knife-edge with respect to corresponding aperture-zone radii allows determining the mirror shape in one direction. Foucault thus obtained an accurate quantitative tool for aspherization control. In collaboration with Secretan and Eichens, a 40-cm aperture silver-on-glass reflector was completed in 1858 and presented before the French Academy [59]. This collaboration continued with an 80-cm aperture - f/5.7 reflector; the Saint-Gobain glass disk, twice as thick at center as at edge, was retouched by Foucault to a quite exact paraboloid shape – i.e. a 4.3 μm asphericity in r^4. The telescope focus was located inside (or near) a total reflection prism at the head-ring center; an f/5.7–f/20 focal expander re-imaged the focus which then was observed at the side of the tube (Fig. 1.11).[6]

• **George Airy** (1801–1892) showed, in 1835 [2], that with perfect seeing conditions, such as close to vacuum conditions, a circular mirror (or a lens) cannot provide an infinitely small image of a point. The observed image at the Gaussian focus is an interference pattern formed of a bright central peak surrounded by concentric rings. Airy calculated the intensity distribution at the diffracted image. Considering a mirror of diameter D, used at wavelength λ, which provides perfect spherical wavefronts converging to a Gaussian focus, the maximum resolution of *diffraction limited images* (cf. Sect. 1.11) is the angle

$$\varphi = 1.22\,\frac{\lambda}{D}\,. \tag{1.2}$$

For instance, a 27-cm aperture telescope used at 0.55-nm wavelength with perfect images just resolves two stars separated of 0.5 arcsec.

• **William Rowan Hamilton** (1805–1865) published, in 1833 [71, 170], the first and famous analysis of the geometrical theory of aberrations by introducing *characteristic functions*. In the case of centered optical systems, he deduced the general form of the aberration function in terms of a power series using three fundamental parameters: the aperture radius, the field radius, and the azimuth angle.

• **Joseph Petzval** (1807–1891) investigated with considerable success the third-order aberrations for the new task of designing large aperture and wide field objectives for photography. Unfortunately, his extensive manuscript on the subject was destroyed by thieves and he never rewrote it; he demonstrated the practical value of his analysis by constructing, around 1840 [17], an unrivaled "portrait lens." Petzval was probably the first to have derived the two coupled coefficients which simultaneously define primary astigmatism and field curvature. In anastigmatic systems, the field curvature is often called *Petzval's curvature*.

• **Ludwig von Seidel** (1821–1896), in 1856 [144], elaborated the first formal analysis of the five monochromatic third-order aberrations by explicitly expressing their amounts introduced by a given surface. Thus, summing them independently through the system, the *Seidel sums* allows deriving its general properties.

6 The telescope life extended from 1864–1965 in the Marseille Observatory with research on nebulae, galaxies, and double stars. Stephan did the first trials to measure stellar diameters from modified fringe patterns as predicted by Fizeau; unfortunately, the two-aperture base allowed by the mirror was too small, and he concluded in 1874 that stellar diameters must be smaller than 0.16 arcsec (Stephan [148]) (the first stellar diameters were resolved in 1922 by Michelson with the Hooker telescope and enlarging base). Later, the first astrophysical images with narrow-band etalons were obtained by Fabry, Perot, and Buisson. In both Foucault's 40- and 80-cm reflectors, the mirror was supported by an *inflatable cushion* in which the observer blew in or let out some air until obtaining a satisfactory image. This supporting concept was next used by Henry Draper in the USA [48].

• **Ernst Abbe** (1840–1905), in 1873 [1], discovered the condition for a system to satisfy both the correction of primary spherical aberration – stigmatism – and primary coma, thus providing *aplanatism*. Using large aperture microscopes, the botanist J. Lister had previously noticed their unexpected image quality and concluded that more than the spherical aberration was corrected. The *Abbe sine condition* is an important theorem in optical design. Considering the axial beam, this can be stated as follows: *if the surface generated by the locus of the intersection points of incident rays and emergent conjugates is a sphere, then the system is aplanatic*.

• **Lord Rayleigh** (1842–1919) showed, in 1879 [123], that if tolerating a 20% light decrease in the intensity at the central peak, then the corresponding departure from the Gaussian reference sphere amounts to a quarter of a wavelength in terms of primary spherical aberration. It was further shown that for primary coma and astigmatism, the peak intensity is less affected by such a quarter wave deformation. Hence, this result became known as *Rayleigh's quarter wave criterion*, one of the simplest and most useful rules among the various tolerance criteria which have been formulated [17]).

• **Georges Ritchey** (1864–1945) was an uncontested expert in optical polishing, in the design and construction of surfacing machines, and in the development and practice of accurate optical testing. He had a remarkable ability to achieve the important features making new giant reflectors a complete success. After completing the optics of the 60-inch reflector for Mount Wilson, he succeeded with those of the 100-inch Hooker Reflector in 1917. His deep understanding and interest in the advantages of aplanatic systems, and his encouragement of Chrétien, led to the Ritchey-Chrétien form, whose second prototype was a 1-m aperture reflector (Fig. 1.12).

• **Karl Schwarzschild** (1873–1916), known for his first-rank achievements in several fields of physics, formulated the complete third-order theory of one- and two-mirror systems in 1905 [142]. His *eikonal method* allowed him to determine the amount of each third-order aberration in a given point of the field. From his general formulation, he derived the two-mirror cases for an object at infinity. He discovered that *for any two-mirror anastigmat telescope, the axial separation of the mirrors must be twice the system focal length*. Most of these anastigmatic systems are described in Sect. 4.1.

• **Henri Chrétien** (1879–1956) elaborated the complete theory of two-mirror telescopes satisfying the sine condition, thus corrected from all-order spherical aberration and all-order linear coma. Apparently before 1910, he derived the theoretical shape of the mirrors by integrations of differential equations including this condition. This led him to formalize the so-called *mirror parametric equations*. From these results, he derived the third- and fifth-order theory for the Cassegrain and Schwarzschild forms. In the Cassegrain aplanatic telescope – known as the Ritchey-Chrétien telescope – Chrétien derived accurate representations of the primary and secondary mirrors as hyperboloids. These studies were published in two articles in 1922 [30]. Aware of all sorts of optical systems and of their key points for aberration correction, Chrétien also invented spectacular *anamorphotic systems* known

Fig. 1.12 The 1-m Ritchey-Chrétien Reflector, primary mirror f/4, Cassegrain focal ratio f/7.3, focal plane scale 28 arcsec/mm. Designed by George W. Ritchey, and completed in 1934, it was originally installed at the US Naval Observatory in Washington D.C. and then relocated to Flagstaff in 1955 (after Chrétien [29])

as Cinemascope for movie panoramic recording and projection. His famous "*Calcul des Combinaisons Optiques*" [29], reissued several times, is a brilliant and profound exposition of all useful properties of optical systems; it also contains many invaluable historical notes. Chrétien is a co-founder of the Institute of Optics which was created in 1920 in Paris.

• **Bernhard V. Schmidt** (1879–1935) invented, in 1929, a new class of reflector called *wide field telescopes* or *wide field cameras*. In Hamburg he built the first such reflector using a 36 cm aperture aspherical corrector plate and obtained perfect image quality on curved photographic films over an entire 7.5° field of the sky. Schmidt telescopes quickly became of primordial use to establish complete cartographies of the sky. Associated with the Palomar 5-m telescope, the Palomar 1.2 m aperture Schmidt was built for this complementary task with 5° field surveys (cf. Chaps. 4 and 5). The situation changed with large-format CCDs which require a flat field of view. FOV of 1.5 or 2° can be provided by a Ritchey-Crétien telescope with two mirrors of the same curvature, equipped with a Gascoigne two-plate astigmatism corrector. With such a 2.5 m telescope, the Sloan Digital Sky Survey (SDSS) uses time-delay-and-integrate imaging – or drift scan imaging as is also used by liquid mirror telescopes (LMTs) – and produces multi-passband surveys and fiber-fed spectrograms. All-reflective Schmidt systems bring an efficient solution for a telescope dedicated to wide-field spectroscopic surveys with 4,000 optical fibers motorized on a 5° FOV, LAMOST is the telescope showing the largest *optical etendue* (cf. Sect. 1.9.3 and Chaps. 4 and 5).

The reflector projects with 1–1.5 m diameter glass mirrors undertaken in France after Foucault and up to 1907 ended in fiascos. Finally, the glass mirror concept was scaled up in the USA for successive 1.5 and 2.5 m reflectors at Mount Wilson using Saint-Gobain blanks. Pyrex glass material was developed for the Palomar 5-m and Caucasus 6-m, and Zerodur vitroceram for the VLT's four 8.2-m blanks. Some historical telescopes built in the period 1900–2008 are listed in Table 1.1.

• **Fritz Zernike** (1888–1966) initially developed, in 1934 [178], the phase-contrast method in order to improve the Foucault test. His important contribution to diffraction theory led him to the invention of the phase-contrast microscope for which he received the Nobel prize. Zernike also invented the *orthogonal polynomials* for representing wavefront surfaces now in general use in optical testing.

• **Maurice Paul** (1890–1981), a pupil of Chrétien, published, in 1935 [170, 178], a general analysis of three-mirror telescopes. He investigated such anastigmatic systems and critically analyzed the cases of aspheric plates and lenses for field correction.

• **Albert Bouwers** (1893–1972) introduced, in 1948 [19], a *concentric meniscus* lens for wide-field catadioptric cameras using a spherical concave mirror. These systems were used in astronomy for panoramic imaging and spectrograph cameras.

• **Dimitri Maksutov** (1896–1964) invented, in 1944 [100], independently of Bouwers, catadioptric cameras using a *null power lens* for correcting the spherical aberration of a concave mirror. His system was widely built for astronomical purposes.

• **André Couder** (1897–1979) invented a two-mirror anastigmat telescope with a concave secondary: *Couder's telescope* (cf. Sect. 4.1) is an outgrowth of the Schwarzschild theory of two-mirror telescopes. Inventing the *null test* method for large mirrors (1927), he made many contributions to the development of reflectors, such as proposing the *Couder law* for mirror support systems and advocating *vase form* metal mirrors plated with enamel (cf. Sects. 7.2 and 8.2).

• **Cecil R. Burch** (1901–1983) described a two-mirror Schwarzschild design which led to the development of reflective microscope objectives. In 1943, Burch published a powerful method for treating the Seidel aberrations of an optical system that he called a "*see-saw diagram*" [23] or diagram of Schmidt plates. Burch's method has been and is still used for searching for aplanatic or anastigmatic solutions.

Table 1.1 Some historical reflectors built in the period 1900–2008

Year	Apert.[m]	Name, Location	Type	f-ratio	Mount	Focus
1908	1.5	Mt Wilson	PH	f/5/16	equat.	Ne, Ca
1917	2.5	Hooker, Mt Wilson	PH	f/5/16	equat.	Ne,Ca,Co
1930	.36	Schmidt, Hamburg	WF	f/1.75	equat.	Schmidt
1934	1.0	Naval O., Flagstaff	RC	f/4/7.3	equat.	Ca
1948	5.0	Hale, Mt Palomar	PH	f/3.3/16	equat.	Pr,Ca,Co
1948	1.2	Sch., Mt Palomar	WF	f/2.5	equat.	Schmidt
1959	3.0	Shane, Lick Obs.	PH	f/5/13.5	equat.	Pr,Co
1960	1.3	Schm., Tautenburg	WF	f/3/10	equat.	Sch,Ca,Co
1971	4.0	KPNO, Arizona	RC	f/2.6/8	equat.	Pr,Ca,Co
1973	1.2	UK Sch., Australia	WF	f/2.5	equat.	Schmidt
1974	6.0	SAO, Caucasus	PH	f/4/30	alt-az	Na
1975	3.9	AAT, Australia	PH	f/3.3/8	equat.	Pr,Ca,Co
1977	2.5	I. du Pont, Chile	RC	f/3/7.5	equat.	Ca
1978	≡.1	Einstein-Xray, USA	W	grazing	space	F_2
1979	3.6	CFHT, Hawaii	PH	f/3.8/8	equat.	Pr,Ca,Co
1983	0.6	IRAS-IR, US Euro.	RC	f/1.5/9	space	Ca
1985	2.4	Hubble-HST, USA	RC	f/2.5/24	space	Ca
1987	4.2	WHT, Canaries	PH	f/2.5/11	alt-az	Pr,Ca,Na
1989	3.5	NTT, ESO Chile	RC	f/2.2/11	alt-az	Na
1990	≡.2	Rosat-Xray, Europe	W	grazing	space	F_2
1993	10.0	Keck-2, Hawaii	PH	f/1.7/15	alt-az	Ca,Na,Co
1994	2.7	LMT, UBC-Laval	P	f/1.9	transit	Pr
1995	0.6	ISO-IR, ESA	RC	f/1.6/9	space	Ca
1998	8.2	VLT-4, Mt Paranal	RC	f/1.8/14	alt-az	Ca,Na,Co
1999	2.5	SDSS, New Mexico	WF	f/2.25/5	alt-az	Ca
1999	9.2	HET, Texas	4M	f/1.5/5	az-track	Gr
1999	≡.3	Chandra-Xray, USA	W	grazing	space	F_2,F_3
1999	≡.4	XMM-Xray, ESA	W	grazing	space	F_2,F_3
1999	8.2	SUBARU, Hawaii	RC	f/1.8/12	alt-az	Pr,Ca,Na
2000	8.1	GEMINI-2, US-Chile	RC	f/1.8/9	alt-az	Pr,Ca,Na
2000	6.5	MAGELLAN-2, USA	PE	f/1.2/11	alt-az	Na
2003	0.5	GALEX-UV, US-EU	RC	f/2/6	space	Ca
2003	0.8	Spitzer-IR, USA	RC	f/2/12	space	Ca
2005	6.0	LMT, UBC, Canada	P	f/1.5	transit	Pr
2005	8.4	LBT-2, Arizona	PE	f/1.2/15	alt-az	Pr,Gr,Co
2005	10.2	SALT, South Africa	4M	f/1.3/4	az-track	Gr
2008	4.0	VISTA, Mt Paranal	WF	f/1/3	alt-az	Ca
2008	10.2	GTC, Canaries	PH	f/1.7/15	alt-az	Ca, Na
2008	4.0	LAMOST, Xinglong	WF	f/5	sideros.	Schmidt

Abbreviations: PH: paraboloid-hyperboloid, WF: wide field, RC: Ritchey-Chrétien, 4M: 4-mirror design, PE: paraboloid-ellipsoid, Ne: Newton, Ca: Cassegrain, Co: coudé or recombined, Pr: prime, Na: Nasmyth, Gr: Gregory, W: Wolter grazing system, F_i: final image focus after i grazing-incidence mirrors.
The sign ≡ in column two stands for equivalent aperture diameter.

• **John D. Strong** (1905–1992) developed the evaporation process of aluminizing glass mirrors under vacuum [150]. This required a vacuum tank, efficient pumps, and tungsten heaters for the aluminum sublimation at 1,460 K. For large mirror reflective coatings, this technique definitively replaced the silver-on-glass chemical method after 1935.

• **Mark U. Serrurier** (1905–1988) invented the compensated flexure tube for the Palomar 5-m telescope. This allows maintaining perfect alignment between the primary and secondary mirrors, thus preventing primary coma de-centering aberrations. He noticed that when the declination axis is fixed on a rigid hub defining the center of the telescope tube, both the bottom ring supporting the primary mirror and the head ring supporting the prime focus cage or the Cassegrain mirror can flex by the same amount if they stay in parallel planes. He found the solution with triangular struts from the hub to each end by deriving the appropriate section ratio between the upper and lower bars. This is known as *Serrurier's truss* [145] and is directly applied, or used as a basic concept, in all designs of large telescope trusses.

• **James G. Baker** (1914–2005) invented several forms of wide field telescopes that were built in large numbers for ground-based and space astronomy. Among them, one may cite the flat-field Baker-Schmidt-Cassegrain telescopes [170], the flat-field Paul-Baker three-mirror telescopes and, in 1945, the Super-Schmidt cameras and the Baker-Nunn satellite tracking cameras [8]. Baker was probably the first person to use a computer in optical design.

• **Norman J. Rumsey** (1922–2007) discovered, in 1969 [137], a flat- and wide-field three-mirror telescope presenting particularly interesting features. His reflector, with three hyperboloid surfaces, is an *anastigmat* which is four-times shorter than the equivalent Schmidt. Furthermore, the de-centering problems of the primary and tertiary mirrors are avoided by figuring these mirrors in the same blank. Sky surveys with detector arrays would benefit from the high quality imaging. Modified-Rumsey telescopes with the primary and tertiary mirrors – designed as a continuous optical surface on a single substrate – can be simultaneously aspherized by *active optics* methods (cf. Chap. 6).

• **Daniel Malacara,** in 1978, edited *Optical Shop Testing* [101] which describes all tests applicable to optical components and systems. This book had become a classic of optical shops. In collaboration with specialists in the field, the book is a review on mountings using interferometers, Foucault, Ronchi and Hartmann tests including the phase modulation tests (Zernike, Lyot, etc.), tests on a star, holographic test, fringe scanning technique, and many useful null testing configurations for compensating the aberrations of the most-common wavefront surfaces.

1.2 Snell's Law and Glass Dispersion

The first proof that the propagation of light is with finite velocity was obtained by Römer (1676), who observed that the time intervals between the epochs of eclipses of Jupiter's first satellite were varying over an annual period. From his measures he derived the first determination of the velocity of light. In 1728, Bradley noticed the annual elliptic variations of the position of the stars which were not at the ecliptic pole – the so-called *stellar aberration* (20.5 arcsec) – and thus derived a new measure of velocity. In 1975, the Fifteenth General Conference on Weights and Measures and BIPM [7] defined the velocity of light in a vacuum as

$$c = 299,792,458 \text{ m/s} . \tag{1.3}$$

The light is the visible part of the *electromagnetic spectrum* which covers wavelengths from gamma rays to radio waves that all propagate at velocity c in a vacuum. The propagation of the electric and magnetic vectors **E** and **H** is represented by Maxwell's equations of electromagnetism (cf. Born and Wolf [17]). In an homogeneous medium (as all media considered hereafter), these equations reduce to the *wave equations*

$$\nabla^2\mathbf{E} - \frac{\varepsilon\,\mu}{c^2}\frac{\partial^2\mathbf{E}}{\partial t^2} = 0\,, \qquad \nabla^2\mathbf{H} - \frac{\varepsilon\,\mu}{c^2}\frac{\partial^2\mathbf{H}}{\partial t^2} = 0\,, \tag{1.4}$$

where ε and μ are the *dielectric constant* (or permittivity) and the *magnetic permeability* of the medium. From these standard equations of wave motion, the velocity v of a wave is

$$v = \frac{c}{\sqrt{\varepsilon\,\mu}} \le c\,, \tag{1.5}$$

where $\mu = 1$ for all non-magnetic media and $\varepsilon \ge 1$ for a transparent medium ($\varepsilon = 1$ in vacuum). Thus, the velocity of light is smaller than c in air or glass; this conclusion was first demonstrated from experiment by propagation of light in water in 1850 by Foucault and Fizeau.

In order to introduce the concept of *geometrical wavefront*, let us assume points in an isotropic medium: a ray from a point source O propagates in a straight line through another point P during a delay Δt. During the same delay, the light propagating from O, via a neighboring ray of P, stands on a *wavefront surface* Σ passing through P. Thus, a wavefront surface contains the notion of both ray package, i.e. *optical pencil* or *optical beam*, and *constant phase* in time over all its front. A wavefront or constant phase surface is generally an *asphere*, i.e. a non-spherical surface because of the geometrical aberrations that are generated from refraction or reflection at the boundary of successive media. However, the wavefront concept is only a useful approximation at large distances from a focus or an obstruction edge. Near the focusing regions, where a *caustic* is the envelope of light rays, the effect of diffraction requires using the electromagnetic theory to correctly describe the light distribution of phase and intensity.

Let us consider a ray or a plane wave Σ that propagates in a medium of refractive index n and reaching at incident angle i a plane surface separating a second medium of refractive index n' in which the ray is refracted at angle i' so that the refracted plane wave is Σ' (Fig. 1.13). The sine *refraction law*, first clearly established from experiment by Snell in 1621 and published by Descartes in 1637 (see Sect. 1.1), or *Snell's law* is

---

[7] Bureau International des Poids et Mesures

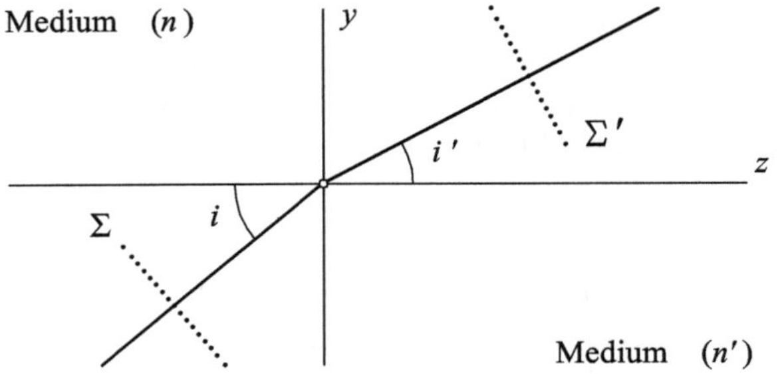

Fig. 1.13 Ray propagation through a plane surface separating two media: Snell's law

$$n' \sin i' = n \sin i\,. \tag{1.6}$$

If $n' < n$ and $\sin i' = 1$, we find $i = \arcsin(n'/n)$, that is $i = 41.8°$ for $n = 1.5$ and $n' = 1$. For larger value of i, (1.6) cannot be satisfied, no refracted ray exists, the ray is reflected at the separating surface in the medium of higher refractive index; this *total internal reflection* has many applications in the design of prisms.

Denoting v and v' the velocity of light in these respective media, the sine ratio of the angles is equal to the velocity ratio (cf. Sect. 1.3), i.e.

$$\frac{\sin i'}{v'} = \frac{\sin i}{v}\,, \tag{1.7}$$

so we obtain $n'v' = n\,v$. By setting $n = 1$ for a vacuum medium, the velocities in each medium are

$$v = \frac{c}{n}\,, \quad \text{and} \quad v' = \frac{c}{n'}\,, \tag{1.8}$$

and the identification with (1.5) gives *Maxwell's formula* for the refractive index: $n = \sqrt{\varepsilon\,\mu}$, where $n \geq 1$ for any transparent medium, and conventionally $n = 1$ in a vacuum.

The dependance of the refractive index of glass on the wavelength has been represented by various possible functions. Some of them – known as Schott or Hertzberger formulas – are directly related to power series of the wavelength λ. The Sellmeier-1 dispersion formula

$$n^2 = 1 + \sum_{i=1,2,3} \frac{K_i\,\lambda^2}{\lambda^2 - L_i}\,, \tag{1.9}$$

with three couples $(K_i,\ L_i)$ of constant parameters, is used by Schott and some other glass manufacturers. Because the Sellmeier-1 formula does not have linear coefficients, the dispersion set up is iterative and thus, takes more computer time to fit the data than for a power series formula. Given a wavelength, an accuracy better than $\Delta n/n = 10^{-5}$ is always achieved into a large spectral range. The Sellmeier-3 formula has a supplementary (K, L) couple if necessary.

Whatever the slope of the normal unit vector $\mathbf{n}$ at the intersection point of the surface boundary, if $\mathbf{r}$ and $\mathbf{r}'$ are the unit vectors along the incident and refracted rays, a vector product representation of Snell's law is

$$n'\,\mathbf{r}' \times \mathbf{n} = n\;\mathbf{r} \times \mathbf{n}\,. \tag{1.10}$$

This relation ensures the local coplanarity between the two rays and the normal. For analytic ray tracing, one generally takes the convention with rays in the first medium coming from left to right towards a positive z axis. An additional sign convention is to consider that incident and refracted rays have their z direction cosines both represented by the same sign, whatever the sign of $x,\ y$ direction cosines. Using a convenient convention, one can also treat the case of a reflected ray with (1.10) by replacing n' by $-n$.

1.3 Fermat's Principle

In 1657, in a letter to Cureau [73], Pierre de Fermat stated that the law of refraction might be deduced from a minimum principle, and this similarly as the remark by Heron of Alexandria (100 AD) for the reflection on a plane. In 1662, considering optical paths between two points in different media, he proved that a ray follows the optical path which takes the least time since this also entails the direct recovering of Snell's law.

In all generality, *Fermat's principle* of *least time* may be enounced as: the time delay for the light to travel from a point A to a point B, through a medium of refractive index $n(x, y, z)$,

$$\Delta t = \int_{\mathrm{A}}^{\mathrm{B}} \frac{ds}{v} = \frac{1}{c}\int_{\mathrm{A}}^{\mathrm{B}} n\,ds\,, \tag{1.11}$$

is minimal or stationary. The ray follows a curvilinear line of elementary length ds, and the local velocity of the light v is a function of the coordinates on this trajectory.

The quantity $\ell = \int n\,ds = c\,\Delta t$ is called the *optical path*. If the light propagates into isotropic – i.e. homogeneous – successive media, the rays follow straight lines with successive deviations at surface boundaries. Postulating that Δt is minimum entails that ℓ is the *shortest* optical path. This is known as the second formulation of Fermat's principle. Let us denote I_k the intersection point at the k-th surface boundary so that the only possible optical ray is A...I_k...B, i.e. the least time ray: the intersection points are such as the variation of the optical path if moving them onto their associated surface boundary must be zero. Following Welford [167] in considering generalized curvilinear coordinates drawn onto the boundary surfaces, and denoting these transversal coordinates p_k and q_k, the stationarity of the optical path means that for the physically possible ray these coordinates must be such that

$$\frac{\partial\,\Delta t}{\partial p_k} = \frac{\partial\,\Delta t}{\partial q_k} = 0 \quad \text{i.e.} \quad \frac{\partial\,\ell}{\partial p_k} = \frac{\partial\,\ell}{\partial q_k} = 0 \quad (k = 1, 2\ldots)\,. \tag{1.12}$$

If A and B are general points of an optical system, for example including a diffraction grating that provides a spectrum for each diffraction order, then the integral of the optical path calculated from (1.12) is not necessarily a minimum but is *stationary*. The term *stationary* implies the notion of returning towards the result of constancy of an integral after zeroing this quantity in a calculus of variations. A detailed treatment on stationarity is given in Born and Wolf [17] including the diffraction theory involving use of electromagnetic equations.

Let us apply Fermat's principle to the simple case of two homogeneous media separated by a plane boundary, $k = 1$, through which the only possible ray – least time ray – passes at point I_1 denoted I (Fig. 1.14). Any ray close to AIB, such as the dashed line ray, would not satisfy Eq. (1.12). Without loss of generality, Snell's law may be derived by setting the incident and refracted rays in the z, y plane of a frame where the x, y plane separates the media of indexes n and n' at its origin. Denoting $(0,0,z_A)$, $(0,y,0)$, and $(0,y_B,z_B)$ the coordinates of points A, I, and B respectively, the optical path is

$$\ell = n\,[\,y^2 + z_A^2]^{1/2} + n'\,[(y_B - y)^2 + z_B^2]^{1/2}. \tag{1.13}$$

The stationarity condition for a virtual displacement of point I along y axis is

$$\frac{\partial\,\ell}{\partial y} = n\,\frac{y}{[\,y^2 + z_A^2]^{1/2}} - n'\,\frac{y_B - y}{[(y_B - y)^2 + z_B^2]^{1/2}} = 0\,. \tag{1.14}$$

The coefficient of n and n' can be identified as the sines of the incidence and refraction angles i and i' which therefore demonstrates Snell's law. Using the time delay Δt, instead of the optical path ℓ, and the similar condition $\partial(\Delta t)/\partial y = 0$, we obtain the other form $(\sin i\,')/v' = (\sin i)/v$, where the light velocities v and v' are respectively c/n and c/n'.

A non-homogeneous medium with a refractive index radially distributed from a pole as $n = n_0/(1 - r^2/a^2)$, where n_0 and a are constants, provides curved rays and leads to an inversion transformation imaging; this has been investigated by Maxwell [105] and is known as Maxwell's *fish-eye*. The determination of the propagation curves in a *gradient index medium*, has recently found several important developments in optical fibers. Given an index distribution, systematic determinations of

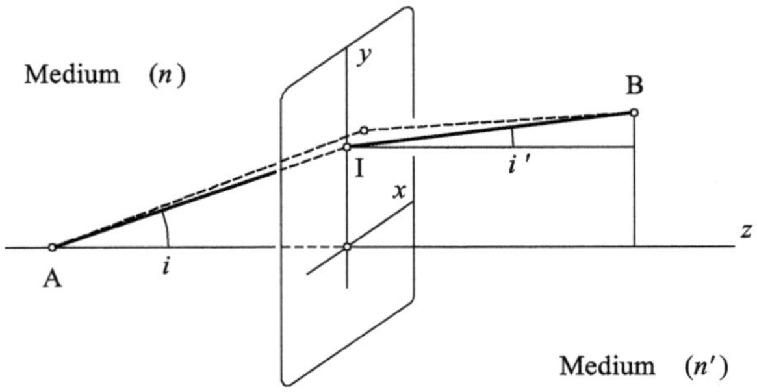

Fig. 1.14 Derivation of Snell's law from Fermat's principle

the optical path lines which satisfy a stationary value of $\int n\, ds$ have been treated by Carathéodory [26].

Calculus of variations is of considerable importance because it allows deriving theorems which have analogue significance in various fields of physics. There is a close analogy between Fermat's principle in geometrical optics and in dynamics with the movement of a particle submitted to a force derived from a potential $U(x, y, z)$. Hamilton demonstrated that the action $\mathcal{A}$ may be expressed by $\int \sqrt{U+a}\, d\ell$ – where a is a constant –, and should be minimal or at least stationary. *Hamilton's principle* of stationary action and his masterful comparison with Fermat's principle became of great value in De Broglie's wave mechanics (1924).

Considering a source point in a homogeneous medium, the emitted rays are all normal to the propagating pencil wavefronts which are concentric spherical surfaces, also said to be *homocentric wavefronts*. The rays are all orthogonal to these surfaces and then form a *normal congruence*. If the rays are now refracted by an optical surface in a second homogeneous medium, then each propagation line remains straight; such a congruence is called a *normal rectilinear congruence*. Malus showed in 1808 [102] that, whatever the surface shape, the refracted or reflected pencil – which generally is no longer homocentric – again forms a normal rectilinear congruence. Dupin (1816), Quetelet (1825) and others generalized this property which is known as *Malus-Dupin theorem*:

→ *In any homogeneous medium rays propagate in normal rectilinear congruence.*

1.4 Gaussian Optics and Conjugate Distances

Considering a curved surface (diopter) separating two optical media of refractive indexes n in the object space and n' in the image space, and an optical axis z taken perpendicular to this diopter with the origin O onto its surface, the *first-order optics theory* based on Snell's law allows determining the axial positions of the focii F and F' and of the conjugate distances of an object and its real or virtual image.

The general form of a local surface is defined by its two principal curvatures. We may set the section plane of these curvatures along x and y directions of the x, y, z frame. The diopter shape is represented by the power series

$$z = \tfrac{1}{2}\, c_x\, x^2 + \tfrac{1}{2}\, c_y\, y^2 + O(x^p, y^q)\,, \tag{1.15}$$

where p and q are integers greater than 2.

The determination of focus positions and conjugate distances in each x, z or y, z plane leads to expressions of similar forms in c_x and c_y (see for instance Sect. 3.5.7). Then, we do not restrict the analysis by considering c_x=c_y and a diopter represented by

$$z = \tfrac{1}{2}\, c\, r^2 + O(x^p, y^q)\,. \tag{1.16}$$

K.F. Gauss [66], in a celebrated memoir of 1841 (see comments by Wilson [171]) and with complete generality, demonstrated that the higher order terms O, do not enter into the determination of the focus positions and conjugate distances. These are obtained from the first-order optics theory or *Gaussian theory*, also called *paraxial theory*. In Gaussian optics, the representation of any surface of an optical system reduces to the quadratic form,

$$z_\ell = z_{0,\ell} + \tfrac{1}{2}\, c_\ell\, r^2, \qquad \ell = 1, 2, 3, \cdots, \tag{1.17}$$

where all the axisymmetric surfaces, individually numbered ℓ, have the same axis. Such a system is usually called a *centered system*.

Hence, considering a refractive or reflective spherical surface, we in fact cannot distinguish it from a paraboloid or another conicoid of the same curvature because its shape only differs by the next order term. The Gaussian theory concerns the analysis with the so-called *paraxial rays*, i.e. rays which are in the immediate neighborhood of the axis. In this theory the sine expansion of Snell's law, which relates the incident angle i in medium n to the conjugate emerging angle i' in medium n',

$$n'\left(i' - \tfrac{1}{3!}\, i'^3 + \cdots\right) = n\left(i - \tfrac{1}{3!}\, i^3 + \cdots\right), \tag{1.18}$$

is accurately approximated by

$$n' i' = n\, i\,. \tag{1.19}$$

• **Sign convention**: Because of the small angles considered, a paraxial drawing may have difficulty showing simultaneously the focal points and the curvature of a diopter. This curvature is schematically represented by a bracket with ends turned towards the center of curvature. We use the Cartesian sign convention: a positive curvature corresponds to a surface whose concavity is turned towards z positive.

1.4.1 Diopter of Curvature $\mathbf{c = 1/R}$

Let us consider a surface – a diopter – which separates two media of refractive indexes n and n' (Fig. 1.15), and set the origin of the z-axis at its vertex. The abscissa $f = \mathrm{OF}$ of the object focus F in the object space is defined as the distance from the surface vertex O to the convergence point F of rays issued from a source point at infinity in the image space.

Conversely, the abscissa $f' = \mathrm{OF}'$ of the image focus F′ in the image space, is defined from a source point at infinity in the object space. These abscissa represent the *focal lengths*

$$f' = \frac{n'}{n'-n}\, R\,, \qquad f = -\frac{n}{n'-n}\, R\,, \tag{1.20a}$$

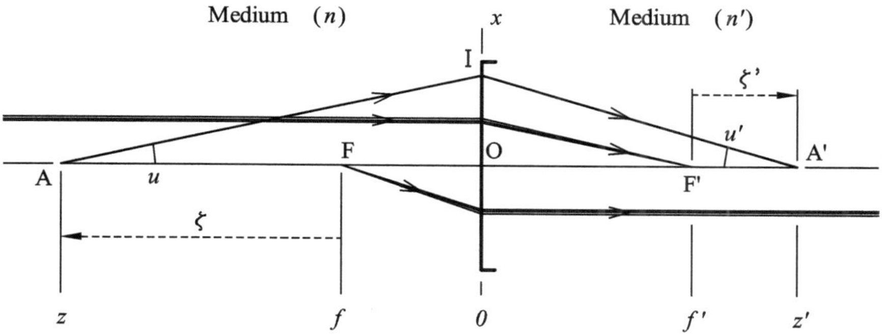

Fig. 1.15 Paraxial focii – also called Gaussian focii – and conjugate distances. Bracket ends towards z show a diopter with curvature $c = 1/R > 0$

from which we deduce

$$\frac{n'}{f'} = -\frac{n}{f} = \frac{n'-n}{R}, \qquad \text{and} \qquad f' + f = R. \tag{1.20b}$$

The quantity

$$K = n'/f' \tag{1.21}$$

is called the *optical power* of the surface or of the optical system. Also in the image space, its reciprocal f'/n' is the *effective focal length* (efl) of the system.

The conjugate distances $z = \mathrm{OA}$ and $z' = \mathrm{OA}'$ of an axial object A and its image A′ are obtained by considering a four-angle equation set in u, i, i', u', expressing the local geometrical transformation at intersection point I. The quantities u, u' are the *aperture angles* of a ray at the object and image points, sometimes called *convergence angles*. Solving this system leads to

$$\frac{n'-n}{R} = \frac{n}{z'-R} - \frac{n'}{z-R}, \tag{1.22a}$$

which may be written

$$\frac{1}{R} = \frac{n'}{n'-n}\,\frac{1}{z'} - \frac{n}{n'-n}\,\frac{1}{z}. \tag{1.22b}$$

Using Eq. (1.20), we obtain the vertex conjugate distance equation

$$\frac{f'}{z'} + \frac{f}{z} = 1. \tag{1.23a}$$

Let us denote $\zeta = \mathrm{FA}$ and $\zeta' = \mathrm{F'A'}$ the distances of the object and its image from their respective focus. Then $z = \zeta + f$ and $z' = \zeta' + f'$. After substitution in the above equation, we obtain the Newton conjugate distance equation

$$\zeta'\,\zeta = f'\,f. \tag{1.23b}$$

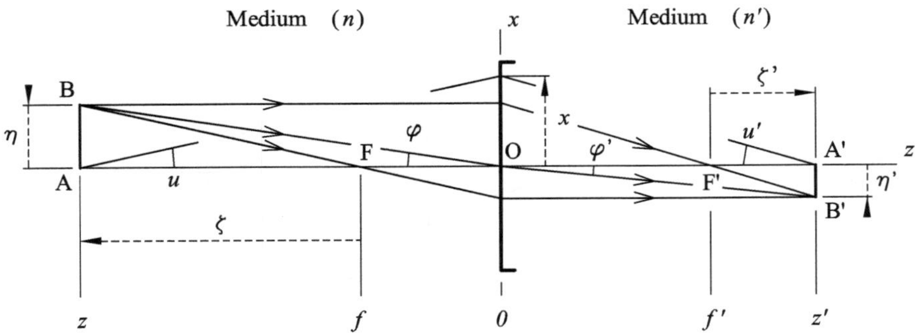

Fig. 1.16 Conjugates and transverse magnification

The conjugate distances from the vertex diopter and those from the location of the focii are related by

$$\frac{\zeta'}{z'} + \frac{\zeta}{z} = 1\,. \tag{1.23c}$$

Another important optics quantity is the *transverse magnification M*. Considering an object AB and its image A′B′, let $\eta = \text{AB}$ and $\eta' = \text{A}'\text{B}'$ be their heights in the x direction (Fig. 1.16). In Gaussian approximation, the transverse magnification can be defined either by the ratio of the *conjugate heights* or by the ratio of the *aperture angles* u, u', as

$$M = \frac{\eta'}{\eta} = \frac{u}{u'}\,, \tag{1.24}$$

which is negative if the image is reverted.

Let φ and φ' be the *field angles*, i.e. the angles of these segments seen from the origin O. We have $\eta = z\,\varphi$ and $\eta' = z'\varphi'$. Since $n\,\varphi = n'\varphi'$ in the Snell's law approximation, the transverse magnification is

$$M = \frac{\eta'}{\eta} = \frac{n}{n'}\frac{z'}{z}\,. \tag{1.25a}$$

Considering homothetic triangles in the object and image spaces, we may also represent the transverse magnification with Newton's conjugates respectively as

$$M = -f/\zeta = -\zeta'/f'. \tag{1.25b}$$

1.4.2 Mirror in Medium n

Setting the refractive index of the image space to $n' = -n$ for a mirror in medium n, from (1.21) the power is

$$K = -n/f' = -n/f = -2n/R\,. \tag{1.26a}$$

The focal lengths, conjugate distance equations and magnification are

$$\frac{1}{f'} = \frac{1}{f} = \frac{2}{R}, \qquad \frac{1}{z'} + \frac{1}{z} = \frac{2}{R}, \qquad \zeta'\zeta = \frac{R^2}{4}, \qquad M = -\frac{z'}{z}, \tag{1.26b}$$

whatever the n index of the medium.

For a mirror in a refractive medium equal to unity, $n = 1 = -n'$, the power is

$$K = -1/f' = -1/f = -2/R\,. \tag{1.26c}$$

In the sign convention of the incident light propagating towards $z > 0$, a concave mirror is with f', f and R all negative and its power K is positive.

With centered systems that are composed of several axially separated diopters, the Gaussian theory provides a powerful method for determining the efl f' of the global system. The concept introduces the notion of object and image principal points and *principal planes*. For a single diopter these planes are both at its vertex. For a global system, the image-side principal plane goes through the intersection point of a ray parallel to the axis in the space object with its conjugate passing through the image focus. Thus, the efl is the distance from the image principal plane to the image focus; this is the rigorous definition of the focal length of an optical system. Its efl calculation uses abscissa transformations of the axial locations of the successive principal planes.

This process provides the paraxial properties of thick lenses and of axially separated surfaces.

1.4.3 Power of Combined Systems

One shows that the resulting power of two systems of respective power K_1, K_2 axially separated by a distance d is

$$K = K_1 + K_2 - \frac{d}{n^*} K_1 K_2\,, \tag{1.27}$$

where n^* is the refractive index of the medium that separate these systems.

1.4.4 Lens in Air or in Vacuum

Let $c_1 = 1/R_1$, $c_2 = 1/R_2$ and d be the curvatures and axial thickness of a lens with refractive index n. From (1.20a) to (1.21) we obtain $K_1 = (n-1)/R_1$ and $K_2 = -(n-1)/R_2$, hence by substitution in the above equation, the power of a lens in air is

$$K = \frac{1}{f'} = -\frac{1}{f} = (n-1)\left(\frac{1}{R_1} - \frac{1}{R_2} + \frac{n-1}{n}\,\frac{d}{R_1 R_2}\right). \tag{1.28a}$$

However, the refractive index of a vacuum is approximatively $n_a = 1.0003$ under normal conditions of pressure and temperature. But it is a universal practice in optical design to specify the refractive indices of glasses with respect to air, therefore no correction of its refractive index is required for the above power of a lens working in air.

Hence, if a lens is designed for working under vacuum, then, in (1.28a), the quantity $(n + n_a - 1)$ must be substituted to $(n - 1)$. This entails

$$K = (n+n_a-1)\left(\frac{1}{R_1} - \frac{1}{R_2} + \frac{n+n_a-1}{n+n_a}\,\frac{d}{R_1 R_2}\right) \tag{1.28b}$$

for the *power* of a *lens in vacuum*.

1.4.5 Afocal Systems

With at least two optical surfaces we may obtain *afocal systems* by setting the resulting power to zero. For instance this is achieved by a thick lens in air if, from (1.28a), its thickness is $d = n(R_1 - R_2)/(n - 1)$. Other basic systems are the four Mersenne two-mirror designs with confocal paraboloids in *L'Harmonie Universelle*, 1636 ([32], see also Sect. 2.3) and the two-lens designs by Galileo of the Cassegrain form and by Kepler of the Gregory form (Fig. 1.17). Without knowing *anastigmatism* (see Sect. 1.9), Mersenne invented such afocal systems, and with its Cassegrain form of an afocal system, he also invented the basis of the first

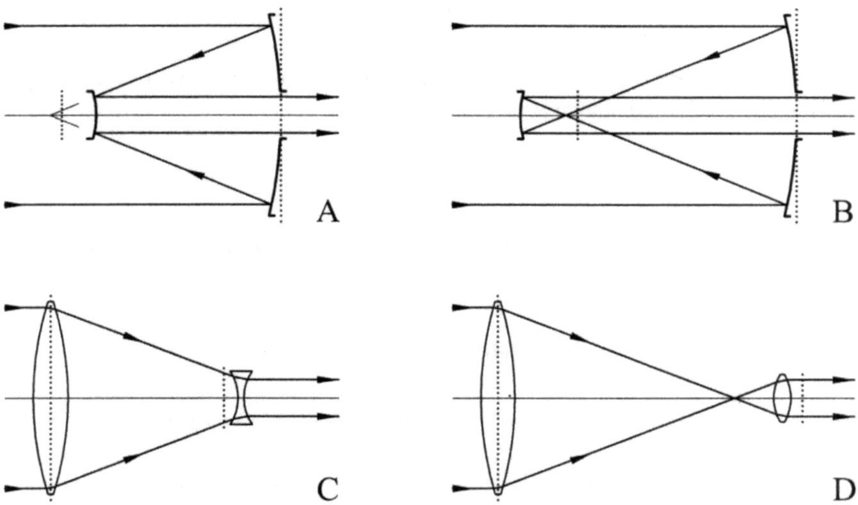

Fig. 1.17 Afocal systems. Two of the four reflective two-mirror systems by Mersenne: (**A**) Cassegrain forms, (**B**) Gregory form. Refractive two-lens designs: (**C**) by Galileo, (**D**) by Kepler. The entrance and exit pupils are in *dotted lines* (adapted from King [85] and Wilson [170])

mirror *telephoto effect* which provides a long focal length by a compact design (see Wilson [170]).

Afocal systems can be characterized by a *compression ratio* or an *expansion ratio* which allows defining the relative change in aperture diameter of the conjugate beams (cf. Sect. 2.4).

1.4.6 Pupils and Principal Rays

It is useful to define an aperture area where all the field beams enter into an optical system; this is called the *entrance pupil*, or *input pupil*. In the next medium we obtain an image pupil. At the system output medium we obtain the *exit pupil*, or *output pupil*, which is the system conjugate of the entrance pupil. An *aperture stop* physically defines the beam area in which all beams are allowed to pass, then defining all the pupils of the system. A *field stop* physically defines the linear size of the field of view. In astronomy, we generally define the entrance pupil – or aperture stop – of a telescope as the first element, the primary mirror.

Given a field angle, the ray passing through the center of the entrance pupil is called the *principal ray* or *chief ray*. The principal ray also passes through the center of the successive pupils. Rays which pass at the edge of the pupil are called *marginal rays*.

If a system is *telecentric*, then the principal rays in the image space are parallel to the axis and the image focal surface is a plane. This is achieved by setting the entrance pupil at the object focal plane thus providing an exit pupil at infinity.

If a system is *homocentric*, then the principal rays in the image space are converging towards a common center and the image focal surface is curved and normal to this central direction. This is achieved by monocentric systems whose entrance and exit pupils are both located at the common center. Then in the image space, the principal rays are all normal to the focal surface.

1.4.7 Aperture Ratio or Focal Ratio

When a system is designed to work with objects at infinity or at large distance, a convenient measure of the light-gathering power is the so-called *aperture ratio* or *focal ratio* or f-*ratio*.

The f-ratio Ω is a *positive* number defined in the image space from an axial point at infinity in the object space. Denoting D the diameter of the parallel beam in the object space and efl the effective focal length – assumed positive – in the image space, $\Omega = \text{efl/D}$. In the Gaussian approximation, the ratio Ω is two times the absolute value of the maximum aperture angle u'_{max},

$$\Omega = \text{efl}\,/\,\text{D} = 2\,|u'_{max}|\,. \tag{1.29}$$

A system with a focal ratio $\Omega = 3$ is usually denoted f/3.

1.5 Lagrange Invariant

A fundamental law of Gaussian optics can be derived from the properties of A, B points and their conjugates A′, B′ (Fig. 1.16). This law which links the transverse magnification $M = \eta'/\eta$ to the aperture angle ratio u/u', is the Lagrange invariant also known as Smith, Lagrange, or Helmholtz relationship.

Let us consider the case of a single diopter, Eq. (1.25a) may be written

$$n\,\frac{1}{z}\,\eta = n'\,\frac{1}{z'}\,\eta' \,. \tag{1.30}$$

Now from the object point A, the ray of aperture angle u meets the x, y plane at the height $x = u\,z$ and goes through the image point A′ from the same height than in object space, then

$$u\,z = u'\,z' \,, \tag{1.31}$$

whatever the case of refractive or reflective diopter for the signs.

Multiplying the above equations together, we obtain

$$n\,u\,\eta = n'\,u'\,\eta' \,. \tag{1.32}$$

Considering a system with a next diopter with next medium of index n'', we could obtain in the same manner $n''\,u''\,\eta'' = n'\,u'\,\eta'$ and so on. The result applies to any intermediate space and to the last space – the final image space – of an optical system, hence

$$n\,u\,\eta = H\text{-invariant} \,, \tag{1.33a}$$

and is called the *Lagrange invariant.*

With afocal systems, the aperture angles are $u = u' = 0$, the object and image locations are at infinity which leads to $\eta \rightarrow \pm\infty$ and $\eta' \rightarrow \mp\infty$, so the above formulation is indeterminate.

In the object space, the ray height at the optical surface is $x = u\,z$, and the field height is $\eta = z\,\varphi$ where φ is the field angle; the substitution in (1.33a) of $u\,\eta$ by $x\,\varphi$ gives

$$n\,x\,\varphi = H\text{-invariant} \,, \tag{1.33b}$$

which is the *Lagrange invariant afocal form.*

This invariance property entails the following consequences for focal and afocal systems having both ends in a same medium ($n' = n$) as in air.

• **Focal ratios and linear fields of view**: Consider focal reducer systems currently used in astronomy, which are instruments working at finite distances with a magnification $M = \eta'/\eta$ such that $M^2 < 1$.

→ *If the output f-ratio is* $1/M$ *times faster than the input f-ratio, then the linear size of the output field of view is* M *times smaller than that of the input field.*

• **Afocal beams and angular fields of view**: Consider an afocal system, such as a Mersenne telescope or a laser beam expander, we may define a number $K = |x'/x|$ expressing the diameter transformation of the conjugate beams.

→ *If the diameter of the output beam is decreased by a factor* $1/K$*, then the field angle of the output beam is increased by a factor* K.

Typical examples of the Lagrange invariant consequence in astronomy are: the increased detection of extended objects by Courtès (1952) with focal reducers, and for the afocal case the well known Galilean form of refracting telescope (Fig. 1.17).

1.6 Étendue Invariant and Lagrange Invariant

1.6.1 Lagrange Invariant

In general, an optical surface introduces aberrations at the focused image which deteriorate its sharpness. At first, this alteration of image quality is due to primary aberrations (cf. Sect. 1.8) and their analytic calculation require using the Lagrange invariant. Summarizing the results in the previous section, the two forms of the *Lagrange invariant* H are

$$H = n\,u\,\eta \quad \equiv n\,x\,\varphi = \text{Lagrange invariant}\,. \qquad (1.34)$$

In ray tracing analysis some other authors, such as for instance Welford [167] and Wilson [170], define the Lagrange invariant with the opposite sign.

1.6.2 Étendue Invariant

Let us consider a source of light – or focal plane – whose area $\bar{\mathcal{A}}$ uniformly emits radiations of intensity I_ν within a frequency band $\delta\nu$. Assuming that the emitting surface is centered and perpendicular to the axis of an optical system, the total amount of energy per second W reaching the entrance pupil of area $\mathcal{A}$, located at distance d from the source, is represented by the *inverse square law*

$$W = \frac{\mathcal{A}\,\bar{\mathcal{A}}}{d^2}\;I_\nu\,\delta\nu\,. \qquad (1.35a)$$

This relationship is well known in spectroscopy and photometry (Pecker and Schatzman [120], Sterken and Manfroid [149]). The solid angle Ω of the field of view is $\Omega = \bar{\mathcal{A}}/d^2$ (Fig. 1.18), so the energy is represented by

$$W = \mathcal{A}\;\Omega\;I_\nu\,\delta\nu\,. \qquad (1.35b)$$

In the expression of the energy W, the product $\mathcal{A}\,\Omega$ of the geometrical terms is called the *optical Étendue* after Jacquinot [81] who formalized this quantity as an invariant through a perfectly transmissive optical system. The optical Étendue is generally denoted

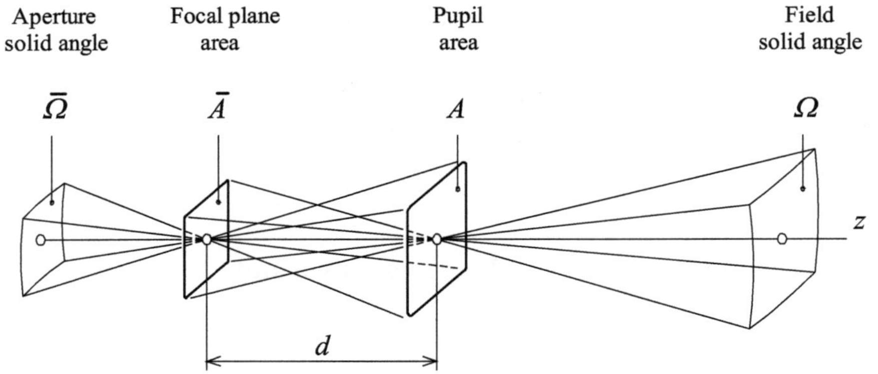

Fig. 1.18 The *optical Étendue* $\mathcal{A}\,\Omega \equiv \bar{\mathcal{A}}\,\bar{\Omega}$ is invariant through a perfect system

$$E = \mathcal{A}\ \Omega = \bar{\mathcal{A}}\ \bar{\Omega} = \text{Étendue invariant}\,, \tag{1.36a}$$

where $\bar{\mathcal{A}}$ and $\bar{\Omega}$ are the focal plane area and the full aperture solid angle at the system input. Therefore, these products may be substituted by the products $\mathcal{A}'\,\Omega' = \bar{\mathcal{A}}'\,\bar{\Omega}'$ $(=\mathcal{A}\,\Omega)$ for the system output.

1.6.3 Equivalence of the Étendue and Lagrange Invariants

Assuming an optical system with both ends in the same medium ($n' = n$) such as in air, an *equivalence* between the Lagrange invariant and the Étendue can be derived from the above relations.

If the entrance pupil is a square aperture, then $\mathcal{A} = 4x^2$, and if the field is a square solid angle, $\Omega = 4\,\varphi^2_{max}$, then we obtain from (1.36a) and (1.34) the equivalence

$$E = 16\,x^2\varphi^2_{max} \ \equiv\ 16\,H^2. \tag{1.36b}$$

If the entrance pupil and field of view are circular, then $\mathcal{A} = \pi x^2$ and $\Omega = 2\pi(1{-}\cos\varphi_{max})$. If the field angle is small then $\Omega = \pi\,\varphi^2_{max}$ and we obtain similarly the equivalence

$$E = \pi^2 x^2 \varphi^2_{max} \ \equiv\ \pi^2\,H^2. \tag{1.36c}$$

S. Carnot (1796–1832), in posthumous papers, noticed that (Bruhat [22]): "in a general thesis, the driving power is an invariable quantity in nature which strictly speaking is neither produced nor destroyed." This was brought to attention by Clapeyron. In 1845, J.R. Mayer gave the first formulation of the general law on energy conservation during a transfer process. These notions were given more accurate mathematical forms by Helmholtz (1847), Clausius (1850), who later introduced the term "entropy," and Lord Kelvin (1853).

Hence, one of the most fundamental theorems derived from Gaussian optics and applying to any optical system, may be stated as:

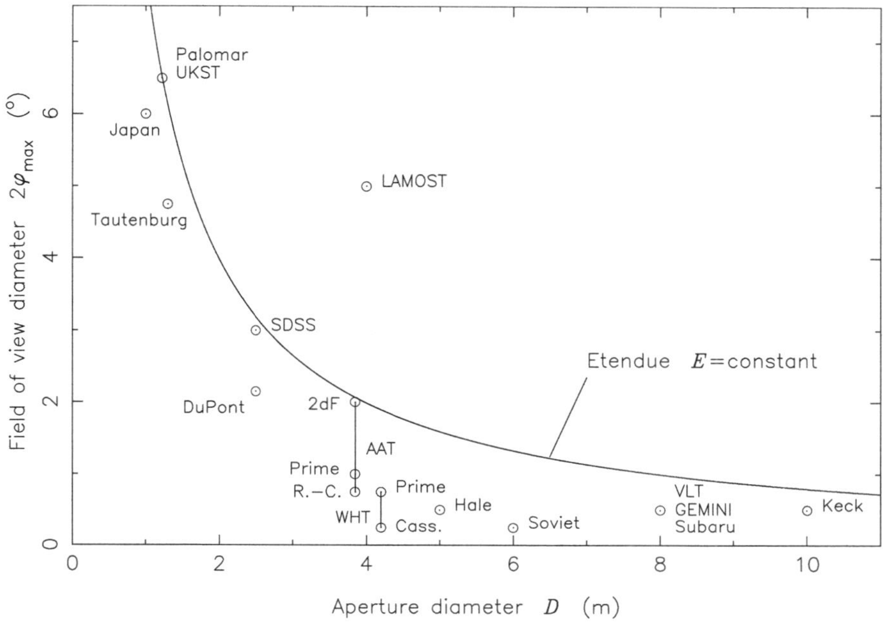

Fig. 1.19 *Optical Étendue* $E = (\pi^2/4)\, D^2\, \varphi_{max}^2$ of various telescopes equipped with multi-object spectroscopic facilities. Compared to the displayed curve, the gain in Étendue is greater by factor 6 with LAMOST

→ *The Étendue invariant E is related to the thermodynamics Carnot's law which expresses that, in a closed system, entropy does not decrease so the total energy transferred through a perfect optical system is conserved.*

→ *The Lagrange invariant H is a 1-dimensional representation of the optical Étendue.*

Denoting $D = 2x$ the input pupil diameter of a telescope, from (1.36c) and for a circular semi-field φ_{max}, the Étendue is

$$E = \mathcal{A}\,\Omega = \frac{\pi^2}{4}\, D^2\, \varphi_{max}^2. \tag{1.36d}$$

For instance, this relation allows comparing the Étendue of various telescopes (Fig. 1.19).

1.7 Analytical Representation of Optical Surfaces

Gaussian optics only consider the curvature term of an optical surface and a tilt term for prisms or mirrors; these terms constitute the two fundamental surfaces of the 1st-order theory. The correction of the optical aberrations (see next sections) requires taking into account the next order terms. Investigating the case of stigmatic singlet lenses, Descartes laid down the first analytic theory leading to ovals which are non-spherical surfaces. In the present terminology, non-spherical surfaces are called *aspherical surfaces* or more briefly *aspherics*.

Representing an optical surface in a cylindrical coordinate frame z, r, θ, we may distinguish between axisymmetric aspherical surfaces for centered systems and non-axisymmetric aspherics for non-centered systems. Their representation is usually given with respect to the tangent plane x, y at the vertex, hence $z\{r=0,\ \theta\} = 0$.

1.7.1 Conicoids

The *conicoids* are rotational symmetry aspherics whose meridian sections belong to the conic family. These shapes are represented by

$$(1+\kappa)\,z^2 \;-\; 2Rz \;+\; r^2 = 0\,, \tag{1.37}$$

where $R = 1/c$ is the radius of curvature and κ the *conic constant*, also called the *Schwarzschild constant*, which characterizes the asphericity since $\kappa = 0$ generates a sphere. This equation may be written

$$z = \frac{c\,r^2}{1+\sqrt{1-(1+\kappa)\,c^2 r^2}}\,, \tag{1.38a}$$

whose first terms of the expansion are

$$z = \frac{1}{2\,R}\,r^2 + \frac{1+\kappa}{8\,R^3}\,r^4 + \frac{(1+\kappa)^2}{16\,R^5}\,r^6 + \frac{5(1+\kappa)^3}{128\,R^7}\,r^8 + \cdots\,. \tag{1.38b}$$

We obtain for the general expansion of a conicoid

$$z = \sum_{n=1,2,3,\ldots} \frac{(2n-2)!\;n}{2^{2n-1}\,(n!)^2}\;\frac{(1+\kappa)^{n-1}}{R^{2n-1}}\;r^{2n}\,, \tag{1.38c}$$

where the indeterminate form $(1+\kappa)^{n-1} \to 1$ when $n \to 1$ and $\kappa \to -1$.
The conicoids are classed into families from the value of their conic constant κ (see below and Fig. 1.20):

$\kappa < -1$	hyperboloid
$\kappa = -1$	paraboloid
$-1 < \kappa < 0$	elongated ellipsoid or prolate ellipsoid
$\kappa = 0$	sphere
$\kappa > 0$	flattened ellipsoid or oblate ellipsoid

One may use the *eccentricity* $e = \sqrt{-\kappa}$ which is imaginary for a flattened ellipsoid.

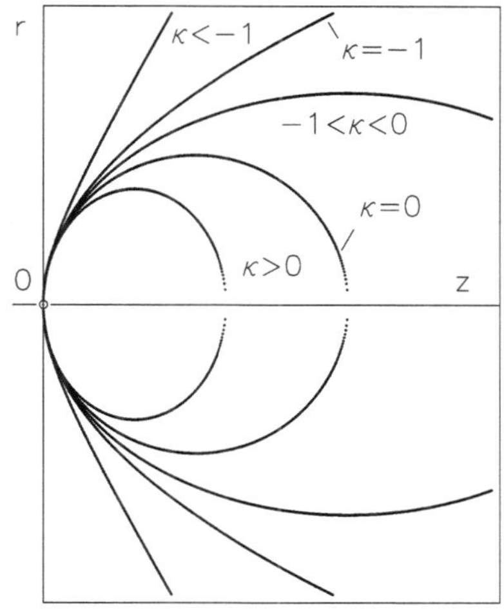

Fig. 1.20 Conicoid sections of same curvature

A particular case is a *biconicoid surface*; the conic sections differ in the x- and y-directions. Denoting $c_x, c_y, \kappa_x, \kappa_y$ the curvatures and conic constants, this surface is represented by

$$z = \frac{c_x\, x^2 + c_y\, y^2}{1 + \sqrt{1 - (1+\kappa_x)\, c_x^2\, x^2 - (1+\kappa_y)\, c_y^2\, y^2}}\,. \tag{1.38d}$$

1.7.2 Spheroids

Axisymmetric surfaces generally differ by the third term from the (1.38b) expansion. In this case the shape is called a *spheroid* and is represented by the power series

$$z = \sum_{n=2,4,6,\ldots} A_n\, r^n\,, \tag{1.39}$$

where if having $A_2 = 1/2R$ and $A_4 = (1+\kappa)/8R^3$, then $A_6/A_4 \neq (1+\kappa)/2R^2$.

If $A_6/A_4 = (1+\kappa)/2R^2$ and if the next term differs from that of a conicoid, then the spheroid may be called a *deformed conicoid* whose shape departs from a conicoid by the 7-th order term (i.e. the $A_8 r^8$ term).

1.7.3 Non-Axisymmetric Surfaces and Zernike Polynomials

Non-axisymmetric optical surfaces are used for non-centered systems. Such surfaces may also represent a wavefront shape including some of the optical aberrations. Taking the origin of the azimuth angle θ in the z, x plane by setting $x = r\cos\theta,\ y = r\sin\theta$, the general shape may be represented by

$$Z = \sum z_{n,m} = \sum_{n,m} (A_{n,m} r^n \cos m\theta \; + \; B_{n,m} r^n \sin m\theta)\,, \qquad (1.40a)$$

$$\text{with} \quad n, m \ \text{positive integers}, \quad n+m \ \text{even}, \quad m \leq n$$

and where $A_{n,m}$, $B_{n,m}$ are coefficients. A particular class of non-centered systems presents a symmetry plane: setting this symmetry in the z, x plane allows studying optical systems with $B_{n,m} = 0$. The surface $Z(r, \theta)$ is generated by successive *optical surface modes* $z_{n,m}$ as

$$\begin{aligned} Z = & A_{00} + A_{20}\, r^2 + A_{40}\, r^4 + A_{60}\, r^6 + \cdots \\ & + A_{11}\, r \cos\theta + A_{31}\, r^3 \cos\theta + A_{51}\, r^5 \cos\theta + \cdots \\ & + A_{22}\, r^2 \cos 2\theta + A_{42}\, r^4 \cos 2\theta + \cdots \\ & + A_{33}\, r^3 \cos 3\theta + \cdots \\ & + \cdots \end{aligned} \qquad (1.40b)$$

where the coefficients $A_{n,m}$ are denoted A_{nm} for simplicity. The general representation of a surface – having a continuous tangent plane – is obtained by co-adding the $B_{nm}\, r^n \sin m\theta$ terms.

In the polynomial representation introduced by Zernike [178, 179] of an optical surface, or a wavefront, each polynomial has certain simple properties of invariance. Let us consider a dimensionless aperture radius $\rho = r/r_{\max}$, and represent a particular surface by

$$Z = \sum R_{n,0}\{\rho\} + \sum R_{n,m}\{\rho\} \cos m\theta + \sum R_{n,m}\{\rho\} \cos m\theta, \qquad (1.40c)$$

where the dimensioning coefficients in front of each term $R_{n,m}\{\rho\}$ have been assumed equal to unit length (they do not appear here). The terms of this expansion are called *Zernike circle polynomials*. The normalization and determination of the radial components $R_{n,m}\{\rho\}$ are such as (cf. Born and Wolf [17])

$$R^2_{n,m}\{\rho\} \leq 1 \quad \forall \rho \in [0,\, 1]\,,$$

and, with $m \leq n$,

$$R_{n,m}\{\rho\} = \sum_{s=0}^{(n-m)/2} \frac{(-1)^s\, (n-s)!}{s!\left[\frac{1}{2}(n+m) - s\right]!\left[\frac{1}{2}(n-m) - s\right]!}\, \rho^{n-2s}. \qquad (1.40d)$$

A c_{nm} dimensioned representation of a surface with the first Zernike circle polynomials is

$$\begin{aligned} Z = & c_{00} + c_{20}(2\rho^2 - 1) + c_{40}(6\rho^4 - 6\rho^2 + 1) + c_{60}(20\rho^6 - 30\rho^4 + 12\rho^2 - 1) + \cdots \\ & + c_{11}\rho \cos\theta + c_{31}(3\rho^2 - 2)\rho \cos\theta + c_{51}(10\rho^4 - 12\rho^2 + 3)\rho \cos\theta + \cdots \\ & + c_{22}\rho^2 \cos 2\theta + c_{42}(4\rho^2 - 3)\rho^2 \cos 2\theta + \cdots \\ & + c_{33}\rho^3 \cos 3\theta + \cdots \\ & + \cdots \end{aligned} \qquad (1.40e)$$

where a similar series including coefficients c'_{nm} for the sinus terms must be added.

1.8 Seidel Representation of Third-Order Aberrations

Geometrical aberrations introduced by centered optical systems have been investigated for improving the performance of lenses used in photography, an invention discovered by Nicéphore Niepce in 1816 and improved by J. Daguerre in 1839. These lenses required large apertures and large fields of view, and thus needed a better correction of on- and off-axis aberrations. The early investigations on optical aberrations were by ray tracing, but J. Petzval obtained considerable results by an analytical study (Sect. 1.1). Although his manuscript was destroyed in a fire by thieves, he demonstrated the importance of his calculations by constructing his four-lens objective in 1840 well-known as "Petzval portrait lens" [17, 29].

1.8.1 The Seidel Theory

Considering centered systems, the first analysis on geometrical aberrations was published by W. Rowan Hamilton, in 1833 [71] (cf. Wilson [170]), where he introduced its *Characteristic function*. He derived the general form of the primary aberrations by using three fundamental parameters: the aperture height of the ray, its azimuth angle, and the field height. L. Seidel published a famous analysis on opical aberrations in 1856 [144]. His theory provides a powerful evaluation of the amount of primary aberration types introduced by each surface of an optical system, which then allows one to calculate their sums for the whole system. In 1895, an analytical study on aberrations led Bruns to consider a function related to the aberration function that he called *eikonal function*. Historical notes on this work are in the book by Herzberger [75]. Schwarzschild introduced the *Schwarzschild eikonal* in 1905, which is closely related to its perturbation function in the motion of planets, and called by him the *Seidel eikonal* (cf. Born and Wolf [17]).

In general the rays from a point source in the object field and emerging from a centered system do not lean against a spherical wavefront. The general shape of the aberrated wavefront or *aberration function* can be defined from three variables which characterize all the aberration terms. We follow Welford ([167] Sect. 8.2) and denote these dimensionless variables as:

ρ: aperture radius at exit pupil plane x, y, normalized from 0 to 1,
θ: azimuth angle at exit pupil plane, polar coordinate with ρ,
$\bar{\eta}$: image height at image focal plane $\mathcal{X}$, $\mathcal{Y}$, normalized from 0 to 1,

where $x = \rho\cos\theta$, $y = \rho\sin\theta$, and the normalized image height $\bar{\eta} = \eta'/\eta'_{\text{max}}$ is here taken in the $\mathcal{X}$-direction which is parallel to the x-axis. In the Gaussian approximation, the normalized field angle $\bar{\varphi} = \varphi'/\varphi'_{\text{max}}$ is equivalent to the normalized image height ($\bar{\varphi} \equiv \bar{\eta}$); to simplify the writing, the prime on $\bar{\eta}$ and $\bar{\varphi}$ have been omitted.

With respect to the Gaussian reference sphere, the wavefront only including *primary aberrations* is represented at the exit pupil level by the *aberration wavefront function*

$$W_{[4]}(\rho,\theta,\bar{\eta}) = \tfrac{1}{8}\,S_I\,\rho^4 + \tfrac{1}{2}\,S_{II}\,\bar{\eta}\,\rho^3\cos\theta + \tfrac{1}{4}\,S_{III}\,\bar{\eta}^2\rho^2\cos 2\theta + \tfrac{1}{4}\,(2S_{III}+S_{IV})\,\bar{\eta}^2\rho^2 + \tfrac{1}{2}\,S_V\,\bar{\eta}^3\rho\cos\theta\,, \tag{1.41}$$

where S_I to S_V are the five *3rd-order Seidel coefficients* which have the dimension of length and represent respectively the amount of primary *spherical aberration, coma, astigmatism, field curvature* and *distortion*. These terms are called *third-order aberrations*. To distinguish them from those of the same family, which could also include higher-order coefficients, we will refer to them as

$$Sphe\,3\,,\quad Coma\,3,\quad Astm\,3\,,\quad Petz\,3,\quad \text{and}\quad Dist\,3\,. \tag{1.42}$$

Explicit relations have been derived to calculate the transverse image size of all third-order aberrations for a single surface, thus providing each of its Seidel coefficients S_i as functions of $\rho,\ \theta,\ \bar{\eta}$ parameters. These relations take also into account the case of an aspheric surface. Since Seidel, many authors have re-formulated the explicit expressions for S_i coefficients. One may refer, for instance, to the books by Conrady [36], Chrétien [29], Kingslake [86], Welford ([167], Appendix A), and Wilson [170] who explicitly gives the Seidel sums for two-mirror telescopes. Of universal practice in raytracing analysis, each of the S_i coefficients of a surface efficiently allows one to calculate, by simple co-addition, the Seidel sum $\sum S_i$ of a whole system.

Given the field height of the source point and its associate principal ray which is normal to the wavefront $W_{[4]}$, (1.41) allows deriving the lateral shift at the local image plane for any aperture ray. The two components of this shift, which define the amount of *transverse aberration* (or *lateral aberration*), are

$$\begin{cases} \Delta\mathcal{X} = -\tfrac{1}{2}\,[\,S_I\,\rho^3\sin\theta + S_{II}\,\bar{\eta}\,\rho^2\sin 2\theta + (S_{III}+S_{IV})\,\bar{\eta}^2\rho\,\sin\theta\,]\,,\\ \Delta\mathcal{Y} = -\tfrac{1}{2}\,[\,S_I\,\rho^3\cos\theta + S_{II}\,\bar{\eta}\,\rho^2(2+\cos 2\theta)\\ \qquad\quad + (3S_{III}+S_{IV})\,\bar{\eta}^2\rho\,\cos\theta + S_V\,\bar{\eta}^3\,]\,. \end{cases} \tag{1.43}$$

This allows one to draw the geometrical shape of the aberrated image for *Sphe* 3, *Coma* 3, *Astm* 3, and its global radial shift *Dist* 3. Given an aberration term, its image pattern may be traced for all rays emerging at a fixed height $\bar{\eta}$ = constant and then varying ρ and θ.

Several important properties may be immediately derived from the aberration function (1.41).

If *Sphe* 3, *Coma* 3, and *Astm* 3 vanish, i.e. $S_I = S_{II} = S_{III} = 0$, we see that the value of S_{IV} is a measure of the field curvature; this curvature is also called the *Petzval curvature*, after Petzval who discovered, in 1840, that the *Petzval condition* $S_{IV} = 0$ is a necessary condition for obtaining a flat image field.

If *Astm* 3 is not corrected, $S_{III} \neq 0$, we may regroup the third and fourth terms of (1.41) in the form

$$\tfrac{1}{4}\,(\,3S_{III}+S_{IV})\,\bar{\eta}^2\,x^2\;+\;\tfrac{1}{4}\,(\,S_{III}+S_{IV})\,\bar{\eta}^2\,y^2, \tag{1.44}$$

which, in the image space, are respectively the x, z and y, z sections of the *tangential focal surface* and *sagittal focal surface* of curvature $1/R_t$ and $1/R_s$. Denoting $1/R_P$ the curvature of the *Petzval focal surface*, from (1.44), these surfaces are related by

$$\frac{3}{R_s} - \frac{1}{R_t} = \frac{2}{R_P}\,. \tag{1.45}$$

When *Astm* 3 is the only third-order aberration, the beam converges on two perpendicular segments which, on the principal ray, are separated by the so-called *astigmatism length*, i.e. the distance between the tangential and sagittal focal surfaces.

If $S_{III} \neq 0$ and if *Sphe* 3 and *Coma* 3 are small, then the least confusion images are obtained on the *mean curvature focal surface* whose curvature is

$$\frac{1}{R_m} = \frac{1}{2}\left(\frac{1}{R_s} + \frac{1}{R_t}\right), \tag{1.46}$$

as represented by the fourth righthand term of (1.41).

If *Astm* 3 vanishes ($S_{III} = 0$), then $R_m = R_s = R_t = R_P$: the four surfaces merge into one, the Petzval surface.

In the general case, depending on the sign of *Astm* 3, an off-axis ray intersects these four surfaces by following equal path intervals between them, and in the subscript order (t, m, s, P) or (P, s, m, t) of their radius (Fig. 1.21).

Similarly as $W_{[4]}$ in (1.41) for the third-order aberrations, other functions may be considered for the analysis of aberrations at any order. The general form of

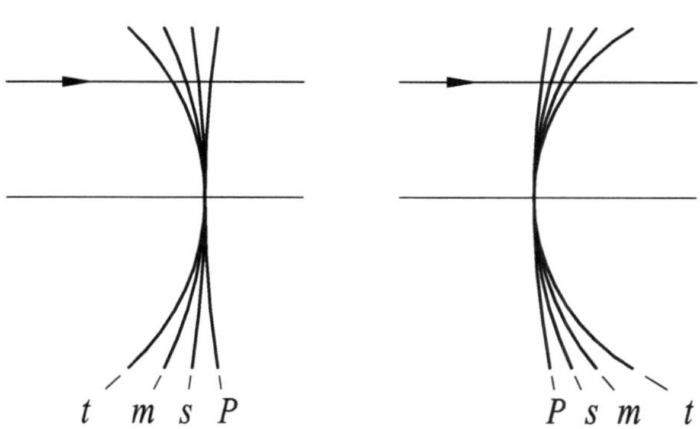

Fig. 1.21 Meridian sections at the Gaussian plane of the equally spaced tangential, least confusion, sagittal and Petzval focal surfaces for opposite signs of *Astm* 3

the wavefront function is defined by the *Hamilton characteristic function* as the expansion

$$W = W_{[0]} + W_{[2]} + W_{[4]} + W_{[6]} + \cdots, \tag{1.47}$$

where each function $W_{[i]}$ involves one or several terms of the form

$$w = a_{l,n,m}\ \bar{\eta}^l\ \rho^n \cos m\theta\,, \quad l,\ n,\ m \text{ integers} \geq 0\,, \quad m \leq n\,. \tag{1.48}$$

The first function $W_{[0]}$ reduces to a constant ($l = n = m = 0$) giving the origin coordinates of the wavefront. $W_{[2]}$ includes the dioptrics first-order terms *curvature* and *tilt* – i.e. the Gaussian terms – that we will refer to as

$$Cv\,1 \quad \text{for } l = 0,\ n = 2,\ m = 0\,, \tag{1.49a}$$

$$Tilt\,1 \quad \text{for } l = 1,\ n = 1,\ m = 1\,. \tag{1.49b}$$

Each term of the power series W is included in an order K_W of the wavefront function which can be defined by

$$K_W = l + n - 1\,. \tag{1.50}$$

The Gaussian terms and the third and fifth-order aberration terms involved in the Hamilton expansion of the characteristic function W are displayed by Table 1.2 where for simplicity the coefficients $a_{l,n,m}$, which have dimension of a length, are not shown.

For each primary aberration the shape of the wavefront can be represented either by a pure term of the type $w_{nm} = A_{nm}\rho^n \cos m\theta$ or by co-adding a first-order term (Fig. 1.22).

Table 1.2 First terms of the Hamilton wavefront function

n	$m = 0$	$m = 1$	$m = 2$	$m = 3$	K_W
1		$\bar{\eta}\,\rho\,\cos\theta$			1
2	ρ^2				
1		$\bar{\eta}^3\rho\,\cos\theta$			3
2	$\bar{\eta}^2\rho^2$		$\bar{\eta}^2\rho^2\cos 2\theta$		
3		$\bar{\eta}\,\rho^3\cos\theta$			
4	ρ^4				
1		$\bar{\eta}^5\rho\,\cos\theta$			5
2	$\bar{\eta}^4\rho^2$		$\bar{\eta}^4\rho^2\cos 2\theta$		
3		$\bar{\eta}^3\rho^3\cos\theta$		$\bar{\eta}^3\rho^3\cos 3\theta$	
4	$\bar{\eta}^2\rho^4$		$\bar{\eta}^2\rho^4\cos 2\theta$		
5		$\bar{\eta}\,\rho^5\cos\theta$			
6	ρ^6				

Fig. 1.22 (*Up*) Wavefronts converging at the Gaussian-, mean confusion-, and marginal-focus in the presence of primary spherical aberration $w = 5\lambda\rho^4$, $5\lambda(\rho^4 - \frac{3}{2}\rho^2)$, $5\lambda(\rho^4 - 2\rho^2)$. (*Center*) Wavefronts of primary coma $w = 4\lambda\rho^3\cos\theta$, $4\lambda(\rho^3 - \rho)\cos\theta$. (*Down*) Wavefronts of primary astigmatism $w = 1.6\lambda\rho^2\cos 2\theta$, $1.6\lambda\rho^2(\cos 2\theta + 1)$, $1.6\lambda\rho^2(\cos 2\theta + 2)$

1.8.2 Seidel Aberration Modes – Elastic Deformation Modes

The active optics methods are used to aspherize mirrors and lenses whose surface shape is generated by elasticity and represented by one or several terms $z_{n,m} = A_{n,m}\,\rho^n\cos m\theta$ of the power series (1.40b).

In this representation, the normalized image height parameter $\bar{\eta}$ – or normalized field angle $\bar{\varphi}$ – and successive powers in l do not explicitly appear. By regrouping in W the aberration terms of same integers n and m, we may obtain the $A_{n,m}$ coefficients. For instance, if a mirror is elastically aspherized for simultaneously correcting $a_{1,3,1}\,\bar{\eta}\,\rho^3\cos\theta$ and $a_{3,3,1}\,\bar{\eta}^3\rho^3\cos\theta$, which represent the first two terms of *linear coma*, of respective orders $K_W = 3$ and 5 in Table 1.2, the coefficient A_{31} should be

$$A_{31} = (a_{1,3,1}\,\bar{\eta} + a_{3,3,1}\,\bar{\eta}^3)\,/\,r_{\max}^3\,.$$

From the two integers n and m, we may define a $z_{n,m}$ term of (1.40b) as an optical surface mode included in the order K_O of *optical surface modes*,

$$K_O = n + m - 1\,, \qquad n + m \text{ even}, \quad m \le n\,. \tag{1.51}$$

For simplicity, we will refer to each optical surface mode by abbreviating the same name as that of the aberration wavefront appearing in the lowest order K_W of the wavefront function, followed by its order number K_O.[8]

From (1.40b), both optics and elastic modes, which are used hereafter and in the next chapters for *active optics*, are denoted by the following *abbreviated modes*,

$$\begin{array}{rlllll} z_{\text{Opt}} = A_{00} + & A_{20}\, r^2 + & A_{40}\, r^4 + & A_{60}\, r^6 + \cdots & \\ & Cv\,1 & Sphe\,3 & Sphe\,5 & \\ & + A_{11}\, r\cos\theta + & A_{31}\, r^3 \cos\theta + & A_{51}\, r^5\cos\theta + \cdots & \\ & Tilt1 & Coma3 & Coma5 & \\ & & + A_{22}\, r^2\cos 2\theta + & A_{42}\, r^4\cos 2\theta + \cdots & \quad (1.52) \\ & & Astm3 & Astm5 & \\ & & & + A_{33}\, r^3\cos 3\theta + \cdots & \\ & & & Tri5 & \\ & & & + \cdots & \end{array}$$

From the Clebsch elasticity theory which concerns constant thickness plates, we will show that the modes belonging to the two lower diagonal lines of this *optics triangular matrix* can be easily generated and co-added by flexure of a circular plate (cf. Sect. 7.2). Also including the *Sphe* 3 mode, we will often refer to them as *Clebsch-Seidel modes*.

1.8.3 Zernike rms Polynomials

For wavefront testing analysis, the rms values of the deviation to a theoretical surface are usually required. From a Zernike circle polynomial, one may build its associated rms polynomial, the *Zernike rms polynomials* $z_{n,m}$ (Noll [117]). From the radial components $R_{n,m}\{\rho\}$ defined by (1.40d), $z_{n,m}$ are given by

$$z_{n,m}\{\rho,\theta\} = k_{n,m}^{-1}\, R_{n,m}\{\rho,\theta\}, \qquad k_{n,m} = \sqrt{n+1} \times \begin{cases} 1 & \text{if } m=0, \\ \sqrt{2} & \text{if } m \neq 0. \end{cases} \tag{1.53}$$

The constant $k_{n,m}$ allows one to obtain the rms polynomial $z_{n,m}$ from the ptv polynomial $R_{n,m}$ (Table 1.3).

[8] **N.B**: We refer to the same abbreviations for the optical surface modes in z_{Opt} as for the wavefront terms in W, but generally their meaning is different. For instance, the aberration wavefront term *Astm* 3 is represented by the two terms $\bar{\eta}^2\rho^2$ and $\bar{\eta}^2\rho^2\cos 2\theta$ (cf. (1.41) and Table 1.2). Now using the abbreviation of the optical surface modes in (1.52), this term is obtained by the co-addition of *Cv* 1 and *Astm* 3 modes. In the context, the words "wavefront aberrations term" and "optical surface mode," or more concisely "aberration term" and "optical mode," allow one to distinguish between them.

Table 1.3 Coefficients $k_{n,m}$ and Zernike polynomials $R_{n,m}$ defining the rms polynomials $z_{n,m}$ up to the order $K_O = n+m-1 = 9$ (after Hugot [78])

i	Mode	$k_{n,m}$	$R_{n,m}(\rho,\theta)$	K_O
1	*Piston*	1	1	
2	*Tilt* 1 *x*	2	$\rho\cos\theta$	1
3	*Tilt* 1 *y*	2	$\rho\sin\theta$	
4	*Cv* 1	$\sqrt{3}$	$2\rho^2-1$	
5	*Astm* 3 *x*	$\sqrt{6}$	$\rho^2\cos 2\theta$	3
6	*Astm* 3 *y*	$\sqrt{6}$	$\rho^2\sin 2\theta$	
7	*Coma* 3 *x*	$\sqrt{8}$	$(3\rho^2-2)\,\rho\cos\theta$	
8	*Coma* 3 *y*	$\sqrt{8}$	$(3\rho^2-2)\,\rho\sin\theta$	
9	*Sphe* 3	$\sqrt{5}$	$6\rho^4-6\rho^2+1$	
10	*Tri* 5 *x*	$\sqrt{8}$	$\rho^3\cos 3\theta$	5
11	*Tri* 5 *y*	$\sqrt{8}$	$\rho^3\sin 3\theta$	
12	*Astm* 5 *x*	$\sqrt{10}$	$(4\rho^2-3)\,\rho^2\cos 2\theta$	
13	*Astm* 5 *y*	$\sqrt{10}$	$(4\rho^2-3)\,\rho^2\sin 2\theta$	
14	*Coma* 5 *x*	$\sqrt{12}$	$(10\rho^4-12\rho^2+3)\,\rho\cos\theta$	
15	*Coma* 5 *y*	$\sqrt{12}$	$(10\rho^4-12\rho^2+3)\,\rho\cos\theta$	
16	*Sphe* 5	$\sqrt{7}$	$20\rho^6-30\rho^4+12\rho^2-1$	
17	*Squa* 7 *x*	$\sqrt{10}$	$\rho^4\cos 4\theta$	7
18	*Squa* 7 *y*	$\sqrt{10}$	$\rho^4\sin 4\theta$	
19	*Tri* 7 *x*	$\sqrt{12}$	$(5\rho^2-4)\,\rho^3\cos 3\theta$	
20	*Tri* 7 *y*	$\sqrt{12}$	$(5\rho^2-4)\,\rho^3\sin 3\theta$	
21	*Astm* 7 *x*	$\sqrt{14}$	$(15\rho^4-20\rho^2+6)\,\rho^2\cos 2\theta$	
22	*Astm* 7 *y*	$\sqrt{14}$	$(15\rho^4-20\rho^2+6)\,\rho^2\sin 2\theta$	
23	*Coma* 7 *x*	$\sqrt{16}$	$(35\rho^6-60\rho^4+30\rho^2-4)\,\rho\cos\theta$	
24	*Coma* 7 *y*	$\sqrt{16}$	$(35\rho^6-60\rho^4+30\rho^2-4)\,\rho\sin\theta$	
25	*Sphe* 7	$\sqrt{9}$	$70\rho^8-140\rho^6+90\rho^4-20\rho^2+1$	
26	*Penta* 9 *x*	$\sqrt{12}$	$\rho^5\cos 5\theta$	9
27	*Penta* 9 *y*	$\sqrt{12}$	$\rho^5\sin 5\theta$	
28	*Squa* 9 *x*	$\sqrt{14}$	$(6\rho^2-5)\,\rho^4\cos 4\theta$	
29	*Squa* 9 *y*	$\sqrt{14}$	$(6\rho^2-5)\,\rho^4\sin 4\theta$	
30	*Tri* 9 *x*	$\sqrt{16}$	$(21\rho^4-30\rho^2+10)\,\rho^3\cos 3\theta$	
31	*Tri* 9 *y*	$\sqrt{16}$	$(21\rho^4-30\rho^2+10)\,\rho^3\sin 3\theta$	
32	*Astm* 9 *x*	$\sqrt{18}$	$(56\rho^6-105\rho^4+60\rho^2-10)\,\rho^2\cos 2\theta$	
33	*Astm* 9 *y*	$\sqrt{18}$	$(56\rho^6-105\rho^4+60\rho^2-10)\,\rho^2\sin 2\theta$	
34	*Coma* 9 *x*	$\sqrt{20}$	$(126\rho^8-280\rho^6+210\rho^4-60\rho^2+5)\,\rho\cos\theta$	
35	*Coma* 9 *y*	$\sqrt{20}$	$(126\rho^8-280\rho^6+210\rho^4-60\rho^2+5)\,\rho\sin\theta$	
36	*Sphe* 9	$\sqrt{11}$	$252\rho^{10}-630\rho^8+560\rho^6-210\rho^4+30\rho^2-1$	

1.9 Stigmatism, Aplanatism, and Anastigmatism

The third-order spherical aberration, coma, and astigmatism – *Sphe* 3, *Coma* 3, *Astm* 3 – are sometimes called *pupil aberrations*, they should preferably be corrected on or near the aperture stop.

1.9.1 Stigmatism

If an optical system is free from *Sphe* 3 aberration, then the first Seidel sum is null,

$$\sum S_I = 0\,, \tag{1.54}$$

and the design is said to be a *stigmatic system* in the third order.

• **Stigmatic single mirror**: The first known stigmatic surface is a paraboloid mirror for an object at infinity. The discovery of this geometrical property belongs to the Greek heritage. Considering a meridian plane, Diocles (cf. Sect. 1.1.1) demonstrated that a parabola provides the convergence of lines that are parallel to its axis at a single point. Ellipsoid and hyperboloid mirrors provide the stigmatic property for an object at finite distances.

Adding the plane and the sphere to these three shapes defines the conicoid family (cf. Sect. 1.7.1).

• **Stigmatic singlet lens**: In Descartes' attempt to develop refractive telescopes with a singlet lens, he invented the ovoid surfaces – known as *Descartes' ovoids* [45] (cf. Sect. 9.1.2) – where the other surface of the lens is a sphere of the same curvature as the incident or refracted wavefront. This provides *stigmatic singlet lenses*. The difficulty encountered with the axil chromatism of a singlet lens stopped these developments although in fact Descartes invented analytical geometry for the purpose of solving the stigmatism problem.

Descartes' ovoids include the subfamily of conicoids.

• **Any centered system**: From the above stigmatic single mirror or lens, we have the following important theorem.

→ *In any dioptric, catoptric, or catadioptric centered system, Descartes' ovoids – which include the conicoid family – provide stigmatism at all orders. Spherical aberration at all orders, Sphe* 3*, Sphe* 5*, etc, can be completely corrected for finite aperture angles U, U' reaching large values.*

The Descartes *theory of stigmatism* is the first theory of abberrations including either lenses, mirrors or both of them. For more than two centuries this theory, where the meridian section of an optical surface can be a fourth degree curve, remained the only one.

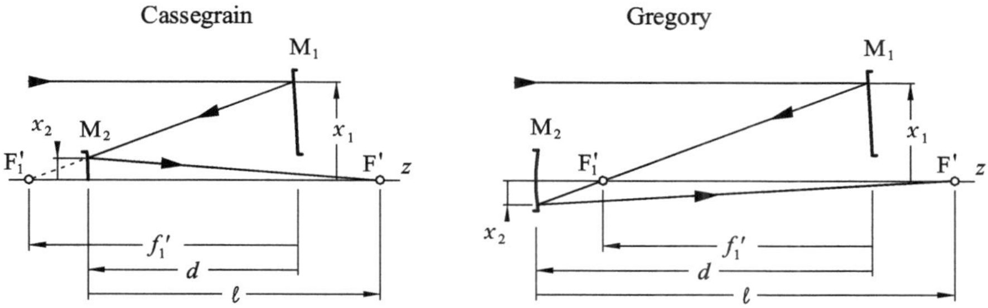

Fig. 1.23 Gaussian parameters of Cassegrain and Gregory forms. These telescopes provide an *inverted image* and an *erect image*, respectively

• **Stigmatic two-mirror telescopes**: Let us consider two-mirror telescopes in all generality, i.e. the Cassegrain and Gregory forms (Fig. 1.23).

Denote d the algebraic distance M_1M_2 from the vertex of the primary to that of the secondary ($d < 0$ for both forms in Fig. 1.23 since opposite to the z-direction), and $R_1 = 2f_1'$, $R_2 = 2f_2'$ their radii of curvature ($R_1 < 0$ and $f_1' < 0$ in both forms). Setting $n = -1$ in (1.28a) for the reflection case, the power of the system is

$$\frac{1}{f'} = 2\left(-\frac{1}{R_1} + \frac{1}{R_2} - \frac{2d}{R_1\,R_2}\right), \quad \begin{cases} > 0 & \text{for Cassegrain,} \\ < 0 & \text{for Gregory,} \end{cases} \tag{1.55}$$

since the sign of $R_2 = 2f_2'$ is opposite in the two forms ($R_2 < 0$ for Cassegrain and $R_2 > 0$ for Gregory). The secondary generates a transverse magnification

$$M = \frac{f'}{f_1'} = \frac{2f'}{R_1}, \quad \begin{cases} < 0 & \text{for Cassegrain,} \\ > 0 & \text{for Gregory,} \end{cases} \tag{1.56}$$

where the sign means that the image is *erect* for Gregory and *inverted* for Cassegrain. Denoting ℓ the distance M_2F' from the secondary to the resulting focus ($\ell > 0$ in both forms). A representation for d/f' with these parameters is

$$\frac{d}{f'} = \frac{1}{M}\left(1 - \frac{\ell}{f'}\right). \tag{1.57}$$

Following Wilson [7] in the theory of two-mirror focal systems, a representation of the first Seidel sum is

$$\sum S_I = \frac{1}{4}\left(\frac{x_1}{f'}\right)^4 \left\{-M^3(1+\kappa_1)\,f' + (M+1)^3\left[\left(\frac{M-1}{M+1}\right)^2 + \kappa_2\right]\ell\right\}, \tag{1.58}$$

where x_1 is the marginal ray height at the primary. The *stigmatism condition*, $\sum S_{\rm I} = 0$, entails

$$1 + \kappa_1 - \left[\left(\frac{M-1}{M+1}\right)^2 + \kappa_2\right]\left(\frac{M+1}{M}\right)^3 \frac{\ell}{f'} = 0 . \tag{1.59}$$

There is an infinity of (κ_1, κ_2) sets satisfying this condition.

A classical form is with a paraboloid primary; hence the conic constants are in this case

$$\begin{cases} \kappa_1 = -1 , & (1.60a) \\ \kappa_2 = -\left(\dfrac{M-1}{M+1}\right)^2 , & (1.60b) \end{cases}$$

which then leads to a hyperboloid secondary for Cassegrain and ellipsoid for Gregory.

If one of the mirrors is a sphere ($\kappa_1 = 0$, or $\kappa_2 = 0$), then the total amount of *Sphe* 3 aberration must be corrected by the other mirror. Thus we obtain from (1.59) two other systems for each form, Cassegrain and Gregory (Table 1.4). Designs with a spherical primary are called *Pressman-Camichel telescopes*, whilst those with spherical secondary are *Dall-Kirkham telescopes*. We include in Table 1.4 the Ritchey-Chrétien (RC) telescopes whose conic constants are given further by Eq. (1.70).

One may define the *telephoto effect*, T, as the following ratio,

$$T = \left| f' / \ell \right| . \tag{1.61}$$

This ratio is usable for comparisons only if the image field is at the vertex of the primary mirror or just behind it towards the light propagation. A high telephoto effect characterizes a compact system. Compared to the Gregory form, T is the highest for the Cassegrain.

Table 1.4 Comparison of the conic constants for stigmatic two-mirror telescopes: Cassegrain, Newton, quasi-flat secondary and Gregory forms

Telescope type	Primary M_1		Secondary M_2	
Cassegrain PH	Paraboloid	$\kappa_1 = -1$	Hyperboloid	$\kappa_2 < -1$
Cass. M_1 spherical	Sphere	$\kappa_1 = 0$	Ellipsoid	$\kappa_2 > 0$
Cass. M_2 spherical	Ellipsoid	$-1 < \kappa_1 < 0$	Sphere	$\kappa_2 = 0$
Cass. Aplanatic RC	Hyperboloid	$\kappa_1 < -1$	Hyperboloid	$\kappa_2 < -1$
Flat M_2 – Newton	Paraboloid	$\kappa_1 = -1$	Plane	$1/R_2 = 0$
Quasi-flat M_2	Spherical	$\kappa_1 = 0$	Aspherical	
Gregory PE	Paraboloid	$\kappa_1 = -1$	Ellipsoid	$-1 < \kappa_2 < 0$
Greg. M_1 spherical	Sphere	$\kappa_1 = 0$	Hyperboloid	$\kappa_2 < -1$
Greg. M_2 spherical	Hyperboloid	$\kappa_1 < -1$	Sphere	$\kappa_2 = 0$
Greg. Aplanatic RC	Ellipsoid	$-1 < \kappa_1 < 0$	Ellipsoid	$-1 < \kappa_2 < 0$

1.9.2 Aplanatism and Abbe's Sine Condition

If an optical system is free from $Sphe\,3$, and $Coma\,3$ aberration terms, the corresponding Seidel sums are

$$\sum S_I = \sum S_{II} = 0\,, \tag{1.62}$$

and the design is said to be an *aplanatic system* in the third order. One may notice that up to 1873 and somewhat later, the meaning of aplanatism applied for stigmatism.

One of the most fundamental laws in optical design implicitly expresses these two conditions. A complete demonstration of this law is given by Welford [166, 167] by using a skew ray – i.e. a ray which is off a meridian plane – and the *skew invariant* which applies to any centered system.

Consider an off-axis pencil emitted by a source point in medium n and at the optical height η (Fig. 1.16) and denote u the paraxial aperture angle formed from this point by the skew ray and the local principal ray. The skew invariant theorem states that

$$n'\eta' u' = n\,\eta\,u\,, \tag{1.63}$$

which means that any skew conjugate planes in which the paraxial aperture angles u, u' lie both intersect their respective pupil conjugate planes at the same azimuth angles ($\theta' = \theta$).

Consider now the *finite* aperture angles U and conjugate U', which are the angle used in ray tracing. If the energy emitted by the object field of area η^2 is *totally* transferred to the image field of area η'^2, then the one-dimensional conservation of the energy, as shown by Clausius in 1864 [33] (an account of the Clausius general law is given by Chrétien [29]), is represented by

$$n'\eta' \sin U' = n\,\eta\,\sin U\,, \tag{1.64}$$

where the finite aperture angles U and U' may be as large as $\pm\pi/2$.

For small field angles η/f and η'/f', the two later equations provide a relation between finite and paraxial aperture angles. In the normal (focal) case

$$\frac{\sin U'}{u'} = \frac{\sin U}{u}\,. \tag{1.65a}$$

This important property is known as the *Abbe sine condition*, a condition where terms in all orders for on-axis spherical aberration and linear-field coma are cancelled. This result was first established in a somewhat different form by E. Abbe [1] in 1873 who achieved major advances in the design of microscope objectives (cf. Sect. 9.1.3). In the modern sense, the terminology *"aplanatic system"* or *"aplanatism condition"* must be strictly used in the third-order theory for stating that both $Sphe\,3$ and $Coma\,3$ are removed, $S_I = S_{II} = 0$.[9]

[9] The word *aplanatic* [from α π λ α ν η τ - ο ς (α privative + π λ α ν α - ϵ ι ν to wander) = free from error] was currently used – long before Abbe – to indicate *only* that the axial image of a point is

When both object and image are at infinity, the sine condition in this afocal case becomes

$$\frac{X'}{x'} = \frac{X}{x}, \tag{1.65b}$$

where X, x and their conjugates are the ray aperture heights replacing the aperture angles. For a telescope with object and image fields in medium $n = n' = 1$, then $x = -u' f'$; the *sine condition* writes

$$\frac{x}{u'} = \frac{X}{\sin U'} = -f' = \text{constant}. \tag{1.65c}$$

This fundamental result may be enounced by the following theorem.

→ *In any centered system, Abbe's sine condition provides aplanatism at all orders. Both spherical aberration at all orders, Sphe 3, Sphe 5, etc., and coma at all orders, Coma 3, Coma 5, etc., can be completely corrected for finite aperture angles U, U' reaching large values.*

The Abbe sine condition implies the following on-axis property (Fig. 1.24).

→ *For an object at infinity and a system satisfying the sine condition, any ray of the axial beam converging at the image focus F' intersects its input conjugate ray on a sphere centered at the focus F', the* "Abbe sphere."

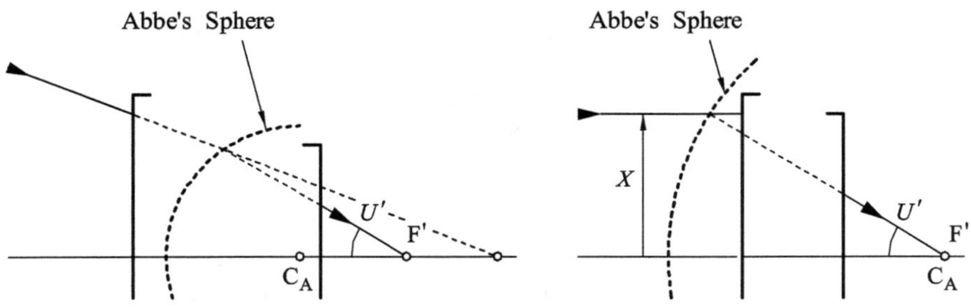

Fig. 1.24 On-axis beam satisfying the Abbe sine condition. The *Abbe sphere* of center C_A is shown by a *dotted line* for a convergent beam (*Left*), and for an object at infinity (*Right*)

free from spherical aberration. It stood for the word *stigmatic*, which did *not* exist in the language of opticians before ~1900. Hence, in his memoir of 1873 and the following ones, Abbe specified that the sine condition allows one to obtain, in addition to an axial "aplanatic" point, any other "aplanatic" points in the field of view provided their distance from the axis is infinitesimally small.

After introducing the term *stigmatic*, the modern significance of the term *aplanatic* was generalized into the literature during the period 1920–40.

The sine condition was brought to the attention of astronomers in stipulations decided by the Paris Congress for the "Carte du Ciel" (1889), mainly on the instigation of H.A. Steinheil acting as German delegate, for the standardization of astrographs. This recommendation was to avoid the dissymmetric coma images of solely stigmatic objectives which caused errors in the position of stars and required use of an inaccurate *magnitude* equation for those corrections.

These notes are based on the historical account by Chrétien [29], Sect. 379.

The Abbe sine condition entails that both *all-order spherical aberration* terms and *linear coma* terms are zeroed. Thus from (1.48) and Table 1.2, the latter condition implies that the terms in linear dependence on field height $\bar{\eta}$ – whose sum is the so-called *linear coma* – are also null,

$$a_{1,n,1}\,\bar{\eta}\,\rho^n\cos\theta = 0\,, \quad \forall n = 3,\,5,\,7,\,\cdots\,. \tag{1.66}$$

As demonstrated by Welford [167], the above condition is derived from the *Staeble-Lihotzky isoplanatism condition*, simultaneously published in 1919 by F. Staeble and E. Lihotzky, which is also a condition for zero linear coma in the presence of spherical aberration.

Since Abbe's sine condition does not involve non-linear terms on the height as $\bar{\eta}^3\rho^3\cos\theta$ coma or $\bar{\eta}^2\rho^4$ spherical aberration, the stationarity of this condition is valid for field image points which have an off-axis distance infinitesimally close to the system axis.

• **Numerical aperture**: The *numerical aperture* (N.A.) allows characterizing the light-gathering power of an optical system in the object space (whilst the f-ratio is preferably used in the image space). This number is defined from the maximum value of the *finite* aperture angle U in medium n by

$$\text{N.A.} = n\,\sin|U_{\max}|\,.$$

The terminology N.A. is a universal use for characterizing the input beam of microscope objectives. Such objectives generally satisfy Abbe's sine condition; their front part may be immersed in oil.

• **Isoplanatic singlet lens**: Any single thick lens with one or two aspheric surfaces can be designed aplanatic for one wavelength. In convergent beams, a single lens with both spherical surfaces also can be. Such lenses are used as microscope objective components (cf. Sect. 9.1.3). If a singlet lens is corrected for coma *only*, then

$$S_I \neq 0\,, \quad S_{II} = 0\,. \tag{1.67}$$

Such a design is called an *isoplanatic system.*

For a lens, the condition $S_{I\,I} = 0$ requires a separate location of the pupil. This remote pupil and lens configuration – *isoplanatic lens mounting* – is sometimes used for spectrograph collimators.

• **Isoplanatic single mirror**: Analytic relations expressing the Seidel aberrations of one- or two-mirror systems are known (cf. for instance Wilson [170]). Considering a single concave mirror and an object at infinity, the two first ones are

$$S_I = -\frac{1}{4}\left(\frac{x}{f'}\right)^4(1+\kappa)\,f'\,,\quad S_{II} = -\frac{1}{4}\left(\frac{x}{f'}\right)^3\varphi\left[\,2f' - (1+\kappa)\,s\,\right], \tag{1.68}$$

where φ is the field angle defined by the principal ray, and s the axial separation of the pupil from the mirror vertex. If a spherical mirror ($\kappa = 0$) is used with a pupil located at $s = 2f'$, i.e. at its center of curvature, then $S_{II} = 0$ and $S_I \neq 0$. This *isoplanatic mirror mounting* is the basic first property of Schmidt telescopes.

• **Aplanatic Ritchey-Chrétien telescopes**: H. Chrétien [30] published, in a second paper of 1922, the theory of two-mirror telescopes satisfying the sine condition, then obtaining the so-called exact *mirror parametric equations.*[10] Within the mirror class represented by conicoids, the Cassegrain aplanatic form have since been built in very large size. In the third-order theory, this requires satisfying the stigmatism condition $\sum S_I = 0$, expressed by (1.59), and also the coma free condition which then reduces, whatever the pupil position (cf. Wilson [170]), to

$$\Big(\sum S_{II}\Big)_{Stig} = -\frac{1}{4}\left(\frac{x_1}{f'}\right)^3 \varphi_1 \left\{2f' + (M+1)^3\left[\left(\frac{M-1}{M+1}\right)^2 + \kappa_2\right]d\right\} = 0\,. \quad (1.69)$$

This determines κ_2, and substituting it in (1.59) we obtain

$$\begin{cases} \kappa_1 = -1 - \dfrac{2}{M^3}\dfrac{\ell}{d}\,, & (1.70a) \\[2ex] \kappa_2 = -\left(\dfrac{M-1}{M+1}\right)^2 - \dfrac{2}{(M+1)^3}\dfrac{f'}{d}\,, & (1.70b) \end{cases}$$

which are the conic constants for the Cassegrain and Gregory aplanatic forms.

Referring to the signs of M, f', d and ℓ as shown in the previous section, $M^{-3}\ell d^{-1}$ is positive for Cassegrain and negative for Gregory. Therefore the primary mirror of aplanatic telescopes is a hyperboloid for the Cassegrain form, and an ellipsoid for the Gregory form (cf. Table 1.4).

• **Parametric representation of two-mirror telescopes satisfying the sine condition**: The exact shapes of the mirror surfaces for a two-mirror telescope satisfying the sine condition are *not* conicoid asphercis. In a certain manner it is fortunate that, even for all usual cases of very large Ritchey-Chrétien telescopes, a *truly* accurate representation of those surfaces can be obtain by use of conicoids.

For instance the classical representation of optical surfaces by even polynomials – such as for conicoids – cannot be use for the raytrace of *grazing incidence telescopes* strictly satisfying the sine condition because high-order coma terms *cannot* be neglected (cf. Sect. 10.1.3).

In his second paper of 1922, introducing the sine condition, Chrétien [30] was the first to derive from integrations a parametric representation of the optical surfaces for the general case of two-mirror telescopes, hence deriving systems that are

[10] G. Ritchey encouraged Chrétien in investigating the case of the coma free two-mirror telescopes. Without knowledge of Schwarzschild's third-order theory of aplanatic two-mirror telescopes (1905), Chrétien derived apparently before 1910 the all-order theory of two-mirror telescopes by integrations of differential equations that he laid down from the sine condition (see also historical note in Sect. 1.1).

completely free from all-order spherical aberrations and linear coma terms. In a cylindrical coordinate frame z, r, Chrétien expressed the mirror surfaces by

$$z_{1,2} = F_{1,2}\left[t(U'),\, \ell/f',\, d/f'\right], \qquad r_{1,2} = G_{1,2}\left[t(U'),\, \ell/f',\, d/f'\right], \tag{1.71}$$

where $t(U')$ is the parametric variable of the finite aperture angle U'.

Also using a similar parametric representation, Lynden-Bell [98] carried out a general study of the two-mirror case for finite aperture angle U' that may reach extremely large values such as $\pm\pi/2$ (or even $\pm\pi$) for the purpose of determining and drawing the shape of the most typical telescopes of this family. In several cases, also including virtual image fields, the mirror surfaces were found to be *cusp-* or *trumpet*-shaped at their vertices.

This was followed up by Willstrop and Lynden-Bell [169] in a classification review of all two-mirror telescopes satisfying the sine condition for varied driver parameters ℓ/f' and d/f'. The results proved that all telescopes of practical interest have already been discovered.

1.9.3 Anastigmatism

If an optical system is free from $Sphe\,3$, $Coma\,3$, and $Astm\,3$ aberration terms, the three Seidel sums are

$$\sum S_I = \sum S_{II} = \sum S_{III} = 0\,, \tag{1.72}$$

and the design is said to be an *anastigmatic system* in the third order.

In this case, the tangential, mean curvature and sagittal surfaces merge into one, the *Petzval surface*.

• **Anastigmatic singlet lens**: Any single thick lens with one or two aspheric surfaces can be designed anastigmatic for one wavelength. A special case of anastigmatism is a *monocentric lens* of index n with front and back spherical radii R and R/n centered in C: if the incident beams converge behind C on a sphere of radius nR centered at C, then all the refracted beams converge on its back surface (cf. Sect. 9.1.3).

• **Anastigmatic two-mirror telescopes**: With the two first Seidel sums set to zero, the astigmatism of an aplanatic telescope is

$$\left(\sum S_{III}\right)_{Aplan} = \frac{1}{2}\left(\frac{x_1}{f'}\right)^2 \varphi_1^2 \left(d + 2f'\right)\frac{f'}{\ell}\,, \tag{1.73}$$

and if this sum is also set to zero, then

$$d + 2f' = 0\,, \tag{1.74}$$

which is the *anastigmatism condition* for two-mirror telescopes.

By substitution in (1.70b), the conic constants of an anastigmatic telescope are

$$\begin{cases} \kappa_1 = -1 + \dfrac{1}{M^3}\dfrac{\ell}{f'}, & (1.75a) \\ \kappa_2 = -\left(\dfrac{M-1}{M+1}\right)^2 + \dfrac{1}{(M+1)^3}. & (1.75b) \end{cases}$$

In an attempt to find a perfect telescope, K. Schwarzschild in his celebrated paper of 1905 [142] formulated the complete third-order theory of one- and two-mirror telescopes including all the above conditions. His eikonal method allowed him to determine telescopes satisfying $\Sigma\, S_I = \Sigma\, S_{II} = \Sigma\, S_{III} = 0$, and particularly the *Schwarzschild telescope* which in addition satisfies $\Sigma\, S_{IV} = 0$ for a flat field. Unfortunately, his telescope was impracticable because of a convex primary.[11]

Various designs of the *anastigmatic telescope family* are shown in Sect. 4.1.

• **All reflective Schmidt telescopes**: The *Schmidt telescope* is with no doubt the most important design in the anastigmat instrument class. It must be considered as an *extrapolate* design from the two-mirror anastigmat family. Although not knowing the Schwarzschild results, B. Schmidt [140] in 1930 indirectly found that the primary mirror could be substituted by a pseudo-flat refracting corrector plate which then is located at the center of curvature of the secondary. This satisfies the aplanatism and anastigmatism conditions, i.e. both (1.65c) and (1.74).

The case of purely *reflective Schmidts* has been investigated (i) for spectrographs with normal diffraction aspherized gratings which hold for the primary, (ii) for spectroscopy sky surveys with a large telescope by tilt of the primary (cf. Chaps. 4, 5 and LAMOST).

Let us consider as an example of interest, a fully obstructed design which is the basic case of reflective Schmidts. From the anastigmatism condition, the focus is midway between the two mirrors (Fig. 1.25).

For all Schmidt telescopes the transverse magnification is $M \simeq 0$ because generally the primary element (corrector plate, mirror, or grating) shows a very low optical power. Assuming that $M = 0$ entails that the primary M_1 is a purely plane-aspheric surface; then from (1.75) the conic constants are $\kappa_1 \to \infty$ (since the primary curvature is null) and $\kappa_2 = 0$ which gives a spherical secondary in the third-order. Starting from the condition $d+2f' = 0$, we will determine the shape of the mirrors with higher-order terms in two different ways:

(C_1) by using a perfectly spherical secondary and stigmatism condition,
(C_2) by using the Abbe sine condition with Abbe's sphere construction.

In order to determine the shape of the mirrors for each condition, let us denote $1/R_2$ the curvature of the secondary, $f' = R_2/2$ the efl, and $\zeta = z/f'$, $\rho = r/f'$

[11] Schwarzschild was widely involved in many important fields of physics. Within a year of Einstein's publication of the theory of general relativity, he discovered in 1916 the first and one of the most important of rare solutions for the metric. This probably explains why Schwarzschild did not focus his attention on the Seidel conditions for aplanatic two-mirror telescopes, conditions which were implicitly included in his theory.

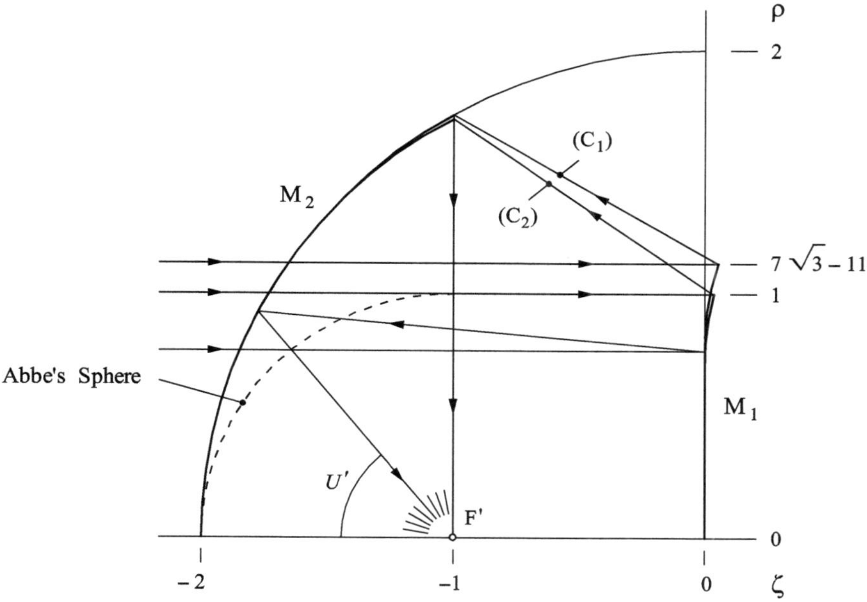

Fig. 1.25 Comparison of two reflective Schmidt telescopes without tilt of the primary and fulfilling the anastigmatism condition $d+2f'=0$. The Abbe sphere - *dotted line* - is shown up to f/0 which corresponds to a finite aperture angle U' varied of $\pm\pi/2$. The designs are with condition (C_1): $\kappa_2=0$ and $\sum S_I=0$, or condition (C_2): $\sum S_I=\sum S_{II}=0$ [93]

dimensionless variables of a cylindrical frame whose origin is at the vertex of the primary. The shapes ζ_1, ζ_2 of the primary and secondary are represented in the form $\zeta=\sum a_n\rho^n$. For condition (C_1) we use A_n coefficients as given in Table 4.1 subsequently converted in dimensionless coefficients a_n. For condition (C_2) we use a parametric representation (Lemaitre [93]) subsequently transformed in the power series $\zeta(\rho)$. The result of a_n calculations for $n = 0, 2, 4, 6, 8$ in the two cases is, for the primary and secondary respectively,

$$\begin{aligned}
(\zeta_1)_{Stig} &= \{\,0,\quad 0,\ 1/2^6,\ 3/2^9,\ 5\times 3^2/2^{14}\,\}\\
(\zeta_1)_{Aplan} &= \{\,0,\quad 0,\ 1/2^6,\ 3/2^9,\quad 3/2^{10}\,\}\\
\\
(\zeta_2)_{Stig} &= \{-2,\ 1/2^2,\ 1/2^6,\ 1/2^9,\quad 5/2^{14}\,\}\\
(\zeta_2)_{Aplan} &= \{-2,\ 1/2^2,\ 1/2^6,\ 1/2^9,\quad 1/2^{11}\,\}
\end{aligned} \tag{1.76}$$

which, comparing these coefficients, shows that whatever (C_1) or (C_2) condition the shape of the mirrors is the same up to the fifth-order included,

$$\begin{cases}
\Delta\zeta_1=\Delta\zeta_2=0 \quad \text{for } n=0,\ 2,\ 4,\ 6, & (1.77a)\\
\Delta\zeta_1(n=8)=\Delta\zeta_2(n=8)=-\,3\times 2^{-14}\rho^8. & (1.77b)
\end{cases}$$

Considering the f-ratio $\Omega = f'/D$, the last equation leads to a difference in the slopes of $\Delta(d\zeta/d\rho) = -3\times 2^{-18}\Omega^{-7}$ which is negligible even for the usual systems with fast f-ratio. Although other high-order field aberrations remain, this result confirms the validity of always designing reflective Schmidts with a concave secondary of perfectly spherical shape.

We will see in Sect. 4.1.4 that the field balance of aberration correction requires a low powered primary whose extremal radial curvatures $d\zeta_1^2/d^2\rho$ are set opposite.

• **Afocal two-mirror telescopes – Mersenne anastigmats**: The two-mirror afocal forms were first investigated by M. Mersenne [108] who showed in 1636 that two *confocal paraboloids* provide beam compressors (or expanders) free from spherical aberration ($\sum S_I = 0$).

Three centuries after Mersenne's publication, his optical systems have become famous – and must be regarded with no doubt as the most important of optical designs – because it was found that also $\sum S_{II} = \sum S_{III} = 0$. A brief historical account on the discovery of the properties of Mersenne systems is given in Sect. 2.3.

The four designs of *Mersenne's afocal telescopes* are discussed in Sect. 2.3.

1.10 Petzval Curvature and Distortion

1.10.1 Petvzal Curvature

If a system is an anastigmat, the tangential, mean curvature, and sagittal focal surfaces merge into the *Petzval surface*. Hence for an object at infinty, the images lie on the Petzval surface of curvature C_P. If the system is only stigmatic or aplanatic, the best images are obtained on the mean curvature focal surface which is generally designed with a flat shape.

The Petzval curvature C_P of an optical system is related to the fourth Seidel sum of $Petz$ 3 terms by

$$C_P = -\frac{n}{H^2}\sum S_{IV}\,, \tag{1.78}$$

where H is the Lagrange invariant, whose dimension is a length, and n the refractive index of the last medium.

The Petzval curvature can be directly derived from the curvature $c_\ell = 1/R_\ell$ of each surface numbered ℓ and from the refractive indexes n_ℓ and n'_ℓ of the media before and after the surface, respectively. Let us denote $n_{\ell+1}$ – instead of n'_ℓ – the second index after surface ℓ, and use the Cartesian sign convention of positive curvature corresponding to a sag variation towards positive z (in concordance with previous Sections).

The *Petzval theorem* provides the determination of the Petzval curvature of a system having number L surfaces,

$$\begin{cases} C_P \equiv \dfrac{1}{R_P} = n_{L+1} \sum_{\ell=1}^{L} \left(\dfrac{1}{n_{\ell+1}} - \dfrac{1}{n_\ell} \right) \dfrac{1}{R_\ell}, \\ \text{refractive index:} \begin{cases} n_{L+1} \text{ for the last medium,} \\ n_{\ell+1} = -n_\ell \text{ for a mirror.} \end{cases} \end{cases} \tag{1.79}$$

The important condition $C_P = 0$ for a flat field is called the *Petzval condition*.

• **Petzval curvature of some basic systems**: The above relations allow derivation of the Petzval curvature in any optical system. For instance, we may consider the following cases:

Refractive diopter: L=1, indexes n_1 and n_2=n_{L+1}, $\quad C_P = \frac{n_1-n_2}{n_2}\, c_1$

Singlet lens: L=2, n_1=1, n_2=n, n_3=1 $= n_{L+1}$, $\quad C_P = \frac{n-1}{n}\,(c_2 - c_1)$

Single mirror: L=1, n_1=1, n_2=−1=n_{L+1}, $\quad C_P = 2\, c_1$

Two mirrors: L=2, n_1=1, n_2=−1, n_3=1=n_{L+1}, $\quad C_P = 2\,(c_2 - c_1)$

Three mirrors: L =3, n_1=n_3=1, n_2=n_4=−1=n_{L+1}, $C_P = 2\,(c_1 - c_2 + c_3)$

Dyson copying catadioptric system: The *Dyson copying system* [49] uses plano-convex singlet lens and a concave mirror whose surfaces are monocentric and both spherical (Fig. 1.26). Working in monochromatic light at magnification $M = -1$, this system is "perfect" since free from the five Seidel sums. The calculation of the Petzval sum is simplified by starting with the object on the glass. After returning through the lens i.e. L=3, the successive indexes are n_1=n, n_2=1, n_3=−1, n_4=−n=n_{L+1}, which leads by setting c_1=c_3 in (1.79) to

$$C_P = -n\left[\left(\tfrac{1}{1} - \tfrac{1}{n}\right)\tfrac{1}{R_1} + \left(\tfrac{1}{-1} - \tfrac{1}{1}\right)\tfrac{1}{R_2} + \left(\tfrac{1}{-n} - \tfrac{1}{-1}\right)\tfrac{1}{R_3}\right] = \tfrac{2n}{R_2} - \tfrac{2(n-1)}{R_1}$$

Hence, the *Petzval condition* of flat field is achieved if $c_2 = \frac{n-1}{n}\, c_1$.

Offner image transport system: The *Offner image transport* system [118] has catoptric equivalence to the Dyson system. All Seidel sums are also nulled. The three mirrors are spherical and monocentric (Fig. 1.26). Since $c_3 = c_1$,

$$C_P = 2\left(c_1 - c_2 + c_3\right) = 2\left(2\, c_1 - c_2\right).$$

Hence the flat fielding condition $C_P = 0$ is obtained by setting $c_2 = 2\, c_1$. Because of their central obstruction, these systems are used off-axis.

• **Telecentric systems**: An important special case in optics is the design of *telecentric systems*: If at a focal plane the principal rays are parallel to the axis of the

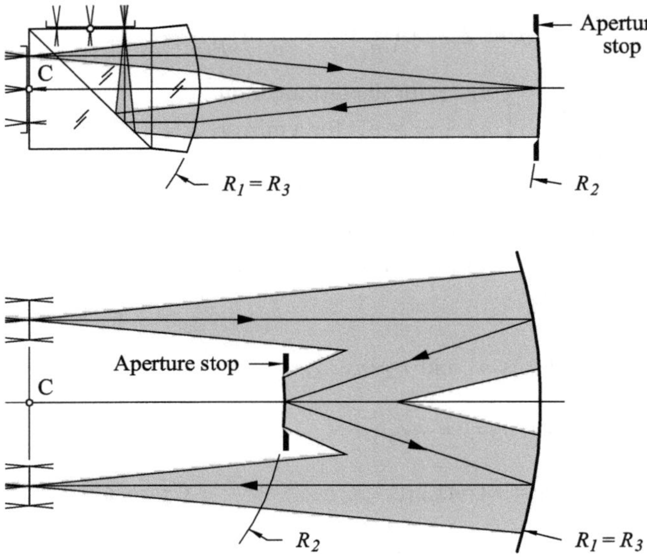

Fig. 1.26 Flat-field telecentric anastigmats working at transverse magnification $M = -1$. *Up*: The Dyson copying system. *Down*: The Offner image transport. Both designs have all Seidel sums nulled and thus are "perfect instruments" in the third-order theory

system, then the system is said to have telecentric beams. Therefore, the lateral position of the barycenters of any image in the field is *invariant* to focusing errors. For instance, a telecentric design is currently used in *astrometric telescopes* for increasing the accuracy in the determination of stellar positions.

In a focal instrument (both object and image are at finite distances), telecentric output beams are obtained by an aperture stop exactly located at the entrance focal plane. Hence the exit pupil is at infinity.

Conversely, if the input beams are telecentric, then the output pupil is at the image focus F'. By using consecutively such a direct system and a conversed one, we obtain a compound *telecentric transfer system* where the input and output pupils are both at infinity. This is achieved in the Dyson and Offner designs. In a dioptric form, telecentric transfers are co-added by pairs in the design of *periscopes*.

1.10.2 Distortion

The fifth Seidel sum $\sum S_V$ is the third-order distortion, abbreviated $Dist\,3$. This aberration gives rise to a radial shift of the image in its focal plane which then entails a non-linear correspondence between the object and the image scales.

In the image plane, the first lateral component of the principal ray height η' is given by the transverse magnification M in (1.24) which defines the paraxial scale.

The second component is derived from S_V in (1.43) which gives the *Dist* 3 radial shift $\Delta\mathcal{Y} = \Delta\eta' = -\frac{1}{2}\, S_V\, \bar{\eta}^3$. Hence the ray height in the image field is represented by the odd expansion of its conjugate

$$\eta' = M\,\eta + k_3\,\eta^3 + k_5\,\eta^5 + \cdots, \tag{1.80}$$

where M is the transverse magnification and $k_3 = -\,S_V/(2\,\eta_{max}^3)$.

If M is negative, the distortion is in *pincushion* form for S_V positive and in *barrel* form for S_V negative (Fig. 1.27).

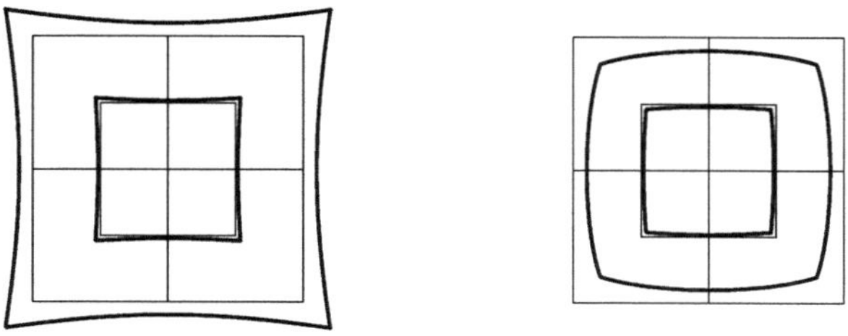

Fig. 1.27 *Left*: Pincushion distortion. *Right*: Barrel distortion

In general, telescopes do not require to be corrected for distortion because this effect does not degrade the image quality and can always be removed by image processing.

However, *liquid mirror telescopes* (LMTs) – which are zenithal and transit instruments – require a distortion correction and, in addition, a sky projection correction when not working at the equator of the Earth. This was first solved by E.H. Richardson [76] who showed that a three-lens correction was inappropriate and thus designed dedicated four-lens correctors for large LMTs (cf. also Sect. 7.7.2).

1.11 Diffraction

The colored lines of a thin oil film lying on water or the colored lines of equal thickness of a very thin lens – which were recognized as interference fringes by T. Young – as well as the image showing oscillations in intensity when produced by a point source through an optical system, are evidence effects of the wave nature of light. Although easily observable, some of these phenomena were first referenced during the Renaissance by Leonardo da Vinci. It was F.M. Grimaldi, in his book (1665), who drew attention to them with accurate descriptions. Geometrical optics – or ray trace – were inadequate for deducing these phenomena, and neither could the later corpuscular theory explain the diffraction effects. In an attempt at explanation, in 1678 Huygens [80] constructed the following concept:

→ *Any point of a wavefront gives rise to secondary disturbances which are spherical wavelets and, at any later instant, the light distribution may be regarded as the sum of these wavelets.*

1.11.1 The Diffraction Theory

Inspired by Huygens' construction, Fresnel in 1816 laid down a celebrated memoir [62] in which he explained the principle of interference known as the *Huygens-Fresnel principle*. Reviewing Fresnel's prize memoir to the French Academy, Poisson deduced (1818) that, following Fresnel's theory, a bright spot should appear at the center of the shadow of a small disk, and concluded that this should be wrong since never observed. However, Arago, who was also a reviewer, carried out the experiment and observed the predicted white spot. In 1882, Kirchhoff [87] gave a more complete and sound mathematical basis to Fresnel's theory; this is known as *Fresnel-Kirchhoff diffraction theory*.

Let us assume a wave propagation in a vacuum. According to the wave equation (1.4), each Cartesian component $W(\mathbf{r}, t)$ of the field vectors $\mathbf{E}$ and $\mathbf{H}$ must satisfy the *scalar wave equation*

$$\nabla^2 W - \frac{1}{c^2}\frac{\partial^2 W}{\partial t^2} = 0 . \tag{1.81}$$

where $\nabla^2 W$ is the Laplacian.

Use of an exponential function instead of trigonometric functions simplifies the calculations with scalar waves. Assuming it is strictly *monochromatic* of wavelength $\lambda = c/\nu$, where the frequency ν is the number of vibrations per second, we may separate the space- and time-components. Denoting $i = \sqrt{-1}$ and $\Re$ for a real part, we have the form

$$W(\mathbf{r}, t) = \Re\{U(\mathbf{r})\, e^{-i(\omega t + \phi_0)}\}, \tag{1.82a}$$

$$U(\mathbf{r}) = U_0(\mathbf{r})\, e^{ik\mathbf{r}}, \tag{1.82b}$$

where U is the complex amplitude, ϕ_0 a phase constant, ω the *angular frequency*, and k the *wave number*. In a vacuum these later quantities are

$$\omega = 2\pi\,\nu \;\equiv\; 2\pi\, c/\lambda\,,$$
$$k = \omega/c \;\equiv\; 2\pi/\lambda\,.$$

By substitution of W into (1.81), we find that the space-dependent part U must satisfy the equation

$$\nabla^2 U + k^2 U = 0 . \tag{1.83}$$

Considering a wave passing through a plane aperture stop (Fig. 1.28) whose area is $\mathcal{A}(x, y)$, a solution of the space-dependent term is expressed by the *Fresnel-Kirchhoff diffraction integral*

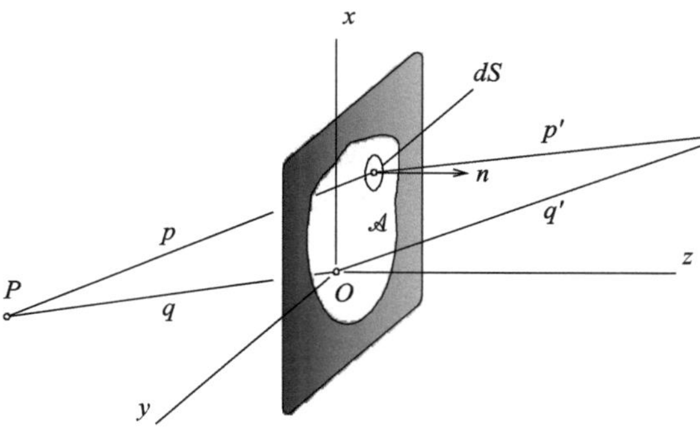

Fig. 1.28 Diffraction by the area $\mathcal{A}$ of a plane aperture

$$U = U_0 \iint_{\mathcal{A}} \frac{e^{ik(p+p')}}{p\,p'} \left[\cos(n, p) - \cos(n, p')\right] dS\,, \tag{1.84a}$$

in which U_0 is a constant, p, p' the respective distances from the source point P and from the point P' – where the disturbance is determined – to the elementary surface dS in the area $\mathcal{A}$, and (n, p), (n, p') the angles with the normal n to the plane $\mathcal{A}$. Setting the origin O of the x, y plane at any point inside the aperture $\mathcal{A}$, we will assume that points P and P' are at large distances from the origin and that the angles generated by the lines PO and OP' with PP' are moderate. Thus, the term pp' may be replaced by qq' and the term into the square bracket replaced by $2\cos\delta$ where δ is the angle between PP' and the z axis. These two terms will not vary appreciably over the aperture, whilst the exponent term $p+p'$ will change by many wavelengths. The diffraction integral of the disturbance at P' then becomes

$$U = U_0 \frac{\cos\delta}{q\,q'} \iint_{\mathcal{A}} e^{ik(p+p')}\, dS\,. \tag{1.84b}$$

Expanding p and p' as functions of q, q' and power series of x/q, y/q, x/q' and y/q', the wave disturbance is

$$\begin{cases} U = U_0 \frac{\cos\delta}{q\,q'} \iint_{\mathcal{A}} e^{ik\,f(x,y)}\, dx\, dy\,, & (1.85a) \\ f(x,y) = r_x\, x + r_y\, y + g_1 x^2 + g_2 xy + g_3 y^2 + \cdots\,, & (1.85b) \end{cases}$$

where, in a displacement of the element dS into the aperture plane, $r_x(\mathrm{P'})$ and $r_y(\mathrm{P'})$ are the Cartesian components of the relative variation of the optical path $p+p'$ with respect to $q+q'$, whilst $g_1(\mathrm{P'})$, g_2, g_3 are functions of the next order relative variations.

The Fresnel-Kirchhoff diffraction theory may lead to integrations by considering only the first two terms of $f(x, y)$ if assuming that the aperture area $\mathcal{A}$ is small; this simplest case is called the *Fraunhofer diffraction*. When the quadratic terms cannot be neglected, such as for the diffraction of an infinite long edge, this is the

Fresnel diffraction case. A detailed account on these two cases is given by Born and Wolf [17].

When observing a diffracted wave near an aperture edge, one can see a broad intensity oscillation of the first fringes at the edge. A useful interpretation of this phenomena, which was investigated by Sommerfeld (1896), is to consider that the observed pattern is the superposition result of the incident wave with a locally cylindric *edge wave* such as was generated from the edge.

1.11.2 Diffraction from a Circular Aperture

Denoting $\rho,\ \theta$ the polar coordinates of a current point into the plane of a circular aperture of radius $\rho = a$, the coordinates of this point are

$$\rho\,\cos\theta = x\,, \qquad \rho\,\sin\theta = y.$$

From the definition of r_x and r_y in (1.85b), it follows that the polar coordinates $r,\ \psi$ of a diffracted ray at the image plane are

$$r\,\cos\psi = r_x\,, \qquad r\,\sin\psi = r_y\,.$$

After substitution of $f(x, y) = r_x\,x + r_y\,y$ into (1.85a), the diffraction integral becomes

$$U(r) = C\int_0^a\!\!\int_0^{2\pi} e^{ik\rho\,r\cos(\theta-\psi)}\rho\,d\rho\,d\theta\,. \tag{1.86a}$$

A well-known integral representation of Bessel's functions (cf. Jahnke and Emde [82]) is

$$J_n(z) = \frac{i^{-n}}{2\pi}\int_0^{2\pi} e^{iz\cos\alpha}e^{in\alpha}d\alpha\,,$$

hence we obtain

$$U = 2\pi C\int_0^a J_0(k\rho\,r)\,\rho\,d\rho\,, \tag{1.86b}$$

where J_0 is the zero-order Bessel function. Since $\int_0^u u'\,J_0(u')\,du' = u\,J_1(u)$, and also $J_1(u)/u \to 1/2$ when $u \to 0$, the wave oscillation at the Gaussian plane is

$$U = \pi a^2 C\,\frac{2J_1(kar)}{kar}\,, \tag{1.86c}$$

and the intensity is represented by

$$I = |\,U\,|^2 = \left[\frac{2J_1(kar)}{kar}\right]^2 I_0\,. \tag{1.87}$$

This is the celebrated formula first derived in 1835 by Airy [2] in a different form.

Now representing its argument by $x = kar$, a perfectly spherical wavefront converges at the Gauss focus by radially distributing the intensity of the light according to the so-called *Airy function* $[2J_1(x)/x]^2$ (Fig. 1.29) which provides a pattern with successive dark rings (Fig. 1.30). This function has a maximum value equal to unity for $x = 0$, i.e. on the axis. The null intensities are given by the zeros of $J_1(x)$, whilst the secondary maxima are given by the zeros of $[2J_1(x)/x]^2$ derivative (Table 1.5). The first dark ring occurs at $x = 1.220\pi$. Since the wave number is $k = 2\pi/\lambda$, the corresponding angular radius is

$$r = 0.610\,\frac{\lambda}{a}\,. \tag{1.88a}$$

For any instrument with an aperture diameter $D = 2a$, the maximal angular resolution due to the diffraction limitation, called the *resolving power*, is defined from the Airy patterns of two object points – or stars – as the angle

$$\varphi = 1.220\,\frac{\lambda}{D} \tag{1.88b}$$

separating their central peaks.

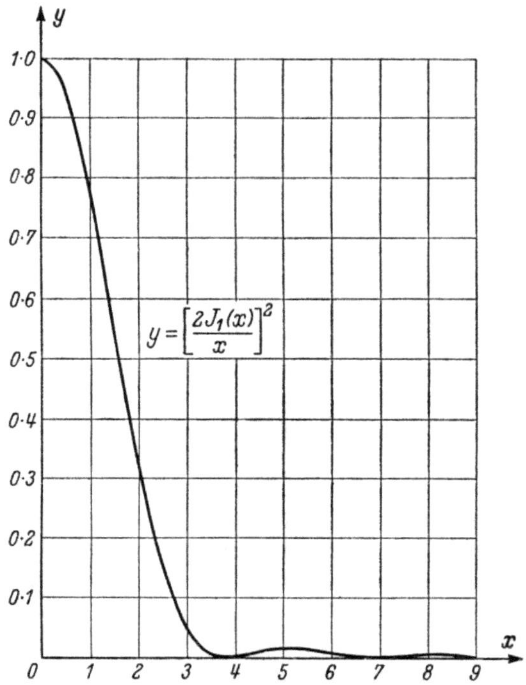

Fig. 1.29 Fraunhofer diffraction of a circular aperture with normalized intensity $y = [2J_1(x)/x]^2$ (after Born and Wolf [17])

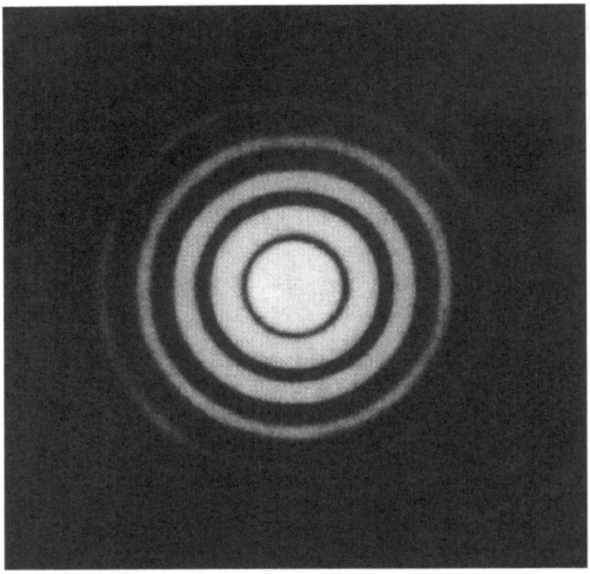

Fig. 1.30 Fraunhofer diffraction – Airy pattern – of a circular aperture 6 mm in diameter, magnification 50 ×, $\lambda = 579$ nm. The central intensity has been overexposed to show the weak subsidiary maxima (after Institute of Optics, Orsay)

Table 1.5 First minima and maxima of the function $[2J_1(x)/x]^2$ (after Born and Wolf [17])

	x	$[2J_1(x)/x]^2$
Central maximum	0	1
First minimum	$1.220\pi = 3.832$	0
Second maximum	$1.635\pi = 5.136$	0.0175
Second minimum	$2.233\pi = 7.016$	0
Third maximum	$2.674\pi = 8.417$	0.0042
Third minimum	$3.238\pi = 10.174$	0
Fourth maximum	$3.669\pi = 11.620$	0.0016

According to Rayleigh's definition, this angle is the radius of the first ring of null intensity and known as the *Rayleigh two-point resolution criterion*. The angular resolution may as well be defined as the diameter of the Airy function at half-maximum which then is $\varphi = 1.04\,\lambda/D$ and sometimes referred to simply as $\varphi = \lambda/D$. If this separation is reduced to $\varphi = 0.947\,\lambda/D$, the dip between the two peaks just disappears; this is known as the *Sparrow two-point resolution criterion* (Wetherell [168]).

For comparisons of a wave obtained from optical testing, it is useful to consider as an absolute reference the encircled energy of a perfectly stigmatic wave (Fig. 1.31).

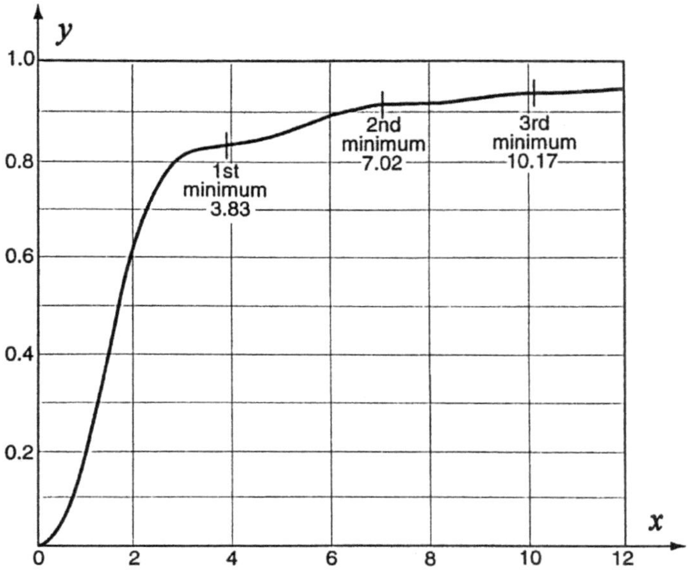

Fig. 1.31 Function $y = 1 - J_0^2(x) - J_1^2(x)$ representing the fractional encircled energy into a radius $r = x/ka$ of Airy's pattern (after Born and Wolf [17])

1.11.3 Diffraction from an Annular Aperture

An important case with telescopes is to consider the diffracted image of a wavefront obstructed at its center, such as caused by a secondary mirror (Born and Wolf [17]). Let us define the boundaries of this aperture area as the radii a and ϵa of two concentric circles, where ϵ is a positive number less than infinity. In this Fraunhofer case, the light distribution at the focal plane is represented by

$$I(r) = \left[\frac{2J_1(kar)}{kar} - \epsilon^2 \frac{2J_1(\epsilon kar)}{\epsilon kar}\right]^2 I_0. \tag{1.89}$$

Compared to a full aperture, this result shows that the resolving power is slightly increased but the maximum intensity of the rings is also increased. With $\epsilon = 1/2$, the first root of (1.89) is for $x = kar = 1.00\pi$ instead of 1.22π for ϵ=0, and the intensity of the second maximum is 0.092 instead of 0.018 for ϵ=0.

1.11.4 Point Spread Function (PSF) and Diffracted Aberrations

The *point spread function* – or PSF – is the relative function $I(x, y)/I_0$ representing the intensity distribution of the light in the image region of the Gaussian focal plane.

• **Spherical wavefront**: In the basic case of a perfect spherical wavefront, the normalized diffraction function representing the intensity distribution over the focal plane is the PSF.

For a full circular aperture, the PSF is the Airy function $I(r)/I_0$ in (1.87).

If the circular aperture is a ring, then the PSF is ratio $I(r)/I_0$ in (1.89).

For a rectangular aperture of sides $2a$ and $2b$, one shows that the PSF is

$$\frac{I(x,y)}{I_0} = \left(\frac{\sin kax}{kax}\right)^2 \left(\frac{\sin kby}{kby}\right)^2.$$

Important cases of diffracted images are, for instance, with a pupil whose transmission is a function (*apodization*), or with a polarized input beam. By introducing convenient parameters, the diffraction theory has been mostly developed in a scalar form, thus avoiding the difficulties of the general vectorial form.

• **Aberrated wavefront**: In the presence of a single primary aberration, many diffraction patterns from an object point have been recorded by K. Nienhuis [116], Françon et al. [61] and others for various aberration amplitudes. The diffraction theory applied to a single primary aberration was treated by F. Zernike and B. Nijboer [180], Nienhuis, Kingslake, who obtained isophotes in meridian sections of the beam and at the focal plane. The theoretical results are in accordance with the diffraction patterns that are observed for single primary aberrations (Fig. 1.32). Details on these analyses and the resulting spatial distributions of intensity are reviewed by Born and Wolf [17].

Hereafter theoretical results on the determination of an optimal diffraction imaging in the presence of primary aberrations allow the definition of optical tolerances.

1.11.5 Diffraction-Limited Criteria and Wavefront Tolerances

An important aspect for highly accurate optical systems is the achievement of a diffraction-limited imaging. A certain freedom in the polished shape and set up of the elements always exists, but these parameters must conform to strict criteria. The raytrace and manufacturing processes have to be matched with respect to criteria.

In the presence of aberrations the point spread function is very complicated in form, but when the aberrations are small the effect on the PSF is that the central intensity decreases, the half-width of the central maximum does not change whereas more light appears in the rings. Thus, we could set a tolerance level on that amount of aberration which produces such a just perceptible change. The first investigations in this field were developed by Rayleigh [123] and more systematically by Strehl [147].

Denoting I_{Aber}/I_0 the point spread function delivered by an optical system with residual aberrations and I_{Th}/I_0 the theoretical PSF if that system would be perfect, Strehl suggested that an appropriate tolerance level for a drop in intensity of the central peak would be such as

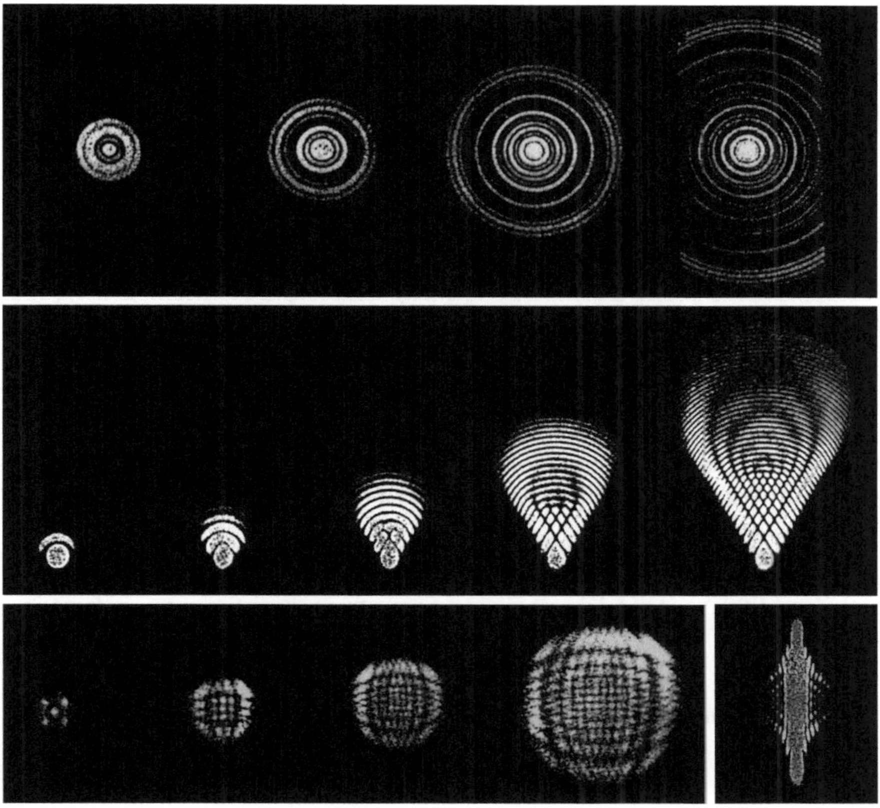

Fig. 1.32 Diffraction patterns in the presence of primary aberrations. (*Up*) Primary spherical aberration at the least confusion focal plane $w = A_{40}(\rho^4 - \frac{3}{2}\rho^2)$ with $A_{40} = 1.4\lambda,\ 3.7\lambda,\ 8.4\lambda,\ 17.5\lambda$. (*Center*) Primary coma in the Gaussian focal plane $w = A_{31}\rho^3\cos\theta$ with $A_{31} = 0.3\lambda,\ \lambda,\ 2.4\lambda,\ 5\lambda,\ 10\lambda$. (*Down*) Primary astigmatism at the least confusion plane $w = A_{42}\rho^2\cos 2\theta$ with $A_{42} = 1.4\lambda,\ 2.7\lambda,\ 3.5\lambda,\ 6.5\lambda$. Image in the plane containing one of the two separated focal lines for $w = 2.7\lambda\rho^2\cos 2\theta$ (after K. Nienhuis [116])

$$S \equiv \frac{I_{\text{Aber}}}{I_{\text{Th}}} \geq 0.8 \quad \text{where } S \text{ is the \textit{Strehl intensity ratio},} \tag{1.90a}$$

a tolerance which is generally accepted.

For instance, if a defocus is considered as an aberration, a Strehl ratio $S = 0.8$ provides a tolerance in agreement with the early result of a $\lambda/4$ defocus by Rayleigh. For *Sphe* 3, this Strehl value also corresponds to the *quarter-wave Rayleigh criterion*. However, the quarter-wave criterion *do not* satisfy $S = 0.8$ for *Coma* 3, *Astm* 3 or higher-order aberrations. Therefore, the Strehl definition has been preferred as an extension definition to the general case including all aberration types.

→ *If the Strehl intensity ratio of an optical system is at least 0.8 or greater, then the system is well corrected and so-called "diffraction limited."*

Considering the general form of the wavefront function, such as defined by (1.47) and (1.48), where (ρ, θ) are the normalized coordinates of a point in a circular pupil, it has been shown by Maréchal [103] (cf. also Born and Wolf [17] and Wetherell [168]) that the value of the *Strehl ratio* is

$$S = 1 - \frac{4\pi^2}{\lambda^2}\left\{\iint W^2(\rho,\theta)\,\rho\, d\rho d\theta - \left[\iint W(\rho,\theta)\,\rho\, d\rho d\theta\right]^2\right\}, \qquad (1.90b)$$

where the integrations are taken over the normalized area of the pupil ($\rho \in [0, 1]$, $\theta \in [0, 2\pi]$). When the aberrations are sufficiently small, this equation may be approximated by the two first terms of an expansion as

$$S \simeq 1 - \frac{4\pi^2}{\lambda^2}\,\Delta W^2_{\rm rms} \equiv 1 - \sigma^2_\phi\,, \qquad (1.90c)$$

where $\Delta W^2_{\rm rms}$ and σ^2_ϕ are the variances of the wavefront and phase aberration, respectively. When $S = 0.8$, the so-called *Maréchal criterion* [103] is

$$\sigma^2_\phi = 0.2 \quad [\text{in rad}^2], \qquad (1.91a)$$

and from (1.90c), we obtain a mean square wavefront aberration, or *wavefront variance*,

$$\Delta W^2_{\rm rms} \simeq \lambda^2/197.4\,, \qquad (1.91b)$$

thus a root mean square wavefront aberration value

$$\Delta W_{\rm rms} \simeq \pm\,\lambda\,/\,14\,, \qquad (1.91c)$$

which is valid whatever the aberration type. Hence, Maréchal's criterion provides an equivalence to the Strehl value $S = 0.8$, and may be enounced as follows.

→ *If the root mean square departure of the wavefront to the best fit reference sphere does not exceed $\lambda/14$, then the system is diffraction limited.*

Now considering the wavefront tolerancing of various aberrations, we give hereafter the results published by Welford [167] in his Section "Optical Tolerances." For example, denoting α, β, and γ some of the first coefficient $a_{l,n,m}\bar{\eta}^l$ of the wavefront function W as defined by (1.47) and (1.48), we may consider defocus as an aberration and write it $\alpha_1\rho^2$ where $\rho = 1$ at the pupil full aperture. For a defocus, the Strehl tolerance $S = 0.8$ leads to a $\Delta W_{\rm ptv}$ wavefront deviation

$$\alpha_1 = \pm 0.25\,\lambda\,. \qquad (1.92a)$$

This is the famous result obtained by Rayleigh, known as *quarter-wave Rayleigh criterion* [123] stating that an optical system can be considered as perfect if the wavefront is included between two concentric surfaces separated by $\lambda/4$.

Similarly, denoting $\beta_1\rho^4$ the primary spherical aberration, the $\Delta W_{\rm ptv}$ tolerance limit is

$$\beta_1 = \pm 0.24\,\lambda\,, \qquad (1.92b)$$

and if this aberration is balanced but choosing the best focus, i.e. considering a $W(\rho, \theta)$ term of the form $\alpha_1 \rho^2 + \beta_1 \rho^4$, the ΔW_{ptv} tolerance limit is

$$\beta_1 = \pm 0.95\,\lambda \quad \text{with} \quad \alpha_1 = -\beta_1\,. \tag{1.92c}$$

In the same way, we can find the tolerance limit for other aberrations and balance them with appropriate terms. Table 1.6 displays some ΔW_{ptv} wavefront tolerance limits satisfying either the Strehl criterion $S = 0.8$ or equivalently the Maréchal criterion $\Delta W_{\text{rms}} = \lambda/14$.

Table 1.6 Best sphere fit wavefront tolerances ΔW_{ptv} for obtaining diffraction limited images, i.e. for $S = 0.8$ or equivalently $\Delta W_{\text{rms}} = \lambda/14$. (after Welford [167])

Aberration	Aberration polynomial	Coefficients at tolerance limit
$Cv\,1$, defocus	$\alpha_1 \rho^2$	$\alpha_1 = \pm 0.25\,\lambda$
$Sphe\,3$, primary spherical aber.	$\beta_1 \rho^4$	$\beta_1 = \pm 0.24\,\lambda$
$Sphe\,3$ with choice of best focus $Cv\,1$	$\alpha_1 \rho^2 + \beta_1 \rho^4$	$\beta_1 = \pm 0.95\,\lambda$ $\alpha_1 = \mp 0.95\,\lambda$
$Sphe\,5$ with choice of best $Sphe\,3$ and $Cv\,1$	$\alpha_1 \rho^2 + \beta_1 \rho^4 + \gamma_1 \rho^6$	$\gamma_1 = \pm 3.47\,\lambda$ $\beta_1 = \mp 5.62\,\lambda$ $\alpha_1 = \pm 2.24\,\lambda$
$Coma\,3$, primary coma	$\beta_2 \rho^3 \cos\theta$	$\beta_2 = \pm 0.20\,\lambda$
$Astm\,3$, primary astigmatism	$\beta_3 \rho^2 \cos 2\theta$	$\beta_3 = \pm 0.17\,\lambda$

• **Note**: Many useful optical systems do not require a diffraction-limited quality in the purpose they have been designed for. With those systems, e.g. photographic objectives, optics of television cameras, etc, the specification is to achieve a given number resolving elements, typically 400–1000 elements in each of x, y directions. Their optical tolerances are based on the concept of the *optical transfer function*.

1.12 Some Image Processor Options

A review on the theoretical and practical limitation that an astronomical system may reach in observational astrophysics is given by Léna, Rouan et al. [96] for the whole electromagnetic spectrum. Restraining hereafter to *image processors* in the range from extreme-ultraviolet to infrared, there are many options for which each one induces a single- or multi-telescope system, a preferred telescope design and an optimal telescope focal length. Some associated imaging systems may have two or more processors in series such as, for instance, an atmospheric dispersion compensator plus an adaptive optics module followed by an imaging spectrograph. We hereafter follow some descriptions by Richardson [131] on image processor options.

1.12.1 Human Eye

If the processor is a human eye, objects laying over a large field can be correctly imaged for distances varying from infinity to *punctum proximum* (25 cm). The optics standard parameters of the human eye generally assume that the low gradient index of the lens may be approximated by a constant index. Le Grand [70] proposed $n_d = 1.44$ or 1.45 for this index; we have found that raytraces are more realistic with $n_d = 1.45$ for the lens (Table 1.7).

Table 1.7 Optical parameters of the standard human eye working at infinity or at *punctum proximum* (after Le Grand [70]) and with some complements from raytrace by the author) (Lengths in [mm])

Surface	Radius of Curv.		Axial Separat.		Index
	∞	p.p.	∞	p.p.	n_d
Cornea 1st	7.8	7.8	0.55	0.55	1.377
······ 2nd	6.5	6.5	3.05	2.65	1.337
Pupil (d=4)			0.00	0.00	1.337
Lens surf. 1	10.2	6.1	4.00	4.50	1.450
··· surf. 2 *	−6.0	−5.5	16.60	16.50	1.336
Retina †	−13.4	−13.4			

* deformed hyperboloids of conic constants $\kappa_{\infty} \simeq -4.6$, $\kappa_{\text{p.p.}} \simeq -4.7$.
† parented to a flattened ellipsoid, $\kappa_{\infty} = \kappa_{\text{p.p.}} \simeq +0.1$.
The axial separation from the cornea first surface to retina is 24.2 mm.

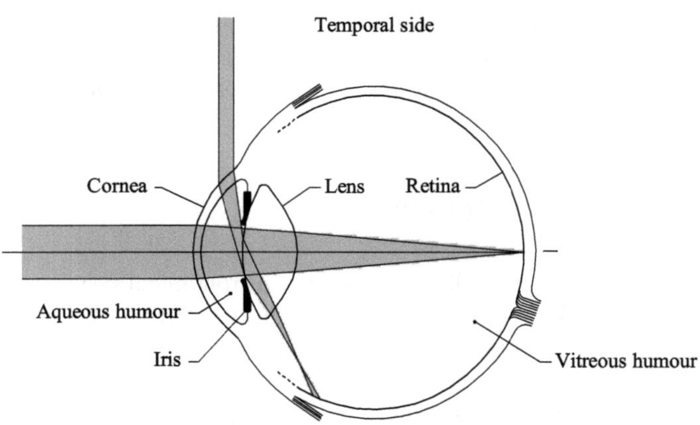

Fig. 1.33 Horizontal section of the human eye

For objects at infinity, the efl is 22.4 mm in the image space, thus equivalent to an efl of 16.7 mm in the air. In the central field of about 5°, the resolution in the object space is 1 arcmin corresponding to 5 μm on the retina. The low resolution imaged field reaches 100° from the axis in the temporal direction and 60–80° in the other directions (Fig. 1.33). The aperture of the iris, which is the input pupil, varies from 2 to 8 mm and takes the value of 4 mm for normal daytime conditions.

1.12.2 Eyepiece

If the field is magnified and observed with the eye, the processor is an *eyepiece*. Eyepieces have a very different task from that of the objectives because of the location of the exit pupil: this pupil must be with some clearance after the last lens of the eyepiece – called the "eye lens," the opposite is the "field lens" – in order for the observer to position his eye's pupil, the iris, on it. This axial clearance varies from 5 to 20 mm and is called *eye relief*. Also, the diameter of the exit pupil shall be small enough for all of the beam of light to enter the iris. Another complication in eyepiece design is the increased field of view which is in inverse proportion to the diameter of the pupil. For example, imaging a (1/3)° field with a 1 m aperture telescope, the apparent field with a 5 mm pupil must be 200-times greater, thus 67°, requiring a complicated design with several elements.

The raytrace design of eyepieces leads to systems of several glass types having up to eight elements with spherical surfaces, or with less elements but using aspherical surfaces. Available eyepieces have an apparent field of view going up to 80° which requires the eye to pivot to accurately observe subfields. Typical Zeiss eyepieces are the Abbe *orthoscopic* and the classical Plössl with four elements, and the Erfle with five elements (Fig. 1.34). Descriptions and comparisons of some eyepieces are given by Taylor [153] and Smith [146].

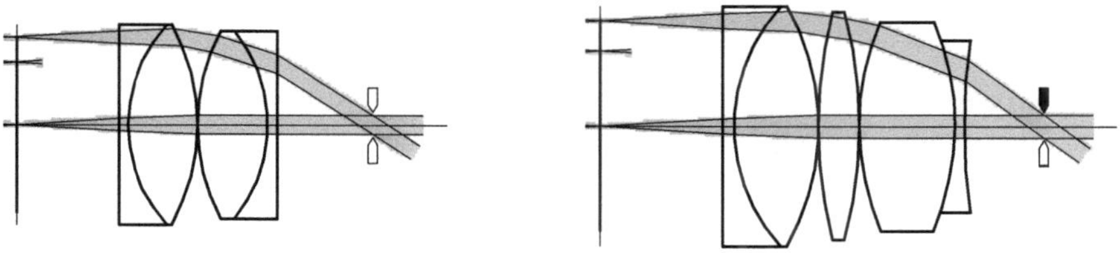

Fig. 1.34 *Left*: Plössl eyepiece (type 2-2). *Right*: Erfle eyepiece (type 2-1-2)

1.12.3 Interferometer

If the image of a same object is issued from two or several separated apertures, or if a wavefront is separated in two wavefronts by a beam splitter or a phase mask, which causes the recombined waves of light to add and subtract, the processor is an *interferometer*.

Examples of interferometers are infrared *Fourier transform spectrometers* (cf. Sect. 2.5.1), *stellar interferometers*, pioneered by E. Stephan and A. Michelson, *telescope arrays* pioneered by A. Labeyrie (cf. Sect. 2.5.2), *wavefront analyzers* and some *coronographs*. For some interferometers requiring a field compensation, *variable curvature mirrors* have been installed at the retro-reflected focü of the mobile carriages (Chap. 2).

1.12.4 Coronograph

If a faint object is detected at the immediate proximity of a bright object, the processor is a *coronograph*. The first coronographs used a spatial filter at the focal plane, a Lyot stop, to block the bright image and transmit the surrounding light to a next imaging system.

An appropriate modification of the spatial transmission function of the pupil, known as *apodization*, allows the suppression of the secondary maxima. In astronomy, this facilitates the detection of a faint object when close to a bright punctual object. New type coronographs use interferometric techniques such as with a phase mask or are based on the incident flux division by pupil reversal and phase shift of one beam. The beam recombination provides destructive interferences at the center of the field. For instance, *four-quadrant phase masks* were proposed and elaborated by D. Rouan et al. (2000). A new concept providing an achromatic destructive interference over one octave at the center of the field was recently introduced by D. Rouan et al. (see details in Lena, Rouan et al. [1.96]). It consists of cellular phase shifter mirrors – called *chessboard mirrors* – introduced in each arm of a nulling interferometer. The number of shifting cells is determined from integer roots of polynomial relations (*diophantine optics*). Exoplanet detection will benefit from this concept.

1.12.5 Polarimeter

If the processor is a polarizer, such as the N-shaped Wollaston prism or a semi-wave plate, that causes the light waves to oscillate in a single plane, this instrument is a *polarimeter*. A usual convention is to describe linear polarization as the orientation of the electric field vector.

1.12.6 Slit Spectrograph

If the processor is a diffraction grating or a prism, the light entering through a slit is angularly dispersed with respect to the wavelengths from a spectrum: this instrument is a *spectrograph*. There exists reflective and transmission gratings which are deposited on plane, spherical, or aspherical substrates. The line distributions generated by holographic processes provide *constant-* or *variable-spacing diffraction gratings*.

Although a concave, toroid, reflective grating with constant-spaced lines, or a concave spherical concave grating with variable-spaced lines, are efficient *single surface slit spectrographs* for ultraviolet studies (cf. Sects. 3.5.7 and 7.8.2), spectrographs are generally designed with a collimator optics before a plane or plane-aspheric grating and a camera optics after it. If a reflective grating reflects the dispersed light back to the collimator so that the collimator also serves as a camera, this arrangement is called a *Littrow's mountings* spectrograph. Some spectrographs require an internal re-imaging of the pupil – as for example for cross dispersion spectrographs using both echelle and classical gratings – which is provided by additional mirrors or lenses. Such a design is called a *white pupil mounting* or *Baranne's mounting* as originated by A. Baranne [11, 12]. One usually distinguishes between slit, long-slit, and multi-slit spectrographs; the latter case is called a *multi-object spectrograph* (MOS).

1.12.7 Slitless Spectrograph

If a dispersive element is located before the focal plane of a telescope, the processor is called a *slitless spectrograph*.

• **Objective grating**: If a concave reflective grating is substituted to the primary mirror of a small telescope, one obtains the basic form of slitless spectroscopy with an *objective grating*. For an object at infinity and given a wavelength, the design with a *paraboloid reflective grating* working at a normal diffraction angle is rigourously stigmatic on the axis; similarly a stigmatic solution is also obtained with a *spherical reflective grating* and variable spacing lines. For an object point at finite distance, there always exists a conicoid grating shape providing a stigmatic image for a wavelength diffracted perpendicularly to the grating vertex, sometimes called *normal diffraction mounting*. Because gratings are not available in very large size, other alternatives have been developed for *objective spectroscopy*.

• **Normal field objective prisms**: If a prism pair is designed to cancel the deviation at wavelength λ_0 but still provides a dispersion, the processor is known as *Fehrenbach prisms* or *normal field objective prisms*. For the determination of stellar radial velocities, such prism pairs have been pioneered by Fehrenbach in 1944 [56] and developed up to 0.6 m aperture diameter (Fehrenbach et al. [57]) as telescope input pupil components. Exposures are taken with the prism-pair at 0° and then with in-plane rotation of 180°, thus providing two counter-dispersed spectra of 5° sky field as for the Schmidt telescope at Haute Provence Observatory [58]. Let $n_{d,1}$ and $n_{d,2}$ be the d-yellow hydrogen line refractive indices of prisms with small wedge angles A_1 and A_2, respectively. If the wedges are opposite such as $A_2 = -A_1 \equiv A$, then the beam deviation angle of the two prisms – which form a parallel plane plate – is

$$\alpha = (n_{d,1} - n_{d,2})\, A\,.$$

By choosing a glass pair such as $n_{d,2} = n_{d,1} \equiv n_d$ the deviation is cancelled, $\alpha = 0$. Introducing the Abbe number $\nu_d = (n_d - 1)/(n_F - n_C)$, the dispersion angle between the F-blue and C-red hydrogen wavelengths is, from the above relation,

$$\Delta\alpha = \big(1/\nu_{d,1} - 1/\nu_{d,2}\big)(n_d - 1)\, A\,. \tag{1.93}$$

From the Schott (n_d, ν_d) diagram of glasses, one can see that the dispersion of these two wavelengths is ~ 2.3-times larger for the glass pair SK14-F5 ($n_d = 1.603$) than for the glass pair BK6-LLF6 ($n_d = 1.531$). The design of Fehrenbach normal field prisms for a given central wavelength λ_0 sometimes requires the glass manufacturer to elaborate a special glass.

It must be noticed that Fehrenbach normal field prisms have the highest throughput compared to any slitless dispersion system. Instead of being installed at the telescope input, they also can be located at a reimaged pupil of a *large* telescope. For galaxies at extremely large distances, this should allow efficient detections of large redshifted spectra – typically $z \simeq 7$ – such as the Lyman cutoff observed in the near infrared region.

• **Grism**: If the processor is a simple Δ-shaped prism, the beam is turned through a deviation angle but the dispersion is less than from a transmission grating. Therefore, if a transmission grating is bonded to the prism such as the resulting deviation cancels at wavelength λ_0, then the beam is still dispersed. Such a combination is called a *grism* or *Carpenter prism* and is usually located before a telescope focus for objective spectroscopy.

• **Grens**: If the surface of a prism – which is not bonded to the grating – is given some power to act as a lens, the processor is sometimes called a *grens*. For instance, wide-field correctors equipped with a grens were designed by Richardson [127, 128] for objective spectroscopy at the prime focus of large telescopes (Fig. 1.35).

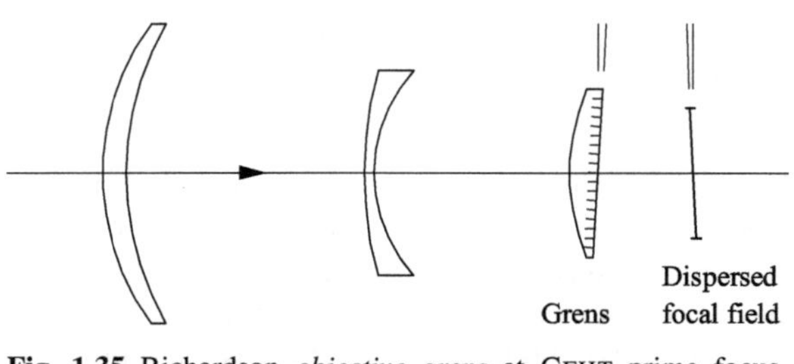

Fig. 1.35 Richardson *objective grens* at CFHT prime focus. The wedge of the third lens is 1.1° with a 45 ℓ/mm grating giving 1,000 Å/mm dispersion on a focal plane tilted 0.38° (after Richardson [127])

1.12.8 Multi-Object Spectroscopy with Slits or Fiber Optics

If a spectrograph simultaneously disperses objects whose positions in the field are *discrete*, the process is called *multi-object spectroscopy*. Two classical alternatives are with multi-slit masks or optics fibers.

• **Multi-slit mask**: From the spectrograph imaging mode, the coordinates of the objects are selected in a rectangular field and a *multi-slit mask* is then made as input of the spectrograph mode. The spectra are generally obtained on a rectangular detector and have homothetical positions to those of the objects in the telescope field. This technique is well adapted for a typical 5-7 arcmin field. Accurate *multi-slit masks* are generated by high-power YAG lasers which, in addition to rectangle slits, allow curvilinear-slit cuts of constant width in the dispersion direction for arch-like object studies (Di Biagio et al. [14]).

• **Fiber optics**: Optics fibers have the advantage of selecting objects whatever their location in a small field (5-7 arcmin) or a large field. The fiber output ends feed the *long slit* of one or several spectrographs. However, optics fibers suffer from *focal ratio degradation* (FRD) so that the collimator of the spectrograph must have a somewhat faster f-ratio than that of the emerging telescope beams. Some examples of large field instruments are the galaxy and quasar surveys 2dF (400 fibers, 2° field at 4 m AAT), SDSS (640 fibers, $2 \times 1.5°$ field at 2.5 m extended field R-C), 6dF (150 fibers, 6° field at 1.2 m UK Schmidt) and next LAMOST (4000 fibers, 5° at 4 m Schmidt) (cf. Table 1.1). For a large sky scale and low object density, each fiber input can be motorized.

1.12.9 Integral Field Spectrographs

If a spectrograph allows the entrance of all N^2 contiguous subareas of a field area $(\delta x, \delta y)$ subdivided into $N \times N$ elements, the arrangement is an *integral field spectrograph* (IFS). The data recording process $(\delta x, \delta y, \lambda)$, then called *3-D spectroscopy*, allows obtaining simultaneous spectra of all the field image elements of the seeing size. The basic constraints for acquiring three-dimensional information – the so-called "data cube" – with two-dimensional detectors was discussed by Monnet [109].

If an arrangement allows integral field spectroscopy when placed in series with a classical spectrograph, this is called an *integral field unit* (IFU).

G. Courtès showed in 1952 [40] that the detection of extended objects is greatly increased by use of fast camera optics. Such systems which then reduce the efl of a telescope are called *focal reducers*. Usually, the dispersor – grism, Fabry-Perot, colored or interference filters – is introduced in the afocal beams before the camera optics. Sometimes it can be very efficient to use an interference filter in convergent beams (Courtès [41]). An interesting instrumental concept – which is not really a spectrograph but a simultaneous *multi-bandpass imager* or *multi-bandpass photometer* – consists of reimaging several spectral zones of a dispersed image of the telescope pupil on individual regions of a detector or on several detectors (Courtès [40]); the optical design is facilitated by use of a *white pupil mounting* [11]. These multi-bandpass photometers are efficient systems for the determination of faint astronomical objects having a large redshift.

We may distinguish several alternatives for IFS, all using collimating optics, disperser or scanning interferometer, and camera optics.

• **Bundle fiber and lenslets**: A *bundle fiber* with the output fiber ends feeding a long slit provide a simple arrangement for an IFU. However, the light losses because of the gaps between the cores of the fibers can be avoided by use of a 2-D microlens array which images the telescope pupil on the fiber input. This requires re-arranging the entrance bundle fiber, also in number N^2, into a cartesian frame such as proposed by Courtès [42] and developed by Allington-Smith et al. [4]. At the fiber output, a 1-D microlens array may also be used for a sharp re-imaging of the field of view elements on one or several *long-slit spectrographs*.

• **Lenslets**: A mostly transparent IFS consists to form the telescopic image on a 2-D *microlens array*, or *lenslet*, which images the telescope pupil into N^2 subpupils on a mask. The mask has N^2 holes, much smaller than the microlenses, which allow all the light to enter the spectrograph. This provides a simple arrangement for an IFU. After the beam collimation a grism must be appropriately oriented about the axis of the camera optics in order to optimize the 2-D filling of the spectra on the detector area (Courtès [43]). Depending on the spectral dispersion, the filling of the N^2 spectra on the detector area requires an optimization involving the microlens size, the detector size, and the grism orientation with respect to the lenslet. The well-defined hole- or slit-function and optical étendue of this system (Jacquinot [81]) provide a uniform spectral resolution for stellar or extended objects.

• **Scanning Fabry-Perot**: For an N^2 high sampling of contiguous subareas, another alternative is to use a *scanning Fabry-Perot* instead of a classical disperser. As shown by Caplan [25] and Bland-Hawthorn [16], this requires the data processing of a more complex point spread function, the Airy function, than with a grating spectrograph. The temporal fluctuation of the sky during the integration makes it difficult to obtain an accurate intensity ratio of two distant spectral lines unless rapid scanning is used (Caplan [25]). Some accounts on these instruments are given by Le Coarer, Amram et al. [35], Georgelin, Comte et al. [67] and Tully [163].

• **Fourier transform spectrometer**: With the advent of 2-D infrared detectors, 3-D or imaging *Fourier transform spectrometers* (FTS) have been designed and built for the near infrared by Maillard [99], who proposed an option for the JW Space Telescope. The state-of-the-art of detector technology only allows a moderate N^2 number for the far infrared, however the SPIRE FTS, with optics designed by Dohlen et al. [47], is one of the two imaging spectrometers installed on the Herschel Space Telescope for observations from the L_2 Lagrange point of the Sun-Earth system.

• **Image slicer**: Bowen [20] introduced the first *image slicer* (IS) to avoid the light loss at the *narrow slit* of high-resolution stellar spectrographs. This uses a 1-D plane-facet array of k facets which by appropriate orientation reflects at 90° the telescope image of a square area N^2a^2 of N^2 elements into k rectangular elements of area $(Na/k) \times (Na)$. Richardson [126, 129] introduced a new IS concept with two concave mirrors of equal curvature whose centers of curvature are on their reciprocal vertices. Each mirror is separated in two halves – the slit mirrors and the aperture mirrors – to form a straight slot, and the two slots are arranged in orthogonal directions. The starlight enters the IS through a cylindric lens which produces two mutually perpendicular line images: the first image line is at the slot of the aperture mirrors and the second line coincides with the slot of the slit mirrors. These later mirrors allow the central part of the light to directly pass through whilst the reflected light goes back to the aperture mirrors. Slight tilts of all four mirrors provides the next sliced subfield elements after even numbers of reflections, thus requiring efficient coatings. A lens located at the slit images the telescope pupil on the spectrograph grating. The sliced subfields are in odd number k and symmetrically distributed end-to-end from the central subfield. The gain in detection is obtained by superposing all the k spectra; this is achieved by a cylindric lens located after the camera optics of the spectrograph (cf. also the account by Hunter [79]). In the Walraven IS [164], the beam propagates by multiple internal reflections into a plane parallel glass plate which is in contact with a triangular area of a prism to allow the light to emerge. Starting from a modified-Walraven IS, Diego [46] added to this system input and output prisms of same wedge, thus obtaining k sliced images all co-focused side-to-side at the same axial position.

Multiple reflections can be avoided with the Lemaitre IS [94] which is designed from two spherical concave mirrors. The slicing transformation of a square field into k rectangular subfields is achieved by uniform rotation of k element areas of each mirror around a lateral axis, so the facets are distributed along helices. If the radii of curvature of the facets are $2R$ for the *field mirror array*, R for the *pupil mirror array*, and their axial separation is R, the sliced subfields are reimaged by the pupil mirror with transverse magnification $M = -1$. In the stellar mode for high spectral resolution, the reimaged subfields are just superposed (Fig. 1.36-*Up*). Therefore, the Étendue or Lagrange invariant expressed by (1.34) immediately entails that a square section of the telescope beam of aperture angle u is transformed into an anamorphosed section whose linear semi-apertures are $u'_x = ku$ and $u'_y = u$ in accordance with the output Étendue $(N^2a^2/k) \times (4ku^2)$. The emerging beam is given a square section at the spectrograph pupil – i.e. the grating – by inserting a cylindrical field lens near the slit. In the integral field spectroscopy mode, the subfields must be arranged end-to-end to feed a long-slit spectrograph. This requires a second rotation of the facets of the pupil mirror array around the other lateral direction (Fig. 1.36-*Down*), so this mirror is much more difficult to build since it is not directly generated by differential rotations of parallel slice elements from a continuous mirror surface. This mode, developed with the Content IS [37] for IFUs, uses an additional 1-D optics array located at the slit array to reimage the pupil at a correct position in the spectrograph. Transverse magnifications differing from -1 are obtained by changing the curvature of the pupil mirror.

When the slice number k becomes large, one may encounter efficiency difficulties due to the amount of diffracted light from the small mirror elements, so a compromise must be adopted by appropriate scaling of the IS.

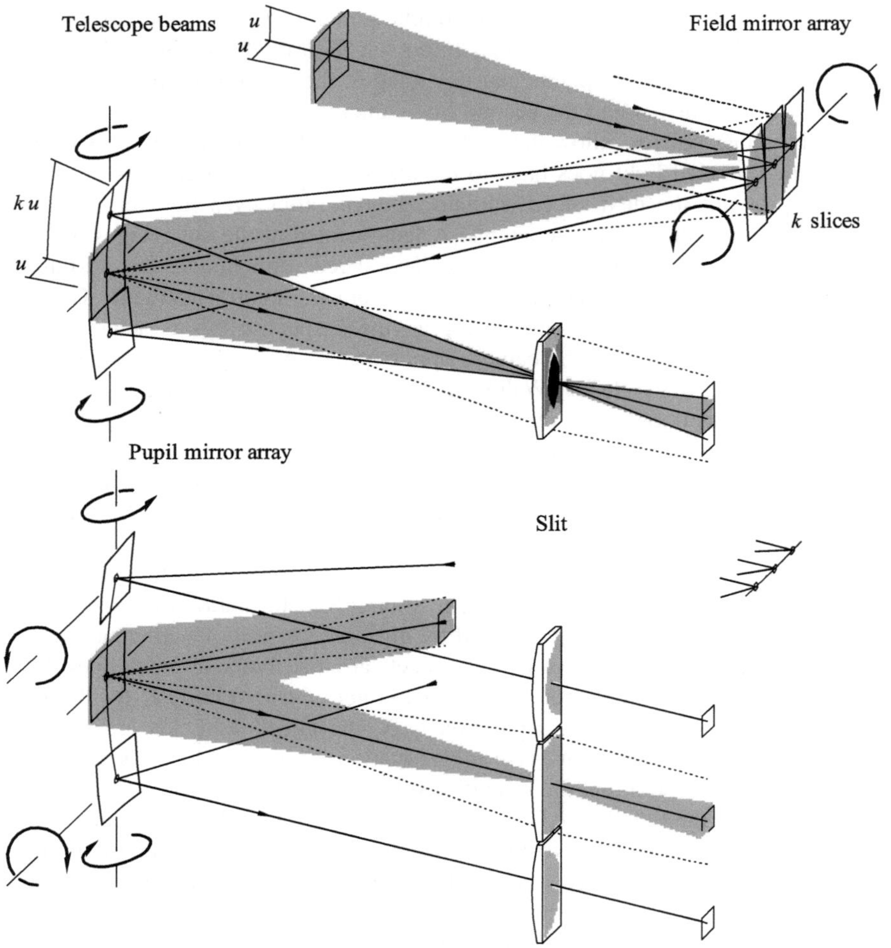

Fig. 1.36 Principle of the helix image slicer. The pupil is imaged by the field mirror array on the pupil mirror array. The reimaging of the sliced subfields by the pupil mirror is here shown with magnification $M = -1$. (*Up*): Classical mode for superposed spectra, (*Down*): Long slit mode for IFUs (after Lemaitre [94] and Content [37])

1.12.10 Back-Surface Mirrors

The large mirrors in telescopes are always front-surface mirrors, but for small mirrors there are advantages with back-surface mirrors. A common example is a right-angle deviation prism. In that case no metal coating is required because the reflection at the diagonal is by internal reflection. If the angle of incidence of the light is lower than the *critical incidence angle* it is necessary to have a reflective coating on the outside of the back surface, such as in a bathroom mirror. This reflective surface can be well protected against corrosion and dust.

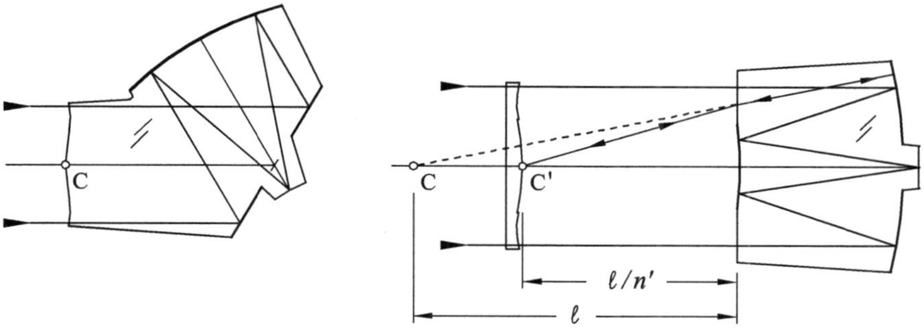

Fig. 1.37 Solid-folded (*Left*) and semi-solid (*Right*) Schmidt spectrograph cameras

If a concave back-surface mirror has a shorter radius on the front than on the back, it is called a *Mangin mirror*. Given a wavelength, the spherical aberration caused when light passes through the front is balanced by that caused by the reflection at the spherical back. Mangin mirrors are used in some spectrographs as the camera mirror.

If the glass is made thick, a very fast f-ratio can be designed such as *solid* or *semi-solid Schmidt spectrograph cameras* (cf. for instance *Astronomical Optics* by Schroeder [141]). For detector access, a solid Schmidt is designed in a folded form whilst a semi-solid Schmidt is designed in the Cassegrain form (Fig. 1.37). Compared to a classical Schmidt in the air where the focal ratio f/D is defined by the curvature of the mirror, the same curved back-surface mirror provides a focal ratio $f/n'D$ which thus is faster by the refractive index n'. Therefore, since the efl is n' times shorter, in a direct imaging mode these systems provides a gain in detection of a factor n'^2.

1.12.11 Field Derotator

If a telescope compensates for the Earth's rotation by altitude and azimuth motions, i.e. an *alt-az mount* telescope such as usually preferred for apertures larger than 5 m, the rotation of the field of view must be compensated by a *field derotator*. This is achieved by three plane mirrors all rotating around the field axis (Fig. 1.38). The processor may be substituted by an axial rotation of the focal instrument.

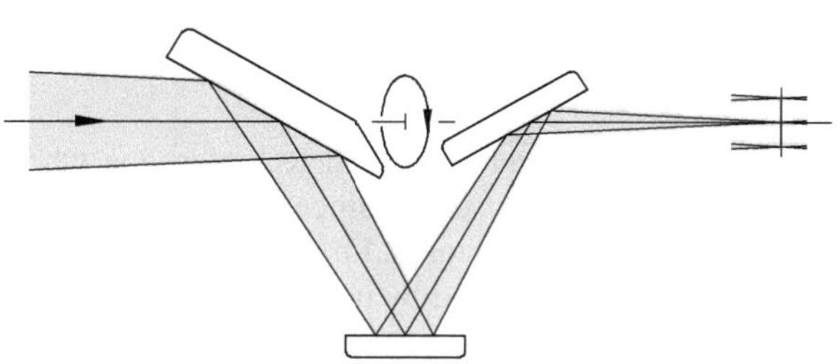

Fig. 1.38 Three-mirror field derotator

1.12.12 Pupil Derotator

If a coronograph is installed in a fixed position on the platform of a Nasmyth focus, the processor is a *pupil derotator*. Its rotation rate is opposite to that of a field derotator, whilst this latter device is not used. Thus, the residual aberrations of the telescope and the diffracted light from the spiders supporting the secondary mirror do not rotate at the detector during the observation.

1.12.13 Telescope Field Corrector

If *Sphe* 3, *Coma* 3, *Astm* 3, and *Petz* 3 and some higher-order aberrations are simultaneously corrected at a telescope focus, the processor is called a field corrector. Then the telescope benefits from an extended field of view. The first correctors were required for the prime focus. One may distinguish between corrector types for paraboloid-hyperboloid (PH) or for Ritchey-Chrétien (RC) telescopes.

• **Prime focus of PH telescopes**: Only considering the correction of *Coma* 3 without introducing *Sphe* 3, F.E. Ross [135, 136] designed the first correctors of paraboloid mirrors for the 60-inch at Mount Wilson in 1933 and later the 200-inch at Palomar. His correctors included a thin low-power meniscus, with a strongly concave first surface, placed at some distance from a thin air spaced afocal doublet, itself at some distance from the focus. A *Ross corrector* provides a limited field extent since *Astm* 3 and *Petz* 3 still remain with also some chromatic differences for *Sphe* 3 and *Coma* 3. Subsequently, A.B. Meinel [106] investigated the field correction by several aspherical plates and showed that *Coma* 3, *Astm* 3 and part of corresponding 5th-order terms can be corrected. Cancelling the four 3rd-order aberrations and some higher-order terms, C.G. Wynne [172] introduced in 1967 a four lens corrector with all lenses of the same glass and where the two first lenses are thin air separated whilst the next ones are two spaced lenses. Considering the two first lenses as equivalent to a positive lens, the power distribution positive-negative-positive provides the best arrangement for the correction of a large field. Faulde and Wilson [55] then showed that an equivalent performance is obtained with three spaced lenses whose central negative lens is with an aspherical surface. Epps et al. [50] introduced, for the Keck Telescope, a second aspherical surface in a three spaced lens corrector which included an option with atmospheric dispersion compensator. The widest field corrector presently built for a paraboloid mirror is $1° \times 1°$ field at the Canada France Hawaii Telescope (CFHT).

• **Prime focus of RC telescopes**: For the prime focus of an RC telescope where the primary is an hyperboloid, the field correction is easier since *Sphe* 3 is available from the mirror to compensate for that of additional lenses. First, Gascoigne [64, 65] showed that both *Sphe* 3 and *Coma* 3 can be corrected by using an aspherical plate located at an appropriate distance from the focus. A *Gascoigne corrector* provides a limited field extent since *Astm* 3 and *Petz* 3 still remain with also some chromatic differences for *Sphe* 3 and *Coma* 3. Next, cancelling the four 3rd-order aberrations and some of the next order, Wynne [173] designed several correctors for large telescopes, showing that three spaced lenses in the same glass, with a power distribution positive-negative-positive, provides a wide field by only using spherical surfaces. Richardson [130] showed that given a field of view and f-ratio, there exists an optimum corrector size that matches the image quality requirement.

• **Prime focus of spherical primary telescopes**: Large segmented telescopes like the Hobby-Eberly Telescope (HET) require correcting the huge amount of *Sphe* 3 and the next aberrations. For a typical field of view of 4–5 arcmin, two holed pairs of concave mirrors provide the correction (Ramsey et al. [122], Sebring et al. [143]). These aspherical mirrors have their concavity in front together, and work as image transports by successively re-imaging the field at a second and third focus.

• **Cassegrain focus of RC telescopes**: The Cassegrain focal surface of an RC telescope can be made flat by equalizing the curvature of the mirrors (cf. the Petzval sum in Sect. 1.10.1). In this special form, the aplanatic combination is called a *flat field Ritchey-Chrétien*. This was realized by I.S. Bowen with the 2.5 m Irénée du Pont de Nemours Telescope (Table 1.1). The next correction is *Astm* 3 which was obtained by Bowen and Vaughan [21] with a single aspherical plate located at some distance before the focus. This requires a slight re-design of the mirrors which thus departs from a strict RC form to provide an *extended field Ritchey-Chrétien*. Because of the remaining chromatic difference of *Astm*3 and other residual aberrations, Wynne [176] further replaced the aspherical plate by two spaced aspheric lenses of different glass. Examples of such extended field telescopes are the ultraviolet space telescope GALEX 1.2° field, and the Sloan Digital Sky Survey Telescope (SDSS) 2° × 1.5° field (Table 1.1).

A detailed account on telescope field correctors is given by Wilson [170].

1.12.14 Atmospheric Dispersion Compensator

If the processor is two prismatic units which can increase or decrease the amount of dispersion to match that of the atmosphere, this system is called an *atmospheric dispersion compensator* (ADC). Two concepts are used for ADCs: (1) two prism pairs which are counter rotated, and (2) two thin single prisms with a variable axial separation.

In a study of refraction, Claudius Ptolemy (∼140 AD in Alexandria) mentioned the effect of *atmospheric refraction*; at that time, the refraction was assumed as proportional to the incidence. Tycho Brahe made some attempts to measure the atmospheric refraction, but his correction of the parallax (3 arcmin) for the determination of the planet position was totally false which postponed its refraction correction. The first model of the atmospheric refraction goes back to J-D. Cassini (1662), who assumed that the atmosphere was a medium with constant index n_a abruptly ending at a fixed elevation h_a before the "æther." Although *Cassini's model* only requires use of Snell's law by appropriate choice of the two constants (n_a, h_a), we may notice

Table 1.8 Refractive index of the air versus wavelength λ for p_0=1013.25×10^2 Pa, T_0=273.15 K, and p_w=0 from Allen and Cox [3]

λ [nm]	300	400	500	600	700	800	1,000	2,000
$(n-1)\,10^6$	307.6	298.3	294.3	292.2	290.9	290.1	289.2	288.0

For a water vapor pressure p_w=550 Pa, subtract 0.2 to $(n-1)\,10^6$; for other values apply a linear correction. For other temperatures and pressures multiply $(n-1)$ by $p\,T_0/p_0T$.

that his resulting refraction angles – difference of two arsin functions – are remarkably accurate for zenith distances up to 45°.

Considering a more realistic model, let us denote n the refractive index of the air and z the zenith distance, the atmospheric diffraction angle is accurately represented for $z \leq 80°$ by

$$\varphi = \big[n(\lambda,\ p,\ T,\ p_w) - 1\big] \tan z\,, \tag{1.94}$$

where λ, p, T, and p_w are the wavelength, air pressure and temperature, and water vapor pressure at the observation place. The values of the refractive index in Table 1.8 are extracted from *Allen's Astrophysical Quantities* [31].

The *atmospheric dispersion* is the variation of the refraction in a given wavelength range (λ_1, λ_2). This transforms a star image into a spectrum, and the purpose of an ADC is to correct this degradation. From (1.94), the angular spectrum caused by the atmospheric dispersion may be expressed as

$$\Delta\varphi = (n_{\lambda 1} - n_{\lambda 2})\ \tan z\,.$$

From the ultraviolet atmospheric cut-off to the infrared limit of sensitivity of a CCD detector, corresponding to the spectral range 300–1,000 nm, one obtains from Table 1.8 the angular dispersions

$$\Delta\varphi\{z{=}45°\} = 3.79\ \text{arcsec}, \qquad \Delta\varphi\{z{=}60°\} = 6.57\ \text{arcsec}.$$

It can be shown from these results that, even for a telescope located at the high elevation of 4,000 m, the image degradation must be corrected for an upgraded imaging.

- **Counter rotated prism pairs**: A first form of ADC is with two N-shaped prisms which form a *counter rotated prism compensator* (Fig. 1.39). Each prism pair does not deviate the beams for the wavelength corresponding to the mean refractive index of the central spectral range. Such designs were developed by Wynne [174], Epps et al. [50] as an option included to a wide-field corrector, Wynne and Worswick [175], D-q. Su [151], Bingham [15], Wang and Su [165] and others.

- **Varying separation thin prism pair**: A second form of ADC is with two thin prisms whose axial separation is varied (Fig. 1.40). This concept proposed by Beckers was designed by Avila and Rupprecht [7] who called it a *linear atmospheric dispersion compensator* (LADC). Compared to the previous concept by counter

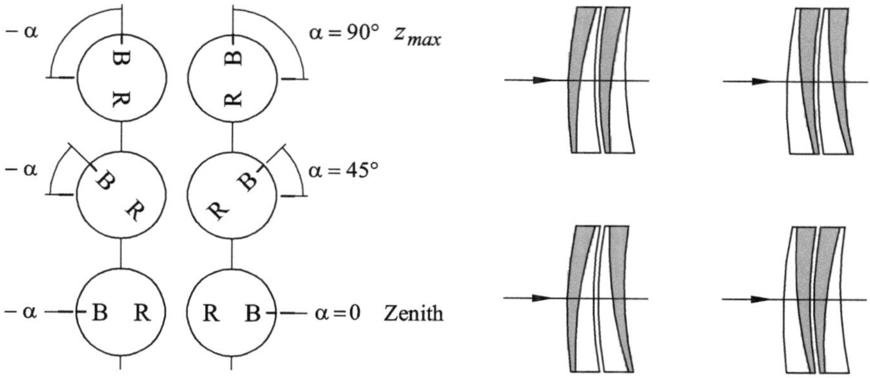

Fig. 1.39 *Left*: Schematic of counter rotations of each prism pair in an ADC. The maximum compensation is when the prism pairs co-add their dispersion, at α=90° (B: blue, R: red). *Right*: The four possible forms of an ADC shown at maximum zenithal compensation z_{max} (most dispersive glass shown in gray)

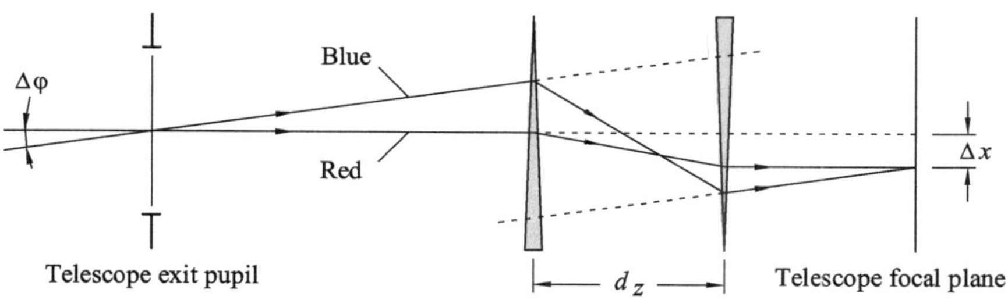

Fig. 1.40 Schematic of an ADC with two thin prisms and varying separation - LADC/VLT (after Avila, Rupprecht and Beckers [7])

rotation, the design is much longer but shows the advantage of avoiding tilting of the pupil axis and, with only one glass (silica), benefits from higher throughputs from ultraviolet to infrared.

In an alt-az telescope, the ADC is usually located in series with the field derotator.

1.12.15 Adaptive Optics

• **Seeing-limited and diffraction-limited imaging**: Ground-based astronomical observations suffer from wavefront degradations caused by the Earth's atmosphere. Thus, a large telescope with *passive optics* can only provide a *seeing-limited* imaging whilst its aperture diameter should theoretically allow the much higher angular resolution of *diffraction-limited* imaging. For extremely short exposures, a seeing-limited star image is made of a collection of speckle elements whose individual size is that of an Airy pattern. Most observation sites located at near-sea-level elevations show a 1-2 arcsec seeing. For exceptionally good sites at 4,200 m elevation (Mauna Kea) the seeing is 0.5–1 arcsec.

If the processor is a telescope mirror whose shape, orientation, and position can be modified by actuators, controlled either by an open- or a closed-loop system working at a low time frequency, say, $f \leq 0.1$ Hz, then the system is called an *active optics system*.

If now the *local* shape and *tip-tilt* positions of a mirror are modified by a closed-loop control at high time frequency, say, $f \geq 50$ Hz, which compensates for the image degradation caused by turbulence in the Earth's atmosphere, then the system is called an *adaptive optics system*.

• **Atmospheric seeing corrections with deformable mirrors**: Use of a telescope with adaptive optics corrections allow obtaining a sharp image of a star, i.e. close to the Airy pattern. This is usually called *high angular resolution* imaging. For this purpose the deformable mirror (DM) of an adaptive optics system must be located at or near a reimaged pupil of the telescope input pupil (Fig. 1.41).

In some recent telescopes, a first DM is the Cassegrain mirror which thus is the telescope input pupil. Present studies of extremely large telescopes (ELTs) include a large pupil DM in their main mirror train.

Realtime wavefront sensing of stars – dichroically separated or at the telescope field edge – and fast algorithms for the wavefront reconstruction allow driving the DM actuators. The closed-loop control also implicitly compensates for low aberration residuals and vibrations of the telescope optics, say, for a fraction of a wavelength. If there are not enough bright stars in the field of view, the light of a laser guide star (LGS) allows wavefront analysis by use, for instance, of the Na doublet lines retro-reflected by the atmosphere.

Because the turbulence effects are mainly due to several dominant atmospheric layers, a single layer wavefront correction only provides an *isoplanatic field* of ∼ 30 arcsec in the near infrared. Using several DMs coupled with several wavefront guide stars allow increasing the isoplanatic field up to 2-3 arcmin. Such a technique is called *multi-conjugate adaptive optics* (MCAO).

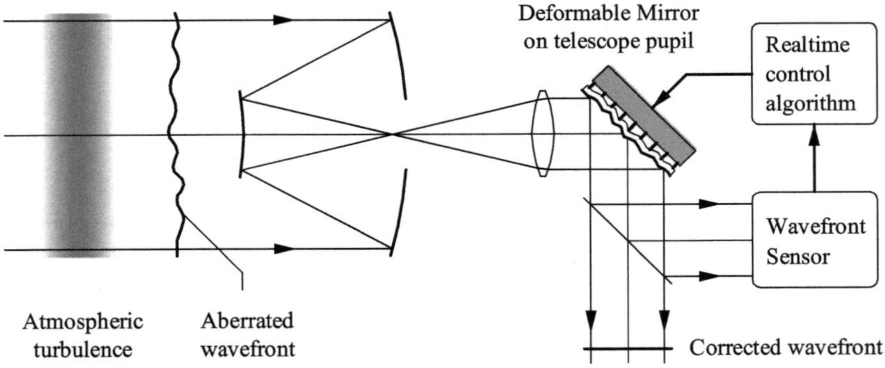

Fig. 1.41 Principle of adaptive optics closed-loop control

Complementary DMs with $\sim 10^4$ actuators under development for the ELT projects are: large and very thin mirrors of 2–3 m aperture directly usable as a component of the telescope main optical train; smaller mirrors of 10–20 cm aperture located after a telescope focus; micro-opto-electro-mechanical mirrors (MOEMS) of 20–30 mm aperture also implemented after a telescope focus.

Whatever the principle of the acting forces – magnetostriction induced strain or piezoelectric bimorph strain – the wavefront correction is actually generated onto a thin and continuous reflective plate. Although in most DM systems the reflective plate is totally of uniform thickness, it is clear that the intermediate regions where no force is applied should be designed with a smaller thickness than that in the regions where the forces must act. A mirror plate made of continuous *tulip-like thickness distributions*, similar to those found in Sect. 2.1.2, could improve the performance in terms of dynamic range and influence function when the tulip profile element has a biaxial or a threefold symmetry.

1.13 Elasticity Theory

The elaboration of elasticity theory is based on infinitesimal formalism and is a more recent science than dioptrics. However, one may notice that for particular cases such as typically with the development of the bow over the millenniums, the mean to store a maximum energy was empirically found by giving the bow cross-sections an appropriate thickness distribution. This problem leads to *equal constraint cantilever bars* of particular thickness distributions whose area of the cross-section may be null at the unclamped end (cf. Galileo [63], Clebsch [34], Saint-Venant [138]), but may also remain of finite value (Lemaitre [95]).

Noticeable accounts on the development of elasticity theory are given by Todhunter and Pearson [161], Love [97] and Timoshenko [156, 158, 159], Timoshenko and Woinowsky-Krieger [155], and Landau and Lifshitz [92]. One briefly summarizes hereafter the first milestones in the early development of elasticity theory.

1.13.1 Historical Introduction

It was Galileo Galilei who introduced the first problem of elasticity in considering the strength of solids to rupture a cantilever beam in *Discorsi e Dimostrazioni Matematiche* (1638). This, known as *Galileo's problem*, consisted of determining the cross-section thickness distribution along the horizontal axis of a beam whose one end is built into a wall whilst the tendency to break it arises from its own weight or a load applied to its other end (Fig. 1.42).

Although not knowing the law linking the elastic displacements to the forces producing them, Galileo nevertheless clearly enounced and emphasized this important question of strength. This led to the fundamental discovery of the proportionality

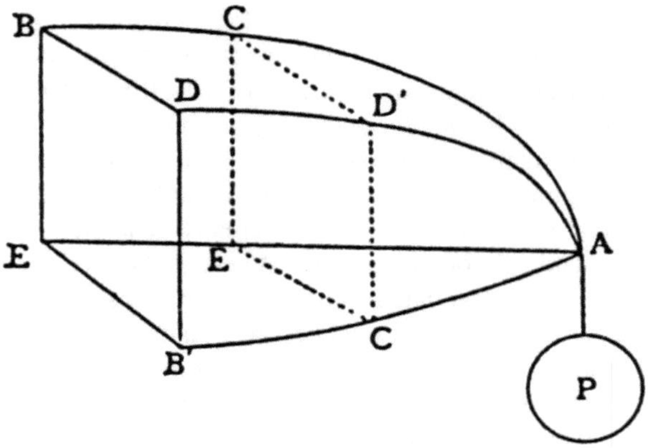

Fig. 1.42 Galileo's problem of *equal strength cantilevers* for a load. His drawing of the thickness variation of a cantilever beam is close to the exact solution (Galileo, *Discorsi e Dimostrazioni Matematiche* [63])

law between stress and strain by Hooke that he published in 1678 in *De Potentia Restitutiva* [77], known as *Hooke's law*.

Without knowing Hooke's law, Mariotte [104] reformulated it in 1680 and remarked that the flexure of a beam arises from parts of its section which are in extension and in contraction (cf. Love [97]). He concluded that the position of the neutral curve, required in the solution of Galileo's problem, is at the center of the section.

• **Equation of equilibrium of rods and beams**: Considering Galileo's problem, Jacob Bernoulli in 1705 also concluded that the resulting elastic bent curve arises from the extension and contraction zones which are separated at the middle of the cross-sections by the neutral surface. This entails that the bending moment M is proportional to the curvature $1/R$ of the rod when bent, a result which was later assumed by Leonhard Euler (1707–1783). This local proportionality is expressed by

$$M = EI\,/\,R\,, \tag{1.95}$$

where E is the *Young modulus* characterizing the material whose first notion was introduced by Young [177] in 1807 – who also introduced the concept of shear as an elastic strain –, and I the *moment of inertia* of the cross-section around the inertia axis which is perpendicular to the flexure plane and passes through the center of the section (this name is because of the analogy with an inertia mass around a straight line).

For a circular section of diameter $2a$, its surface element $d\mathcal{A}$ is delimited by the contour and two parallel lines to the inertia axis. Denoting d the distance of $d\mathcal{A}$ to this axis, the moment of inertia is

$$I = \int_{\mathcal{A}} d^2\, d\mathcal{A} = \tfrac{1}{4}\,\pi\, a^4\,. \tag{1.96a}$$

For a beam of rectangular cross-section with sides $2a$, $2b$, the principal axes of inertia pass through its center and are parallel to the sides. The principal moments of inertia are

$$I_1 = \tfrac{4}{3}\, a^3 b\,, \qquad I_2 = \tfrac{4}{3}\, ab^3\,. \tag{1.96b}$$

It was early known that the work done by a moment is the product $M\,\delta\theta$ of moment and the angle of rotation. Thus from (1.95), the strain energy per unit length of the rod may be denoted $\frac{1}{2}\,EI/R^2$ where the varying quantity is the square of the curvature. In 1742, Daniel Bernoulli [13] suggested to Euler that the differential equation of the flexure could be found by making the integral of the square of the curvature along the rod a *minimum*. Euler, acting on this suggestion, considered *thin* rods of a constant circular section and straight in a unstressed state on which opposite forces F and moments M are applied to each end. Using curvilinear coordinates s, θ along the rod where θ is the angle between the tangent to the rod and the z-axis and where $dy = ds \sin\theta$, $dz = ds \cos\theta$, the general differential equation of the flexure is

$$\tfrac{1}{2}\, EI\,(d\theta/ds)^2 + F\,\cos\theta = c_1\,, \tag{1.97}$$

where c_1 is a constant. Euler [51] derived this equation in 1744 and gave some solutions of *elastica* including zero, one or several loops (Fig. 1.43). He also classified them noticing that the curves may or may not include inflectional points.

Considering thin rods or thin beams, (1.97) leads to the curvilinear length

$$s = \pm\sqrt{\tfrac{1}{2}\, EI}\int_{\theta_1}^{\theta} \frac{d\theta}{\sqrt{c_1 - F\cos\theta}} + c_2\,, \tag{1.98a}$$

so the function $\theta(s)$ can be obtained in terms of elliptic functions (cf. for instance Landau and Lifshitz [92]). The parametric equations representing the flexure of the central line of the thin rod are

$$y = \pm\sqrt{2EI\,(c_1 - F\cos\theta)/F^2} + c_3, \tag{1.98b}$$

$$z = \pm\sqrt{\tfrac{1}{2}\, EI}\int_{\theta_1}^{\theta} \frac{\cos\theta\, d\theta}{\sqrt{c_1 - F\cos\theta}} + c_4. \tag{1.98c}$$

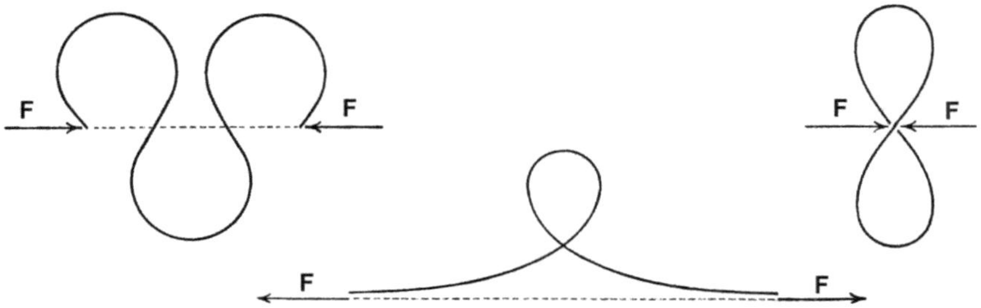

Fig. 1.43 Plane flexures of a thin circular rod. These curves – also called *elastica* – were obtained by Euler [51] who classified the various shapes (after Love [97])

• **Equal curvature bar of constant cross-section**: If the force F is nulled and a moment M per unit length around the x-axis is applied to the free end, this moment is introduced in (1.97) by setting $c_1 = M$. Hence $1/R = \sqrt{2M/EI}$ = constant and we obtain an *equal curvature bar*. The curvilinear length is

$$s = \sqrt{EI/2M}\ (\theta + \text{constant})\,. \tag{1.99}$$

• **Cantilever beam**: Restricting to the case without loops i.e $0 \le \theta \le \pi/2$ or $-\pi/2 \le \theta \le 0$, Eq. (1.98) allows solving the Galileo problem of a cantilever of constant-section clamped at one end along the y-axis, where $s = 0$, $\theta = \pi/2$, and deformed at the other free end by a force F parallel to the z-axis where $M = 0$ i.e. $d\theta/ds = 0$ (Fig. 1.44-*Left*). Denoting ℓ the total length of the bar and $\theta(\ell) = \theta_0$ its rotation at the free end, these conditions entail

$$c_1 = F\cos\theta_0, \quad c_2 = 0, \quad c_3^2 = 2EIc_1/F^2, \quad c_4 = 0, \tag{1.100a}$$

where the negative sign is taken for the square root term of (1.98b) and the positive for the two other equations. The equation for θ_0 is

$$\ell = \int ds = \sqrt{\frac{EI}{2F}} \int_{\theta_0}^{\pi/2} \frac{d\theta}{\sqrt{\cos\theta_0 - \cos\theta}}\,. \tag{1.100b}$$

We may consider now the case of a small deformation, i.e. where θ_0 is close to $\pi/2$. Using the complement angle $\varphi = \pi/2 - \theta$ and approximating $\sin\varphi$ by φ, the integrations of ℓ and z provide respectively at the free end of the cantilever

$$\varphi_0 = \frac{F\ell^2}{2EI}, \qquad z(\varphi_0) = \frac{F\ell^3}{3EI}\,. \tag{1.100c}$$

The flexure $z(\varphi)$ may be represented by a polynomial form $z(y)$ with coefficients defined by the boundaries. Hence, from (1.98b) and (1.98c), we obtain

$$z = \frac{F}{6EI}\,(3\ell y^2 - y^3)\,, \tag{1.100d}$$

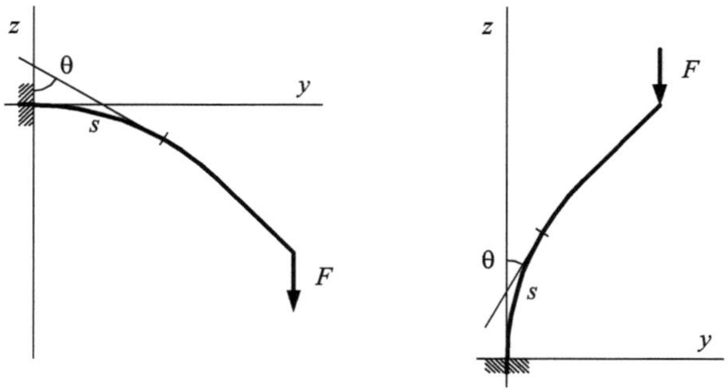

Fig. 1.44 Cantilever bar (*Left*). Compressed bar (*Right*)

which gives $[d^2z/dy^2]_{y=\ell} = 0$, accordingly to a null curvature at the free end.

If the force F is suppressed and a moment M around the x-axis is applied to the free end, the flexure of the central line is approximated by a parabola $z = (M/2EI)\,y^2$; in fact this result formally corresponds to the previous case of equal curvature bars. For a cantilever where the force is applied at some distance from its end, or for cases differing from a cantilever such as bars with both free ends, similar representations of φ_0, $z(\varphi_0)$ and $z(y)$ are of classical use (cf. for instance Roark and Young [134]).

• **Variable cross-section cantilevers – Equal constraint – Parabolic flexure**: Beside determining the strength of a cantilever, Galileo's problem implicitly included the determination of the variable cross-section of a bar of equal strength, a case which we now call *of equal constraint*. He intuitively produced a drawing close to a correct solution (cf. Fig. 1.42).

Galileo's problem of a variable cross-section cantilever of equal constraint was first solved by Clebsch [34], in 1862, for a beam of constant width and a force concentrated at the unclamped end. If a beam is of width a=constant in the x-direction, the thickness solution in the z, y plane (i.e. the vertical plane if the force is a weight) is a *parabola* whose vertex is at the beam end.

Denoting M_y the bending moment along the y-axis, and Q_y the shearing force acting in the bar, the statics equilibrium is represented by

$$\frac{dM_y}{dy} + Q_y = 0\,. \tag{1.101}$$

Considering both cases of *equal constraint* and *parabolic flexure* cantilevers, the maximum stress σ and the curvature of the neutral line $1/R$ in the y, z plane of a bar are related by $\sigma = Et/2R$, where $t(y)$ is the thickness in this plane and E the Young modulus. A parabolic flexure is expressed by $z = 1/2R\ y^2$. Hence the bending moment is expressed by

$$M_y = EI_x\,\frac{d^2z}{dy^2} = I_x\,\frac{E}{R} = 2I_x\,\frac{\sigma}{t}\,, \tag{1.102}$$

where, from (1.96b), the inertia moments are $I_x = \pi t^4/64$ for a rod, and $I_x = at^3/12$ for a beam. The shearing force is determined at distance y as the sum of the external forces acting from the clamped origin of the bar to this distance. Denoting ℓ the length of the cantilever, for a rod of total weight P, the shearing force is

$$Q_y = P - \frac{\pi\mu g}{4}\int_0^y t^2 dy = \frac{\pi\mu g}{4}\int_y^\ell t^2 dy\,. \tag{1.103}$$

where μ and g are the density and the gravity.

Considering bars with loading cases as a concentrated force F at the unclamped end, a line force f per unit length applied along the bar, and a flexure due to own weight, these shearing forces are

$$Q_y = \begin{cases} \text{rod:} \quad F, \quad (\ell - z)f, \quad \frac{1}{4}\pi\mu g \int_y^\ell t^2 dy\,, \\ \text{beam:} \quad F, \quad (\ell - z)f, \quad a\mu g \int_y^\ell t\, dy\,. \end{cases} \tag{1.104}$$

where F, f, and g are negative since oriented towards negative, z.

After substitution of the moment and shearing force in (1.101), we obtain a set of 12 differential equations which can all be expressed in the general form

$$t^p \frac{dt}{dy} - \alpha G = 0\,, \qquad G = \int_y^\ell t^q\, dy\,, \tag{1.105}$$

where p, q are positive integers given in Table 1.9 together with coefficient α and function $G(t, y, \ell)$.

Table 1.9 Exponent p and terms α, G of the differential equation (1.105) of rod and beam cantilevers of *equal constraint* (σ) or *parabolic flexure* (y^2) flexed by various loads, with beam width a constant (after Lemaitre [95])

Bar	Condition	Exp.	Point force F		Line load f		Own weight	
type		p	α	G	α	G	α	G
Rod	$\sigma =$const.	2	$\frac{32F}{3\pi\sigma}$	1	$\frac{32f}{3\pi\sigma}$	$\ell - y$	$\frac{8\mu g}{3\sigma}$	$\int_y^\ell t^2 dy$
Beam	$\sigma =$const.	1	$\frac{3F}{a\sigma}$	1	$\frac{3f}{a\sigma}$	$\ell - y$	$\frac{3\mu g}{\sigma}$	$\int_y^\ell t\, dy$
Rod	$z = \frac{1}{2R}\, y^2$	3	$\frac{16RF}{\pi E}$	1	$\frac{16Rf}{\pi E}$	$\ell - y$	$\frac{4R\mu g}{E}$	$\int_y^\ell t^2 dy$
Beam	$z = \frac{1}{2R}\, y^2$	2	$\frac{4RF}{aE}$	1	$\frac{4Rf}{aE}$	$\ell - y$	$\frac{4R\mu g}{E}$	$\int_y^\ell t\, dy$

NB: F, f, and g are negative since towards the z-axis negative.

From the general equation (1.105) and the data in Table 1.9, we can derive the distribution $t(y)$ when the bar is with null or *finite* thickness at its end $y = \ell$. When the bar is of finite thickness we refer to the *truncated bar class*. Its length is given as the same length ℓ as that of an untruncated bar.

Denoting t_0 the thickness at $y = 0$, and t_ℓ the thickness at $y = \ell$, we briefly list hereafter the cantilever geometries resulting from integration.

1. *Equal constraint flexure of a rod and concentrate force F*: We obtain $t^2\frac{dt}{dy} - \frac{32F}{3\pi\sigma} = 0$. The solution

$$t = t_0\left\{1 - \left[1 - \left(\frac{t_\ell}{t_0}\right)^3\right]\frac{y}{\ell}\right\}^{1/3} \quad \text{with} \quad 1 - \left(\tfrac{t_\ell}{t_0}\right)^3 = -\frac{32\ell F}{\pi t_0^3 \sigma} \tag{1.106a}$$

is a *truncated cubical parabola* whose vertex is outside the rod, at $y_v = \ell/(1 - t_\ell^3/t_0^3)$. If the rod is with a null thickness at its end ($t_\ell = 0$) the result, first derived by Clebsch [34], is a *cubical parabola*, term that he introduced because of the even symmetry.

2. *Equal constraint flexure of a beam and concentrate force F*: We obtain $t\,\frac{dt}{dy} - \frac{3F}{a\sigma} = 0$. The solution

$$t = t_0 \left\{1 - \left[1 - \left(\frac{t_\ell}{t_0}\right)^2\right]\frac{y}{\ell}\right\}^{1/2} \quad \text{with} \quad 1 - \left(\tfrac{t_\ell}{t_0}\right)^2 = -\frac{6\,\ell F}{a\,t_0^2 \sigma} \tag{1.106b}$$

is a *truncated parabola* (Fig. 1.45). If $t_\ell = 0$, then $t^2 = t_0^2(1 - y/\ell)$ is a parabola.
3. *Parabolic flexure of a rod and concentrate force F*: We obtain $t^3\frac{dt}{dy} - \frac{16RF}{\pi E} = 0$. The solution

$$t = t_0 \left\{1 - \left[1 - \left(\frac{t_\ell}{t_0}\right)^4\right]\frac{y}{\ell}\right\}^{1/4} \quad \text{with} \quad 1 - \left(\tfrac{t_\ell}{t_0}\right)^4 = -\frac{64\ell RF}{\pi t_0^4 E} \tag{1.106c}$$

is a *truncated biquadratic* whose vertex is with null curvature.
4. *Parabolic flexure of a beam and concentrate force F*: We obtain $t^3\frac{dt}{dy} - \frac{4RF}{aE} = 0$. The solution

$$t = t_0 \left\{1 - \left[1 - \left(\frac{t_\ell}{t_0}\right)^3\right]\frac{y}{\ell}\right\}^{1/3} \quad \text{with} \quad 1 - \left(\tfrac{t_\ell}{t_0}\right)^3 = -\frac{12\ell RF}{a\,t_0^3 E} \tag{1.106d}$$

is a *truncated cubical parabola*.
5. *Equal constraint and parabolic flexure by a line load f*: One shows that for the four cases in Table 1.9, the general solution is of the form (Lemaitre [95])

$$t = t_0 \left\{1 - \left[1 - \left(\frac{t_\ell}{t_0}\right)^{p+1}\right]\left(2 - \frac{y}{\ell}\right)\frac{y}{\ell}\right\}^{1/(p+1)}, \tag{1.107a}$$

and the geometries for the truncated bar class are given by $p = 1, 2,$ or 3.

If the thickness at the edge becomes $t_\ell = 0$, these distributions degenerate and become more simply

$$t = t_0 \left[1 - \frac{y}{\ell}\right]^{2/(p+1)}, \tag{1.107b}$$

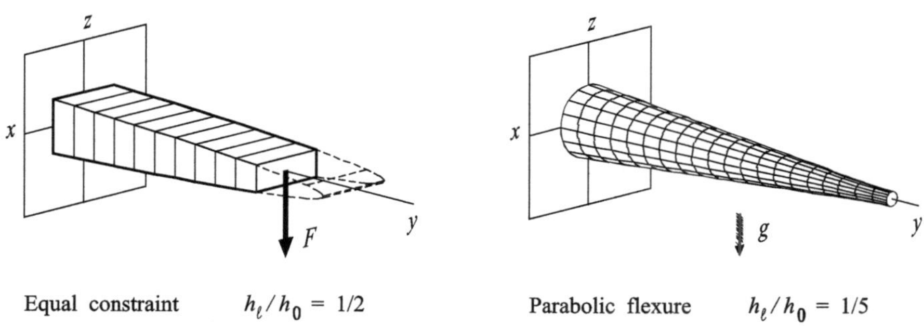

Fig. 1.45 (*Left*) Beam of equal constraint flexures by a concentrate force F. The edge thickness ratio is t_ℓ/t_0=1/2. Its thickness section in the y, z plane is a *truncated parabola*. (*Right*) Rod with a parabolic flexure by its own weight. The edge thickness ratio is t_ℓ/t_0=1/5. Its thickness differs slightly from a truncated cone (after Lemaitre [95])

which, for $p = 1$ (beam σ =const.) gives a *wedge* distribution with a *cutting* line end, for $p = 2$ (rod σ =const., and beam R =const.) gives another kind of *cubic parabola* with a *round* end at $y = \ell$, and for $p = 3$ (rod R =const.) gives a *parabola*.

6. *Equal constraint and parabolic flexure due to own weight*: One shows [95] that for the four cases in Table 1.9, the differential equation for the general cases with truncated bars is

$$\frac{dt}{t_0} = -\frac{2\beta}{p-q+1}\left[\left(\frac{t}{t_0}\right)^{p+q+1} - \left(\frac{t_\ell}{t_0}\right)^{p+q+1}\right]^{1/2}\left(\frac{t}{t_0}\right)^{-p}\frac{dy}{\ell} \tag{1.108a}$$

where q is an integer and β a dimensionless constant. Starting from a given value of the relative thickness t_ℓ/t_0 at the edge, a numerical integration allows determining the constant β and then the thickness distribution t/t_0. This resolution has been carried out by computer for the case $p = 3$, $q = 2$, of an equal curvature rod; the solution is almost a *truncated cone* (cf. Fig. 1.45).

The degenerated cases of null thickness at the edge, $t_\ell = 0$, correspond to $\beta = 1$. After integration we obtain the general form

$$t = t_0\left[1 - \frac{y}{\ell}\right]^{2/(p-q+1)}, \tag{1.108b}$$

which, for rods and beams with σ=constant ($p-q = 0$) gives a *parabolic thickness*, and for rods and beams with parabolic flexure ($p-q = 1$) gives a *cone* and a *wedge* respectively.[12]

7. *Special cases of beam cantilevers with constant thickness*: Beams of equal constraint or with parabolic flexure can be obtained, when the thickness t=constant, if now the width $a(y)$ in the x-direction varies with the position y of the cross section. We assume that $a(y)|_{\max} \ll \ell$. The inertia moment $I_x = at^3/12$ is varying with a and, for the loading cases considered in Table 1.9, the shearing forces are

$$Q_y = F, \quad (\ell - z)f, \quad t\mu g\int_y^\ell a\,dy\,. \tag{1.109}$$

After substitutions of the bending moment M_y and of Q_y into the equilibrium equation (1.101), we obtain, for the equal constraint cases,

---

[12] For a null thickness end, $t_\ell = 0$, Clebsch [34] derived the solution of equal constraint when a bar is flexed by its own weight by considering homothetic cross-sections. In the cases of both rod and beam he found *parabolas*. For a beam of square cross-sections, his result can be easily demonstrated: the inertia moment and shearing force become $I_x = t^4/12$ and $Q_y = \mu g\int_y^\ell t^2\,dy$ which, after substitution into the equilibrium equation (1.105), entails $p = q = 2$. Hence $p - q = 0$, so the Clebsch *parabola* is the exact answer to Galileo's problem and to his drawing of Fig. 1.42 which shows homothetic cross-sections.

In his annotations of Clebsch's book, Saint-Venant ([138] p. 359, cf. Thodhunter and Pearson [161] vol.II, part II, pp. 263–164) comments that the "paradox of a null thickness" at the suspension point leads to an infinite shearing force, but this can be solved by giving the vertex a finite thickness. He did not consider Clebsch's integration constant that was set to zero and which would have led to cantilevers with finite thickness, the above truncated class.

$$\frac{da}{dy} = -\frac{6F}{t^2\sigma}, \qquad -\frac{6f}{t^2\sigma}(\ell - y), \qquad -\frac{6\mu g}{t\sigma}\int_y^\ell a\,dy, \tag{1.110}$$

and since $R = Et/2\sigma$, the substitutions of σ give expressions of the *same* form for parabolic flexure beams.

Considering the first case of a concentrate force F and denoting a_0 and a_ℓ the widths at $y = 0$ and ℓ, the solution is represented by

$$a = a_0\left[1 - \left(1 - \frac{a_\ell}{a_0}\right)\frac{y}{\ell}\right], \quad \text{with } 1 - \frac{a_\ell}{a_0} = -\frac{6\,\ell\,F}{t^2 a_0\,\sigma} \equiv -\frac{12\,\ell\,RF}{t^3 a_0 E}. \tag{1.111}$$

Hence for constant thickness beams of equal constraint or of parabolic flexure, the geometry is the same: a *trapeze* which degenerates into a *triangle* when $a_\ell = 0$ (Fig. 1.46).

From (1.111), whether a *truncated* or a *sharp* end, the constant constraint and parabolic flexure are satisfied if the angle formed by the lateral faces is

$$2\arctan\left|\frac{6F}{t^2\sigma}\right| \equiv 2\arctan\left|\frac{12RF}{t^3E}\right|. \tag{1.112}$$

In practice, these solutions have been developed for improving suspension systems by adding to the cantilever a symmetric part with respect to the x, z plane, so a force applied at $y = 0$ is in reaction with two opposite forces at the ends $\pm\ell$. The trapeze plate is sliced into parallel strips which are symmetrically rearranged on the central strip with freedom to slide. This arrangement is known as a *multiple-leaf spring system* (cf. Fig. 1.46).

• **Compressed beams and buckling critical load**: Euler [52, 53] investigated the important case of a *compressed beam* (or rod) of circular and constant section. Considering one end clamped at the origin whilst the other is free (Fig. 1.44-*Right*), the force F must be substituted by $-F$ in Eq. (1.98), the boundary conditions are $\theta = 0$ for $s = 0$, and $d\theta/ds = 0$ for $s(\theta_0) = \ell$. This entails

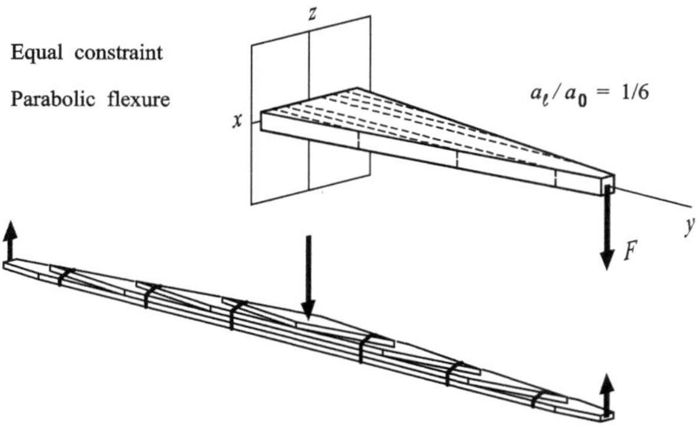

Fig. 1.46 (*Up*) Constant thickness beams providing both equal constraint and parabolic flexure. For a finite width end the shape is a *trapeze* and otherwise a *triangle*. (*Down*) From segment cuts, as shown in *dashed lines*, one obtains the classical rearrangement for suspension systems

$$c_1 = -F\cos\theta_0, \quad c_2 = 0, \quad c_3^2 = 2EI(F + c_1)/F^2, \quad c_4 = 0, \qquad (1.113a)$$

where the negative sign is taken for the square root term of (1.98b) and the positive for the two other equations. The equation for θ_0 is

$$\ell = \sqrt{\frac{EI}{2F}} \int_0^{\theta_0} \frac{d\theta}{\sqrt{\cos\theta - \cos\theta_0}}. \qquad (1.113b)$$

For a small flexure, $\theta_0 \ll 1$, the integration leads to $\arcsin\theta/\theta_0$

$$\ell = \sqrt{\frac{EI}{F}} \int_0^{\theta_0} \frac{d\theta}{\sqrt{\theta_0^2 - \theta^2}} = \frac{\pi}{2}\sqrt{\frac{EI}{F}}, \qquad (1.113c)$$

hence θ_0 vanishes. This shows that the flexure only exists if θ_0 exists, i.e. if

$$F \geq \pi^2 EI/4\ell^2. \qquad (1.113d)$$

This important result by Euler defines the so-called *buckling critical load* – here the force $F_{\text{Cr}} = \pi^2 EI/4\ell^2$ – where the beam suddenly ceases to be straight. This introduced the first mathematical formalism in the general notion of *elastic stability*.

Equations (1.100b) and (1.113b) are nearly identical. They are different only by the integration range. One is for the cantilever bars, while the second is for the compressed beams and buckling limit. Other buckling limits of straight beams can be determined from Eqs. (98) and taking into account the boundaries, e.g. for beams with ends either clamped or free to rotate (Fig. 1.47).

Another case of any importance, that of a beam ceasing to be straight under its own weight, was also treated by Euler and Lagrange. In 1773, Lagrange [90] determined a variable cross-section distribution of columns providing the strongest resistance to buckling.

• **Torsion of beams of various cross sections**: The torsion of rods of constant diameter was first investigated by Coulomb [39] in 1784 for the determination of the attraction laws in electrostatics and electromagnetism. Developing torsion balances for accurate measures of those forces and discovering the inverse square law

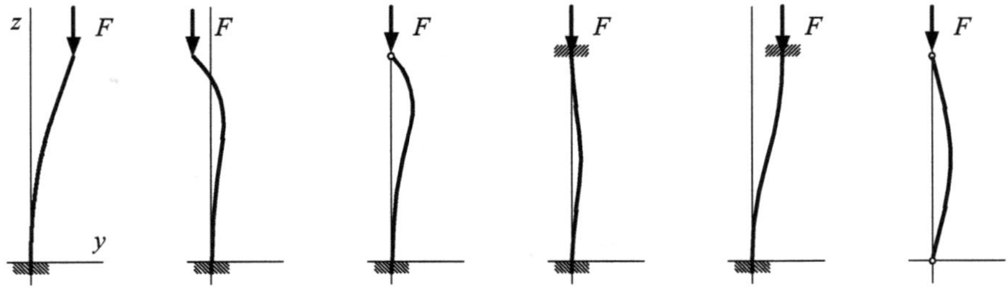

Fig. 1.47 Buckling of beams for various boundary conditions. The two beams on the left show the first and second buckling modes

q_1q_2/d^2, Coulomb showed that the torsion moment M_t around the z-axis of the rod is proportional to its torsion angle τ per unit length. Also for cylinders of any uniform cross-section, the total rotation is the product $\theta = \tau L$ of the angle τ and the length L of the bar. The constant c linking the quantities M_t and τ, so-called *torsional rigidity* and of dimension [F][L^2], is (cf. for instance Timoshenko [158])

$$c = \frac{M_t}{\tau} = \frac{G\mathcal{A}^4}{4\pi^2 I_p}, \qquad I_p = \int_{\mathcal{A}} r^2 \, d\mathcal{A}, \tag{1.114}$$

where G is the *shear modulus* of the material, $\mathcal{A}$ the area of the cross-section, and I_p the *polar moment of inertia* around the center of the cross-section for which r is taken as the polar distance to the element area $d\mathcal{A}$.

For a rod of diameter $2a$, the polar moment of inertia and the torsional rigidity are

$$I_p = \tfrac{1}{2}\,\pi a^4 = 2I\,, \qquad c = \tfrac{1}{2}\,\pi G a^4. \tag{1.115a}$$

For a cylinder of elliptical cross-section whose lengths of principal axes are $2a$, $2b$, the polar moment of inertia and the torsional rigidity are

$$I_p = \tfrac{1}{4}\,\pi ab\,(a^2 + b^2), \qquad c = \pi G\,\frac{a^3 b^3}{a^2 + b^2}\,. \tag{1.115b}$$

Saint-Venant [91] noticed that the assumption of cross-sections remaining plane during the torsion is perfectly valid only if Coulomb's theory is restricted to the case of a rod. For any other case, he devised the *large torsion theory* of various cross-section cylinders in the celebrated memoir *La Torsion des Prismes* (1855). Investigating bars of square (Fig. 1.48), rectangular, elliptical, and various cross-sections expressed in a polynomial form, Saint-Venant derived the axial and transversal displacements. For instance for an elliptical cylinder, he showed that the shape of any cross-section is represented by the local equation

$$z = \frac{M_t(b^2 - a^2)}{\pi G\, a^3 b^3}\, xy\,, \tag{1.116}$$

where the x, y coordinates coincide with the principal axes of the elliptic cross-section when distorted. The level curves of this surface are hyperbolas whose asymptotes are the principal axes of the ellipse.

Whatever the magnitude of the torsion angle per unit length, if *no* axial force is applied to the prism ends, then the deformation of any volume element made of two cross-sections is with a *constant* volume and is said to be a *pure shear deformation* (cf. Sect. 1.13.3).

1.13.2 Elasticity Constants of Isotropic Materials

The elasticity theory of isotropic media involves two *elasticity constants* characterizing the elastic properties of the material which thus is sometimes called a

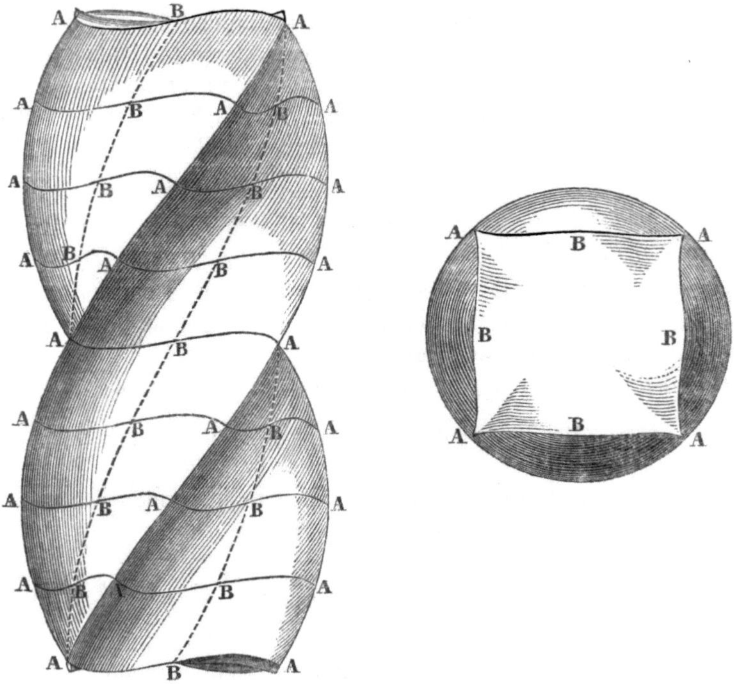

Fig. 1.48 Torsion of a beam of square cross-section in the large deformation theory (from *La Torsion des Prismes* by Saint-Venant [91])

bi-constant theory. However, four different constants are usually measured from various experimental tests, so that any one of the four can be represented as a function of two others.

Whatever the sign convention involved in their measurement, the four elasticity constants may be determined as follows:

• **Young modulus** E: Considering a rod elongated of δz along a z-axis by a uniform stress σ_z applied on each end, the *Young modulus* is the ratio $E = |\sigma_z/(\delta z/z)|$. Its dimension is [F][L^{-2}].

• **Shear modulus** G: Considering a rod of diameter $2a$ distorted of the angle τ per unit length by a torsion moment M, the *shear modulus* is the ratio $G = |2M/\pi\tau a^4|$. Its dimension is [F][L^{-2}].

• **Poisson ratio** ν: Considering a rod elongated of $\delta z/z$ along a z-axis by a uniform stress and whose cross-section is then changed of $\delta r/r$, the *Poisson ratio* is the ratio $\nu = -(\delta r/r)/(\delta z/z)$.

• **Isotropic compression modulus**: Considering a sphere of radius a compressed by a pressure $p > 0$, the *isotropic compression modulus* is the ratio $K = -\frac{1}{3}\, p\,/(\delta a/a)$. Its dimension is [F][L^{-2}].

These quantities are linked by the formulas

$$E = \frac{9KG}{3K+G}\,, \qquad \nu = \frac{3K-2G}{2(3K+G)}\,, \tag{1.117a}$$

$$K = \frac{E}{3(1-2\nu)}\,, \qquad G = \frac{E}{2(1+\nu)}\,. \tag{1.117b}$$

The two *Lamé coefficients* [91], $\lambda = \nu E/[(1-2\nu)(1+\nu)]$ and $\mu \equiv G$, are sometimes used.

Although from these quantities, the Poisson ratio may have values such as $-1 < \nu < 1/2$, we do not know in nature a *continuous* and *isotropic* material with $\nu < 0$, i.e. which would laterally expand when longitudinally elongated. For all materials, the Poisson ratio is in the domain $0 \le \nu \le 1/2$. For rubber, the Poisson ratio is close to 1/2, hence the isotropic compression modulus K is close to infinity; this means that for rubber the volume change in the deformation is quasi null.

Glass or vitro-ceramic materials show perfectly invariant quantities E, G, ν, and K up to the rupture. For instance in the measure of E, the axial stress σ_z is exactly proportional to the strain $\delta z/z$ up to rupture which corresponds to the *ultimate strength* σ_{ult}. However many substances, such as most metal alloys, show a deviation from linearity between stress and strain. Because of this deviation, the experimental measures of these elastic constants by dedicated testing machines are generally carried out in the low stress level of their elastic domain.

As is customary in elasticity analysis, the elastic constants E and ν are generally used instead of G and K. For some typically linear materials used in active optics, the values of these constants and the tensile maximum stress, $\sigma_{T\max}$, are given in Table 1.10.

Table 1.10 Young modulus E, Poisson's ratio ν, and tensile maximum strength $\sigma_{T\max}$ of some linear strain-stress materials at 20°C

Material	E [GPa]	ν	$\sigma_{T\max}$ (*) [MPa]
Borofloat glass, Pyrex, Duran	64.0	0.200	7
Fused silica ULE $SiO_2 + TiO_2$	68.8	0.170	20
Fused silica SiO_2 Suprasil	77.5	0.165	19
BK7 optical glass	82.0	0.206	10
U-BK7 optical glass	90.0	0.206	12
Zerodur vitroceram SCHOTT	90.2	0.243	22
VO2 vitroceram XINHU	91.0	0.247	22
Titanium alloy Ti90 Al6 V4	122	0.333	900
ZPF CERATECH	150	0.280	60
Stainless steel Fe87 Cr13	201	0.315	1,000
Stainless steel Fe87 Cr13 quenched	201	0.315	1,400
Beryllium pure VHP	287	0.110	400
Silicon carbide polycrystal CVD	430	0.210	150
Sapphire polycrystal Al_2O_3	440	0.300	100

* $\sigma_{T\max}$ is the limit linear range for metal alloys. For brittle materials, $\sigma_{T\max} = \frac{1}{4}\,\sigma_{10^3\,\mathrm{sec}}$ from 10^3 seconds load duration (cf. Table 5.2 in Sect. 5.2.5).

1.13.3 Displacement Vector and Strain Tensor

Let us consider any point of a solid whose coordinate is x, y, z when in state without stress, and denote $\mathbf{r}$ its position vector from the origin. When the solid is deformed by external forces, or internal forces such as those exerted by a gravity field, the position vector becomes $\mathbf{r}'$. The *displacement vector* $\mathbf{u}$ representing the flexure is

$$\mathbf{u} = \mathbf{r}' - \mathbf{r}\,. \tag{1.118}$$

where its cartesian components $u_x(x, y, z)$, $u_y(x, y, z)$ and $u_z(x, y, z)$ may be denoted u_i,

If dl is the length separating two adjacent points, $dl^2 = dx^2 + dy^2 + dz^2 = dx_i^2$, this length becomes dl' in the deformation, $dl'^2 = (dx_i + du_i)^2$. Using the subscripts i, k, l, the simplifications in the summations lead to

$$d\,l'^2 - d\,l^2 = 2\,\varepsilon_{ik}\,dx_i\,dx_k\,, \tag{1.119}$$

where ε_{ik} is the *strain tensor*. Following Landau and Lifschitz [92], one shows that for *small* deformations the high-order terms can be neglected so the strain tensor becomes symmetric, $\varepsilon_{ik} = \varepsilon_{ki}$. Hence the strain tensor is defined by

$$\varepsilon_{ik} = \frac{1}{2}\left(\frac{\partial u_i}{\partial x_k} + \frac{\partial u_k}{\partial x_i}\right), \qquad \varepsilon_{ik} = \varepsilon_{ki}\,. \tag{1.120}$$

The strain components include the relative axial displacement ε_{ii} or *normal strain* in the direction x_i, and the associated relative lateral displacements ε_{ik} and ε_{il} or *shear strains* in the directions x_k and x_l respectively.

Let us introduce now $u,\ v,\ w$ as a quite usual nomenclature for representing the components of the displacement vector; in a cartesian coordinate system we have the equivalent notations

$$u \equiv u_x, \quad v \equiv u_y, \quad w \equiv u_z,$$

where only the $u,\ v,\ w$ notation will be used hereafter. From (1.120), the strain components are

$$\begin{aligned} &\varepsilon_{xx} = \frac{\partial u}{\partial x}, \quad \varepsilon_{yy} = \frac{\partial v}{\partial y}, \quad \varepsilon_{zz} = \frac{\partial w}{\partial z}, \\ 2\varepsilon_{yz} = \frac{\partial w}{\partial y} + \frac{\partial v}{\partial z}, \quad &2\varepsilon_{zx} = \frac{\partial u}{\partial z} + \frac{\partial w}{\partial x}, \quad 2\varepsilon_{xy} = \frac{\partial v}{\partial x} + \frac{\partial u}{\partial y}\,. \end{aligned} \tag{1.121a}$$

In a cylindric coordinate system, let us denote by the subscript t a tangential direction which is perpendicular both to the z-axis and a radial direction r so the components of the displacement vector are denoted $u \equiv u_r$, $v \equiv u_t$, and $w \equiv u_z$ respectively. The strain components are

$$\varepsilon_{rr} = \frac{\partial u}{\partial r}, \quad \varepsilon_{tt} = \frac{1}{r}\frac{\partial v}{\partial \theta} + \frac{u}{r}, \quad \varepsilon_{zz} = \frac{\partial w}{\partial z},$$
$$2\varepsilon_{tz} = \frac{1}{r}\frac{\partial w}{\partial \theta} + \frac{\partial v}{\partial z}, \quad 2\varepsilon_{zr} = \frac{\partial u}{\partial z} + \frac{\partial w}{\partial r}, \quad 2\varepsilon_{rt} = \frac{\partial v}{\partial r} - \frac{v}{r} + \frac{1}{r}\frac{\partial u}{\partial \theta}. \quad (1.121b)$$

From a general form strain tensor ε_{ik}, let $\varepsilon^{(i)}$ be the three diagonal strains taken in the principal directions at a given point. In a first approximation, the strains or relative elongations of the solid element in these principal directions are, from (1.119),

$$\sqrt{1+2\varepsilon^{(i)}} - 1 \simeq \varepsilon^{(i)}, \quad (1.122a)$$

so the resulting volume when the displacements occur is

$$dV' = (1+\varepsilon^{(1)})\,(1+\varepsilon^{(2)})\,(1+\varepsilon^{(3)})\,dV \simeq (1+\varepsilon^{(1)}+\varepsilon^{(2)}+\varepsilon^{(3)})\,dV. \quad (1.122b)$$

Since the sum of the principal values $\varepsilon^{(1)}+\varepsilon^{(2)}+\varepsilon^{(3)}$ of a tensor is an invariant, it is also equal to the sum $\varepsilon_{ii} = \varepsilon_{11}+\varepsilon_{22}+\varepsilon_{33}$ of its diagonal terms. Hence, in any coordinate system, the relative volume change is

$$(dV' - dV)\,/\,V = \varepsilon_{ii}\,. \quad (1.122c)$$

If $\varepsilon_{ii} = 0$, then the volume element does not change. It can be shown that this corresponds to the case of strains generated by a torsion. Except for substances whose Poisson's ratio is $\nu = 1/2$ (such as rubber) and which then conserve a constant volume, a result is the following.

→ *Whatever a Poisson's ratio* $\nu \in [0,\ 1/2]$ *, if the elastic deformation of a small element is with constant volume, then the geometric transformation is a torsion.*

1.13.4 The Stress-Strain Linear Relations and Strain Energy

Consider A volume element dV of a perfectly elastic and isotropic substance characterized by E, ν. One defines stress components σ_{ik} for representing the stress vectors arising at the facets of the element (Fig. 1.49). Since $\sigma_{kl} = \sigma_{lk}$, the three *axial stress* components σ_{ii} and three *shear stress* components σ_{kl} are expressed in force per unit surface area. These stresses are functions of the strains such as

$$\sigma_{xx} = \frac{E}{(1+\nu)(1-2\nu)}\,[(1-\nu)\,\varepsilon_{xx} + \nu\,(\varepsilon_{yy}+\varepsilon_{zz})]\,,$$
$$\sigma_{yy} = \frac{E}{(1+\nu)(1-2\nu)}\,[(1-\nu)\,\varepsilon_{yy} + \nu\,(\varepsilon_{zz}+\varepsilon_{xx})]\,,$$
$$\sigma_{zz} = \frac{E}{(1+\nu)(1-2\nu)}\,[(1-\nu)\,\varepsilon_{zz} + \nu\,(\varepsilon_{xx}+\varepsilon_{yy})]\,,$$
$$\sigma_{yz} = \frac{E}{1+\nu}\,\varepsilon_{yz}, \quad \sigma_{zx} = \frac{E}{1+\nu}\,\varepsilon_{zx}, \quad \sigma_{xy} = \frac{E}{1+\nu}\,\varepsilon_{xy}. \quad (1.123a)$$

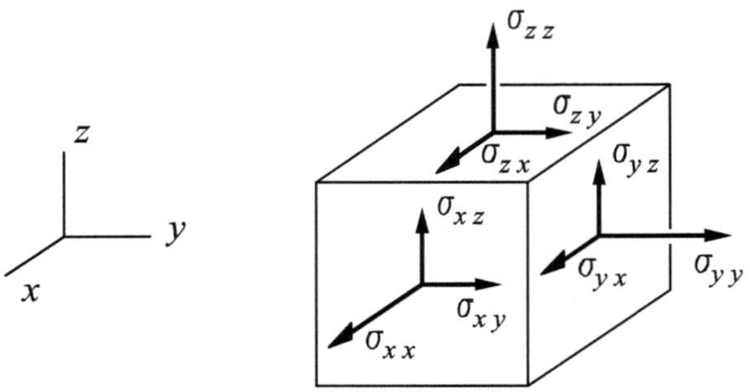

Fig. 1.49 Axial and shear stress components acting on a volume element ($\sigma_{kl} = \sigma_{lk}$)

The inverse relations are

$$\begin{aligned}
\varepsilon_{xx} &= \tfrac{1}{E}\left[\sigma_{xx} - \nu\left(\sigma_{yy} + \sigma_{zz}\right)\right],\\
\varepsilon_{yy} &= \tfrac{1}{E}\left[\sigma_{yy} - \nu\left(\sigma_{zz} + \sigma_{xx}\right)\right],\\
\varepsilon_{zz} &= \tfrac{1}{E}\left[\sigma_{zz} - \nu\left(\sigma_{xx} + \sigma_{yy}\right)\right],\\
\varepsilon_{yz} = \tfrac{1+\nu}{E}\,\sigma_{yz}, \quad \varepsilon_{zx} &= \tfrac{1+\nu}{E}\,\sigma_{zx}, \quad \varepsilon_{xy} = \tfrac{1+\nu}{E}\,\sigma_{xy}.
\end{aligned} \tag{1.123b}$$

These so-called *stress-strain relations*, also known as Hooke's law or Navier's relations, were derived in above general form by Navier [113] in 1820.[13]

If both the solid and the deforming loads have a *rotational* symmetry around the z-axis, the tangential displacement component $v \equiv u_t = 0$, and also $\partial u/\partial\theta = \partial w/\partial\theta = 0$; hence, from (1.121b), $\varepsilon_{tz} = \varepsilon_{rt} = 0$. In this case, the so-called *thick plate theory* involves the stress-strain relation

$$\begin{pmatrix}\sigma_{rr}\\ \sigma_{tt}\\ \sigma_{zz}\\ \sigma_{zr}\end{pmatrix} = \frac{E}{(1+\nu)(1-2\nu)}\begin{pmatrix}1-\nu & \nu & \nu & 0\\ \nu & 1-\nu & \nu & 0\\ \nu & \nu & 1-\nu & 0\\ 0 & 0 & 0 & 1-2\nu\end{pmatrix}\begin{pmatrix}\varepsilon_{rr}\\ \varepsilon_{tt}\\ \varepsilon_{zz}\\ \varepsilon_{zr}\end{pmatrix} \tag{1.124a}$$

Conversely the strain components are

$$\begin{pmatrix}\varepsilon_{rr}\\ \varepsilon_{tt}\\ \varepsilon_{zz}\\ \varepsilon_{zr}\end{pmatrix} = \frac{1}{E}\begin{pmatrix}1 & -\nu & -\nu & 0\\ -\nu & 1 & -\nu & 0\\ -\nu & -\nu & 1 & 0\\ 0 & 0 & 0 & 1+\nu\end{pmatrix}\begin{pmatrix}\sigma_{rr}\\ \sigma_{tt}\\ \sigma_{zz}\\ \sigma_{zr}\end{pmatrix} \tag{1.124b}$$

[13] Navier's relations were finalized with the shear components by A. Cauchy in 1827–29. Navier, Poisson, Cauchy and Lamé assumed that all materials were with $\nu = 1/4$ which then led to the so-called *uni-constant theory*. A bi-constant representation was later upheld by G. Green who showed that the number of elastic moduli goes up to 21 for crystals.

From the stress components, it is useful to derive the energy that occurs in the deformation of the volume element. The total work in the deformation of the element, or *free energy*, is represented per volume unit by

$$d\mathcal{F} = \frac{1}{2}\,\sigma_{ik}\varepsilon_{ik} = \frac{E}{2(1+\nu)}\left(\varepsilon_{ik}^2 + \frac{\nu}{1-2\nu}\,\varepsilon_{ll}^2\right), \tag{1.125}$$

where ε_{ik}^2 is the sum of the square of all the components of the ε_{ik} symmetric tensor, and ε_{ll}^2 is the square of the sum of its diagonal components.

$$\varepsilon_{ik}^2 = \varepsilon_{11}^2 + \varepsilon_{22}^2 + \varepsilon_{33}^2 \;+\; 2\,(\varepsilon_{12}^2 + \varepsilon_{23}^2 + \varepsilon_{31}^2)\,,$$
$$\varepsilon_{ll}^2 = (\varepsilon_{11} + \varepsilon_{22} + \varepsilon_{33})^2.$$

The problem of determining the displacement vector, i.e. its flexure components u, v, w, consists of finding the form of integral functions which satisfies the symmetry of the stresses or of the strains, if any, and to set up their limit values at the boundaries. In general, the three-dimensional problem is highly complex and is solved by finite element analysis and computer.

In active optics methods the mirror-plate is generally designed with rotational symmetry. The z-component w of the displacement vector is the starting feature; it represents the flexure to achieve. Hence the problem consists of determining the stress components. This leads to the determination of the resulting external forces and moments at the boundaries and to the associated determination of an appropriate thickness distribution.

1.13.5 Uniform Torsion of a Rod and Strain Components

The 3-D problem of the elastic deformation in torsion of a rod is a classical example where the analytical displacements can be easily obtained (cf. for instance Timoshenko [157]). Although the torsion angle τ per unit length is assumed to be small, the total rotation $\theta_{\rm m} = \tau\ell$ of one end may reach one or several times 2π for a rod whose length-to-radius ratio ℓ/a is large. From (1.114), the torsion angle per unit length is $\tau = M/c$ where M and c are the twisting moment and the torsional rigidity. Considering the shear modulus G and a rod of radius a, from (1.115a) we obtain

$$\tau = 2M\,/\,(\pi G a^4)\,. \tag{1.126}$$

In the first approximation of a small τ-value, the study of the torsion of a rod leads to the conclusion that all the strain components ε_{ik} are null except for the shear strain ε_{tz} so the volume of the rod remains *unchanged* in a torsion since $\sigma_{rr} + \sigma_{tt} + \sigma_{zz} = 0$.

However and as shown hereafter, when a *large torsion* occurs so the $\tau^2 a^2$ terms cannot be neglected, the dimensions of a rod are modified *unless* opposite axial forces that increase its length are added to the opposite twisting moments generating the torsion.

• **Torsion generated by twisting moments only**: Let us consider a straight rod where *only* opposite twisting moments M_t are applied to its ends around the z-axis. A generatrix at a radial distance $r = a$ from this axis is deformed into a helix (Fig. 1.50). At any point of this helix, the inclination angle with respect to a direction parallel to z is a constant $a\tau$ whatever the z ordinate. When twisted, the two facets $z =$ constant of a volume element remain parallel and their axial separation is $dz' = dz \cos \tau a$. Hence the axial strain component is

$$\varepsilon_{zz} = (dz' - dz)/dz = \cos \tau a - 1 = -\tfrac{1}{2}\,\tau^2 a^2 \,. \tag{1.127}$$

Denoting the components of the displacement vector as

$$u \equiv u_r\,, \quad v \equiv u_t \quad \text{and} \quad w \equiv u_z\,,$$

any cross-sections perpendicular to the z-axis remain *plane* ($\partial w/\partial\theta = \partial w/\partial r = 0$) so the axial displacement w is a function of z only. From (1.121b),

$$w = \varepsilon_{zz} z + \text{constant} = -\tfrac{1}{2}\,\tau^2 a^2 z\,, \tag{1.128}$$

which shows that the axial length of the rod is *shortened*.

These cross-sections remain circular, thus $\partial v/\partial\theta = \partial u/\partial\theta = 0$ in (1.121b), so the strain components become

$$\begin{aligned} \varepsilon_{rr} &= \frac{\partial u}{\partial r}\,, & \varepsilon_{tt} &= \frac{u}{r}\,, & \varepsilon_{zz} &= -\tfrac{1}{2}\,\tau^2 a^2, \\ 2\varepsilon_{tz} &= \frac{\partial v}{\partial z}\,, & 2\varepsilon_{zr} &= \frac{\partial u}{\partial z}\,, & 2\varepsilon_{rt} &= \frac{\partial v}{\partial r} - \frac{v}{r}\,. \end{aligned} \tag{1.129}$$

The diameter of the distorted cross-sections does not vary with z, hence the radial component of the displacement is a function of the radius only, $u = u(r)$. Assuming

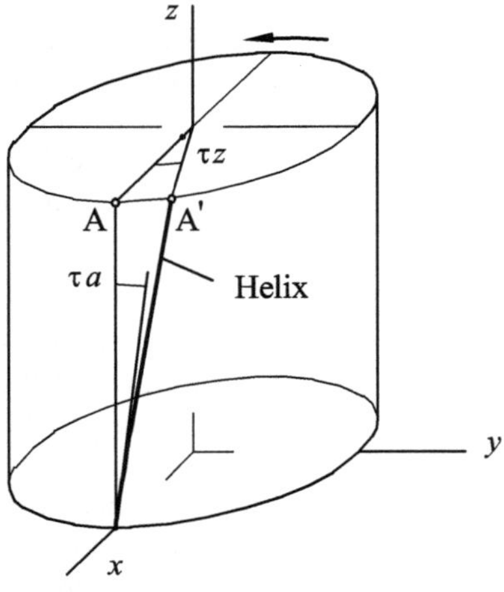

Fig. 1.50 Torsion of a rod by opposite twisting moments applied at its ends

that in the torsion the element volume does not change even for strain terms of quadratic form such as $\tau^2 a^2$, from (1.122c) $\varepsilon_{rr} + \varepsilon_{tt} + \varepsilon_{zz} = 0$. From (1.121b), this entails

$$\frac{\partial u}{\partial r} + \frac{u}{r} - \tfrac{1}{2}\,\tau^2 a^2 = 0\,. \tag{1.130a}$$

We obtain $u = \frac{1}{4}\,\tau^2 a^2 (r + c_1/r)$. Setting the constant $c_1 = 0$ for a plain rod,

$$u = \tfrac{1}{4}\,\tau^2 a^2 r\,, \tag{1.130b}$$

which shows that the radius of the cross-section is *increased*.

Any point located in a cross-section rotates by the same angle $\theta = \tau a$; therefore a straight radial line of the cross-section is assumed to remain straight. At the distance z from the origin, the length of the curvilinear displacement is $v = \tau a z$, and for any point of the rod this displacement is represented by

$$v = \tau\, r\, z\,. \tag{1.131}$$

The shear strain ε_{tz} purely applies along circles centered on the z-axis which, for any point at distance $r = a$, leads to $\varepsilon_{tz} = \tau a/2$. Finally the strain and displacement components are

$$\begin{cases} \varepsilon_{rr} = \frac{1}{4}\,\tau^2 a^2, & \varepsilon_{tt} = \frac{1}{4}\,\tau^2 a^2, \quad \varepsilon_{zz} = -\frac{1}{2}\,\tau^2 a^2, \\ \varepsilon_{tz} = \frac{1}{2}\,\tau\, a\,, & \varepsilon_{zr} = 0\,, \qquad\quad \varepsilon_{rt} = 0\,, \\ u = \frac{1}{4}\,\tau^2 a^2 r\,, & v = \tau\, r\, z\,, \qquad w = -\frac{1}{2}\,\tau^2 a^2 z\,. \end{cases} \tag{1.132}$$

The strain-stress relations (1.111a) with the subscripts x, y, z substituted by r, t, z provide the determination of the stress components at a current point $r = a$. Using only the shear modulus instead of the Young modulus and Poisson's ratio [cf. Eq. (1.117)], we obtain after simplification

$$\sigma_{zz} = -\,G\,\tau^2 a^2, \qquad \sigma_{tz} = G\,\tau\, a\,. \tag{1.133}$$

One may notice that the stress component σ_{zz} varies with the square power of the torsion angle. This entails that the effect of this stress with respect to the shearing stress component σ_{tz} becomes important when the τ-value is *large*.

Let now a be the external radius of the rod. If the central length of the rod were unchanged, the axial stress would be null at the center of the cross-section. But from (1.128) the shortening $w = -\tau^2 a^2 z/2$ of the rod axis is due to a compression of the central zone. Hence the axial stress distribution or stress function $\sigma_z(r)$ must vary as r^2/a^2 over the cross-section and must have a finite value on the axis. An appropriate representation is of the form

$$\sigma_z(r) = \sigma_{zz}\left(C_1 - r^2/a^2\right), \tag{1.134}$$

where C_1 is a constant. Since no axial force is applied at the rod ends, the resultant of the stress elements taken over the cross-section must be zero in this direction,

$$\int_{\mathcal{A}} \sigma_z \, d\mathcal{A} = 2\pi \int_0^a \sigma_{zz} \big(C_1 - r^2/a^2\big)\, r\, dr = 0\,, \tag{1.135}$$

which gives $C_1 = 1/2$. The stress distribution is

$$\sigma_z(r) = G\tau^2 a^2 \left(\frac{r^2}{a^2} - \frac{1}{2} \right), \tag{1.136a}$$

and takes a null value at the distance $r/a = 1/\sqrt{2}$ from the axis. The axial stresses at the center of the cross-section is a compression and is opposite at the edge,

$$\sigma_z(0) = -\sigma_z(a) = -G\tau^2 a^2/2\,. \tag{1.136b}$$

Since only the moments are applied to the ends of the rod, we assume that the rod is long enough so the stress distribution $\sigma(r)$ arises in the main part of the rod and locally vanishes at the ends.

• **Torsion with constant length**: From the above result, we conclude that the diameter and length of the rod may be made unchanged for a *large* torsion angle if, in addition to the opposite twisting moments M_t, opposite external forces F_z are applied to the ends (cf. Fig. 1.50).

The intensity of the stretching force F_z to apply towards the positive z-end is derived from $\sigma_z(0)$ in (1.136b),

$$F_z = -2\pi \int_0^a \sigma_z(0)\, r\, dr = \tfrac{1}{2}\pi\, G\tau^2 a^4\,, \tag{1.137}$$

therefore the rod is elongated and recovers its *initial* length.

1.13.6 Love-Kirchhoff Hypotheses and Thin Plate Theory

The bending of plates is a complex three-dimensional problem which can only be accurately solved from finite element analysis. For the basic and classical case where the thickness t of the plate is small compared to the typical overall dimension ℓ of its two-dimension surface area, the so-called *thin plate theory* allows reducing the problem to a two-dimension one by introducing some conditions. These conditions are usually referred to as *Love-Kirchhoff hypotheses*.

Assuming a plane plate and setting an x, y plane that lies at its middle plane and z-axis normal to this plane, Love-Kirchhoff hypotheses are as follows.

1. For points originally at the mid-plane, the components of the displacement vector reduce to

$$u\,|_{z=0} = v\,|_{z=0} = 0\,, \qquad w\,|_{z=0} = w(x, y)\,. \tag{1.138a}$$

2. Any point located on normal lines to the mid-plane remains on normal lines to the deformed mid-surface (known as *Euler-Bernoulli hypothesis*). No stress exists along these normal lines,

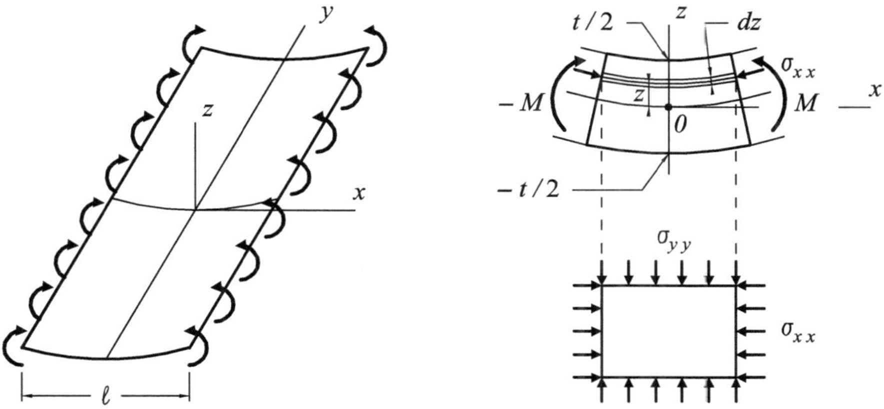

Fig. 1.51 *Left*: Cylindrical bending of a long plate. *Right*: Unit stresses acting in a volume element

$$\sigma_{zz} = \sigma_{yz} = \sigma_{xz} = 0\,. \tag{1.138b}$$

3. Consequently the stress-stain relations reduce to three stresses and three strains such as

$$\begin{pmatrix} \sigma_{xx} \\ \sigma_{yy} \\ \sigma_{xy} \end{pmatrix} = \frac{E}{(1+\nu)(1-2\nu)} \begin{pmatrix} 1-\nu & \nu & 0 \\ \nu & 1-\nu & 0 \\ 0 & 0 & 1-2\nu \end{pmatrix} \begin{pmatrix} \varepsilon_{xx} \\ \varepsilon_{yy} \\ \varepsilon_{xy} \end{pmatrix} \tag{1.138c}$$

Love-Kirchhoff hypotheses are accurately achieved when the thickness ratio t/ℓ of the plate satisfies, say, $t/\ell < 1/10$. In addition, hypothesis **2** also implies that the middle surface of the plate is *free* from any "in-plane" stress, $\sigma_{xx}\,|_{z=0} = \sigma_{yy}\,|_{z=0} = \sigma_{xy}\,|_{z=0}$.

This not necessarily implies that the flexure is small compared to the thickness. For instance, the case of a cylindrically bent plate treated hereafter may lead to a large flexure with low stress level.

1.13.7 Bending of Thin Plates and Developable Surfaces

From Love-Kirchhoff hypotheses, the components u, v, and w of the displacement take a greatly simplified form if we consider the basic case of a *thin* plate. Let us also consider a plate geometry and loads such as the shape of the flexure is a *developable surface*. The simplest of such surfaces is a cylinder. A cylindrical flexure may be obtained by opposite bending moments uniformly distributed on the long edges in the y-direction of a rectangular plate (Fig. 1.51).

We assume that the long edges allow considering a plate strip element of width dy as taking a strict cylindrical deformation when bent, so the shear strain ε_{xy} can be neglected, $\varepsilon_{xy} = 0$. This could be achieved by maintaining the long edges as straight lines, so the curvature in any section x = constant is null ($1/R_y = 0$). If no "in-plane" force is applied in the x-direction on the long edges, there is no stretching of the *middle surface* during bending; the stresses σ_{xx} and σ_{yy} in the strip element are only generated by the curvature $1/R_x$ due to the moments. Denoting t the thickness of the plate, we will show that for thin plate i.e. $t/|R_x| \ll 1$, these stresses are null at the middle surface or *neutral surface* of the plate and linearly increase up to maximum opposite values on the faces where $z = \pm t/2$.

Restricting hereafter to the case of a thin plate, the stress component σ_{zz} in the volume element of a strip-plate can be neglected when no external load is applied, thus $\sigma_{zz} = 0$. Assuming that the long y-length is constant during bending entails that the component ε_{yy} is null. The stress-strain linear relations (1.123b) provide

$$\varepsilon_{xx} = \frac{1}{E}\left(\sigma_{xx} - \nu\,\sigma_{yy}\right), \qquad \varepsilon_{yy} = \frac{1}{E}\left(\sigma_{yy} - \nu\,\sigma_{xx}\right) = 0\,, \tag{1.139a}$$

from where

$$\sigma_{xx} = E\,\varepsilon_{xx}/(1-\nu^2)\,, \qquad \sigma_{yy} = \nu\,\sigma_{xx}\,. \tag{1.139b}$$

For a thin plate, the displacement or flexure component $w \equiv u_z$ is the same at any level z of the plate and does not vary along y, thus $w = w(x)$. Denoting $1/R = d^2w/dx^2$ the curvature, the strain ε_{xx} at a distance z from the middle surface is $\varepsilon_{xx} = -z/R$. Hence from (1.139b),

$$\sigma_{xx} = -\frac{E\,z}{1-\nu^2}\frac{d^2w}{dx^2}\,. \tag{1.140}$$

Depending on the boundary conditions at the long edges, the plate may also be submitted to the action of tensile or compression forces acting in the x-direction. The corresponding induced stress must be added to the above stress.

From the expression of σ_{xx}, the bending moment in the strip-plate element per unit length in the y-direction is

$$M = -\int_{-t/2}^{t/2} \sigma_{xx}\,z\,dz = \frac{E\,t^3}{12(1-\nu^2)}\frac{d^2w}{dx^2}\,. \tag{1.141}$$

The quantity linking the bending moment M to the curvature d^2w/dx^2 is usually denoted

$$D = \frac{E\,t^3}{12(1-\nu^2)}\,, \tag{1.142}$$

and called the *flexural rigidity* or rigidity.

The representation of the z-displacement or flexure curve w of the strip-plate element is a solution of

$$\frac{d^2w}{dx^2} - \frac{M}{D} = 0\,. \tag{1.143}$$

Compared to the flexure of rods or beams such as represented by (1.95), the flexural rigidity D is similar to the quantity EI where now the dimension is decreased of a length because M in (1.143) is a moment per unit length.

If the origin of the coordinates is taken at the center of the neutral surface of the plate which is only submitted to the bending moments $\pm M_0$ at the edges $x = \pm\ell/2$ of the plate, the two integration constants are null. This leads to the flexure

$$w = \frac{M_0}{2D}\, x^2, \tag{1.144}$$

which is a *parabola*. In fact the true flexure is a circular arc and the parabola must be considered as the result of a first-order approximation for a constant curvature.
Sign Convention: We have adopted a positive sign convention for the bending moment when the curvature of the deformation is also positive. Since $M_0 > 0$ for $x > 0$, and $D > 0$, then $w(\pm\ell/2) > 0$.[14]

In the general case of cylindrical bending of a long plate, we may consider the following various loads

q: uniform load per unit length applied to the element strip,
F: force applied to the positive edge in the x-direction,
M_0: bending moment applied to the positive edge,

whose resulting effect at any cross section of the strip is represented by the bending moment

$$M = -\frac{q}{2}\left(\frac{\ell^2}{4} - x^2\right) + Fw + M_0\,, \quad x^2 \le \ell^2/4\,. \tag{1.145}$$

After substitution, the differential equation of the bending is

$$\frac{d^2w}{dx^2} - \frac{F}{D}\,w = -\frac{q}{2D}\left(\frac{\ell^2}{4} - x^2\right) + \frac{M_0}{D}\,. \tag{1.146}$$

• **Loaded plate without edge force in x-direction** ($F = 0$): Let us consider a uninform load q applied all over the plate and bending moments $\pm M_0$ applied to the long edges. If no force acts in the x-direction at the long edges, then the first corresponding boundary at those edges is a freedom to move in the x-direction. Various boundary cases are shown in Fig. 1.52.
The successive integrations of (1.146) with $F = 0$ leads to

$$w = \frac{q}{24D}\,x^4 + \left(\frac{M_0}{2D} - \frac{q\ell^2}{16D}\right)x^2 + C_1\,x + C_2\,. \tag{1.147a}$$

[14] Many authors prefer using an opposite sign convention for the bending moment M in (1.143) (cf. for instance Timoshenko and Woinowsky-Krieger [155]); this leads to an opposite sign in the representation of the flexure. For the applications of active optics methods, we always use here the more natural sign convention of a positive flexure for a positive curvature i.e. a positive second derivative.

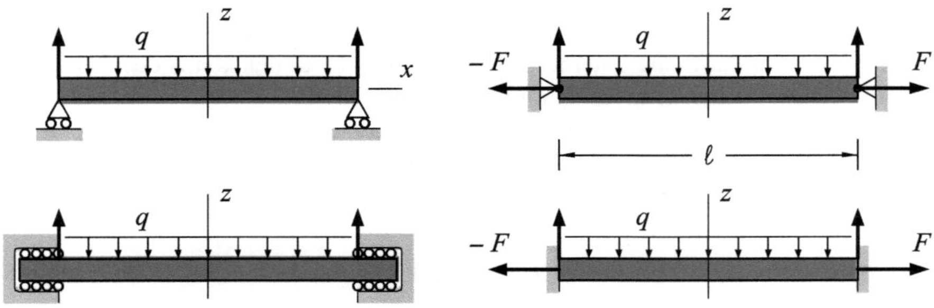

Fig. 1.52 Four boundary conditions in the cylindrical bending of a long plate. The uniform load q is negative in the four cases

Taking the origin of the z, y frame at the middle of the neutral surface and since the load is symmetrical, the ordinate and slope are null at the origin. So the constants C_1=C_2 =0 and the flexure is

$$w = \frac{q}{24D}\,x^4 + \left(\frac{M_0}{2D} - \frac{q\ell^2}{16D}\right)x^2. \tag{1.147b}$$

If the plate is with *simply supported edges* free to move in x, the bending moment at the edge is null (cf. Fig. 1.52). Setting $M_0 = 0$ entails

$$w = \tfrac{q}{48D}\,(2x^2 - 3\ell^2)\,x^2, \qquad w\{\pm\ell/2\} = -5q\ell^4/384D\,, \tag{1.148a}$$

where if $q < 0$, then $w\{\pm\ell/2\} > 0$.

If the plate is with *built-in edges* free to move in x, the slope is null at the edge (cf. Fig. 1.52). Setting $dw/dx|_{\pm\ell/2} = 0$ entails $M_0 = q\ell^2/12$ and

$$w = \tfrac{q}{48D}\,(2x^2 - \ell^2)\,x^2, \qquad w\{\pm\ell/2\} = -q\ell^4/384D\,, \tag{1.148b}$$

where if $q < 0$, then $w\{\pm\ell/2\} > 0$.

• **Loaded plate with opposite edge forces $F, -F$, in x-direction**: Other load configurations of interest are similar to the two previous cases with the load q, moments $M_0, -M_0$, but now including opposite forces $F, -F$, applied to the long edges at $x = \pm\ell/2$ and in the x-direction (cf. Fig. 1.52). Set $F > 0$ at $x = \ell/2$ and denote μ a dimensionless quantity defined by

$$\frac{F}{D} = \frac{4\mu^2}{\ell^2}\,, \tag{1.149}$$

so the the differential equation of the flexure (1.144) becomes

$$\frac{d^2w}{dx^2} - \frac{4\mu^2}{\ell^2}\,w = -\frac{q}{2D}\left(\frac{\ell^2}{4} - x^2\right) + \frac{M_0}{D}\,. \tag{1.150}$$

The particular solution w_1 is of the form $w_1 = A_1 x^2 + A_2$. After identification of A_1 and A_2 coefficients, we obtain

$$w_1 = -\frac{q\ell^2}{8D\mu^2}\,x^2 - \frac{q\ell^4}{32D}\,\frac{2-\mu^2}{\mu^4} - \frac{M_0\ell^2}{4D\mu^2}\,. \tag{1.151}$$

The solution without the right-hand side of (1.150) is

$$w_2 = C_1 \cosh\frac{2\mu x}{\ell} + C_2 \sinh\frac{2\mu x}{\ell}\,, \tag{1.152}$$

Since the loading of the plate is symmetrical with respect to the z, y plane, then $C_2 = 0$. The general solution is of the form $w = w_1 + C_1 \cosh(2\mu x/\ell)$. Setting the origin of the flexure on the neutral surface and midway between the edges, $w(0) = 0$ is achieved if

$$C_1 - \frac{q\ell^4}{32D}\,\frac{2-\mu^2}{u^4} - \frac{M_0\ell^2}{4D\mu^2} = 0 \quad \text{i.e.}\quad C_1 = -w_1(0)\,, \tag{1.153}$$

which entails

$$w = w_1(0)\left(1 - \cosh\frac{2\mu x}{\ell}\right) - \frac{q\ell^2}{8D\mu^2}\,x^2. \tag{1.154a}$$

Substituting $1 - \cosh 2\mu x/\ell = -2\sinh^2 \mu x/\ell$, we can represent the flexure as a function of the force F (or μ) and moment M_0 by

$$w = \left(\frac{q\ell^4}{16D}\,\frac{2-\mu^2}{\mu^4} + \frac{M_0\ell^2}{2D\mu^2}\right)\sinh^2\frac{\mu x}{\ell} - \frac{q\ell^2}{8D\mu^2}\,x^2. \tag{1.154b}$$

The most interesting case is when the opposite forces $F, -F$, are generated by the deformation of the plate *itself* when the edges are not allowed to move in the x-direction. During the bending, the plate elongates and its curvilinear width – ℓ when not stressed – is increased by $\delta\ell$ which represents the difference of the length between the arc along the flexure curve and the length ℓ of the cord.

We have seen with (1.148b) that the (movable) built-in edge condition entails $dw/dx|_{\pm\ell/2} = 0$ and $M_0 = q\ell^2/12$. Now the intensity of the forces $\pm F$ due to the *unmovable built-in edges* can be obtained. This was solved by Timoshenko and Woinowsky-Krieger ([155], Chap. 1), who showed that for small flexures, an element strip elongates of

$$\delta\ell = \frac{1}{2}\int_{-\ell/2}^{\ell/2}\left(\frac{dw}{dx}\right)^2 dx\,. \tag{1.155a}$$

This elongation is only caused by the extension of the plate from the blocked edges. Still assuming that the lateral strain ε_{yy} of the strip is unchanged in the long plate the co-added component of constant stress is, from (1.139b), $\sigma'_{xx} = -E\,\delta\ell/(1-\nu^2)\ell \equiv -F/t$. Also using (1.149) and (1.142), we obtain

$$\delta\ell = (1-\nu^2)\,\frac{F\ell}{Et} = \frac{t^2\mu^2}{3\,\ell}\,. \tag{1.155b}$$

Equalizing the latter two expressions and substituting the flexure represented by (1.154b), we thus obtain for *unmovable built-in edges* a relation which now contains the unknown μ only,

$$(1-\nu^2)^2\,\frac{q^2\ell^8}{E^2t^8}\,\Psi(\mu) = \mu^2\,, \tag{1.156}$$

where $\Psi(\mu)$ is derived from (1.155a) and includes finite term polynomials in μ^n and $\tanh^m\mu$. Any numerical solving in function of the dimensionless ratio $(1-\nu^2)^2(q/E)^2(\ell/t)^8$ allows us to obtain the intensity of force F in (1.149) and the flexure from (1.154b).

1.13.8 Bending of Thin Plates and Non-developable Surfaces

Let us consider hereafter a thin plate of thickness t and in-plane typical dimension ℓ, such as $t/\ell \ll 1$, and assume that the absolute value of the maximum curvature $|1/R_{\min}|$ occurring in the bending is $\ell/|R_{\min}| \ll 1$. The latter condition is equivalent to state that the maximum value $w_{\max}$ of the flexure taken over all the surface remains small compared the thickness of the plate. Hence the conditions are respectively

$$t/\ell \ll 1\,, \qquad |w_{\max}|/t \ll 1\,. \tag{1.157}$$

Under these conditions, the bending introduces a negligible strain along the middle surface and the perpendicular stress distribution is null on it; this surface is then called the *neutral surface*.

One generally distinguishes between plates bent into an *synclastic surface* and into a *anticlastic surface*. The principal radii of curvature R_x and R_y of a synclastic surface are $R_xR_y > 0$, whilst for an anticlastic surface $R_xR_y < 0$ (Fig. 1.53).

Let us assume that the plate is flat when not bent, and set a coordinate frame x, y, z with the origin at the neutral surface and the z-axis normal to the surface. The above hypotheses of thin plate and small deformation entail that the flexure w is a function of x and y only. Since the middle surface is also the neutral surface, the components of the displacement vector at this surface are

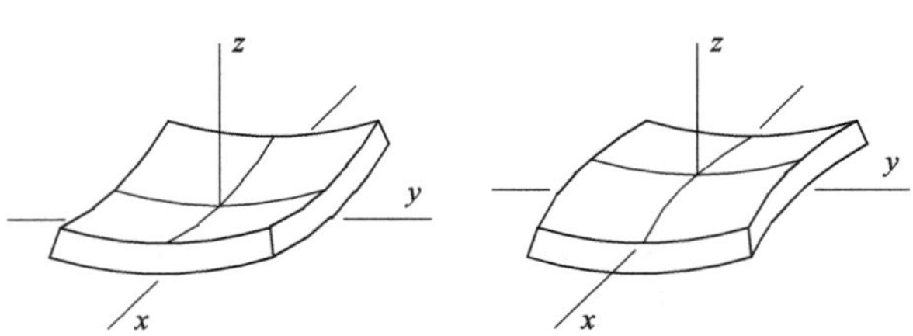

Fig. 1.53 *Left*: Synclastic surface, $R_xR_y > 0$. *Right*: Anticlastic surface, $R_xR_y < 0$

$$u\,|_{z=0} = v\,|_{z=0} = 0\,, \qquad w\,|_{z=0} = w(x,y)\,. \tag{1.158a}$$

For a thin plate and a small deformation, the second of Love-Kirchhoff hypotheses states that the stresses

$$\sigma_{zz} = \sigma_{yz} = \sigma_{zx} = 0 \tag{1.158b}$$

at the outer surface of the plate as well as at the inner volume of it. The flexure of all points originally located at any distance z from the middle surface can be derived by orthogonal and equal lengths from the flexure of the middle surface.

The third equation of the stress-strain relations (1.123a) leads to

$$\sigma_{zz} = \frac{E}{(1+\nu)(1-2\nu)}\big[(1-\nu)\,\varepsilon_{zz} + \nu\,(\varepsilon_{xx}+\varepsilon_{yy})\big] = 0\,. \tag{1.159}$$

Hence, the Love-Kirchhoff conditions (1.158) entail

$$\varepsilon_{zz} = -\frac{\nu}{1-\nu}\,(\varepsilon_{xx}+\varepsilon_{yy})\,, \qquad \varepsilon_{yz} = 0\,, \qquad \varepsilon_{zx} = 0\,. \tag{1.160}$$

The sections that are perpendicular to the x- and y-axes when the plate is flat at rest, become tilted around the y- and x-axes respectively in the bending but they remain flat and normal to the mid-surface. Since the tilt angle is equal to the slope of w, the two other components of the displacement vector are

$$u = -z\,\frac{\partial w}{\partial x}\,, \qquad v = -z\,\frac{\partial w}{\partial y}\,, \tag{1.161}$$

which satisfy the conditions (1.158a) of null value at the neutral surface. From u and v, the determination of all the components (1.121a) of the strain tensor gives

$$\varepsilon_{xx} = -z\,\frac{\partial^2 w}{\partial x^2}\,, \quad \varepsilon_{yy} = -z\,\frac{\partial^2 w}{\partial y^2}\,, \quad \varepsilon_{zz} = \frac{\nu\, z}{1-\nu}\left(\frac{\partial^2 w}{\partial x^2} + \frac{\partial^2 w}{\partial y^2}\right),$$
$$\varepsilon_{yz} = \varepsilon_{zx} = 0\,, \qquad \varepsilon_{xy} = -z\,\frac{\partial^2 w}{\partial x\,\partial y}\,, \tag{1.162a}$$

and, from the stress-strain relations (1.123a), the associated stresses are

$$\sigma_{xx} = -\frac{Ez}{1-\nu^2}\left(\frac{\partial^2 w}{\partial x^2} + \nu\frac{\partial^2 w}{\partial y^2}\right), \quad \sigma_{yy} = -\frac{Ez}{1-\nu^2}\left(\frac{\partial^2 w}{\partial y^2} + \nu\frac{\partial^2 w}{\partial x^2}\right),$$
$$\sigma_{zz} = \sigma_{yz} = \sigma_{zx} = 0, \qquad \sigma_{xy} = -\frac{Ez}{1+\nu}\,\frac{\partial^2 w}{\partial x \partial y}\,. \tag{1.162b}$$

• **Energy of a bent plate**: These components allow us to determine the free energy of the volume element $d\mathcal{V}$. Substituting them into the representation (1.125) of the elementary free energy $d\mathcal{F}$, we obtain after simplification

$$d\mathcal{F} = \frac{E\,z^2}{1+\nu}\left[\frac{1}{2(1-\nu)}\left(\frac{\partial^2 w}{\partial x^2} + \frac{\partial^2 w}{\partial y^2}\right)^2 + \left(\frac{\partial^2 w}{\partial x\,\partial y}\right)^2 - \frac{\partial^2 w}{\partial x^2}\,\frac{\partial^2 w}{\partial y^2}\right] d\mathcal{V}. \tag{1.163a}$$

The total free energy of the plate is obtained by integration over its total volume. For a *constant* thickness plate whose thickness varies from $-t/2$ to $t/2$, and denoting the rigidity $D = Et^3/12(1-\nu^2)$ such as defined by (1.142), the total free energy or *flexural energy* of the plate is

$$\mathcal{F} = D\iint_{\mathcal{A}}\left\{\frac{1}{2}\left(\frac{\partial^2 w}{\partial x^2}+\frac{\partial^2 w}{\partial y^2}\right)^2+(1-\nu)\left[\left(\frac{\partial^2 w}{\partial x\,\partial y}\right)^2-\frac{\partial^2 w}{\partial x^2}\frac{\partial^2 w}{\partial y^2}\right]\right\}d\mathcal{A}. \tag{1.163b}$$

where $d\mathcal{A} = dx\,dy$ and the integration is taken over the surface area $\mathcal{A}$ of the plate. In deriving this latter equation, we have assumed a thin plate and a small deformation, so the w-displacement of any point of the plate is the same as that of a point with identical coordinates x, y belonging to the neutral surface.

• **Equation of deformation of a plate**: The free energy due to the internal stresses and strains can be used to obtain the general form of the flexure $w(x, y)$ of constant thickness plates. First, one must remark that the sum of the free energy and of the potential energy is a *constant*. The potential energy is the opposite of the work of the external load applied to the surface of area $\mathcal{A}$ in the normal direction; for small deformations this normal is the z-axis so this work taken over the whole plate is $\iint q\,w\,d\mathcal{A}$, where q is the external load of dimension FL^{-2}. Second, we state that the total energy is *minimal* in any variation δw of the displacement. This condition is written as

$$\delta\mathcal{F} - \iint_{\mathcal{A}} q\,\delta w\,d\mathcal{A} = 0. \tag{1.164a}$$

The variation $\delta\mathcal{F}$ of the free energy is carried out by considering the boundary conditions that are applied at the contour line $\mathcal{C}$ of the plate. In the general case, this involves using curvilinear coordinates. A complete analysis is given by Landau and Lifshitz [92] which, starting from (1.163b), leads to the following expression

$$\iint_{\mathcal{A}}\nabla^2\nabla^2 w\,\delta w\,d\mathcal{A}+\oint_{\mathcal{C}}P_1\delta w\,d\mathcal{C}+\oint_{\mathcal{C}}P_2\,\delta w\,d\mathcal{C}-\frac{1}{D}\iint_{\mathcal{A}}q\,\delta w\,d\mathcal{A}=0, \tag{1.164b}$$

where $\nabla^2\cdot = \partial^2\cdot/\partial x^2+\partial^2\cdot/\partial y^2$ is the Laplacian and the P_1, P_2 terms are functions of w, ν and of the shape of the contour. The two corresponding integrals must be taken equal to zero over the contour, thus providing two conditions, each of them defining the force and the moment applied at the contour. Satisfying the two conditions, this relation becomes

$$\iint_{\mathcal{A}}\left[\nabla^2\nabla^2 w - \frac{q}{D}\right]\delta w\,d\mathcal{A} = 0. \tag{1.164c}$$

Since the variation of the flexure δw is arbitrarily done, the principle of minimal energy is satisfied if the factor into brackets is null. Therefore, we obtain the equilibrium equation

$$\nabla^2\nabla^2 w - q(x, y)/D = 0, \tag{1.164d}$$

or explicitly

$$\frac{\partial^4 w}{\partial x^4} + \frac{2\,\partial^4 w}{\partial x^2 \partial^2 y} + \frac{\partial^4 w}{\partial y^4} - \frac{q}{D} = 0\,. \tag{1.164e}$$

This biharmonic equation, also known as *Poisson's equation*, gives the solutions for small deformations of a thin and *constant* thickness plate ($D =$ constant) provided no in-plane force exists at the middle surface. This relation was first derived by S. Denis Poisson [121] who found the complete general integral solution for circular unholed plates (1828).[15]

• **Bending moments, twisting moments, shearing forces, and net shearing forces**: The *bending moments* M_x and M_y act tangentially around the y- and x-axes respectively on the y, z and x, z faces of a plate element of volume $t\,dx\,dy$. The *twisting moments* M_{xy} and M_{yx} act on the same faces but perpendicularly to them i.e. around the x- and y-axes. From the stress components (1.124b), these moments are, per unit length of the element side,

$$M_x = -\int_{-t/2}^{t/2} \sigma_{xx}\, z\, dz = D\left(\frac{\partial^2 w}{\partial x^2} + \nu \frac{\partial^2 w}{\partial y^2}\right), \tag{1.165a}$$

$$M_y = -\int_{-t/2}^{t/2} \sigma_{yy}\, z\, dz = D\left(\frac{\partial^2 w}{\partial y^2} + \nu \frac{\partial^2 w}{\partial x^2}\right), \tag{1.165b}$$

$$M_{xy} = M_{yx} = -\int_{-t/2}^{t/2} \sigma_{xy}\, z\, dz = D(1-\nu)\left(\frac{\partial^2 w}{\partial x\, \partial y}\right), \tag{1.165c}$$

accordingly to the sign convention (cf. the previous subsection).

We may remark that combining these moments entails

$$\frac{1}{D(1+\nu)}\left(M_x + M_y\right) = \nabla^2 w\,, \tag{1.166a}$$

[15] Sophie Germain [68] established a biharmonic equation in memoirs to the French Academy, the last in 1815, as attempts to explain the vibration modes of an elastic plate. This problem was brought to her attention by the fascinating experiments by Chladni [28] showing various patterns formed by the sand which moved on a vibrating plate, when excited by acoustic strings of different tones, until reaching the two-dimensional positions of the nodes. After seeing *Chladni's patterns* – or nodal curves – in 1810, Napoleon funded the Institute of France for an extraordinary prize to be awarded for a theory providing a mathematical explanation. At the issue of the contest announcement, the examining committee noticed some flaws in Germain's last memoir but deduced that her approach was correct, and the prize was awarded to her in 1816. In fact a full analysis of *Germain's problem* would have required first knowing a boundary condition involving the twisting moment M_{rt} at the plate edge. This condition for non-axisymmetrical deformations was found much later by Kirchhoff, in 1850 (see hereafter).

Turning to the case of the vibrations of an elastic membrane, only involving second derivatives, the nodal curves of a circular membrane – like drum heads – were first derived by Clebsch [34], in 1862, where the radial part of the solutions are Bessel functions.

Germain's problem for a circular plate was finally solved by Rayleigh [124, 125], in 1873, by using the Kirchhoff boundary condition of the thin plate theory. He also solved this problem for the nodal curves of a square plate. Ritz [133] later generalized Rayleigh's extremals method of the total energy (potential and kinetic), classically known as the Rayleigh-Ritz method.

and

$$\frac{1}{D}\left(\frac{\partial^2 M_x}{\partial x^2} + \frac{2\,\partial^2 M_{xy}}{\partial x \partial y} + \frac{\partial^2 M_y}{\partial y^2}\right) = \nabla^2\nabla^2 w\,. \tag{1.166b}$$

Hence, from (1.164d), this second derivative sum of the moments must be zero if there is no transverse load ($q = 0$).

The *shearing forces* Q_x and Q_y per unit length act on the sections of the element in the z-direction. The equilibrium equations of the statics, obtained from the moments around the y- and x-axes, and from the forces in the z-directions are respectively

$$\frac{\partial M_x}{\partial x} + \frac{\partial M_{xy}}{\partial y} + Q_x = 0\,, \qquad \frac{\partial M_y}{\partial y} + \frac{\partial M_{yx}}{\partial x} + Q_y = 0\,, \tag{1.167}$$

and

$$\frac{\partial Q_x}{\partial x} + \frac{\partial Q_y}{\partial y} + q = 0\,. \tag{1.168}$$

From (1.167), we obtain the expression of the shearing forces

$$Q_x = -\,D\frac{\partial}{\partial x}\,\nabla^2 w\,, \qquad Q_y = -\,D\frac{\partial}{\partial y}\,\nabla^2 w\,. \tag{1.169}$$

Substituting them into (1.168) entails

$$\frac{\partial^2}{\partial x^2}\,\nabla^2 w + \frac{\partial^2}{\partial y^2}\,\nabla^2 w - \frac{q}{D} = 0,$$

which is the biharmonic Poisson equation.

Finally Kirchhoff [88, 89] defined, in 1850, the *net shearing forces* V_x and V_y per unit length as

$$V_x = Q_x - \frac{\partial M_{xy}}{\partial x}\,, \qquad V_y = Q_y - \frac{\partial M_{yx}}{\partial y}\,, \tag{1.170}$$

which are the resulting axial forces acting into the plate. Theses forces enter in the formulation of a boundary condition, known as *Kirchhoff's condition* at the edge.[16]

For a plate with a free edge, the net shearing forces become null at the contour.

---

[16] The Kirchhoff definition of the net shearing force is an important formulation for a boundary condition at a plate edge. This attracted the attention of Rayleigh and permitted him to elaborate the theory of the vibrations of plates.

Its physical significance entailed a clarified statement of the boundaries as was explained by Kelvin and Tait [83]. Independently the same question was explained by Boussinesq [18]. A detailed account on Kirchhoff's condition is given by Love [97], p. 460, and by Timoshenko and Woinowsky-Krieger [155], p. 84.

1.13.9 Bending of Rectangular Plates of Constant Thickness

• **Plate distorted by bending moments**: Let us consider a flat rectangular plate of lengths a, b in the x, y directions only submitted to opposite bending moments M_a and M_b, i.e. where no transverse load is applied to its surface ($q = 0$). We assume that the bending moments are in the principal directions of the frame, positive for x and y, and that the frame origin remains at the center of the middle surface when at rest and when bent (Fig. 1.54).

Setting the origin at the center of the mid-surface, and for symmetry reasons,

$$w\{0,0\} = 0\,, \qquad \left.\frac{\partial w}{\partial x}\right|_{x=0} = \left.\frac{\partial w}{\partial y}\right|_{y=0} = 0\,. \tag{1.171a}$$

The Kirchhoff conditions for a plate with free edges is that the net shearing forces V_x and V_y are zeroed on this contour,

$$V_x\{a, y\} = V_y\{x, b\} = 0\,. \tag{1.171b}$$

Since the x- and y-axis are in the principal directions of the bending moments, the twisting moments M_{xy} and M_{yx} vanish whatever x and y which entails that $Q_x = Q_y = 0$ and $V_x = V_y = 0$. The bending moments M_x and M_y are constant moments whatever x and y. Equating them to the moments M_a and M_b applied to the edges, we obtain from (1.165a) to (1.165b),

$$\frac{\partial^2 w}{\partial x^2} = \frac{M_a - \nu M_b}{D(1-\nu^2)}\,, \qquad \frac{\partial^2 w}{\partial y^2} = \frac{M_b - \nu M_a}{D(1-\nu^2)}\,.$$

From the origin and boundary conditions, all the integration constants are equal to zero. The solution is

$$w = \frac{1}{2D(1-\nu^2)}\left[(M_a - \nu M_b)\,x^2 + (M_b - \nu M_a)\,y^2\right]. \tag{1.172a}$$

Introducing the principal curvatures $1/R_x$ and $1/R_y$, the flexure can be represented by

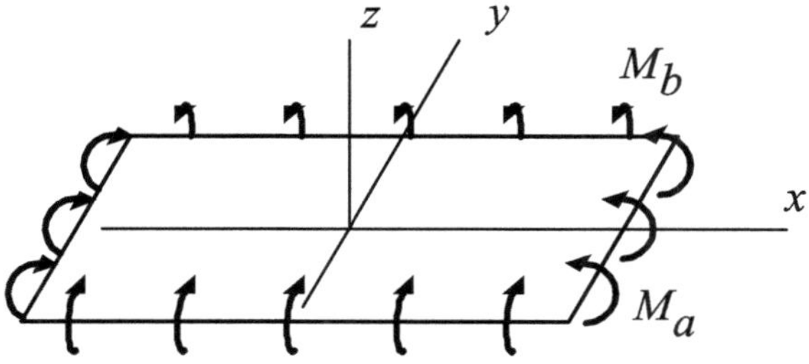

Fig. 1.54 Rectangular plate only submitted to bending moments M_a and M_b at the edges

$$w = \frac{x^2}{2R_x} + \frac{y^2}{2R_y}, \tag{1.172b}$$

with

$$\frac{1}{R_x} = \frac{M_a - \nu M_b}{D(1-\nu^2)}, \qquad \frac{1}{R_y} = \frac{M_b - \nu M_a}{D(1-\nu^2)}. \tag{1.173}$$

If $M_b = M_a$, then $R_y = R_x = D(1+\nu)/M_a$, the flexure is a synclastic surface which is a *paraboloid*,

$$w = \frac{M_a}{2D(1+\nu)}\left(x^2 + y^2\right). \tag{1.174}$$

If $M_b = -M_a$, then $R_y = -R_x = -D(1-\nu)/M_a$, the flexure is an anticlastic surface or *saddle*,

$$w = \frac{M_a}{2D(1-\nu)}\left(x^2 - y^2\right). \tag{1.175}$$

If $M_b = 0$, we can compare the flexure (1.172a) with that of a long plate whose long edges are simply supported on straight lines and bent by uniform moments while the long length is kept constant (cf. Sect. 1.13.6), i.e. where the flexure is in x^2 only. For a rectangular plate only submitted to the bending moments M_a and $-M_a$ on two opposite sides, a flexure component arises in the y-direction,

$$w = \frac{M_a}{2D(1-\nu^2)}\left(x^2 - \nu\, y^2\right). \tag{1.176}$$

With respect to the initial plane of the plate when at rest, the level lines of the deflected surface are *hyperbolas* separated by asymptotes. From (1.175) and denoting α the half-value of the largest angle between the asymptotes, Poisson's ratio is expressed by

$$\nu = -\frac{R_x}{R_y} = \tan^{-2}\alpha\,. \tag{1.177}$$

This result, showing that the angle α is simply related to ν, was proposed by Cornu [38] for the determination of Poisson's ratio (Fig. 1.55). Since $\nu \le 1/2$, we always have $\arctan\alpha \ge \sqrt{2}$ i.e. $\alpha \ge 54.735°$.

Because the four edges of the plate do not remain straight lines in the bending, for a more accurate application of the boundaries the optical interferences are generally obtained with a long plate where the moments M_a and $-M_a$ are applied to the short edges. Cornu's method with a He-Ne laser source has become of classical use for accurate *Poisson's ratio measurements*.

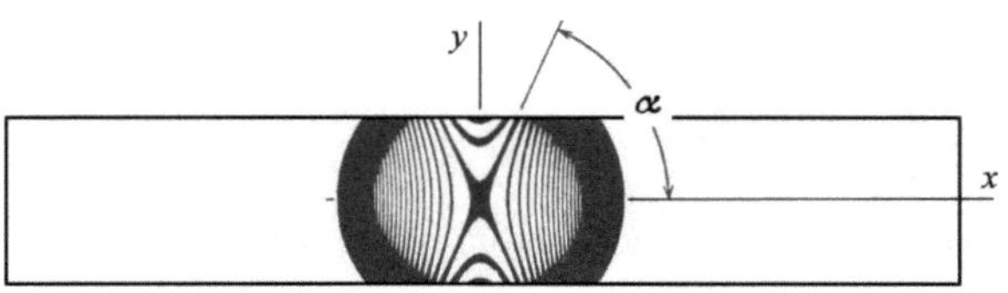

Fig. 1.55 Interferogram of a bent rectangular plate for the determination of Poisson's ratio. This method was proposed by Cornu [38] (after Timoshenko [158])

• **Plate bent by a load**: Because of the discontinuity at the boundaries due to the corners of the plate, the flexure of rectangular plates submitted to a load q is a complex problem. The general solution of Poisson's equation without second member cannot be dissociated from that of the particular solution.

For a simply supported rectangular plate where the edges remain straight lines, a first and classical solution of the problem is due to Navier [114] who showed that if both the load q and the flexure w are represented by a double trigonometric series of the form

$$\sum_{m=1}^{\infty}\sum_{n=1}^{\infty} \sin\frac{m\pi x}{a}\ \sin\frac{m\pi y}{b}\,, \tag{1.178}$$

then, Poisson's equation can be satisfied.

An extended account of the analytical developments of rectangular plate flexure using the double series representation and other representations is given in Timoshenko and Woinowsky-Krieger [155], who also consider cases with various edge conditions and treat continuous floor-slab plates. For active optics methods, the rectangular plate case is of very limited interest.

1.13.10 Axisymmetric Bending of Circular Plates of Constant Thickness

The flexure w of the mid-surface of a thin circular plate resulting from an axisymmetric load is derived from Poisson's equation. This equation is now only relative to the radial distance r from the center and becomes, from (1.164b),

$$\nabla^2\nabla^2 w \;\equiv\; \frac{1}{r}\frac{d}{dr}\left\{r\,\frac{d}{dr}\left[\frac{1}{r}\frac{d}{dr}\left(r\,\frac{dw}{dr}\right)\right]\right\} = \frac{q(r)}{D}\,, \tag{1.179}$$

where $q(r)$ is the load per unit surface area and D the constant rigidity.

Considering hereafter a uniform load, $q =$ constant, distributed all over the surface of the plate, the successive integrations can be carried out directly. As first demonstrated by Denis Poisson [121] in his famous memoir of 1829, we find

$$w = \frac{q}{64D}\,r^4 + C_1 r^2 + C_2 r^2 \ln r + C_3 \ln r + C_4\,, \tag{1.180}$$

where C_i are unknown constants that are determined from the boundaries. The terms $r^2 \ln r$ and $\ln r$ allow treating cases with a ring force or a central force, and holed plates.

Starting from the stress-relations (1.124b), and similarly as for rectangular plates, we derive the three stress components that differ from zero, namely σ_{rr}, σ_{tt} and σ_{zr}. This allows determining the bending and twisting moments, per unit length, as

$$M_r = D\left(\frac{d^2w}{dr^2} + \frac{\nu}{r}\frac{dw}{dr}\right),\quad M_t = D\left(\nu\,\frac{d^2w}{dr^2} + \frac{1}{r}\frac{dw}{dr}\right),\quad M_{rt} = 0\,. \tag{1.181}$$

The shearing forces Q_r, Q_t and net shearing force V_r, V_t are

$$Q_r = -D\frac{d}{dr}\nabla^2 w\,, \qquad V_r = Q_r\,, \qquad Q_t = V_t = 0\,. \tag{1.182}$$

The total shearing force acting along the circle of radius r is $2\pi r Q_r$. This force is in static equilibrium with the force resulting from the total load applied inside this area. For instance, with a uniform load applied all over the surface, $2\pi r Q_r + \int_0^r 2\pi q\, r\, dr = 0$, so that $Q_r = -qr/2$ in this loading case. Hence, the shearing force Q_r can always be determined from the loading configuration.

It is always advantageous to derive the flexure from the shearing force in (1.182). The integration of Poisson's equation is then directly operated from the integro-differential equation

$$\frac{d}{dr}\left(r\frac{dw}{dr}\right) = -\frac{r}{D}\int Q_r\, dr \tag{1.183}$$

by substituting the expression of Q_r as a function of the known load.

1.13.11 Circular Plates and Axisymmetric Loading Manifolds

• **Sign convention**: In all Chapters, the *sign convention* for the flexure w in the z-direction, sometimes denoted u_z or z, is that with a positive flexure – which means towards z positive – when the curvature term $w(r^2)$ is positive, this term being considered as the first-order mode of the flexure. Hence the sign of an applied force or load must be conveniently chosen.

- If a force F or a load q is positive, then it acts towards z positive.
- If $w(r^2) = 0$, we apply the convention to the next-order term $w(r^4)$.
- A positive moment M_r at outer edge generates a positive curvature.

In the thin plate theory of small deformations one uses this to set the origin of the flexure at the mid-plane vertex of the plate, however the flexure is the same at its external surfaces. We list hereafter the flexure for various loads and boundaries, the associated shearing force $Q_r = V_r$, and the maximum flexure $w\{a\}$ at the edge. The sign of the load is given for a flexure with positive curvature (term $w(r^2) > 0$) or, if this term is null, for $w(r^4) > 0$.

1. Free Edge and Uniform Bending Moment at Edge: If $M > 0$ is a constant bending moment at the edge and no other force acts on the plate, then $Q_r = 0$, and

$$w = \frac{Ma^2}{2(1+\nu)D}\frac{r^2}{a^2}\,, \qquad w\{a\} = \frac{M\,a^2}{2(1+\nu)D}\,. \tag{1.184a}$$

2. Simply Supported Edge and Uniform Load: If $q < 0$ is a uniform load applied over all the surface, then $M_r\{a\} = 0$, $Q_r = -qr/2$, and

$$w = \frac{qa^4}{64D}\left(\frac{r^2}{a^2} - 2\frac{3+\nu}{1+\nu}\right)\frac{r^2}{a^2}, \qquad w\{a\} = -\frac{5+\nu}{1+\nu}\frac{qa^4}{64D}. \tag{1.184b}$$

3. Built-in (or Clamped) Edge and Uniform Load: If $q < 0$ is a uniform load, then the slope at edge is $dw/dr|_{r=a} = 0,\ Q_r = -qr/2$, and

$$w = \frac{qa^4}{64D}\left(\frac{r^2}{a^2} - 2\right)\frac{r^2}{a^2}, \qquad w\{a\} = -\frac{qa^4}{64D}. \tag{1.184c}$$

4. Simply Supported Edge and Concentrated Force at Center: If $F < 0$ is a force applied at center, then $M_r\{a\} = 0$, $Q_r = -F/2\pi r$, and

$$w = \frac{Fa^2}{16\pi D}\left(\ln\frac{r^2}{a^2} - \frac{3+\nu}{1+\nu}\right)\frac{r^2}{a^2}, \qquad w\{a\} = -\frac{3+\nu}{1+\nu}\frac{Fa^2}{16\pi D}. \tag{1.184d}$$

5. Built-in Edge and Concentrated Force at Center[17] : If $F < 0$ is a force applied at center, then $dw/dr|_a = 0$, $Q_r = -F/2\pi r$, and

$$w = \frac{Fa^2}{16\pi D}\left(\ln\frac{r^2}{a^2} - 1\right)\frac{r^2}{a^2}, \qquad w\{a\} = -\frac{Fa^2}{16\pi D}. \tag{1.184e}$$

6. Free Edge and Opposite Central Force and Load: If $F < 0$ is a central force such as $F + \pi a^2 q = 0$, then $M_r\{a\} = 0$, $Q_r = -(F/2\pi)(1/r - r/a^2)$, and

$$w = \frac{Fa^2}{64\pi D}\left(4\ln\frac{r^2}{a^2} - 2\frac{3+\nu}{1+\nu} - \frac{r^2}{a^2}\right)\frac{r^2}{a^2}, \qquad w\{a\} = -\frac{7+3\nu}{1+\nu}\frac{Fa^2}{64\pi D}. \tag{1.184f}$$

7. Bent and Supported Edge with Uniform Load for r^4 Flexure: If $q > 0$ is a load and by generating an edge moment $M = (3+\nu)\,qa^2/16 > 0$, then $Q_r = -qr/2$, and

$$w = \frac{qa^4}{64D}\frac{r^4}{a^4}, \qquad w\{a\} = \frac{qa^4}{64D}. \tag{1.184g}$$

8. Plate Bent by a Concentric Ring Force: When a ring force acts on a centered circle of radius b and the plate is freely supported or clamped at the edge, the determination of the flexure must be dissociated into an inner zone, $r \leq b$, and an outer zone $b \leq r \leq a$. This problem was first solved by Saint-Venant [138] (cf. [158] p. 64).

9. Plate Bent by an Off-Center Force Point: When a force point is applied off-center of the plate, the determination of the flexure has been solved by Clebsch

[17] The flexures of cases **1–5** were first derived by Poisson in his famous and detailed memoir [121] of 1828 where he used the uni-constant theory $\{E,\ \nu = 1/4\}$ and thus created the *thin plate theory* of elasticity. See also the comments in Love's book [97], p. 489.

[34]. Starting from Poisson's equation in polar coordinates (r, θ), one shows that the flexure is represented by *Clebsch's polynomials* (cf. Chap. 7).

1.13.12 Deformation of a Plate in a Gravity Field

The volume forces generated by a gravity field act on all the elements of a body. For instance the length of a bar is not the same when placed vertically or horizontally on the ground. The flexure of a solid due to the gravity is sometimes called *own weight flexure*.

When a flat and horizontal plate is supported on the edge, its flexure under gravity can be easily derived. Denoting μ the density of the plate, the sum of the volume forces per unit surface area of the plate over the thickness t is equivalent to the pressure $t\mu g$. In the sign convention, the gravity vector $\mathbf{g}$ is opposite to the z-axis and a uniform load q is positive towards z positive. Hence, the substitution

$$q \rightarrow t\,\mu\,g\,, \qquad g < 0\,, \tag{1.185}$$

straightforwardly provides the flexure of a plate under gravity.

For instance, this substitution in (1.184b), (1.184c) or (1.184f) gives the flexure of a simply supported plate at the edge, of a built-in plate or of a plate suspended from its vertex, respectively.

1.13.13 Saint-Venant's Principle

The small deformation theory of circular plates allows us to simply express the component $w(r)$ of the displacement vector $u,\ v,\ w$ of the middle surface and any other point departing from this surface is displaced by the same amount $w(r)$ at the same radius r. The large deformation theory of thin circular plates (cf. Chap. 2) considers that the radial strain ε_{rr} becomes a function of both u and w. The thick plate theory of small deformations takes into account the shear strains that lead to cross-sections over the plate thickness which are no longer orthogonal to the middle surface and become *S-shaped*. With these improvements of the basic thin plate theory, the difficulties involved for finding the mathematical expressions of the displacements do not generally allow us to obtain explicitly represented solutions.

In the general case of a solid or plate where both thickness and flexure are not small, the complexity is such as it is out of purpose to search for the functions satisfying a partial derivative equation set, this even if the boundary conditions are particularly simple.

In other respects from the practical point of view, it is experimentally illusory to consider that for all loading cases we could exactly apply given surface force distributions $\mathbf{F}$ over a given area $\delta\mathcal{A}$. Although a uniform load can be accurately distributed by a pressure difference whatever the flexure is, in most cases local deformations only arise due to the application of concentrated forces as generally

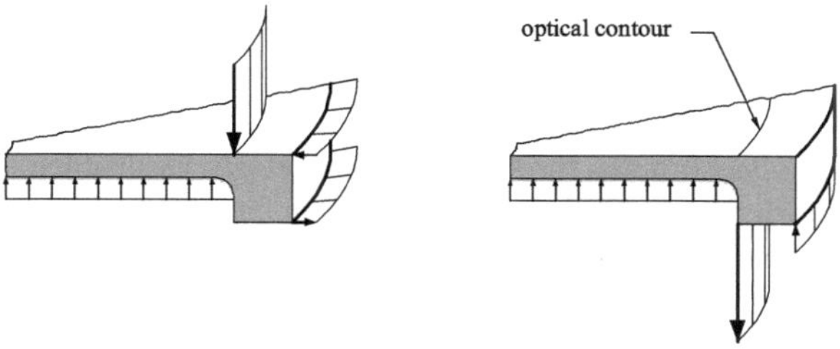

Fig. 1.56 *Saint-Venant's principle of equivalence*: example of two quasi-equivalent load configurations applied at the boundary of an axisymmetric mirror

happens at the boundaries of the solid. Of course, these local deformations can be determined, for instance by using Hertz's contact formulas [74] (cf. Landau and Lifshitz [92] p. 42) or Dirac's function, but when the main purpose is to derive the whole displacements of the solid, it is clear that these local deformations do not substantially affect them.

These remarks led Saint-Venant to enounce a useful principle which introduced some flexibility in the practical application of the boundary conditions.[18] We recall that a set of forces define a torsor which, at any given point, is globally represented by a resultant force and a resultant moment. An excellent statement of *Saint-Venant's principle of equivalence* has been given by Germain and Muller [69] as follows:

→ *If one substitutes a first distribution of given surface forces* **F**, *acting on a part* $\delta\mathcal{A}_B$ *of a boundary area, by a second one acting on the neighborhood and determining the same torsor whilst the other boundary conditions on the complementary parts of* $\mathcal{A}_B$ *relatively to* $\mathcal{A}$ *remain unchanged, then, in all regions of* $\mathcal{A}$ *sufficiently distant from* $\mathcal{A}_B$ *the stress and strain components are practically unchanged.*

The application of Saint-Venant's principle allows determining several quasi-equivalent loading configurations at the contour of a solid (Fig. 1.56).

In active optics methods, the application of Saint-Venant's principle allows finding external force configurations at the boundaries which minimize the local deformations of the optical surface near the clear aperture contours. We will often use it in the next chapters such as, for instance, with the monomode and multimode deformable mirrors in Chap. 7.

1.13.14 Computational Modeling and Finite Element Analysis

Computational modeling – sometimes called the "third branch of science" for bridging analytical theory and experimentation – is the ultimate method to accurately

[18] Saint-Venant first enounced the equivalence principle in *Sur la Torsion des Prismes* [139] pp. 298–299.

solve any sort of equilibrium or time-dependent problems. *Finite element analysis* allows determining the elastic deformations of a solid in static equilibrium.

Developed for more than three decades, evolute software for finite element analysis are now plainly efficient to solve complex three-dimensional elasticity problems. Finite element analysis can be briefly summarized as follows. For each finite volume element, the three-dimensional equations of elasticity allow writing the continuity conditions from three equilibrium equations that use the six stresses σ_{ik} associated to this element. The loads acting at the boundary of concerned volume elements determine the stresses of all elements. Navier's stress-strain relations (see (1.123b) in Sect. 1.13.4) allow us to derive the strains ε_{ik} for all finite elements, thus providing the component

$$u(r,\theta,z), \quad v(r,\theta,z), \quad w(r,\theta,z) \tag{1.186}$$

of the displacement vector for each element (cf. (1.121b) in Sect. 1.13.3). Iteration algorithms allow repeating the solving process until no variation occurs in the displacement vectors, which thus corresponds to the static equilibrium. A convenient accuracy is reached when increasing the number of finite elements entails quasi-equivalent displacements.

1.14 Active Optics

1.14.1 Spherical Polishing

The sphere is the *natural* shape obtained in the surfacing with abrasive grains of two rigid blanks of the same size that are brought into contact in a relative movement with three degrees of freedom. These movements are three rotations which reduce for a plane surface to a rotation and two translations. By progressively decreasing the size of the abrasive grains, this process provides extremely accurate spheres as was known by mankind in the "polished stone age" for the elaboration of hatchets and low reflection mirrors.

For astronomical optics, the finishing process generally uses square segments in a soft material like pitch – originally, a hardened pine resin – which are thermally sealed on the tool substrate. The spherical polishing within a diffraction limited criteria is *naturally* achieved by a *rigid* tool of the *same* diameter as the optical surface.

Let d be the diameter of the tool or optical surface. In a cylindrical frame z, r, θ, some appropriate rules are as follows: (i) the duration of a z-rotation 2π between the two surfaces must be at least 7 times greater than that of a full loop relative displacement in a z, r plane, (ii) the full displacement in the z, r plane must be $\Delta r \simeq d/3$, (iii) the lateral off centering ℓ of the two surfaces in contact must be varied such as $\ell/d \in [0, 1/7]$.

Useful information on grinding abrasives, polishing oxides, optical cements, polishing pitches, and cinematics of surfacing machines can be found in [5, 154].

1.14.2 Optical Surfaces Free from Ripple Errors

It has been sometimes said that the most important tool for the aspherization of a surface is an efficient optical testing. This affirmation implicitly admits the principle of the conventional method by zonal retouches with conveniently small polishing tools to obtain an optical surface with the required peak-to-valley (ptv) or root mean square (rms) tolerance. However, the small size of these polishing tools makes it difficult to avoid generating extremely local footprints on the optical surface whose number increases when the tool size decreases. This effect of slope discontinuities, known as *ripple errors* or extremely *high spatial frequency errors*, provides a scattering light which may become difficult to measure even with strict diffraction limited tolerance criteria on encircled energy (cf. Sect. 1.11.2). The *stress lap polishing* is an alternative with controlled flexible tools to partly avoid the ripple errors, but is not fully satisfactory however.

Active optics methods directly applied by elastic deformation of the optical surface are of particular interest because the surface can be figured as a sphere by full-aperture grinding and polishing tools which, therefore, naturally provide the advantages of *continuity, smoothness*, and *accuracy*. Compared to the conventional method of generating aspherics, active optics allows avoiding the zonal defects of slope discontinuities due to inherent local polishing tools. Then, optical surfaces generated from active optics are *free* from "*ripple errors*" and "*high spatial frequency errors*."

Active optics methods allow us to generate an aspherical surface from spherical polishing, but also allow us to generate shape *variations* of the surface.

1.14.3 Active Optics and Time-Dependence Control

Optical surfaces can be obtained from "*active optics*" in the three following cases: (i) after spherical *stress polishing* when in an elastically relaxed state, (ii) during *in situ stressing* after a spherical polishing, or, (iii) by a combination of the latter two cases. The flexure may reach a 10 mm range or more, without time dependence.

Some optical systems require an "*in situ active optics*" control, such as a telescope mirror, a variable curvature mirror for field compensation in two-arm interferometers, etc. Generally, these systems use a *low frequency bandpass control.*

In contrast, *"adaptive optics"* is a *high frequency bandpass control* essentially concerned with wavefront corrections of the atmospheric seeing, thus cannot compensate for more than a *few* wavelengths range, i.e. 1 or 1.5 μm in the visible.

1.14.4 Various Aspect of Active Optics

Active optics methods provide both axisymmetrical surfaces and non-axisymmetrical surfaces. Current research and developments in active optics may be subdivided into the following aspects:

- Large amplitude aspherization by *stress figuring* and/or by *in situ stressing*.
- *In situ correction* of shape and drift due to the optics orientation in field gravity.
- Available *variable asphericity* for various focii selected by mirror interchanging.
- Field compensation by *variable curvature mirrors* in telescope interferometers.
- Optics and diffraction gratings by replication techniques from *active submasters*.
- Diffractive corrections by photosensitive recording with *active compensators*.
- Mirror concepts for optics modal corrections with *adaptive optics systems*.

In 1965, the International Astronomical Union held a symposium in Tucson on the construction of large telescopes. On this occasion, the second aspect above for improving telescope imaging quality was discussed. As found in the proceedings, it was stated that "the primary mirror would be actively aligned with the relay optics" and it was also observed that: [19]

> *[...] Active optics is a sophistication that astronomers haven't worried about to date, but I'm afraid when we consider very large optical systems the tolerances of alignments of the optical elements will require us to consider actively controlling collimation as well as the figure.*
>
> Aden B. Meinel [107]

This same year, active optics allowed the complete aspherization of a telecope corrector plate using the stress figuring method suggested in 1930 by Bernard Schmidt:

> *[...] The method easily yields zone-free plates. [...] Although the theory is elementary and the process is not difficult, the method appears to have been long neglected.*
>
> Edgar Everhart [54]

Various theoretical aspects and application fields of active optics are described throughout the next chapters. Active optics methods constitute the main subject of the book.

References

1. E. Abbe, in Schultze's *Archiv für Mikroskopische Anatomie*, **IX**, 413–468 (1873)
2. G.B. Airy, Trans. Camb. Phil. Soc., **5**, 283 (1835)
3. C.W. Allen, *Allen's Atrophysical Quantites*, Fourth issue by A.N. Cox, Springer, 263 (2000)
4. J. Allington-Smith, R. Content, R. Haynes, I. Lewis, in *Optical Telescopes of Today and Tomorrow*, SPIE Proc., **2871**, 1284–1294 (1996)

[19] Thanks to Marc Ferrari for this reference.

5. *Amateur Telescope Making*, A.G. Ingalls ed., Scientific American Inc. publ., Book one (1953) and two (1954)
6. J.M. Arnaudiès, P. Delezoide, Nombres(2,3)-constructibles, Advances in Mathematics, 158 (2001), Constructions géométriques par intersection de coniques, APMEP bull. **446**, 367–382 (2003), and www.apmep.asso.fr/BV446Som.html
7. G. Avila, G. Rupprecht, J. Beckers, in *Optical Telescopes of Today and Tomorrow*, SPIE Proc., **2871**, 1135 (1996)
8. J.G. Baker, U.S. Patent No. 2 458 132 (1945)
9. W.W. Rouse Ball, *Short Account of the History of Mathematics*, Sterling Publ., London, (2002)
10. A. Baranne, F. Launay, Cassegrain: un célèbre inconnu de l'astronomie instrumentale, J. Opt., **28**, 158–172 (1997)
11. A. Baranne, Un nouveau montage spectrographique, Comptes Rendus Acad. Sc., **260**, 2383 (1965)
12. A. Baranne, White pupil story, in *Very Large telescope and their Instrumentation*, ESO Proc., Garching, Vol. II, 1195–1206 (1998)
13. D. Bernoulli, 26th letter to Euler (Oct. 1742), in *Correspondance Mathématique et Physique* by Fuss, t. 2, St Petersburg (1843). Also cf. Love (loc. cit.)
14. B. Di Biagio, E. Le Coarer, G.R. Lemaitre, in *Instrumentation in Astronomy VII*, SPIE Proc., **1235**, 422–427 (1990)
15. R.G. Bingham, in *Very large Telescopes and their Instrumentation*, ESO Proc., Garching, **2**, 1157 (1988)
16. J. Bland-Hawthorn, in *Tridimensional Optical Spectroscopic Methods in Astrophysics*, IAU Proc., G. Comte & M. Marcelin eds., ASP Conf. Ser., **71**, 72–84 (1995)
17. M. Born, E. Wolf, *Principles of Optics*, Cambridge Univ. Press (1999)
18. J. Boussinesq, J. Math., Paris, ser. 2, **16**, 125–274 (1871), and ser. 3, **5**, 329–344 (1879)
19. A. Bouwers, *Achievements in Optics*, Elsevier edt., New York (1948)
20. I.S. Bowen, Astrophys. J., **88**, 113 (1938)
21. I.S. Bowen, A.H. Vaughan, The optical design of the 40-in. telescope and the Irenee DuPont telescope at Las Campanas, Appl. Opt., **12**, 1430–1434 (1973)
22. G. Bruhat, *Thermodynamique*, Masson edt., issue 8, p. 74, (1968)
23. C.R. Burch, On aspheric anastigmatic systems, Proc. Phys. Soc., **55**, 433–444 (1943)
24. J. Caplan, private communication, and Marseille Observatory Inventory No. IM13000003 in www.oamp.fr/patrimoine/museevirtuel-lunettes.html
25. J. Caplan, in *Tridimensional Optical Spectroscopic Methods in Astrophysics*, IAU Proc., G. Comte & M. Marcelin eds., ASP Conf. Ser., **71**, 85–88 (1995)
26. C. Carathéodory, *Geometrische Optik*, Springer, Berlin (1937)
27. L. Cassegrain, Journal des Sçavans, J.-B. Denis edt., **7**, 71–74, April 25 issue (1672)
28. E.F.F. Chladni, *Die Akustik*, Leipzig (1802)
29. H. Chrétien, *Calcul des Combinaisons Optiques*, 5th ed., Masson edt., Paris, 422 (1980)
30. H. Chrétien, Le télescope de Newton et le télescope aplanétique, Rev. d'Optique, **1**, 13–22 and 51–64 (1922)
31. J.A. Church, Refractor designs: Clairaut's forgotten legacy, Sky & Telescopes, **66**(3), 259–261 (1983)
32. A.C. Clairaut, Mémoires sur les moyens de perfectionner les lunettes d'approche, Mém. Acad. Roy. Sc., 380–437 (1756), 524–550 (1757) and 378–437 (1762). Due to the Seven Years' War, the publications of these memoirs were somewhat delayed; they appear in the 1761, 1762 and 1764 Memoirs of the ARS.
33. R. Clausius, Poggendorf Ann., **121**, 1–44 (1864)
34. A.R.F. Clebsch, *Theorie der Elasticität fester Körper*, Teubner edt., Leipzig (1862)
35. E. Le Coarer, P. Amram et al., Astron. Astrophys., **257**, 289 (1992)
36. A.E. Conrady, *Applied Optics and Optical Design*, Oxford (1929), reissued by Dover Publ., New York (1957)

37. R. Content, in *Optical Telescopes of Today and Tomorrow*, SPIE Proc., **2871**, 1295–1307 (1997)
38. A. Cornu, Méthode optique pour l'étude des déformations élastiques, Comptes Rendus Acad. Sc. Paris, vol. 69, 333–337 (1869)
39. C. Coulomb, *Histoire de l'Académie* for 1784, 229–269 (1787)
40. G. Courtès, Comptes Rendus Acad. Sc., **234**, 506 (1952)
41. G. Courtès, Astron. J., **69**, 325–333 (1964)
42. G. Courtès, An integral field spectrograph (IFS) for large telescopes, Proc. IAU Conf., C.M. Humphries ed., Reidel Publ. Co., 123–128 (1982)
43. G. Courtès, in *Tridimensional Optical Spectroscopic Methods in Astrophysics*, IAU Proc., G. Comte & M. Marcelin eds., ASP Conf. Ser., **71**, 1–11 (1995)
44. A. Danjon, A. Couder, *Lunettes et Télescopes* (1933), (reissued: Blanchard edt., Paris, 1979)
45. R. Descartes, *La Géometrie Livre II* and *La Dioptrique* in *Discours de la Méthode*, Adam & Tannery edt., 389–441 (1637), reissue Vrin edt., Paris (1996)
46. F. Diego, Appl. Opt., **31**(31), 6284 (1993)
47. K. Dohlen, A. Origne, D. Pouliquen, B.M. Swinyard, in *UV, Optical and IR Space Telescopes and Instrumentation*, SPIE Proc., **4013**, 119–128 (2000)
48. H. Draper, *Smithsonian Contributions to Knowledge*, **14**(article 4), (1864), (reissued 1904)
49. J. Dyson, Unit magnification system without Seidel aberrations, J. Opt. Soc. Am., **49**, 713 (1946)
50. H.W. Epps, J.P.R. Angel, E. Anderson, in *Very Large Telescopes, their Instrumentation and Programs*, IAU Proc., **79**, 519 (1984)
51. L. Euler, *Methodus Inveniendi Lineas Curvas Maximi Minimive Proprietate Gaudentes* in the *Additamentum* of *De Curvis Elasticis*, Lausanne (1744)
52. L. Euler, Sur la Force des Colonnes, Mémoires Acad. Sc. Berlin, t. XIII, 252–281 (1758)
53. L. Euler, Sur la Force des Colonnes, *Acta Acad. Petropolitanæ*, Pars prior edt., 121–193 (1778)
54. E. Everhart, Making corrector plates by Schmidt's vacuum method, Appl. Opt., **5**(5), 713–715 (1966)
55. M. Faulde, R.N. Wilson, Astron. Astrophys., **26**, 11 (1973) (1934)
56. C. Fehrenbach, Principes fondamentaux de classification stellarie, Annales d'Astrophysique, **10**, 257 (1947)
57. C. Fehrenbach, R. Burnage, Comptes Rend. Acad. Sc. Paris, **281**-B, 481–483 (1975)
58. C. Fehrenbach, *Des Hommes, des Télescopes, des Étoiles*, CNRS edt., ISBN 2-222-04459-6 (1991) and 2nd issue with complements, Vuibert edt., ISBN 978-2711740383 (2007). Note: This book gives an historical account on the French development of astronomical telescopes and instrumentation during the first half of the 20th century.
59. L. Foucault, Essai d'un nouveau télescope en verre argenté, C.R. Acad. Sc., **49**, 85–87 (1859)
60. L. Foucault, C.R. Acad. Sc., **44**, 339–342 (1857)
61. M. Françon, M. Cagnet, J.-C. Thrierr, Institut d'Optique de Paris, in *Atlas of Optical Phenomena*, Springer-Verlag edt. (1962)
62. A. Fresnel, Ann. Chim. et Phys, **1**(2), 239 (1816) (cf. also Mém. Acad. Sc. Paris, Vol. 5, 338–475 (1821–22))
63. G. Galilei, *Discorsi e Dimostrazioni Matematiche Intorno á due Nuove Scienze*, Leiden, Elsevier edt. (1638)
64. S.C.B. Gascoigne, The Observatory, **85**, 79 (1965)
65. S.C.B. Gascoigne, Recent advances in astronomical optics (p. 1419–1429), Appl. Opt., **12**(4), 1419 (1973)
66. K.F. Gauss, *Dioptrische Untersuchungen*, Göttingen, Memories from 1838 to 1841 (1841)
67. Y.P. Georgelin, G. Comte et al., in *Tridimensional Optical Spectroscopic Methods in Astrophysics*, IAU Proc., G. Comte & M. Marcelin eds., ASP Conf. Ser., **71**, 300–307 (1995)
68. S. Germain, *Recherches sur la Théorie des Surfaces Élastiques*, Mme. V. Courcier edt., Paris (1821)

69. P. Germain, P. Muller, *Introduction à la Mécanique des Milieux Continus*, Masson edt., Paris, 2nd issue, 140 (1995)
70. Y. Le Grand, in *Optique Physiologique*, Revue d' Optique edt., Paris, 3rd issue, Vol. I, 68, 74 and 103 (1965)
71. Sir W.R. Hamilton, Report Brit. Assoc., **3**, 360 (1833)
72. H. Harting, Zur Theorie der Zweitheiligen Verkitteten Fermrohrobjective, Z. Instrum., **18**, 357–380 (1898)
73. C. Henry, P. Tannery, *Oeuvres de Fermat* – 5 Vol., Gauthier-Villars edt., Paris. Vol. 2, 354 (1891)
74. H. Hertz, Über die Berührung fester elastischer Körper (1881), English translation in *H. Hertz Miscellaneous Papers*, Macmillan edt., New York, 146–183 (1896)
75. M. Herzberger, *Modern Geometrical Optics*, Interscience Publ., New York (1958)
76. P. Hickson, E.H. Richardson, A curvature-compensated corrector for drift-scan observations, Publ. Astron. Soc. Pac., **110**, 1081–1086 (1998)
77. R. Hooke, *De Potentia*, or *Of Spring Explaining the Power of Springing Bodies*, London (1678)
78. E. Hugot, Performance evaluation of toroid mirrors generated by elasticity, Phase B study on PLANET-FINDER/VLT, report Obs. Astron. Marseille Provence (2006)
79. D.M. Hunter, in *Methods of Experimental Physics*, L. Marton edt., Academic, New York, Vol. 12, Part 4, 193 (1974)
80. C. Huygens, Traité de la Lumière, (completed in 1678), Leiden (1690)
81. P. Jacquinot, The luminosity of spectrometers with prisms, gratings or Fabry-Perot etalons, J. Opt. Soc. Am., **44**, 761–765 (1954)
82. E. Jahnke, F. Emde, *Tables of Functions*, Dover Publ., 4th issue, 149 (1945)
83. W. Kelvin (Lord, Thomsom), G.P. Tait, *Treatise of Natural Philosophy*, vol. 1, part 2, 188 (1883)
84. A. Kerber, Ueber di chromatische Korrektur von doppelobjektiven, Central-Zeitung für Optik und Mechanik, t. **8**, p. 145 (1887). Chrétien also refers to Kerber's paper in Central Ztg. f. Opt. u. Mech., t. **10**, p. 147 (1889)
85. H.C. King, *The History of the Telescope*, C. Griffin Co. edt., London (1955)
86. R. Kingslake, *Lens Design Fundamentals*, Academic Press, New York (1978)
87. G. Kirchhoff, Berl. Ber., 641 (1882), Ann. d Physik, **18**(2), 663 (1883)
88. G.R. Kirchhoff, Uber das Gleichewicht und die Bewegung iener elastischen Scheibe, Journ. Crelle, **40**, 51 (1850)
89. G.R. Kirchhoff, Vorlesungen über Mathematische Physik, *Mechanik*, 450 (1877)
90. J.L. Lagrange, *Miscellanea Taurinensia*, vol. 5 (1773)
91. G. Lamé, *Leçons sur la Théorie de l'Élasticité des Corps Solides*, Ecole Polytechnique, Paris (1852)
92. L.D. Landau, E.M. Lifshitz, *Theory of Elasticity* in *Course of Theoretical Physics* – Vol. 7, USSR Acad. of Sc., Butterworth & Heinemann eds, 3rd edition (1986)
93. G.R. Lemaitre, Reflective Schmidt anastigmat telescopes and pseudo-flat made by elasticity, J. Opt. Soc. Am., **66**, No. 12, 1334–1340 (1976)
94. G.R. Lemaitre, private communication to P. Connes (1978) and G. Monnet (1978)
95. G.R. Lemaitre, Equal curvature and equal constraint cantilevers: Extension of Euler (sic) and Clebsch formulas, Meccanica, **32**, 459–503 (1997). In fact Euler did not do any research on this subject and only Clebsch must be credited with the first advances on this problem.
96. P. Léna, D. Rouan, F. Lebrun, F. Mignard, D. Pelat, in *L'Observation en Astrophysique*, edt. EDP Science/CNRS (2008).
97. A.E.H. Love, *A Treatise on the Mathematical Theory of Elasticity*, Dover Publ., 4th edition (1944), reissue (1980)
98. D. Lynden-Bell, Exact optics: a unification of optical telescopes, Mont. Not. R. Astron. Soc., **334**, 787–796 (2002)
99. J.-P. Maillard, in *Tridimensional Optical Spectroscopic Methods in Astrophysics*, IAU Proc., G. Comte & M. Marcelin eds., ASP Conf. Ser., **71**, 316–327 (1995)

100. D. Maksutow, New catadioptric menicus systems, Journ. Opt. Soc. Am., **34**, 270 (1944)
101. D. Malacara, *Optical Shop Testing*, John Wiley & Sons edt., New York, 2nd edition (1992)
102. E. Malus, Optique Dioptrique, Journ. École Polytechn., **7**, 1–44, 84–129 (1808)
103. A. Marechal, Rev. Optique, **26**, 257 (1947)
104. E. Mariotte, *Traité du Mouvement des Eaux*, Paris, 1886
105. J.C. Maxwell, Cambridge and Dublin Math. J., **8**, 188 (1854)
106. A.B. Meinel, Astrophys. J., **118**, 335–344 (1953)
107. A.B. Meinel, in *The Construction of Large Telescopes*, Proc. of IAU Symposium No. 27, D.L. Crawford ed., Section: optical design, 31 (1966)
108. M. Mersenne, *L'Harmonie Universelle*, Paris (1636)
109. G. Monnet, in *Tridimensional Optical Spectroscopic Methods in Astrophysics*, IAU Proc., G. Comte & M. Marcelin eds., ASP Conf. Ser., **71**, 12–17 (1995)
110. O.F. Mossotti, Nuova theoria degli instrumenti ottici, Anal. Univ. Toscana, Pisa, **4**, 38–165 (1853)
111. O.F. Mossotti, Nuova theoria degli instrumenti ottici, Anal. Univ. Toscana Pisa, **5**, 5–95 (1858)
112. O.F. Mossotti, *Nuova Theoria Strumenti Ottici*, Casa Nistri edt., Pisa, 171–191 (1859)
113. C.L. Navier, *Sur les Lois de l'Équilibre et du Mouvement des Corps Solides Elastiques*, Mém. Acad. Sc. Paris, Vol. 7, 375–393 (1827). (The memoir was read in 1821. In this memoir, Navier refers to *Mécanique Analytique*, vol. 1, which seems to have been published by him in 1815 probably as a revised version of Lagrange's book)
114. C.L. Navier, *Mémoire sur la Flexion des Plans Élastiques*, lithographic edition, Paris, 38 pages (1820) and issued in Bulletin de la Société Philomatique, Paris (1823). This paper was presented to the French Academy in 1820. The original manuscript is in the library of the École des Ponts et Chaussées.
115. I. Newton, Phil. Trans., **7**, 4006–4007 (1672)
116. K. Nienhuis, Thesis, University of Groningen (1948)
117. R.J. Noll, Zernike polynomials and atmospheric turbulence, J. Opt. Soc. Am., **66**, 207–211 (1976)
118. A. Offner, New concept in projection mask aligners, Opt. Eng., **14**, 131 (1975)
119. M. Paul, Systèmes correcteurs pour réflecteurs astronomiques, Rev. d' Optique **14**(5), 169–202 (1935)
120. J.-C. Pecker, E. Schatzman, *Astrophysique Générale*, Masson edt., Paris, 121–122 (1959)
121. S.D. Poisson, *Mémoire sur l'Équilibre et le Mouvement des Corps Solides*, Mém. Acad. Sc. Paris, vol. 8 (1829)
122. L.W. Ramsey, T.A. Sebring, C. Sneden, in *Advanced Technology Telescopes V*, SPIE Proc., **2199**, 31 (1994)
123. Lord Rayleigh, Phil. Mag., **8**(5), 403 (1879)
124. J.W. Rayleigh (Lord, Strutt), Proc. London Mathematical Society, No. 86, **20** (1873)
125. J.W. Rayleigh (Lord, Strutt), *The Theory of Sound*, London, vol. 1 (1877), vol. 2 (1878)
126. E.H. Richardson, The spectrographs of the Dominion Astrophysical Observatory, J. Royal Astron. Soc. Canada, **62**, 313 (1968)
127. E.H. Richardson, Canadian J. Phys., **57–9**, 1365–1369 (1979)
128. E.H. Richardson, Proc. SPIE Conf. on *Instrumentation in Astronomy IV*, **331**, 253 (1982)
129. E.H. Richardson, J.M. Fletcher, W.A. Grundman, in *Very Large Telescopes, their Instrumentation and Programs*, IAU Proc., **79**, 469 (1984)
130. E.H. Richardson, C.F.H. Harmer, W.A. Grundmann, MNRAS, **206**, 47–54 (1984)
131. E.H. Richardson, in *Encyclopedia of Astronomy and Astrophysics* (2003)
132. R. Riekher, *Fernrohre und ihre Meister*, Verlag Tecknik edt., Berlin (1957), (reissued 1990)
133. W. Ritz, *Gessamelte Werke* (1911)
134. R.J. Roark, W.C. Young, *Formulas for Stress and Strain*, McGraw-Hill Book Co., 5th issue (1975)
135. F.E. Ross, Astrophys. J., **77**, 243 (1933)
136. F.E. Ross, Astrophys. J., **81**, 156 (1935)

137. N.J. Rumsey, A compact three-reflection astronomical camera, in *Optical Instruments and Techniques*, ICO 8 Meeting, London, Home Dickson edt., Oriel Press Newcastle, 514–520 (1969)
138. A. Saint-Venant (Barré de), Flamant *Théorie de l'Élasticité des Corps Solides de Clebsch*, Dunod edt., Paris, 858–859 (1881). (This is a French translation of Clebsch's book including many important annotations and complements. This book is often referred to as "Clebsch Annoted Version")
139. A. Saint-Venant (Barré de), *La Torsion des Prismes*, Mémoires des Savants Étrangers, Acad. Sc., Paris, vol. 14 (1855)
140. Schmidt, B., Mitteilungen der Hamburger Sternwarte, R. Schorr edt., 10 (1930)
141. D.J. Schroeder, *Astronomical Optics*, Academic Press edt. (1987)
142. Schwarzschild, K., Untersuchungen zur geometrischen Optik, I, II, III, Göttinger Abh, Neue Folge, Band IV, No. 1 (1905)
143. T.A. Sebring, J.A. Booth, J.M. Good, V.L. Krabbendam, F.B. Ray, in *Advanced Technology Telescopes V*, SPIE Proc., **2199**, 565 (1994)
144. L. von Seidel, Astronomische Nachrichten, **43**, Nos 1027 p289, 1028 p305 and 1029 p321 (1856)
145. M. Serrurier, *Structural structure of 200-inch telescope for Mount Palomar Observatory*, Civil Engineering, **8**, 524 (1938)
146. P.J. Smith, http://www.users.bigpond.com/PGIFL/INDEX.html
147. K. Srehl, Z. f. Instrumkde, **22**, 213 (1902)
148. J.-M.E. Stephan, C.R. Acad. Sc., **78**, 1008–1012 (1874)
149. C. Sterken, J. Manfroid, *Astronomical Photometry*, R. Boyd edt., Kluwer Acad. Publ. Dordrecht (1992)
150. J. Strong, *Procedure in Experimental Physics*, Prentice-Hall edt., Englewood Cliffs, N.J., 24th issue (1966)
151. D.-q. Su, Astron. Astrophys., **156**, 381 (1986)
152. A. Szulc, Improved solution for the cemented doublet, Appl. Opt., **35**(19), 3548–3558 (1996)
153. E.W. Taylor, The inverting eyepiece and its evolution, J. Sci. Instrum., **22**(3), 43 (1945)
154. J. Texereau, *How to Make a Telescope*, Willmann-Bell Inc., 2nd issue (1998)
155. S.P. Timoshenko, S. Woinowsky-Krieger, *Theory of Plates and Shells*, McGraw-Hill edt. (1959)
156. S.P. Timoshenko, *Theory of Elastic Stability*, McGraw-Hill edt. (1961)
157. S.P. Timoshenko, *Elements of Strength of Materials*, Wadsworth Publ., 5th issue, Sect. 55 (1968)
158. S.P. Timoshenko, *Theory of Elasticity*, McGraw-Hill edt. (1970)
159. S.P. Timoshenko, *History of Strength of Materials*, McGraw-Hill edt. (1983)
160. W. Tobin, in *Léon Foucault*, Cambridge Univ. Press (2002)
161. I. Todhunter, K. Pearson, *A History of the Theory of Elasticity*, Dover Publ., Vol. I and Vol. II, reissue (1960)
162. G.J. Toomer, *DIOCLES – On Burning Mirrors*, Sources in the History of Mathematics and the Physical Sciences 1, Springer-Verlag, New York (1976). See also the improved translation of Diocles' book and of Anthemius' book in *Les Catoptriciens Grecs*, Vol. 1, by R. Rashed, Les Belles Lettres edt., Paris, 2nd issue (2002)
163. B. Tully, Astrophys. J., **27**, 415 (1974)
164. T. Walraven, J.H. Walraven, in *Auxiliary Instrumentation for Large Telescopes*, ESO Proc., Garching, 175 (1972)
165. Y.-n. Wang, D.-q. Su, Astron. Astrophys., **232**, 589 (1990)
166. W.T. Welford, A note on the skew invariant in optical systems, Optica Acta, **15**, 621 (1968)
167. W.T. Welford, *Aberrations of Optical Systems*, Adam Hilger edt., 4th edition (2002)
168. W.B. Wetherell, in *Applied Optics and Optical Engineering*, R.R. Shannon & J.C. Wyant eds., Academic Press, London, Vol. VII, Chap. 6, 214 (1980)
169. R.V. Willstrop, D. Lynden-Bell, Exact optics – II: Exploration of designs on- and off-axis, Mont. Not. R. Astron. Soc., **342**, 33–49 (2003)
170. R.N. Wilson, *Reflecting Telescope Optics I*, Springer edt. (1996)

171. R.N. Wilson, Karl Schwarzschild and Telescope Optics, Karl Scharzschild Lecture to the German Astronomical Society, Bochum, 1993, published in *Review of Modern Astronomy*, **7**, 1 (1994)
172. C.G. Wynne, Appl. Opt., **6**(7), 1227 (1967)
173. G.G. Wynne, Astrophys. Jour., **152**(3), 675 (1968)
174. C.G. Wynne, The Observatory, **104**, 140 (1984)
175. C.G. Wynne and S.P. Worswick, Mont. Not. R. Astron., **620**, 657 (1986)
176. C.G. Wynne, MNRAS, **253**, 160–166 (1991)
177. T. Young, *A Course of Lectures on Natural Philosophy and the Mechanical Arts*, London, Lecture XIII (1807)
178. F. Zernike, Diffraction theory of the knife-edge test and its improved form, the phase-contrast method, Mont. Not. R. Astron. Soc., **94** 377–384 (1934)
179. F. Zernike, *Physica*, **1**, 689 (1934)
180. F. Zernike, B.R.A. Nijboer, Contribution to *La Théorie des Images Optiques*, Revue d'Optique, Paris (1949)

Chapter 2
Dioptrics and Elasticity – Variable Curvature Mirrors (VCMs)

The elastic deformation modes corresponding to the first-order modes of the optical matrix characterizing the wavefront shape are the curvature ($Cv\,1$) and tilt ($Tilt\,1$). These are the two fundamental modes involved in Gaussian optics. Because a tilt is easily obtained by a global rotation of a rigid substrate, this chapter only reduces to mirrors generating a $Cv\,1$ mode. Such *variable curvature mirrors* (VCM) are also sometimes called *zoom mirrors*.

Let us denote $z(r)$ – instead of $w(r)$, because z is usual for representing an optical surface – the optical figure achieved by flexure of a circular plate which is flat at rest. In the thin plate theory, a curvature mode $Cv\,1$ is represented by

$$z \equiv w = A_{20}\,r^2 \equiv \frac{1}{2R}\,r^2\,, \tag{2.1}$$

where R is the radius of curvature of the bent optical surface.

Two classes of substrate thicknesses provide the curvature mode as investigated hereafter: Constant thickness distribution (CTD) and Variable thickness distribution (VTD).

2.1 Thin Circular Plates and Small Deformation Theory

2.1.1 Plates of Constant Thickness Distribution – CTD

Let us consider a possibly holed plane circular plate with a constant thickness t and rigidity $D = Et^3/[12(1-\nu^2)] = constant$, where E and ν are the Young modulus and Poisson's ratio, respectively. If an external pair of concentric circle forces or a bending moment are applied to the perimeter region without surface load ($q = 0$), then bilaplacian Poisson's equation representing the flexure z of the plate reduces to

$$\nabla^2\nabla^2 z = 0\,, \tag{2.2}$$

whose general solution

$$z = B_{20} + C_{20}\,\ln r + D_{20}\,r^2 + E_{20}\,r^2\ln r \tag{2.3}$$

G.R. Lemaitre, *Astronomical Optics and Elasticity Theory,* Astronomy and Astrophysics Library, DOI 10.1007/978-3-540-68905-8_2,

contains the curvature term D_{20}. The external forces taken into consideration by (2.2) can be a central one, a uniform distribution of forces onto a concentric circle, and a uniform radial moment onto the inner and outer edges. Choosing the deformation origin at the plate vertex leads to set $B_{20} = 0$.

We search for external forces and/or radial moments leading to the conditions $C_{20} = E_{20} = 0$. The Laplacian is represented by

$$\nabla^2 z = \frac{1}{r}\frac{d}{dr}\left(r\,\frac{dz}{dr}\right) = 4\,(D_{20} + E_{20} + E_{20}\,\ln r) \tag{2.4}$$

so the radial shearing force per unit length [cf. (1.187)] is

$$Q_r = -\,D\,\frac{d}{dr}\,(\nabla^2 z) = -\,4\,E_{20}\,D\,\frac{1}{r}\,. \tag{2.5}$$

Condition $E_{20} = 0$ implies a null shearing force all over the plate, $Q_r = 0$. Therefore, the zeroing of $r^2 \ln r$ coefficient in (2.3) implies that no central force or circle-line force can be applied onto the plate to achieve a $Cv\,1$ mode. From conditions $B_{20} = E_{20} = 0$, the radial bending moment per unit length is

$$M_r = D\left(\frac{d^2z}{dr^2} + \frac{\nu}{r}\frac{dz}{dr}\right) = D\left[\,2(1+\nu)D_{20} - (1-\nu)C_{20}\,\frac{1}{r^2}\,\right]. \tag{2.6}$$

Coefficients C_{20} and D_{20} are determined from the boundaries, i.e. the bending moments $M_r\{a\}$ at the plate edge and $M_r\{b\}$ at the edge of its central hole. After solving the system, these coefficients are

$$C_{20} = \frac{1}{1-\nu}\,\frac{a^2b^2\,[M_r\{a\} - M_r\{b\}]}{(a^2-b^2)\,D}\,, \tag{2.7a}$$

$$D_{20} = \frac{1}{2(1+\nu)}\,\frac{a^2M_r\{a\} - b^2M_r\{b\}}{(a^2-b^2)\,D}\,. \tag{2.7b}$$

A pure curvature mode $Cv\,1$ is obtained if the logarithm term vanishes, i.e. if $C_{20} = 0$. From (2.7a), this leads to two solutions: one is with $b = 0$ and the other with $M_r\{b\} = M_r\{a\}$. The coefficient D_{20} is identical for the two solutions which are expressed by

$$B_{20} = C_{20} = E_{20} = 0 \quad \text{and} \quad D_{20} = \frac{M_r\{a\}}{2(1+\nu)D}\,. \tag{2.8a}$$

Finally, by identifying the A_{20} coefficient in (2.1) with D_{20} coefficient, the curvature of the flexure mode $Cv\,1$ is

$$\frac{1}{R} \equiv 2A_{20} = \frac{M_r\{a\}}{(1+\nu)D} = 12\,(1-\nu)\,\frac{M_r\{a\}}{E\,t^3}\,. \tag{2.8b}$$

• **Conclusion for CTD solutions with bending moments**: → *A plate or a slightly curved meniscus generates a* $Cv\,1$ *deformation mode* $z = A_{20}r^2$ *if its thickness* $t = \mathcal{T}_{20}\, t_0$ *is a constant such as*

$$\mathcal{T}_{20} = 1 \quad \textit{with} \quad \frac{t_0}{a} = \left[6(1-\nu)\frac{M_r\{a\}}{A_{20}Ea^3} \right]^{1/3}, \tag{2.9}$$

and if a bending moment M_r *is applied only. The two solutions providing such a mode for variable curvature mirrors are*:

1. *A plain plate,* $b = 0$*, with uniform bending moment* $M_r\{a\}$ *applied to its edge,*
2. *A holed plate with uniform bending moments* $M_r\{b\} = M_r\{a\}$ *applied to edges.*

From the stress-strain relation one shows that the radial and tangential stresses σ_{rr} and σ_{tt} are equal. Their value are maximum at the surfaces of the plate, i.e. for $z = \pm\, t/2$, and expressed by

$$\sigma_{rr} = \sigma_{tt} = \pm \frac{E\,t}{2(1-\nu)R} = \pm \frac{6M_r\{a\}}{t^2}. \tag{2.10}$$

The basic solution for a plain plate providing a curvature mode $Cv\,1$ is displayed by Fig. 2.1. Two designs using an outer ring built-in at the perimeter plate are equivalent and allow generating the bending moment by mean of axial forces (Fig. 2.1). The intensity of these forces can be derived from the *vase form* study in Chap. 7.

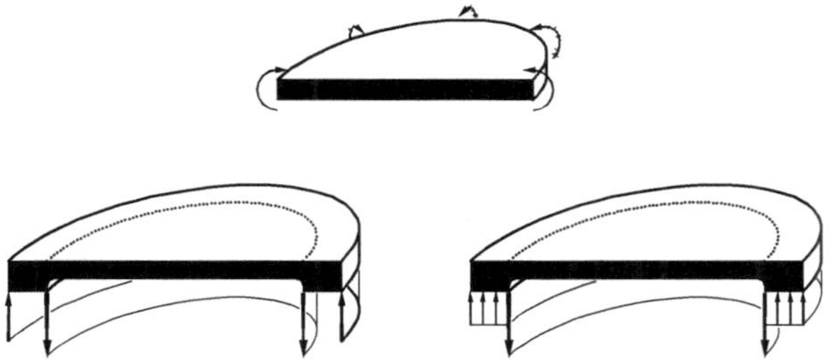

Fig. 2.1 Variable Curvature Mirrors derived from the CTD class. *Up*: Basic solution – uniform bending moment applied along the perimeter, *Down*: Axial forces on a *vase form* providing equivalent bending moments

2.1.2 *Plates of Variable Thickness Distribution – VTD – Cycloid-Like form – Tulip-Like Form*

Considering variable thickness distributions (VTDs), it is possible to find several configurations which actively generate the first-order curvature mode $Cv\,1$ (Lemaitre [35, 36]). We will see below, from the thin plate theory, that the possible

VTD geometries depend on the load distributions and associated reactions at the substrate boundaries. The radial and tangential bending moments M_r and M_t are represented by

$$M_r = D\left(\frac{d^2z}{dr^2} + \frac{\nu}{r}\frac{dz}{dr}\right), \qquad M_t = D\left(\nu\frac{d^2z}{dr^2} + \frac{1}{r}\frac{dz}{dr}\right), \tag{2.11}$$

where $D(r) = Et^3(r)/[12(1-\nu^2)]$ is the variable rigidity. The static equilibrium between the components M_r, M_t of the bending moments and the shearing force Q_r acting in a plate element is derived about a local tangential axis. This equilibrium writes

$$M_r + r\frac{dM_r}{dr} - M_t + r\,Q_r = 0\,. \tag{2.12}$$

After substitution of $M_r, dM_r/dr$, M_t and division by $r\,D$, this differential equation becomes

$$\frac{d^3z}{dr^3} + \left(\frac{1}{D}\frac{dD}{dr} + \frac{1}{r}\right)\frac{d^2z}{dr^2} + \left(\frac{\nu}{rD}\frac{dD}{dr} - \frac{1}{r^2}\right)\frac{dz}{dr} = -\frac{Q_r}{D},$$

that is,

$$D\frac{d}{dr}(\nabla^2 z) + \left(\frac{d^2z}{dr^2} + \frac{\nu}{r}\frac{dz}{dr}\right)\frac{dD}{dr} = -\,Q_r\,. \tag{2.13}$$

Notation (2.1) of a flexural curvature mode leads to $\nabla^2 z = 4A_{20}$. After substitutions, the first derivative of the rigidity is

$$\frac{dD}{dr} = -\frac{Q_r}{2(1+\nu)A_{20}} \equiv -\frac{R}{1+\nu}\,Q_r\,, \tag{2.14}$$

thus a direct function of the shearing force.

Three loading configurations and boundary reactions on the substrate are of interest for practical applications. Each of them is associated with a particular shearing force.

• **VTD Type 1 – Uniform loading and reaction at edge**: A uniform load q is applied all over the surface of the substrate in reaction at the edge $r = a$. At a current radius r of the substrate, the shearing force is defined by the equilibrium $\pi r^2 q + 2\pi r\,Q_r = 0$ of the inner element to r, that is

$$Q_r = -\frac{1}{2}\,q\,r\,, \tag{2.15}$$

and after substitution in (2.14), we obtain the rigidity

$$D = -\frac{q\,R}{4(1+\nu)}\,(constant - r^2)\,.$$

The interest of VTDs is to avoid the application of moments at the boundaries. We can select a null bending moment at the edge, $M_r(a) = 0$. From (2.11), this is satisfied by taking the rigidity $D(a) = 0$. Therefore, the rigidity is

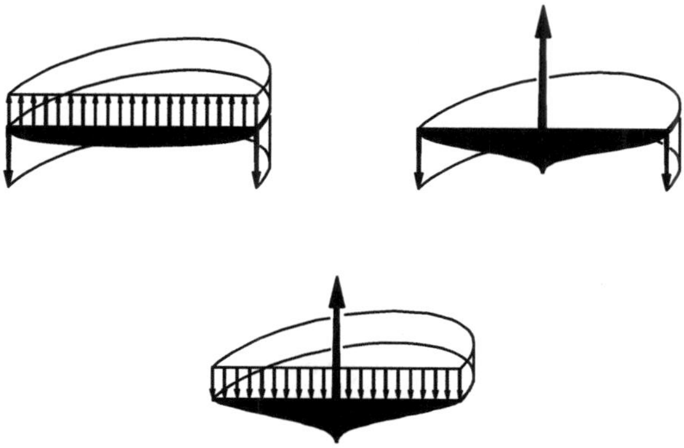

Fig. 2.2 Variable Curvature Mirrors derived from the VTD class. Dimensionless thicknesses $\mathcal{T}_{20}$ with $\rho = r/a$ and $\rho \in [0, 1]$ (after Lemaitre [35]). *Up-left*: Uniform loading and reaction at the edge, $\mathcal{T}_{20} = (1-\rho^2)^{1/3}$. *Up-right*: Axial force at center and edge reaction, $\mathcal{T}_{20} = (-\ln \rho^2)^{1/3}$. *Down*: Uniform loading and central reaction, $\mathcal{T}_{20} = (\rho^2 - \ln \rho^2 - 1)^{1/3}$

$$D = -\frac{q\,a^2 R}{4(1+\nu)}\left(1-\frac{r^2}{a^2}\right),$$

and the thickness distribution is

$$t = -\left[3(1-\nu)\frac{q\,R}{E\,a}\left(1-\frac{r^2}{a^2}\right)\right]^{1/3} a\,.$$

Because of the smoothly decreasing profile for low values of r and vertical tangents at the substrate edge $r = a$, we call this thickness distribution a *cycloid-like form* (Fig. 2.2).

• **Conclusion for VTD solution Type 1**: → *Variable curvature mirrors are obtained by uniform loading q and reaction at the edge provided a cycloid-like thickness $t = \mathcal{T}_{20}\, t_0$ is such as*

$$\mathcal{T}_{20} = \left(1-\frac{r^2}{a^2}\right)^{1/3} \quad with \quad \frac{t_0}{a} = -\left[3(1-\nu)\frac{q\,R}{E\,a}\right]^{1/3}, \tag{2.16}$$

where $1/R = 2A_{20}$ is the curvature of the deformation, and the product qR negative.[1]

At mirror surfaces, the maximum stresses are derived from substitution of the rigidity in the definition of M_r and M_t in (2.11). Their determinations are the following

[1] We will obtain the same VTD for generating the third-order astigmatism mode $Astm\,3$, i.e. $\mathcal{T}_{22} = [1-r^2/a^2]^{1/3}$ where a $\cos 2\theta$ shearing force is now applied to the edge (cf. Sect. 3.5.2). Hence the superposition of both modes $Cv\,1$ and $Astm\,3$ allows, for instance, to generate cylindric-like flexures on mirrors with circular contour (cf. Sect. 3.5.2).

$$\sigma_{rr} = \pm \frac{6M_r}{t^2} = \pm \frac{3}{2}\frac{a^2}{t_0^2}\, q\, \mathcal{T}_{20} = \pm \left[\frac{3}{8(1-\nu)^2}\frac{a^2}{R^2}\, qE^2\right]^{1/3} \mathcal{T}_{20}\,, \tag{2.17a}$$

$$\sigma_{tt} = \pm \frac{6M_t}{t^2} = \sigma_{rr}\,, \tag{2.17b}$$

showing that the radial and tangential stresses are identical $\forall\, r$ – as for a CTD [see (2.10)] – and maximal at the center of the substrate.

• **VTD Type 2 – Axial force at center and reaction at edge**: The substrate is deflected by an axial force F applied to its center that gives rise to a reaction $-F$ at the edge. If we consider an equivalent uniform load q applied to all its surface, we can define the central force by $F = \pi a^2 q$, and the associated shearing force by $\pi a^2 q + 2\pi r Q_r = 0$, thus

$$Q_r = -\frac{q\, a^2}{2r}\,. \tag{2.18}$$

After substitution and integration of (2.14), the rigidity is represented by

$$D = -\frac{q\, R\, a^2}{2(1+\nu)}\,(\,constant - \ln r\,)\,.$$

Similarly as in the previous configuration, we select a null bending moment at the edge by taking $D_r(a) = 0$ in (2.11). Therefore, the *constant* = $\ln a$, and the rigidity is

$$D = -\frac{q\, a^2 R}{4(1+\nu)}\left(-\ln\frac{r^2}{a^2}\right).$$

Because of an infinite thickness at $r = 0$ and vertical tangents at the substrate edge, we call this thickness distribution a *tulip-like form* (Fig. 2.2).

• **Conclusion for VTD solution Type 2**: → *Variable curvature mirrors are obtained by axial force F at center and reaction at the edge provided a tulip-like thickness $t = \mathcal{T}_{20}\, t_0$ such as*

$$\mathcal{T}_{20} = \left(-\ln\frac{r^2}{a^2}\right)^{1/3} \quad with \quad \frac{t_0}{a} = -\left[3(1-\nu)\,\frac{F\,R}{\pi\, E\, a^3}\right]^{1/3}, \tag{2.19}$$

where $1/R = 2A_{20}$ is the curvature of the deformation, and the product FR negative.[2]

---

[2] In the region where the radius r is near the edge radius a, say $0.75 < r/a \le 1$, we may compare the asymptotic expansion of this VTD with that of Type 1, $\mathcal{T}_{20} = [1 - r^2/a^2]^{1/3}$, which also allows generating the *Astm* 3 mode. We know that

$$\ln x = x - 1 - \tfrac{1}{2}(x-1)^2 + \tfrac{1}{3}(x-1)^3 - \cdots, \quad 0 < x \le 2,$$

which, denoting $\rho = r/a$, entails for a VTD Type 2

$$(-\ln \rho^2)^{1/3} = \left[1 - \rho^2 + \tfrac{1}{2}(1-\rho^2)^2 + \tfrac{1}{3}(1-\rho^2)^3 + \cdots\right]^{1/3}.$$

• **VTD Type 3 – Uniform loading and reaction at center**: The substrate is deflected by a uniform load q and in reaction at its center by a force $F = -\pi a^2 q$. The associated shearing force at current radius r is defined by the static equilibrium $F + \pi r^2 q + 2\pi r Q_r = 0$, thus

$$Q_r = \frac{q}{2}\left(\frac{a^2}{r} - r\right). \tag{2.20}$$

After substitution and integration of (2.14), the rigidity is represented by

$$D = \frac{q\,R\,a^2}{2(1+\nu)}\left(constant + \frac{r^2}{2\,a^2} - \ln r\right).$$

Similarly as in the previous configurations, we select a null bending moment at the edge by taking $D_r(a) = 0$ in (2.11). Therefore, the *constant* = $1/2\,(\ln a^2 - 1)$, and the rigidity is

$$D = \frac{q\,a^2 R}{4(1+\nu)}\left(\frac{r^2}{a^2} - \ln\frac{r^2}{a^2} - 1\right).$$

Because of an infinite thickness at $r = 0$ and vertical tangents at the substrate edge, this thickness distribution is also a *tulip-like form* (Fig. 2.2).

• **Conclusion for VTD solution Type 3**: → *Variable curvature mirrors are obtained by uniform loading and reaction at the center provided a tulip-like thickness* $t = \mathcal{T}_{20}\,t_0$ *such as*

$$\mathcal{T}_{20} = \left(\frac{r^2}{a^2} - \ln\frac{r^2}{a^2} - 1\right)^{1/3} \quad with \quad \frac{t_0}{a} = \left[3(1-\nu)\,\frac{q\,R}{E\,a}\right]^{1/3}, \tag{2.21}$$

where $1/R = 2A_{20}$ *is the curvature of the deformation, and the product* qR *positive.*

Hence we obtain

$$\lim\left\{\frac{\mathcal{T}_{20\,\text{Type 2}}}{\mathcal{T}_{20\,\text{Type 1}}}\right\}_{\rho\to 1} = \lim\left\{\left[1 + \tfrac{1}{2}(1-\rho^2) + \tfrac{1}{3}(1-\rho^2)^2 + \cdots\right]^{1/3}\right\} = 1,$$

so the two types are asymptotically the same near the edge. For $\rho = 0.85$, the ratio in the above limit is $\simeq 1.05$.

Since a Type 2 only requires a central force reacting at the edge, when superposing a $Cv\,1$ mode with an $Astm\,3$ mode the number of actuators is lower than for the case of a Type 1 which requires generating an edge-moment, a more difficult condition to achieve in practice. Hence, although theoretically less perfect than for a Type 1, the superposition of $Cv\,1$ and $Astm\,3$ modes with a Type 2 has been proposed by Hugot [24] for telescope integral field units (cf. Sect. 1.12.9). In this development, one may find a more appropriate VTD of the form $\mathcal{T} = \alpha\mathcal{T}_{20\,T1} + (1-\alpha)\mathcal{T}_{20\,T2}$, where $0 \le \alpha \le 1$, which allows us to match the accuracy of both generated modes relative to the wavefront tolerances.

2.1.3 Optical Focal-Ratio Variation

From the three VTDs, we can determine the optical f-ratio variation, i.e. the zoom-range, generated by the $Cv\,1$ deformation of the VCM. Assuming a flat mirror when in an unstressed state, let

$$\Omega = |f/2a| = |R/4a| = |1/8aA_{20}| \tag{2.22}$$

be this f-ratio variation. After substitution, all three VTDs can be expressed by

$$\frac{t}{a} = \left[12(1-\nu)\Omega\,\frac{q}{E}\right]^{1/3} \mathcal{T}_{20}\,. \tag{2.23}$$

For these distributions, the radial and tangential stresses are identical, $\sigma_{rr} = \sigma_{tt}$. In practical applications, these stresses must be evidently lower than the tensile maximum stress $\sigma_{T\max}$ of the mirror substrate. Therefore, the maximal value of the stresses σ_{rr} or σ_{tt}, derived from Eqs. (2.17a) and (2.17b), must satisfy

$$\left[\frac{3}{128(1-\nu)^2\Omega^2}\,qE^2\right]^{1/3} \left|\mathcal{T}_{20}\right|_{\max} < \sigma_{T\max}\,. \tag{2.24}$$

With the tulip-like VCMs, because of the point forces applied at the center, $\mathcal{T}_{20}(0) \rightarrow \infty$ in (2.19) and (2.21), and also for the stresses (2.24). In fact, the stem of the profile thickness is very narrow because of its logarithmic nature in $(-\ln\rho^2)^{1/3}$ that comes from the infinite pressure due to the central point-force. For practical applications it is always possible to limit the central thickness to a finite value. The stem truncation is done with respect to the Rayleigh quarter-wave criterion applied to the central area; the axial force is not applied on a point but on a small area, say of typical radius $a/50$.

With the cycloid-like VCM, $|\mathcal{T}_{20}|_{\max} = \mathcal{T}_{20}(0) = 1$ from (2.16).

2.1.4 Buckling Instability

A *self-buckling instability* may happen during a curvature change. This is similar to the meniscus shell "jumping toy," in polymer material, which is manually brought, temporarily, to opposite curvature. Avoiding buckling instability requires taking into account the radial tension N_r existing at the middle surface and showing that the maximum compression value of N_r remains small compared to a critical value. This self-buckling instability is avoided by restricting curvatures to always having the same sign during zooming. Furthermore, all three VTDs $\mathcal{T}_{20}$ are decreasing to zero at the edge which also prevents from this instability.

2.2 Thin Plates and Large Deformation Theory – VTD

In the previous Section, the radial and tangential stresses in the middle surface of the plate have not been considered, so that the results are valid only if the sags $a^2/2R$ are small compared to the mean thickness $< t >$ of the substrates. In order to design VCMs generating a large zoom range with the best accuracy, the analysis is deepened by taking into account the strain of the middle surface. As for constant thickness plates (cf. Timoshenko and Woinowsky-Krieger [58]) in the axisymmetric case, the displacement of a point of the middle surface can be resolved into two components: Assuming a plane middle surface before loading, let us denote $z,\ u$ the axial and radial displacements (instead of using the notation $w,\ u$ which should be more appropriate for a curved surface at rest). Then, considering the *large deformation theory*, the relative elongations, or strains, in the radial and tangential directions are defined by

$$\varepsilon_{rr} = \frac{du}{dr} + \frac{1}{2}\left(\frac{dz}{dr}\right)^2, \qquad \varepsilon_{tt} = \frac{u}{r}, \tag{2.25}$$

where the second term of ε_{rr} takes into account the large deformation case, as can be compared with the low deformation strains in (1.109b). The corresponding radial and tangential tensile forces in the middle surface, N_r and N_t per unit length, are defined by

$$\varepsilon_{rr} = \frac{1}{Et}(N_r - \nu N_t), \qquad \varepsilon_{tt} = \frac{1}{Et}(N_t - \nu N_r),$$

which leads to

$$N_r = \frac{Et}{1-\nu^2}(\varepsilon_{rr} + \nu\varepsilon_{tt}) = \frac{Et}{1-\nu^2}\left[\frac{du}{dr} + \nu\frac{u}{r} + \frac{1}{2}\left(\frac{dz}{dr}\right)^2\right], \tag{2.26a}$$

$$N_t = \frac{Et}{1-\nu^2}(\nu\varepsilon_{rr} + \varepsilon_{tt}) = \frac{Et}{1-\nu^2}\left[\nu\frac{du}{dr} + \frac{u}{r} + \frac{\nu}{2}\left(\frac{dz}{dr}\right)^2\right]. \tag{2.26b}$$

Taking into account the tensile forces (Fig. 2.3) and considering the equilibrium of a segment of dimension dr, $r\,d\theta$ and thickness t, the sum of the projection of theses forces onto the radial direction, after division by $r\,dr\,d\theta$, is

$$N_r - N_t + r\frac{dN_r}{dr} = 0\,. \tag{2.27}$$

After calculation of the N_r derivative with respect to z, u, and t, substitution in (2.27) and division by $Etr/(1-\nu^2)$, we obtain the first equation of equilibrium

$$\frac{d^2u}{dr^2} + \left(\frac{1}{t}\frac{dt}{dr} + \frac{1}{r}\right)\frac{du}{dr} + \left(\frac{\nu}{t}\frac{dt}{dr} - \frac{1}{r}\right)\frac{u}{r} + \frac{1}{2}\left(\frac{1}{t}\frac{dt}{dr} + \frac{1-\nu}{r}\right)\left(\frac{dz}{dr}\right)^2 + \frac{d^2z}{dr^2}\frac{dz}{dr} = 0. \tag{2.28}$$

The second equation is given by the equilibrium of the bending moments. This is (2.13) in which Q_r stands for the total shearing force Q_r^* that takes also into

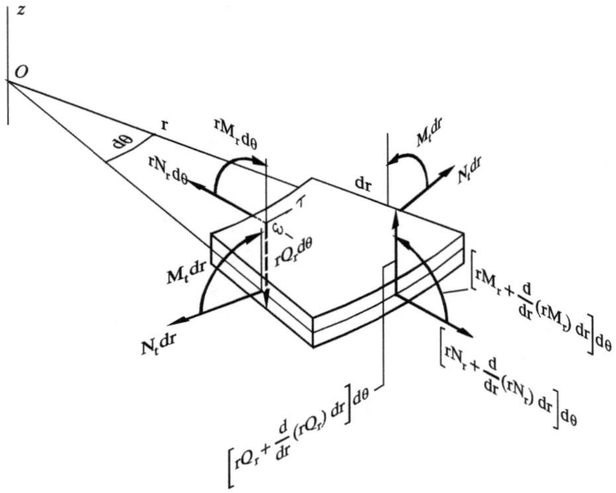

Fig. 2.3 Forces and moments providing the equilibrium of a plate segment

account the axial component of the radial force N_r in the middle surface. The total shearing force Q_r^* is expressed by

$$Q_r^* \;=\; -\,N_r\,\frac{dz}{dr} + Q_r\,, \tag{2.29}$$

where the shearing force Q_r is defined by the external loading cases in (2.15), (2.18), or (2.20). From the expression of N_r in (2.26a), we obtain

$$\frac{Q_r^*}{D} \;=\; -\,\frac{1}{t^2}\left[\frac{du}{dr} + \nu\frac{u}{r} + \frac{1}{2}\left(\frac{dz}{dr}\right)^2\right]\frac{dz}{dr} + \frac{Q_r}{D}\,. \tag{2.30}$$

After substitution in (2.13), the second equation of equilibrium becomes

$$\frac{d}{dr}(\nabla^2 z) + \left(\frac{d^2 z}{dr^2} + \frac{\nu}{r}\frac{dz}{dr}\right)\frac{1}{D}\,\frac{dD}{dr} - \frac{1}{t^2}\left[\frac{du}{dr} + \nu\frac{u}{r} + \frac{1}{2}\left(\frac{dz}{dr}\right)^2\right]\frac{dz}{dr} + \frac{Q_r}{D} = 0. \tag{2.31}$$

With the hypothesis (2.1) of paraboloid flexure generating a curvature $1/R = 2A_{20}$ when external forces are applied, and since $dD/D = 3dt/t$, the substitutions in (2.28) and (2.31) lead to the system (Ferrari [17, 18])

$$\left\{\begin{aligned}
&\frac{d^2u}{dr^2} + \left(\frac{1}{t}\frac{dt}{dr} + \frac{1}{r}\right)\frac{du}{dr} + \left(\frac{\nu}{t}\frac{dt}{dr} - \frac{1}{r}\right)\frac{u}{r} + \left(\frac{1}{t}\frac{dt}{dr} + \frac{3-\nu}{r}\right)\frac{r^2}{2R^2} = 0\,, && \text{(2.32a)}\\
&\frac{du}{dr} + \nu\frac{u}{r} - 3(1+\nu)\,\frac{t}{r}\frac{dt}{dr} + \frac{r^2}{2R^2} + 12(1-\nu^2)\,\frac{R}{r}\frac{Q_r}{E\,t} = 0\,, && \text{(2.32b)}
\end{aligned}\right.$$

requiring a numerical integration. This was carried out for the design of some VCM with large zoom range.

For each of three VTD types studied in Sect. 2.1.2, the associated shearing force Q_r is expressed by (2.15), (2.18), or (2.20). The boundary conditions are defined by a null thickness at the edge and a finite radial elongation ε_0 at the center

$$\Big(t\Big)_{r\,=\,a} = 0\,, \qquad \Big(\frac{du}{dr}\Big)_{r\,=\,0} = \varepsilon_0\,. \tag{2.33}$$

Given a VTD type, the integration is carried out by use of dimensionless variables

$$\rho = \frac{r}{a}\,, \quad U = \frac{u}{r}\,, \quad \mathcal{T}_{20} = \frac{t}{t_0}\,, \quad \kappa = \frac{q}{E}\,, \quad \Omega = \left|\frac{R}{4a}\right|,$$

where Ω is now the design optimal f-ratio of the zoom range (cf. Sect. 2.6.2). Noticing that $u(0) = 0$ and considering one of the thicknesses obtained in (2.16), (2.19), or (2.21) from the small deformation theory, we start the integration with $\mathcal{T}_{20}\{\rho_i = 1\}$

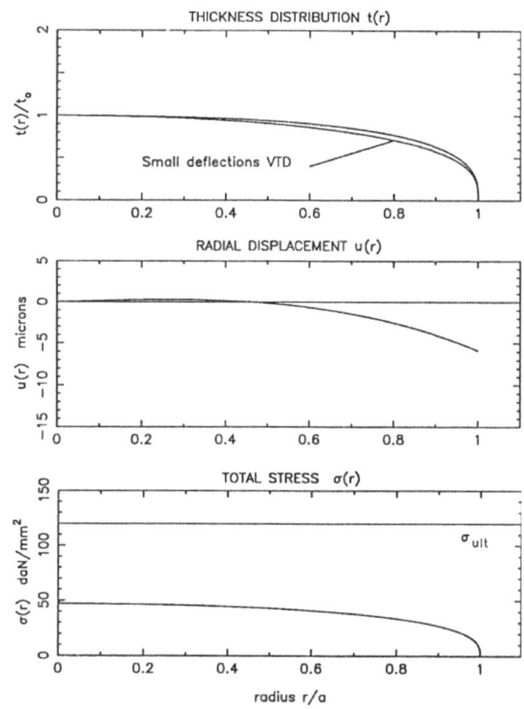

Poisson's ratio	ν = 0.315	Young mod.	E = 2.05 10^6 daN.cm^{-2}
Diameter	$2a$ = 16.00 mm	Uniform load	q = 5.280 daN.cm^{-2}
Central thick.	t_0 = 0.300 mm	Flexural sag	z_0 = − 0.300 mm
Rad. of curv.	R = −106.6 mm	Max. stress	σ_r = ± 50.02 daN.mm^{-2}

Fig. 2.4 Type 1: Uniform loading and edge reaction – Cycloid-like VCM. Results of integration from large deformation theory. Zoom: [f/∞–f/2.5]. The substrate material is Fe87 Cr13 stainless steel alloy in a quenched state (after Ferrari [18])

from one of those equations and inject a value $(dU/d\rho)_{\rho_1} = \varepsilon_0$ unknown, where ρ_1 is small. Equations (2.32b) and (2.32a) provide $d\mathcal{T}_{20}/d\rho$ and $d^2U/d\rho^2$, respectively, the latter gives U_{i+2}. Thus, all elements are known to increment for the next step $\rho_2 = \rho_1 + \delta\rho$ with a very small $\delta\rho$. We can continue the integration in the radial direction by successive increases $\rho_{i+1} = \rho_i + \delta\rho$. Thus, by changing the starting values of the radial elongation ε_0, the numerical process is repeated up to satisfy at edge $\mathcal{T}_{20}\{1\} = 0$.

The maximum radial stresses σ_{rr} on each surface of the substrate is the sum of two components

$$\sigma_{rr} = N_r / t \pm 6 M_r / t^2 . \tag{2.34}$$

Figures 2.4 and 2.5 display the reduced thickness $\mathcal{T}_{20} = t/t_0$, the radial deformation u and the maximum stress σ_{rr} resulting from the integration for a cycloid-like form (VTD Type 1) and a tulip-like form (VTD Type 2). The air pressure load generates convex flexures all over the zoom range. The basic sag used for the integration is

$$z_0 = A_{20}a^2 = a^2/2R = -2a/\Omega , \tag{2.35}$$

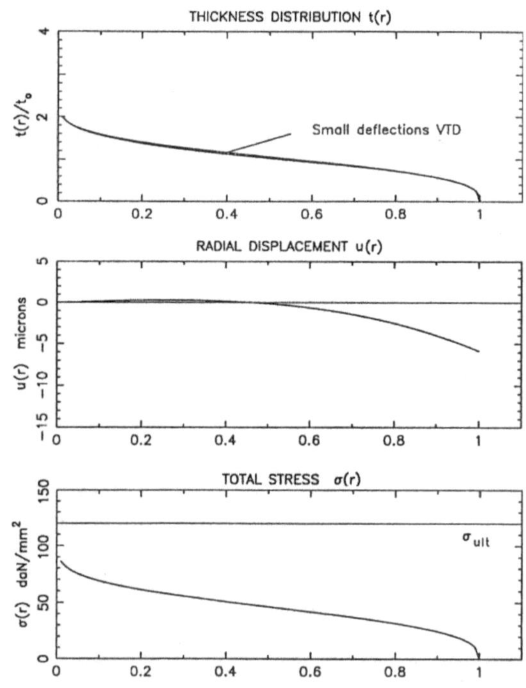

Poisson's ratio	$\nu = 0.315$	Young mod.	$E = 2.05\ 10^6$ daN.cm^{-2}
Diameter	$2a = 16.00$ mm	Uniform load	$F = 9.700$ daN
Central thick.	$t_0 = 0.600$ mm	Flexural sag	$z_0 = -0.400$ mm
Rad. of curv.	$R = -106.6$ mm	Max. stress	$\sigma_r = \pm 116.4$ daN.mm^{-2}

Fig. 2.5 Type 2: Axial force at center and edge reaction – Tulip-like VCM. Results of integrations from large deformation theory. Zoom: [f/∞ – f/2.5]. The substrate material is Fe87 Cr13 stainless steel alloy in a quenched state (after Ferrari [18])

and negative in both cases ($R < 0$, $q > 0$ and $F > 0$). In the integrations for VTDs Type 1 and 2, the thickness distributions are determined for the *basic* f-ratio value f/3.33 considered as the mean value of the zoom range, thus determining z_0. At the limit of the zoom range, the maximum *deformation-ratio* in Type 1 reaches $z_0/t_0 = -1.33$ at f/2.5, i.e. a flexure sag larger than the thickness.

The forces N_r and N_t are both positive at the central zone of the substrate. At the outer part, the force N_r decreases to zero at the perimeter but the force N_t becomes negative. For much larger deformations, this could entail an elastic instability traduced by multiple wavelets along the edge similarly as analytically developed by Casal [7] in his theory of membranes.

The relations between the load q – or central force F – and the deformation-ratio z/t_0 over the zoom ranges [f/∞ – f/2.5] have been determined by Ferrari [20]. These curves show the important non-linearity as well for a VTD Type 1 as for a VTD Type 2 (Fig. 2.6).
From these results, the load q or F can be represented in an odd power series of the deformation ratio z/t_0 by

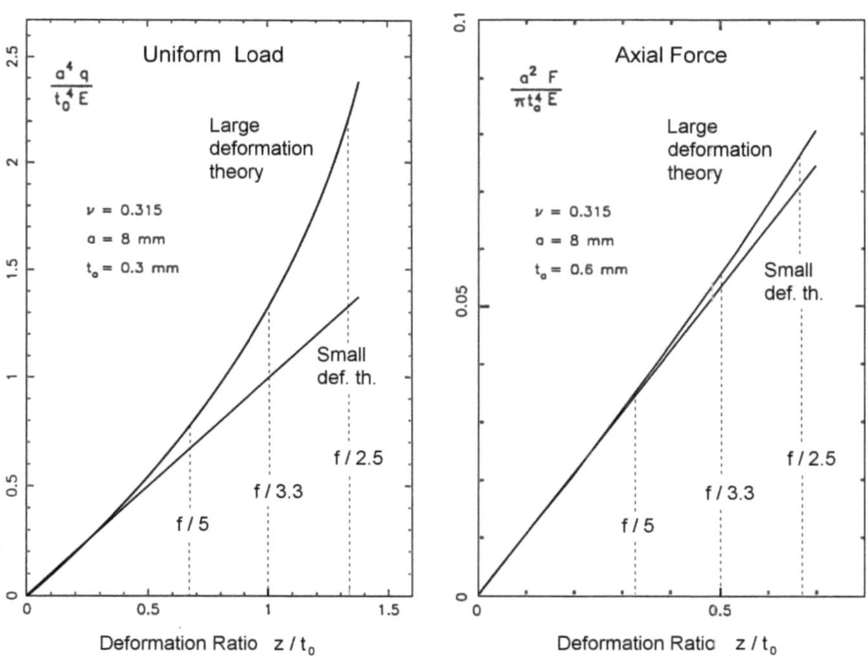

Fig. 2.6 Comparison of dimensionless load-flexure relations. *Left*: VTD Type 1 – Uniform load in reaction at the edge. *Right*: VTD Type 2 – Central force in reaction at the edge

$$\frac{a^4}{t_0^4}\frac{q}{E} \equiv \frac{a^4}{\pi t_0^4}\frac{F}{a^2 E} = \sum_{i\,=\,1,3,5,\ldots} \alpha_i \left(\frac{z}{t_0}\right)^i , \tag{2.36}$$

where α_i are dimensionless coefficients.

2.3 The Mersenne Afocal Two-Mirror Telescopes

The *Mersenne two-mirror telescopes*, published in 1636 [45, 46], are made of two confocal paraboloid mirrors thus providing *afocal systems*.

Let us consider a concave paraboloid primary mirror M_1 of curvature $1/R_1$ and two paraboloid secondary mirrors M_{2a} and M_{2b}, all having their focus located at the origin of a cylindric coordinate frame. Their shape is represented by

$$z_i = -\frac{R_1}{2k_i} + \frac{k_i}{2R_1}\, r^2 , \tag{2.37}$$

where the suffix $i = 1,\ 2a$ and $2b$ characterizes each mirror, and k_i are dimensionless parameters defined by

$$k_1 = 1 , \quad k_{2a} = \frac{R_1}{R_{2a}} > 0 , \quad k_{2b} = -\,k_{2a} < 0 , \tag{2.38}$$

so that the surfaces of M_{2a} and M_{2b} have opposite curvatures $1/R_{2b} = -\,1/R_{2a}$.

Assuming that each side of M_{2a} and M_{2b} mirrors may be used, Mersenne obtained *four* distinct afocal forms (Fig. 2.7).

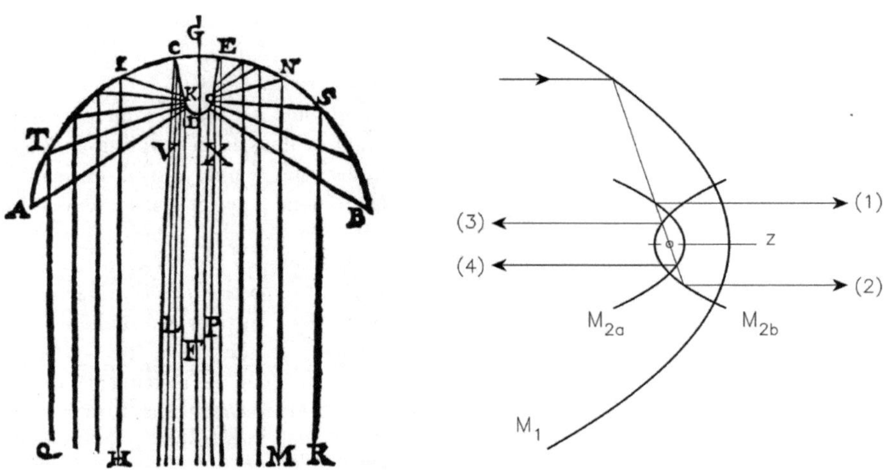

Fig. 2.7 Mersenne afocal two-mirror telescopes. *Left*: One of the four forms proposed by Mersenne in 1636 [45]. *Right*: The four forms

- FORM 1 uses the convex side of M_{2a} mirror (Cassegrain form),
- FORM 2 uses the concave side of M_{2b} mirror (Gregory form),
- FORM 3 uses the convex side of M_{2b} mirror (retro-reflective form),
- FORM 4 uses the concave side of M_{2a} mirror (retro-reflective form).

Forms 1 and 2 (see also Fig. 1.6) may be used in the paraxial zone as well as in grazing incidences for both mirrors. Retro-reflective forms 3 and 4 preferably applies to rays with large heights at the primary corresponding to conjugates with low heights at the secondary and conversely; however the third-order aberration theory that uses the classical Hamilton/Seidel formulation [53] may not be able to correctly model those two latter forms.

Denoting in a general formulation

$$k = R_1/R_2 \tag{2.39}$$

the algebraic radius ratio of the two mirrors, and h the height of an incident ray, it can readily be shown that the conjugate height h' of the system emerging ray is a solution of

$$h'^2 - \left(h - \frac{R_1^2}{h}\right)\frac{h'}{k} - R_2^2 = 0\,, \tag{2.40}$$

from where the roots may be written as

$$\begin{cases} \dfrac{h}{h'} = k = \mathit{constant} \ \text{ for } \begin{cases}\text{Form 1}\\ \text{Form 2}\end{cases} & \text{(2.41a)} \\[2ex] h\,h' = -\dfrac{R_1^2}{k} = \mathit{constant} \ \text{ for } \begin{cases}\text{Form 3}\\ \text{Form 4}\end{cases} & \text{(2.41b)} \end{cases}$$

With the two first forms, the heights of emerging rays vary in accordance with (2.41a) as an *homothetic transformation congruence* of the Gaussian optics type. With the two latter *retro-reflective systems*, the emerging rays get closer to the telescope axis if the incident ray moves away from the axis. These heights vary in accordance with (2.41b) that we call an *inversion transformation congruence* of the conjugate heights; this latter case was certainly remarked by Marin Mersenne who displayed numerous rays in his figure – actually Fig. 2.7–*Left* – of *Harmonie Universelle* (1636).

The four Mersenne forms are free from all order spherical aberration; thus, in the third-order Seidel theory, the first sums relative to *Sphe* 3 are

$$\mathbf{\Sigma}_1\, S_I = \mathbf{\Sigma}_2\, S_I = (\mathbf{\Sigma}_3\, S_I) = (\mathbf{\Sigma}_4\, S_I) = 0\,, \tag{2.42}$$

where the parentheses point out that, because of the inversion transformation congruence between the input and output conjugate rays, actually these sums cannot be properly derived from this theory for Mersenne forms 3 and 4; in some optical design codes, the selection of the second alternative intersection point of an optical surface is achieved by a special option as "alternate even" in Zemax. For instance even for an extremely small field of view – typically of 1 or 10 arcsec – raytraces with Mersenne form 3, $k = -2$, and linear central obstructions of 0.3 or 0.4, show *huge* field aberrations; when anastigmatically focalized, the residual blur images show a remarkable *rotational* symmetry.

• **Quasi-perfect two-mirror systems**: Restraining only to the Mersenne afocal forms 1 and 2, we must mention that for three and a half centuries, *all* of the additional properties of these systems remained unknown. We now know that they are remarkable *anastigmatic systems*, i.e. also free from *Coma* 3 and *Astm* 3,

$$\boldsymbol{\Sigma}_1 S_{II} = \boldsymbol{\Sigma}_2 S_{II} = \boldsymbol{\Sigma}_1 S_{III} = \boldsymbol{\Sigma}_2 S_{III} = 0\,, \tag{2.43}$$

and that *Dist* 3 can be cancelled, $\boldsymbol{\Sigma}\, S_V = 0$, if the input pupil is conveniently located.

Astonishingly on may notice, first, that the aplanatic property of these two simple systems has *not* been clearly understood, say, up to the 1960s. Second, the demonstration of the anastigmatic property came much later.[3]

The *Petzval theorem* or Seidel sums $\boldsymbol{\Sigma}\, S_{IV}$ allow deriving the Petzval curvature (cf. Sect. 1.10.1). For Mersenne form 1 or 2, the *Petz* 3 curvature is also the field curvature. Whatever the position of the pupil and mirrors, for a two-mirror system this curvature is

$$1/R_P = -2/R_1 + 2/R_2\,, \tag{2.44}$$

which entails that the absolute value of the Petzval curvature is the smallest for any Cassegrain form, thus for Mersenne form 1.

A generalized Schwarzschild theorem formalized by Wilson [62] states that

→ *n separated aspheric mirrors or lenses in any geometry allow the correction of n conditions.*

It must be added that some particular geometries are more favorable than others.

For instance, the particular confocal geometry of *Mersenne forms* 1 *and* 2 provides the correction of the three conditions $\boldsymbol{\Sigma}\, S_I = \boldsymbol{\Sigma}\, S_{II} = \boldsymbol{\Sigma}\, S_{III} = 0$ with only two aspherical surfaces and then are absolutely fundamental optical systems. Because also correcting the spherical aberration and linear coma at *all* aberration orders, the Mersenne two-mirror systems are important designs for the development of 3-, 4- and 5-mirror telescopes.

---

[3] Although the Seidel theory has been well established since the 1860s, the *aplanatism* property of a confocal Mersenne mirror-pair seems to have been only remarked in the 1960s probably from raytrace designs of long-slit spectrographs for telescope prime focus. The complete demonstration of the *anastigmatism* property seems to have been first given by Martin Krautter [29] in 1986 by use of coefficients first formalized by Schwarzschild in his general aberration theory of a two-mirror telescope (1905). Using the Seidel sums, other demonstrations of the anastigmatism were later given by D. Korsch [28] and with a more classical formalism by R. Wilson [62].

It is surprising for those who know the work of M. Paul in 1935 [51] that he did not at that time, in setting up his well-known anastigmatic three-mirror telescope, depart from the principle of the Mersenne telescope. Instead, he derived his telescope form from a paraboloidal primary corrected by an afocal field corrector consisting of two separated mirrors. It is clear he was referring in no way to the properties of Mersenne telescopes.

Another special case is the remarkable geometry of the *flat-field anastigmat three-mirror Rumsey telescope* [52] where M_1-M_2 is a quasi-confocal pair allowing the system to satisfy the four conditions $\Sigma\, S_I = \Sigma\, S_{II} = \Sigma\, S_{III} = \Sigma\, S_{IV} = 0$ with only three aspherical mirrors. In a modified design, a Rumsey-Lemaitre telescope [39] was built by active optics methods starting from the *same* sphere for M_1 and M_3 mirrors that are both on a single substrate, thus aspherizing the three surfaces of the telescope by only figuring two stressed *spherical* surfaces (see Sect. 6.6.7 for the aspherization of the M_1-M_3 double vase form mirrors and Sect. 3.3.5 for that of the M_2 tulip form mirror).

2.4 Beam Compressors, Expanders and Cat's Eyes – Active Optics Pupil Transfers

Two-arm interferometers developed for high spatial or high spectral resolutions, require the use of *retro-reflective systems*. Because retro-reflective Mersenne telescopes (forms 3 and 4) cannot be used with all rays in the paraxial space [cf. Sect. 2.3 (2.41b)], then considering the Mersenne anastigmat of Gregory type (form 2) having real intermediate focus, we may obtain a *retro-reflective system* by addition of a mirror M_2 at the common focus of M_1 and M_3 paraboloids (Fig. 2.8). When M_2 deviates from a flat shape, the system is no longer anastigmatic but remains *aplanatic*.

Such a three-mirror *beam compression system* or a *beam expander system* can also be characterized by its compression ratio. The form corresponding to a compression ratio equal to unity is called a *cat's eye system*. If mirror M_2 is a VCM, then the system allows an *active pupil transfer*.

Let us define the beam compression ratio by $k = R_1/R_3$. Considering retro-reflective forms for which always $k > 0$, a coordinate system with its origin at the vertex of the M_1 mirror – $f_i = R_i/2$ are all negative in Fig. 2.8 – and an object pupil at abscissa p_1, the conjugate distance equation

$$(p_1' - f_1)(p_1 - f_1) = f_1^2 \,, \tag{2.45}$$

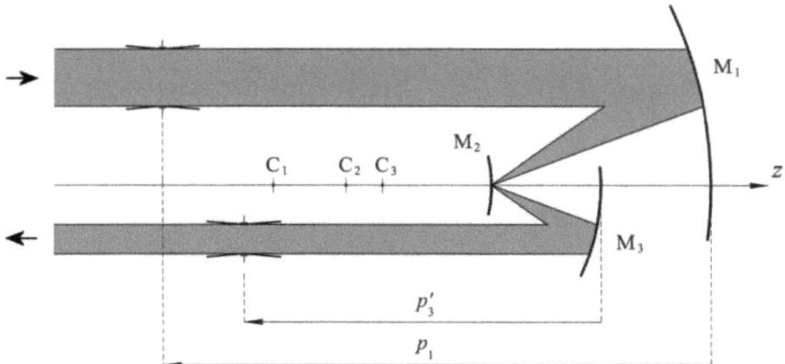

Fig. 2.8 Active pupil transfer by VCM-M_2 retro-reflective systems. These systems are aplanats if M_1 and M_3 are confocal paraboloids

provides the abscissa of the pupil image p'_1. Translating the system origin to the vertex of M_2, this image pupil corresponds to an object pupil of abscissa p_2. The passing equation is $p_2 = p'_1 - f_1$ which provides the image pupil p'_2 by using the conjugate distance equation of M_2

$$p'_2 = \frac{f_2^2}{\dfrac{f_1^2}{p_1 - f_1} - f_2} + f_2 \,.$$

Finally, from the passing equation $p_3 = p'_2 + f_3$, the abscissa p'_3 of the output pupil with respect to p_1, in the M_3 frame, is

$$p'_3 = \frac{f_3^2}{\dfrac{f_2^2}{\dfrac{f_1^2}{p_1 - f_1} - f_2} + f_2} + f_3 \,,$$

thus after transformation,

$$p'_3 = -\left(\frac{f_3}{f_1}\right)^2 p_1 + \left(1 + \frac{f_3}{f_1} + \frac{f_3}{f_2}\right) f_3 \,, \qquad (2.46)$$

where $f_3/f_1 = 1/k$ is the reciprocal beam compression ratio.

For *retro-reflective beam compressors*, the pupil conjugate distances are with $k > 1$.

For *retro-reflective beam expanders*, the pupil conjugate distances are with $0 < k < 1$.

For *retro-reflective cat's eyes*, $k = 1$, the sections of input and output beams have an identical size since paraboloids M_1 and M_3 are coinciding, $R_3 = R_1$. Thus, the conjugate distances of the pupils p_1 and p'_3 with respect to the vertex of the M_1 mirror are linked by

$$p'_3 = -p_1 + \left(2 + \frac{f_1}{f_2}\right) f_1 \,. \qquad (2.47)$$

When the VCM M_2 has a plane shape, $R_2 = \infty$, this formula reduces to $p_1 + p'_3 = 2f_1$. In this case, if $p_1 = f_1$, then $p'_3 = f_1$: → Both input and output pupils are at focus abscissa.

The aplanatic properties of these retro-reflective systems are useful in high-resolution interferometry which always requires a small field of view.

2.5 VCMs as Field Compensators of Interferometers

A VCM installed at the focal plane of a cat's eye system allows us to control the output pupil distance and to maintain it at a convenient location where the beams are recombined. At the same time, the pupil conjugation provides the convenient correction of optical paths in the field of view, so that these systems are also *field compensators*.

Fourier transform spectrometers and telescope array interferometers require VCMs that are used by *active optics control systems*.

2.5.1 Fourier Transform Spectrometers

Two-arm interferometers are particularly useful for the spectral analysis in the infrared and far-infrared - typically in the range 1–20 μm. A plane parallel plate splits the collimated beam into two channels as with a Michelson-type interferometer. The translation of one arm provides a variation of the optical path difference. After recombination, the beams are imaged on a monopixel detector. The spectral recording is achieved by measuring the intensity of the resulting interferences during the arm translation; then, the spectrum is obtained by calculations using a Fourier transform analysis. The largest Fourier transform spectrometers (FTSs) were built by the Laboratoire de Physique Moleculaire et Applications (LPMA) of the University Pierre and Marie Curie, Paris (Fig. 2.9).

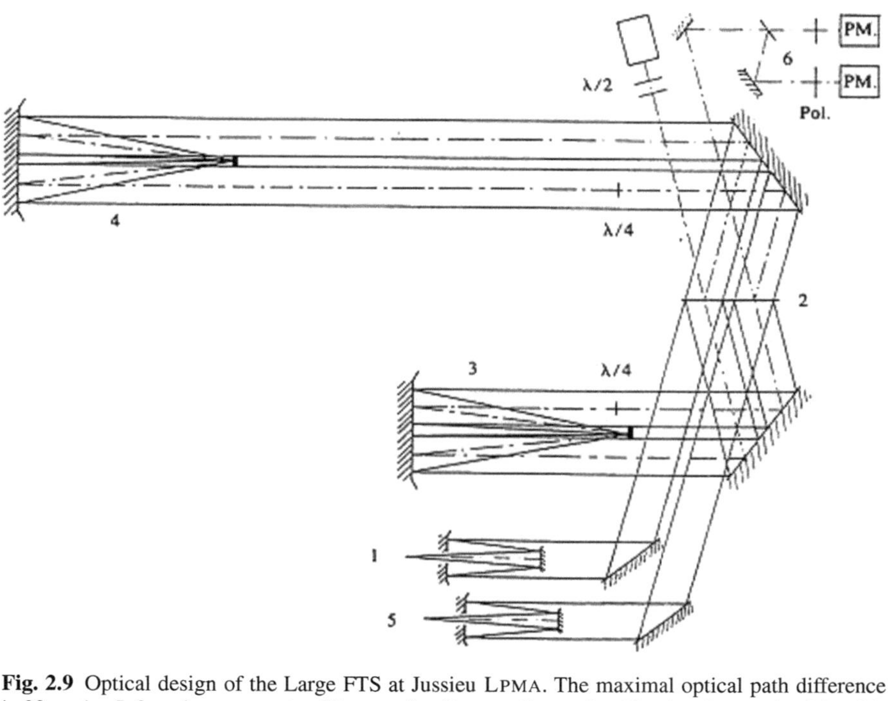

Fig. 2.9 Optical design of the Large FTS at Jussieu LPMA. The maximal optical path difference is 22 m. 1 – Infrared source and collimator, 2 – Beam splitters, 3 – Fixed cat's eye, 4 – Movable cat's eye with VCM, 5 – Camera optics and detector, 6 – Control of path difference by He-Ne laser (– · – · –) (after Valentin [60])

FTSs have been proposed and developed by Jacquinot [25, 26] in order to achieve high spectral resolutions without requiring use of diffraction gratings. Although some FTS concepts not requiring use of on-line computation – by recording stationary interferences on holographic plates – have been proposed and built (Stroke [56]), most instrument concepts are based on a multiplex computational analysis.

In the astrophysical case, high-resolution FTSs have been built for infrared studies of the Sun, Jupiter, and Saturn at Meudon, at McMath-Pierce Solar Telescope of the National Solar Observatory (NSO) and other institutes (Connes [9], Brault [5]), and lower-resolution FTSs for infrared studies of stars (Maillard [41]). In the case of molecular physics studies at the laboratory, very high resolution FTSs have been built (Valentin [60]). Recent developments of FTS methods have led to a large variety of instruments, each of them being specialized to the physics study of a particular problem (see Bracewell [4], Mertz [47], Thorne [57], Smith [54], Davis [11], Christy et al. [8]).

Compact FTSs for broadband studies have proved useful by use of 1D and 2D detectors. A FTS concept with a 1D detector and slightly tilted mirrors (Dohlen [14]) can be used without translation of a mirror; the same design with a 2D detector provides the long slit mode (Dohlen [15]). Replacing the Michelson mirrors by diffraction gratings (Dohi and Suzuki [13]), a 2D detector also allows the fast recording of a temporal chemical reaction, similar to the conventional long slit mode. These instruments do not need to vary the optical path and all optical pieces are statics.

In using 2D infrared detectors and varying the optical path, cat's eye FTSs provide 3D super-imaging and become integral field spectrometers similar to integral field spectrographs in the visible (cf. Sect. 1.12.9), but with the advantage of a free spectral resolution, i.e. which is not preselected by the optical design. 3-D or super-imaging FTSs have been developed at CFHT by Maillard [42] for observations in the spectral range 1–5 μm.

VCM and cat's eye systems were originally developed by Lemaitre in 1975 (cf. Connes and Michel [10]) for the mobil arm of the FTS of Aimé Cotton Laboratory in Bellevue. Such a system was recently implemented on the Large FTS of the Molecular Physics Laboratory and Applications (LPMA) at University P. and M. Curie, Paris-Jussieu, by Valentin and Henry [61] (see Fig. 2.9).

For large optical path variations ($\Delta\ell = 22\,\text{m}$ at Jussieu LPMA) in dual arm interferometers, thus providing high spectral resolutions in the infrared, a *VCM and cat's eye system* allows a substantial increase in luminosity and S/N ratios. This is due to the gain on the *optical Étendue* which is typically increased by a factor of 100.

2.5.2 Stellar Interferometers and Telescope Arrays

Following Young's and Fizeau's ideas of observing interferences from the light of a single wavefront passing through a pair of holes, interferometric methods have been developed in astronomy for the measurement of stellar diameters. The first approach used two sub-apertures (lunules) oppositely separated at a 65 cm distance on the 80-cm Foucault telescope; although the small size of the subapertures overcame the degradation of the atmospheric seeing, Stephan [55] deduced in 1873 that his optical baseline was too small to observe a fringe pattern which related to an angular star diameter and rightly concluded that the diameters of proximate bright stars are smaller than 0.15 arcsec. The first measurements of stellar diameters – 0.05–0.04 arcsec – where obtained by Michelson [48] by using two aperture mirrors of larger baseline mounted and driven by the 100-inch Mount Wilson Telescope. This class of interferometers has been developed to resolve stellar diameters up to 0.02 arcsec.

Another class of stellar interferometers has been developed by recombining the beams of a pair of telescopes, thus allowing a longer baseline and corresponding higher resolutions. Progress in pointing accuracy and cell sensitivity led Hanbury Brown [6] to resolve star diameters up to 0.7 10^{-3} arcsec with two radio-telescopes and intensity correlation methods. Progress in 2D detectors for the visible and laser positioning control allowed Labeyrie [32, 33] to coherently recombine two optical telescopes (I2T), thus, opening the way to telescope arrays.

By using a sufficiently large number of unit telescopes, a telescope array allows to us recover the high-resolution imaging that should be provided by a monolithic telescope of the same aperture as the baseline. By increasing the amount of light with large unit telescopes, telescope arrays joined to adaptive optics compensations allow the study of much fainter objects. Giant baselines are planned for use in space where diffraction limited images are immediately available. For ground based astronomy, the remarkable advance realized with GI2T by Vakili and Percheron [59] and Mourard et al. [49] has led astronomers to develop the array concept in symbiosis with those of very large telescopes.

The construction by the European Southern Observatory (ESO) of the Very Large Telescope (VLT) and its associated Interferometer (VLTI) realizes a symbiosis approach in combining deep field programs, high-resolution imaging, and interferometric programs by implementation of four 8 m unit telescopes and several auxiliary telescopes. The main features of the ESO interferometric array have been developed and described in Woltjer et al. [63], Beckers [1–3], Merkle [44], von der Lühe et al. [30, 31], Mariotti et al. [43], Glindeman et al. [23], Koehler and Flebus [27] (Fig. 2.10).

The VLTI was designed for recombining the light of four 8 m unit telescopes and four 1.8 m auxiliary telescopes. This is the first telescope array allowing a *field compensation*, sometimes called *field cophasing*. Eight delay lines equipped with cat's eye carriages by Derie [12], allow us to maintain the equality of optical paths during the interferometric acquisitions at the recombining laboratory (Fig. 2.11).

Starting to operate in 2006, the VCMs provide a compensation over 2-3 arcsec sky field diameter (Ferrari et al. [19]). Each VCM is installed at the Cassegrain focus of a translating f/1.6–f/6.5 Ritchey-Chrétien telescope of 550 mm clear aperture. Such cat's eye systems open the way to dual object studies with large format detectors (Leinert and Graser [34], Paresce et al. [50]). Each VCM provides the control of the pupil transfer to the recombination laboratory. With a large zoom range [f/∞–f/2.5], the convex curvatures of the VCMs are in closed-loop control with

Fig. 2.10 View of the VLT Interferometer at Cerro Paranal (courtesy ESO)

Fig. 2.11 Views of VLTI delay lines and translation cat's eye (courtesy ESO). The cat's eyes are Ritchey-Chrétien telescopes with a VCM at focus (after F. Derie [12])

the location of the cat's eye carriages. A carriage translation can compensate for optical-path differences up to 120 m.

2.6 Construction of VCMs with VTDs

2.6.1 Elastic Deformability and Choice of Material Substrate

In active optics applications, an important choice is that of the material. This material must be extremely linear, in the sense of Hooke's law of a linear stress-strain relationship. Glass, vitro ceramics and some metal alloys possess this linear characteristic. If large deformations are required, an additional important feature is the *elastic deformability ratio.*

To define this quantity, one may express the parabolic flexure which corresponds to the curvature mode $Cv\,1$ as a function of physical quantities depending on the substrate material. For this, one may use either constant thickness distribution (CTD) or variable thickness distribution (VTD) mirrors. For instance, from (2.1), and (2.8b) (2.10) for CTD, the flexure can be represented as

$$z = \frac{1}{2R}\,r^2 = \pm\,(1-\nu)\,\frac{|\sigma|}{E\,t}\,r^2, \tag{2.48a}$$

where the positive sign is for $R > 0$ and from which the sag at mirror full aperture is

$$z\{a\} = \pm\,(1-\nu)\,\frac{|\sigma|}{E}\,\frac{a^2}{t}, \tag{2.48b}$$

where $\sigma = \sigma_{rr} = \sigma_{tt}$ are the stresses arising at the surfaces of the plate. We would obtain, for instance, an identical result for a Type 1 VTD mirror which shows a continuous thickness t_0 at center so that now $t \to t_0$ in (2.48b), the stress σ arising in the plate is for $r = 0$ and is the same as before in CTD.

From material mechanical testing, it is well known that the *tensile* ultimate stress is much lower than the *compression* ultimate stress. Let $\sigma_{T\max}$ be the *tensile maximum stress* of a material as an acceptable limit that must not be exceeded to avoid rupture or plastic strain. Since a tension σ is positive in our sign convention, from (2.48b), and given a plate of diameter $2a$ and thickness t, the *elastic deformability ratio* of a material may be characterized by the dimensionless quantity

$$(1-\nu)\,\frac{\sigma_{T\max}}{E}\,. \tag{2.48c}$$

As most materials used in active optics have a Poisson's ratio $\nu \in [\,0.11,\ 0.33\,]$, it is more convenient to simply define the elastic deformability as

$$\text{elastic deformability} = \frac{\sigma_{T\max}}{E}\,. \tag{2.48d}$$

For instance from Table (1.10), comparing a Zerodur vitroceram (where $\sigma_{t\max} = 22$ MPa and $E = 90.2$ GPa), with a quenched stainless steel Fe87 Cr13 (where $\sigma_{t\max} = 1.4 \times 10^3$ MPa and $E = 201$ GPa), the gain in elastic deformability is $\sim$28-times in favor of this metal alloy; taking into account the Poisson ratio would lead to a gain $\sim$25.

Other linear alloys usable for mirrors, such as Ti90Al6V4 or Be95Cu5, could also provide large gains in elastic deformability, but their practical machining to the convenient geometry of a VTD seems more difficult. For large zoom range VCMs, stainless steel Fe87 Cr13 in a quenched state has been found preferable.

2.6.2 Zoom Range and Choice of a Thickness Distribution

For a small zoom range, the design of a VCM using a VTD is straightforward. With the examples in Fig. 2.6, if the zoom range is corresponding to a variation of [f/∞–f/7] with VTD Type 1 or a variation of [f/∞–f/5] for VTD Type 2, the stress-strain relations are quasi-linear and any thickness profile $\mathcal{T}_{20}$ provides accurate curvatures which are *affine paraboloids*.

For a large zoom range such as [f/∞–f/2.5], it is not possible to obtain affine paraboloids all over the range. Assuming that the VCM is polished flat or slightly convex at an f-ratio Ω_0 close to infinity when at rest, and that the zoom range varies down to Ω_{min}, a balance of the surface deviation to a paraboloid can be obtained if the thickness $\mathcal{T}_{20}$ is determined for an Ω-value such as $\Omega \in [\,\Omega_0, \Omega_{min}\,]$. In this range the full sag variation is subdivided into four equal segments and the optimal design f-ratio Ω for the calculation of $\mathcal{T}_{20}$ is determined for the junction of the last two segments by using the balance criterion

$$\frac{1}{\Omega} = \frac{1}{\Omega_0} + \frac{3}{4\,\Omega_{min}}\,. \tag{2.49}$$

Therefore with a flat or quasi-flat VCM at rest, i.e. $\Omega_0 \simeq \infty$, this criterion for the determination of $\mathcal{T}_{20}$ means that the mirror is a paraboloid when at f/Ω, while maximum spherical aberration residuals of the optical surface occur with opposite signs at ∼f/2Ω_{min} and f/Ω_{min}.

2.6.3 Achievement of Boundary Conditions

Considering the VTD class, the boundary condition at the VCM perimeter is a free supported edge for Type 1 and Type 2, and free edge for Type 3. From (2.34), no radial bending moment and no radial tension must be applied,

$$M_r(a) = 0 \quad \text{and} \quad N_r(a) = 0\,.$$

With metal mirrors, taking into account that only the axial reaction acts at the edge of Type 1 and Type 2, these conditions can be approximated by using a small cylindric collaret that links the VCM to an outer thick ring (Fig. 2.12). Convenient axial length and radial thickness of the collaret can be defined with respect to the thickness $\mathcal{T}_{20}$ at the edge proximity, for instance from the value of $\mathcal{T}(0.99)$.

Because of the instable equilibrium of the acting forces in Type 3, the boundaries only require the VCM to be radially maintained into an outer ring. Since there is no reaction exerted to the edge, a possible solution is to use a very thin linking plate at the edge proximity and compensate for the sag by axial displacement of the zone of central contact during the deformation. Although these boundaries are difficult to fulfill, the comparison of the dimensionless thicknesses displayed by Fig. 2.2 shows

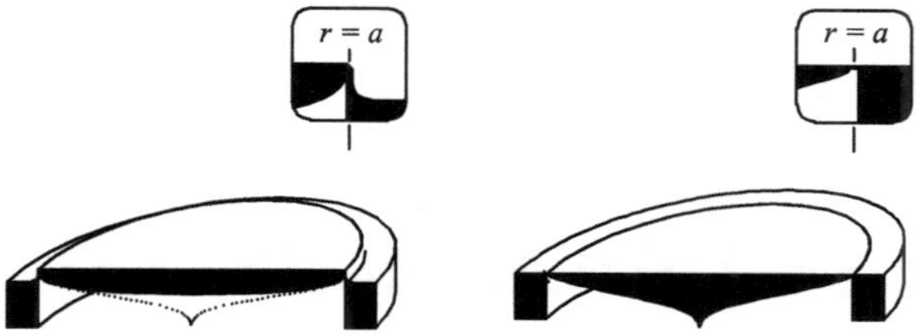

Fig. 2.12 Holosteric solutions linking active substrate and outer ring. *Left*: VTDs Type 1 and 2: Boundaries achieved by a thin cylinder. *Right*: VTD Type 3: Boundaries achieved by a thin plane plate

that Type 3 VCMs are much thinner towards the edge. Therefore compared to Type 1 and 2, Type 3 VCMs are less sensitive to surface deviations from the paraboloid shape and should theoretically provide largest zoom ranges.

2.6.4 Design and Results with VTD Type 1 – Cycloid-Like Form

Type 1 VCMs have been built for optical telescope array interferometers and Fourier transform spectrometers (Ferrari et al. [21]). The following results are obtained with the design parameters of Fig. 2.4 and a thickness distribution $\mathcal{T}_{20}$ determined for f/Ω = f/3.33 allows optimizing the zoom range [f/∞–f/2.5]. An air pressure load generates convex shapes. The substrate is a stainless steel Fe87 Cr13 quenched at a Brinell hardness BH = 330. The integration of system (2.32) provides the dimensionless thickness $\mathcal{T}_{20}$ and then thickness t. A small positive lens-like thickness is added to t in order to compensate for the plastic deformation resulting from the prestressing. The resulting thickness $t^*(r)$ was executed by a numeric command lathe on the rear face of the substrate (Table 2.1).

The prestressing was carried out by slightly overpassing the maximum curvature of the zoom-range (cf. next Section). After prestressing and final plane resurfacing, the mean aspect-ratio of these VCMs may be approximated by $< t > /D \simeq 1/60$. He-Ne interferometric tests were carried out by use of Fizeau lenses of various curvatures and an accurate pressure gauge of resolution 10^{-4} (Fig. 2.13).

Table 2.1 Thickness $t^*(r)$ of a Type 1 VCM before plane surfacing. Eso Vlti and Jussieu Lpma: Zoom range [f/∞–f/2.5]. At VCM edge, the radial thickness of the cylinder-collaret is 25 μm. t^* is with a 18 μm extra-thickness at center which includes a 14 μm positive lens-like shape for plasticity correction (see Sect. 2.7) (Loom)

r	0	1	2	3	4	5	6	7	7.9	8^+	12
t^*	318	316	311	301	286	265	235	188	150	5,000	5,000

[r: mm, t^*: μm]

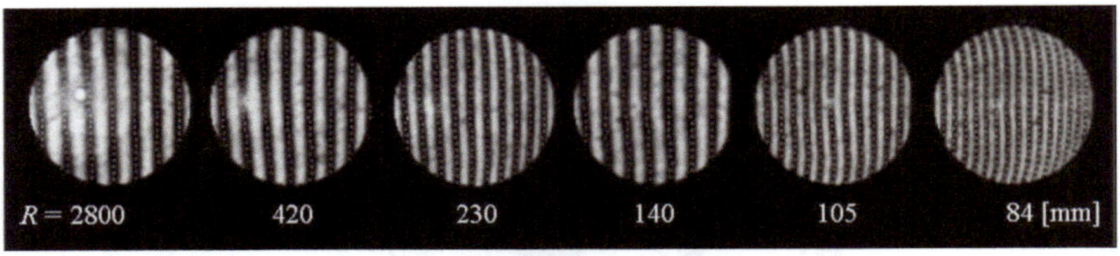

Fig. 2.13 Optical tests of a Type 1 VCM with respect to concave calibers. Fizeau interferograms for a clear aperture $2a = 5$ mm (LOOM)

2.6.5 Design and Results with a VTD Type 2 – Tulip-Like Form

Type 2 VCMs have been built for particular applications where discrete curvatures are preferred to a continuous variation of the curvature (Lemaitre [37]). The central force is generated by a motorized lead screw. These actuators are less complex to use than air pressure units with controller but may provide some vibrations during the zooming. Tulip-like VCMs are useful for pre-positioning the first pupil of telescope arrays such as with the Auxiliary telescopes of VLTI. Figure 2.6 shows that, for a given zoom range, the compared VTDs from large and small deformation theories present less deviations than for Type 2. In the following example with a VCM in quenched stainless steel Fe87 Cr13. A positive central force generated convex shapes. The thickness t was derived by solving $\mathcal{T}_{20}$ in system (2.32) for the design value f/Ω = f/6 and a zoom-range [f/∞–f/4.5]. Before optical surfacing, the theoretical thickness $t(r)$ was increased by a constant extra-thickness and the rear side of the mirror executed by a numeric command lathe (Table 2.2).

Table 2.2 Thickness $t^*(r)$ of a Type 2 VCM before plane surfacing. Zoom range [f/∞ – f/4.5]. At the VCM edge, the radial thickness of the cylinder-collaret is 25 μm. t^* includes an extra-thickness of 5 μm (LOOM)

r	0.1	1	2	3	4	5	6	7	7.9	8^+	12
t^*	515	410	359	320	286	253	217	171	135	5,000	5,000

[r: mm, t^*: μm]

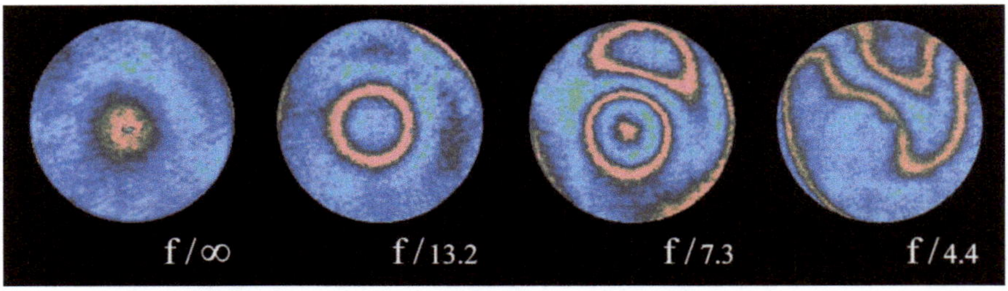

Fig. 2.14 Optical tests of a Type 2 VCM with respect to concave calibers. Zoom range f/∞–f/4.5. He-Ne patterns for full aperture $2a = 16$ mm. The central force is varied from 0 to 5.85 daN (LOOM)

For this zoom-range, the plastic deformation remains negligible and does not requires a compensation of the thickness distribution. After prestressing and plane surfacing, interferometric tests were carried out by use of Fizeau lenses of discrete curvatures and a ball-screw actuator driven by an encoded motorized system (Fig. 2.14).

2.7 Plasticity and Hysteresis

Because of the very large zoom range developed with "cycloid" type VCMs for 8 m aperture recombined telescopes (VLTI), it was found necessary to take under consideration the *plastic deformation* as well as the *hysteresis deformation loop* of the metal substrate. While plasticity goes back to the dawn of time, the discovery of hysteresis is due to J.A. Ewing in the 1880s.

A compensation of the plastic deformation and a hysteresis loop model can be determined in order to increase (i) the geometrical accuracy of the optical curvatures and (ii) the resolution of the curvature control. The plastic deformation error is corrected in the mirror figuring process, while hysteresis errors are compensated by the *closed-loop control system.*

2.7.1 Stress-Strain Linearization and Plasticity Compensation

• **Extended linear range:** For metallic alloys, the *Ewing-Muir process* [16] allows one to extend the linear range of the stress-strain relation. This *plastic tightening*, in French *raidissement plastique*, applies to the VCM substrates in quenched Fe87 Cr13 alloy. The process consists of prestressing the substrate at $\sigma_{\rm p.s}$ slightly higher than the working tensile maximum stress $\sigma_{t\,\max}$. Figure 2.15 displays the process in the case of an elongated rod. After applying a pre-stressing at $\sigma_{\rm p.s}$, the new length at rest becomes permanently increased, but for next loadings laying under $\sigma_{\rm p.s}$ the stress-strain law has been extended while remaining linear.

The prestressing applied to Type 1 VCMs (design in Table 2.1) was typically $\sigma_{\rm p.s} = 70\,{\rm daN.mm^{-2}}$ corresponding to a loading $q = 8.25$ daN.cm^{-2}. After final polishing, the VCMs operate up to a tensile maximum stress $\sigma_{t\,\max} = 68.4\,{\rm daN.mm^{-2}}$ corresponding to a loading $q_{\max} = 8.05\,{\rm daN.cm^{-2}}$ for the maximum permissible curvature $C_{\max}$.

Let C_0 be the initial curvature of a never stressed VCM and C_0^* its final curvature at rest after prestressing. The curvature difference due to *plasticity* is

$$\Delta C_{\rm Plas} = C_0^* - C_0\,. \tag{2.50}$$

He-Ne interferograms in Fig. 2.16 display the VCM shapes during the prestressing cycle of a previously unstressed VCM. The optical figures are recorded with

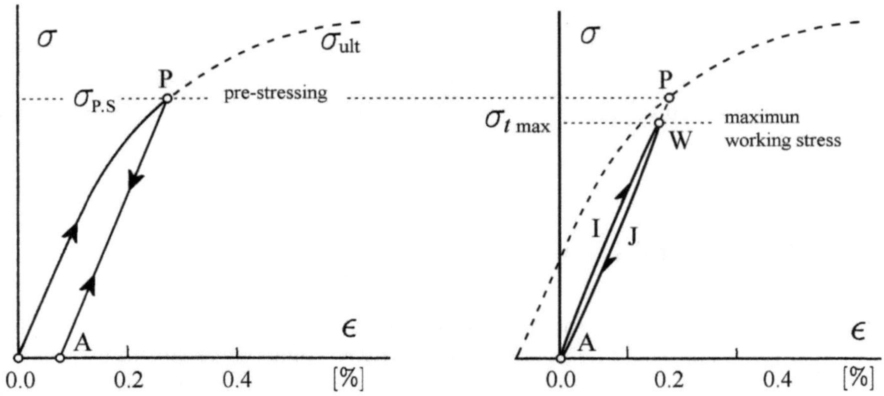

Fig. 2.15 Plasticity and hysteresis of quenched Fe87 Cr13 alloys. *Left*: Stress-strain diagram, Ewing-Muir linearization by prestressing. *Right*: Hysteresis loop in extended elastic domain after prestressing (the width of the loop is exagerated for clarity of the figure)

respect to reference lens calibers that are concave spheres of discrete curvature. These are mounted onto a wheel facing the VCM in a Fizeau mounting.

In a first approximation, the difference between the initial and final interferograms shows that the plastic deformation is of quadratic form

$$z_{\rm Plas} \propto r^2 . \qquad (2.51)$$

From Type 1 VCMs built with the design parameters in Fig. 2.4, the maximum tested elastic sag was $z_{\rm Elas} = -381\,\mu$m for the last caliber $R = 84$ mm; the mean value of the measured plastic sags was $z_{\rm Plas} = -14\,\mu$m. Thus, the typical *plasto-elastic deformation ratio* is

$$z_{\rm Plas}/z_{\rm Elas} \equiv (C_0^* - C_0)/(C_{\rm max} - C_0) = 3.67 \pm 0.15\,\% . \qquad (2.52)$$

These results can be used for analysis with a plasticity theory (Lubliner [40]) for model investigations. Given the stress distribution of σ_r (Fig. 2.4), that is also of cycloid-like form, the plastic deformation appears near the substrate faces from mirror axis to a radius about half the aperture, $r \approx a/2$.

In order to compensate for the effect induced by the plastic deformation, the following conditions apply (Lemaitre et al. [38]):

• **Plasticity Compensation:** *Assuming a VCM figured quasi-flat with curvature* C_0 *at rest, which becomes* $C_0^\star$ *at rest after prestressing, and denoting* $\Delta_{\rm Plas} = C_0^\star - C_0$,

1 → *if the optical figurings are always executed at same curvature before and after pre-stressing,*

2 → *and if the rear side of the substrate* $Z_{\rm RS}(r)$ *is defined by the co-addition of the thickness distribution (elasticity term) and of a lens shape (plasticity term) following the sign of loading* q *as*

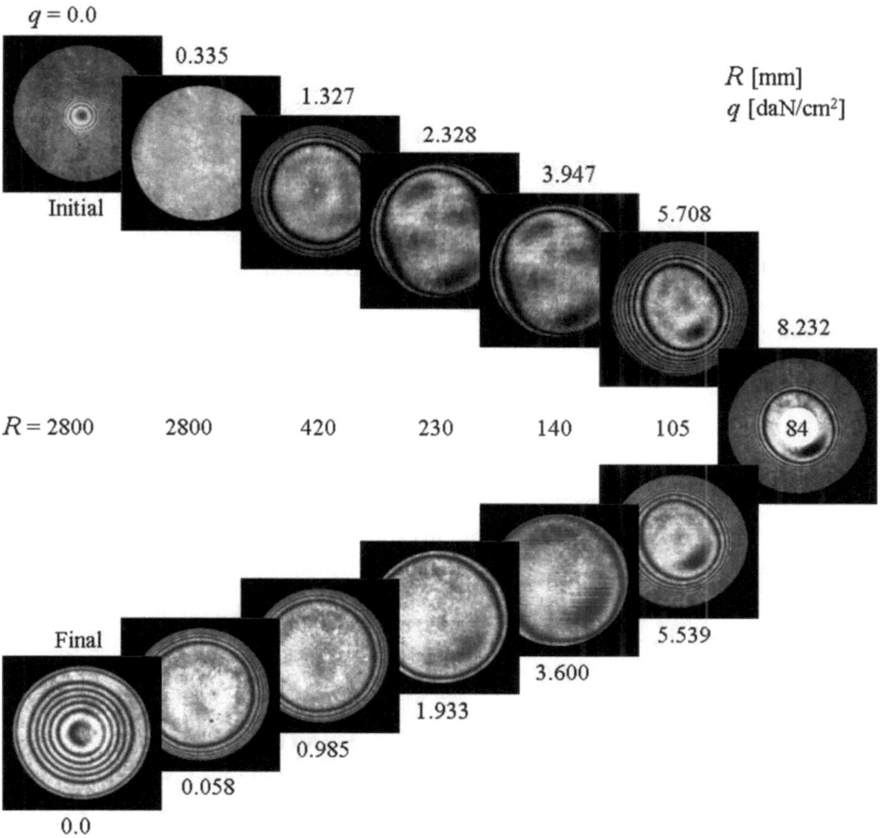

Fig. 2.16 He-Ne fringes with respect to calibers (R) and loads (q) of a VCM – Type 1 VTD-form – during *prestressing loop*. Plastic deformation is derived from the two patterns on left. The mirror is then resurfaced to a plane figure at $q = 0$. These VCMs of zoom range ($f/\infty - f/2.5$) are the core of cat's eye carriages of the VLTI eight delay lines (LOOM)

$$Z_{\mathrm{RS}} = t(r) - \Delta C_{\mathrm{Plas}}(a^2 - r^2)/2 \quad for \quad q > 0 \; i.e. \forall C < 0, \; or \qquad (2.53a)$$

$$Z_{\mathrm{RS}} = t(r) + \Delta C_{\mathrm{Plas}}\, r^2/2 \qquad for \; q < 0 \; i.e. \forall C > 0, \qquad (2.53b)$$

then the optimal properties of the VCM design with the large deformation theory are recovered.

These conditions can be realized by the construction and prestressing tests of preliminary prototypes allowing the determination of the lens shaped correction. Hence, the *plasticity correction* was set up in the final design of Type 1 VCMs, following (2.53a), which generates a zoom range with all negative curvature ($\forall C <$

0). This correction is included into the thicknesses $t^*(r)$ of Table 2.1 with $C_0 = 0$, $C_0^* = -0.44\,10^{-3}\,\text{mm}^{-1}$ and $\Delta C_{\text{Plas}} = C_0^\star - C_0 = C_0^\star$.

2.7.2 *Hysteresis Compensation and Curvature Control*

In Sect. 2.2, we have shown from the large deformation theory that the representation of the loading as a function of the flexure ratio is not linear (2.36) but expressed by an odd series in z/t_0. For Type 1 VCMs of zooming range [f/∞–f/2.5] deformed by air pressure, this representation provides a convenient accuracy by limiting the power series development up to $i = 5$. Considering the curvature C instead of the flexure ratio, the *load-curvature* relationship derived from (2.36) can be represented by (Ferrari et al. [22])

$$q = \beta_1 (C - C_0) + \beta_3 (C - C_0)^3 + \beta_5 (C - C_0)^5 \,, \tag{2.54}$$

where β_i are coefficients, C_0 the mirror curvature at rest and $\forall\, C < 0$.

For large deformations, metal substrates show a *flexural hysteresis*:

1 → *During the de-loading, the same curvatures as during loading are obtained by lower applied loads.*
2 → *After the loading and subsequent de-loading sequence, the initial and final curvatures are identical.*

The largest of the hysteresis loops is the path AIWJA (Fig. 2.14 *Right*), where the extremal working point W of maximum load q_{max} and curvature C_{max} is reached. Considering a loading sequence up to q_{seq}, and provided $q_{\text{seq}} \leq q_{\text{max}} < q_{\text{p.s}}$ such as defined from prestressing, the above load-curvature relationship is only valid for increasing pressures; when decreasing the load from a q_{seq} loading, the β_i coefficients become slightly different. Let $\overline{\beta}_i\big|_{\text{seq}}$ be those coefficients. Given a curvature C, the load difference Δq between the increasing and decreasing pressures is a function of the maximum pressure q_{seq} or of the associated curvature C_{seq} reached in the going up sequence. The *hysteresis amplitudes* Δq, increasing with higher deformation sequences, are also represented by fifth-order odd polynomials

$$\Delta q\,\big|_{\text{seq}} = \beta_1^* (C - C_0) + \beta_3^* (C - C_0)^3 + \beta_5^* (C - C_0)^5 \,, \tag{2.55}$$

with

$$\beta_i^* = \beta_i - \overline{\beta}_i\big|_{\text{seq}} \,.$$

Remaining under the maximum working stress defined by the prestressing (Sect. 2.6.1), measurements have been carried out on 12 cycloid-type VCMs having a zoom-range [f/∞–f/2.5]. Hysteresis amplitudes were determined by Shack-Hartmann optical tests for loading sequences $q_{\text{seq}} < q_{\text{max}}$. Considering a representation of the hysteresis Δq in function of the load q instead of the curvature, we obtain the form

$$\Delta q\,\big|_{\rm seq} = \delta_1\, q + \delta_3\, q^3 + \delta_5\, q^5\,, \qquad q \le q_{\rm seq} \le q_{\rm max}\,, \tag{2.56}$$

where δ_i coefficients are deduced from series (2.54) and (2.55).

From the (2.56) form, we can model all the hysteresis loops from simple properties. By definition, given a sequence, the hysteresis is null at the maximum load of the sequence, i.e. at $q = q_{\rm seq}$. In addition, the results from Shack-Hartmann tests show that the slopes of the hysteresis loops at $q = 0$ and $q = q_{\rm seq}$ are *opposite* (Ferrari [22]). Then, we have the two conditions

$$\Big[\Delta q\,\big|_{\rm seq}\Big]_{q\,=\,q_{\rm seq}} = 0\,, \quad \Big[\frac{d}{dq}\,\Delta q\,\big|_{\rm seq}\Big]_{q\,=\,q_{\rm seq}} = -\Big[\frac{d}{dq}\,\Delta q\,\big|_{\rm seq}\Big]_{q\,=\,0}\,,$$

which entails

$$\delta_3 = -\,\frac{3}{2q_{\rm seq}^2}\,\delta_1 \quad \text{and} \quad \delta_5 = \frac{1}{2q_{\rm seq}^4}\,\delta_1\,. \tag{2.57}$$

The third condition to determine δ_1 is provided by the coordinates of the maximum amplitude. The results from Shack-Hartmann tests show that the slope at the origin is of the form

$$\delta_1 = a_1\, q_{\rm seq} + a_3\, q_{\rm seq}^3\,, \tag{2.58}$$

where a_1 and a_3 are coefficients. This allows one to construct a hysteresis grid model from the locus of the maximums, i.e. the coordinates of antinode A points. Therefore, the *hysteresis amplitude grid* for any sequence up to $q_{\rm seq}$ is defined by

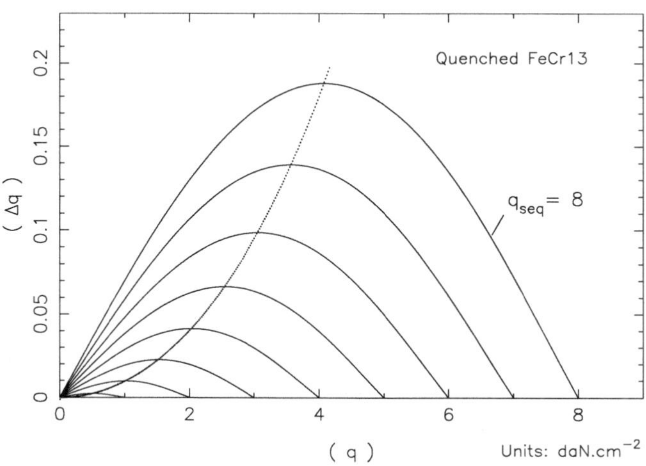

Fig. 2.17 Hysteresis $\Delta q(q)$ vs loading sequences $q_{\rm seq}$. Cycloid-like VCM – VTD Type 1 – actuated by air pressure. At $q_{\rm max} = 8$ daN.cm^{-2}, the radial stress is $\sigma = 68$ daN.mm^{-2} and the hysteresis amplitude $\Delta q/q_{\rm sec}$ reach 2.3% [38]

$$\Delta q\,\Big|_{\rm seq} = \frac{1}{2}\left(a_1\, q_{\rm seq} + a_3\, q_{\rm seq}^3\right)\left(2 - 3\frac{q^2}{q_{\rm seq}^2} + \frac{q^4}{q_{\rm seq}^4}\right) q\,, \qquad (2.59a)$$

$$\text{with Antinodes}\quad \begin{cases} q_{\rm A}\Big|_{\rm seq} = \kappa_1\, q_{\rm seq}\,, \\ \Delta q_{\rm A}\Big|_{\rm seq} = \kappa_1\,\kappa_2\left(a_1\, q_{\rm seq}^2 + a_3\, q_{\rm seq}^4\right), \\ \kappa_1 = \left[(9-\sqrt{41})/10\right]^{1/2} = 0.5095, \\ \kappa_2 = (13+3\sqrt{41})/50 = 0.6441. \end{cases} \qquad (2.60a)$$

The hysteresis curves are represented by the first positive part of five real-root curves (Fig. 2.17). The locus of the antinodal A-points is a biquadratic curve (Fig. 2.17 *Dotted line*). All these curves are determined from (2.59a) and set (2.59b) with coefficient values

$$a_1 = 7.65\ 10^{-3}\,\mathrm{cm}^2\cdot\mathrm{daN}^{-1} \quad \text{and} \quad a_3 = 2.05\ 10^{-5}\,\mathrm{cm}^6\cdot\mathrm{daN}^{-3}.$$

In order to achieve the best resolution of curvature control, this *hysteresis compensation model* is included in the VCM curvature *closed-loop control system* at the same level as the positioning control of the cat's eye translation.

References

1. J.M. Beckers, Field of view considerations for telescope arrays, SPIE Proc., **628**, 255 (1986)
2. J.M. Beckers et al., The VLTI implementation plan, ESO/VLT Report No. 59b (1989)
3. J.M. Beckers, Interferometric imaging with the VLTI, J. Optics (Paris), **22**, 73 (1991)
4. R.N. Bracewell, in *The Fourier Transform and Its Applications*, McGraw-Hill ed., New York (1965)
5. J.W. Brault, Fourier transform spectroscopy, in *High Resolution in Astronomy*, 15th Advanced Course, SAAS-FEE, 1–62 (1985)
6. R. Hanbury Brown, Measurement of stellar diameters, Ann. Rev. Astron. Astrophys., **6**, 13–38 (1968)
7. P. Casal, C. Fayard, B. Authier, Calculations of the elastic deformations of revolution membranes, Appl. Opt., **20**(11), 1983–1989 (1981)
8. A.A. Christy, Y. Ozaki, V.G. Gregoriou, in *Modern Fourier Transform Infrared Spectroscopy*, Elsevier ed., ISBN 0444 500 448, (2001)
9. P. Connes, Astronomical Fourier spectroscopy, Ann. Rev. Astron. Astrophys., **8**, 209–230 (1970)
10. P. Connes, G. Michel, Astronomical Fourier spectrometers, Appl. Opt., **14**, 9, 2067–2084 (1975)
11. S.P. Davis, M.C. Clark, J.W. Brault, in *Fourier Transform Spectroscopy*, Academic Press, San Diego ISBN 0120 425 106 (2001)
12. F. Derie, VLTI delay lines: Design, development and performance requirements, in *Interferometry in Optical Astronomy*, SPIE Proc., **4006**, 25–30 (2000)
13. T. Dohi and T. Suzuki, Appl. Opt., **10**, 1137 (1971)
14. K. Dohlen, Design of a FTS for environmental surveillance, Doctoral dissertation, University of London (1994)

15. K. Dohlen, Interferometric spectrometer for liquid mirror survey telescopes, in *Optical Telescopes of Today and Tomorrow*, SPIE/ESO Proc., **2871**,1359–1364 (1997)
16. J.A. Ewing, *The Strength of Materials*, Cambridge Univ. Press, 2nd ed., (1906). See also S.P. Timoshenko, *Résistance des Matériaux*, Dunod edt., Paris, 365–366 (1968)
17. M. Ferrari, G.R. Lemaitre, Analysis of large deflection zoom mirrors for the ESO VLTI, Astron. Astrophys. **274**, 12–18 (1993)
18. M. Ferrari, Optique active et grandes déformations élastiques, Doctoral dissertation, University Aix-Marseille I (1994)
19. M. Ferrari, F. Derie, B. Delabre, J.-M. Mariotti, VLTI's VCMs – Pupil transfer inside the delay line cat's eye, ESO Report No VLT-TRE-ESO-15220-1509 (1997)
20. M. Ferrari, Development of variable curvature mirrors for the delay lines of the VLTI, Astron. Astrophys. Suppl. Ser., **128**, 221–227 (1998)
21. M. Ferrari, G.R. Lemaitre, S. Mazzanti, P. Lanzoni, F. Derie, VLTI pupil transfer: Variable curvature mirrors [10], Final results and performances, in *Astronomical Telescopes and Instrumentation*, SPIE Proc., **4006**, 104–116 (2000)
22. M. Ferrari, S. Mazzanti, G.R. Lemaitre, J. Lemerrer, P. Lanzoni, P. Dargent, F. Derie, A. Huxley, A. Wallanders, Variable curvature mirrors - Implementation in the VLTI delay lines for field compensation, in *Interferometry for Optical Astronomy* II, SPIE Proc., **4838**, 1155–1162 (2002)
23. A. Glindeman et al. The VLT Interferometer, in *Interferometry in Optical Astronomy*, SPIE Proc., **4006**, 2–12 (2000)
24. E. Hugot, Optique astronomique et élasticité – Ph. D., Université de Provence – Aix Marseille I, Chap. 4 (2007)
25. P. Jacquinot, C. Dufour, J. Rech. CNRS, Lab. Bellevue (Paris), **6**, 91 (1948)
26. P. Jacquinot, J. Opt. Soc. Amer., **54**, 761 (1954)
27. B. Koehler, C. Flebus, VLT auxiliary telescopes, in *Interferometry in Optical Astronomy*, SPIE Proc., **4006**, 13–24 (2000)
28. D. Korsch, in *Reflective Optics*, Academic Press Inc., 173 (1991)
29. M. Krautter, Aplanatic two-mirror surfaces, in *Optical System Design, Analysis, and Production for Advanced Technology Systems*, SPIE Proc., **655**, 127–137 (1986)
30. O. von der Lühe, J.M. Beckers, R. Braun, The configuration of the VLTI on the Paranal site, in *High Resolution by Interferometry* **II**, ESO Conf. Proc., 959–968 (1991)
31. O. von der Lühe et al., A new plan for the VLTI, The Messenger, ESO ed., **87**, 8–14 (1997)
32. A. Labeyrie, Stellar interferometry methods, Ann. Rev. Astron. Astrophys., **16**, 77–102 (1978)
33. A. Labeyrie, Interferometry with arrays of large-aperture ground based telescopes, Proc. KPNO Conf. on *Optical and Infrared Telescopes for the 1990s*, **II**, 786–796 (1980)
34. C. Leinert, U. Graser, MIDI – The Mid-Infrared interferometer instrument for the VLTI, in *Astronomical Interferometry*, SPIE Proc., **3350**, 389–402 (1998)
35. G.R. Lemaitre, Élasticité et miroirs à focale variable, C. R. Acad. Sc. Paris, **282 B**, 87–89 (1976)
36. G.R. Lemaitre, French patent No 2343262 (1976), US patent No 4119 366 (1976)
37. G.R. Lemaitre, S. Mazzanti, M. Ferrari, P. Montiel, P. Lanzoni, Tulip-form variable curvature mirrors, in *Astronomical Interferometry*, SPIE Proc., **3350**, 373–379 (1998)
38. G.R. Lemaitre, M. Ferrari, S. Mazzanti, P. Lanzoni, P. Joulié, VLTI pupil transfer: Variable curvature mirrors [2], Plasticity, hysteresis and curvature control, in *Astronomical Telescopes and Instrumentation*, SPIE Proc., **4006**, 192–197 (2000)
39. G.R. Lemaitre, P. Montiel, P. Joulié, K. Dohlen, P. Lanzoni, Active optics and modified-Rumsey wide-field telescopes: MINITRUST demonstrators with vase- and tulip-form mirrors, Appl. Opt., **44**(34), 7322–7332 (2005)
40. J. Lubliner, in *Plasticity Theory*, MacMillan Publ. ed., New York (1991)
41. J.-P. Maillard, Seismology with Fourier transform spectrometer, Appl. Opt., **35**, 16, 2734–2746 (1996)
42. J.-P. Maillard, A super-imaging FTS for the VLTI, in *Scientific Drivers for* ESO *Future* VLT/VLTI *instrumentation*, ESO Conf., 193 (2002)

43. J.-M. Mariotti et al., The VLTI program: A status report, in *Astronomical Interferometry*, SPIE Proc., bf 3350, 800–806 (1998)
44. F. Merkle, Synthetic-aperture imaging with the VLT, Journ. Opt. Soc. Am., **A 5**(6), 904 (1989)
45. M. Mersenne, *Traité de l'Harmonie Universelle*, Paris (1636)
46. M. Mersenne, *L'Optique et la Catoptrique*, posthume publication, in *La Perspective Curieuse* by J.-F. Niceron, F. Langlois alias Chartres edt., Paris (1652)
47. L. Mertz, in *Transformation in Optics*, John Wiley and Sons ed., New York (1965)
48. A.A. Michelson, On the application of interference methods to astronomical measurements, Ap. J., **51**, 257–262 (1920)
49. D. Mourard et al., The GI2T/REGAIN interferometer, in *Astronomical Interferometry*, SPIE Proc., **3350**, 517–525 (1998)
50. F. Paresce, F. Delplancke, F. Derie, A. Glindemann, A. Richichi, M. Tarrenghi, Scientific objectives of ESO's PRIMA facility, in *Interferometry for Optical Astronomy* I, SPIE Proc., **4838**, 486–495 (2002)
51. M. Paul, Systèmes correcteurs pour réflecteurs astronomiques, Rev. Opt., **14**(5), 169–202 (1935)
52. N.J. Rumsey, A compact three-reflection astronomical camera, in *Optical Instruments and Techniques*, ICO 8 Meeting, London, Home Dickson edt., Oriel Press Newcastle, 514–520 (1969)
53. L. Seidel, Astron. Nachr., **43**, 289 (1856)
54. B.C. Smith, in *Fundamentals of Fourier Transform Infrared Spectroscopy*, CRC Press, Florida ISBN 0849 324 610, (1996)
55. E. Stephan, Sur les franges d'interférence observées avec de grands instruments, C. R. Acad. Sc., **76**, 1008–1010 (1873)
56. G.W. Stroke and A.T. Funkhauser, Physics Letters, **16**, 272 (1965)
57. A.P. Thorne, in *Spectrophysics*, Chapman and Hall ed., London (1988)
58. S.P. Timoshenko, S. Woinowsky-Krieger, *Theory of plate and shells*, McGraw-Hill ed., 396 (1976)
59. F. Vakili, I. Percheron, Beam combination and coherence tracking with diluted arrays, in *High Resolution by Interferometry* **II**, ESO Conf. Proc., 1247–1256 (1991)
60. A. Valentin, Fourier spectroscopy with a very long optical path length, Spectrochim. Acta, **51**-A(7), 1127–1142 (1995)
61. A. Valentin, A. Henry, The Fourier infrared spectrometer of Laboratoire de Physique Moléculaire et Applications, Paris-Jussieu, Internal report of LPMA/CNRS, Univ. P. & M. Curie, Paris (2002)
62. R.N. Wilson, Karl Schwarzschild and Telescope Optics, Karl Schwarzschild Lecture given to the German Astronomical Society, Bochum (1993), published in *Review of Modern Astronomy*, **7**, 1 (1994). See also *Reflecting Telescope Optics I*, Springer edt., 88 (1996)
63. L. Woltjer et al., Proposal for the construction of the 16-m VLT, ESO/VLT Report no. 57 (1987)

Chapter 3
Active Optics and Correction of Third-Order Aberrations

3.1 Elasticity Theory with Constant Thickness Distributions – CTD Class

The elastic deformations of a circular plate belonging to the CTD class leads to solving the bi-laplacian equation $\nabla^2\nabla^2 z - q/D = 0$, where D is the plate rigidity and q the external load function applied to the surface of the plate. Axisymmetric or non-axisymmetric deformations generated by point forces or uniform loads distributed on the plate are derived from this fourth derivative equation also known as Poisson's equation.

Rigorous elasticity configurations can be found for obtaining by flexure particular optical modes such as 3rd-order aberrations - $Sphe\,3$, $Coma\,3$, $Astm\,3$ - or some higher-order aberrations. Restraining to the cases where only external loads $q = 0$ or $constant$ are applied over all the optical aperture, these deformations - that we have called *Clebsch-Seidel modes* (see definition in Chap. 7) - require an azimuth modulation in $m\theta$ of the bending moment applied along the circular contour.

Such flexural single modes and their co-addition capability allowed by CTD plates are investigated with meniscus forms and vase forms in Chap. 7.

3.2 Elasticity Theory with Variable Thickness Distributions – VTD Class

The elastic deformations of a circular substrate belonging to the VTD class can provide particularly simple mirror configurations for obtaining flexures exactly corresponding to the 3rd-order aberration modes and other optical modes [14, 19]. For instance, a VTD applied to the curvature mode $Cv\,1$, i.e. the flexure $z = A_{20}r^2$, is achieved by direct action-reacting forces (Chap. 2); this avoids generating a uniform bending moment at the contour as required with a CTD. Since line forces are easier to generate than bending moments, mirror designs with a VTD and simply supported contours are preferred when this is possible. In the present chapter we mainly investigate VTDs using simply supported contours as well as CTDs using clamped (i.e. built-in) edges.

G.R. Lemaitre, *Astronomical Optics and Elasticity Theory*, Astronomy and Astrophysics Library, DOI 10.1007/978-3-540-68905-8_3,

Considering a VTD where the rigidity can be represented by $D(r, \theta)$ in a cylindrical coordinate system, the radial and tangential bending moments and twisting moment, per unit length, are defined by

$$M_r = D(r,\theta)\left[\frac{\partial^2 z}{\partial r^2} + \nu\left(\frac{1}{r}\frac{\partial z}{\partial r} + \frac{1}{r^2}\frac{\partial^2 z}{\partial \theta^2}\right)\right], \tag{3.1a}$$

$$M_t = D(r,\theta)\left[\frac{1}{r}\frac{\partial z}{\partial r} + \frac{1}{r^2}\frac{\partial^2 z}{\partial \theta^2} + \nu\frac{\partial^2 z}{\partial r^2}\right], \tag{3.1b}$$

$$M_{rt} = M_{tr} = (1-\nu)\, D(r,\theta)\left[\frac{1}{r^2}\frac{\partial z}{\partial \theta} - \frac{1}{r}\frac{\partial^2 z}{\partial r \partial \theta}\right]. \tag{3.1c}$$

As used in Chap. 1 and the next Chapters, this representation of the flexural moments entails a positive flexure if a positive radial bending moment M_r is applied at $r = a$ for generating the fundamental mode z_{20} (curvature mode). For z_{nm} modes with $m = n$, we may also verify that M_r is positive in the x, z section i.e. if $y = \theta = 0$. This sign convention is natural and in agreement with the generally used optics convention: a curved surface z_{20} is of positive curvature if $\forall\, r$, $z_{20}(r) > z_{20}(0)$.

Since the Laplacian is

$$\nabla^2 z = \frac{\partial^2 z}{\partial r^2} + \frac{1}{r}\frac{\partial z}{\partial r} + \frac{1}{r^2}\frac{\partial^2 z}{\partial \theta^2},$$

the bending moments satisfy

$$M_r + M_t = (1+\nu)\, D(r,\theta)\, \nabla^2 z\,. \tag{3.2}$$

The determination of radial and tangential components of the shearing forces Q_r and Q_t [14], as functions of the moments, is derived from the equilibrium of a segment $t\, r\, d\theta\, dr$ around tangential axis $\mathrm{O}\tau'$ parallel to $\omega\tau$ and around radial axis $\mathrm{O}\omega$, respectively (Fig. 3.1).

For the radial shearing force Q_r, the resulting components around $\mathrm{O}\tau'$ are

$$Q_r\, r\, d\theta\, dr + \frac{\partial}{\partial r}(r M_r)\, d\theta\, dr - M_t\, d\theta\, dr - \frac{\partial M_{rt}}{\partial \theta}\, d\theta\, dr = 0\,,$$

where the third term in M_t is the sum of two components tilted of $\pm\, d\theta/2$ from the radial axis $O\omega$. After simplification, the radial shearing force is represented by

$$Q_r = -\frac{\partial M_r}{\partial r} - \frac{1}{r}\left(M_r - M_t - \frac{\partial M_{rt}}{\partial \theta}\right). \tag{3.3}$$

The tangential shearing force Q_t is derived from the moments around $O\omega$. The resulting components are

$$Q_t\, r\, d\theta\, dr + \frac{\partial M_t}{\partial \theta}\, d\theta\, dr - M_{rt}\, d\theta\, dr - \frac{\partial}{\partial r}(r M_{rt})\, d\theta\, dr = 0\,,$$

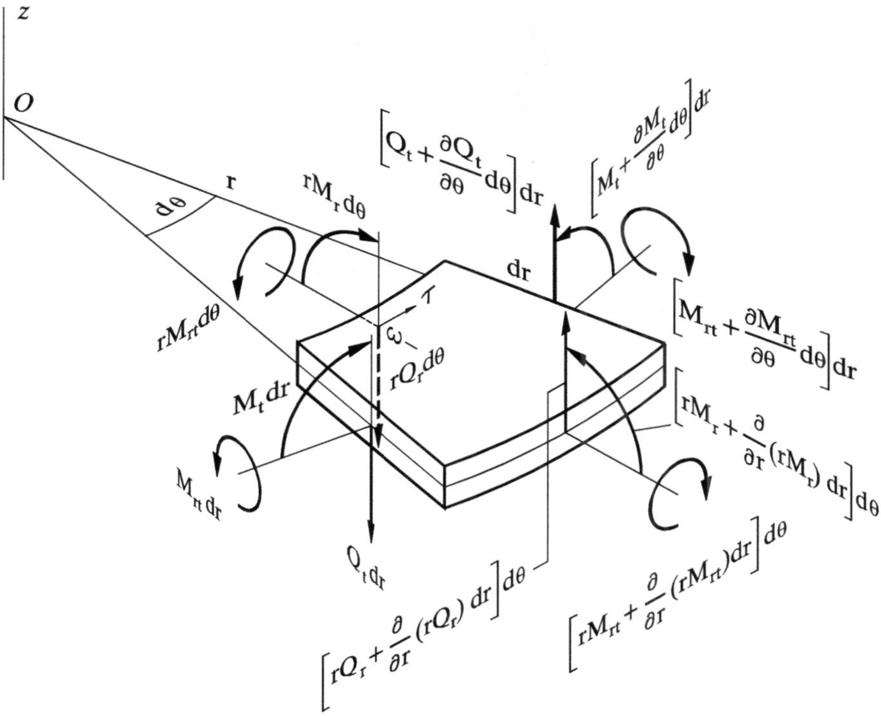

Fig. 3.1 Bending moments, twisting moment, and shearing forces providing the equilibrium of a plate element

and after simplification, the tangential shearing force is

$$Q_t = -\frac{1}{r}\left(\frac{\partial M_t}{\partial \theta} - 2\,M_{rt}\right) + \frac{\partial M_{rt}}{\partial r}\,. \tag{3.4}$$

Finally the external load q applied per unit area onto the surface of the elementary segment is in static equilibrium with the shearing forces (Fig. 3.2). After dividing the terms of the equilibrium equation by element area $r\,d\theta\,dr$,

$$\frac{1}{r}\left[\frac{\partial}{\partial r}(r\,Q_r) + \frac{\partial Q_t}{\partial \theta}\right] + q = 0\,. \tag{3.5}$$

This partial derivative equation linking the shearing forces to external load q is the general relation of variable thickness plates [14]. This equation degenerates into Poisson's equation $D\nabla^2\nabla^2 z - q = 0$ if $D = constant$ (cf. Sect. 7.2). The VTD $t(r,\theta)$ is defined by

$$t^3(r,\theta) = 12\,(1-\nu^2)\,D(r,\theta)\,/\,E\,.$$

Let us consider flexures having the shape of wavefront aberration modes. These modes belong to a circular polynomial series and are represented by

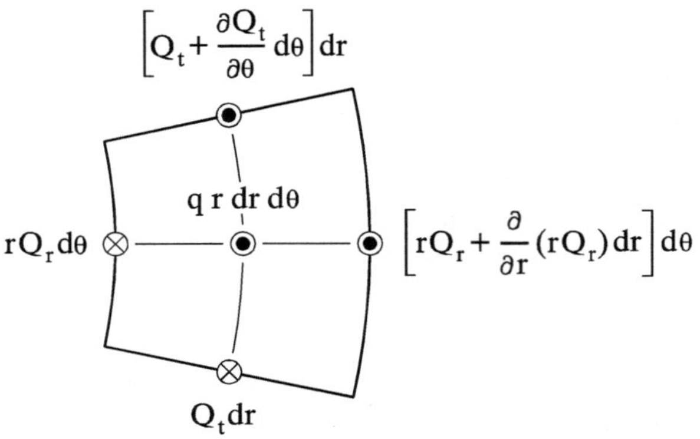

Fig. 3.2 External load and shearing forces providing the equilibrium of a plate element in z-direction

$$z \equiv z_{nm} = A_{nm}\, r^n \cos m\theta\,, \qquad \text{with} \;\; m \le n\,, \;\; m+n \text{ even} \ge 2\,, \tag{3.6}$$

where n and m are integers. Given the condition $m \le n$, the series development of such optics modes generates the terms of a triangular matrix. For low-order modes we use the simple suffix denotation nm instead of n, m. With $m+n=4$, the three terms coming after the dioptrics are the third aberration modes z_{40}, z_{31}, and z_{22}, i.e. spherical aberration, coma, and astigmatism respectively, also denoted $Sphe\,3$, $Coma\,3$, and $Astm\,3$.

The following derivative elements

$$\begin{aligned}
\frac{\partial^2 z}{\partial r^2} &= n(n-1)A_{nm}r^{n-2}\cos m\theta\,,\\
\frac{1}{r}\frac{\partial z}{\partial r} + \frac{1}{r^2}\frac{\partial^2 z}{\partial \theta^2} &= (n-m^2)A_{nm}r^{n-2}\cos m\theta\,,\\
\frac{1}{r}\frac{\partial^2 z}{\partial r \partial \theta} - \frac{1}{r^2}\frac{\partial z}{\partial \theta} &= -m(n-1)A_{nm}r^{n-2}\sin m\theta\,,
\end{aligned} \tag{3.7}$$

allow one to derive from (3.1a, b, c) the bending and twisting moments as

$$M_r = \Big[\, n(n-1) + \nu(n-m^2)\Big]\; D(r,\theta)\; A_{nm}\; r^{n-2}\cos m\theta\,, \tag{3.8a}$$

$$M_t = \Big[\, n - m^2 + \nu n(n-1)\Big]\; D(r,\theta)\; A_{nm}\; r^{n-2}\cos m\theta\,, \tag{3.8b}$$

$$M_{rt} = m\,(n-1)(1-\nu)\,D(r,\theta)\,A_{nm}\,r^{n-2}\sin m\theta\,. \tag{3.8c}$$

Restricting to VTDS of rotational symmetry, which is an interesting case for practical applications, let us consider a rigidity $D(r,\theta) \equiv D(r)$ of the following analytical form [14]

$$D = A_0 \ln r + \sum_{i=1}^{j} A_i\, r^{-\alpha_i}\,, \tag{3.9}$$

where the unknown constant coefficients A_0, A_i, and α_i are only functions of the external loading. Denoting

$$A'_0 = A_0\, A_{nm} \quad \text{and} \quad A'_i = A_i\, A_{nm}\,, \tag{3.10}$$

and substituting the rigidity in Eq. (3.9), the moments are represented by

$$M_r = \left[n(n-1) + \nu(n-m^2)\right]\left[A'_0\, r^{n-2}\ln r + \sum A'_i\, r^{n-2-\alpha_i}\right]\cos m\theta\,, \tag{3.11a}$$

$$M_t = \left[n - m^2 + \nu n(n-1)\right]\left[A'_0\, r^{n-2}\ln r + \sum A'_i\, r^{n-2-\alpha_i}\right]\cos m\theta\,, \tag{3.11b}$$

$$M_{rt} = m\,(n-1)(1-\nu)\left[A'_0\, r^{n-2}\ln r + \sum A'_i\, r^{n-2-\alpha_i}\right]\sin m\theta\,. \tag{3.11c}$$

After calculation, the radial shearing force, derived from the equilibrium (3.3), is

$$\begin{aligned} Q_r = & -(n-2)(n^2-m^2)\, A'_0\, r^{n-3}\ln r\ \cos m\theta \\ & -\left[n(n-1) + \nu(n-m^2)\right] A'_0\, r^{n-3}\ \cos m\theta \\ & +\sum_{i=1}^{j}\{-(n-2)(n^2-m^2) \\ & +\left[n(n-1) + \nu(n-m^2)\right]\alpha_i\}\, A'_i\, r^{n-3-\alpha_i}\ \cos m\theta\,. \end{aligned} \tag{3.12}$$

and the tangential shearing force derived from (3.4) is

$$\begin{aligned} Q_t = & \; m\,(n^2-m^2)\, A'_0\, r^{n-3}\ln r\ \sin m\theta \\ & +m\,(n-1)(1-\nu)\, A'_0\, r^{n-3}\ \sin m\theta \\ & +\sum_{i=1}^{j} m\left[n^2-m^2-(n-1)(1-\nu)\,\alpha_i\right] A'_i\, r^{n-3-\alpha_i}\ \sin m\theta\,. \end{aligned} \tag{3.13}$$

Finally, the equilibrium equation (3.5) provides the external load applied to the surface of the plate. After calculation, the load is represented by

$$\begin{aligned} q = & \;(n^2-m^2)\left[m^2-(n-2)^2\right] A'_0\, r^{n-4}\ln r\ \cos m\theta \\ & -\left[n(n-2)(2n-1+\nu) - m^2(2n-3-\nu)\right] A'_0\, r^{n-4}\ \cos m\theta \\ & +\sum_{i=1}^{j}\left(\begin{array}{c} +\left[n(n-1)+\nu(n-m^2)\right]\alpha_i^2 \\ -\left[n(n-2)(2n-1+\nu) - m^2(2n-3-\nu)\right]\alpha_i \\ +(n^2-m^2)\left[(n-2)^2-m^2\right] \end{array}\right) A'_i\, r^{n-4-\alpha_i}\ \cos m\theta\,. \end{aligned} \tag{3.14}$$

For a given mode z_{nm}, these results allow the determination of the rigidity and of the associated external forces to apply to the mirror surface and also the forces and moments to its contour. Hereafter and for reasons of practical application, we will mainly consider the two cases $q = 0$ and a uniform load $q = constant$.

As previously found for variable curvature mirrors i.e. the $Cv\,1$ mode, the choice of the external load $q = 0$ or $q = constant$ in (3.14) allows one to recover the single solution with $D\{a\} = constant$ and $M_r\{a\} = constant$ that belongs to the CTD class, as well as the three solutions with $D\{a\} = 0$ and $M_r\{a\} = 0$ that belong to the VTD class and characterized by the two forms called cycloid-like and tulip-like. The four rigidities described in Chap. 2, appear from the coefficients A_0 (logarithm term), A_1, A_2, and the roots $\alpha_1 = 0, \alpha_2 = -2$.

The net shearing force V_r represents the resultant acting in z direction into the plate at a radius r. This force was first derived by Kirchhoff [11, 12, 27] when a twisting moment M_{rt} exists into the plate i.e. for the case of non-axisymmetric deformations. This force is defined by[1]

$$V_r = Q_r - \frac{1}{r}\,\frac{\partial M_{rt}}{\partial \theta}\,. \tag{3.15}$$

After substitution, the net shearing force is

$$\begin{aligned} V_r = &-\Big[(n-2)n^2 + m^2 - \nu(n-1)m^2\Big]\, A_0'\, r^{n-3} \ln r\, \cos m\theta \\ &-\Big[n(n-1) + \nu(n - m^2)\Big]\, A_0'\, r^{n-3}\, \cos m\theta \\ &+\sum_{i=1}^{j} \begin{pmatrix} -\,[(n-2)n^2 + m^2 - \nu(n-1)m^2] \\ +\,[n(n-1) + \nu(n-m^2)]\,\alpha_i \end{pmatrix} A_i'\, r^{n-3-\alpha_i}\, \cos m\theta \end{aligned} \tag{3.16}$$

• **Boundaries and continuity conditions**: In this chapter on the thin plate theory we assume that the "in-plane" forces at the middle-surface are negligible so the plate is free to radially move at the boundaries. Hence the four cases describing an edge boundary condition, or the link continuity condition between adjacent zones, at a contour of radius $r = a$ are the following.

1. *Built-in or clamped edge boundary*

$$\left(\frac{\partial z}{\partial r}\right)_{r=a} = 0\,. \tag{3.17a}$$

2. *Free edge boundary*

$$V_r\{a,\theta\} = 0 \quad \text{and} \quad M_r\{a,\theta\} = 0\,. \tag{3.17b}$$

---

[1] The present definition of the net shearing force V_r is consistent with the positive sign convention of the three flexural moments in the above Eq. (3.1).

There is an error in *Theory of Plates and Shells* by Timoshenko and Woinowsky-Krieger at Eq. (j) p. 284: Their convention uses a negative sign convention in the definition of the three moments M_r, M_t and M_{rt} while the sign of their shearing forces Q_r and Q_t with respect to the Laplacian term is as above Eqs. (8a) and (8b), so that the two equations in (8c) are the same. Hence the correctly associated representation of the net shearing force should be $V_r = Q_r + \partial M_{rt}/(r\,\partial r)$, with their notation.

Several other authors as well use TWK's negative sign convention or the present positive sign convention in defining the two bending moments, but the torsion moment M_{rt} appears with an opposite sign whatever convention is used. In order to respect the equilibrium equations of statics, the sign before M_{rt} is also changed in those equations so the Poisson biharmonic equation is satisfied. However, there is an error in the sign before $\partial M_{rt}/(r\,\partial r)$.

3. *Simply supported edge boundary*

$$M_r\{a,\theta\} = 0\,. \tag{3.17c}$$

4. *Linked edges of adjacent zones 1 and 2 – continuity*

$$z\{a,\theta\}\big|_1 = z\{a,\theta\}\big|_2\,, \qquad \left(\frac{\partial z}{\partial r}\right)_{r=a}\Big|_1 = \left(\frac{\partial z}{\partial r}\right)_{r=a}\Big|_2\,,$$

$$M_r\{a,\theta\}\,\big|_1 = M_r\{a,\theta\}\,\big|_2 \quad \text{and} \quad V_r\{a,\theta\}\,\big|_1 = V_r\{a,\theta\}\,\big|_2\,. \tag{3.17d}$$

In the next sections are presented the rigidities and associated force configurations providing the third-order aberration modes. First, for a given mode n, m, the rigidity is derived from the choice of the loading q in (3.14) which provides solutions for the coefficients α_i and the existence or not of the A_0 term. Then, the global form of the rigidity is known from (3.9) by combining all A_0 and A_i terms. The bending moment M_r and net shearing force V_r, given by (3.11a) and (3.16), define the possible associated configurations of external forces and boundaries allowing one to select compatible arrangements of A_0 and A_i terms, therefore, providing the final rigidity for each distribution.

3.3 Active Optics and Third-Order Spherical Aberration

The third-order spherical aberration, *Sphe* 3 mode, is defined by $n = 4$ and $m = 0$. This axisymmetric wavefront function is represented by

$$z_{40} = A_{40}\, r^4\,. \tag{3.18}$$

The solutions α_i to consider must be of the simplest form as possible, that is, for instance, a configuration of two loads in equilibrium, defined by a combination of two of the three following elements: an axial force at center, a uniform load onto the mirror surface, a constant force along the contour.

Considering (3.14), the two first terms in $\ln\{r\}$ and in r^0 cannot provide simultaneously $q = 0$ or $q = constant$ except if $A_0' = A_0 = 0$ and $q = 0$, so that no component in $A_0 \ln r$ is possible for the rigidity. From (3.11a), (3.16), and (3.14), the remaining part of M_r, V_r, and q are respectively

$$M_r = 4\,(3+\nu)\,A_{40} \sum_{i=1}^{j} A_i\, r^{2-\alpha_i}\,, \tag{3.19a}$$

$$V_r = Q_r = 4\,A_{40} \sum_{i=1}^{j} [\,-8 + (3+\nu)\,\alpha_i\,]\,A_i\, r^{1-\alpha_i}\,, \tag{3.19b}$$

$$q = 4\,A_{40} \sum_{i=1}^{j} [\,(3+\nu)\,\alpha_i^2 - 2\,(7+\nu)\,\alpha_i + 16\,]\,A_i\, r^{-\alpha_i}\,. \tag{3.19c}$$

The roots are

$$\begin{cases} \text{for } q = 0 \ \rightarrow \ \alpha_1 = 8/(3+\nu) \ \text{ and } \alpha_2 = 2\,, \\ \text{for } q = constant \ \rightarrow \ \alpha_3 = 0\,. \end{cases}$$

Therefore, substituting these α_i in (3.9), all the distributions of flexural rigidity are contained in the representation

$$D = A_1\, r^{-8/(3+\nu)} + A_2\, r^{-2} + A_3\,. \tag{3.20}$$

Let us write the rigidity as

$$\frac{D}{D_0} = C_1 \left(\frac{a^2}{r^2}\right)^{4/(3+\nu)} + C_2\,\frac{a^2}{r^2} + C_3\,, \tag{3.21}$$

with

$$D_0\, C_i\, a^{\alpha_i} = A_i\,, \tag{3.22}$$

where the rigidity D_0 is a constant used for normalization, and C_1, C_2, C_3 are dimensionless coefficients. Since $D = Et^3/12(1-\nu^2)$, a dimensionless thickness $\mathcal{T}_{40}$ can be similarly defined from a constant t_0, the *scaling thickness*, by

$$\mathcal{T}_{40} = \frac{t}{t_0} = \left(\frac{D}{D_0}\right)^{1/3}. \tag{3.23}$$

Substituting the roots α_i in (3.19a) and (3.19b), the bending moment and net shearing force are

$$M_r = 4(3+\nu)A_{40}\left[A_1\, r^{2-8/(3+\nu)} + A_2 + A_3\, r^2\right], \tag{3.24a}$$

$$V_r = -8A_{40}\left[(1-\nu)A_2\,\frac{1}{r} + 4A_3\, r\right]. \tag{3.24b}$$

Depending on the possible arrangements, selection and value of coefficients A_i, various configurations allow generating the *Sphe* 3 mode in the CTD and VTD classes.

3.3.1 *Configurations in the CTD Class ($A_1 = A_2 = 0$)*

Solutions in the CTD class are obtained if the coefficients $A_1 = 0$ and $A_2 = 0$: these are plates or moderately curved meniscus when deformed by a uniform load applied to their surface. With $\alpha_3 = 0$, (3.19c) provides

$$A_3 = \frac{q}{64\, A_{40}}\,. \tag{3.25}$$

At the contour, the bending moment $M_r\{a\}$ and the axial reaction V_r to the uniform load are given by (3.24a) and (3.24b). Choosing $D_0 = A_3$, we obtain the results:

→ *A plate or a meniscus provides a Sphe 3 deformation mode* $z = A_{40}\, r^4$, *if a uniform load q is applied to its surface and a bending moment and shearing force to its edge. The characterizing features are*

$$\mathcal{T}_{40} = 1\,, \tag{3.26a}$$

$$M_r\{a\} = \frac{3+\nu}{16}\, q\, a^2\,, \quad V_r\{a\} = -\frac{1}{2}\, q\, a\,, \quad t_0 = \left[\frac{3(1-\nu^2)}{16\, A_{40}}\, \frac{q}{E}\right]^{1/3}, \tag{3.26b}$$

where the product $q\, E_{40}$ *is positive.*

The uniform bending moment at the contour, $M_r\{a\}$, exactly cancels the curvature component of the flexure (see also Sect. 7.3). Examples of such configurations are displayed by Fig. 3.3.

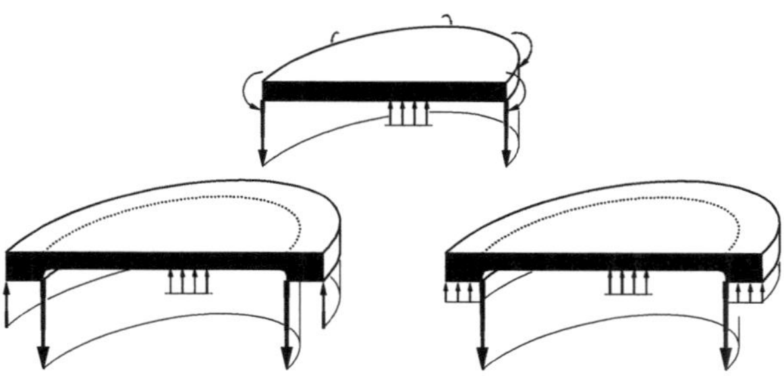

Fig. 3.3 CTD class: Mirrors providing the *Sphe* 3 mode $z_{40} = A_{40}\, r^4$. *Up*: Basic solution. *Down*: Bending moment generated by a *vase form*

3.3.2 Configurations in the VTD Class

Interesting solutions in the VTD class can be derived by assuming a null rigidity at the contour, $D\{a\} = 0$. This condition directly entails that $M_r\{a\} = 0$ which is easy to satisfy in practice and corresponds to a simply supported edge or to a free edge. From (3.21), the condition $D\{a\} = 0$ is realized if

$$C_1 + C_2 + C_3 = 0\,. \tag{3.27}$$

After substitution of C_3, these configurations are all of the form [13]

$$\frac{D}{D_0} = C_1 \left(\frac{a^2}{r^2}\right)^{4/(3+\nu)} + C_2\, \frac{a^2}{r^2} - (C_1 + C_2)\,. \tag{3.28}$$

With setting $M_r\{a\} = 0$ and depending on the A_i selection, we obtain three solutions (Lemaitre [14]).

• **VTD 1 - Uniform load and simply supported edge** ($A_2 = 0$): With $A_2 = 0$, the substitution of A_1, α_1 and A_3, α_3 in the bending moments (3.24a) provides $A_1 = -a^{8/(3+\nu)} A_3$ since $M_r\{a\} = 0$. From (3.22), this result becomes $C_3 = -C_1$ which is compatible with (3.27) only if $C_2 = 0$. Choosing $D_0 = -q/64A_{40}$, we derive the net shearing force from (3.24b) as $V_r = -qrC_1/2$. This is a uniform load in reaction to edge if $C_1 = 1$, then $A_3 = -D_0$. This case is summarized as follows:

→ *A tulip-shaped mirror provides a Sphe 3 deformation mode $z = A_{40} r^4$, if a uniform load q applied to its surface is in reaction at its edge. The characterizing features are*

$$\mathcal{T}_{40} = \left[\left(\frac{a^2}{r^2} \right)^{4/(3+\nu)} - 1 \right]^{1/3}, \tag{3.29a}$$

$$M_r\{a\} = 0, \quad V_r = -\frac{1}{2} q r, \quad t_0 = -\left[\frac{3(1-\nu^2)}{16 A_{40}} \frac{q}{E} \right]^{1/3}, \tag{3.29b}$$

where the product $q A_{40}$ is negative.

• **VTD 2 - Central force and simply supported edge** ($A_3 = 0$): With $A_3 = 0$, the substitution of A_1, α_1 and A_2, α_2 in the bending moments (3.24a) provides $A_1 = -a^{2(1-\nu)/(3+\nu)} A_2$ since $M_r\{a\} = 0$. From (3.22), this result becomes $C_1 + C_2 = 0$;
the third term of the rigidity in (3.28) vanishes. Choosing $D_0 = -q/64A_{40}$, we derive the net shearing force from (3.24b) as $V_r = -((1-\nu)qa^2C_2 \,/\, 8r$. This is obtained by central force F in reaction to the edge if $C_2 = -4/(1-\nu)$ and $F = \pi a^2 q$, then $A_2 = -4D_0a^2/(1-\nu)$. The mirror and its deforming configuration are summarized as follows:

→ *A tulip-shaped mirror provides a Sphe 3 deformation mode $z = A_{40} r^4$, if a central force F applied to its surface is in reaction at its edge. The characterizing features are*

$$\mathcal{T}_{40} = \left(\frac{4}{1-\nu} \right)^{1/3} \left[\left(\frac{a^2}{r^2} \right)^{4/(3+\nu)} - \frac{a^2}{r^2} \right]^{1/3}, \tag{3.30a}$$

$$M_r\{a\} = 0, \quad V_r = -\frac{F}{2\pi r}, \quad t_0 = -\left[\frac{3(1-\nu^2)}{16 A_{40}} \frac{q}{E} \right]^{1/3}, \tag{3.30b}$$

where $F = \pi a^2 q$ and the products $q A_{40}$ and $F A_{40}$ are negative.

• **VTD 3 - Uniform load and free edge**: The input of A_1, A_2, and A_3 in the load (3.19c) provides $A_3 = q/64A_{40}$ and after substituting the A_i in (3.24b), the force V_r is

$$V_r = -\frac{1}{2} q r - 8(1-\nu) A_{40} A_2 \frac{1}{r}.$$

If the second term on the right is considered as $qa^2/2r$ or $F/2\pi r$ with $F = \pi a^2 q$, the expression of V_r represents the net shearing forces generated by a uniform load q reacting with a central force $-F$ which corresponds to a free edge condition $V_r\{a\} = 0$; so, retaining this case, we set $A_2 = -qa^2/16(1-\nu)A_{40}$. Substituting A_2 and A_3 in (3.24a), we obtain $M_r\{a\} = 0$ if $A_1 = (3+\nu)qa^{8/(3+\nu)}/64(1-\nu)A_{40}$. Normalizing with $D_0 = q/64A_{40}$ as for the previous configurations, the C_i coefficients, derived from (3.22), are finally

$$C_1 = \frac{3+\nu}{1-\nu}, \quad C_2 = -\frac{4}{1-\nu}, \quad C_3 = 1.$$

→ *A tulip-shaped mirror provides a Sphe 3 deformation mode $z = A_{40}r^4$, if a uniform load q applied to its surface is in reaction at its center. The characterizing features are*

$$\mathcal{T}_{40} = \left[\frac{3+\nu}{1-\nu}\left(\frac{a^2}{r^2}\right)^{4/(3+\nu)} - \frac{4}{1-\nu}\,\frac{a^2}{r^2} + 1\right]^{1/3} \tag{3.31a}$$

$$M_r\{a\} = 0, \quad V_r = \frac{q}{2}\left(\frac{a^2}{r} - r\right), \quad t_0 = \left[\frac{3(1-\nu^2)}{16\,A_{40}}\,\frac{q}{E}\right]^{1/3}, \tag{3.31b}$$

where the product qA_{40} is positive.

These three configurations show a central thickness $\mathcal{T}_{40}(0) \to \infty$ which locally respects a null variation of mirror curvature at its center. For practical applications, the mirror central thickness will be given a finite value by considering an enough purity of the diffraction images, i.e. the central cut could provide a slight amount of curvature deformation but laying within the quarter-wave Rayleigh criteria or the Maréchal criteria [3(a)]. For comparisons, the normalized thickness distributions are displayed by Fig. 3.4 and associated external forces by Fig. 3.5.

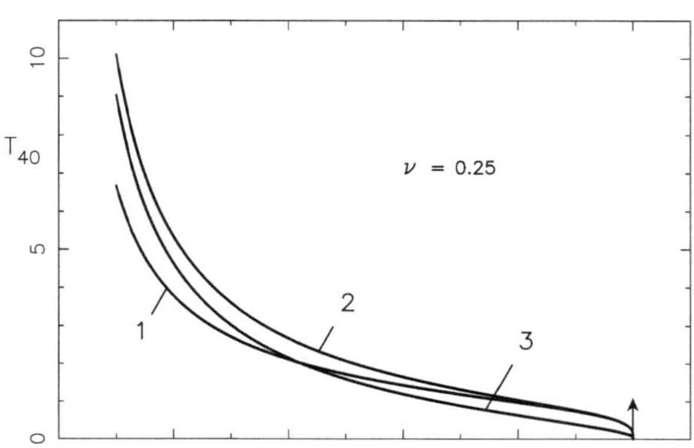

Fig. 3.4 Normalized dimensionless thicknesses $\mathcal{T}_{40}$ in the VTD class, generating a flexure $z_{40} = A_{40}r^4$ i.e. the *Sphe* 3 mode. $\rho = r/a \in [0, 1]$

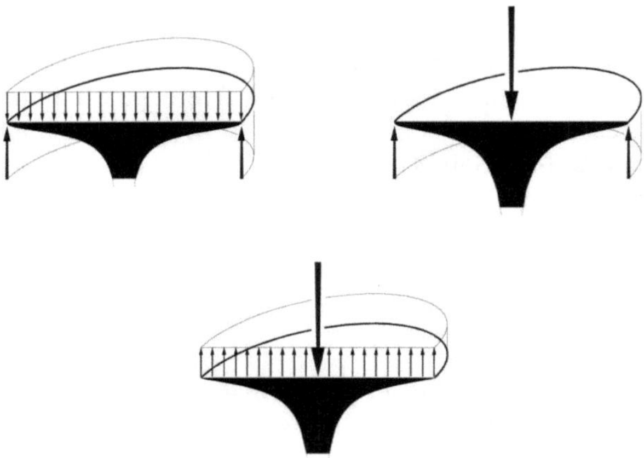

Fig. 3.5 VTD class: Cases **1**, **2**, and **3** providing *Sphe* 3 mode $z = A_{40}\, r^4$

1. Uniform loading and reaction at edge $\mathcal{T}_{40} = [\rho^{-8/(3+\nu)} - 1\,]^{1/3}$
2. Axial force at center and edge reaction $\mathcal{T}_{40} = [\,\frac{4}{1-\nu}\,\rho^{-8/(3+\nu)} - \frac{4}{1-\nu}\,\rho^{-2}\,]^{1/3}$
3. Uniform loading and reaction at center $\mathcal{T}_{40} = [\,\frac{3+\nu}{1-\nu}\,\rho^{-8/(3+\nu)} - \frac{4}{1-\nu}\,\rho^{-2} + 1\,]^{1/3}$

The radial and tangential maximum stresses must be lower than the ultimate stress σ_{ult} of the material. The maximum stresses are represented by

$$\sigma_r = \frac{6\,M_r}{t^2} < \sigma_{\text{ult}} \quad \text{and} \quad \sigma_t = \frac{6\,M_t}{t^2} < \sigma_{\text{ult}}\,. \tag{3.32}$$

Considering the case of *Sphe* 3, the bending moment M_r and M_t are derived from (3.1a) to (3.1b). For the radial bending moment, we obtain

$$M_r = 4(3+\nu)A_{40}\; r^2 D\,.$$

With the three previous rigidities and after calculation and simplifications, the radial maximum stress is

$$\sigma_r = \frac{3(3+\nu)\,a^2 q}{8\,t_0^2}\;\mathcal{S}_r \quad \text{with} \quad \mathcal{S}_r = \frac{r^2}{a^2}\,\mathcal{T}_{40}\,, \tag{3.33}$$

where $\mathcal{S}_r$ is a dimensionless maximum stress (Fig. 3.6).

For identical dimensions, loads, and flexures, the configuration VTD 3 provides the smallest radial maximum stress. For these configurations the tangential maximum stress $\mathcal{S}_t$ is smaller than $\mathcal{S}_r$.

3.3.3 Hybrid Configurations

With the CTD configuration such as displayed by Fig. 3.3, the *Sphe* 3 mode is purely achieved up to $r \leq a$, i.e. the optical clear aperture of the mirror is limited to the inner plate or meniscus which is clamped into the ring. The ring allows one to

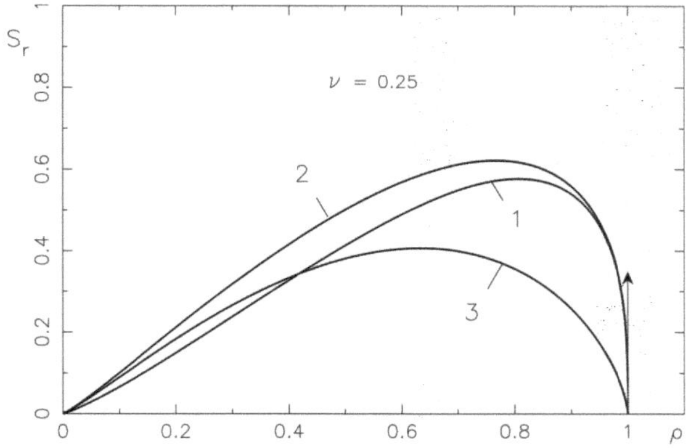

Fig. 3.6 VTD class: Dimensionless maximum stresses $\mathcal{S}_r = \rho^2 \mathcal{T}_{40}$ in configurations **1**, **2**, and **3** when generating *Sphe* 3 flexure mode $z_{40} = A_{40}\, r^4$. Dimensionless radius $\rho = r/a \in [\,0, 1\,]$

generate the bending moment $M_r\{a\}$ by applying an external pair of ring forces, but its flexure is not purely of the fourth degree. However, from the latter results, we can obtain also an exact *Sphe* 3 flexure of the zone outer to $r = a$.

Let us consider a configuration composed of two concentric zones. The first zone is a CTD such as defined by Eq. (3.26) and applying for $0 \le r \le a$; the second zone extending on $a \le r \le b$ is a VTD 3 defined by Eq. (3.31) where a is replaced by b. After this substitution, the bending moment of the outer zone is represented by

$$M_r = \frac{3+\nu}{16}\, q \left[\frac{3+\nu}{1-\nu}\, r^2 \left(\frac{b^2}{r^2} \right)^{4/(3+\nu)} - \frac{4}{1-\nu}\, b^2 + r^2 \right],$$

which satisfies $M_r\{b\} = 0$.

If the clamping of the inner VTD 3 with the outer CTD is done at $r = a$ with the same thicknesses, we then set the same uniform loadings q in intensity and sign. These two conditions are written by

$$\mathcal{T}_{40}(a)\big|_{\mathrm{VTD\,3}} = \mathcal{T}_{40}(a)\big|_{\mathrm{CTD}} \quad \text{and} \quad q\,\big|_{\mathrm{VTD\,3}} = q\,\big|_{\mathrm{CTD}}\,, \tag{3.34}$$

the first equality providing the radial location of the link between the two distributions as, from (3.26a) to (3.31a),

$$\frac{3+\nu}{1-\nu} \left(\frac{b^2}{a^2} \right)^{4/(3+\nu)} - \frac{4}{1-\nu}\, \frac{b^2}{a^2} + 1 = 1\,.$$

Therefore, the radius a of the rigidity junction, where is also acting in the ring-force reaction, is

$$\frac{a^2}{b^2} = \left(\frac{3+\nu}{4} \right)^{\frac{3+\nu}{1-\nu}} \tag{3.35}$$

After substitution of this ratio in the bending moment of the VTD 3, we obtain $M_r\{a\} = (3+\nu) q a^2/16$ which is identical to $M_r\{a\}$ of the CTD [see (3.26b)], so that the condition (3.35) is satisfied.

→ *A mirror with two concentric zones provides a Sphe 3 deformation mode* $z = A_{40}\, r^4$ *on all its surface* $r \in [\,0, b\,]$, *if a uniform load* q *applied to all its surface is in reaction with a ring force at the rigidity junction* $r = a$. *The thickness* $t = \mathcal{T}_{40}\, t_0$ *and the geometric parameters are represented by*

$$\begin{aligned} \mathcal{T}_{40} &= 1\,, && 0 \le r \le a\,,\\ \mathcal{T}_{40} &= \left[\frac{3+\nu}{1-\nu}\left(\frac{b^2}{r^2}\right)^{4/(3+\nu)} - \frac{4}{1-\nu}\,\frac{b^2}{r^2} + 1\right]^{1/3}, && a \le r \le b\,, \end{aligned} \tag{3.36}$$
$$\text{with} \quad \frac{a^2}{b^2} = \left(\frac{3+\nu}{4}\right)^{\frac{3+\nu}{1-\nu}}, \qquad t_0 = \left[\frac{3(1-\nu^2)}{16\,A_{40}}\,\frac{q}{E}\right]^{1/3},$$

where the product $A_{40}q$ *is positive.*

The junction ratio a/b is varying slowly with the Poisson's ratio. For $\nu \in [\,0,\ 1/2\,]$, we obtain $a/b \in [\,0.6495,\ 0.6266\,]$, respectively. This hybrid configuration is represented in Fig. 3.7-*Left* for a Schott Zerodur vitroceram where $\nu = 1/4 \;\to\; a/b = 0.6377$. We can derive other hybrid configurations by modifying the outer rigidity and its associated load. This is achieved if the ratios t_0^3/q are identical for the two zones (Fig. 3.7-*Right*).

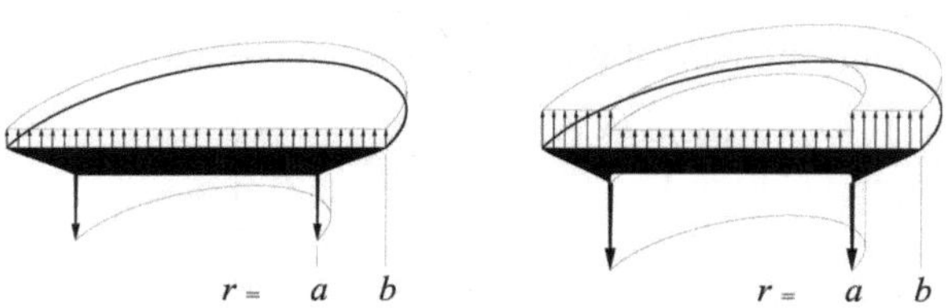

Fig. 3.7 Hybrid CTD-VTD configurations providing the *Sphe* 3 mode $z = A_{40}\, r^4$ by uniform loads q applied all over the clear aperture diameter $2b$. (*Left*) Same load q for the two zones. (*Right*) q (outer zone) $>$ q (inner zone)

NB: Hybrid configurations using the CTD and VTD classes can be similarly derived for the $Cv\,1$ mode. In this case, the uniform load applied to the central zone is suppressed for $0 \le r \le a$.

3.3.4 Balance with a Curvature Mode

The best focussing of a wavefront containing a spherical aberration, i.e. the *least confusion image*, is located between its paraxial (Gauss) and marginal focii. Let $Z(\rho)$ be a wavefront of biquadratic form whose dimensionless representation is $\zeta = Z/R$ with respect to its radius of curvature R. The sum of the two modes $Cv\,1$ and *Sphe* 3 may be written as

$$\zeta = 2\,\rho_0^2\,\rho^2 \; - \; \rho^4\,, \quad \text{with} \quad \rho \in [0,\ 1\,]\,,$$

where $\rho = 1$ and ρ_0 are the clear aperture radius and the null powered radius, respectively.

The algebraic balance of the first derivative extremals is provided by $\rho_0 = \sqrt{3}/2$, and define the location of the *least confusion focus*. This is corresponding to slopes $d\zeta/d\rho = \pm 1$ at $\rho = 1/2$ or 1.

The algebraic balance of the second derivative extremals is provided by $\rho_0 = \sqrt{3/2}$, and is used as mirror best shape for the *field balance of aberrations* in all-reflective systems (see Chaps. 4 and 5). This corresponds to curvatures $d^2\zeta/d\rho^2 = \pm\,6$ at $\rho = 0$ or 1.

These particular shapes are displayed by level fringes in Fig. 3.8.

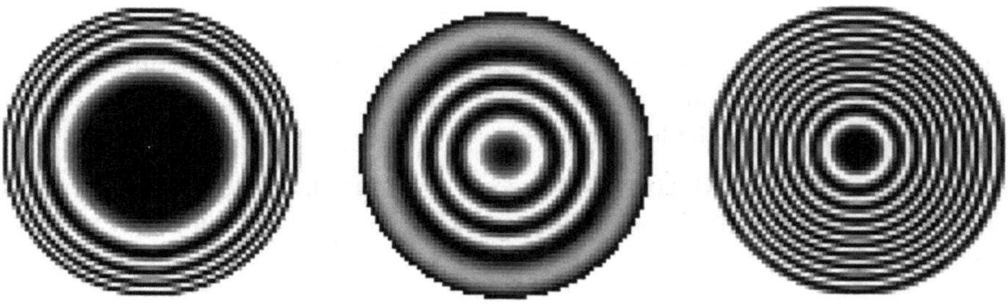

Fig. 3.8 Co-addition of *Sphe* 3 with *Cv* 1 modes: $\zeta = 2\,\rho_0^2\,\rho^2 \; - \; \rho^4$. (*Left to right*) $\rho_0 = 0 \leftrightarrow$ flat center ; $\rho_0 = \sqrt{3}/2 \leftrightarrow$ slope balance ; $\rho_0 = \sqrt{3/2} \leftrightarrow$ curvature balance

Somewhat similarly, the elastic deformation of a plate in basic boundary cases, such as clamped edge or simply supported edge, naturally provides a flexure combining the two modes z_{20} and z_{40}.

Chapters 5, 6, and 7 include the study of several configurations providing combined flexure z_{20} and z_{40} modes by using the CTD class, quasi-CTD class or the hybrid rigidity class. We will see in these chapters that active optics configurations can be derived in an easier way than for the case where the *Sphe* 3 mode, z_{40}, is pure ($z_{40} \propto r^4$) as treated herafter.

3.3.5 Examples of Application

• **Elasto-optical design parameters**: Using the Schwarzschild notation, the representation of the quadric surface of an axisymmetric mirror is

$$z = \frac{1}{2R}\,r^2 + \frac{1+\kappa}{8R^3}\,r^4 + \frac{(1+\kappa)^2}{16R^5}\,r^6 + \cdots\,,$$

where κ is the conic constant. Limiting to the development of the two first terms of this series, the deviation to the sphere of curvature $1/R$, i.e. the aspheric term, is the part $\kappa\, r^4/8R^3$ of the second right-hand term. The asphericity to realize by flexure is $z = A_{40}\, r^4$; then, by identification,

$$A_{40} = \frac{\kappa}{8\,R^3} = \frac{\kappa}{512\,a^3\,\Omega^3} \quad \text{with} \quad \Omega = \frac{\mathrm{f}}{\mathrm{d}} = \left|\frac{R}{4a}\right|, \tag{3.37}$$

where Ω is the mirror f-ratio and $2a$ its clear aperture.

Considering the *scaling* of the thickness t_0 given by (3.26b) for a CTD, which is also the same for the three previous VTD classes (3.29b),(3.30b), and (3.31b), as for the hybrid class (3.36), the substitution of A_{40}, leads in all cases to the aspect ratio

$$\frac{t_0}{a} = \left[\frac{3\,(1-\nu^2)}{2\,\kappa}\,\frac{q}{E}\right]^{1/3} 4\,\Omega\,. \tag{3.38}$$

This ratio provides the set up of the execution conditions: The scaling thickness (t_0) of the mirror is fully defined from the mirror semi-aperture (a), the optics (Ω, κ), the material (ν, E), and the load (q).

• **Metal mirror designed with VTD 2 configuration**: The design and construction of a mirror aspherized by a central force in reaction at its simply supported edge has been carried out. Because of the edge support condition without bending moment, a metal mirror has been preferred. The design of the *tulip-shaped* mirror includes, in one piece, the thickness profile simply supported to an outer ring via a cylindric collar of thin radial thickness [16, 17].

A back support linked to the ring allows one to generate a central force $F \equiv \pi a^2 q$. A view of the mirror, the resulting interferogram of the *Sphe* 3 deformation obtained in-situ, and the design parameters derived from (3.38) are displayed by Fig. 3.9.

• **Vitroceram mirror designed with VTD 3 configuration**: The design and construction of a convex mirror hyperbolized by a uniform load in reaction at its center

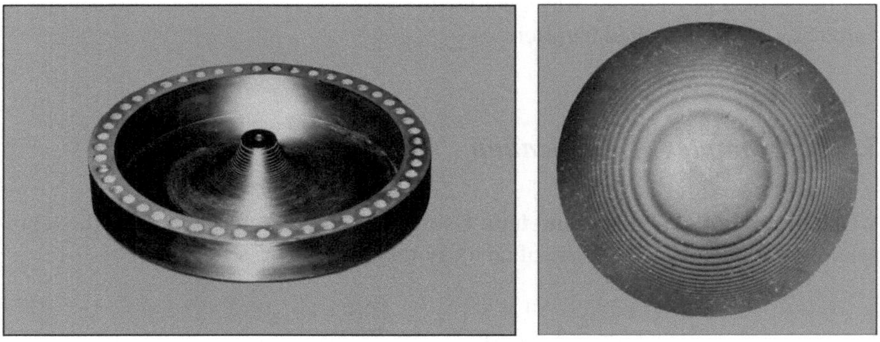

Fig. 3.9 (*Left*) View of a metal mirror providing the flexure $z = A_{40}\, r^4$. (*Right*) He-Ne interferogram of *Sphe* 3 mode obtained by a central force. Design parameters: Stainless steel alloy Fe87 Cr13-quenched, $a = 90$ mm, $\kappa = -1$ (paraboloid), $\Omega = 3/2$, $\nu = 0.305$, $E/q = -2.5\,10^6$, $a/t_0 = 20.41$

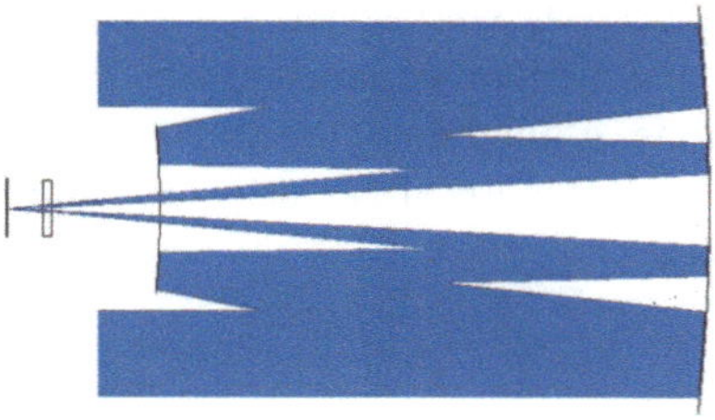

Fig. 3.10 (*Left*) MINITRUST flat-field three-reflection anastigmat telescope. (*Right*) View of the holed secondary mirror hyperbolized by stress polishing (LOOM)

has been carried out. This mirror is the secondary of the flat field three reflection telescope MINITRUST [23].

This modified-Rumsey optical system (Fig. 3.10) is a very compact anastigmat - four-times shorter than a Schmidt system - and should provide new capabilities with large format detectors in wide field astronomy for ground-based observations as well as for sky-surveys in space. During the stress polishing, the free edge condition - edge without bending moment or axial force - has been ensured by a soft paste non-soluble into water. Applied between the mirror edge and its surrounding removable ring, the paste is sucked into the gap by the partial vacuum, thus ensuring airtightness. The central hole, required for this mirror, provides support of convenient lateral stability for the stress polishing. At the telescope, other advantages of a mirror supported at its central hole are a substantial gain in weight, and a low deformation to own weight in the gravity field [15].

Because of the relatively large diameter of the central hole, the inner built-in condition is not exactly satisfied; this gives rise to a slight rotation of the inner ring, thus generating a small curvature $Cv\,1$ added to $Sphe\,3$. In addition, a $Sphe\,5$ mode necessary for this application was simultaneously obtained from a slight change of the VTD towards the edge. So, the determination of the real VTD has been carried

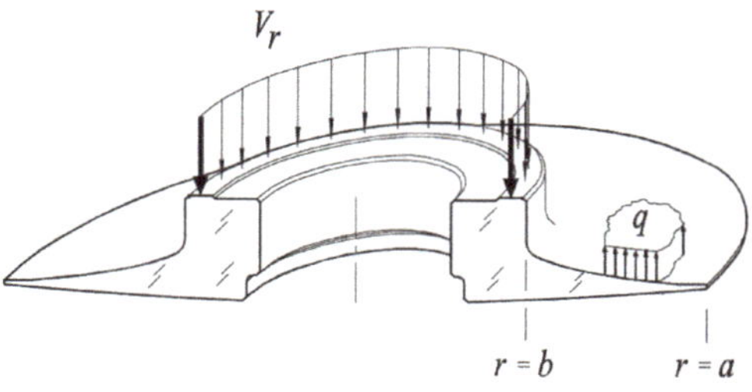

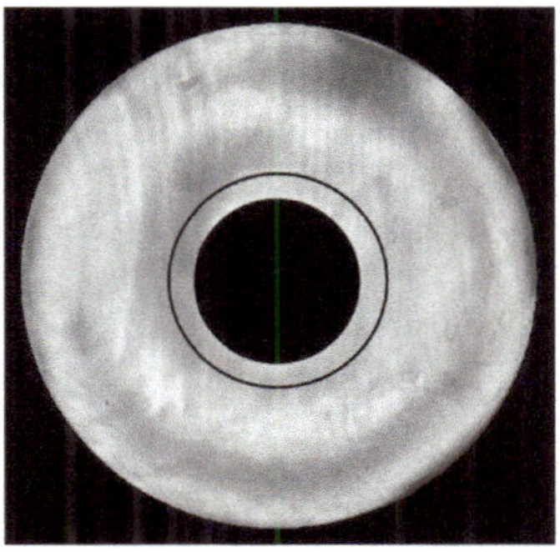

Fig. 3.11 (*Left*) *Tulip-form* secondary mirror and stress polishing loads. (*Right*) He-Ne interferogram with respect to a sphere when under stressing. Design parameters: Zerodur-special vitroceram from Schott, $a = 103$ mm, $\kappa = -3.917$ (hyperboloid), $\Omega = -8/3$, $\nu = 0.240$, $E/q = 8.89\ 10^5$, $a/t_0 = 18.31$ (MINITRUST [23]) (LOOM)

out by analytical integration with the three modes *Cv* 1, *Sphe* 3, and *Sphe* 5. However, the effects of *Cv* 1 and *Sphe* 5 are small, and the final parameters are close to those directly derived from (3.38) for *Sphe* 3 alone. A diagram of the mirror with its applied loads, the resulting interferogram of the deformation after elastic relaxation, and the design parameters are displayed by Fig. 3.11.

3.4 Active Optics and Third-Order Coma

The third-order coma, *Coma* 3 mode, is defined by $n = 3$ and $m = 1$. This asymmetric wavefront function, or mirror shape, is represented by

$$z_{31} = A_{31}\, r^3 \cos\theta\,. \tag{3.39}$$

The solutions α_i of this mode are derived from (3.14) by stating that the load q is null. For the *Coma* 3 mode, the substitution of n, m leads to zero for the coefficient of the first term in $r^{-1} \ln r \cos\theta$ whatever A'_0 is, while the coefficient in $r^{-1} \cos\theta$ of the second term is $-4(3+\nu)A'_0$ which requires $A'_0 = 0$ only if $q = 0$. Therefore, after simplification, the *Coma* 3 configurations are all generated by the third and remaining term of (3.11a), (3.16), and (3.14). These terms are

$$M_r = 2(3+\nu)A_{31} \sum_{i=1}^{j} A_i\, r^{1-\alpha_i} \cos\theta\,, \tag{3.40a}$$

$$V_r = -\,2A_{31} \sum_{i=1}^{j} [\,5-\nu - (3+\nu)\,\alpha_i\,]\, A_i\, r^{-\alpha_i} \cos\theta\,, \tag{3.40b}$$

$$q = 2(3+\nu)A_{31} \sum_{i=1}^{j} (\alpha_i - 2)\,\alpha_i\; A_i\, r^{-1-\alpha_i} \cos\theta\,. \tag{3.40c}$$

Since a 2-D prismatic load of the form $q = q_0 r \cos\theta$ is to be rejected as extremely difficult to achieve practically, the roots to retain are those corresponding to a null uniform load

$$q = 0 \;\rightarrow\quad \alpha_1 = 2 \;\text{ and }\; \alpha_2 = 0\,.$$

Substituting the α_i in (3.9), the distributions of flexural rigidity are contained in the representation

$$D = A_1\, r^{-2} + \,A_2\,. \tag{3.41}$$

With the notation of (3.22) and (3.23), let us write the rigidity and a dimensionless thickness as

$$\frac{D}{D_0} = C_1\,\frac{a^2}{r^2} + \,C_2\,, \qquad \mathcal{T}_{31} = \left(\frac{D}{D_0}\right)^{1/3}\,. \tag{3.42}$$

From (3.40a) and (3.40b), the bending moment and net shearing force are

$$M_r = 2(3+\nu)A_{31}\left[\frac{A_1}{r} + A_2\, r\right]\cos\theta\,, \tag{3.43a}$$

$$V_r = 2A_{31}\left[(1+3\nu)\frac{A_1}{r^2} - (5-\nu)A_2\right]\cos\theta\,. \tag{3.43b}$$

Given the form (3.41) of the rigidity, two configurations can be derived such as described hereafter.

3.4.1 *Configuration in the CTD Class* ($A_1 = 0$)

A solution in the CTD class is obtained if the coefficient $A_1 = 0$: this is a plate or a moderately curved meniscus when deformed both by a bending moment and a net shearing force applied to its contour. Let us define the amplitude V_0 of the net shearing force acting per unit length where the mirror perimeter $r = a$, that is, a prismatic ring-force

$$V_r\{a\} = V_0\cos\theta\,.$$

From (3.43b), we deduce the amplitude of this force as

$$V_0 = -2(5-\nu)\, A_{31}\, A_2\,.$$

By setting $A_2 = 4D_0/(5-\nu)$, and since, with (3.22), $D_0 C_i a^{\alpha_i} = A_i$, the C_i and dimensionless thickness are

$$C_1 = 0\,,\qquad C_2 = \frac{4}{5-\nu}\,,\qquad \frac{D}{D_0} = \frac{4}{5-\nu}\,.$$

Similarly, the bending moment $M_r\{a\}$ is derived from (3.43a). The parameters of this configuration are summarized as follows [19]:

→ *A plate or a slightly curved meniscus generates a* Coma 3 *deformation mode* $z = A_{31}r^3\cos\theta$, *if its thickness* $t = \mathcal{T}_{31}\, t_0$ *is a constant, and if a prismatic bending moment* M_r *and a prismatic ring-force* $V_r = V_0\cos\theta$ *are applied to its perimeter* $r = a$*:*

$$\mathcal{T}_{31} = \left(\frac{4}{5-\nu}\right)^{1/3}, \tag{3.44a}$$

$$M_r\{a,\theta\} = -\frac{3+\nu}{5-\nu}\, a\, V_r\,,\qquad V_0 = -8A_{31}D_0\,,\qquad t_0 = -\left[\frac{3(1-\nu^2)}{2\,A_{31}}\,\frac{V_0}{E}\right]^{1/3}, \tag{3.44b}$$

where the product $V_0 A_{31}$ *is negative.*

This configuration is shown in Fig. 3.12-*Left*.

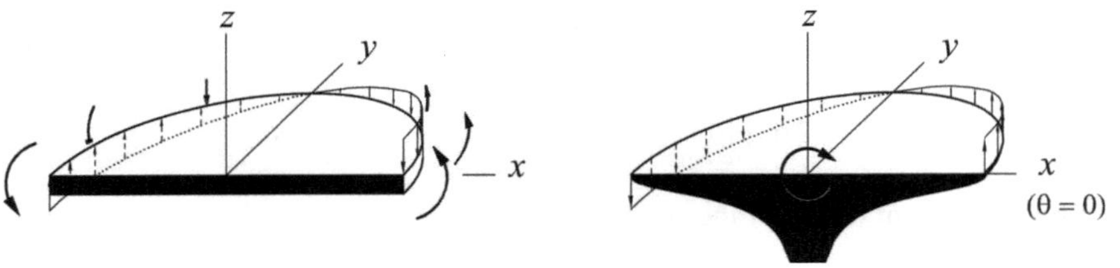

Fig. 3.12 Mirrors providing *Coma* 3 mode $z_{31} = A_{31} r^3 \cos\theta$. (*Left*) Solution in CTD class. (*Right*) Solution in VTD class

3.4.2 Configuration in the VTD Class

A solution in the VTD class is obtained for the case of a null rigidity at the edge, $D\{a\} = 0$. This is provided in (3.42) if $C_2 = -C_1$. Setting the C_1 coefficient to $2/(3+\nu)$, we obtain

$$C_1 = -C_2 = \frac{2}{3+\nu}, \qquad \frac{D}{D_0} = \frac{2}{3+\nu}\left(\frac{a^2}{r^2} - 1\right),$$

and from (3.22), $A_1 = 2D_0 a^2/(3+\nu)$ and $A_2 = -2D_0/(3+\nu)$. After substitution in Eq. (3.43), the bending moment is

$$M_r = 4\, A_{31} D_0 \left(\frac{a^2}{r} - r\right) \cos\theta\,,$$

and the net shearing force is

$$V_r = 4 A_{31} D_0 \left(\frac{1+3\nu}{3+\nu}\,\frac{a^2}{r^2} + \frac{5-\nu}{3+\nu}\right) \cos\theta\,.$$

Considering M_r and V_r at the edge $r = a$, and near the center for $r = b$ small, we have

$$M_r\{a,\,\theta\} = 0\,, \qquad M_r\{b,\,\theta\} \propto A_{31} D_0\, a^2 \cos\theta\,/b\,,$$
$$V_r\{a,\,\theta\} = 8 A_{31} D_0 \cos\theta\,, \qquad V_r\{b\,,\theta\} \propto A_{31} D_0\, a^2 \cos\theta\,/b^2\,,$$

which corresponds to a prismatic external ring-force $V_r = V_0 \cos\theta$ applied per unit length to a simply supported edge. This configuration is summarized as follows (Lemaitre [14]):

→ *A tulip-shaped mirror provides a Coma 3 deformation mode* $z = A_{31} r^3 \cos\theta$, *if a prismatic ring-force* $V_r = V_0 \cos\theta$ *applied along its simply supported edge is in reaction to a moment applied near the center. Its thickness* $t = \mathcal{T}_{31}\, t_0$ *and characterizing features are*

$$\mathcal{T}_{31} = \left[\frac{2}{3+\nu} \left(\frac{a^2}{r^2} - 1 \right) \right]^{1/3}, \tag{3.45a}$$

$$M_r\{a,\theta\} = 0\,, \qquad V_0 = 8A_{31}D_0\,, \qquad t_0 = \left[\frac{3(1-\nu^2)}{2\,A_{31}} \frac{V_0}{E} \right]^{1/3}, \tag{3.45b}$$

where the product $V_0 A_{31}$ *is positive.*

The configuration is displayed by Fig. 3.12-*Right*, and dimensionless distribution $\mathcal{T}_{31}(\rho)$ by Fig. 3.13.

The central part of the mirror can be considered as infinitely rigid, so that the only reaction to the prismatic ring-force $V_r\{a\}$ can be generated by a central moment $\mathcal{M}_y$ around the y-axis determined by

$$\mathcal{M}_y = -4 \int_0^{\pi/2} a \cos\theta \; V_r\{a,\,\theta\}\; a\, d\theta = -\pi\, V_0\, a^2$$

For practical reasons, the reacting moment can be applied at a small distance from the mirror center where $b \ll a$.

The radial and tangential maximum stress σ_r and σ_t must be lower than the ultimate stress $\sigma_{\rm ult}$ of the material. Substituting the *Coma* 3 flexure mode in (3.1a), we obtain the radial bending moment

$$M_r = 2(3+\nu)A_{31}D\, r\cos\theta\,.$$

The radial maximum stress $\sigma_r = 6M_r/t^2$, derived with the scaling thickness t_0 in (3.45b) and the definition of the rigidity, is

$$\sigma_r = \frac{3(3+\nu)\,a\,V_0}{2\,t_0^2}\,\mathcal{S}_r \qquad \text{with} \qquad \mathcal{S}_r = \frac{r}{a}\,\mathcal{T}_{31}\cos\theta\,,$$

where S_r is a dimensionless maximum stress (Fig. 3.13).

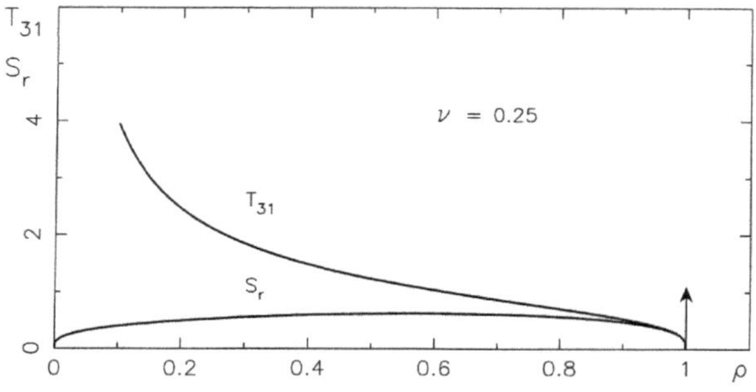

Fig. 3.13 VTD class: *Coma* 3 flexure mode $z_{31} = A_{31}r^3\cos\theta$. Dimensionless thickness $\mathcal{T}_{31} = \left[\frac{2}{3+\nu}\left(\rho^{-2} - 1\right)\right]^{1/3}$ with $\rho = r/a$. Dimensionless maximum stress $\mathcal{S}_r = \rho\,\mathcal{T}_{31}\cos\theta$ plotted for $\theta = 0$

3.4.3 Hybrid Configurations

Hybrid configurations use a combination of the rigidity classes and can provide a *Coma* 3 mode flexure over the full diameter of a mirror. A simple combination is achieved with a CTD class for a central zone $0 \le r \le a$, and a VTD class for the outer zone $a \le r \le b$. Since the dimensionless thicknesses have been normalized in the two rigidity cases (3.44) and (3.45) with identical amplitude flexures A_{31} and scaling thicknesses t_0, we can directly link the two distributions at their thickness intersection by writing

$$\mathcal{T}_{31}(a)\,|_{CTD} = \mathcal{T}_{31}(a)\,|_{VTD}\,.$$

→ *A mirror with two concentric zones provides a* *Coma* 3 *deformation mode* $z = A_{31}\, r^3 \cos\theta$ *on all its surface* $r \in [\,0, b\,]$, *if a prismatic ring-force* $V_r = V_0 \cos\theta$ *is applied to its edge* $r = b$ *in reaction with a ring-force at its rigidity junction* $r = a$. *The thickness* $t = \mathcal{T}_{31}\, t_0$ *and the geometric parameters are represented by*

$$\mathcal{T}_{31} = \left(\frac{4}{5-\nu}\right)^{1/3}, \qquad 0 \le r \le a\,, \tag{3.46a}$$

$$\mathcal{T}_{31} = \left[\frac{2}{3+\nu}\left(\frac{b^2}{r^2} - 1\right)\right]^{1/3}, \qquad a \le r \le b\,, \tag{3.46b}$$

$$\textit{with}\quad \frac{a^2}{b^2} = \frac{5-\nu}{11+\nu}\,, \qquad V_0 = 8A_{31}D_0\,, \qquad t_0 = \left[\frac{3(1-\nu^2)}{2\,A_{31}}\,\frac{V_0}{E}\right]^{1/3} \tag{3.46c}$$

where the product $V_0 A_{31}$ *is positive.*

For $\nu \in [\,0,\ 1/2\,]$, we obtain $a/b \in [\,0.6742,\ 0.6255\,]$, respectively. This hybrid configuration is displayed in Fig. 3.13 for a stainless steel alloy Fe87 Cr13 where $\nu = 0.305 \ \rightarrow\ a/b = 0.6444$. We can derive other hybrid configurations by modifying the outer rigidity and the intensity of the prismatic ring forces. This is achieved with the constancy of the ratio t_0^3/V_0. Two advantages of these configurations are a *natural* setting of the reacting moment at the link $r = a$ and a *full-aperture* solution for *Coma* 3 (Fig. 3.14).

3.4.4 Balance with a Tilt Mode

Geometrical properties of a coma wavefront: One defines a coma mode of order n by the form $z_{n1} = A_{n1}\, r^n \cos\theta$, with $n = 3, 5, 7, \ldots$ Considering tilted planes containing a y-axis, i.e. of the form $z_{11} = A_{11}\, r \cos\theta$, the intersection of the two surfaces are ellipses belonging to the circular cylinder $r = (A_{11}/A_{n1})^{1/(n-1)}$.

If we consider now balanced wavefronts resulting from the co-addition of a *Tilt* 1 mode and *Coma* n mode, i.e. of the form $Z = A_{11} r \cos\theta - A_{n1} r^n \cos\theta$, the sections in the $z = 0$ plane are *circles* $\forall\, n$ having the radius of the previous cylinder.

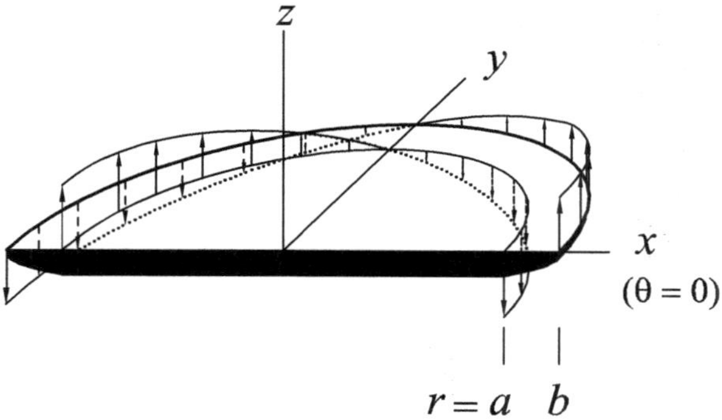

Fig. 3.14 Hybrid CTD-VTD configurations providing *Coma* 3 mode $z = A_{31}\, r^3 \cos\theta$ by two prismatic ring forces applied at $r = a$ and $r = b$

Since the co-addition of a *Tilt* 1 mode does not modify the coma image and neither the bending moment and net shearing force generating any elastic mode, this entails the following results:

→ *The boundaries of a Coma* 3 *distorted mirror can be achieved by a rigid ring linked to it via cylindric collars of thin radial thickness:*

• VTD class: *a clamped center reacting with a perimeter single prismatic ring-force, applied via a thin collar, generate only* $V_r\{a, \theta\}$, *thus providing a simply supported edge.*
• CTD and hybrid classes: *a pair of opposite prismatic ring-force, applied at* $r = a$ *and* $r = a' > a$ *via thin collars, generate* $M_r\{a, \theta\}$ *and* $V_r\{a, \theta\}$, *thus providing to a linked edge at* $r = a$ [see (3.17d)].

These properties simplify the realization of boundary conditions in practical applications. Level fringes of the co-addition of *Coma* 3 mode with varying *Tilt* 1 mode are displayed by Fig. 3.15.

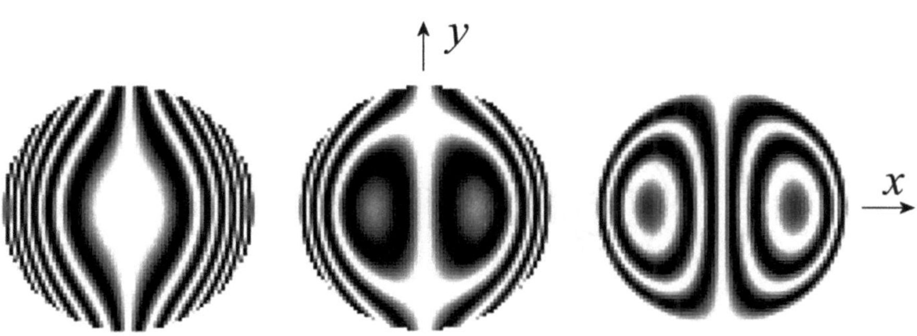

Fig. 3.15 Co-addition of *Coma* 3 and *Tilt* 1 modes: $\zeta = (3\rho_0^2\,\rho - \rho^3)\cos\theta$. (*Left to right*) $\rho_0 = 0 \leftrightarrow$ flat center. $\rho_0 = 1/\sqrt{6}$ and $1/\sqrt{3} \leftrightarrow$ circles at $1/\sqrt{2}$ and 1 respectively

Note: In the representation of optical modes radial Zernike polynomials [3(b), 6, 32(a)], a given order polynomial $Z_n^m(\rho)$ is built by mode co-addition of the same subfamily up to n order, and coefficients are normalized such as $Z_n^m(0)=Z_n^m(1)=1$. For instance, since $Z_1^1 = \rho$ and $Z_3^1 = 3\rho^3 - 2\rho$, the Zernike representation of *Coma* 3 mode with dimensionless radius is $z_{31}=A_{31}\rho^3\cos\theta=\frac{1}{3}A_{31}\,(Z_3^1+2Z_1^1)$ $\cos\theta$.

3.4.5 Coma from a Pupil and Concave Mirror System

Let us consider an object at infinity providing an incident and parallel beam passing through a pupil and reflected by a concave mirror of curvature $1/R$ and conic constant κ. Denote s the axial separation between the pupil and the mirror, i the incident angle of the principal ray with respect to the mirror axis, $a_{\rm Tel}$ the semi-aperture radius of the telescope mirror and $\Omega = {\rm f/d} = R/4a_{\rm Tel}$ the aperture number. If the incident principal ray belongs to the (x,z) plane, the wavefront representation of *Coma* 3, derived from the Wilson equations [32(b)] of Seidel sums, is

$$z_{31} = \frac{i}{16\,\Omega^2 a_{\rm Tel}^2}\left[1 - \frac{s}{R}(1+\kappa)\right] r^3\cos\theta\,. \tag{3.47}$$

This relation contains some fundamental properties of a pupil and mirror system, in particular:

→ *A spherical mirror* $\kappa = 0$ *provides a coma-free system if the pupil is located at its center of curvature* $s = R$ (Schmidt system).

→ *If the pupil is on the mirror* ($s = 0$), *the size of Coma* 3 *is the same whatever the conic constant is.*

At the focal plane of a concave mirror and in this latter case ($s = 0$), the linear size of the *Coma* 3 aberration is $\ell_x = 3iR/32\Omega^2$ and $\ell_y = 2iR/32\Omega^2$ (Fig. 3.16), and the coma spike is directed towards the mirror axis.

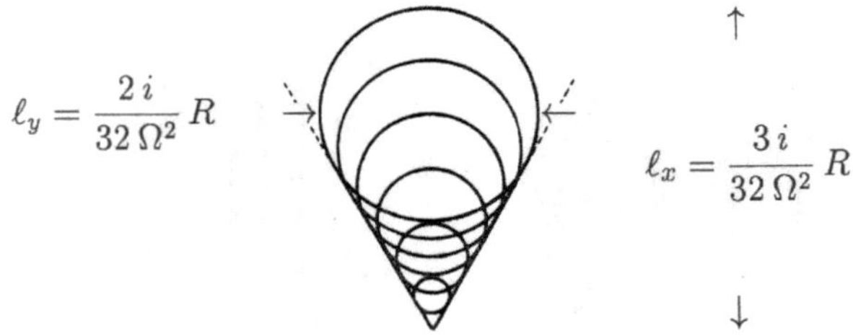

Fig. 3.16 Linear size of the *Coma* 3 image from a concave and axisymmetric pupil mirror; object at infinity, $s = 0$ and $\forall\,\kappa$

3.4.6 Examples of Active Optics Coma Correction

• **Local correction of** *Coma* 3 **by "acomatic mirror" – Star tracking systems**: Considering that the pupil is on the concave mirror, i.e. $s = 0$ in (3.47), which is the case of a telescope primary mirror, star tracking or guiding imager systems are often used in an off-axis area of the focal surface. These systems generally use an optical transport re-imaging the selected sky area on a detector. In the case of non-aplanatic telescopes, the improvement of image quality can be achieved by implementing an "*acomatic mirror*" into the re-imaging optical transport. For an incidence angle i, the shape of the equivalent telescope mirror that would compensate for this aberration is, from (3.47),

$$z_{31} = -\frac{i}{32\,\Omega^2 a_{\mathrm{Tel}}^2}\, r^3 \cos\theta\,. \tag{3.48}$$

Stating that the sum of the shapes expressed by (3.48) and by the acomatic mirror (3.39), must be zero, i.e.

$$z_{31}\{a_{\mathrm{Tel}},\,\theta\}\Big|_{\mathrm{Tel.}} + z_{31}\{a,\,\theta\}\Big|_{\mathrm{A.M.}} = 0\,,$$

we obtain the A_{31} coefficient of the acomatic mirror,

$$A_{31} = \frac{i}{32\,\Omega^2}\,\frac{a_{\mathrm{Tel}}}{a^3}\,, \tag{3.49}$$

which compensates for *Coma* 3 of the telescope mirror at i angle.

Substituting this coefficient in the representation of the scaling thickness t_0 given by (3.44b), (3.45b), or (3.46b), depending on the selected rigidity class - CTD, VTD, or hybrid -, we obtain the aspect ratio

$$\frac{t_0}{a} = 4\left[\frac{3(1-\nu^2)\Omega^2}{4\,i}\,\frac{V_0}{a_{\mathrm{Tel}}E}\right]^{1/3}. \tag{3.50a}$$

This aspect ratio provides the setting of the execution conditions of the acomatic mirror involving the scaling thickness (t_0), the semi-aperture (a), the telescope optics ($a_{\mathrm{Tel}}, \Omega, i$), the elastic constants (ν, E), and the ring-force intensity (V_0) per unit length.

From the geometrical properties of *Coma* 3 wavefront (Sect. 3.4.4), the boundary conditions can be realized by axially thick cylinders linked via thin radial collars to the mirror substrate. This mirror-cylinder link allows one to generate a prismatic ring-force $V_r = V_0 \cos\theta$ by only applying two opposite point-force $\mathcal{F} = \pm a \int V_r\, d\theta = \pm 2a V_0$ at $\theta = 0$ and $\theta = \pi$ on the cylinder rear side. Thus, the cosine modulation is naturally achieved by the cylinder axial thickness. In function of the point-force $\mathcal{F}$, the design relation becomes

$$\frac{t_0}{2a} = \left[\frac{3(1-\nu^2)\Omega^2}{i}\,\frac{\mathcal{F}}{a\,a_{\mathrm{Tel}}E}\right]^{1/3}. \tag{3.50b}$$

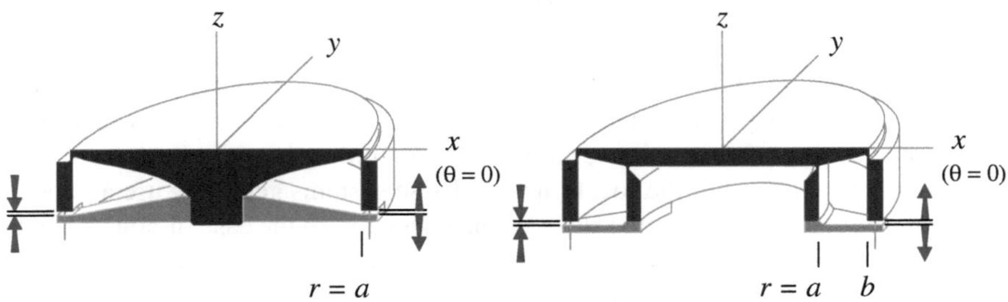

Fig. 3.17 Configurations providing a *Coma* 3 mode deflexure from two point-force $\mathcal{F} = \pm 2aF_0$ applied at $\theta = 0$ and $\theta = \pi$, usable in a star tracking system. (*Left*) VTD mirror. (*Right*) Hybrid mirror. Design parameters: telescope aperture $2a_{\rm Tel} = 3.6$ m, acomatic mirror $2a = 0.16$ m, $t_0/a = 1/10$, $\nu = 0.305$, $\Omega = $ f/d$ = 3.8$, $i = 0.005$ rad, $\mathcal{F}/a\, a_{\rm Tel} E = 7.95\, 10^{-9}$

The thin collars are readily obtained with a metal substrate such as quenched stainless steel Fe87 Cr13. Designs of *Coma* 3 mode mirrors deformed by a couple of point-force $\mathcal{F}$ and $-\mathcal{F}$ respectively applied at $\theta = 0$ and $\theta = \pi$, in the VTD and hybrid classes are displayed by Fig. 3.17.

In the CTD class, solutions for *Coma* 3 with meniscus or vase form distorted by clamped radial arms are given in Chap. 7. Radial arms are of preferable use for glass or vitro-ceram acomatic mirrors.

• **Spectroscopy in convergent beams and acomatic gratings**: Some astrophysical studies in slitless spectroscopy and low spectral resolution use a transmission grating located at a convenient distance before the telescope focal plane and detector (typical dispersion $\simeq$ 1,000 Å/mm). Associated with a multi-passband filter, theses systems provide quick spectral selections of stellar-like objects, some of them being subsequently analyzed with slit spectrographs.

A plane transmission grating used in a convergent beam generates a coma aberration in the dispersed image. At a given dispersion order, this aberration can be corrected for the central wavelength of the spectra by using an aspherized grating. The design and realization of an *acomatic grating* for the Cassegrain focus at f/8 of CFHT has been carried out [19, 20, 33] with a 75 ℓ/mm grating working in first-order and providing a mean dispersion of 750 Å/mm. After achieving the *Coma* 3 correction at $\lambda_0 = 500$ nm, the next residual aberration is *Astm* 3 which does not debase the spectral resolution; the remaining and dominating blur effect is the tilt of individual spectra with respect to the focal plane (Fig. 3.18).

Acomatic gratings are obtained by replication from an active optics submaster under stress. The submaster is designed in accordance with the CTD class (Fig. 3.12-*Left*) by using a vase form in quenched stainless steel Fe87 Cr13. A plane grating is deposited on the plane surface of the submaster when not stressed. A set of opposite ring-force at $r = a$ and $r = b$, is generated, via two cylinders and thin collars, from an axially acting push-screw and pull-screw. These screws operate the point forces $\mathcal{F} = \pm 2aV_0$ in $\theta = 0$ and $\theta = \pi$ azimuths. The acomatic grating replica is obtained on Schott UBK7 optical glass with a transparent resin of the same refractive index (Fig. 3.19).

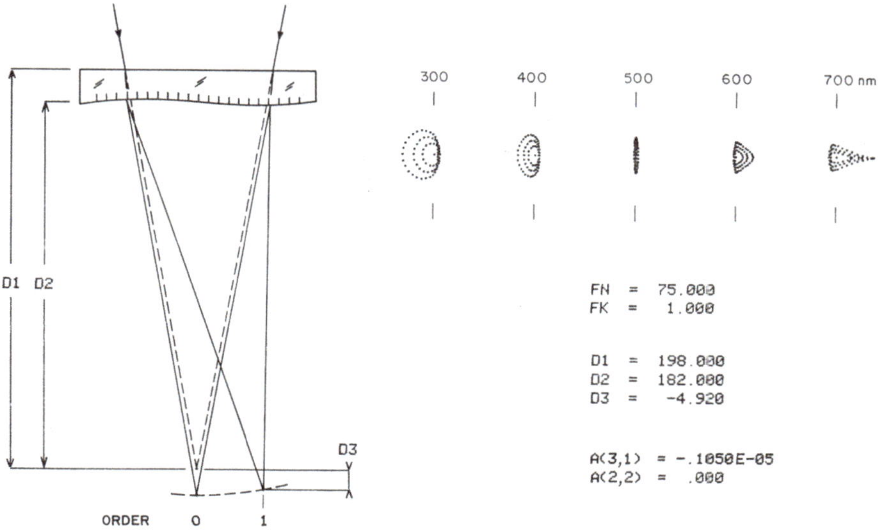

Fig. 3.18 Acomatic transmission grating working in first diffraction order. Focussing on residual tangential astigmatism at central wavelength 500 nm

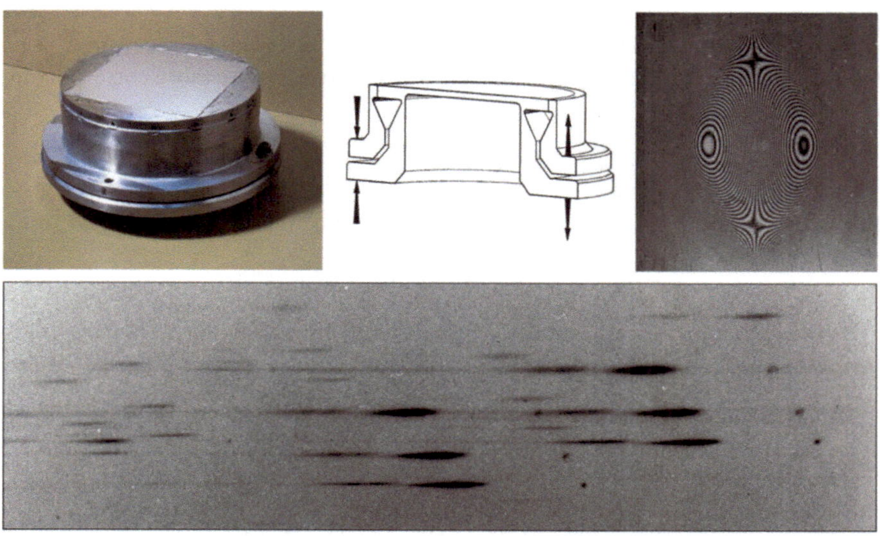

Fig. 3.19 (*Left*) Vase form metal submaster in the VTD class generating acomatic gratings by replication under stress. (*Right*) He-Ne interferogram of the grating when collimated in zero-order with respect to a plane surface. (*Down*) Markarian 382 set-up spectra from a 75 ℓ/mm *acomatic grating* with the wide-field electronic camera at CFHT f/8 Cassegrain focus after Wlérick, Cayrel, Lelièvre, and Servan [33] (LOOM)

3.5 Active Optics and Third-Order Astigmatism

The third-order astigmatism, *Astm* 3 mode, is defined by $n = 2$ and $m = 2$. This bisymmetric wavefront function, or mirror shape, is represented by the quadric surface

$$z_{22} = A_{22}\, r^2 \cos 2\theta\,, \tag{3.51}$$

which is a *hyperbolic paraboloid* (saddle).

The solutions α_i of this mode are derived from the representation of the load q by (3.14). In this relation, the first term in $\ln r$ is null whatever is A'_0, but the second term leads to a surface load $q \propto A'_0 r^{-2} \cos 2\theta$ which would be difficult to generate in practice, so that we consider hereafter $A'_0 = 0$. Then, the distributions for M_r, V_r, and q are all included by the third and remaining term of (3.11a), (3.16), and (3.14), respectively. After simplification, these terms are

$$M_r = 2(1-\nu)A_{22} \sum_{i=1}^{j} A_i\, r^{-\alpha_i} \cos 2\theta\,, \tag{3.52a}$$

$$V_r = -\,2A_{22} \sum_{i=1}^{j} (\,2-\alpha_i\,)\, A_i\, r^{-1-\alpha_i} \cos 2\theta\,, \tag{3.52b}$$

$$q = 2(1-\nu)A_{22} \sum_{i=1}^{j} \alpha_i\,(2+\alpha_i)\, A_i\, r^{-2-\alpha_i} \cos 2\theta\,. \tag{3.52c}$$

Since 2-D prismatic loads of the form $q = q_0 r^{\beta} \cos\theta$ are to be rejected as extremely difficult to achieve practically, the α_i roots to retain are those corresponding to a null uniform load

$$q = 0 \;\rightarrow\; \alpha_1 = 0 \;\text{ and }\; \alpha_2 = -2\,.$$

Substituting α_i in (3.9), the distributions of flexural rigidity are contained in the representation

$$D = A_1 + A_2\, r^2. \tag{3.53}$$

With the notation of (3.22) and (3.23), let us write the rigidity and a dimensionless thickness as

$$\frac{D}{D_0} = C_1 + C_2 \frac{r^2}{a^2}\,, \qquad \mathcal{T}_{31} = \left(\frac{D}{D_0}\right)^{1/3}. \tag{3.54}$$

From (3.52a) and (3.52b), the bending moment and net shearing force are

$$M_r = 2(1-\nu)A_{22}\,(\,A_1 + A_2\, r^2\,) \cos 2\theta\,, \tag{3.55a}$$

$$V_r = -4(1-\nu)A_{22} \left(\frac{A_1}{r} + 2A_2\, r\right) \cos 2\theta\,. \tag{3.55b}$$

Given the form (3.54) of the rigidity, two configurations can be derived such as described hereafter.

3.5.1 Configuration in the CTD Class ($A_2 = 0$)

A solution in the CTD class is obtained when the coefficient $A_2 = 0$ in (3.53): this solution is a plane or a moderately curved meniscus when deformed both by bending moment and net shearing force applied to its contour. Let V_0 be the maximum amplitude of the net shearing force acting per unit length at the mirror perimeter $r = a$, so the ring force is

$$V_r\{a\} = V_0 \cos 2\theta\,.$$

From (3.55b), we deduce the amplitude of this force as

$$V_0 = -4(1-\nu)\, A_{22}\, A_1/a\,.$$

By setting $A_1 = D_0$, and since, with (3.22), $D_0 C_i a^{\alpha_i} = A_i$, the C_i and dimensionless thickness are

$$C_1 = 1\,, \qquad C_2 = 0\,, \qquad \mathcal{T}_{22} = 1\,.$$

Similarly, the bending moment $M_r\{a\}$ is derived from (3.55a). The parameters of this configuration are summarized as follows [19]:

→ *A plate or a slightly curved meniscus generates an* $Astm\,3$ *deformation mode* $z = A_{22}r^2\cos 2\theta$, *if its thickness* $t = \mathcal{T}_{22}\, t_0$ *is a constant, and if a bending moment* M_r *and a ring-force* $V_r = V_0 \cos 2\theta$ *are applied to its perimeter* $r = a$*:*

$$\mathcal{T}_{22} = 1, \tag{3.56a}$$

$$M_r\{a,\theta\} = -\frac{a\,V_r}{2}, \qquad V_0 = -4(1-\nu)\frac{A_{22}D_0}{a}, \qquad t_0 = -\left[3(1+\nu)\frac{a\,V_0}{A_{22}\,E}\right]^{1/3}, \tag{3.56b}$$

where the product $V_0 A_{22}$ *is negative.*

This CTD configuration is displayed in Fig. 3.20-*Left*.

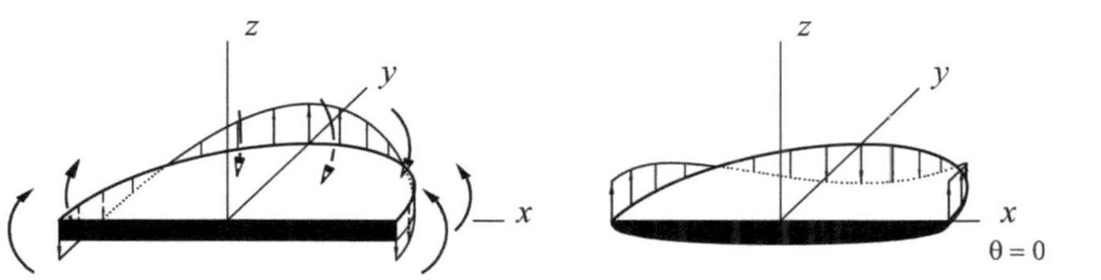

Fig. 3.20 Mirrors providing $Astm\,3$ mode $z_{22} = A_{22}r^2\cos 2\theta$. (*Left*) Solution in CTD class. (*Right*) Solution in VTD class

3.5.2 Configuration in the VTD Class

A solution in the VTD class is obtained for the case of a null rigidity at the edge, $D\{a\} = 0$. From (3.53), this is provided whenever $A_2 = -A_1/a^2$. Setting the coefficient $A_1 = D_0$, we obtain $A_2 = -D_0/a^2$ and

$$C_1 = 1\,, \quad C_2 = -1\,, \quad \frac{D}{D_0} = 1 - \frac{r^2}{a^2}\,,$$

After substitution in Eq. (3.55), the bending moment and and the net shearing force are respectively

$$M_r = 2(1-\nu)\,A_{22} D_0 \left(1 - \frac{r^2}{a^2}\right) \cos 2\theta\,,$$
$$V_r = -4(1-\nu)\,A_{22} D_0 \left(\frac{1}{r} - \frac{2r}{a^2}\right) \cos 2\theta\,.$$

At the edge, this provides $M_r\{a,\theta\} = 0$; the net shearing force distribution V_r is fully generated from applying to the edge

$$V_r\{a,\,\theta\} = 4(1-\nu)\frac{A_{22} D_0}{a} \cos 2\theta\,.$$

This simple configuration corresponds to an external ring-force in $\cos 2\theta$ applied per length unit to a simply supported edge summarized as follows (Lemaitre [14]):

→ *A cycloid-like shaped mirror provides an Astm 3 deformation mode* $z = A_{22}\, r^2 \cos 2\theta$*, if a ring-force* $V_r = V_0 \cos 2\theta$ *is applied along its simply supported edge. This force distribution is self-reacting. The thickness* $t = \mathcal{T}_{22}\ t_0$ *and characterizing features are*

$$\mathcal{T}_{22} = \left(1 - \frac{r^2}{a^2}\right)^{1/3}, \tag{3.57a}$$

$$M_r\{a,\theta\} = 0\,, \quad V_0 = 4(1-\nu)\frac{A_{22} D_0}{a}, \quad t_0 = \left[3(1+\nu)\frac{a\,V_0}{A_{22} E}\right]^{1/3}, \tag{3.57b}$$

where the product $V_0 A_{22}$ *is positive.*

This VTD configuration is displayed in Fig. 3.20-*Right*.
The radial and tangential maximum stress σ_r and σ_t must be lower than the ultimate stress $\sigma_{\rm ult}$ of the material. The radial maximum stress $\sigma_r = 6M_r/t^2$, derived with the scaling thickness t_0 and the definition of the rigidity, is

$$\sigma_r = \frac{3\,a\,V_0}{t_0^2}\,\mathcal{S}_r \quad \text{with} \quad \mathcal{S}_r = \mathcal{T}_{22} \cos 2\theta\,,$$

where the dimensionless maximum stress $\mathcal{S}_r$ in (x, z) section, i.e. $\theta = 0$, appears to be identical to the thickness $\mathcal{T}_{22}$ (Fig. 3.21).

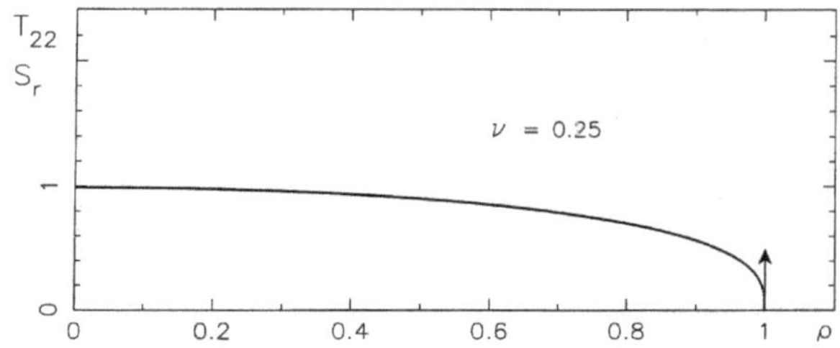

Fig. 3.21 VTD class: *Cycloid like form.* Thickness and stresses of *Astm* 3 flexure mode $z_{22} = A_{22}\,r^2 \cos 2\theta$: $\mathcal{T}_{22} = (1 - r^2/a^2)^{1/3}$, $\mathcal{S}_r = \mathcal{T}_{22} \cos 2\theta$ for $\theta = 0$

One notices that VTDs for *Astm*3 mode and *Cv*1 mode type 1 (Sect. 2.1.2) are identical and that, from (3.14), if $m = n$, then $\mathcal{T}_{nn} = (1 - r^2/a^2)^{1/3}$.

3.5.3 Hybrid Configurations

Similarly as for *Sphe* 3 and *Coma* 3 modes, hybrid configurations can be derived for an *Astm* 3 mode providing this deflection over the full mirror diameter. With a CTD as the central zone and a VTD for the outer zone $a \le r \le b$, we can link the two distributions at $r = a$, and for instance, assume that their thicknesses are locally equal by writing $t(a)\,|_{CTD} = t(a)\,|_{VTD}$. However, two features must be taken into account:

- the VTD $\mathcal{T}_{22}$ is of finite thickness at the center for the *Astm* 3 case,
- the external force applied to the edge of a VTD mirror is self-reacting.

From these properties, hybrid configurations will not provide interesting simplifications for generating *Astm* 3 deformation modes.

3.5.4 Balance with a Curvature Mode and Cylindric Deformations

The co-addition of an *Astm* 3 mode $z = A_{22}\,r^2 \cos 2\theta$ with particular values of the curvature mode *Cv* 1 can generate cylindric surfaces $z \propto x^2$ or $z \propto y^2$ (level fringes in Fig. 3.22). This can be achieved as well with a CTD or a VTD but the boundaries are simplest with a VTD since the application of bending moments at the edge is not required.

Noticing that the *Cv* 1 mode generated by a uniform load in reaction to the edge leads to the VTD (2.16) which is fully identical to the VTD providing the *Astm* 3 mode, the superposition of these modes can apply. Therefore, we can generate *cylindrical mirrors* by deformation of a thickness profile $\mathcal{T}_{22} \equiv \mathcal{T}_{20} = (1 - r^2/a^2)^{1/3}$. Let us write

$$Z = z_{20} + z_{22} = A_{20}\,r^2 + A_{22}\,r^2 \cos 2\theta$$

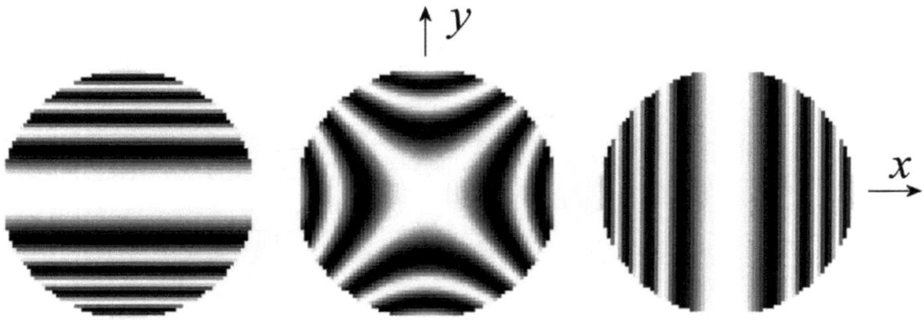

Fig. 3.22 Co-addition of *Astm* 3 mode with varying *Cv* 1 mode. The curvature modes are respectively $z_{20} = -A_{22}r^2,\ 0,\ +A_{22}r^2$

the co-addition of *Cv* 1 with *Astm* 3 . If we consider curvature amplitudes A_{20} of the same sags as those of astigmatism amplitudes at $\theta = 0$ and $\pi/2$, i.e. $\pm A_{22}$, the cylindric deformations are represented by

$$Z_{\mathrm{Cyl}} = A_{22}\, r^2(\pm 1 + \cos 2\theta) \quad \text{with} \quad A_{20} = \pm A_{22}\,. \tag{3.58}$$

The associated loads are derived from (2.16) and (3.57b). From the first,

$$\frac{t_0}{a} = -\left[\, 3(1-\nu)\,\frac{q\,R}{E\,a}\,\right]^{1/3},$$

and from the second,

$$\frac{t_0}{a} = \left[\, 3(1+\nu)\frac{V_0}{A_{22}E\,a^2}\,\right]^{1/3}.$$

By equalizing and setting $A_{20} = \pm A_{22}$, we obtain the coupling relation of the net shearing force with the uniform load

$$q = \pm 2\,\frac{1+\nu}{1-\nu}\,\frac{V_0}{a}\,, \tag{3.59}$$

so that q and the perimeter external force

$$V_r(a) = \left(\cos 2\theta \pm \frac{1+\nu}{1-\nu}\right) V_0 \quad \text{with} \quad V_0 = \frac{A_{22}\,E\,t_0^3}{3(1+\nu)\,a} \tag{3.60}$$

fully define each of the two cases providing cylindric deformations.

3.5.5 Sagittal and Tangential Ray Fans in Mirror Imaging

• **Conjugation relations in the general case**: Let us consider the general case of an astigmatic incident beam reflected by a bi-axisymmetric surface of curvatures $1/R_s$

and $1/R_t$ in its sagittal and tangential sections respectively [4, 30a]. Let s and t be the distances of an astigmatic object from the surface vertex to the sagittal and tangential focii and, s' and t' the distances from the vertex to the conjugated focal images. The relations of conjugate imaging are

$$\frac{1}{s}+\frac{1}{s'}=\frac{2\cos i}{R_t}, \quad (3.61a)$$

$$\frac{1}{t}+\frac{1}{t'}=\frac{2}{R_s\cos i}. \quad (3.61b)$$

→ *Sagittal ray fans* provide tangential focii that are on the *tangential focal surface*,
→ *Tangential ray fans* provide sagittal focii that are on the *sagittal focal surface*.

• **Astigmatism of an axi-symmetric concave mirror around its center**: A concave mirror or a singlet lens illuminated off-axis provide a third-order astigmatism if having a rotational symmetry. For instance, let us consider the case of a concave spherical mirror of curvature $1/R$ with its vertex V corresponding to a mechanical center, and C its center of curvature (Fig. 3.23).

After reflection on the mirror, an object point-source O taken sideways at distance h from C, and into a P plane normal in C to VC, gives a conjugated image showing $Astm\,3$. The reflected beam leans against two distant perpendicular segments centred on the principal ray in T' and S' points. These segments are the *tangential* and *sagittal focii*, respectively. Denoting $i=h/R$ the incident angle of the principal ray at the mirror vertex, the *astigmatism length* – axial distance of these segments – is a function of i. Since the intersection of the input beam with the output beam is a *sphere* – the mirror –, the design satisfies the Abbe sine condition of *aplanatism* [1] i.e. $Sphe\,3=0$ and $Coma\,3=0$.

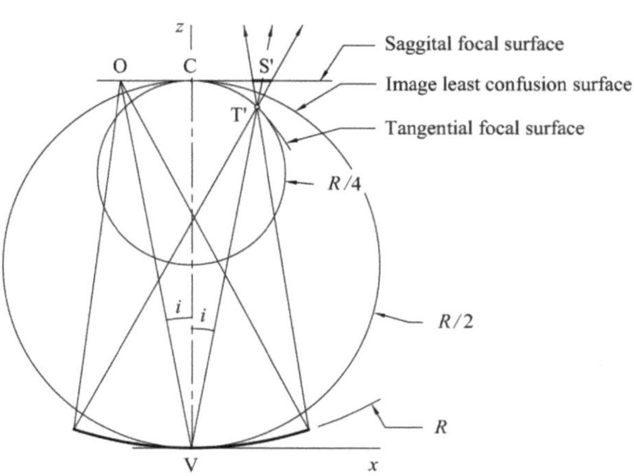

Fig. 3.23 Third-order astigmatism of a spherical mirror around its center of curvature. The mounting is *aplanatic*, i.e. $Sphe\,3=0$ and $Coma\,3=0$

Since the mirror is spherical, $R_s = R_t = R$, and since the input beam is stigmatic

$$s = t = R\,/\cos i\,.$$

The distances from the vertex to the conjugated focal images are

$$s' = s \;\simeq\; R\,(1 + i^2/2) \quad \text{and} \quad t' \;\simeq\; R\,(1 - 3\,i^2/2)\,.$$

Denoting $\ell = s' - t'$, $2a$ the clear aperture diameter of the mirror and $\Omega = R/2a$ the beam f-ratio, the *astigmatism length* and the diameter of the transverse least confusion image are respectively

$$\ell_{M\simeq -1} = 2\,R\,i^2 \quad \text{and} \quad d = R\,i^2/\,\Omega\,. \tag{3.62}$$

The radii of curvature of the tangential, least confusion, and sagittal focal surfaces are respectively

$$\mathcal{R}_t = R/4\,, \qquad \mathcal{R}_{l.c} = R/2\,, \qquad \mathcal{R}_s = \infty\,. \tag{3.63}$$

These results apply also if the mirror is replaced by a diffraction grating of spherical shape. In the symmetry plane of the mounting, spectroscopists move the input slit along the principal ray from $s = R/\cos i$ to $s = R\cos i$, i.e. onto the circle of radius $R/2$; then, the tangential image also moves to reach the same sphere. This *Rowland circle* [3(c)] is currently used in optical mounting of concave gratings because it is free from coma.

• **Astigmatism of an axi-symmetric concave mirror with object at infinity** ($M = 0$): In this case, the distances of the object focii are $s = t = \infty$, and for a pupil at the mirror, conjugation relations become

$$\frac{1}{s'} = \frac{2\cos i}{R_t} \quad \text{and} \quad \frac{1}{t'} = \frac{2}{R_s\cos i}\,.$$

If the mirror is axisymmetric $R_t = R_s = R$, and whatever is its aspherical shape, the resulting astigmatism length $\ell = s' - t'$ is

$$\ell_{M=0} = R\,i^2/2 = \ell_{M=-1}/4\,. \tag{3.64}$$

→ *Astigmatism lengths are four-times larger at magnification* $M = -1$ *than at* $M = 0$.

• **Shape of a mirror correcting** $Astm\,3$ **at any magnification** (M): With the sign convention of (3.61), let us define the magnification as $M = -s/s' = -t/t'$. Thus, the main curvatures are

$$R_t = \frac{2\,\cos i}{1 - M}\,s' \quad \text{and} \quad R_s = \frac{2}{(1 - M)\cos i}\,t'\,.$$

The correction provides stigmatic images if $s' = t'$. Then, M vanishes and we obtain

$$R_t = R_s \cos^2 i \,, \tag{3.65}$$

whatever is the magnification. Let us generate an elastic deformation in $r^2 \cos 2\theta$ (saddle) which is co-added to an axi-symmetric mirror of curvature $1/R$. We set the latter as the mean of sagittal and tangential curvatures, hence

$$\frac{2}{R} = \frac{1}{R_s} + \frac{1}{R_t} \,. \tag{3.66}$$

So, by solving this equation system,

$$\frac{1}{R_s} = \left(1 - \frac{1-\cos^2 i}{1+\cos^2 i}\right)\frac{1}{R} \quad \text{and} \quad \frac{1}{R_t} = \left(1 + \frac{1-\cos^2 i}{1+\cos^2 i}\right)\frac{1}{R}\,, \tag{3.67}$$

where the radius of curvature R_s belongs to the sagittal plane ($R_x \equiv R_s$).

For a concave mirror where both principal curvatures have the same sign, this result leads to a less well-known property.

→ *At given incidence angle* i, *Astm* 3 *corrected mirrors have an identical shape whatever the magnification* M.[2]

The mirror shape is the co-addition of a sphere and a saddle, that is represented by the expanded series

$$Z = \frac{1}{2R}\, r^2 - \frac{1-\cos^2 i}{1+\cos^2 i}\,\frac{1}{2R}\, r^2 \cos 2\theta + \frac{1}{8R^3}\, r^4 - \frac{3\,i^2}{16R^3}\, r^4 \cos 2\theta + \ldots\,, \tag{3.68}$$

where the θ origin is the x-axis.

• **Optical design of an aplanatic image transport mirror** ($M = -1$): Given an incidence angle i, the mirror shape of the image-transport is represented by (3.68). Neglecting the fifth-order astigmatism - *Astm* 5 mode - which is of small saddle sag, the *Astm* 3 mode to be generated by flexure is

$$z_{22} = -\frac{1-\cos^2 i}{1+\cos^2 i}\,\frac{1}{2R}\, r^2 \cos 2\theta \;\simeq\; -\frac{i^2}{4R}\, r^2 \cos 2\theta \,. \tag{3.69}$$

Compared to Fig. 3.23, the optical design of the mounting must be modified as follows. We set $s = t = s' = t'$ which provides a magnification $M = -1$. From the expressions of R_s and R_t in (3.67), we obtain the mounting geometry

$$s = t = s' = t' = \left(1 + \frac{i^4}{8} + \cdots\right) R \;\simeq\; R \,. \tag{3.70}$$

---

[2] This property may be also demonstrated by considering an oblique pencil reflected at the surface of a conicoid which principal ray passes through one of the geometrical focii: there is no astigmatism for conjugates at any position along the principal ray (cf. for instance in Chap. 1 [16], p. 161).

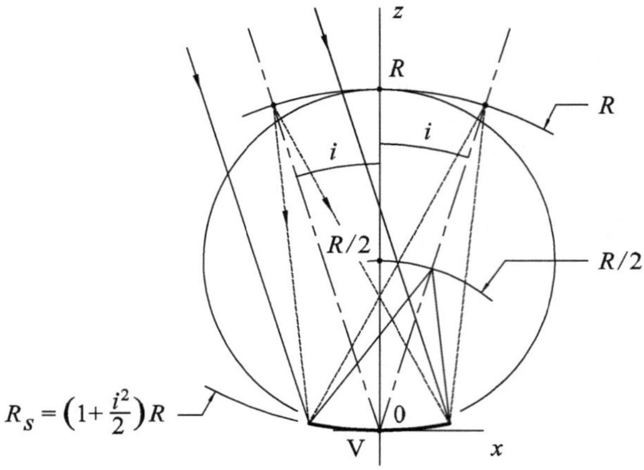

Fig. 3.24 Location of stigmatic images with the same *Astm* 3 corrected mirror. $M = -1$: object and image on the sphere R. $M = 0$: image on the sphere $R/2$

Therefore, after elastic deformation, the stigmatic object and image now belong to the sphere of radius R centered on the mirror vertex (Fig. 3.24 - *dotted lines*). Compared to the initial position of point source O in Fig. 3.23, its axial displacement is $\Delta z = -R\,i^2/2$ towards the mirror assuming small i ($R_s \simeq (1 + i^2/4)R$). In addition, to obtain a best imaging, the conjugate fields of view must be conveniently tilted about the y-axis.

The above Gaussian approximation in the quasi-normal incidence case is equivalent to assume that the shape of the mirror results from the co-addition of two *quadric surfaces* – a paraboloid and a hyperbolic paraboloid – which gives a local region of a *toroidal surface* [see also (1.38d)].

• **Object at infinity** ($M = 0$): The mirror is of the same shape as for $M = -1$ [see (3.68)], and the distance of the stigmatic image from the mirror vertex is

$$s' = t' = \left(1 + \frac{i^2}{4} + \cdots\right)\frac{R}{2}\,. \tag{3.71}$$

Thus, in first approximation, the locus of the stigmatic images belongs to a sphere of radius $R/2$, centered at the mirror vertex (Fig. 3.24).

3.5.6 *Aspherization of Concave Mirrors – Examples*

• **Off-axis astigmatism correction by VTD concave mirrors**: *Image transport systems* designed with a single optical surface – a mirror used off-axis – are useful in ultraviolet, visible, and infrared. For instance, such a system has been proposed by the author as a star tracker at telescope focus by viewing the reflected field at the spectrograph slit input [18]. For multi-object spectroscopy, reflective slits cannot be tilted for the off-axis guiding; but now using YAG lasers for the 2-D mask cuts of rectilinear or curvilinear micro-slits [5], it should be possible to replace the slit tilted plane by normal incidence mask-gratings with multi-aperture cuts, thus providing an output beam clearance. The low line density of such field gratings allows the transport mirror to reimage all order beams, thus recovering the field of view without light loss due to the reflective slit-mask-grating.

The design of a concave-saddle mirror requires the adaptation of the f-ratio and of the scaling of its mean curvature R in order to maintain the transported imaging quality of the field. The sagittal and tangential curvature of the mirror define the elastic deformation to generate by stress polishing or in situ. From (3.69) we obtain the A_{22} coefficient, and substituting in (3.57b) we obtain the elasto-optics coupling that gives the perimeter force $V_r = V_0 \cos 2\theta$. For a moderate i angle, these are

$$A_{22} = -\frac{i^2}{4R}, \qquad V_0 = -\frac{i^2}{12\,(1+\nu)}\,\frac{E\,t_0^3}{a\,R}. \tag{3.72}$$

Defining the f-ratio of the image-transport beams by Ω

$$\Omega = \frac{R}{2a}, \qquad V_0 = -\frac{\Omega\,i^2}{6\,(1+\nu)}\,\frac{E\,t_0^3}{R^2}. \tag{3.73}$$

For a given optical design $(R,\ \Omega,\ i)$ and material $(E,\ \nu)$, the latter relation defines the amplitude of the perimeter force per unit length with respect to the central thickness of the *cycloid-like mirror* (Fig. 3.25-*Right*).

The final design is achieved by coupling the mirror to a circular outer ring via a collar of thin radial thickness, hence ensuring a *simply supported edge* boundary (3.17c). This allows using two pairs of opposite point-force and obtaining smooth angular deformations. Mirrors with outer collar and ring are built in to a *holosteric piece*. Since the ring rigidity is larger than the inner profile, the applied forces and stress level are given by the ring. Developing a half ring into a straight bar and assuming clamped ends, the theory of bars and *cantilevers* (Lemaitre [21]) provide a convenient strain-stress equivalence if the ring radial thickness is small compared

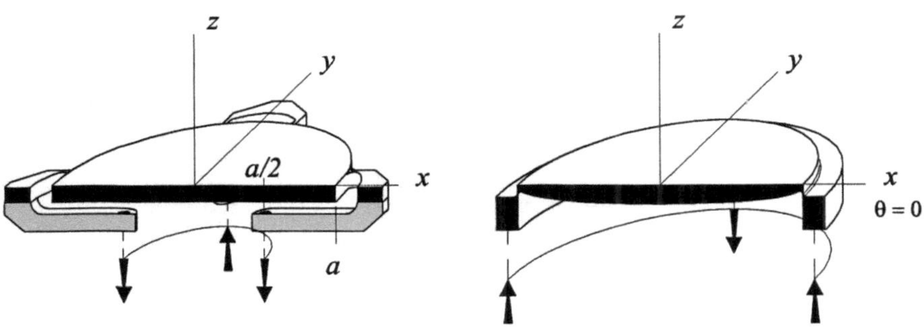

Fig. 3.25 Configuration providing *Astm* 3 mode $z_{22} = A_{22}\,r^2 \cos 2\theta$ by only requiring four axial forces applied at $\theta = 0,\ \pi/2,\ \pi$ and $3\pi/2$. (*Left*) CTD class - meniscus form. (*Right*) VTD class – cycloid-like form

to a. Let t_2 be the axial thickness of the outer ring, the resulting maximum stress is

$$\sigma_{\max} = \frac{1}{1+\nu} A_{22} E\, t_2 \; < \sigma_{\mathrm{ult}}\,, \tag{3.74}$$

where σ_{ult} is the ultimate stress.

Metal substrates conveniently allow the construction of a single piece mirror including the perimeter collar and ring. Several stainless steel mirrors have been built in Fe87 Cr13 quenched alloy. The optical tests are in accordance with the analysis and provide diffraction-limited images (Fig. 3.26).

• **Off-axis astigmatism correction by CTD concave mirrors**: Vase form and meniscus form mirrors can be designed to provide the *Astm* 3 mode. Degenerated solutions using only four external forces have been found [22] (Sect. 7.4). With vase forms, a smooth modulation in cos 2θ is also achieved by the axial thickness of the outer ring (Fig. 3.27).

Compared to the VTDs, these solutions are well adapted for glass or vitroceram substrates. Of the two CTDs, the vase form is preferred because the loading geometrical configurations are more compact than with meniscus form (Lemaitre [22]) (Fig. 3.25-*Left*). The folded arms in vase form can be shorter than for meniscuses.

Four-force configurations of vase form have been built and tested (see interferogram Fig. 7.4). The determination of the location and of the intensity of the four forces are given by (7.22a) and (7.23a).

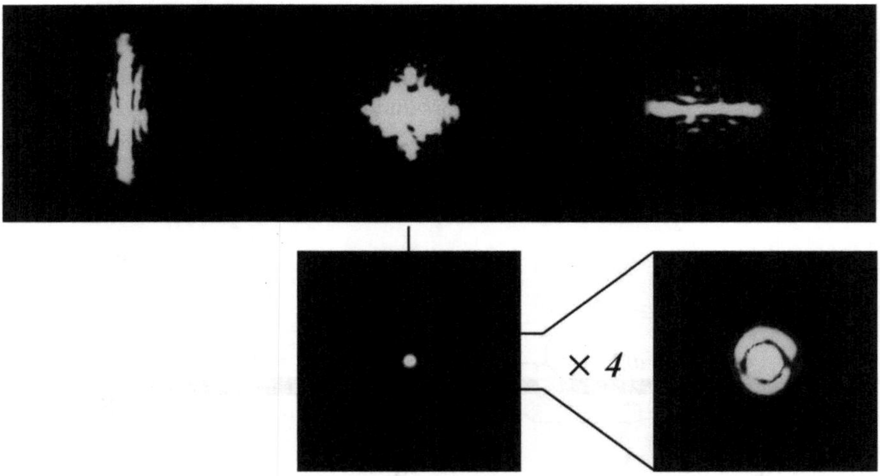

Fig. 3.26 VTD class: Saddle aspherization of a concave mirror. Design: cycloid like – collar – ring as in Fig. 3.25-*right*. Fe87 Cr13 quenched alloy. Aperture $2a = 80$ mm, $R = 640$ mm, f/8 beams, $i = 8.6°$, $A_{22} = -3.51\,10^{-5}$ mm^{-1}, $t_0 = 4$ mm, $t_2 = 12$ mm. *Astm* 3 mode is generated by four axial forces applied to outer ring at $\theta = 0,\ \pi/2,\ \pi,\ 3\pi/2$. (*Up*) Astigmatic focii and least confusion image before flexure. (*Down*) Airy disc after flexure (LOOM)

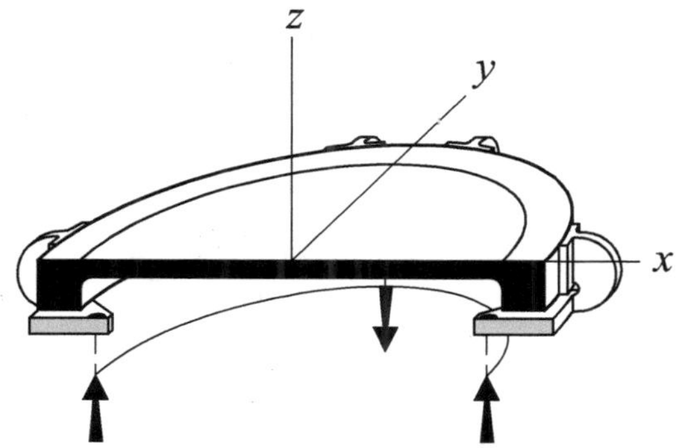

Fig. 3.27 CTD – case 2: *Vase-form* configuration providing $Astm\,3$ mode z_{22} with only four axial forces applied at $\theta = 0,\ \pi/2,\ \pi\ ,\ 3\pi/2$

3.5.7 Concave Diffraction Gratings and Saddle Correction

• **Aberrations of constant-spaced-line concave gratings**: A reflective concave grating simultaneously provides the imaging and dispersion modes. Since this is achieved without need of collimator and camera optics, the aberrations are expected to be much larger than those using plane grating systems. The theory has been developed by many authors, among them Zernike [34], Beutler [2], Namioka [24, 25], Welford [31, 30(b)], and is based on the determination of the length of optical path ℓ from the stigmatic point source P to its image P' via a grating point G. Similarly to demonstrate Fermat's principle, the series expansion of ℓ shall be stationary whatever the location of G at the grating surface (Fig. 3.28).

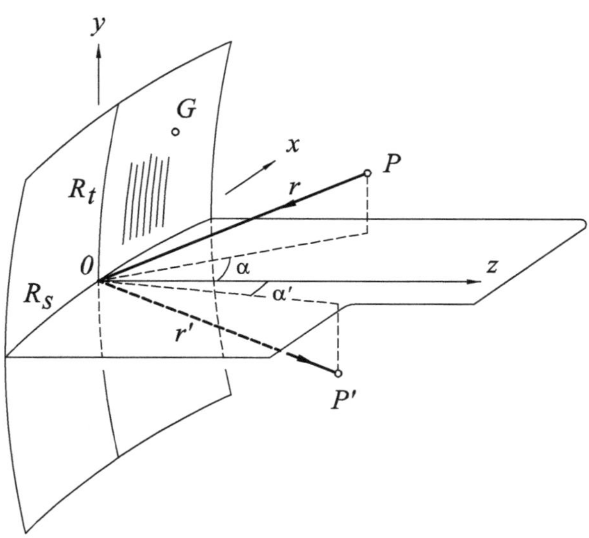

Fig. 3.28 Concave diffraction grating with two main curvatures

Let us consider a concave grating of principal curvatures $1/R_s$, $1/R_t$ in sagittal (x,z) and tangential (y,z) sections respectively, where the line projections on the tangent plane (x,y) at the grating vertex O are parallel to the y-axis and have a constant spacing. Let (x,y,z) be the coordinates of G; δ and δ' the y-distances of P and P' to the (x,z) plane; r and r' the lengths OP and OP'; and α, α' the incidence and diffraction angles of the principal ray projection in the (x,z) plane. The first terms of the optical path length $\ell =$ GP + GP' are

$$\begin{aligned}\ell = r + r' - (\sin\alpha + sin\alpha')\,x + \frac{1}{2}\left(\frac{\delta^2}{r^2}\sin\alpha + \frac{\delta'^2}{r'^2}\sin\alpha'\right)x \\ -\left(\frac{\delta}{r}+\frac{\delta'}{r'}\right)y + \frac{1}{2}\left[\left(\frac{\cos^2\alpha}{r} - \frac{\cos\alpha}{R_s}\right) + \left(\frac{\cos^2\alpha'}{r'} - \frac{\cos\alpha'}{R_s}\right)\right]x^2 \\ +\frac{1}{2}\left[\left(\frac{1}{r} - \frac{\cos\alpha}{R_t}\right) + \left(\frac{1}{r'} - \frac{\cos\alpha'}{R_t}\right)\right]y^2 + \cdots .\end{aligned} \tag{3.75}$$

In order to get ℓ stationary with respect to the G point moving in x, y on the grating surface, all coefficients of the series must be set to zero whatever is the term power ≥ 1, except for terms in x providing the dispersion by a linear displacement in x. Then, the variations of ℓ are

$$\frac{\partial\ell}{\partial x} = constant\,, \qquad \text{and} \qquad \frac{\partial\ell}{\partial y} = 0\,,$$

which simultaneously entails the conditions

$$\frac{\delta}{r} + \frac{\delta'}{r'} = 0\,, \qquad \text{and} \qquad \sin\alpha + \sin\alpha' = \frac{k\lambda}{\sigma}\,. \tag{3.76}$$

The latter relation is the *grating equation*, k is the diffraction order and σ the line spacing. When substituting it in (3.75), the product $\lambda\,x$ is constant.

By denoting

$$T = \frac{\cos^2\alpha}{r} - \frac{\cos\alpha}{R_s}\,, \qquad S = \frac{1}{r} - \frac{\cos\alpha}{R_t}\,, \qquad \text{etc...,} \tag{3.77}$$

a concise writing of the optical path length is

$$\ell = const. - \frac{k\lambda}{2\sigma}\left(1 - \frac{\delta'^2}{r'^2}\right)x + \frac{1}{2}(T+T')\,x^2 + \frac{1}{2}(S+S')\,y^2 + \cdots \tag{3.78}$$

The sagittal sections define the tangential focii, then the best spectral resolution is obtained by tangential lines parallel to the grating lines. This is achieved by the condition

$$T + T' = 0\,. \tag{3.79}$$

In (3.78), the second component of $Astm\,3$ is represented by $(S+S')y^2/2$. At a given wavelength λ, the component in $\delta'^2 x$ of the second righthand term corresponds to the curvature of the spectral lines with respect to a rectilinear input slit.

The next terms of the series expansion are in y^2x and y^3 and correspond to *Coma* 3. One derives from these coefficients that if

$$T = T' = 0\,, \tag{3.80}$$

this aberration is zeroed. Then, the projection of P and P', on the symmetry plane normal to the lines, belongs to the *Rowland circle*. This circle of radius $R_s/2$ is tangent to the grating vertex.

• **Saddle deformed concave gratings**: If the grating shape is an outer cut of a toroidal surface, Haber [7] pointed out that exact stigmatism can be obtained for two images in the spectrum located symmetrically from the grating normal. Let us move P and P' onto the Rowland circle, hence form (3.80), $r = R_s \cos\alpha$ and $r' = R_s \cos\alpha'$. From (3.78), the correction of astigmatism is achieved if

$$S + S' = 0\,, \tag{3.81}$$

and after substitution of r and r' and for a given value $\alpha' = \alpha'_0$, the condition

$$R_t = \cos\alpha \,\cos\alpha'_0 \, R_s\,. \tag{3.82}$$

provides stigmatic points at diffraction angles $\pm\alpha'_0$.

Let us generate an elastic deformation in $r^2 \cos 2\theta$ (saddle) which is co-added to an axi-symmetric grating of curvature $1/R$. We set the latter as the mean of sagittal and tangential curvatures, hence

$$\frac{2}{R} = \frac{1}{R_s} + \frac{1}{R_t}\,,$$

and solving this equation set,

$$\frac{1}{R_s} = \left(1 - \frac{1 - \cos\alpha \,\cos\alpha'_0}{1 + \cos\alpha \,\cos\alpha'_0}\right)\frac{1}{R}\,, \quad \frac{1}{R_t} = \left(1 + \frac{1 - \cos\alpha \,\cos\alpha'_0}{1 + \cos\alpha \,\cos\alpha'_0}\right)\frac{1}{R}\,, \tag{3.83}$$

where the sagittal curvature $1/R_s$ belongs to the sagittal plane, i.e. the dispersion plane (x, z) with $\theta = 0$ on x-axis $(R_x \equiv R_s)$ (Fig. 3.29).

The mean curvature $1/R$ allows us to represent the grating ellipsoidal shape by expanding the co-addition modes z_{20}, z_{22}, z_{40}, of a sphere with a saddle. The grating shape

$$Z = \frac{1}{2R}\,r^2 - \frac{1 - \cos\alpha \,\cos\alpha'_0}{1 + \cos\alpha \,\cos\alpha'_0}\,\frac{1}{2R}\,r^2 \cos 2\theta + \frac{1}{8R^3}\,r^4 - \cdots \tag{3.84}$$

corresponds to two stigmatic points at diffraction angles $\alpha' = \pm\alpha'_0$.

For a detector imaging the two stigmatic points, the field balance of the lateral astigmatism variation on the Rowland circle defines the semi-field size $\alpha'_{max} = \sqrt{2}\,\alpha'_0$ (Fig. 3.29).

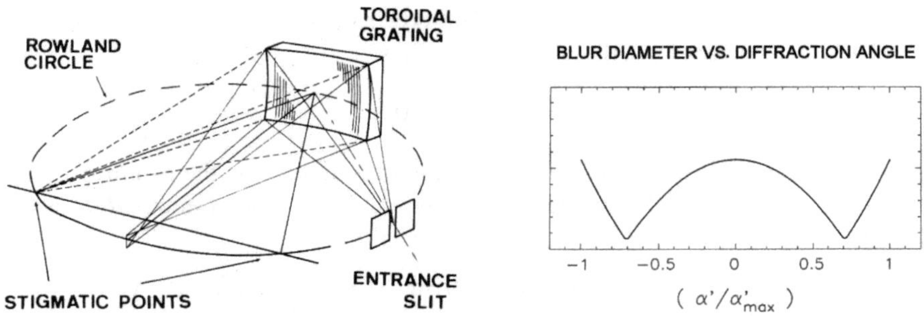

Fig. 3.29 Toroid grating with two stigmatic points on Rowland circle. (*Left*) Stigmatism at $\pm\alpha'_0$. (*Right*) Field imaging balance $\alpha'_{\max} = \sqrt{2}\,\alpha'_0$; (Courtesy Huber et al. [9]) (SOHO Mission)

3.5.8 *Aspherization of Single Surface Spectrographs – Example*

Because of the low reflectivity of available coatings, extreme ultraviolet instrumentation require the use of a minimal number of optical surfaces. In this wavelength range $\lambda\lambda\,[\,100-2,000\,\text{Å}]$, the highest efficiency is obtained by *single surface spectrographs.*

The aspherization of toroidal gratings is obtained by replication methods and active optics submasters co-adding the saddle mode. This method was developed by the author (cf. Huber et al. [8, 9]). The replication technique is of basic use by grating manufacturers. Although concave submasters can be used, the choice of convex submasters minimize the number of replicas for making concave toroidal gratings. In a first stage, a spherical grating is deposited on a submaster without stress. During a controlled loading, the saddle aspherization of the final grating is obtained by a second replication on a rigid substrate.

Let us denote

$$\psi^2 = 2\,\frac{1-\cos\alpha\,\cos\alpha'_0}{1+\cos\alpha\,\cos\alpha'_0} \simeq 2\,\frac{\alpha'^2_0+\sin^2\alpha}{(1+\cos\alpha)^2}\,, \tag{3.85}$$

if α'_0 is small.

Considering a saddle z_{22} generated by a VTD submaster, i.e. of thickness $\mathcal{T}_{22} = 1-\rho^2$ (Sect. 3.2.5), the A_{22} coefficient and the associated perimeter force $V_r = V_0\cos 2\theta$ in (3.57b), are respectively

$$A_{22} = -\frac{\psi^2}{4R}\,, \qquad V_0 = -\frac{\psi^2}{12\,(1+\nu)}\,\frac{E\,t_0^3}{a\,R}\,, \tag{3.86}$$

which define the elasto-optics coupling. For a given optical design $(R,\ a,\ \psi)$ and material $(E,\ \nu)$, these relations define the astigmatism amplitude and the perimeter force amplitude per unit length with respect to the central thickness of the *cycloid-like submaster.*

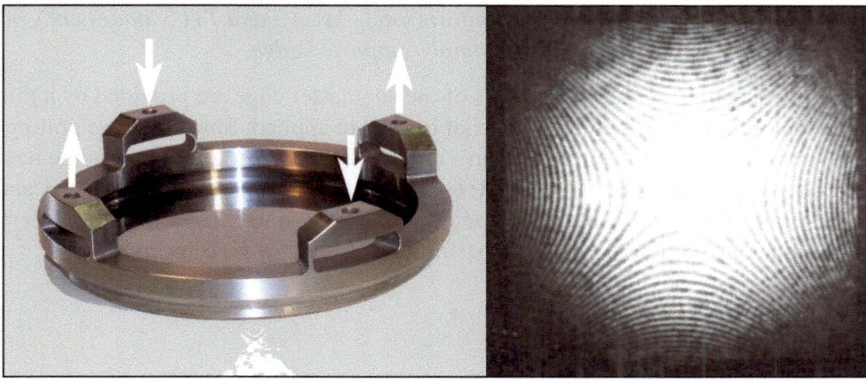

Fig. 3.30 Toroid aspherization of concave gratings by replication. (*Left*) Active submaster of thickness $\mathcal{T}_{22} = (1-\rho^2)^{1/3}$. (*Right*) Twyman fringes of final replica in 0-order. Spectral range [520–630Å], diag. aperture $2a = 0.1$ m, $R = 1$ m, f/10, $1/\sigma$ =3600 ℓ/mm, $\alpha = 11.947°$, $\alpha'_0 = \pm 0.825$. Toricity $R_s/R_t = 1.0222$ [9] (LOOM, ETH Zurich and BACH Research Corp.)

Within the class of constant spacing gratings (in the vertex tangent plane), the first developments of ellipsoid gratings aspherized from active optics submasters were carried out for the CDS and UVCS single surface spectrographs of the Solar Heliospheric Observatory (SOHO) orbiting around Lagrangian point L_1 [10, 26]. These cycloid-like submasters were built in quenched Fe87 Cr13 alloy (Fig. 3.30). Toricity $R_s/R_t = 1.010$ (CDS), 1.022 and 1.058 (UVCS). The toric gratings were produced by BACH Research Corp.

3.5.9 *Higher-Order Aspherizations of Single Surface Spectrographs*

Investigations by Wang [28, 29] concluded that the cycloid-like thickness $\mathcal{T} = (1-\rho^2)^{1/3}$, which is the unique VTD solution for *Astm* 3 mode (3.57a), is also a solution for *Tri* 5 mode $z = A_{33}\, r^3 \cos 3\theta$. The demonstration is as follows: If no loading is applied on the active surface, i.e. $q = 0$ for $0 \le r \le a$, the substitution of $n = m = 3$ in (3.14) entails A'_0=0, and the roots α_1=0, α_2=−2. From (3.9), the rigidity is $D \propto A_1 + A_2\, r^2$. For a simply supported edge, the null bending moment required by condition (3.17c) entails a null rigidity at r=a. Therefore, the rigidity of the unique VTD solution for *Tri* 5 mode is of the form $D \propto 1-\rho^2$.

The cycloid-like thickness distribution $\mathcal{T} = (1-\rho^2)^{1/3}$ allows the co-addition of

- *first-order curvature $z = A_{20}\, r^2$, i.e. Cv 1 mode,*
- *third-order astigmatism $z = A_{22}\, r^2 \cos 2\theta$, i.e. Astm 3 mode, and*
- *fifth-order triangle $z = A_{33}\, r^3 \cos 3\theta$, i.e. Tri 5 mode.*

Except for $Cv\,1$ mode generated by uniform load, $Astm\,3$ and $Tri\,5$ modes are generated by axial displacements of the simply supported edge.

In practice, the axial displacements of the submaster edge are provided by a thin collar and ring on which at least six axial forces are applied. Since concave gratings are used on the Rowland circle that provides the coma correction, $Tri\,5$ is the next aberration to compensate for after $Astm\,3$. Then, the co-addition of $Astm\,3$ and $Tri\,5$ modes on active submasters used for grating aspherization (Fig. 3.30-*Left*) allows the construction of enhanced performance single surface spectrographs.

References

1. E. Abbe, Beitrage zur Theorie des Mikroskops und der mickroskopischen Wahrnehmung, in *Schultzes Arch. f. mikr. Anat.*, Iena, **9**, 413–468 (1873)
2. H.G. Beutler, J. Opt. Soc. Am., **35**, 311 (1945)
3. M. Born, E. Wolf, in *Principles of Optics*, Cambridge University Press, Cambridge 7th Edition, (a) 329, (b) 524, (c) 459 (1999)
4. H. Chrétien, in *Calcul des Combinaisons Optiques*, Masson edt., Paris, 475 (1980)
5. B. Dibiagio, E. Le Coarrer, G.R. Lemaitre, 2D mask generation by pulsed YAG lasers for multi-object spectroscopy, in *Instrumentation in Astronomy VII*, SPIE proc. **1235**, 422–427 (1990)
6. P. Dierickx, ESO internal report for the VLT (1989), and J. Mod. Optics, **39** (3), 569–588 (1992)
7. H. Haber, The torus grating, J. Opt. Soc. Am., **40**, 153 (1950)
8. M.C.E. Huber, G. Tondello, Stigmatic performance of an EUV spectrograph with a single toroidal grating, Appl. Opt., **18** (23), 3948–3953 (1979)
9. M.C.E. Huber, E. Jannitti, G.R. Lemaitre, G. Tondello, Toroidal grating obtained on an elastic substrate, Appl. Opt., **20** (20), 2139–2142 (1981)
10. M.C.E. Huber, J.G. Timothy, J.S. Morgan, G.R. Lemaitre, G. Tondello, Imaging extreme ultraviolet spectrometer employing a single toroidal diffraction grating: The initial evolution, Appl. Opt., **27** (16), 3503–3510 (1988)
11. G.R. Kirchhoff, Uber das gleichgewicht und die bewegung eiver elastischen scheibe, Journ. Crelle, **40**, 51 (1850)
12. G.R. Kirchhoff, Vorlesungen über Mathematische Physik, *Mechanik*, 450 (1877)
13. G.R. Lemaitre, Sur les dioptres asphériques en Optique astronomique, Comptes Rendus Acad. Sc. Paris, **276**, 145–148 (1973)
14. G.R. Lemaitre, Compensation des aberrations par élasticité, Nouv. Rev. Optique, **5** (6), 361–366 (1974)
15. G.R. Lemaitre, Sur la flexion des miroirs secondaires de télescopes, Nouv. Rev. Optique, **7** (6), 389–387 (1976)
16. G.R. Lemaitre, Reflective Schmidt anastigmat telescope and pseudo-flat made by elasticity, J. Opt. Soc. Am., **66**, 12, 1334–1340 (1976)
17. G.R. Lemaitre, J. Flamand, Spectrographic development of diffraction gratings aspherized by elastic relaxation, Astron. Astrophys., **59**, 249–253 (1977)
18. G.R. Lemaitre, L. Wigroux, All-reflective aspherized grating spectrographs at the prime focus of CFHT, in *Instrumentation for Ground-Based Optical Astronomy*, Springer-Verlag, New York, 275–295 (1987)
19. G.R. Lemaitre, Various aspects of active optics, SPIE Proc. on *Active Telescope Systems*, **1114**, 328–341 (1989)

20. G.R. Lemaitre, Off-axis metal mirrors aspherized by Active optics, Proc. Intl. Workshop CNRS/NSO-Tucson on *Mirror Substrate Alternatives*, Publ. OCA-CERGA Grasse - France, 217–226 (1995)
21. G.R. Lemaitre, Equal curvature and equal constraint cantilevers: Extensions of Euler and Clebsch formulas, *Meccanica*, Kluwer Acad. Publ., **32**, 493–503, (1997)
22. G.R. Lemaitre, Active optics and aberration correction with multimode deformable mirrors, in *Laser Optics 2003: Wavefront Transformation and Laser Beam Control*, SPIE Proc. **5481**, 70–81 (2004)
23. G.R. Lemaitre, P. Montiel, P. Joulié, K. Dohlen, P. Lanzoni, Active optics and modified-Rumsey wide-field telescopes: MINITRUST demonstrators with vase- and tulip-form mirrors, Appl. Opt., **44** (34), 7322–7332 (2005)
24. T. Namioka, Theory of ellipsoidal concave gratings I, J. Opt. Soc. Am, **51**, 4 (1961)
25. T. Namioka, Theory of ellipsoidal concave gratings II, J. Opt. Soc. Am, **51**, 13 (1961)
26. NASA-ESA, Four years of SOHO discoveries, ESA Bulletin, **102** (2002) (http://sohowww.nascom.nasa.gov)
27. S.P. Timoshenko, S. Woinowsky-Krieger, in *Theory of Plates and Shells*, McGraw-Hill edit., New York 282 (1959)
28. M. Wang, G.R. Lemaitre, Aspherized concave grating by active submaster for high resolution spectroscopy, ESO Conf. Proc. on *Progress in Telescope and Instrumentation Technologies*, Garching, 729–732 (1992)
29. M. Wang, G.R. Lemaitre, Active optics and deformed toroid concave gratings: Higher-order aspherizations, Astron. Astrophys., **271**, 365–372 (1993)
30. W.T. Welford, Aberration theory of gratings and grating mountings, *Progress in Optics* IV, North Holland Publ., p. 243 (1965)
31. W.T. Welford, in *Aberrations of Optical Systems*, The Adam Hilger Series edt., Bristol, (a) 189, (b) 214 (2002)
32. R.N. Wilson, in *Reflecting Telescope Optics* I, Springer edt., New York (a) 281, (b) 77 (1997)
33. G. Wlérick, R. Cayrel, G. Lelièvre, B. Servant, Internal CFHT Report on the 80 mm wide field electronic camera (1983)
34. F. Zernike, in *Festschrift Pieter Zeeman*, Martinus Nijhoff edt., The Hague, 323 (1935)

Chapter 4
Optical Design with the Schmidt Concept – Telescopes and Spectrographs

4.1 The Schmidt Concept

4.1.1 The Class of Two-Mirror Anastigmatic Telescopes

The basic principle of the wide-field telescope invented by the Estonian optician and astronomer Bernhard Schmidt in 1928 ([67–70], E. Schmidt [72]), is that a single concave and spherical mirror used with a pupil stop at its center of curvature has no unique axis and therefore yields equal size images at all points of its field of view. In the third-order theory, the mounting is free from $Coma\,3$ and astigmatism $Astm\,3$; all images have the same amount of spherical aberration, $Sphe\,3$, coming from the spherical mirror. By using a refractive corrector plate at the mirror center of curvature, one therefore yields equally good images in the whole field of view. In the historical context in Europe, three scientists had previously developed the theoretical analysis on aplanatic telescopes in the two-mirror class, but none of them found or realized that the primary mirror could be used off-axis or could be replaced by an on-axis refractive element. Kellner who patented in 1910 [25] several designs using a corrector lens, locates the plate in a wrong position for wide-field compensations. Schmidt placed the aspherical plate at the mirror center of curvature and emphasized the importance of this location for the entrance pupil of the telescope. The curved focal surface is a monocentric sphere with the mirror. In 1930–31, he succeeded in constructing the first wide-field telescope, 36 cm clear aperture at f/1.75, with which he demonstrated the wide-field performance on 7.5 arc degrees during the two subsequent years. He obtained with Wachmann, about two hundred exposures onto curved films showing perfect images. Such astronomical object densities were never seen before. In 1932, Schmidt published his famous article "*Ein Lichtstarkes Komafreies Spiegelsystem*" [70] and photographies [71]. In fact, his coma-free i.e. aplanatic telescope is also free from third-order astigmatism: nowadays, this is called an anastigmatic telescope. Review papers on B. Schmidt's work were published by Schorr [73], Mayall [47], Wachmann [84, 85], Kross [27] and more recently by E. Schmidt [72], his nephew.

Within the class of two-mirror aplanats (Schwarzschild 1905 [76], Chrétien 1922 [16]) satisfying the Abbe sine-condition, which has been reconsidered with

G.R. Lemaitre, *Astronomical Optics and Elasticity Theory,* Astronomy and Astrophysics Library, DOI 10.1007/978-3-540-68905-8_4,

parametric equation mirrors (Popov [61], Lynden-Bell [44]), the two-mirror anastigmats are particular solutions. Into this sub-class, the Schmidts possess the advantage, over the Schwarzschild or Couder designs (Couder 1926 [18]), of requiring only one aspherical surface (Linfoot [41], Wynne [95]). The Schwarzschild is flat fielded but requires a convex primary mirror which then is not convenient for moderate or large size instruments.

Let us briefly review the class of two-mirror anastigmats, as a function of a dimensionless Petzval curvature factor p, and describe seven designs:

Primary mirror curvature:	c_1
Secondary mirror curvature:	c_2
System focal length [>0]:	f'
Mirror separation [<0]:	$d = \mathrm{M_1M_2}$
Anastigmatism condition:	$d = -2f'$
Power:	$1/f' = 2\,(c_2 - c_1 - 2d\,c_1c_2)$
Petzval:	$C_p = -2\,(c_1 - c_2) = -p/f'$

Curvatures c_1, c_2 are solution of

$$c_1 - c_2 = p/2f',\ c_1c_2 = (1+p)/8f'^2.$$

$$c_1 = [\pm\sqrt{(1+p)^2+1} + p\,]/4f'$$

$$c_2 = [\pm\sqrt{(1+p)^2+1} - p\,]/4f'$$

#	p	c_1	c_2	κ_1	κ_2
1	$-\infty$	$\frac{-1}{4f}$	∞	–	–
2	-2	$\frac{\sqrt{2}-2}{4f}$	$\frac{2+\sqrt{2}}{4f}$	-18.025	-0.5181
3	-1	0	$\frac{1}{2f}$	asph.	0
4	$-\frac{1}{2}$	$\frac{\sqrt{5}-1}{8f}$	$\frac{\sqrt{5}+1}{8f}$	52.989	0.1666
5	0	$\frac{1}{2\sqrt{2}f}$	$\frac{1}{2\sqrt{2}f}$	5.807	0.1722
6	$\frac{1}{3}$	$\frac{1}{2f}$	$\frac{1}{3f}$	2.001	0.1254
7	1	$\frac{\sqrt{5}+1}{4f}$	$\frac{\sqrt{5}-1}{4f}$	0	0

A schematic view of these seven designs of two-mirror anastigmats is displayed by Fig. 4.1.

The construction of Schmidt telescopes with a refractive corrector plate rapidly developed up to the largest instruments of Palomar and Tautenburg observatories having 1.2 and 1.4-m aperture plates, respectively (Ross [65]). Modifications for shortening the design have been proposed by also aspherizing the mirror (Wright [92], Väisälä [83]), then changing its optical properties from anastigmat to aplanat.

Various extensions of the basic Schmidt design have been proposed and built:

(i) Schmidt-Cassegrain form designs with a mirror pair having equal curvatures for a resulting flat field and internal focal surface (Baker [2]), or with a mirror pair having different curvatures for a resulting curved field and external focal surface (Burch [13]) such as developed on the Mariner and Skylab probes (Momtgomery et al. [53], Courtès [19]) or with monocentric Cassegrain mirrors (Linfoot [39], Sigler [77]) for obtaining an enlarged scale of sky.

(ii) Extended field of view has been achieved with an additional refractive Maksutov [45] lens and/or meniscus (Baker [3], Wynne [93, 94]).

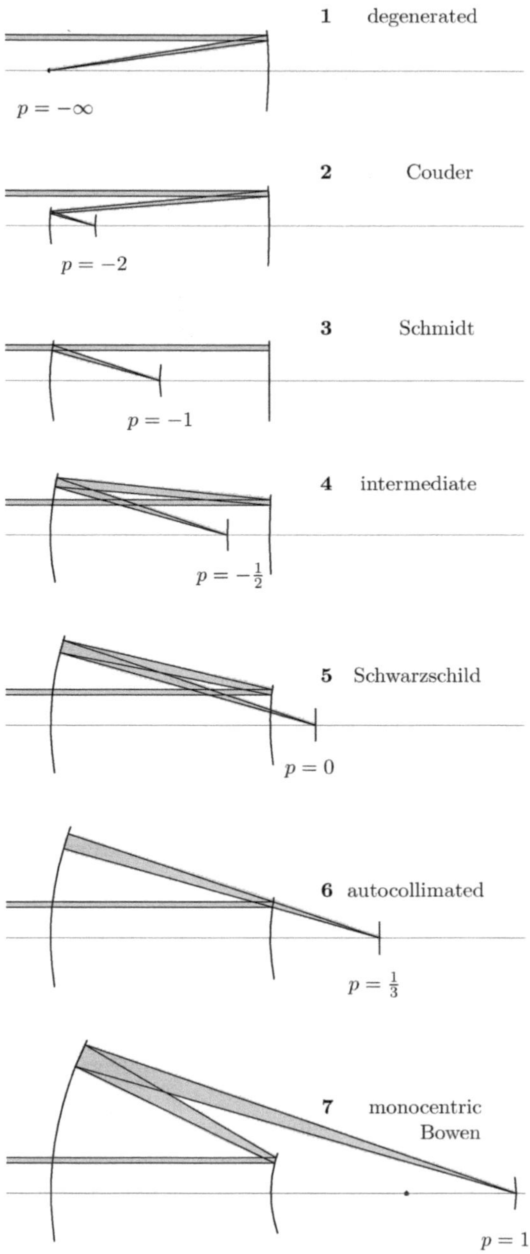

Fig. 4.1 The class of two-mirror anastigmatic telescopes. The mirror curvatures c_1, c_2, are obtained with the positive sign before the square root. The negative sign provides virtual image systems. The Petzval curvature is $C_p = -p/f'$, where f' is the effective focal length. In design (**1**), the secondary mirror degenerates into an infinitely small shell

(**iii**) Objective spectroscopy with a single prism (Schmidt and Wachmann [70]), and with a Fehrenbach normal dispersion prism (Fehrenbach [20, 21]), where the prism is located just in front of the corrector.

(**iv**) Catadioptric Schmidt cameras for spectrographs originally used curved films and are now widely equipped with flat field flatteners. Among the extremely large diversity of these designs, one may cite for instance special arrangements by Baranne [5] who invented the white pupil transfer [4] and some designs by Richardson [64]. Particular camera designs were elaborated with solid or semi-solid forms (Hendrix [24], Schulte [75]) which provide, for the same geometry, faster f-ratio instruments in the proportion of the index of refraction N (these fast systems have been reviewed by Schroeder [74] and by Wilson [88]). Imaging focal reducers using a folded mirror in a refractive Schmidt have also been proposed with flat field designs (Su et al. [80]).

(**v**) Catoptric Schmidt telescope forms have been investigated and built (Lemaitre [29]) leading to a particular correcting mirror shape for obtaining the optimal angular resolution (Lemaitre [31]) over the field and whose monolithic corrector can be aspherized by active optics (Lemaitre [32]). In 2008, this all-reflective form led to the achievement of the giant Schmidt LAMOST (Wang, Su et al. [86, 87], Su and Cui [82]) whose input pupil is 4 m in diameter and whose primary mirror is made of 24 active segments.

(**vi**) Quasi-catoptric spectrographs using aspherized reflective gratings – and a singlet lens field flattener as the only refractive element – have been designed and built for faint object spectrographs (Lemaitre and Flamand [30], Lemaitre [33, 34], Lemaitre and Kohler [36], Lemaitre et al. [37], Lemaitre and Richardson [38]).

A third-order aberration theory of refractive Schmidts has been carried out by Schmidt [67] and Strömgren [79] for minimizing the chromatism, by Carathéodory [14, 15] for a more accurate determination of the on-axis stigmatism, and by Linfoot [40] for the Cassegrain form.

We develop in the next Section a high-order wavefront analysis of Schmidt systems. The basic design of Schmidt telescopes with refractive plates is displayed by Fig. 4.2 and the first built telescope by Fig. 4.3.

The length of a Schmidt telescope is twice that of the focal length. In order to avoid the field vignetting the size of the spherical mirror is the size of the corrector increased by two times the size of the field of view. Ratios between f/3.5 and f/2.7 and 4–5° fields of view were mainly adopted for 0.6–1.4 m refractive corrector plates. For large field sky surveys such as performed with Palomar, ESO and UK telescopes using the pure Schmidt form, the size of the residual aberrations and seeing degradation due to the atmosphere was matched with the resolution element of the detector. The sphero-chromatism of the corrector plate limits the optical performance of the system for broad wavelength ranges. Solutions for reducing this effect will be presented. New developments, such as all-reflective designs and spectrographic cameras using aspherical reflective gratings, need the treatment of a more elaborate theory of aberrations than the third-order one.

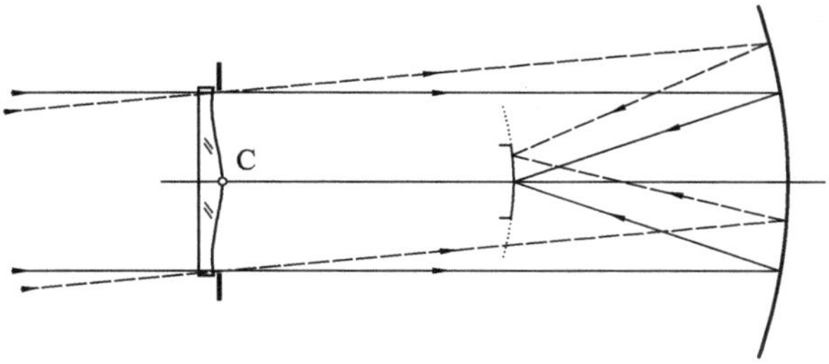

Fig. 4.2 Basic Schmidt telescope with a refractive aspheric plate and stop at the center of curvature C of a spherical mirror. From Fig. 4.1 - design **3**, the M_1 mirror is substituted by a corrector plate (above asphericity of the plate accentuated)

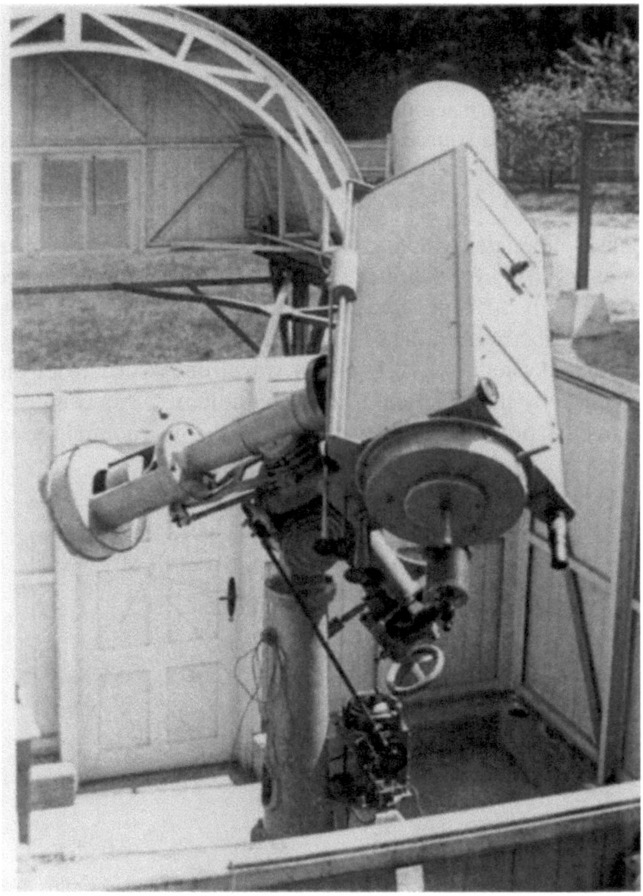

Fig. 4.3 The first Schmidt telescope, 36-cm clear aperture, f/1.75, 7.5° field of view. The corrector plate and mirror were figured by Bernhard Schmidt at the Bergedorf Observatory (Hamburg) during 1930–31

Active methods for obtaining refractive correctors, reflective correctors and aspheric reflective gratings by in-situ stressing or elastic relaxation are described in Chap. 5.

4.1.2 Wavefront Analysis at the Center of Curvature of a Spherical Mirror

In order to determine the optical shape $Z_{\rm Opt}$ to be provided by corrective optics when located at the center of curvature C of a spherical mirror, it is useful to determine, at this location, the shape $Z_{\rm w}(r)$ of a wavefront emitted by a point source S after its reflection by the spherical mirror (Fig. 4.4).

If we consider now that the light propagates in the opposite direction, then the point source S becomes the stigmatic image of the Schmidt system. If the spherical mirror is used alone and if the object point is at infinity, the Gaussian focus G of the mirror is then located midway between segment CV where V is the mirror vertex.

Returning to the case where S is a source point, let R be the radius of curvature of the mirror, and denote

$$\mathrm{GS} = M \times R/\,2$$

where M is a dimensionless parameter which is not necessarily small but may take any negative value between 0 and -1 for Schmidt cameras working at finite distance (Fig. 4.4). For an object at infinity, M is positive and relatively small; for reasons of field balancing of the residual aberrations, the M-value generally differs from zero.

In a cylindrical coordinate system, let $Z_{\rm w}(r)$ be a wavefront issued from the source point S after reflection by the spherical mirror and holding at its center C. The wavefront surface can be expressed by the even terms of the polynomial series

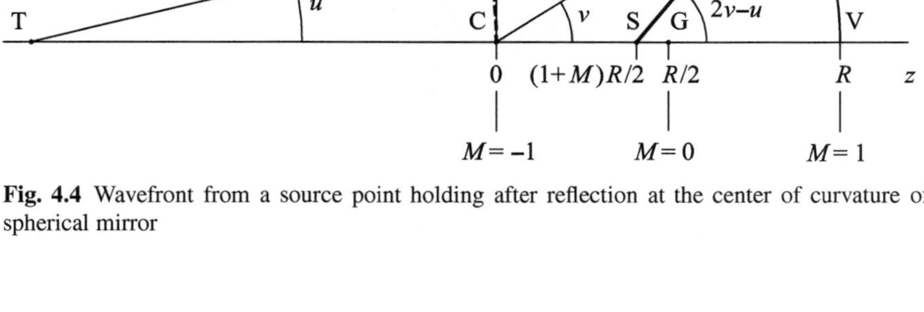

Fig. 4.4 Wavefront from a source point holding after reflection at the center of curvature of a spherical mirror

$$Z_{\rm w}(r) = \sum_{n=2,4,6,...} \mathcal{A}_n \, r^n / R^{n-1} \,, \tag{4.1}$$

where $\mathcal{A}_n$ coefficients are dimensionless. For large spectrographic cameras having an f-ratio as fast as f/1.5 or f/1, the higher-order terms of Eq. (1) have to be determined for an accurate optical correction. This determination is also required for solving the elasticity problem of the design of deformable optics that allows obtaining aspherical plates, mirrors, and gratings.

The shape of the wavefront $Z_{\rm w}(r)$ can be obtained from the condition of a constant light path which is defined on the z-axis by $2 \times \mathrm{SV} + \mathrm{VC} = (3 - M)R\,/2$, so that for a point M at the mirror the sum of positive lengths $\mathrm{SM} = \ell_1$ and $\mathrm{MZ} = \ell_2$ is equal to this path. A set of relations can be deduced from the geometrical properties of Fig. 4.4, where u and v are respectively the angles of the segments ZM and CM with the z-axis. These relations are

$$\ell_1 + \ell_2 = (3 - M)R\,/2 \tag{4.2a}$$

$$Z_{\rm w} = R\cos v - \ell_2 \cos u \tag{4.2b}$$

$$r = R\sin v - \ell_2 \sin u \tag{4.2c}$$

$$\ell_1 \cos(2v - u) = R\cos v - (1 + M)\;R\,/2 \tag{4.2d}$$

$$\ell_1^2 = (5 + 2M + M^2)R^2/4 - (1 + M)R^2 \cos v \tag{4.2e}$$

where the unknowns are $Z_{\rm w}$, u, v, ℓ_1, and ℓ_2. Let us denote some dimensionless quantities with respect to the mirror radius of curvature R, as follows

$$\mathcal{Z}_{\rm w} = Z_{\rm w}/R = \sum \mathcal{A}_n \rho^n, \quad \mathcal{L}_1 = \ell_1/R, \quad \mathcal{L}_2 = \ell_2/R, \quad \rho = r/R. \tag{4.3}$$

The equation set (4.2) becomes

$$\mathcal{L}_1 + \mathcal{L}_2 = (3 - M)\,/2 \tag{4.4a}$$

$$\mathcal{Z}_{\rm w} = \cos v - \mathcal{L}_2 \cos u \tag{4.4b}$$

$$\rho = \sin v - \mathcal{L}_2 \sin u \tag{4.4c}$$

$$\mathcal{L}_1 \cos(2v - u) = \cos v - (1 + M)\,/2 \tag{4.4d}$$

$$\mathcal{L}_1^2 = (5 + 2M + M^2)/4 - (1 + M)\cos v \tag{4.4e}$$

Each $\mathcal{A}_n$ coefficient can be obtained by remarking that

$$\mathcal{A}_n = \frac{1}{n!}\left(\frac{d^n \mathcal{Z}_{\rm w}}{d\rho^n}\right)_{\rho=0} \tag{4.5}$$

so the successive derivatives of $\mathcal{Z}_{\rm w}$ for $\rho = 0$ have to be determined. This is obtained by successive derivations of the unknowns by respect to ρ in each equation of set (4.4), writing and solving them for $\rho = 0$ at orders $n = 0,1,2,...$ For instance, one of these equations is $\ddot{\mathcal{Z}}_{\rm w} = -\dot{v}^2 + \mathcal{L}_2\,\dot{u}^2 - \ddot{\mathcal{L}}_2$ for $\rho = 0$. The resulting determinations of $\mathcal{L}_1$, $\mathcal{L}_2$ optical paths, u, v angles and $\mathcal{A}_n$ coefficients are listed in Table 4.1 in the form of series expansions (Lemaitre, 1985, unpublished).

Table 4.1 Series expansions in ρ^n of dimensionless optical lengths $\mathcal{L}_1$ and $\mathcal{L}_2$, angles $v{-}u$, v and u, and dimensionless coefficients $\mathcal{A}_n$ of the reflected wavefront $\mathcal{Z}_{\mathrm{w}}$ represented by (4.1) with respect to the magnification M. Issued from point source I, the reflected wavefront is passing by the mirror center of curvature ($\mathcal{A}_0 = 0$)

Power n	0	1	2	3	4	5	6
$\mathcal{L}_1/\rho^n$	$\frac{1-M}{2}$	0	$\frac{1-M}{1+M}$	0	$3\left[\left(\frac{2}{1+M}\right)^3 - 1\right]$	0	$45\left[\left(\frac{2}{1+M}\right)^5 + 8\frac{M(1-M)^3}{(1+M)^4} - 1\right]$
$\mathcal{L}_2/\rho^n$	1	0	$-\frac{1-M}{1+M}$	0	$-\mathcal{L}_1^{[4]}/\rho^4$	0	$-\mathcal{L}_1^{[6]}/\rho^6$
$(v-u)/\rho^n$	0	1	0	1	0	$3\left[20\frac{M(1-M)^2}{(1+M)^3} + 3\right]$	0
v/ρ^n	0	$\frac{1-M}{1+M}$	0	$\left(\frac{2}{1+M}\right)^3 - 1$	0	$3\left[3\left(\frac{2}{1+M}\right)^5 + 20\frac{M(1-M)^3}{(1+M)^4} - 3\right]$	0
u/ρ^n	0	$-\frac{2M}{1+M}$	0	$\left(\frac{2}{1+M}\right)^3 - 2$	0	$3\left[3\left(\frac{2}{1+M}\right)^5 - 40\frac{M^2(1-M)^2}{(1+M)^4} - 6\right]$	0
$\mathcal{A}_n$	0	0	$\frac{M}{1+M}$	0	$-\frac{1}{4} + \frac{M}{1+M} - \frac{M^2}{(1+M)^3}$	0	$-\frac{3}{8} + \frac{2M}{(1+M)^4} + \frac{3M^2}{2(1+M)^2} - \frac{2M^4}{(1+M)^5}$

The third-order coefficient *Sphe* 3, such as given in the Table, can also be written $\mathcal{A}_4 = -\frac{1}{4} + \frac{M(1-M^3)}{(1-M)(1+M)^3}$.

The determination of *Sphe* 5 is fully completed. $\mathcal{A}_6$ does not seem to have a simple compact form however.

Higher-order coefficients are respectively $\mathcal{A}_8 = -\frac{45}{64} + 4M + \cdots$, $\mathcal{A}_{10} = -\frac{193}{128} + \cdots$, $\mathcal{A}_{12} = -\frac{1761}{512} + \cdots$.

4.1.3 Wavefront Equation Including the Magnification Ratio M

The dimensionless parameter M is the transverse *magnification ratio* provided the spherical mirror is as usually defined by

$$M = -\mathrm{CS/CT} = -\mathrm{VS/VT}.$$

From Eq. (4.3) and $\mathcal{A}_n$ coefficients in Table 4.1, the *dimensioned* equation of the wavefront surface at the center of curvature C is

$$\begin{aligned} Z_{\mathrm{w}} = & \frac{M}{1+M}\frac{r^2}{R} - \left[\frac{1}{4} - \frac{M(1-M^3)}{(1-M)(1+M)^3}\right]\frac{r^4}{R^3} \\ & - \left[\frac{3}{8} - \frac{2M(1+M-M^3)}{(1+M)^5} - \frac{3M^2}{2(1+M)^2}\right]\frac{r^6}{R^5} \\ & - \left[\frac{45}{64} - 4M + \cdots\right]\frac{r^8}{R^7} - \left[\frac{193}{128} + \cdots\right]\frac{r^{10}}{R^9} - \cdots \end{aligned} \quad (4.6)$$

This equation may be checked when both conjugates are at a finite distance.[1]

When the source point I is close to the Gaussian focus G, then M is small. When $M = 0$ and given a mirror f-ratio $\Omega = f/d = R/4r_{\mathrm{max}}$, the amplitude of the r^{10} coefficient corresponding to the Rayleigh criterion $\lambda/4$ may be characterized by a couple $(\Omega\,;d)$ where $\Omega = [193\,d/2^{24}\lambda]^{1/9}$. For instance at $\lambda = 550\,\mathrm{nm}$, such couples are (f/1.2; 0.25), (f/1.4; 1.00) or (f/1.6; 3.30), where the second number in the parenthesis is the beam aperture diameter d at $z = 0$ expressed in meters.

4.1.4 Optical Design of Correctors – Preliminary Remarks

On-axis refractive correctors as well as tilted reflective correctors – providing or not diffraction limited beams for spectroscopy or imaging – have been found useful for wide-field studies. From the previous wavefront equation $Z_{\mathrm{w}}(r, M)$, a formal analysis could be carried out for the determination of the best corrector shape Z_{Opt}. In the case where the magnification M is of low value, optical ray tracings show that such an analysis is unnecessary. Let us consider the case of a 100% obstruction reflective corrector having its figure straightforwardly defined by $Z_{\mathrm{Opt}} = Z_{\mathrm{w}}/2$: at f/1.5 and with $M \in [0; 0.03]$ – which includes the possibility of having a null power zone outside of the clear aperture –, the result from ray tracing shows that the size of the axial blur image is smaller than 1/50 arcsec at the focal position defined by $\mathcal{L}_1(M)$. The size of the axial blur image is found similarly sharp with a refractive corrector such as $Z_{\mathrm{Opt}} = -Z_{\mathrm{w}}/(N-1)$ at the wavelength corresponding to the refractive index N. This result can be extended to all practical designs based on a *Schmidt concept*.

[1] As a verification, the first coefficients of the wavefront (4.6) give a spherical shape for the particular cases where $M = 1$ and $M = -1$, corresponding to point S at V and point S at C, respectively.

→ *For correctors located at the center of curvature of a spherical mirror up to* $f/1.7 - f/1.5$ *acting in the range* $M \in [0; 0.03]$, *i.e. for objects at infinity, the magnification* M *becomes a positioning parameter for the wavefront null-power zone, and*

(1) *all types of axisymmetric optical correctors can be accurately derived from an affinity of the wavefront equation, i.e.* $Z_{\mathrm{Opt}} = \mathrm{constant} \times Z_{\mathrm{w}}(r, M)$,
(2) *all types of tilted optical correctors can be accurately derived from an affinity of the wavefront equation, i.e.* $Z_{\mathrm{Opt}} = \mathrm{constant} \times Z_{\mathrm{w}}(x, y, M)$.

These affinity rules allow one to avoid the optical path analytical problem of determining the corrector optical shape Z_{Opt} in each case – refractive or reflective, and diffractive or not – from the wavefront Z_{w}. Furthermore, these rules allow a fine set out of aspherical coefficients for iterative optical codes. With reflective Schmidts, these rules also allow the set up of convenient over or under corrections in order to control the image balance in the field.

4.1.5 Object at Infinity – Null Power Zone Positioning

In this Section and the following, only the case where point T now becomes an object at infinity (see Fig. 4.4) will be considered, so that its conjugate image point S is close to the spherical mirror Gaussian focus G. For S at G, the magnification parameter is $M = 0$ and coefficient $\mathcal{A}_2 = 0$, i.e. the aspherical wavefront does not present any curvature at point C ($z = 0$). For instance, when optimizing the corrector for reducing off-axis aberrations or when minimizing the on-axis sphero-chromatism variation of refractive elements, one needs a coefficient $\mathcal{A}_2$ of sign opposite to that of higher coefficients. These wavefronts show an inflexion point, and for a radial distance r_0, the propagation is parallel to the z-axis. If r_{m} is the radius of the clear aperture beam, the radius r_0 for which a ray propagates in a parallel direction can be characterized by the *null-power zone ratio* $\sqrt{k}$ defined from

$$k = r_0^2 \,/\, r_{\mathrm{m}}^2 \,. \tag{4.7}$$

Let us define the f-ratio f/d by $\Omega = |R/4r_{\mathrm{m}}|$ where R is still the radius of curvature of the spherical Schmidt mirror. Since parameter M is necessarily small, a first approximation from the derivation of the two first terms of (4.6) provides the link between M and k,

$$M \simeq k \,/\, 2^5 \Omega^2 \quad \text{with} \quad \Omega = |R \,/\, 4r_{\mathrm{m}}| \,. \tag{4.8}$$

For $k = 1$, the rays are parallel to the z-axis at the height $r_0 = r_{\mathrm{m}}$ i.e. at the full aperture edge. As will be shown, the optical optimizations lead mainly to study cases where $0 \leq k \leq 2$. With the two previous relations, and denoting $\rho = r/r_{\mathrm{m}}$ a reduced radius with respect to the full aperture, a first third-order approximation of the wavefront lying at the mirror center of curvature is $\mathcal{Z}_{\mathrm{w}} = Z_{\mathrm{w}}/R = (2k\rho^2 - \rho^4)/2^{10}$. The various meridian profiles by respect to k are displayed by Fig. 4.5.

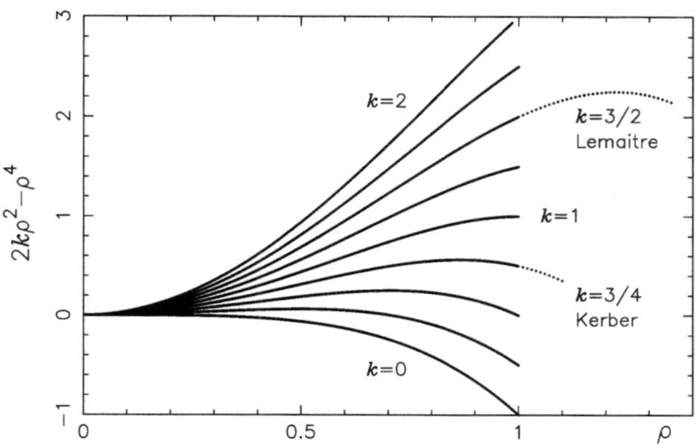

Fig. 4.5 Function $2k\rho^2 - \rho^4$ representing the wavefront profiles with respect to parameter $k = r_0^2/r_m^2$ varying from 0 to 2 by increments of 1/4. The Kerber condition, $k = 3/4$, applies for optimizing refractive Schmidts whereas Lemaitre's condition, $k = 3/2$, applies for optimizing reflective Schmidt telescopes and spectrographs

4.1.6 Optical Equation of Various Corrective Elements

Either refractive correctors, reflective correctors, or self-corrective gratings are Schmidt design manifolds. Let z, r, θ be a cylindrical frame linked to the corrector where the r, θ plane is tangent to its vertex; Set the z-axis as positive toward the spherical mirror of curvature $1/R$, and $\theta = 0$ in the symmetry plane (z, x) of the system, so $x = r\cos\theta,\ y = r\sin\theta$. The general figure $Z_{\mathrm{Opt}}(r, \theta)$ of the corrective optical element is of the form

$$Z_{\mathrm{Opt}} = \sum B_{n,m}\, r^n \cos m\theta\,, \quad n = 2, 3, 4, \ldots, \quad m \le n\,, \tag{4.9}$$

where $B_{n,m}$ coefficients are derived from the $\mathcal{A}_n(M)$ wavefront coefficients and from the inclination angle i of the axial incident ray around the y axis for a mirror or from the incident angle α if the mirror is a reflective grating . These coefficients can be expressed by the form

$$B_{n,m} = s\,\mu\, T_{n,m}(i \text{ or } \alpha)\, \mathcal{A}_n /\, R^{n-1}, \tag{4.10}$$

where s is an under- or over-correction factor for field optimization – factor close to unity – and μ characterizes the corrector type. Coefficient $T_{n,m}$ are all equal to unity for centered system correctors. Parameter μ and first coefficients $B_{n,m}$ are listed in Table 4.2 [34].

For a refractive plate, the rays are assumed to emerge from the aspherical surface that faces the concave mirror. The case of a non-tilted reflective corrector leads to an on-axis full obstruction but the system is usable off-axis; this family belongs to the class of centered systems. The case where the reflective corrector is tilted by an angle i has been listed in Table 4.2 with $\tau = 1/2\sin^2 i$ (terms in τ^2 omitted) and this representation is also valid for Littrow-mounted reflective gratings; these

Table 4.2 μ and $B_{n,m}R^{n-1}$ for various corrector types. Columns 2, 3, and 5 are for *centered systems* [*C. S.*]. Column 4 is for *non-centered systems* [*N-C. S.*]

	Refractive Plate [*C. S.*] correction for index N	Mirror [*C. S.*] working off-axis	Mirror [*N-C. S.*] tilt angle $= i$ $(\tau = \frac{1}{2}\sin^2 i)$	Reflective Grating [*C. S.*] incidence $= \alpha$ diffract. $\beta_0 = 0$
μ	$-1/(N-1)$	$1/2$	$1/2\cos i$	$1/(1+\cos\alpha)$
$B_{2,0}R$	$-s\mathcal{A}_2/(N-1)$	$s\mathcal{A}_2/2$	$s(1-\tau)\mathcal{A}_2/2\cos i$	$s\mathcal{A}_2/(1+\cos\alpha)$
$B_{2,2}R$	0	0	$-s\tau\mathcal{A}_2/2\cos i$	0
$B_{3,1}R^2$	0	0	$-8s^2\mathcal{A}_2^2\tan i$	0
$B_{4,0}R^3$	$-s\mathcal{A}_4/(N-1)$	$s\mathcal{A}_4/2$	$s(1-2\tau)\mathcal{A}_4/2\cos i$	$s\mathcal{A}_4/(1+\cos\alpha)$
$B_{4,2}R^3$	0	0	$-2s\tau\mathcal{A}_4/2\cos i$	0
$B_{5,1}R^4$	0	0	$-24\,s^3\mathcal{A}_2^3\tan i$	0
$B_{6,0}R^5$	$-s\mathcal{A}_6/(N-1)$	$s\mathcal{A}_6/2$	$s(1-3\tau)\mathcal{A}_6/2\cos i$	$s\mathcal{A}_6/(1+\cos\alpha)$
$B_{6,2}R^5$	0	0	$-3s\tau\mathcal{A}_6/2\cos i$	0
$B_{7,1}R^6$	0	0	$8s^2\mathcal{A}_4^2\tan i$	0

cases belong to the class of non-centered system where the figure of the corrector can be approximated by a bi-axial symmetry because of very low coma terms. In the interesting case of spectrograph cameras working in normal diffraction ($\beta_0 = 0$), the reflective gratings are of rotational symmetry.

4.1.7 Under or Over Correction Factor s

If *no* under- or over-correction factor s is applied to the equation Z_{Opt} of the optical surface, such as defined by (4.9) and (4.10), then $s = 1$.

- In the case of telescopes using a refractive plate of circular aperture, (1) at the edge of the field, the corrective element gives a stronger correction than when working on-axis; this is due to the inclination φ of the incident beams, (2) the cross-section of the off-axis beams is elliptical. Therefore, the corrective element increases the correction in the sagittal section of the beams. Because of these two off-axis effects, when optimizing telescope images all over the field of view, it is necessary to choose an s-value slightly smaller than unity, i.e. to under-correct the aspherical element ($s < 1$).
- In the case of all-reflective spectrographs using white pupil gratings illuminated by circular beams, the anamorphosis is inverted to the previous case. It is necessary to choose an s-value slightly larger than unity, i.e. to over-correct the aspherical element ($s > 1$).

It will be shown hereafter that for all types of Schmidt designs, the resolving power is limited by fifth-order astigmatism $Astm\,5$. The angular size of this aberration is $\propto \varphi^2/\Omega^3$. In addition, a refractive corrector provides an on axis chromatic variation of spherical aberration $\propto f(\Delta N, N)/\Omega^3$.

4.2 Refractive Corrector Telescopes

4.2.1 Off-axis Aberrations and Chromatism of a Singlet Corrector

The optical figure of the refractive plate facing the spherical mirror is derived from $B_{n,m}$ coefficients in Table 4.2 – column 2, that is

$$Z_{\mathrm{Opt}} = -\frac{s}{N-1} \sum_{2,4,6,\ldots} \frac{\mathcal{A}_n(M)}{R^{n-1}}\, r^n\,, \tag{4.11}$$

with mean refractive index N, and $\mathcal{A}_n$ coefficients from Table 4.1.

The determination of field aberrations of a Schmidt working with a refractive plate is first investigated in the monochromatic case. Spot diagrams have been traced in order to determine the size of off-axis images at a field angle φ from the axis. At first, one has considered $s = 1$ and various values of parameter M i.e. of the null power position zone k from (4.8). Off-axis spot-diagrams are shown on Fig. 4.6 together with the variation of l_{r}, l_{t} radial and tangential sizes, d mean circular diameter, and Δf refocusing from the sphere of center C and radius $(1{+}M)R/2$, with respect to the k parameter in (4.7) and as functions of the quantities Ω, φ, and R. The minimum chromatism is achieved by $k = 3/4$ (see hereafter), i.e. a null power zone at $r_0 = 0.866\, r_{\mathrm{m}}$ for which the diameter of the largest image is

$$d_\varphi = 0.020\, \varphi^2/\Omega^3, \quad s = 1, \quad k = 3/4 \;\text{ i.e. }\; M = 3/128\,\Omega^2, \tag{4.12a}$$

(cf. point A, Fig. 4.6). For $k = 3/4$ the null power zone is at $r_0 = 1.155\, r_{\mathrm{m}}$, out of the clear aperture and $d_\varphi = 0.0185\, \varphi^2/\Omega^3$ (cf. point B, Fig. 4.6) which is the minimum for $s = 1$.

In the case of purely monochromatic images, it is possible to improve the resolution by under-correcting the plate. If φ_m is the maximum semi-field angle, the under-correction parameter $s = cos^2\, \varphi_{\mathrm{m}}$ provides the best s-balanced optimization. After focusing, the size of the off-axis blur at φ_{m} is shown by the dotted-line of Fig. 4.6. The best resolution is for $k = 4/3$ (cf. point B in Fig. 4.6)

$$d_\varphi = 0.011\, \varphi_{\mathrm{m}}^2/\Omega^3, \quad s = \cos^2\varphi_{\mathrm{m}}, \quad k = 4/3 \;\text{ i.e. }\; M = 1/24\,\Omega^2 \tag{4.12b}$$

Monochromatic spot-diagrams of this optimization are shown on a spherical surface for different field values $\varphi/\varphi_{\mathrm{m}}$ at the bottom line of Fig. 4.6. For instance, at f/3 and

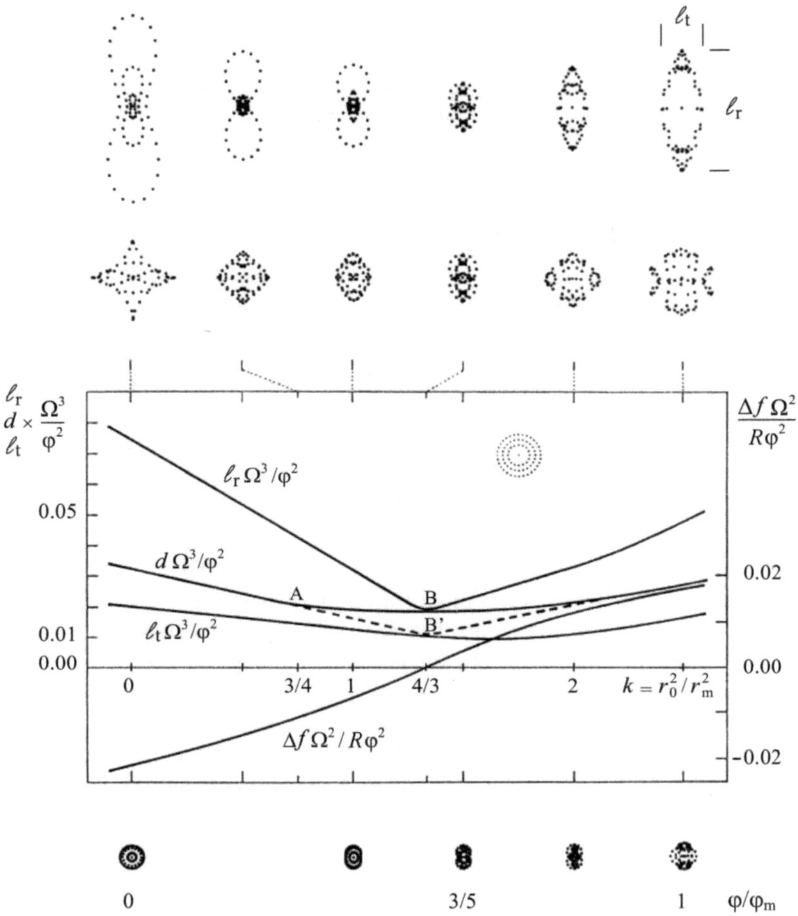

Fig. 4.6 Monochromatic off-axis residual aberrations of refractive corrector Schmidts. With parameter $s = 1$, the off-axis image is plotted, as a function of the k-ratio in (4.7), onto the sphere of center C and radius $(1+M)R/2$ which gives the on-axis stigmatism. The 2-D size of these images (*top line spots*) is denoted ℓ_r and ℓ_t in the radial and tangential directions. Except for $k = 4/3$, each image can be improved to a minimal size of diameter d (*second line spots*) by a refocusing Δf which is positive towards the mirror. For a minimal chromatism, one must set $k = 3/4$, i.e. a null power zone at $r_0 = 0.866\,r_m$, from which the diameter of the best image is $d_\varphi = 0.020\,\varphi_m^2/\Omega^3$ (cf. point A). For purely monochromatic images, the best fit is with $k = 4/3$, so the null power zone is outside the clear aperture at $r_0 = 1.155\,r_m$, and $d_\varphi = 0.0185\,\varphi_m^2/\Omega^3$ (cf. point B)

for $2\,\varphi_m = 5°$, the diameter of the images according to (4.12a) and (4.12b) do not exceed 0.29 and 0.16 arcsec, respectively.

Unfortunately, for usual spectral ranges, the chromatic variation of spherical aberration from the plate is the *dominating* residual aberration. To reduce this effect, the corrector slope must be chosen as weak as possible.

From the optical design of dioptric doublet lenses (cf. $ 1.1.4), a classical condition stated in 1886 by Kerber [26] that minimizes the spherochromatic variation consists of setting the *null-power zone ratio* $\sqrt{k}$ such as $k = 3/4$. The condition also applies to a refractive Schmidt plate.

Kerber's condition: *The profile of a refractive corrector must provide the balance of first derivative extremals. Therefore, the local slope at the half-aperture must be opposite to that of the edge, a condition which is satisfied by a null power ratio* $r_0/r_\mathrm{m} = \sqrt{k} = \sqrt{3}/2 = 0.866.$[2]

The plate being designed for correcting the wavelength λ_0 with refractive index N_0 (perfect on-axis stigmatism), at the wavelength λ corresponding to refractive index N, the angular diameter of the on-axis image (Bowen [11, 12]) is

$$d_\lambda = 1/128\,\Omega^3\nu \quad \text{where} \quad \nu = (N_0 - 1)/(N_0 - N). \tag{4.13}$$

For an f/3 telescope, a plate in fused silica correcting at $\lambda_0 = 405$ nm, and $\lambda = 320$ nm or 656 nm, one has $|\ \nu\ | = 36$ and (4.13) gives $d_\lambda = 1.65$ arcsec. Thus, onto the spherical focal surface:

---

[2] Considering the most significant part of the first two terms of (4.11), a demonstration of *Kerber's condition* (see also Sect. 1.1.4) is the following. We set $s = 1$, and from Table 4.1, $\mathcal{A}_2 = M/(1 + M) \simeq M$ and $\mathcal{A}_4 \simeq -1/4$, so the aspherical surface of the plate is approximated as

$$Z_\mathrm{Opt} = -\frac{1}{N-1}\left(M\frac{r^2}{R} - \frac{r^4}{4R^3}\right), \qquad r \in [\,0,\ r_m\,].$$

Introducing the radius $r_0 = \sqrt{2}\,MR$ and using a dimensionless radius $\rho = r/r_m$, where r_m is the maximum semi-aperture of the plate,

$$Z_\mathrm{Opt} = -\frac{r_m^4}{(N-1)R^3}\,\zeta(\rho), \qquad \text{with} \quad \zeta = \tfrac{1}{2}\,\rho_0^2\,\rho^2 - \tfrac{1}{4}\,\rho^4, \quad \rho \in [\,0,\ 1\,].$$

The equation $\zeta' = 0$ for the first derivative gives $\rho = \rho_0$ for the height with a null optical power. The equation $\zeta'' = 0$ gives $\rho_i = \rho_0/\sqrt{3}$ for the inflexion height. The spectral dispersion is a function of the local wedge of the plate. Therefore, the sphere-chromatism is *minimal* when the slope at the inflexion height is algebraically opposite to that of the edge, i.e. when $\zeta'(\rho_i) = -\zeta'(1)$. After substitution, this condition entails

$$\frac{2}{3\sqrt{3}}\,\rho_0^3 + \rho_0^2 - 1 = 0\,,$$

from where the only acceptable root is $\rho_0 = r_0/r_m = \sqrt{3}/2 = 0.8660...$

From (4.8) and neglecting the paraxial power of the plate, the focal ratio is $\Omega = f\,/\,D \simeq R\,/\,4r_m$. Hence a representation of the aspherical surface of the corrector plate is

$$Z_\mathrm{Opt} = -\frac{r_m}{256(N-1)\Omega^3}\left(\frac{3}{2}\,\rho^2 - \rho^4\right),$$

where the power of the plate is positive in its paraxial region.

→ *The resolving power of a telescope with a refractive corrector plate is*

$$\begin{cases} d_{\lambda,\varphi} \leq d_\lambda + d_\varphi = 1/128\,\Omega^3\nu + 0.020\,\varphi^2/\,\Omega^3, \\ s = 1, \qquad k = 3/4 \quad \text{i.e.} \quad M = 3/128\,\Omega^2. \end{cases} \tag{4.14}$$

Considering singlet plate Schmidts designed at f/3 with $2\varphi_{\rm m} = 5°$ in the wavelength range 320–656 nm, the residual aberration at the field edge is $d_{\lambda,\varphi} = 1.95$ arcsec. Improving this resolution by a factor of two would lead, under the same conditions, to a telescope at f/3×$2^{1/3} \simeq$ f/3.8, thus having a prohibitively long tube. The dominating blur image is the chromatism one, since $d_\lambda \simeq 6\,d_\varphi$.

With an under-correction factor $s = cos^2\,\varphi_m$ and a slight refocusing, the off-axis blur image at the field edge $\varphi = \varphi_m$ decreases substantially (*dotted-line* at point B). The spot-diagram of this s-balanced optimization is plotted for reduced field values (*bottom line spots*) $\varphi/\varphi_{\rm m} = 0$, 2/5, 3/5, 4/5 and 1 (same scale as *top line spots*). For $k = 4/3$, the diameter of the monochromatic residual images reduces to $d_\varphi = 0.011\,\varphi_{\rm m}^2/\,\Omega^3$

4.2.2 *Achromatic Doublet-Plate Corrector*

One way of correcting the *spherochromatism variation* generated by a single plate is to design a doublet-plate corrector. If λ and λ' are the two wavelengths for which the instrument is corrected and N and N' the refractive indices, one can define the Abbe number $\nu = (N_0 - 1)/(N - N')$ for each glass as ν_1 and ν_2. Let us suppose that ν_2 corresponds to the most dispersive glass. If ψ_1 and ψ_2 are the powers of the plates for a given value of the k ratio and ψ that of the doublet corrector at index N_0, the conditions for achromatism are

$$\psi_1/\nu_1 + \psi_2/\nu_2 = 0 \quad \text{and} \quad \psi_1 + \psi_2 = \psi, \tag{4.15a}$$

giving the power of each component as

$$\psi_1 = \frac{\nu_1}{\nu_1 - \nu_2}\,\psi \quad \text{and} \quad \psi_2 = \frac{-\,\nu_2}{\nu_1 - \nu_2}\,\psi\,, \tag{4.15b}$$

the correcting plate of most dispersive glass being divergent on-axis. Assuming that the secondary chromatism is negligible, one can choose the ratio $k = 4/3$ for a better off-axis correction of the two plates [cf. resolution given by (4.12b)], and then the null power zones are out of the clear aperture. The figure of the two plates can be written by using the parameter s of (4.10) as

$$s_1 = \frac{\nu_1}{\nu_1 - \nu_2}\,\cos^2\varphi_m \quad \text{and} \quad s_2 = \frac{-\,\nu_2}{\nu_1 - \nu_2}\,\cos^2\varphi_m. \tag{4.16}$$

For instance a Schmidt at f/3 achromatized at $\lambda = 365$ nm and $\lambda' = 588$ nm by using Schott glass crown UBK7 ($\nu_1 \simeq 27$) and light flint LLF1 ($\nu_2 \simeq 18$), the plates have equivalent asphericities of f/2.06 for the crown and opposite to f/2.35 for the flint. Such correcting plates were built for the 1.2 m, f/3 UK Schmidt in Siding Spring (Australia) and, around 1987, for the 1 m, f/3 ESO Schmidt in La Silla (Chile). Descriptions of achromatic correctors can be found in Su [81], Schroeder [74], or Wilson [88].

4.2.3 Singlet Corrector in Blue and Additional Monocentric Filters in Red

With Schmidt telescopes, astronomers use them to obtain sky surveys in various spectral ranges such as in U, B, V, R bands. Instead of elaborating an expensive achromatic doublet-plate corrector, a low cost way of reducing the *spherochromatic variation* of a singlet corrector plate – which is the dominant aberration of refractive corrector Schmidts – is to share this aberration by use of *monocentric filter meniscuses* (Lemaitre, 1986). The asphericity of the singlet corrector plate is optimized for working alone in a blue range. A filter meniscus is added just before the focus only when working in a dedicated red region and provides a cut-off of the shorter wavelengths. These meniscuses are band-pass filters for an intermediate range and ordinary colored filters for the last infra-red range.

The full spectral range of the telescope can then be shared into blue, red and infra-red, each range corresponding to an equal variation ΔN of the plate refractive index. A Schmidt plate optimized with $k = 3/4$ for the blue shows a *Sphe* 3 under-correction in the red which can be fully compensated by a monocentric meniscus of appropriate thickness. Both mirror and meniscus surfaces have a common center of curvature, so that *no* off-axis aberrations are added by a meniscus. This requires several plate-holders of slightly different curvature and a change of focusing. The red or infra-red sky scales are slightly smaller than that of the blue scale.

Let N_{pl} and N_{me} be the mean refractive indices of the plate and of the meniscus in a red spectral range. The thickness t_{me} of the meniscus is deduced from the *Sph* 3 amount to be introduced; this can be first approximated from that of a plane plate in a convergent beam. The filter meniscus provides a small defocussing Δf. These quantities are

$$t_{\mathrm{me}} = \frac{1}{8}\,\frac{\Delta N_{\mathrm{pl}}}{N_{\mathrm{pl}}-1}\,\frac{N_{\mathrm{me}}^3}{N_{\mathrm{me}}^2-1}\,R\,, \quad \text{and} \quad \Delta f = \left(1-\frac{1}{N_{\mathrm{me}}}\right) t_{\mathrm{me}}\,, \tag{4.17}$$

where $1/R$ is the mirror curvature and ΔN_{pl} is the mean index difference of the plate between two adjacent spectral ranges.

The resolving power is still given by (4.14) where now $d_\lambda \propto 1/\nu$ is divided by the amount of used spectral ranges. The share of d_λ should be three times for

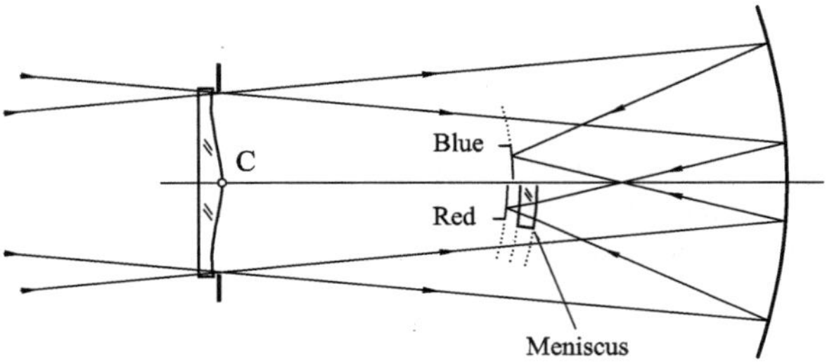

Fig. 4.7 Enhanced performance Schmidt telescope. The asphericity of the plate is optimized for a *blue* spectral range. For the *red* spectral range, an additional monocentric filter-meniscus of convenient thickness provides a similar enhanced performance as that of the blue range. Three spectral ranges may also be defined by use of two meniscuses, thus reducing again the sphero-chromatism residuals

two meniscuses; then the meniscus for the intermediate spectral range could be also as band-pass filter. For instance, an f/3 telescope of 1-meter clear aperture having a plate in BK7 glass designed for 380–510 nm in the blue at $N_{\mathrm{pl}}(415\,\mathrm{nm})$, and a meniscus designed for 510–1,000 nm in the red with $N_{\mathrm{me}}(675\,\mathrm{nm}) = 3/2$, leads to a blue over red sky scale ratio of 1.0058. The meniscus is of moderate thickness, $t_{\mathrm{me}} = 51.85\,\mathrm{mm}$, and the refocusing between the blue to red is $\Delta f = t_{\mathrm{me}}/3$ (Fig. 4.7). A thin low-pass filter is required when working in the blue range.

This meniscus option design does not seem to have been known or practiced for blue and red sky surveys during the area of curved photographic plates. Advantages are (1) higher transparency in the ultraviolet than with a doublet-plate corrector, (2) low cost since there is only one aspherical plate to figure in comparison to an expensive doublet-plate corrector whose positive component is of higher asphericity, and (3) gain in throughput by using curved filters deposited onto a corresponding meniscus or by making each meniscus directly with the filter glass, thus avoiding additional air-glass-air interfaces.

4.3 All-Reflective Telescopes

All-reflecting – or catoptric – Schmidts (Fig. 4.8) have been proposed for carrying out ultraviolet or infrared sky surveys from space (Henize [23]) by initially using a Lallemand-type detector [28] by magnetic focalization. Another approach for direct imaging is to correct the field curvature with flattener lenses for CCD arrays. For extensive spectroscopy observations with numerous optical fibers, each fiber position can be controlled onto the curved focal surface such as with LAMOST [87] onto the 5° field of view. The optical performance of the basic all-reflective Schmidts will be compared between centered systems working off-axis and non-centered systems. The best result is in favor of the latter.

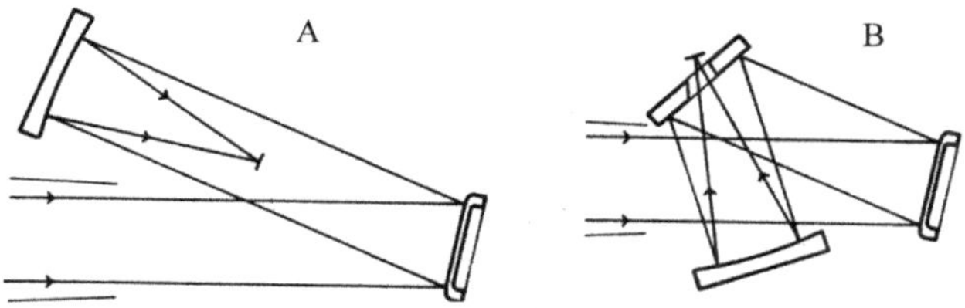

Fig. 4.8 All-reflective Schmidts: (**A**) two-mirror, (**B**) folded three-mirror

4.3.1 Centered Optical Systems used Off-axis

Let us consider, at first, a full obstructed centered system. The optical figure of the aspherical mirror is obtained from $B_{n,m}$ coefficients in Table 4.2 – column 3, that is

$$z_{\mathrm{Opt}} = \frac{s}{2} \sum_{2,4,6,\ldots} \frac{\mathcal{A}_n(M)\, r^n}{R^{n-1}} = \frac{s}{2} \left\{ \frac{M}{1+M} \frac{r^2}{R} - \left[\frac{1}{4} - \frac{M(1-M^3)}{(1-M)(1+M)^3} \right] \frac{r^4}{R^3} - \ldots \right\} \tag{4.18}$$

with $\mathcal{A}_n$ coefficients of Table 4.1.

Similarly as with refractive plate telescopes, the images in the field have been plotted onto the sphere of center C and radius $(1{+}M)R/2$ that provides the on-axis stigmatism ($s = 1$). These images are a function of the positioning ratio $k = r_0^2/r_{\mathrm{m}}^2 = 2^5\Omega^2 M$ [cf. (4.7)]. Onto this sphere, their size is denoted by ℓ_r and ℓ_t in the radial and tangential directions, respectively. A refocussing Δf allows obtaining smallest and circular-like images of diameter d. These residual images and size variations are displayed in Fig. 4.9. The best images are obtained with a *null-power zone ratio* $k = 3/2$, i.e. a null power zone outside of the clear aperture. This is Lemaitre's condition [31] of minimizing the field aberrations which are dominated by fifth-order astigmatism $Astm\,5$. The condition applies to all types of reflective correctors including gratings:

Lemaitre's condition: *The profile of a reflective corrector must provide the balance of the second derivative extremals. Therefore, the local curvature at the center must be opposite to that of the edge, a condition which is satisfied by a null-power zone ratio* $r_0/r_{\mathrm{m}} = \sqrt{k} = \sqrt{3/2} \simeq 1.225$.

The condition applies also to non-centered and diffractive systems investigated in the next Sections. If φ_{m} is the maximum semi-field angle to be optimized, the best parameter for under-correction is $s = \cos\varphi_{\mathrm{m}}$. After a refocussing, the best resolution is $d = 0.011\ \varphi_{\mathrm{m}}^2/\Omega^3$ again with $k = 3/2$. Spot-diagrams of this s-balanced optimization are obtained on a spherical surface for different field values of the reduced field position $\varphi/\varphi_{\mathrm{m}}$ (Fig. 4.9).

To make a two-mirror telescope practicable, the obstruction of the spherical secondary must be avoided. This is achieved if the incidence angle of the principal

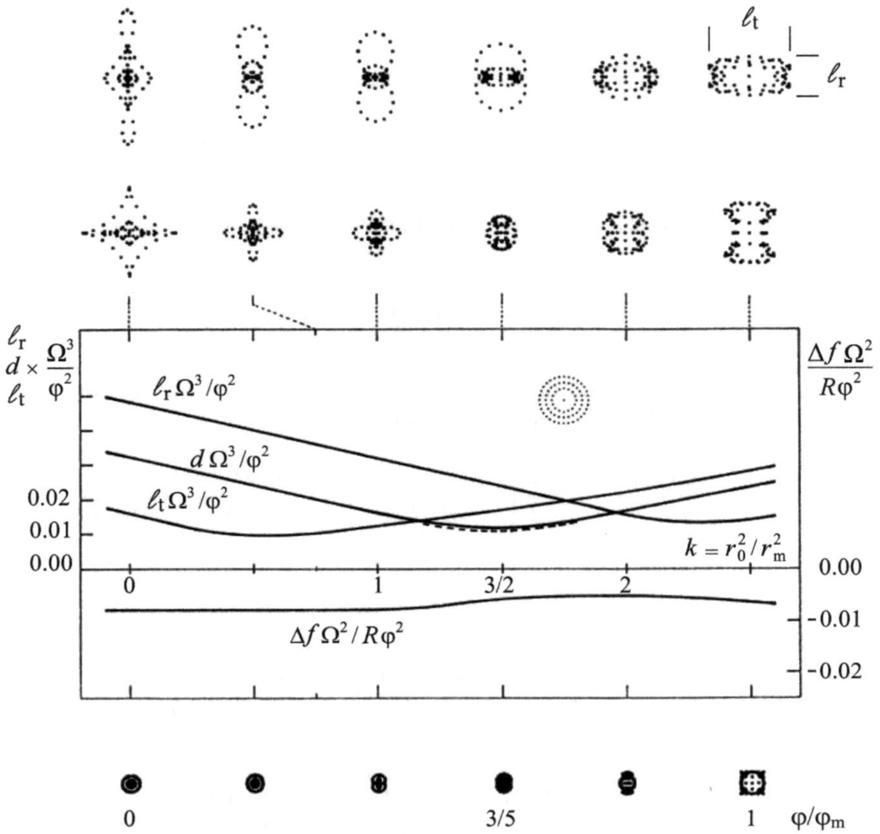

Fig. 4.9 Off-axis residual aberrations of centered reflective Schmidts – 100% obstruction. With parameter $s = 1$, the off-axis image is plotted, as a function of the k-ratio, onto the sphere of center C and radius $(1+M)R/2$ which gives the on-axis stigmatism. The 2-D size of these images (*top line spots*) is denoted ℓ_r and ℓ_t for the radial and tangential directions. The diameter d of the best circular off-axis image is obtained from a refocussing Δf; then, the size variation of the off-axis circular image with respect to k (*second line spots*) shows that $k = 3/2$, i.e. a null power zone at $r_0 = 1.225\, r_m$, provides the minimal blur image. With an under-correction factor $s = \cos\varphi_m$, $k = 3/2$ and slight refocussing, the off-axis blur at the field edge $\varphi = \varphi_m$ is improved (*dotted line*). The spot-diagram of this s-balanced optimization is plotted for various reduced field values (*lower line spots*) $\varphi/\varphi_m = 0$, 1/5, 2/5, 3/5, 4/5 and 1 (same scale as for *top lines*). At any wavelength, the diameter of the residual images is $d = 0.011\,\varphi_m^2/\Omega^3$

beam is at least

$$i = \varphi_m + 1/4\,\Omega\,. \tag{4.19a}$$

When adding a holed flat folding mirror, the three-reflection design provides a better detector access to the focus, and the incidence angle must be at least

$$i = \varphi_m + 7/16\,\Omega\,. \tag{4.19b}$$

Relations (4.19a,b) hold for incident optical beams having a circular cross-section. In these designs, the optical figure of the primary mirror is defined by $B_{n,0}$ coefficients such as given in Table 4.2–column 3. The M-value allows deriving the best fit spherical focal surface – centered at the M_1 mirror vertex – and the diameter d_C of the maximum residual aberration which is dominated by fifth-order astigmatism ($Astm\,5$).

→ *The resolving power of all-reflective centered system Schmidt telescopes working off-axis is*

$$\begin{cases} d_C = 0.011\,(i+\varphi_m)^2/\,\Omega^3, \\ s = \cos\,(i+\varphi_m), \quad k = 3/2 \quad \text{i.e.} \quad M = 3/64\,\Omega^2. \end{cases} \tag{4.20}$$

The under-correction factor s allows one to balance the dominating residual $Astm\,5$ with a small $Sphe\,3$ amount (Fig. 4.10). This resolution formula is similar to that found for a *monochromatic* refractive plate telescope (4.12b), but the gain for large wavelength ranges is considerable compared to (4.14). The axisymmetric primary mirror can be readily figured by the elastic relaxation technique (cf. Sect. 5.3.2).

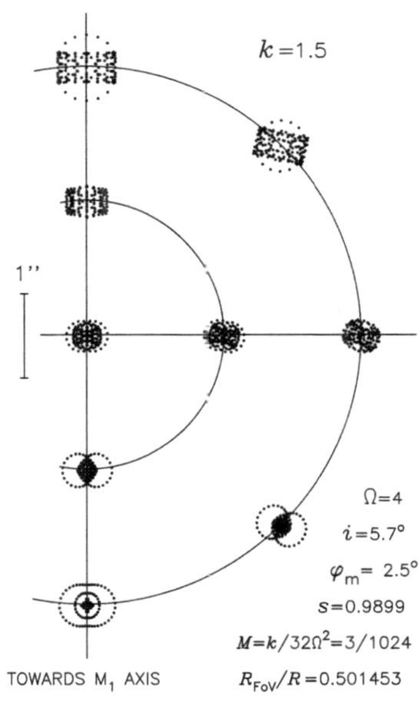

Fig. 4.10 Best residual aberrations of an f/4 - 5° FoV *centered system* reflective Schmidts used off-axis. Mirror M_1 is with rotational symmetry and clear aperture diameter $2r_m$. The spot diagram is for $k = 3/2$ corresponding to a null power zone at $r_0 = 1.224\,r_m$, i.e. outside the clear aperture. Such a mirror can be actively aspherized from a circular plate *semi-built-in* at r_0, i.e. by use of a moderately thicker ring than the plate

4.3.2 Non-Centered Optical Systems

Considering non-centered systems, the basic mounting inclinations i of the primary mirror are also given by (4.19a) for two-mirror systems or by (4.19b) with an additional flat folding mirror. Theoretically, the exact shape of the primary mirror has only one symmetry plane. The mirror $B_{n,m}$ coefficients in Table 4.2–column 4 provide a perfect stigmatism at the center of the field.

In the case of moderate f-ratios up to f/1.7–f/1.5, the coma terms of $B_{n,m}$ with n odd are negligible. Therefore, the mirror surface assimilates to a biaxial symmetric shape generated by quasi-homothetical ellipses obtained by coupling $B_{n,m}$ with $B_{n,0}$ for n even. The first term pair expressing the mirror shape is

$$Z_{\text{Opt}} = B_{2,0}\, r^2 + B_{2,2}\, r^2 \cos 2\theta + \ldots \tag{4.21a}$$

From coefficients in Table 4.2, column 4, we obtain

$$Z_{\text{Opt}} = \frac{s\mathcal{A}_2}{4R\cos i}\left[(2-\sin^2 i)\, r^2 + \sin^2 i\; r^2 \cos 2\theta\right] + \cdots \tag{4.21b}$$

With $x = r\cos\theta,\ y = r\sin\theta$ and the next coupled terms $B_{4,0}$ and $B_{4,2}$

$$Z_{\text{Opt}} = \frac{s}{2\cos i}\left[\frac{\mathcal{A}_2}{R}(x^2\cos^2 i + y^2) + \frac{\mathcal{A}_4}{R^3}(x^2\cos^2 i + y^2)^2 + \cdots\right], \tag{4.21c}$$

leading to the series form

$$Z_{\text{Opt}} = \sum_{n=2,4,6,\ldots} C_n \left(h^2 x^2 + y^2\right)^{n/2}, \qquad C_n = \frac{s\mathcal{A}_n}{2R^{n-1}\cos i}, \qquad h^2 = \cos^2 i\,. \tag{4.21d}$$

By substitution of $\mathcal{A}_2$ and $\mathcal{A}_4$ (Table 4.1), the series becomes

$$Z_{\text{Opt}} = \frac{s}{2\cos i} \times \left\{\frac{M}{1+M}\,\frac{h^2x^2+y^2}{R} - \left[\frac{1}{4} - \frac{M(1-M^3)}{(1-M)(1+M)^3}\right]\frac{\left(h^2x^2+y^2\right)^2}{R^3} - \cdots\right\} \tag{4.21e}$$

Considering now the free parameter $k = r_0^2/r_{\text{m}}^2$, the best resolution in the field is obtained for a *null-power zone ratio* $\sqrt{k} = \sqrt{3/2}$, as for centered systems. Then, from (4.8), $M = k/2^5\Omega^2 = 3/2^6\Omega^2$ where $\Omega = R/4y_{\text{m}}$ is the telescope f-ratio and the y-axis is perpendicular to the telescope symmetry plane. Hence we may represent the optical surface of the primary mirror, with its *elliptic* pupil receiving cylindrical incident beams, by the first approximation

$$Z_{\text{Opt}} \simeq \frac{s}{\cos i}\left[\frac{3}{2^7\Omega^2 R}\left(h^2x^2+y^2\right) - \frac{1}{8R^3}\left(h^2x^2+y^2\right)^2\right]. \tag{4.21f}$$

An under-correction $s < 1$ does not improve the performance. Of the two field edge points in the symmetry plane of the telescope, the largest blur is corresponding to that with the largest deviation angle. The corresponding sideways field residuals have an averaged size of the two opposite residuals in the symmetry plane. The M-value allows deriving the best fit spherical focal surface – which is now

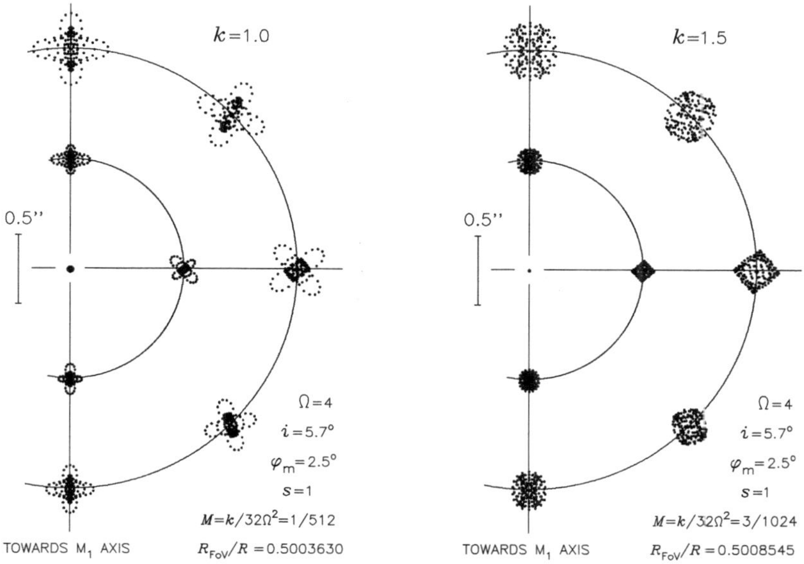

Fig. 4.11 Best residual aberrations of an f/4–5° FoV reflective Schmidt in the *non-centered system* form. The primary mirror, of biaxial symmetry, is generated by homothetical ellipses having principal lengths in a cos i ratio. With $k = 1$, the null-power zone is located at the aperture edge so the mirror is readily aspherized from active optics by use of a *pure built-in* elliptical plate at the clear aperture edge. The best images are obtained for a *null-power zone ratio* $\sqrt{k} = \sqrt{3/2}$, i.e. an elliptical null power zone defined from the circular incident beam by $r_0 = 1.224\, r_{\mathrm{m}}$, thus outside the clear aperture of $2r_{\mathrm{m}}$ diameter

slightly tilted – and the diameter d_{C} of the maximum residual aberration which is also $Astm\,5$ dominated as for the centered system (Lemaitre [31, 32]).

→ *The resolving power of all-reflective non-centered system Schmidt telescopes is*

$$\begin{cases} d_{\mathrm{NC}} = 0.012\, \varphi_{\mathrm{m}}\ (\frac{3}{2}\, i + \varphi_{\mathrm{m}})\ /\ \Omega^3, \\ s = 1, \quad k = 3/2 \ \text{ i.e. } \ M = 3/64\, \Omega^2. \end{cases} \tag{4.22}$$

Spot diagrams in the field for $k = 1$ and $k = 3/2$ show that the latter k-value is the best (Fig. 4.11). The primary mirror shape can be readily figured via elastic relaxation techniques (cf. Sects. 5.3.3 and 5.3.4).

4.3.3 *Gain of Non-Centered Systems Over Centered Designs*

For residual images of larger size than that of the detector pixel, the gain in magnitude of the non-centered design over the centered one is readily determined from

(4.20) and (4.22). Considering the folded three-reflection design with (4.19b), one obtains a gain in limiting magnitude

$$\Delta_{\text{mag.}} = 2.5 \log \left(d_{\text{C}}^2 / d_{\text{NC}}^2 \right) = 5 \log \left[\frac{11}{18} \, \frac{(32\,\Omega\,\varphi_{\text{m}} + 7)^2}{\Omega\,\varphi_{\text{m}}(40\,\Omega\,\varphi_{\text{m}} + 7)} \right]$$

For instance, at f/3 with a semi-field $\varphi_m = 2.5°$, the gain in magnitude is $\Delta_{\text{mag.}} = 2.4$. At the edge of field and with a mirror designed with (4.21d), the non-centered design provides a 100% encircled energy image equal or smaller to 1.2 arcsec corresponding to rms blur images better than 0.6 arcsec.

In the 1980s, this performance allowed the proposal of a 1 m diameter telescope for UV surveys in space (Wray [90, 91]). The all-reflective concept now leads to the very large non-centered telescope LAMOST (China) for spectroscopic surveys.

4.3.4 LAMOST: A Giant Non-Centered Schmidt with Active Optics

A remarkable example of a telescope based on the *non-centered system* case is the Large Area Multi-Object Spectroscopy Telescope (LAMOST) (Wang, Su et al. [87], Su and Cui [82]). Installed at Xinglong Station, this all-reflective Schmidt is completely dedicated to spectroscopic surveys. Starlight beams of 4 m diameter are reflected by a plane-aspheric mirror towards the concave spherical mirror which provides midway between the mirrors an f/5 curved focal surface of 5° sky field of view, i.e. a focal surface of 1.75 m in diameter; these features confer to the instrument an unprecedent *optical étendue* (cf. Sect. 1.6.3). Its optical axis – which is not a polar axis – is fixed in a meridian plane and tilted by 25° to the horizon. Astronomical objects are observed for declination angles $\delta \in [-10°, 90°]$ selected by a convenient inclination of the primary mirror for durations up to 1.5 hour as they pass through the meridian region (Fig. 4.12).

The primary and secondary mirrors, M_1 and M_2, are composed of 24 and 37 segment mirrors respectively. All segments are hexagons in Zerodur from Schott and in VO2 vitroceram from Xinhu Glass, with a diagonal length of 1.1 m. The thickness of the primary sub-mirrors is 25 mm and the active optics system provides a maximum deformation of $\sim 7.5\,\mu$m with respect to the closest tilted plane. The central part of the building holds the rotating focal surface and the controlled connection to 4,000 motorized optical fibers. Each optical fiber has a 3.3 arcsec aperture on the sky, i.e. 320 μm in diameter. Sixteen low resolution spectrographs and some high resolution spectrographs receive 250 fibers each. The spectrographs are located in dedicated rooms under the focal surface.

The M_1 segments, flat when at rest, are all *in-situ stressed* for generating a continuous aspheric surface correcting the spherical aberration of M_2. The declination angle δ and the tracking component angle from the Earth meridian plane determine

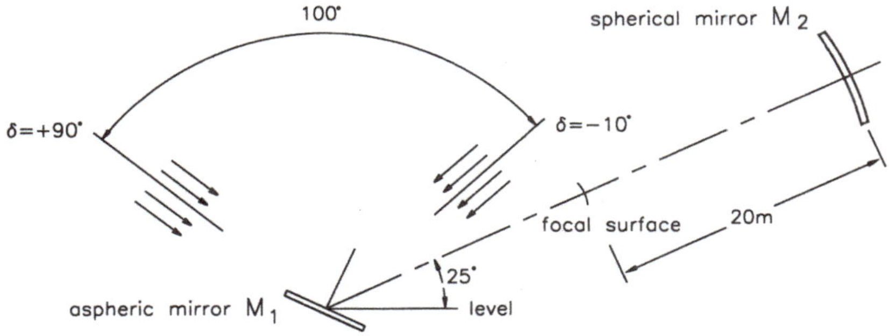

Fig. 4.12 LAMOST non-centered optical system. *Up*: The in-situ aspherized primary mirror is a $5.7 \times 4.4\,\mathrm{m}^2$ aperture made of 24 active segments. The fixed spherical secondary mirror is a $6.7 \times 6.0\,\mathrm{m}^2$ aperture made of 37 segments (after Wang, Su et al. [87], Su, Cui et al. [82] – NIAOT/CAS). *Down*: General view of LAMOST at Xinglong Station (courtesy NAOC/CAS)

the M_1 shape, its inclination angle i and the alt-az angles of its supporting mount; all these quantities vary during tracking.

While the telescope is observing, the variable figure of the M_1 is maintained such that the null-power zone ratio $\sqrt{k}$ is a constant, i.e. always satisfies condition (4.22) $k = r_0^2/r_{\mathrm{m}}^2 = 3/2$ for any direction θ at the M_1 "in-plane." This aspherical figure, elastically generated from actuators, is made of homothetical ellipses in the form $Z_{\mathrm{Opt}} = \Sigma\, C_n(h^2 x^2 + y^2)^{n/2}$, where the coefficients C_n and *pupil anamorphosis ratio* $h = \cos i$ are determined by (4.21d). In a two-term approximation, the M_1 figure is represented by (4.21f) where, from (4.22), the correction factor s is set to unity. During tracking, the magnitudes of i, C_n, h, and alt-az rotations of the primary mirror are determined for the open-loop control. Starlight sensors provide corrections for the closed-loop control.

4.4 All-Reflective Spectrographs with Aspherical Gratings

Catoptric spectrographs based on the use of aspherical reflective gratings located at the center of curvature of spherical mirrors provide astonishingly good performances (Lemaitre [30]). Two problems have to be solved, (i) their optical design requires dedicated codes able to take into account both the asphericity and the local equation to calculate the cosine directors of the diffracted ray, and (ii) they also require dedicated active optic methods using replications for the asph-erization from plane available gratings. Athough such codes are now available, these difficulties explain the reasons of only recent developments of these spectrographs starting in the 1980s.

4.4.1 Comparison of Reflective Grating Spectrograph Designs

With conventional catadioptric spectrographs using reflective gratings, one of the difficulties encountered is the obstruction of the incident beam by the camera optics. To avoid this, the camera optics are designed to be placed at a distance of several times the collimated diameter onto the gratings. It is then necessary to provide camera optics having an aperture substantially larger than that of the collimator beam if severe losses due to vignetting are to be avoided at the edges of spectra. These optics, much larger than the grating size, have increased asphericities, and for cameras of fast f-ratios the figuring of the corrective element becomes critical. For comparison, four spectrograph arrangements having the same focal length, f-ratio, dispersion and spectral range are shown in Fig. 4.13 [30].

The first mounting in Fig. 4.13A is obtained with a classical refractive Schmidt camera. In Fig. 4.13B, a doublet corrector camera is used with all spherical surfaces. The doublet corrector can be replaced by a Maksutov [45] or a Bouwers [9] meniscus, but their optical positioning is closer to the grating giving some obstruction to the collimated beam. A way of avoiding the difficulty of a large and highly aspherical corrector such in Fig. 4.13A, is to place a corrector plate directly onto the grating (Bowen [10]) such as shown by Fig. 4.13C. The light then passes the corrector plate twice, once before and once after diffraction at the grating. In this case, the figure of the twice-through plate is $(1+\cos\alpha)/\cos\alpha \simeq 2$ times lower than that of the once-through plate of the same pupil size, where α is the incidence angle at the grating.

With aspherical gratings, the chromatic aberrations do not exist and the problem of light loss by four refractions (Fig. 4.13C) also vanishes thus providing a performing design for wide spectral ranges and for UV studies as well. Table 4.2 shows that the asphericity of the corrective grating is $(1+\cos\alpha)/(N-1) \simeq 4$-times lower than that of a once-through plate of the same aperture.

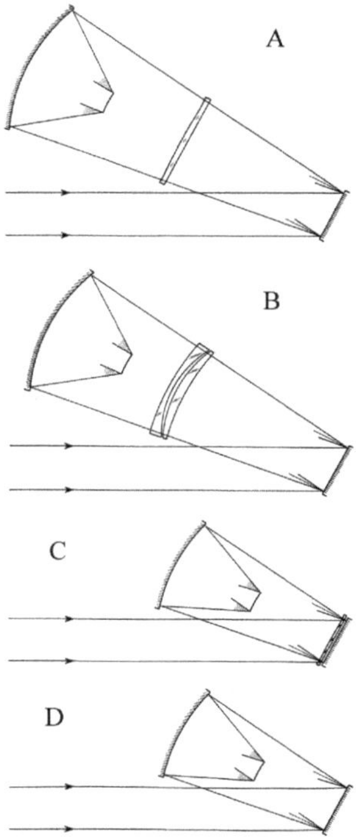

Fig. 4.13 Comparison of four spectrograph camera designs having the same focal length, f-ratio, dispersion, and spectral range

(**A**) Conventional Schmidt with refractive aspheric corrector and plane grating.
(**B**) Doublet lens corrector with all spherical surfaces. The doublet lens corrector can be replaced by a Maksutov or a monocentric Bouwers meniscus but this would provide obstruction difficulties for the incident beam.
(**C**) Bowen proposal with a twice-through refractive aspherical corrector located just before a plane grating.
(**D**) All-reflective or catoptric camera with Lemaitre aspherical grating. The design uses only two surfaces and a *minimal* asphericity correction.

The telescope pupil transfer to the camera optics is naturally obtained very near the grating. This provides an increased capability to achieve fast f-ratios for designs C and D. Furthermore, the normal diffraction mounting in camera D is often practicable and allows use of axisymmetric gratings – as the 100% obstruction two-mirror centered system – thus working with the high performance of a *two-mirror anastigmat* centered system

4.4.2 Diffraction Grating Equation

Let α and β be the incidence and diffraction angles respectively, $\delta = \alpha + \beta$ the deviation angle at the grating, and δ_0 the mounting angle between the collimator and camera axes. These parameters are related by the *grating equation* and the *mounting angle equation*,

$$\sin\alpha - \sin\beta = K\,\lambda/\,\ell \quad \text{and} \quad \delta_0 = \alpha + \beta_0\,, \tag{4.23}$$

where K, λ, and ℓ are the diffraction order, the wavelength, and the groove spacing respectively; β_0 is the diffraction angle at the center of the spectrograph field. The maximum semi-field angle $\varphi_{\rm m}$ at a spectrum edge is half the total variation of the diffraction angle β over the spectral range $[\lambda_1, \lambda_2]$.

In general, for moderate dispersion spectrographs, one can select the incidence angle α, the groove density and the camera size to get a grating mounting working closely to *normal diffraction angle* $\beta_0 = 0$. Assuming an incident beam of circular section that illuminates the grating, the aperture ratio Ω and null-power zone ratio $\sqrt{k}$ are defined in the y-direction from the clear length of the central line grating. In normal diffraction, the half-size (x, y) of the elliptic pupil at the grating is $(r_{\rm m}/\cos\alpha,\ r_{\rm m})$.

Two types of grating optical figure can be distinguished depending on whether the central wavelength is diffracted at $\beta_0 = 0$ or $\beta_0 \neq 0$.

4.4.3 Axisymmetric Gratings ($\beta_0 = 0$)

A rotational symmetry is useful to simplify the grating construction. In this case, the diffraction angle must be $\beta_0 = 0$ or small for the central wavelength. Coefficients (4.9) and (4.10) defining the grating surface are given in Table 4.2 – column 5 for which $B_{n,m}$ reduce to $B_{n,0}$. The residual aberrations are derived from (4.20) of centered system reflective Schmidts with the addition of an *Astm* 5 component in α^2/Ω^3. All gratings must satisfy the condition of *null-power zone ratio* $r_0/r_{\rm m} = \sqrt{k} = \sqrt{3/2}$.

→ *The resolving power of all-reflective spectrographs working in normal diffraction,* $\beta_0 = 0$*, on the best spherical focal surface and in the symmetry plane perpendicular to the grating lines is* $(\varphi_{\rm m} \equiv \beta_{\rm m})$

$$\left\{\begin{array}{l} d_{\rm C} = 0.00043\ \sin^2\!\alpha\ /\ \Omega^3\ +\ 0.011\,\varphi_{\rm m}^2/\ \Omega^3, \\ s = \cos^2\!\varphi_{\rm m}\,, \quad k = 3/2 \ \text{ i.e. } \ M = 3/64\,\Omega^2. \end{array}\right. \tag{4.24}$$

From Table 4.2 – column 5, the shape of the grating is represented by

$$Z_{\rm Opt} = \sum_{2,4,6,\ldots} B_{n,0}\,r^n = \sum C_n\,r^n \quad \text{with} \quad C_n = \frac{s\mathcal{A}_n}{(1+\cos\alpha)R^{n-1}}\,. \tag{4.25}$$

Because of the fact that large size spectrographs have been built with f-ratios as fast as f/1.2, it is useful to develop the power series expressing the grating shape. This allows determining the VTD of vase form active submasters generating aspherized gratings by replication. With a field flattener lens adding a small amount of *Sphe* 3, the field optimization is efficiently carried out with an s-ratio becoming larger than unity for over-correction. Without a field flattener lens, the best s-ratio gives the under-correction $s = \cos^2\varphi_{\rm m}$, and the grating shape is represented by the expansion

$$Z_{\rm Opt} = \frac{\cos^2\varphi_{\rm m}}{1+\cos\alpha}\left\{\frac{M}{1+M}\frac{r^2}{R} - \left[\frac{1}{4} - \frac{M(1-M^3)}{(1-M)(1+M)^3}\right]\frac{r^4}{R^3} - \left[\frac{3}{8} - \frac{2M(1+M-M^3)}{(1+M)^5} - \frac{3M^2}{2(1+M)^2}\right]\frac{r^6}{R^5} - \left[\frac{45}{64} - 4M + ...\right]\frac{r^8}{R^7} - \left[\frac{193}{128} + ...\right]\frac{r^{10}}{R^9} - ...\right\} \tag{4.26}$$

with $M = k/2^5\Omega^2 = 3/2^6\Omega^2$ and $r \le r_{\rm m} = R/4\Omega$.

Considering the various spectrograph designs based on the Schmidt concept, the reflective design with axisymmetric aspherized gratings mounted in normal diffraction is of the highest interest because of the simple and compact design also presenting a high throughput and several other important *advantages* listed in Sect. 4.4.8.

4.4.4 Bi-Axial Symmetric Gratings ($\beta_0 \neq 0$)

For cases where the diffraction angle overpass the range, say, $\beta_0 = \pm 5°$, *bi-axial symmetry* gratings provide more efficient aberration corrections. This allows recovering residual images similar to the case of axisymmetric gratings and a resolving power such as given by (4.24).

→ *The resolving power of all-reflective spectrographs working with bi-axisymmetric gratings out from normal diffraction onto the best-fit spherical focal surface and in the symmetry plane perpendicular to the grating lines is:*

$$\left\{\begin{array}{l} d_{\rm NC} = 0.00043\,\sin^2(\alpha+\beta_0)/\,\Omega^3 \;+\; 0.011\,\varphi_{\rm m}^2/\,\Omega^3, \\ s = \cos^2\varphi_{\rm m}, \quad k = 3/2 \quad \text{i.e.} \quad M = 3/64\,\Omega^2. \end{array}\right. \tag{4.27}$$

Such as with non-centered reflective Schmidts telescopes (Sect. 4.3.2), the grating surface is approximated by generating homothetical ellipses. Thus, the optical surface of the grating is

$$Z_{\rm Opt} = \sum_{2,4,6,...} C_n\left(h^2x^2+y^2\right)^{n/2}, \tag{4.28}$$

with

$$C_n = \frac{s\,\mathcal{A}_n}{(\cos\beta_0+\cos\alpha)\,R^{n-1}} \quad \text{and} \quad h = \cos\beta_0\,. \tag{4.29}$$

Biaxial symmetry gratings also apply to the design of all-reflective Schmidt-Littrow mountings ($\beta_0 = -\alpha$) but do not seem to have been developed up to now. Generally, *Littrow mountings* concern high-resolution spectroscopy requiring long focal lengths with large incidence angles and slow f-ratios. This would lead to moderate asphericity gratings.

For spectrograph camera f-ratios up to f/2, in order to obtain a more accessible focus if necessary, a folding mirror may be added between the grating and the camera mirror. This holed mirror may fold either both incident and diffracted beams or only the latter. For cameras faster than f/2, the detector is fitted at the inner focal surface then requiring large size gratings (Sect. 4.4.11).

4.4.5 *Flat Fielding of All-Reflective Aspherized Grating Spectrographs*

Given the state-of-the-art of present detectors, it is quite impossible to have disposal of a sensitive surface other than a flat shape. Therefore, the cryostat window of the detector has to be replaced by a flattener lens. This operation requires, a small increase of the grating asphericity to compensate for $Sphe\,3$ mode from the field flattener and a small repositioning of the camera optics towards the grating for $Coma\,3$ compensation.

The flattener lens is of positive power. From Petzval's condition in Sect. 1.10.1 (see also Chrétien [17], Born and Wolf [7]), setting the refractive index $N_{i+1} = -N_i$ for a reflection at a surface numbered i of curvature c_i, and also N_{I+1} for the refractive index of the last medium, a null curvature is achieved if

$$C_P \equiv N_{I+1} \sum_{i=1}^{I} \left(\frac{1}{N_{i+1}} - \frac{1}{N_i} \right) c_i = 0\,, \tag{4.30}$$

Thus, with respect to the mirror curvature $1/R$ and curvatures c_1, c_2 of the singlet lens, and after substitutions, a flat field is obtained if

$$(N-1)(c_1 - c_2)/N - 2/R = 0\,, \tag{4.31}$$

where N is the lens refractive index of the central wavelength.

Although it is possible, in convergent beams, to design a singlet lens free from axial chromatism by a convenient balance of its c_1, c_2 curvatures, thickness, and back focal distance to the detector (Wynne [96]), ray-tracing optimizations show that the best field corrections are obtained with an almost convexo-plane singlet lens, i.e. $c_2 \simeq 0$, having the lowest possible thickness. For wide spectral range spectrographs, one of the most often used materials is fused silica, because of its low dispersion and high ultimate strength for a safe use as a cryostat window. Assuming a convexo-plane lens whose first surface of curvature c_1 is passed by the light, a first approximation gives

$$c_1 = 2N\,/\,(N-1)R \qquad \text{and} \qquad c_2 = 0\,. \tag{4.32}$$

A very slight change of c_1, c_2 somewhat improves the performance by giving c_2 a small curvature that minimizes $Astm\,3$ of the lens whilst its power $c_1 - c_2$ is unchanged.

The elastic deformation of the lens due to the cryostat vacuum does not significantly change in its thickness distribution and no elasto-optical effect modifies the image quality in the field. In the case of a lens flattener used as a cryostat window and thus submitted to the atmospheric pressure $q \simeq 10^5$ Pa, the thickness of the lens must be carefully determined. Denoting a the half diameter of the lens, its axial and edge thicknesses must satisfy the following bending and shearing conditions

$$t_0 \geq \sqrt{\frac{3(3+\nu)}{8}\frac{q}{\sigma}}\;a \qquad \text{and} \qquad t_{\text{edge}} \geq \frac{1}{2}\frac{q}{\sigma}\,a\,, \tag{4.33}$$

where the stress σ must be taken two or three times smaller than the ultimate stress σ_{ult} (cf. Glass rupture and loading time in Sect. 5.2.5). For fused silica, $\sigma_{\text{ult}} = 700 \times 10^5$ Pa and Poisson's ratio $\nu = 0.16$.

4.4.6 Examples of All-Reflective Aspherized Grating Spectrographs

The hereafter presented spectrographs are not always strictly all-reflective systems since some of them use a singlet lens for field flattening and as a detector cryostat window. Nevertheless, the terminology "all-reflective" is almost correct because a singlet lens located just before the focal surface does not introduce a significant amount of chromatism but only slightly modifies the correction of the spherical aberration of the system.

• **Laboratory spectrograph with k = 0 geometry gratings**: Although a $k = 3/2$ geometry is preferred for the best aberration correction, the construction of the first aspherized grating was carried out with a geometry $k = 0$ by a *tulip-form submaster* (Lemaitre and Flamand [30]) for a laboratory spectrograph that, thus, used a plane-aspheric grating. A stainless steel submaster was polished flat when not stressed and a standard plane grating from Jobin-Yvon was replicated on it surface. In a second step, the sub-master was elastically aspherized by a central force in reaction with its free edge and then replicated on a Schott Zerodur rigid substrate. This replica is the final aspherized grating (see Chap. 5, Sect. 5.4.4).

The thickness of the tulip-form submaster was close to $\mathcal{T}_{40} \propto [\,\rho^{8/(3+\nu)} - 1/\rho^2]^{1/3}$ [cf. Eq. (3.30)], a form that generates a pure ρ^4-flexure and corresponds to $M = 0$, $\mathcal{A}_2 = 0$ and $\mathcal{A}_4 = -1/4$ in Table 4.1. This entails $k = r_0^2/r_{\text{m}}^2 = 0$, thus a null-powered zone located at the submaster vertex. In fact this form was slightly modified for also correcting the fifth-order spherical aberration. The thickness distribution generated the $Sphe\,3$ and $Sphe\,5$ modes but did not provide any curvature flexure $Cv\,1$ although a small amount arose for this latter mode, without any drawback, caused by the finite thickness of the central part of the tulip form. For $\rho =$

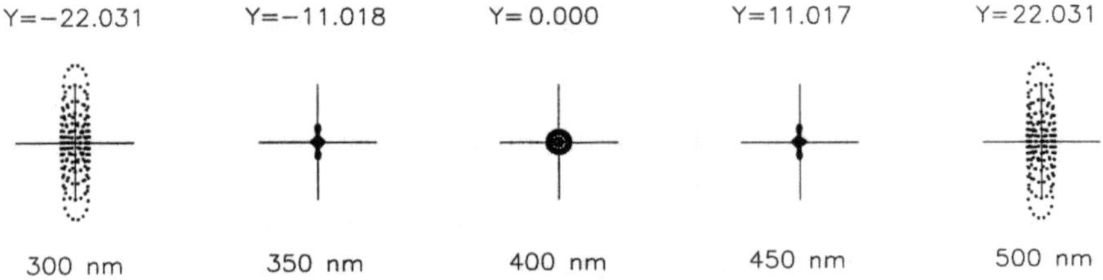

Fig. 4.14 Theoretical residuals of a 3,000–5,000 Å spectrum with a plane-aspheric grating $k = 0$ and $\beta_0 = 0$ onto a curved field 44 mm in length or 15° field. Camera beam f/2×f/1.8 at λ_0. Although a much better aberration correction would occur with $k = 3/2$, this design provides a good correction in the direction of dispersion with residuals smaller than 25×5 μm. The residual field balance leads to *Sphe* 3 at center and *Astm* 5 at edge

1, the corresponding simplysupported submaster edge was $r = 85$ mm. The raytrace best fit of the design under-correction factor was found by setting $s = \cos\varphi_m = 0.991$ [instead of setting $s = \cos^2\varphi_m$ for $k = 3/2$ in (4.24)] and the theoretical shape including the fifth-order submaster active deformation was that of (5.90) in Sect. 5.4.4.

This first aspherized grating, 1,200 ℓ.mm^{-1}, 102×102 mm, was installed in an experiment spectrograph designed as in Fig. 4.13d for optical evaluation. A slit and collimator mirror provided a parallel beam $2r_m = 90$ mm in diameter illuminated the grating mounted in a normal diffraction arrangement, $\beta_0 = 0°$ at $\lambda_0 = 4{,}000$ Å corresponding to the incident angle $\alpha = \delta_0 = 28.7°$. The 90×100 mm elliptical stop at the grating leads to camera beam focal ratios of f/2×f/1.8 for λ_0 and in the two principal directions. Fe- and Cd-spectra were obtained on curved films over the ultraviolet range 3,000–5,000 Å with a 42 Å/mm dispersion and a camera semi-field $\varphi_m = \beta_m = \pm 7.5°$. The results from optical testing were found in full agreement with the theoretical image blurs (Fig. 4.14).

• **Astronomical spectrographs with k=3/2 geometry gratings**: The first aspherized grating built by the *active optics replication process* for astronomical observations was for the UV prime focus (UVPF) spectrograph at CFHT in Hawaii and used an axisymmetric aspheric grating working in the order at diffraction angle $\beta_0 = 0$ for $\lambda = 4,000$ Å.

The UVPF spectrograph made the first run of the CFHT instrumentation in early 1980. It was originally observing with photographic ultraviolet enhanced films (Lemaitre [33]) with a nebular capability of 4 arcmin on the sky. The previous laboratory spectrograph (Fig. 4.14) served as a model for the development of aspherized gratings. The f/3.8 ratio of prime focus was reduced by the camera mirror to f/1.26 in the nebular direction. The dispersion was 55 Å.mm^{-1} at 4,000 Å. This spectrograph was updated for use with a CCD (Lemaitre and Vigroux [35], Boulade et al. [8]). This required introducing a convexo-plane field flattener lens as the cryostat window (Fig. 4.15).

The aspherization of the axisymmetric grating was carried out from replication of a *vase-form active submaster* stressed by air pressure (see Sect. 5.4.3).

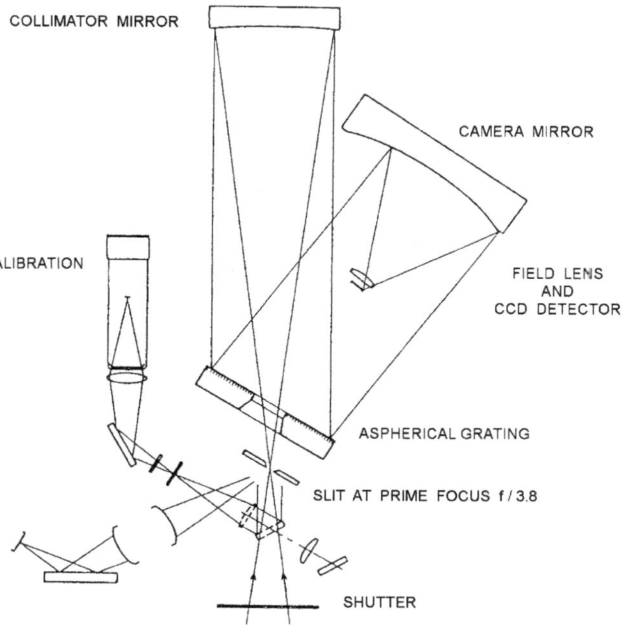

Fig. 4.15 Optical design of the UVPF spectrograph of CFHT. The 1,200ℓ.mm^{-1} axisymmetric aspherized grating provided a spectral range $\lambda\lambda$ [3,000–5,200 Å]. The normal diffraction at 4,000 Å provides an f/1.10 camera beam in the dispersion direction. Originally, all the spectral range was covered by curved UV films 36 mm in length. Updating was performed in 1985 with a thinned CCD covering 890 Å in one frame. Discrete slidings of the CCD and field flattener onto a cylinder allowed covering all the spectral range with three positions (Lemaitre and Vigroux [35])

The submaster was designed to generate the grating geometry $k = r_0^2/r_m^2 = 3/2$ which provides the best balance of the field aberration residuals. Its $r_0 = 71$ mm built-in radius allowed the grating replication for use of an elliptical aperture area. This area is determined by the diameter of the collimator cross-section beam as $2r_m = 2r_0/\sqrt{k} = 116$ mm which corresponds to the grating width. In the other direction and for a normal diffraction angle, the collimator incidence angle $\alpha = 28.7°$ corresponds to a 132 mm grating length. Hence, the useful aperture of the grating – aspherized on a rigid Zerodur holed substrate – was an ellipse 116×132 mm in size.

The UVPF spectrograph design parameters for the CCD version are displayed in Table 4.3, and residual images in Fig. 4.16.

All-reflective spectrographs are also efficient for infrared studies and, of course, for broad-band spectral coverage. Having a number of optical surfaces reduced to a *minimum*, these designs are particularly appropriate for medium and low-resolution spectroscopy, i.e. faint object studies. A set of interchanged gratings provides various dispersions in the same instrument. Such pseudo-plane aspherical gratings having rotational symmetry and in the form $k = 3/2$, have been built with the

Table 4.3 UVPF spectrograph optical parameters f/1.26 × f/1.10. Unit [mm]

Element	Radius of curvature	Axial separation	Figure
Primary mirror CFHT	27, 066.	13,533.	parabola
Slit	∞	437.	——
Collimator mirror	−874.	-372.	parabola
Reflective grating	Z_{Opt} (*)	225.	asphere
Camera mirror	−282.5	–129.495	sphere
Field lens window	−50.	–4.	sphere
in fused silica	∞	–4.	sphere
Focal surface	∞		plane

(*) $k = 3/2$, $r_m = 58$ mm, $\alpha = 28.7°$, $\beta_0 = 0$ at 4,000 Å. Over-correction $s = 1.042$ with flattener lens. Grating optical figure: $Z_{Opt} = 6.021\ 10^{-5}\,r^2 - 5.42\ 10^{-9}\,r^4 - 9.8\ 10^{-14}\,r^6 - 2\ 10^{-18}\,r^8$. Aspherized grating LOOM/Jobin-Yvon, 1,200 ℓ.mm^{-1}, 1st-order, 116×132 mm. Camera-collimator angle $\delta_0 = 28.7°$.

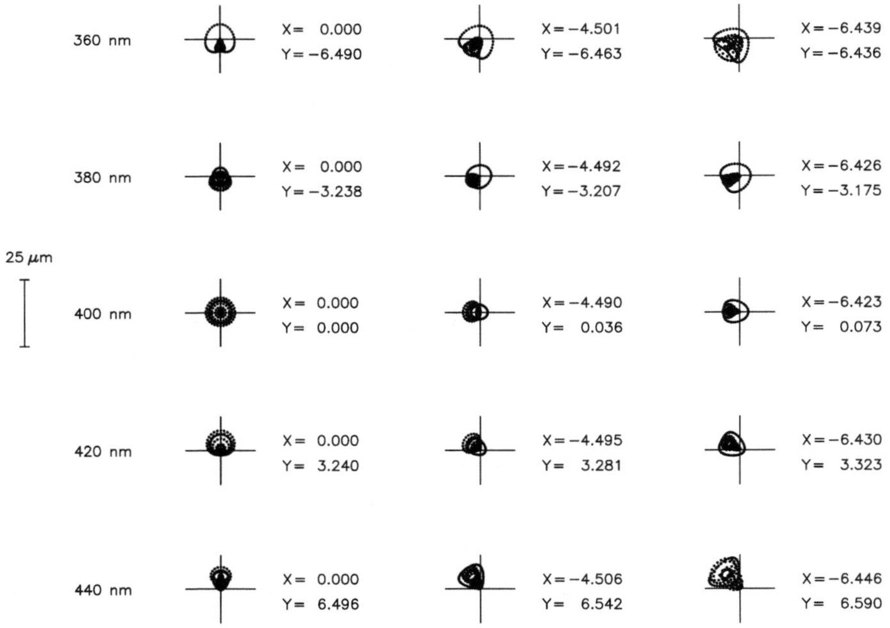

Fig. 4.16 UVPF spectrograph of CFHT – Residual aberrations on the thinned CCD when centered at 4,000 Å. The two other positions, obtained by sliding the field lens with CCD each side on a cylinder, show a similar amount of residual aberrations without refocussing in all the spectral range 3,000–5,200 Å

MARLY spectrographs (Lemaitre and Kohler [36]) of Haute Provence and Nanjing observatories, and CARELEC spectrograph (Lemaitre et al. [37]) of the 2 m f/5–f/15 telescope in Haute Provence. The optical designs of these instruments have shown that axisymmetric gratings can provide excellent image quality with positions deviating substantially from normal diffraction $\beta_0 = 0$ up to $\beta_0 = 12$

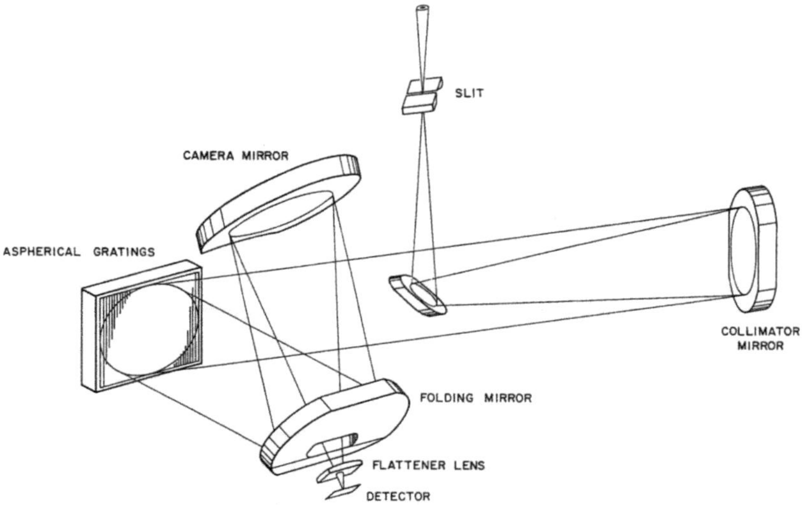

Fig. 4.17 Schematic view of catoptric spectrographs MARLY f/8–f/3.2 and CARELEC f/15–f/2.5, using axisymmetric aspherized gratings of 66 and 100 mm central groove length respectively. The mounting angle between the collimator and camera axes is $\delta_0 = \alpha + \beta_0 = 30°$ [37]

or 15°. The addition of a holed folding mirror (Fig. 4.17) in the camera optics provides better access to the detector and its cryostat. Several vase-form active sub-masters allowed aspherizing by air pressure plane standard gratings from Jobin-Yvon with the optimal profile geometry $k = 3/2$. CARELEC is equipped with five axisymmetric aspherized gratings designed for various pair angles α, β_0 (Table 4.4).

Table 4.4 Axisymmetric gratings of CARELEC Spectrograph f/2.5×f/2.15

J-Y #	$1/\ell$ mm^{-1}	K	λ_B [Å]	α [deg.]	β_0 [deg.]	λ_{min} [Å]	λ_{max} [Å]	$\mathcal{R}$ $\lambda_B/\Delta\lambda$
51006	1,200	1	4,000	29.4	0.6	3,000	6,500	3,750
51009	1,200	1	7,500	42.8	−12.8	6,000	10,500	7,000
		2	3,750	42.8	−12.8	3,000	4,950	7,000
51013	600	1	5,000	23.9	6.1	3,800	9,000	2,000
51049	150	1	5,000	17.7	12.3	3,800	8,500	600
510SP	150	1	8,500	18.8	11.2	5,750	11,000	1,000

Gratings aspherized with $k = 3/2$. Mounting angle $\delta_0 = 30°$. Displayed: Jobin-Yvon grating references, line density $1/\ell$, grating order K, blazed wavelength λ_B, incident angle α, diffraction angle β_0 at center of field, spectral ranges λ_{min}, λ_{max} and spectral resolutions $\mathcal{R}$.

4.4.7 All-Reflective Spectrographs Without Central Obstruction

All-reflective spectrographs without central obstruction can be obtained by use of an off-axis region of the design case shown in Fig. 4.13-D. Thus, if the f-ratio of the starting design is f/n, then the unobstructed design should be, say, f/(2.2n). This latter design would show residual aberrations of an f/n spectrograph. Hence a design without central obstruction is only practicable if derived from a central obstruction design where the focal-ratio is not too fast.

For such unobstructed spectrographs, active optics provides *off-axis plane-aspheric gratings* by replication of an off-axis region of a deformable axisymmetric grating submaster – flat at rest – when stressed by air pressure (up to $q = 25$ Atm.).

These developments were carried out for the design and construction of the ISARD and OSIRIS spectrographs (Lemaitre and Richardson [38]). ISARD is an f/5–f/25 *imager-spectrograph* dedicated to the 2-meter Bernard Lyot telescope at Pic du Midi observatory for faint object studies. OSIRIS is an f/5 telescope and f/2.2 spectrograph on board the ODIN satellite launched in 2001.

4.4.8 Advantages of Quasi-all-Reflective Spectrographs

Several advantages of the above quasi-all-reflective spectrographs render these systems attractive for moderate brightness and faint object studies.

Apart from the field lens, the all-reflective form of the imaging mode is simply obtained by interchanging the aspheric reflective grating by an aspheric mirror; if necessary, the addition of a slightly cylindric lens before the cryostat window field lens flattener somewhat improves the imaging quality.

All-reflective multi-object spectrographs and *imager spectrographs* benefit from:

(1) a *uniform dispersion law* of all spectra over the field of view,
(2) a *quasi-constant linear dispersion* over the full spectral range,
(3) an almost *distortion free* instrument in the nebular direction,
(4) a *high throughput* from the atmospheric cut-off at 320 nm to the far-infrared.

These features are of substantial interest to increase the accuracy in the data reduction process, particularly for instruments dedicated to 3-D spectroscopy.

4.4.9 Diffraction Gratings and Electromagnetic Theoretical Models

The manufacturers of relief-surface diffraction gratings have achieved a high degree of performance in the construction of the facet angle called *blaze angle*. In a given diffraction octave, this technology controls the diffracted energy and is in close agreement with electromagnetic theoretical models. For such broad-band spectral coverage, reflective gratings are more efficient than transmission gratings. For line-density gratings $1/\ell \leq 900\,\mathrm{mm}^{-1}$ used in the visible, the absolute reflectance measured at or near the blaze angle reaches the metal theoretical reflectivity; for instance, Jobin-Yvon classically obtains an efficiency of 90% for aluminum (see hereafter).

For relief-surface diffraction gratings having a groove spacing smaller than five wavelengths ($\ell/\lambda \leq 5$), electromagnetic effects of light propagation appear and modify the reflectance of the polarization components with λ. Local discontinuities such as the Wood anomalies [89] are well known by experimenters since the beginning of the last century. Preliminary explanations were investigated by Rayleigh [62, 63] and later by Meecham [52], Stroke [78] and Maréchal and Stroke [46]. A new formalism was obtained by Petit and Cadilhac [58] and Pavageau and Bousquet [57].

Theoretical analyses have been investigated and developed by Petit [59, 60], Maystre [48–50], Nevière [54–56], McPhedran [51], that allow predicting the reflectance curves $\mathcal{R}_{\perp,\parallel}(\lambda)$ and important local variations with λ. These models allow consideration of many cases of groove profiles such as triangular semi-rectangle *echelette*, holographic *sinusoidal*, rectangular *lamellar*, or low groove density *echelle*. For example, in the case of transmission gratings such as used in some astronomical instruments with so-called Carpenter gratings or grisms – i.e. a grating glued on a prism which refractive index provides a compensated deviation –, it has been found that the transmitted efficiency could be substantially enhanced by deposition of a metallic thin layer onto the small "side-facet" of the grooves. There is no doubt that manufacturing these gratings would present technological difficulties. The case of reflective gratings allows one to obtain the highest efficiency.

The degree of achievement of the electromagnetic models allows prediction of where the grating manufacturers can obtain progress of some percent on the reflectance of a given grating type. A theoretical review on the efficiencies of reflective gratings in function of the blaze and deviation angles was carried out by Loewen et al. [42, 43]. This survey provides comparisons between semi-rectangle and sinusoidal grooves. Figure 4.18 displays an example of these theoretical results for

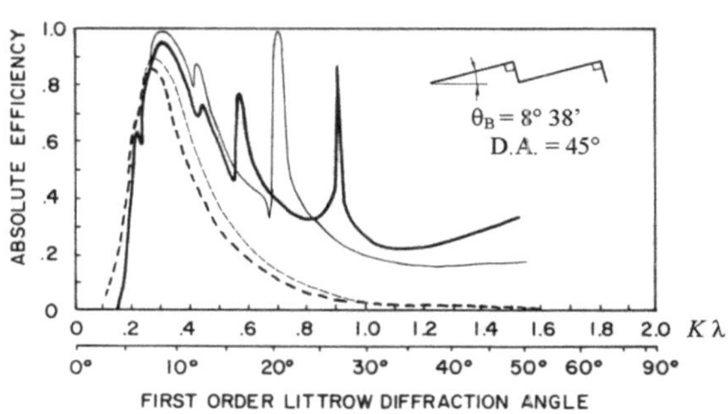

Fig. 4.18 First-order efficiency curves for a 8° 38' blaze angle echelette grating, perfectly conducting, used with 45° deviation angle. *Solid-line*: in the *S*-polarization plane. *Dashed-line*: is the *P*-polarization plane. The *light lines* give the results for Littrow mount as reference (Loewen, Nevière & Maystre [43])

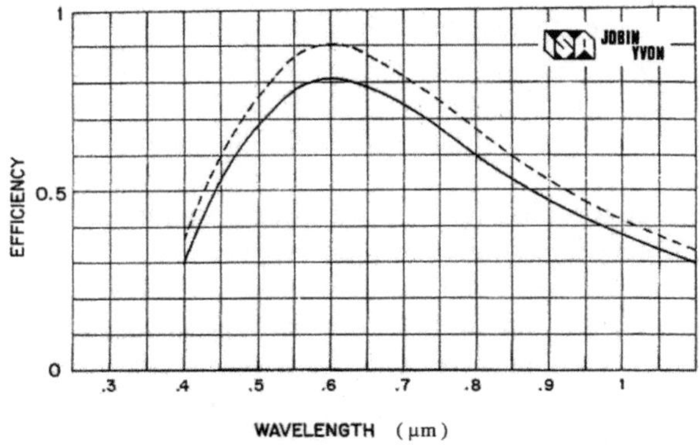

Fig. 4.19 Typical absolute reflectance of an aluminum grating 300 ℓ.mm^{-1}, ref. Jobin-Yvon 51019, measured in the 1st-order under Littrow conditions ($\beta = -\alpha$). Blaze angle $\theta_B = 5.16°$ for $\lambda = 600$ nm, $\lambda/\ell = 2 \sin\theta_B = 5.5$ in Littrow. *Solid-line*: mean values from S and P polarization planes. *Dotted-line*: Expected reflectance with Ag + MgF_2 coating (from J. Famand, Horiba Jobin Yvon Corp.)

the case of an infinite conductor surface with echelettes (semi-rectangle grooves). Figure 4.19 displays the behavior from experiment of the same grating type under unpolarized light of which the two components are denoted S and P.

It must be noticed that holographic and ruled relief-surface diffraction gratings with a line-density $\leq 900\,\text{mm}^{-1}$ have a high efficiency over a large spectral range.

→ *For the low and medium blaze angle region i.e.* $\theta_B = 5$–$18°$ *and deviation angle* $\delta =$ D.A. $= 0$–$45°$, *one obtains unpolarized grating efficiencies* $(P + S)/2$ *higher than 50% over a wavelength region from 0.66 to 1.80-times the blaze wavelength* ($\lambda_B = 2\ell \sin\theta_B$ in Littrow), *i.e. much more than an octave.*

4.4.10 Grating Manufacturing Methods

• **Type A: Diamond ruled gratings**: The first and classical method of making diffraction gratings was pioneered by Joseph von Fraunhofer, in 1813, by ruling of a metal substrate with a diamond translation machine and a precision screw controlling the pitch of the grooves. *Ruled gratings* led him to the discovery and the measure of the main absorption lines of the solar spectrum, known as Fraunhofer lines. Such ruling engines were later improved by H.R. Rowland, who obtained 7.5-inch gratings and the first concave-shaped gratings, by A.A. Michelson and others. G.R. Harrisson and his collaborators introduced a laser interferometric feedback control of the pitch in 1955 which allowed one to build 30-cm gratings. It was found that the limited lifetime of the diamond during the ruling process prevents obtaining more than a ~15-km ruling length.

• **Type B: Holographic surface relief gratings**: Since 1960 and with the appearance of lasers, the ruling process was progressively substituted by the holographic recording of interferometric fringes in a photosensitive film. Starting in the 1980s, blazed holographic gratings by the ion beam method were developed by J. Flamand et al. [22] at Jobin-Yvon. The method allows one to obtain *holographic surface relief gratings* which are efficiently blazed. For most manufacturers, the available standard gratings are limited in size to 12×14 or $15\times20\,\text{cm}^2$ which in fact is the old standard sizes issued from the classical diamond ruling method. Prototype plane gratings in a format as large as $30\times40\,\text{cm}^2$ can presently be obtained with line densities from 150 to $1{,}200\,\ell.\text{mm}^{-1}$.

• **Type C: Volume phase holographic gratings**: Another type of grating called *volume phase holographic gratings* has been developed more recently. Rather than being diffracted by a surface-relief structure, as for previous types A and B, the light undergoes Bragg diffraction as it passes through the volume of a photosensitive film of thickness from a few to some hundred microns. In recent developments the optical accuracy has been improved by sandwiching this film between two flat windows (Arns [1]). In the holographic recording process of plane gratings, the refractive index of the photosensitive film is modulated by slanted and parallel phase planes (Barden [6]). Large size gratings can be obtained for transmission or reflection grating geometries. However, aberration corrections on such gratings suffer from chromatic aberrations.

4.4.11 Towards Large Size Asphérized Reflective Gratings

A reflective relief surface grating generating an aberration correction operates in an achromatic form which, then, is efficient for the widest possible spectral range. A volume phase holographic grating can only provide an aberration correction at a given wavelength λ_0; such gratings suffer from an important chromatic variation error for $\lambda \neq \lambda_0$ which then restrains the spectral range. A holographic surface relief grating – *aspherized reflective grating* – avoids this difficulty and hence must be preferred for *faint object spectrographs.*

With extremely large telescope projects in development, astronomers could expect to soon obtain much larger *holographic surface relief reflective gratings* such as 20×25 or even $40\times50\,\text{cm}^2$. As pointed out in the above sections, only this grating type (Type C) efficiently provides various achromatic aberration corrections. Their asperization requires use of active optics and replication techniques (cf. next Chapter).

4.4.12 Large All-Reflective Aspherized Grating Spectrographs

Given a spectrograph camera mirror f-ratio and a telescope central obstruction ratio, a large size grating is a natural way for the optimal use of a large detector

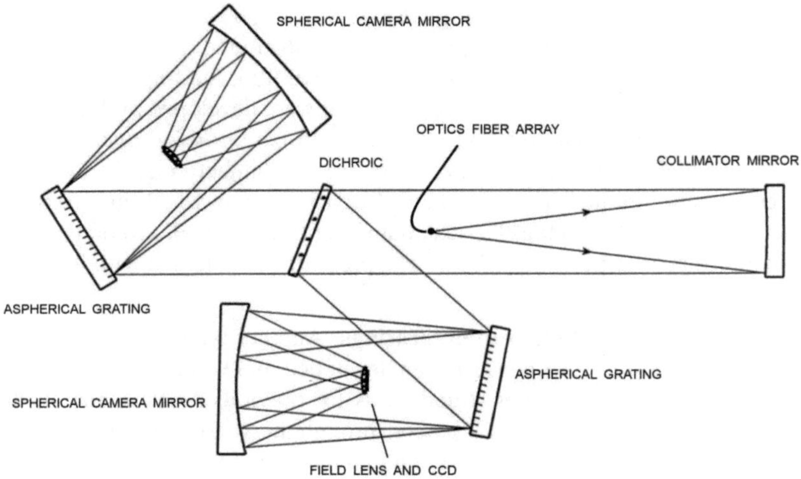

Fig. 4.20 Optical design of LAMOST spectrographs for $\mathcal{R} = 2{,}000$. This uses aspherized reflective gratings of 200 mm groove length and an aperture ratio of f/1.5 in this direction. A dichroic provides the two spectral ranges λλ [370–590 nm] and λλ [570–900 nm]. The CCD detectors have 2048×2048 pixels of 24×24 μm. The mounting angle between the collimator and camera axes is $\delta_0 = 35°$. The normal diffraction angle, $\beta \simeq 0$, allows a design with axisymmetric aspherized gratings (Zhu and Lemaitre [97])

without adding central obstruction. For the second-generation spectrographs of the giant Schmidt LAMOST (cf. Sect. 4.3.4), the spectral resolution will be set two-times larger, i.e. at $\mathcal{R} = 2{,}000$ for both blue and red channels (instead of $\mathcal{R} = 1{,}000$). For such instruments, the throughput will be optimal by use of axisymmetric surface relief gratings working with diffraction angles close to $\beta_0 = 0$ (Fig. 4.20 and Table 4.5). As for the previous examples, the aspherization of such reflective gratings provides an *achromatic* correction of the spherical aberration and then the largest possible spectral range.

Other advantages with aspherized reflective gratings are to *avoid* the difficult task of making huge aspherical refractive Schmidt plates and then to cancel its two associate ghost images. In addition, one should mention that the two *ghost* images generated by the glass substrates of volume phase transmission gratings (generally sandwiched between two plates) are also cancelled by use of reflective gratings. The optical performance of quasi-all-reflective designs – only including a field-flattener cryostat-window lens as refractive element – are similar for both blue and red channel (Fig. 4.21).

The output of the optics fiber array – spectrograph slit – is a parabolic line in a X, Z plane. At λ_0, the lateral shift of the spectra shows a sag of $\Delta Y \simeq 0.495$ mm over the field; the red channel shows a quite similar lateral sag $\Delta Y \simeq 0.505$ mm for $\lambda_0 = 735$ nm at the detector. This X, Y locus of the λ_0 spectral elements can be straightened by rearranging the output of the fiber array – spectrograph slit input – in a lateral circle arc in the Y-direction with an opposite lateral sag of ratio $f_{Col}/f_{Cam} = 4/1.5$.

Table 4.5 Optical parameters of LAMOST spectrographs $\mathcal{R}$ = 2,000. Blue channel [370–590 nm] – Red channel [570–900 nm]. Unit [mm]

Element	Radius of curvature	Axial separation	Figure
Aperture stop	——	802.374	
Slit (optics fibers output)	−802	797.626	on sphere
Collimator mirror	−1, 600	−1,600.000	sphere
Reflective grating	$Z_{\rm Opt}$ (*)	536.000	asphere
Camera mirror	−600	−283.020	sphere
Field lens window	$R_{1\,\rm Lens}$ (†)	−9.000	sphere
in fused silica	3240	−4.000	sphere
Focal surface	∞		plane

(*) Gratings 1,200 and 800 ℓ.mm^{-1}, $\alpha = 35°$, $\beta_0 = 0°$, $Z_G = \Sigma\, C_n\, r^n$: $C_2 = 2.0150\ 10^{-5}$, $C_4 =-6.068\ 10^{-10}$, $C_6 =-2.34\ 10^{-15}$, $C_8 =-1.2\ 10^{-20}$.
(†) First surface lens $R_{1\,\rm Lens}$ =–105 mm (blue) and –103.2 mm (red).
Clear length of groove passing by grating center $2r_m$ = 200 mm; Gratings null-power zone at $r_0/r_m = y_0/y_m = \sqrt{3/2}$. $\delta_0 = 35°$. Collimator f/4. Camera f/1.5 × f/1.22. Fiber feed 250 per unit. Diagonal field angle $\varphi_m = \pm 6.3°$.

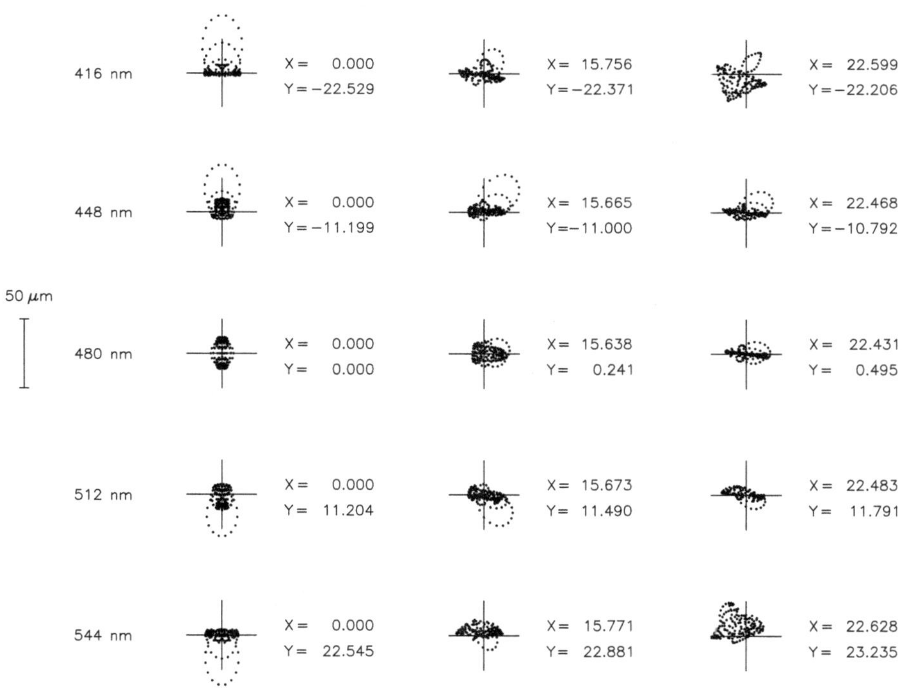

Fig. 4.21 Residual aberrations of LAMOST spectrographs for a resolution $\mathcal{R}$ = 2,000 in blue. The spots show five spectral elements with a central wavelength λ_0 = 480 nm, at heights 0, 0.7, 1 of a 49 × 49 mm FOV for a 2,048 × 2,048 CCD with 24 × 24 μm pixel size. Two positions of the grating cover the blue range $\lambda\lambda$ [370–590 nm]. The X,Y field coordinates of each spot show the position of the principal ray (Zhu and Lemaitre [97])

In accordance with the properties listed in Sect. 4.4.8 one may verify that, from the X, Y field coordinates of each principal ray in Fig. 4.21, the dispersion law is quasi-linear and also *uniform* in the perpendicular direction.

References

1. J.A. Arns, W.S. Colburn, S.C. Barden, Volume phase gratings for spectroscopy, in *Current Developments in Optical Design VIII*, SPIE Proc., **3779**, 313–323 (1999)
2. J.G. Baker, A family of flat-field cameras, equivalent in performance to the Schmidt cameras, Proc. Am. Philos. Soc., **82**, 339 (1940)
3. J.G. Baker, US Patent No 2 458 132 (1945)
4. A. Baranne, Un nouveau montage spectrographique, Comptes Rendus, **260**, 3283–3286 (1965)
5. A. Baranne, M. Mayor, J.-L. Poncet, Coravel – A new tool for radial velocity measurements, Vistas Astro., **23**, 279–316 (1979)
6. S.C. Barden, J.A. Arns, W.S. Colburn, Volume-phase holographic gratings, in *Current Developments in Optical Design VIII*, SPIE Proc., **3779**, 313–323 (1999)
7. M. Born, E. Wolf, *Principles of Optics*, Cambridge University Press, Cambridge 253 (1999)
8. O. Boulade, G.R. Lemaitre, L. Vigroux, UV prime focus spectrograph for the CFHT, Astron. Astrophys., **163**, 301–306 (1986)
9. A. Bouwers, in *New Optical Systems – Achievements in Optics*, Elsevier Edit., Amsterdam (1946)
10. I.S. Bowen, Spectrographic equipment of the 200-inch Hale telescope, Ap. J. **116**, 1–7 (1952)
11. I.S. Bowen, Schmidt cameras, in *Stars and stellar Systems* I – *Telescopes*, G.P. Kuiper and B.M. Middlehurst edts., Univ. of Chicago Press, Chicago 43–61 (1960)
12. I.S. Bowen, Astronomical optics, *Annual Review of Astronomy and Astrophysics*, L. Goldberg ed., **5**, 45–70 (1967)
13. C.R. Burch, On the optical see-saw diagram, MNRAS **102**, 159–165 (1942)
14. C. Carathéodory, *Geometische Optik*, Springer edt., Berlin (1937)
15. C. Carathéodory, *Elementare Theorie des Spiegeltelescops von B. Schmidt*, B.G. Teubner edt., Leipzig u. Berlin, 1–36 (1940) and Hamburg. Math. Einzelschr., 28 (1940)
16. H. Chrétien, Le télescope de Newton et le télescope aplanétique, Rev. d'Opt., **1**, 13 et 51 (1922)
17. H. Chrétien, *Le Calcul des Combinaisons Optiques*, Sennac Edit., Paris, 346–350 (1958)
18. A. Couder, Sur un type nouveau de télescope photographique, Comptes Rendus, **183**, 1276–1279 (1926)
19. G. Courtès, *New Techniques in Space Astronomy*, IAU Conf. Paris, Labuhn & Lüst edt. (1971)
20. C. Fehrenbach, Principes fondamentaux de classification stellaire, Ann. Astrophys. **10**, 257–306 (1947), and **11**, 35 (1948)
21. C. Fehrenbach, R. Burnage, Vitesses radiales mesurées au prisme objectif de 620-mm de l'Observatoire de Haute Provence, Astron. Astrophys. Suppl. Series, **43**, 297 (1981)
22. J. Flamand et coll., *Diffraction Gratings Ruled and Holographic*, Horiba Jobin-Yvon handbook (2003)
23. K.G. Henize, The role of surveys in space astronomy, *Optical Telescopes Technology*, NASA-SP-233, US Gov. Print. Off., Washington DC (1970)
24. D.O. Henrix, W.H. Christie, Some applications of the Schmidt principle in optical design, Sci. Am., **8**, 161, 118–123 (1939)
25. G.A.H. Kellner, American Patent No 969 785, Fig. 3 (1910)
26. A. Kerber, Central Zeitg. f. Opt. and Mech., **8**, 157 (1886) (see Chrétien's book)
27. J. Kross, in *L'oeil du grand Tout*, Life story of B. Schmidt from its invention to the Palomar Schmidt, Laffont edt., Paris (1997)

28. A. Lallemand, N. Duchesne, G. Wlerick, *Advance in Electronics and Electron Physics*, **12**, 5 (1960)
29. G.R. Lemaitre, Reflective Schmidt anastigmat telescopes and pseudo-flat made by elasticity, J. Opt. Soc. Am., **66**(12), 1334–1340 (1976)
30. G.R. Lemaitre, J. Flamand, Spectrographic development of diffraction gratings aspherized by elastic relaxation, Astron. Astrophys., **59**(2), 249–253 (1977)
31. G.R Lemaitre, Sur la résolution des télescopes de Schmidt de type catoptrique, Comptes Rendus Acad. Sc., **288 B**, 297 (1979)
32. G.R. Lemaitre, Asphérisation par relaxation élastique de miroirs astronomiques dont le contour circulaire ou elliptique est encastré ou semi-encastré, Comptes Rendus, **290 B**, 171 (1980)
33. G.R. Lemaitre, Combinaisons optiques à réseaux asphériques: Le spectrographe UV Prime Focus CFHT, Astron. Astrophys. Letters, **103**(2), L14–L16 (1981)
34. G.R. Lemaitre, Optical design with the Schmidt concept, *Astronomy with Schmidt Type Telescopes*, IAU Coll. Proc., M. Capaccioli edt., Asiago, Reidel publ., 533–548 (1984)
35. G.R. Lemaitre, L. Vigroux, All-reflective aspherized grating spectrographs at the prime focus of the CFHT, *Instrumentation for Ground-based Astronomy*, L.B. Robinson ed., Springer-Verlag, New York, 275–295 (1987)
36. G.R. Lemaitre, D. Kohler, Spectrographes à réseaux asphériques par réflexion: Les Marlys des observatoires de Haute-Provence et de Nanjing, Comptes Rendus Acad. Sc., **308 II**, 381–387 (1989)
37. G.R. Lemaitre, D. Kohler, D. Lacroix, J.-P. Meunier, A. Vin, All reflective aspherized grating spectrographs for Haute-Provence and Nanjing observatories: Marlys and Carelec, Astron. Astrophys. **228**, 546–558 (1990)
38. G.R. Lemaitre, E.H. Richardson, Ground-based and orbital off-axis aspherized grating imager-spectrographs: ISARD/Pic-du-Midi and OSIRIS/Odin-Orbiter, *Optical Astronomical Instrumentation*, SPIE Proc., **3355**, 682–695 (1998)
39. E.H. Linfoot, The optics of the Schmidt camera, M.N.R.A.S. **109**, 279–297 (1949)
40. E.H. Linfoot, E. Wolf, On the corrector plates of Schmidt cameras, J. Opt. Soc. Am. **39**, 752 (1949)
41. E.H. Linfoot, Two-mirror systems, *Recent Advances in Optics*, Clarendon edt., Oxford, **277**, Chap. 3, 176–183 (1955)
42. E.G. Loewen, D. Maystre, R.C. McPhedran, Correlation between efficiency of diffraction gratings and theoretical calculations over a wide range, Japan J. Appl. Phys., **141**, 143–152 (1975)
43. E.G. Loewen, M. Nevière, D. Maystre, Grating efficiency theory as it applies to blazed and holographic gratings, Appl. Opt., **16**(10), 2711–2721 (1977)
44. D. Lynden-Bell, Exact optics: A unification of optical telescope design, MNRAS, **334**, 4, 787–796 (2002)
45. D. Maksutov, New catadioptric menicus systems, J. Opt. Soc. Am., **34**, 270 (1944)
46. A. Maréchal, G.W. Stroke, Sur l'origine des effets de polarisation et de diffraction dans les réseaux optiques, Comptes Rendus Acad. Sc., **248**, 2042–2044 (1959)
47. N.U. Mayall, Bernhard Schmidt and his coma-free reflector, PASP **58**, 282–290 (1946)
48. D. Maystre, R. Petit, Détermination du champ diffracté par un réseau holographique, Optics Communications, **2**(7), 309–311 (1970)
49. D. Maystre, R. Petit, Sur l'efficacité du réseau échelette, Nouv. Rev. Opt. Appliquée, **2**(2), 115–120 (1971)
50. D. Maystre, Rigorous vector theories of diffraction gratings, in *Progress in Optics* XXI, E. Wolf edt., Elsevier Sciences publ. (1984)
51. R.C. McPhedran, D. Maystre, A detailed theoretical study of the anomalies of a sinusoidal diffraction grating, Optica Acta, **21**(5), 413–421 1974)
52. W.C. Meecham, Variational method for the calculation of the distribution of energy reflected from a periodic surface. I., J. Appl. Phys., **27**, 361 (1956)
53. D.R. Montgomery, L.A. Adams, Optics and the Mariner imaging instrument, Appl. Opt., **9**, 277 (1970)
54. M. Nevière, M. Cadilhac, Opt. Commun., **4**, 13 (1971)

55. M. Nevière, P. Vincent, R. Petit, Nouv. Rev. Opt. Appliquée, **5**(2), 65–67 (1974)
56. M. Nevière, D. Maystre, J.-P. Laude, Perfect blazing for transmission gratings, J. Opt. Soc. Am., A-**7**(9), 1736–1739 (1990)
57. J. Pavageau, J. Bousquet, Optica Acta, **17**, 469 (1970)
58. R. Petit, M. Cadhilac, Sur la diffraction d'une onde plane par un réseau infiniment conducteur, Comptes Rendus Acad. Sc., B-**262**, 468–471 (1966)
59. R. Petit, Optica Acta, **14**, 3, 301–310 (1967)
60. R. Petit, Electromagnetic theory of gratings, in *Topics in Current Physics*, **22**, Springer-Verlag ed. Berlin (1980)
61. G.M. Popov, New two-mirror systems for astrophysics, *Instrumentation in Astronomy*, SPIE Proc. **2198**, 559–569 (1994)
62. J.W. Rayleigh (Lord Strutt), On the dynamical theory of gratings, Proc. R. Soc. London, A-**79**, 349–416 (1907)
63. J.W. Rayleigh (Lord Strutt), Phil. Mag., **14**, 60 (1907)
64. E.H. Richardson, D. Salmon, The CFHT Herzberg spectrograph, CFHT Bull., Hawaii, **13** (1985)
65. F.E. Ross, The 48-inch Schmidt Telescope, Ap. J., **92**, 400–407 (1940)
66. B. Schmidt, Original manuscript (1929), conserved by The Academy of Estonia, University of Tallin (and by Erik Schmidt). Several aspherical plate profiles are considered which include a Kerber profile, (1929)
67. B. Schmidt, Mitteilungen der Hamburger Sternwarte, R. Schorr edt., 10 (1930)
68. B. Schmidt, A.A. Wachmann, Mitteilungen der Hamburger Sternwarte, R. Schorr edt., 6 and Plate I (1931)
69. B. Schmidt, Y. Wachmann, Mitteilungen der Hamburger Sternwarte, R. Schorr edt., 11 and Plate I (1932)
70. B. Schmidt, Ein Lichtstarkes Komafreies Speigelsysten, Central Zeitung für Optik und Mechanik, **52**, Heft 2, 25 (1932)
71. B. Schmidt, Mitteilungen der Hamburger Sternwarte, R. Schorr edt., 10 and Plates I and II (1936)
72. E. Schmidt, in *Optical Illusions*, Estonian Academy Publishers (Life story of Bernhard Schmidt), 124–125 (1995)
73. R. Schorr, Astronomische Nachrichten, Berlin **259**, 45 (1936) This review paper on B. Schmidt works was translated into English by N.U. Mayall (see also Wachmann, A.A., Kross, J., Schmidt, E.)
74. D.J. Schroeder, *Astronomical Optics*, Academic Press, London (1987)
75. D.H. Schulte, Auxiliary optical systems for the Kitt Peak Observatory, Appl. Opt., **2**(2), 141–151 (1963)
76. K. Schwarzschild, Untersuchungen zur geometrischen Optik, I, II, III, Göttinger Abh, Neue Folge, Band IV, No. 1 (1905) This article is a general investigation of aplanatic systems obtained with two centred mirrors.
77. R.D. Sigler, Compound Schmidt telescope designs with nonzero Petzval curvature, Appl. Opt. **14**, 2302–2305 (1975)
78. G.W. Stroke, Revue d'Optique, **39**, 350 (1960)
79. B. Stromgren, Das Schmidtsche Spiegelteleskop, Viert. Astron. Gessellsch, Leipzig, **70**, 65–86 (1935)
80. D.-q. Su, C. Cao, M. Liang, Some new ideas of the optical system of large telescopes,*Avanced Technology Optical Telecopes* III, SPIE Proc. **628**, 498–503 (1986)
81. D.-q. Su, Researches on Schmidt and achromatic Schmidt telescopes, Acta Astronomica Sinica, **29**(4), 384–395 (1988)
82. D.-q. Su, X. Cui, Active optics in LAMOST, Chin. J. Astron. Astrophys., **4**(1), 1–9 (2004)
83. Y. Väisälä, Uber Spiegelteleskope mit grossem Gesichtsfeld, Astr. Nach., **259**, 197–204 (1936)
84. A.A. Wachmann, From the life of Bernhard Schmidt, Sky and Telescope, November, 4–9 (1955)

85. A.A. Wachmann, Private communications to Erik Schmidt (1985–90). It appears that B. Schmidt probably did not use the stress polishing technique for making its corrector plate although the plate thickness was thin enough for this to work.
86. S.-g. Wang, D.-q. Su, Q.-q. Hu, Two telescope configurations for China, *Advanced Technology Optical Telescopes* IV, SPIE Proc. **2199**, 341–351 (1994)
87. S.-g. Wang, D.-q. Su, Y.-q. Chu, X. Cui, Y.-n. Wang, Special configuration of a very large Schmidt telescope for extensive astronomical spectroscopic observations – LAMOST, Appl. Opt., **35**, 25, 5155–5161 (1996)
88. R.N. Wilson, *Reflecting Telescope Optics* I, Springer-Verlag edt. New York (1996)
89. R.W. Wood, Phil. Mag., **4**, 396 (1902)
90. J.D. Wray, F.G. O'Callaghan, Folded all-reflective Schmidt, *Space Optics*, SPIE Proc., Santa Barbara (1969)
91. J.D. Wray, H.J. Smith, K.G. Henize, G.R. Carruthers, SPIE Proc. **332**, 141 (1982)
92. F.B. Wright, An aplanatic reflector with a flat field related to the Schmidt telescope, Publ. Astron. Soc. Pac., **47**, 300–304 (1935)
93. C.G. Wynne, MNRAS, **107**, 356 (1947a)
94. C.G. Wynne, Chromatic correction of wide-aperture catadioptric systems, Nature, **160**, 91 (1947b)
95. C.G. Wynne, Two-mirror anastigmats, J. Opt. Soc. Am., **59**, 572–580 (1969)
96. C.G. Wynne, Shorter than a Schmidt, MNRAS, **180**, 485–490 (1977)
97. Y.-t. Zhu, G.R. Lemaitre, LAMOST multi-object spectrographs with aspherized gratings, in *Instrument Design for Ground-based Telescopes*, SPIE Proc. **4841**, 1127–1133 (2002)

Chapter 5
Schmidt Correctors and Diffraction Gratings Aspherized by Active Optics

5.1 Various Types of Aspherical Schmidt Correctors

Active optics methods were originally applied to the figuring of refractive Schmidt correctors. Further developments followed with the development of reflective correctors and reflective-diffractive correctors. The various ways to obtain each of the three corrector types are different.

1. The methods developed for refractive plate correctors have been called *elastic relaxation figuring* or more commonly *stress figuring*.
2. Two different methods are practicable for obtaining reflective correctors, namely *stress figuring* and *in-situ stressing*.
3. Starting from a plane or a spherical grating, there are also two methods of obtaining reflective aspherized gratings: duplication on an active submaster and then stressing and second duplication (more commonly used), or simple duplication and stressing.

The main feature of these methods is the inherent *smoothness* of the aspherized surface that is obtained directly from spherical or plane surfaces. This avoids the high spatial frequency errors which are a signature of the classical zonal retouch process. Active optics methods allow obtaining telescopes and astronomical instruments of the highest *intrinsic* image quality.

5.2 Refractive Correctors

5.2.1 Third-Order Optical Profile of Refractive Correctors

The exact aspherical shape of the refractive corrector plate is represented by (4.11) and Table 4.1. The Kerber condition (4.14) of minimizing the spherochromatic variation, $k = 3/4$, gives the location of the null power zone at $r_o/r_m = \sqrt{k} = \sqrt{3}/2$. The chromatic dispersion is the highest at the radial zones $r_m/2$ and r_m, corresponding to maximal local slopes of opposite sign (Fig. 5.1).

G.R. Lemaitre, *Astronomical Optics and Elasticity Theory*, Astronomy and Astrophysics Library, DOI 10.1007/978-3-540-68905-8_5,

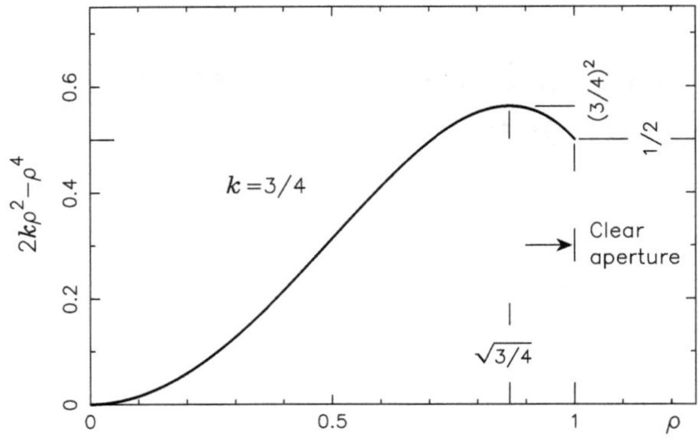

Fig. 5.1 Aspheric surface of a refractive Schmidt corrector plate. Kerber's condition $k = 3/4$ minimizes the axial spherochromatism. Optical shape $Z_{\mathrm{Opt}} \propto 2k\rho^2 - \rho^4 = 3\rho^2/2 - \rho^4$ where $\rho = r/r_{\mathrm{m}} \in [\,0, 1\,]$. The power of the plate is positive on-axis and null at $\rho = \sqrt{k} = 0.866$. This condition provides the balance of the *first derivative extremals* in setting opposite maximum slopes at half-aperture $\rho = 1/2$ and clear aperture edge $\rho = 1$

Let us represent the aspherical shape in third-order with $\mathcal{A}_2$, $\mathcal{A}_4$ coefficients in (4.11) reduced to their main part, namely $\mathcal{A}_2 \simeq M$, $\mathcal{A}_4 \simeq -1/4$ and under-correction parameter also reduced to $s = 1$. Denoting $\Omega = \mathrm{f/d} = R/4r_{\mathrm{m}}$ the telescope f/ratio and $\rho = r/r_{\mathrm{m}}$ a dimensionless radius, the Kerber condition $k = 3/4$ entails $M = 3/2^7\Omega^2$ from (4.8). In this approximation, the refractive surface is

$$Z_{\mathrm{Opt}} \simeq -\frac{1}{2^8(N-1)\Omega^3}\left(\frac{3}{2}\,\rho^2 - \rho^4\right) r_{\mathrm{m}} \quad \text{with} \quad 0 \le \rho \le 1. \tag{5.1}$$

5.2.2 *Elasticity and Circular Constant Thickness Plates*

With the case of refractive corrector plates, it is appropriate to consider the bending of constant thickness plates. The elasticity theory of small deformations of a circular plate when stressed by a uniform load q is due to Poisson [28]. The flexure Z_{Elas} is derived from Poisson's fourth-derivative equation

$$\nabla^2\nabla^2 Z_{\mathrm{Elas}} - q/D = 0\,, \tag{5.2}$$

with the Laplacian

$$\nabla^2\mathbf{.} = \frac{1}{r}\frac{d}{dr}\left(r\,\frac{d\mathbf{.}}{dr}\right), \tag{5.3}$$

and the flexural rigidity

$$D = Et^3/12(1-\nu^2)\,, \tag{5.4}$$

where t, E, and ν are the plate thickness, Young modulus and Poisson's ratio, respectively.

The general solution is represented by

$$Z_{\mathrm{Elas}} = (q/64D)\, r^4 + C_1\, r^2 \ln r + (C_2 - C_1)\, r^2 + C_3 \ln r + C_4\,, \tag{5.5}$$

where $C_{1,2,3,4}$ are constants depending upon boundaries.

The radial stress σ_r and tangential stress σ_t are derived from the corresponding bending moments M_r and M_t

$$M_r = D\left(\frac{d^2Z}{dr^2} + \frac{\nu}{r}\frac{dZ}{dr}\right), \tag{5.6a}$$

$$M_t = D\left(\frac{1}{r}\frac{dZ}{dr} + \nu\,\frac{d^2Z}{dr^2}\right), \tag{5.6b}$$

$$\sigma_r = \pm 6\, M_r/t^2 \quad \text{and} \quad \sigma_t = \pm 6\, M_t/t^2\,. \tag{5.7}$$

The maximum values of stresses σ_r and σ_t allow comparisons with the ultimate strength σ_{ult} of the material for the validity in the application of active optics methods.

5.2.3 Refractive Correctors and the Spherical Figuring Method

B. Schmidt used the classical method by zonal retouch for making his corrector plate (in Chap. 4 [85]).[1] The handwritten elasticity formulas found in Schmidt's personal papers all concern the flexure of beams.[2] Though these formulas are not directly applicable to circular plates, he clearly emphasized that a much smoother profile could be obtained by surfacing a plate bent by partial vacuum while supported at its edge. Proposed by Schmidt as an aspherization concept for obtaining the best profile *continuity*, the elastic relaxation method – or stress figuring method – is now widely known as the basic method of active optics.

The active optics aspherization proposed by Schmidt only requires use of a *full-size* spherical tool and therefore must reject any local retouch. The theoretical problem was fully solved by Couder in 1940 [6], although he did not apply the method. The method was re-suggested for development by Chrétien [3] who pointed out the advantage of accurate optical figures. Finally, *Schmidt's spherical figuring method* seems to have been first applied by Clark in 1964 [4] for a small plate and by Everhart in 1966 [8] in aspherizing a 29-cm clear aperture plate. In 1971, an

[1] From A.A. Wachmann private communications to Erik Schmidt, it appears that B. Schmidt probably did not use the stress figuring technique for making his corrector plate although the plate thickness was thin enough for this to work.

[2] B. Schmidt's personnal scientific papers were given to his nephew E. Schmidt in Palma de Mallorca by A.A. Wachmann, astronomer of Hamburg Observatory. By courtesy of E. Schmidt these papers were consulted by the author.

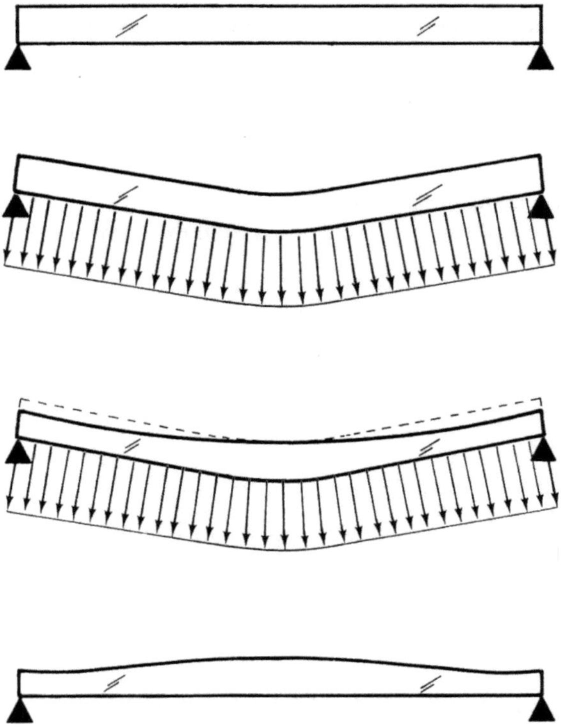

Fig. 5.2 Active optics aspherization of a refractive corrector plate by the Schmidt-Couder-Everhart method for spherical figuring

attempt was made for a 53-cm diameter plate [29]. However, this method presents some difficulties on realizing an accurate edge support and on controlling the tool curvature. Another active optics method which uses a plane figuring tool is now widely preferred (see next section).

The long delay between Schmidt's idea and its application is probably due to technical difficulties such as, for instance, the availability of finding an accurate enough pressure controller. The very simple process for aspherizing the corrector is to apply a partial vacuum under all the surface of the plate, which is in reaction at its edge, while figuring it with a tool of convenient curvature (Fig. 5.2).

The plate of thickness t is supported in reaction around the edge by an optically flat rim belonging to a bowl that can be rotated. The air is partially evacuated to a pressure p under the plate which is subjected to a uniform load $q = p_o - p$, where p_o is the atmospheric pressure. For a plate without hole on which a uniform loading is applied from center to edge, the constants C_1 and C_3 vanish in (5.5), as for C_4 by choosing the coordinate origin at the plate vertex. Assuming that the plate edge r_m is simply supported, and using the dimensionless radius $\rho = r/r_m$, the elastic deformation is

$$Z_{\mathrm{Elas}} = \frac{3(1-\nu^2)}{16} \frac{q\, r_{\mathrm{m}}^3}{E\, t^3} \left(2\, \frac{3+\nu}{1+\nu}\, \rho^2 - \rho^4 \right) r_{\mathrm{m}}\,, \quad 0 \le \rho \le 1. \qquad (5.8)$$

The accessible surface of the plate is ground and polished using a convex spherical tool.

Denoting $\varpi = r_{\mathrm{m}}/2R_{\mathrm{Tool}}$, where R_{Tool} is the radius of curvature of the tool, the equation of the just polished surface – still deformed by depression – is now that of the sphere

$$Z_{\mathrm{Sphe}} = \varpi\, (\rho^2 + \varpi^2 \rho^4)\, r_{\mathrm{m}} \quad \text{with} \quad 0 \le \rho \le 1. \qquad (5.9)$$

When the plate is elastically relaxed by opening the bowl to the atmosphere, the rear surface will revert to a plane. The two surfaces of the plate have to be polished, so that one or both may be aspherized. Denoting $\wp = 1$ for one aspherized face and $\wp = 2$ for two surfaces having half asphericity on each, the front surface will become aspherical in a Kerber profile obtained from the *active optics co-addition law*

$$Z_{\mathrm{Opt}} \;+\; \wp\, (Z_{\mathrm{Elas}} - Z_{\mathrm{Sphe}}) = 0 \quad \text{with} \quad \wp = 1 \text{ or } 2\,. \qquad (5.10)$$

From identification of the coefficients in ρ^2 and ρ^4, the two parameters, R_{Tool} and t can be solved, leading for ϖ, to the following third-degree equation

$$2^{10}\, \frac{3+\nu}{1+\nu}\, \wp\, \varpi^3 \;+\; 2^9\, \wp\, \varpi \;-\; \frac{9+\nu}{(1+\nu)(N-1)}\, \frac{1}{\Omega^3} = 0. \qquad (5.11)$$

This equation always has a unique and positive real root. This root is much smaller than unity since $\Omega^2 > 2$ and $0 < \nu < 1/2$, thus, ϖ^3 is negligible. This is equivalent to saying that the coefficient of ρ^4 in the expansion of the sphere is negligible relative to the coefficient of ρ^4 of the elastic deformation. Since $R_{\mathrm{Tool}}/R = 1/8\varpi\,\Omega$, the radius of the spherical tool and the thickness of the plate can be expressed as functions of the mirror radius of curvature R

$$\begin{cases} R_{\mathrm{Tool}} = 64\, \dfrac{1+\nu}{9+\nu}\, (N-1)\, \wp\, \Omega^2\, R\,, & (5.12a) \\[2ex] t = \left[\dfrac{3}{4}\, \wp\, (1-\nu^2)(N-1)\, \dfrac{q}{E} \right]^{1/3} R. & (5.12b) \end{cases}$$

These relationships fully define the aspherization conditions. For example, with $\nu = 1/5$ and $N = 3/2$, the first one gives $R_{\mathrm{Tool}} = 4.174\, \wp\, \Omega^2 R$.

In order to avoid the breakage – implosion –, the radial and tangential stresses must be compared to the ultimate stress σ_{ult}. These stresses are maximal and equal at the center of the plate,

$$\sigma_r(0) = \sigma_t(0) = \frac{3+\nu}{96\,\Omega^2} \left[\frac{9}{2\wp\,(1-\nu^2)(N-1)}\, \frac{E}{q} \right]^{2/3} q \;\le\; \begin{cases} \sigma_{\mathrm{ult}}/2 \\ \text{or} \\ 2\sigma_{\mathrm{ult}}/3, \end{cases} \qquad (5.13)$$

showing that the highest asphericity is provided by the lowest possible loading. In a basic practical case of loading at $q = 0.85$ bar, i.e. allowing *Sphe* 3 corrections by under or over pressure during the end of the polishing, a conservative rule is to work at a stress value of half or two-thirds of σ_{ult}. For instance, with fused silica ($N = 1.48$, $\nu = 0.17$, $E = 77.5$ GPa, $\sigma_{ult} \simeq 76$ MPa), a $\sigma_{ult}/2$ working stress allows obtaining plates at an f-ratio $\Omega \simeq 1.78$ if only one face is aspherized ($\wp$=1) or at f/1.77$\times 2^{-1/3}$ = f/1.41 if both faces are figured with half asphericity on each ($\wp = 2$).

The imperative condition is that the supporting rim must define a plane to within a quarterwave geometry as well as for the edge of the plate which is the most difficult to achieve. In practice, the plate has to be somewhat larger than the rim size which would lead to a small correction on R_{Tool} in the above formula. The problem can be perfectly treated taking into account the small bending moment at r_m due to the plate extra-radius exceeding the rim radius. This extra-radius of the plate introduces only a small and pure variation of curvature in the ρ^2 term of Z_{Elas} i.e. has no effect on the ρ^4 term. The previous theoretical curvature mode $Cv\,1$ can be recovered by applying a small uniform load to the zone outside the rim.

5.2.4 Refractive Correctors and the Plane Figuring Method

In the method proposed by Schmidt, the main inconvenience is the control of the tool curvature in order to achieve a correct positioning of the null powered zone r_o. Another difficulty is obtaining an axisymmetric geometry within a quarterwave criterion on the very edge of the plate. A final drawback comes from the figuring tool which has a different curvature each time one makes a corrector plate of different f-ratio. All these difficulties vanish with Lemaitre's *two-zone stress figuring method* [11–13, 15] where *flat* tools of full aperture size are used and can be easily controlled by a pressing process onto a flat reference blank during the night. Compared to (5.10), the *active optics co-addition law* is now trivial,

$$R_{Tool} = \infty, \qquad Z_{Sphe} \equiv 0 \quad \text{and} \quad Z_{Opt} + \wp\, Z_{Elas} = 0. \tag{5.14}$$

The plate of outer radius $r = b$ is supported on a metal ring at the clear aperture radius $r = a \equiv r_m$, which divides the surface into two concentric zones. A uniform load q_1 is exerted on the inner zone, and a higher load q_2 – close to the atmospheric load – is applied on the outer zone in the same direction as q_1 (Fig. 5.3).

While the plate is under load, its outer surface is figured with a flat full-aperture tool. A peripheral ring connected to the support aids in centering the plate and also assures an air-tight seal by means of an O-ring or a modelling paste. This seal touches neither surface of the plate but only its edge, exerting a negligible radial force, so the edge of the plate is free to move axially.

Denoting $a \equiv r_m$ the radius of clear aperture, $\rho = r/a$ the dimensionless current radius, the Z_{Elas} solutions of inner zone z_1, and outer zone z_2, are derived from the general solution (5.5),

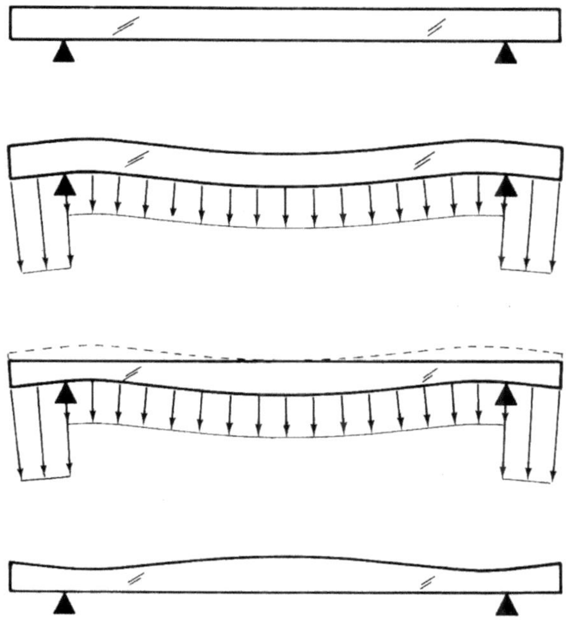

Fig. 5.3 Active optic aspherization of a refractive corrector plate by Lemaitre's two-zone method for plane figuring

$$z_1 = (q_1 a^4/64D)\rho^4 + X_1\,\rho^2 + X_2\,, \qquad 0 \le \rho \le 1, \tag{5.15a}$$

$$z_2 = (q_2 a^4/64D)\rho^4 + X_3\,\rho^2 \ln\rho + (X_4 - X_3)\,\rho^2 + X_5\,\ln\rho + X_6\,, \quad 1 \le \rho \le b/a, \tag{5.15b}$$

where the integration constants $X_{1,\ldots 6}$ are determined from the continuity and boundary conditions.

Continuity conditions at the ring support:

$$\rho = 1 \begin{cases} z_1 = 0 & \text{origin of deformation} & (5.16\text{a})\\ z_2 = 0 & \text{origin of deformation} & (5.16\text{b})\\ dz_1/d\rho = dz_2/d\rho & \text{local slope} & (5.16\text{c})\\ d^2 z_1/d\rho^2 = d^2 z_2/d\rho^2 & \text{local curvature} & (5.16\text{d}) \end{cases}$$

Free edge boundary conditions:

$$\rho = b/a \begin{cases} d\,\nabla^2 z_2/d\rho = 0 & \text{shearing force} & (5.16\text{e})\\ d^2 z_2/d\rho^2 + (\nu/\rho)\,dz_2/d\rho = 0 & \text{bending moment} & (5.16\text{f}) \end{cases}$$

When solving the unknowns, the three parameters, Poisson's ratio ν, radius ratio b/a and load pressure ratio $\eta = q_2/q_1$, define the profile geometry. For instance, with $\nu = 1/5$ and $\eta = 1$, a flexure array of $Z_{\text{Elas}}(b/a)$ represents z_1 and z_2 for various b/a ratios (Fig. 5.4).

Since X_1 is directly obtained from identification with (5.14), the plate thickness is not depending on q_2 but only of q_1, so that (5.12) becomes

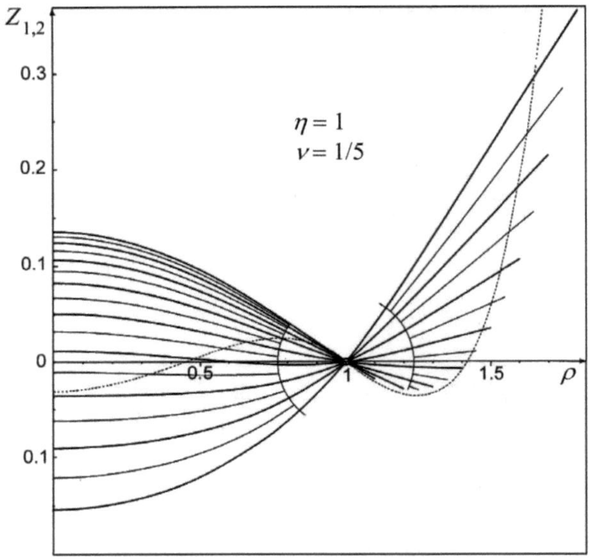

Fig. 5.4 Grid showing the flexure Z_{Elas} of a plate in the two-zone method for $\nu = 1/5$, $\eta = q_2/q_1 = 1$ and various b/a radius ratios. All plates are supported onto a radius $\rho = r/a = 1$. Kerber's profile is obtained for $b/a = 1.4903$. The *dotted line* is the locus of inflexion points

$$R_{\mathrm{Tool}} = \infty \quad \text{and} \quad t = \left[\frac{3}{4} \wp \, (1 - \nu^2)(N - 1) \, \frac{q_1}{E} \right]^{1/3} R. \tag{5.17}$$

Poisson's ratio ν being imposed by the glass type, there is an infinity of pairs $(b/a, \eta)$ that formally satisfy a Kerber profile. A search for pairs $(b/a, \eta)$ giving these profiles has been carried out for $1/6 < \nu < 1/2$, corresponding to most optical glasses. One notices that for $\eta = 1$, the surface area of the outer zone – which is not optically useful – is the largest. For $\eta > 1$, the area of the outer zone decreases. For large values of η, the plate thickness becomes too small for practicable applications. Thus, *a fortiori*, the compromise $3 \leq \eta \leq 6$ has to be chosen for making large diameter correcting plates. The resulting determinations of the ratios b/a, as functions of η and ν, are displayed by Table 5.1.

Table 5.1 Radius ratios b/a providing Kerber-profile plates of clear aperture $2r_{\mathrm{m}} = 2a$ as a function of loading ratio η and Poisson's ratio ν

$\eta = q_2/q_1$	1	2	3	4	5	6
$\nu = 1/6$	1.4888	1.3527	1.2909	1.2536	1.2279	1.2088
$\nu = 1/5$	1.4903	1.3537	1.2917	1.2542	1.2285	1.2093
$\nu = 1/4$	1.4926	1.3552	1.2929	1.2552	1.2293	1.2101
$\nu = 1/3$	1.4962	1.3576	1.2948	1.2569	1.2308	1.2114
$\nu = 1/2$	1.5030	1.3623	1.2985	1.2600	1.2335	1.2138

For comparison with the ultimate strength, the maximum stresses at the surface of the plate are derived from $X_1 = -3a/2^9(N-1)\wp\,\Omega^3$ in (5.15a) and the bending moments M_r and M_t in (5.6a,b). Hence, the highest of the maximum stresses σ_r, arising for $r = a = r_m$, is

$$\sigma_r(a) = \frac{9+\nu}{384\,\Omega^2}\left[\frac{9}{2\wp(1-\nu^2)(N-1)}\,\frac{E}{q_1}\right]^{2/3} q_1 \le \begin{cases}\sigma_{\rm ult}/2\\ \text{or}\\ \sigma_{\rm ult}/3,\end{cases} \tag{5.18}$$

The choice of $\eta = 1$ is preferable for making small corrective plates. For medium to large plates, a uniform loading at $q_2 = 0.85$ bar with $\eta = 4$ i.e. $q_1 = 0.2125$ bar allows reducing the amount of glass outside the clear aperture. The *Sphe* 3 adjustment is achieved by slight under or over pressure during the end of the polishing while maintaining the constancy of the η-ratio. Vacuum-pressure controllers allows maintaining the right pressures q_1, q_2 and the constancy of their ratio.

For aspherizing plates up to 1 m in diameter, a figuring machine has been built together with a melted fine grain steel table rotating onto a hydrostatic polyethylene-oil-steel pad. The Schmidt plates are directly supported on a ring emerging from a rotating table and accurate load controllers operate during the stress figuring process (Fig. 5.5).

For instance, with fused silica, a working stress $\sigma_{\rm ult}/2 \equiv \sigma_{10^3\,{\rm s}}/2$ (see Table 5.2) allows obtaining plates at the fast f-ratio of $\Omega = 1.2$ if only one face is aspherized ($\wp = 1$) or at f/1.2$\times 2^{-1/3}$ = f/0.95 if both faces are figured with half asphericity on

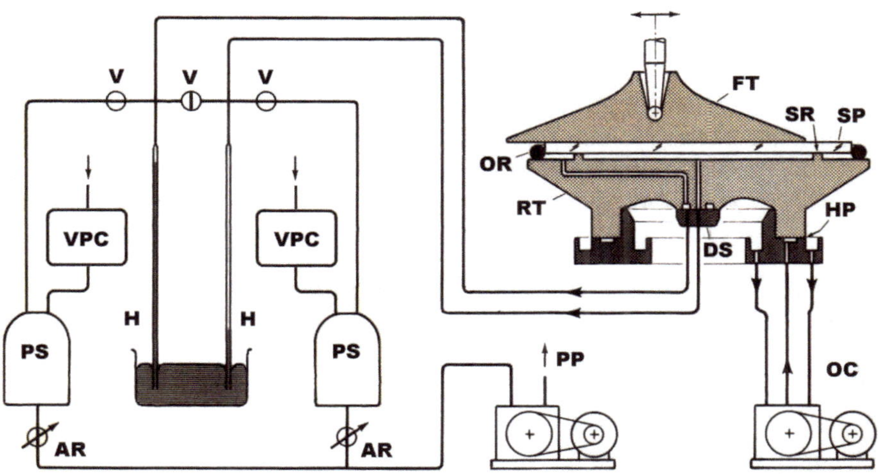

Fig. 5.5 Aspherization of large plates with two partial vacuum zones and full-aperture flat tools. [SP] Schmidt plate, [FT] figuring rigid tool, [SR] merging support ring, [OR] O-ring and radial reaction blocks, [RT] rotating table, [DS] double rotating seal, [HP] hydrostatic polyethylene-oil-steel pad, [OC] oil compressor and regulator, [PP] primary vacuum pump, [VPC] vacuum-pressure controllers, [H] mercury absolute gauges, [V] valves, [PS] air pressure stabilizers, [AR] air flow regulators

Table 5.2 Tensile rupture stresses – or tensile ultimate strengths – σ of some brittle materials versus loading time [MPa]

Loading delay	[*]	[†]	Stress polish.		Permanent	
	1 s	10^3 s	10 h	1 month	5 years	10 years
SCHOTT BK7	82.0	50.0	38.7	28.6	21.3	20.3
KODAK Photo-plates	96.6	59.0	45.7	33.6	25.1	23.8
HERAEUS Fused SiO_2	122.8	75.0	58.0	42.8	31.9	30.4
SCHOTT Zerodur	147.4	90.0	69.6	51.3	38.3	36.4
UGINE Sapphire Al_2O_3	655.1	400.0	309.5	228.1	170.3	162.1
MORTON SiC CVD	974.5	595.0	460.6	339.4	253.3	241.0

The various loading delays are concerned with two cases: (A) stress figuring with removing and without removing the loading each day, (B) quasi-permanently stressed optics i.e. *in situ* active optics. All stresses have been calculated from $\sigma_{10^3\mathrm{sec}}$ using the law (5.19b) with $p = 14$.
[†] Tensile ultimate strengths from manufacturers and corresponding to $\sigma_{10^3\,\mathrm{s}}$.
[*] These values are $\sigma_{1\,\mathrm{s}}$ in (5.19b).
N.B.: For safety in active optics methods, in long duration processes it is customary to not overpass the tensile maximum stress $\sigma_{T\,\max} = \frac{1}{3}\sigma_{10^3\,\mathrm{s}}$ or $\frac{1}{5}\sigma_{10^3\,\mathrm{s}}$.

each ($\wp = 2$). More than 50 corrector plates of 4–10 cm clear apertures at f-ratios from f/2 to f/1.1 have been realized with $\eta = 1$ and outer zone loadings smaller than 0.3 bar for the fastest ones, working also with the less conservative rule $\sigma_{\mathrm{ult}}/2$. This was mainly for space experiments in the UV and conventional ground-based spectrographs, such as, for example, with the Scap flights, Janus flights, Skylab mission cameras by Courtès et al. [7] and semi-solid Schmidt PEDISCOU spectrograph of Haute Provence observatory. The aspheric correction is analyzed from a He-Ne plate-test interferogram formed by the light reflected from each plate face (Fig. 5.6).

At the instigation of Fehrenbach [9], the method has been developed for making the 62-cm clear aperture plate of the f/3.3 Franco-Belgium Schmidt telescope at Haute Provence observatory which was further extensively used with a 62-cm Fehrenbach normal dispersion biprism for radial velocity measurements in the

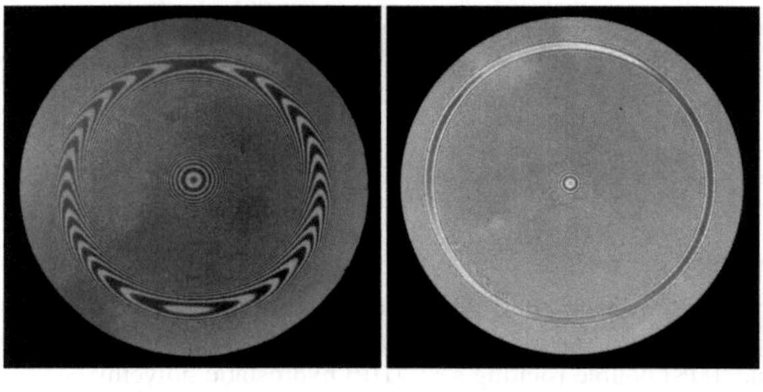

Fig. 5.6 Fizeau interferograms of 5-cm clear aperture fused silica corrector plates aspherized by plane figuring and two-zone partial vacuum. An f/1.1 corrector plate is shown on the *right* (Lemaitre [13]) (LOOM)

slitless spectroscopy mode (cf. Sect. 1.12.7). In Fehrenbach's method the radial velocity of a star is derived from the relative shift of two spectra recorded on the same plate with opposite dispersions. This is obtained by a 180° in-plane rotation of the biprism and a slight lateral telescope depointing avoiding fuzzy spectra that would be caused by a full superposition.

The active optics aspherization of the plate required the construction of a dedicated surfacing machine with a rigid steel table rotating onto a hydrostatic pad. A metal-oil-plastic pad provided a low friction rotating table avoiding ball bearing deformation errors. A preliminary result with this 1 m surfacing machine was the stress aspherization of a 50-cm BK 7 plate for the f/2.7 Schmidt of Lyon observatory (Lemaitre [15]). After a fine grinding of the two sides by use of a central suction disc, the plate was directly supported by a finely ground ring that emerged from the table after an in-situ lathe-type operation removing 0.3–0.4 mm on each side of the ring. For the Haute Provence observatory, similar preparations were carried out with a UBK 7 glass plate and rotating table support ring. The design parameters were $2a = 2r_m$ =620 mm, $b/a = 1.2542$, $\eta = 4$, $q_2 = 0.85$ bar, $\nu = 1/5$, $E = 82$ GPa, R =4,170 mm, $\wp = 2$, and t =24.14 mm. This plate replaced the previous one made by the classical method of zonal tool polishing. Off-focus star images showed that the optical quality is clearly in favor of the active method (Fig. 5.7).

The previous analysis has been carried out by considering constant thickness plates, but this is not exactly the case since the plates are slightly thinner at the clear aperture edge than at their center. In fact, a finer analysis would show that the active methods provide, at least, a *Sphe* 5 coefficient of the *right* sign for higher-order corrections of the spherical aberration.

During the period 1968–2005, the *two-zone stress figuring method* was extensively used at LOOM where more than 60 corrector plates have been produced this way for space- and ground-based telescopes and spectrographs.

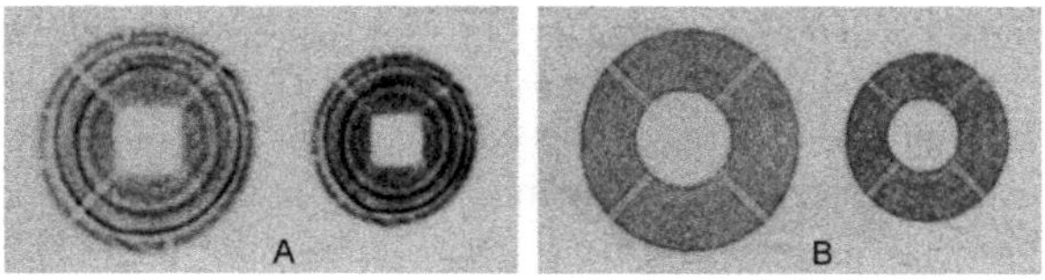

Fig. 5.7 Comparison of the same intrafocus patterns of a star image before and after replacement of the 62-cm corrector plate of the f/3.3 Franco-Belgium Schmidt telescope in Haute Provence observatory. (**A**) Former plate made by the classical method of zonal tool figuring. (**B**) New plate made by two-zone stress figuring method and full-size tool. The latter images show smooth and uniform patterns conferring the telescope with an excellent optical quality (LOOM)

5.2.5 Glass Rupture and Loading Time Dependance

During the figuring of small and fast aperture corrector plates – which is useful with the flat tool method – a breakage may happen. This was the case for some of the

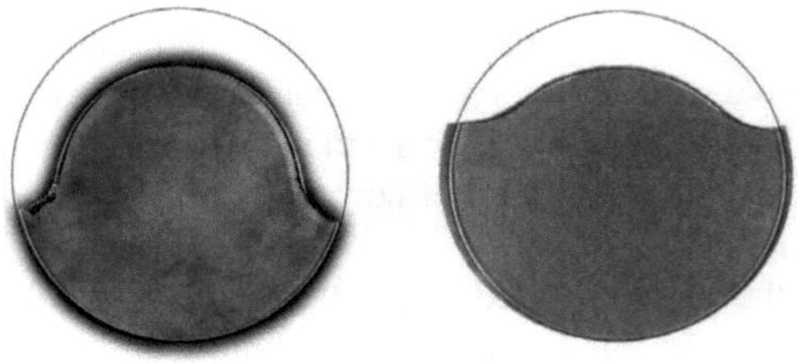

Fig. 5.8 Fracture figure of two corrector plates at f/1 and f/1.2 that arose during the flat polishing of the second face ($\wp = 2$, $\nu = 1/6$, $b/a = 1.4888$, fused silica). The rupture starts from the support ring of radius $r = a \equiv r_{\rm m}$, and propagates along this circle before reaching the edge, thus forming an Ω-shaped line (LOOM)

4–10 cm aperture plates when aspherizing them at f/1. This allows verifying that the maximal stress formula, as given by $\sigma_r(a)$ in (5.18), is valid. The rupture line starts at the supporting circle $r = a$, and continues along this circle line before joining the edge (Fig. 5.8).

The roughness being higher during the grinding, one could suppose, at first, that the breakage occurs during this part of the figuring. In fact this is not so and the breakage may also appear during the polishing. The grinding process is always relatively fast compared to the polishing delay, and the rupture stress of glass is very dependent upon the loading time. This phenomenon of glass mechanics is known by glass manufacturers but poorly described in the literature.

Rupture tests using rectangular glass plates bent onto cylinders of various curvature have been carried out. About 40 rectangular samples were cut from Kodak photographic plates that come polished on each side and are of the borosilicate type, probably in a somewhat quenched state. The elastic constants of the glass, such as given by Kodak, were $E = 7.5$ GPa and $\nu = 0.24$. The thicknesses t of the samples were within a 0.60–0.75 mm range. Each sample was bent for two hours onto successive cylinders of curvature radii $R_x = 414$, 377 and 350 mm. The latter curvature provided the rupture for all samples having passed the two previous curvatures for two hours each. The rupture stress $\sigma_{\rm rupt}$ was calculated with the formula $\sigma_{\rm rupt} = Et/2(1-\nu^2)R_x$. The result of these tests can be modelled by asymptotic-type laws (Fig. 5.9).

Let us consider two possible rupture laws. The first is of the form

$$\sigma_{\rm rupt} = \sigma_{\rm ult} + \sigma_{\rm dyna}\, {\rm e}^{-\mu\tau}, \tag{5.19a}$$

where μ is a constant and τ the loading time. From our rupture tests with Kodak plates we obtain $\sigma_{\rm ult} = 57.22$ MPa, $\sigma_{\rm dyna} = 18.77$ MPa and $\mu = 654 \times 10^6$. This curve is denoted [28] in Fig. 5.9. For $\tau = 0$, $\sigma_{\rm rupt}$ could be defined as an instantaneous stress which is substantially larger than $\sigma_{\rm ult}$. When the loading time $\tau \to \infty$, $\sigma_{\rm rupt}$ becomes equal to $\sigma_{\rm ult}$ which is too optimistic with respect to our results on ruptures happening after 1–2 hours loading.

The second law, mentioned by Haward [10], is more realistic. It may be written as

$$\sigma_{\rm rupt} = \sigma_{1\,{\rm s}}\, \tau^{-1/p}, \tag{5.19b}$$

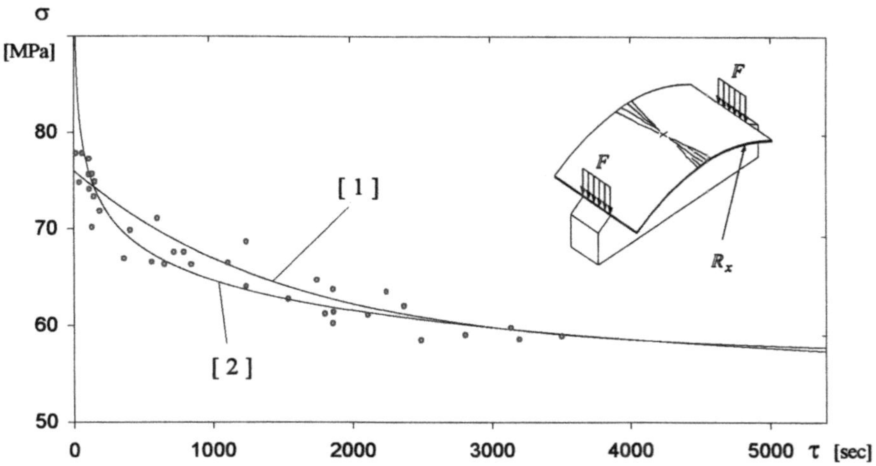

Fig. 5.9 Rupture tensile stresses of cylindric bent glass samples in function of the stressing time τ. The samples were ultra-flat photographic plates from Kodak (LOOM)

where $p=14$ and $\sigma_{1\,\mathrm{s}} = 96.6\,\mathrm{MPa}$ provides the rupture of the Kodak plates in a one second delay. This curve is denoted [6] in Fig. 5.9. This law gives $\sigma_{\mathrm{ult}}(10^3\,\mathrm{s})=$ 59 MPa for Kodak slightly quenched plates, which is in good agreement with the value of 50 MPa given by Schott for BK7 optical glass (not quenched). For fused silica, our measures give $\sigma_{\mathrm{ult}}(10^3\,\mathrm{s})=75\,\mathrm{MPa}$.

Use of the experiment values $\sigma_{10^3\,\mathrm{s}}$ and law (5.19b) with $p=14$, allowed us to determine the *rupture tensile stresses* – or *ultimate tensile strengths* – of some brittle materials for various loading delays (Table 5.2).

Depending on whether the material is stressed just for the figuring – each day or for a month – or if in a permanent *in situ* aspherized state, the time-dependent values σ in Table 5.2 allow us to adopt a design value $\sigma_{T\,\max}$ of the tensile maximal stress for the active optics process, thus minimizing the breaking risk.

From the results in Table 5.2 we obtain the following conclusions.

→ *Fused silica and polycrystalline sapphire are the most convenient isotropic refractive materials to achieve high deformations in active optics aspherization.*

Except when the ultimate tensile stress σ_{ult} is expressly given by glass manufacturers as corresponding to a loading time of 10^3 s, manufacturers generally indicate a recommended stress σ which includes a safety factor for an indefinitely long load duration; thus σ_{ult} for 10^3 s is not known by the user. For instance, with a BK7 glass working under permanent stress, Schott recommend to not exceed a tensile stress of 10 MPa. However, it has been proved as possible to aspherize several 20-cm BK7 corrector plates under a tensile stress of 20 MPa.

Some clarifications concerning the maximum tensile stress and also the time-dependent law itself ought to be given by the glass manufacturers.

5.3 Reflective Correctors

5.3.1 *Optical Figure of the Primary Mirror*

The basic optical parameters defining the shape of the reflective corrector – or primary mirror – are obtained from Tables 4.1, 4.2, and Eq. (4.9). In addition, Lemaitre's condition of minimizing the residual aberrations in the field of view (Sect. 4.3.1) entails a position of the null power zone set by $k = 3/2$, i.e. $r_o/r_m = \sqrt{k} \simeq 1.225$. Thus, the mirror local powers are extremals at the radial zones $r = 0$ and $r = r_m$, corresponding to local curvatures of opposite sign (Fig. 5.10).

Summarizing the results of Sect. 4.3 for centered or non-centered reflective telescopes, the best images are obtained with design parameters

$$k = 3/2\,, \quad \text{i.e.} \quad M = 3/64\,\Omega^2 \quad \text{with} \quad \Omega \equiv \text{f/d} = R/4r_m\,. \tag{5.20a}$$

It may be useful to derive simplified representations of the primary mirror shape. In the third-order theory, if the two first coefficients of Z_w in (4.6) are approximated by their main part, namely $\mathcal{A}_2 \simeq M$ and $\mathcal{A}_4 \simeq -1/4$, and denoting $\rho = r/r_m$ the dimensionless radius, we obtain the following *simplified* optical surface representations.

• **Axisymmetric circular mirrors**: For a *centered system* used off-axis ($s = \cos(i + \varphi_m)$),

$$Z_{Opt} \simeq \frac{\cos(i+\varphi_m)}{2^9\,\Omega^3}\left(3\,\rho^2 - \rho^4\right) r_m \qquad \text{with} \qquad 0 \le \rho \le 1\,. \tag{5.20b}$$

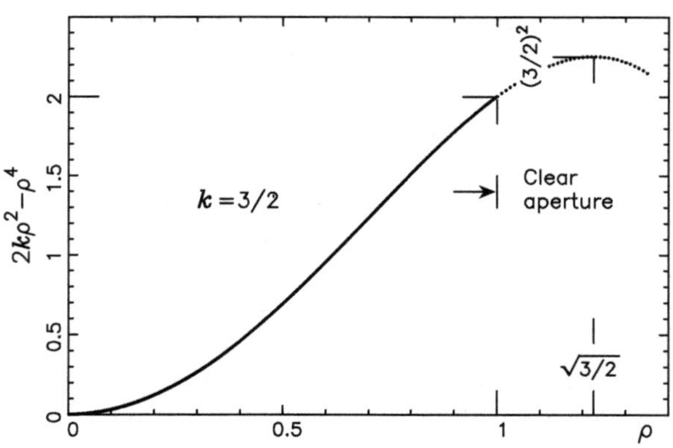

Fig. 5.10 Aspheric section of a reflective Schmidt primary mirror. Lemaitre's condition $k = 3/2$ provides the best balance of the field residual aberrations. An axisymmetric corrector mirror, $Z_{Opt} \propto 2k\rho^2 - \rho^4 = 3\rho^2 - \rho^4$ with $\rho = r/r_m \in [0,1]$, shows a positive power all over its aperture. The null powered zone is at $\rho = \sqrt{k} \simeq 1.225$, thus located outside the clear aperture. This condition realizes the balance of the *second derivative extremals* in setting opposite maximum curvatures at the vertex $\rho = 0$ and clear aperture edge $\rho = 1$ (see Chap. 4)

• **Bisymmetric circular mirrors**: For a *non-centered system* [$s = 1$, deviation angle $2i$ at the primary, $\tau = \frac{1}{2}\sin^2 i$ (see Table 4.2)],

$$Z_{\mathrm{Opt}} \simeq \frac{1}{2^9\,\Omega^3\cos i}\big[\,3(1-\tau)\,\rho^2 - 3\tau\,\rho^2\cos 2\theta \\ -(1-2\tau)\,\rho^4 + 2\tau\,\rho^4\cos 2\theta\,\big]\,r_{\mathrm{m}}\,, \quad 0 \le \rho \le 1\,, \tag{5.20c}$$

where the origin of θ is in the symmetry plane $(z,\ x)$ of the telescope ($x = r\cos\theta$).

• **Bisymmetric elliptic mirror**: For a non-centered system ($s = 1$) and a primary illuminated by circular beams of diameter $2r_{\mathrm{m}}$, an equivalent representation from (4.21c) is

$$Z_{\mathrm{Opt}} \simeq \frac{1}{2^9\Omega^3\cos i}\left[\,3\,\frac{x^2\cos^2 i + y^2}{y_{\mathrm{m}}^2} - \left(\frac{x^2\cos^2 i + y^2}{y_{\mathrm{m}}^2}\right)^2\,\right] y_{\mathrm{m}}, \tag{5.20d}$$

where $y_{\mathrm{m}} = r_{\mathrm{m}}$ is the mirror half clear aperture in perpendicular direction to the telescope symmetry plane, and f-ratio $\Omega = R/4y_{\mathrm{m}} = R/4r_{\mathrm{m}}$. The semi-axes of the clear aperture elliptic contour are $(x_{\mathrm{m}},\ y_{\mathrm{m}})$ with $x_{\mathrm{m}} = y_{\mathrm{m}}/\cos i$. The representation (5.20d) applies with $x \le x_{\mathrm{m}},\ y \le y_{\mathrm{m}}$.

• **Accurate representations**: Use of the exact expressions for $\mathcal{A}_n(M)$ and $B_{n,m}$ coefficients in Tables 4.1 and 4.2 provides a precise shape of the primary mirror (cf. Chap. 4).

If a positive singlet lens is implemented before the focus for flat fielding, then a slight homothetic increase of the coefficients allows maintaining the mirror geometry with $k = constant$. This overcorrection of spherical aberration is easily achieved by setting the s-factor in (4.10) slightly larger than unity.

5.3.2 *Axisymmetric Circular Primaries with k = 3/2 – Vase Form*

• **Centered systems**: From Sect. 4.3.1, the mirror figure for best images in the field is obtained for $k = 3/2$. Although the optical surface can be approximated by (5.20b), let us consider the shape Z_{Opt} of the axisymmetric primary such as represented by (4.18) with exact coefficients $Cv\,1$ and $Sphe\,3$.

For a constant thickness plate bent by uniform load q, whatever is the perimeter boundary – simply supported or built-in edge –, the particular solution of Poisson's equation always appears in the general solution Z_{Elas}, in (5.5), as showing the same fourth-degree term; this term is $(q/64D)\,r^4$. The aspherization process can be directly carried out from a flat polished surface ($Z_{\mathrm{Sphe}} = 0$). Then the *active optics co-addition law* writes

$$R_{\mathrm{Tool}} = \infty\,, \qquad Z_{\mathrm{Sphe}} \equiv 0 \quad \text{and} \quad Z_{\mathrm{Opt}} + \wp\, Z_{\mathrm{Elas}} = 0\,, \tag{5.21}$$

where $\wp = \pm 1$ is depending on the choice of stress figuring or in-situ stressing.

The identification of the elasticity $Z_{\rm Elas}(r^4)$ term with the optical $Z_{\rm Opt}(r^4)$ term in (4.18) provides the first part of the opto-elastic coupling. Let $\Omega = R/4r_{\rm m}$ be the f-ratio and t_1 the optical mirror thickness, after simplifications, the uniform load is

$$q = \frac{\wp \cos(i+\varphi_{\rm m})}{24\,(1-\nu^2)\,\Omega^2}\left[\frac{1}{4} - \frac{M(1-M^3)}{(1-M)(1+M)^3}\right]\frac{t_1^3}{r_{\rm m}^3}\,E\,. \tag{5.22a}$$

Since $M = k/2^5\Omega^2$ and $k = r_0^2/r_{\rm m}^2 = 3/2$ entails $M = 3/2^6\Omega^2$, the mirror constant thickness up to the clear aperture edge is

$$t_1 = 4\,\Omega\left[\frac{3\,\wp\,(1-\nu^2)}{2\cos(i+\varphi_{\rm m})}\,\frac{q}{E}\right]^{1/3}\left(1 + \frac{1}{16\,\Omega^2} + \cdots\right) r_{\rm m}\,, \tag{5.22b}$$

which determines the execution parameter for the *Sphe* 3 correction.

The set up of the *Cv* 1 term for obtaining the right slope at the perimeter $r_{\rm m}$ can be achieved by a *vase form*. This leads to determining the rigidity ratio $\gamma = D_1/D_2 = (t_1/t_2)^3$ of the two rigidity zones with respect to Poisson's ratio and also the outer-inner radii ratio $b/r_{\rm m} \equiv b/a$ of the ring. The geometrical condition balancing *Cv* 1 and *Sphe* 3 modes for $k = 3/2$ is demonstrated in Sect. 7.6.1 (Lemaitre [25]). This condition, [see (7.47)], is

$$\frac{t_2^3}{t_1^3} = \frac{16\left[2\dfrac{a}{b} - 1 - (1+\nu)\ln\dfrac{a}{b}\right] - (5+\nu)\left[1+\nu+(1-\nu)\dfrac{a^2}{b^2}\right]}{(1-\nu^2)\left(1-\dfrac{a^2}{b^2}\right)}\,. \tag{5.22c}$$

Various $(t_2/t_1,\ \nu,\ b/a)$ solution sets for a vase-form geometry satisfying (5.22c) are given in Table 5.3.

The uniform load q is applied to the inner part of the vase form and its reaction arises at the contour $r = r_{\rm m} \equiv a$. For a best validity of this CTD, the junction between the two thicknesses t_1 and t_2 is realized with a radius of curvature equal to thickness t_1 (Fig. 5.11) (see also Fig. 7.9-*Up*).

Spherical aberration corrections including *Sphe* 5, *Sphe* 7 and higher-order terms can be achieved with a *variable thickness vase form*. The thickness distribution of the clear aperture can be determined similarly as given in Sect. 5.4 for the case of making axisymmetric gratings with $k = 3/2$. The result is a thickness distribution

Table 5.3 Thickness ratios t_2/t_1 defining the vase form geometry for $k = 3/2$ with respect to Poisson's ratio ν and radius ratio b/a

ν	1/2	1/3	1/4	1/5	1/6	1/7
$b/a = 1.05$	4.127	3.972	3.937	3.926	3.922	3.921
$b/a = 1.10$	3.341	3.204	3.169	3.157	3.152	3.150
$b/a = 1.15$	2.982	2.848	2.811	2.797	2.790	2.787
$b/a = 1.20$	2.772	2.636	2.597	2.580	2.571	2.566

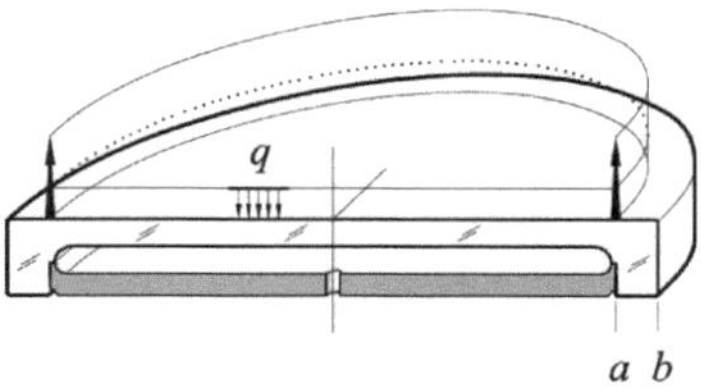

Fig. 5.11 In-situ aspherization of a vase-form mirror by uniform partial vacuum. For obtaining $k = 3/2$ with a mirror in Zerodur ($\nu = 1/4$), from Table 5.5, possible design parameters are for instance $b/a = 1.15$ and $t_2/t_1 = 2.811$

slightly decreasing from center to edge. Reflective Schmidt telescopes are usually with focal ratios lower than f/2.5 or f/3, so the *Sphe* 5 term remains small.

5.3.3 Bisymmetric Circular Primaries with k = 3/2 – MDM

• **Non-centered systems**: In the third-order theory, the optical profile is represented by (5.20c), that is a bi-axial symmetry figure where the astigmatism correction mode is included. In this case, *vase-form multimode deformable mirrors* – or vase-form MDMs – derived from the above vase-form but equipped with radial arms provide convenient solutions by active optics.

These MDMs, mainly discussed in Chap. 7, allow the co-addition of *Cv* 1, *Sphe* 3, *Astm* 3, and *Astm* 5 modes by means of a uniform loading and axial forces applied to the radial arms (cf. Table 7.2 and Fig. 7.9-*Down*).

5.3.4 Bisymmetric Circular Primaries with k = 0 – Tulip Form

• **Non-centered systems**: Although $k = 3/2$ provides the best balance of field aberrations, let us consider $k = 0$ for which the primary mirror is perfectly flat in its paraxial zone. Also in this case, active optics methods provide interesting solutions with *tulip-form mirrors*. For 100% encircled energy, the field residual images will be about three times larger than for $k = 3/2$ (see residual image variation with k in Fig. 4.9 - *solid lines*), but this could be a small handicap for the design of moderate field reflective Schmidt telescopes of fast f-ratio as first priority. With $k = 0$, the two first terms in (5.20c) vanish, so the primary mirror shape is

$$Z_{\mathrm{Opt}} = \frac{1}{2^9\,\Omega^3\,\cos i}\left[-(1-2\tau)\,\rho^4 + 2\tau\,\rho^4\cos 2\theta + \ldots\right] r_{\mathrm{m}}\,, \qquad (5.23)$$

where the second term is *Astm* 5.

Tulip-form mirrors (see Sect. 3.3.2) can provide the compensation of *Sphe* 3 represented by the first term of (5.23). Avoiding the above configuration which requires

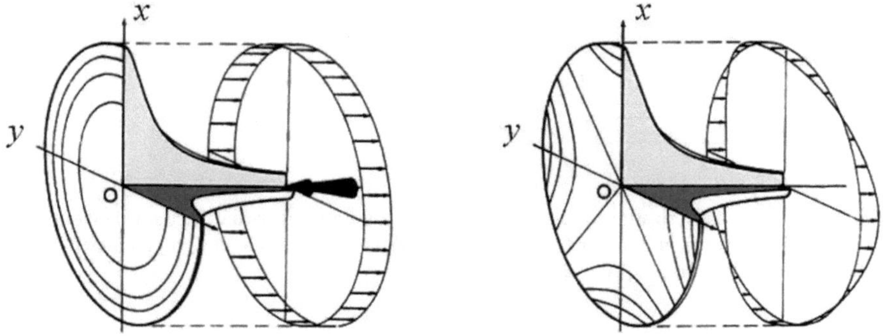

Fig. 5.12 *Tulip-form* thickness distribution of a circular primary mirror in the special case $k = 0$. (*Left*) Thickness $\mathcal{T}_{40}$ and central force F_{40} in reaction to the mirror edge providing the *Sphe* 3 correction. (*Right*) $\mathcal{T}_{42} \equiv \mathcal{T}_{40}$ and perimeter force $F_{42} \cos 2\theta$ generated – via a ring – by two orthogonal force pairs which provide the next *Astm* 5 correction [18]

a uniform external loading q, one can select a configuration using a central external force F_{40} in reaction to the edge $\rho = 1$ of the mirror. We have seen that with $q = 0$ the two roots found for A'_{40} in Sect. 3.3.2 (VTD-2) are $\alpha_1 = 8/(3+\nu)$ and $\alpha_2 = 2$. This leads to the thickness distribution (3.30a),

$$t = \mathcal{T}_{40}\, t_0 \quad \text{with} \quad \mathcal{T}_{40} = \left(\frac{4}{1-\nu}\right)^{1/3} \left[\, 1/\rho^{8/(3+\nu)} - 1/\rho^2 \,\right]^{1/3}. \tag{5.24a}$$

The infinite slope at the null thickness of the mirror edge allows use of a simply supported edge as the boundary condition, while the infinite thickness at the center is due to the imposed null paraxial curvature. The central thickness can be made finite (Fig. 5.12) in agreement with Rayleigh's quarterwave criterion (Lemaitre [14, 16]).

Tulip-form mirrors of the VTD class in Sect. 3.2 can exactly generate an *Astm* 5 mode. Setting $q = 0$, $n = 4$, and $m = 2$ in (3.14), the roots obtained for generating the *Astm* 5 mode, i.e. coefficient A'_{42}, are $\alpha_1 = 3 + \nu$ and $\alpha_2 = 0$; this leads to the variable thickness

$$\mathcal{T}_{42} \propto \left(1/\rho^{3+\nu} - 1 \right)^{1/3} \tag{5.24b}$$

where the flexure is achieved by applying an angular modulated force $\mathcal{F} = F_{42} \cos 2\theta$ at the mirror perimeter. The thickness profile difference between $\mathcal{T}_{40}$ in (5.24a) and above $\mathcal{T}_{42}$ is within a few percent for practical applications. Hence, the simultaneous corrections of the two modes *Sphe* 3 and *Astm* 5 can be achieved by superposition of the two loading systems by use of the $\mathcal{T}_{40}$ thickness profile. This is quite valid because *Astm* 5 is much smaller than *Sphe* 3.

The aspherization is directly achieved from a plane mirror shape. Then, starting conditions are

$$R_{\text{Tool}} = \infty, \qquad Z_{\text{Sphe}} \equiv 0, \quad \text{and} \quad Z_{\text{Opt}} + \wp\, Z_{\text{Elas}} = 0\,,$$

where $\wp = \pm 1$ is depending on the choice of stress figuring or in-situ stressing. From (5.23), the coefficient of *Sphe* 3 mode is

$$A_{40} = -\frac{\cos i}{2^9 \Omega^3} \frac{1}{r_{\rm m}^3}\,.$$

After substitution of A_{40} in the associated thickness $t = t_0\, \mathcal{T}_{40}(\rho)$, and referring to (3.30a), we obtain

$$t_0 = -\left[\frac{3(1-\nu^2)}{16\pi\, A_{40}} \frac{F}{r_{\rm m}^2 E}\right]^{1/3} = 4\,\Omega \left[\frac{3\,\wp\,(1-\nu^2)}{2\pi\,\cos i} \frac{F_{40}}{r_{\rm m}^2 E}\right]^{1/3} r_{\rm m} \qquad (5.24c)$$

which links the mirror thickness $t(r)$ to the central force F_{40}.

For instance, the all-reflective Schmidt telescope FAUST at f/1.5 has been built for an ultraviolet space survey of extended objects. A stainless steel Fe87 Cr13 primary of 18 cm clear aperture has been built and polished flat at rest. The thickness distribution was dimensioned by $t_0 = 5.6$ mm. Its edge was simply supported by a radially thin cylindrical collar linked to a rigid ring of thicknesses $t_z = 25$ mm and $t_r = 14$ mm. Interferograms of Fig. 5.13 display the result of flexure for each mode and their superposition. Figure 5.14 displays the theoretical images of the FAUST rocket program (Monnet et al. [27], Cohendet [5]) and the improvement in image quality during the aspherization process.

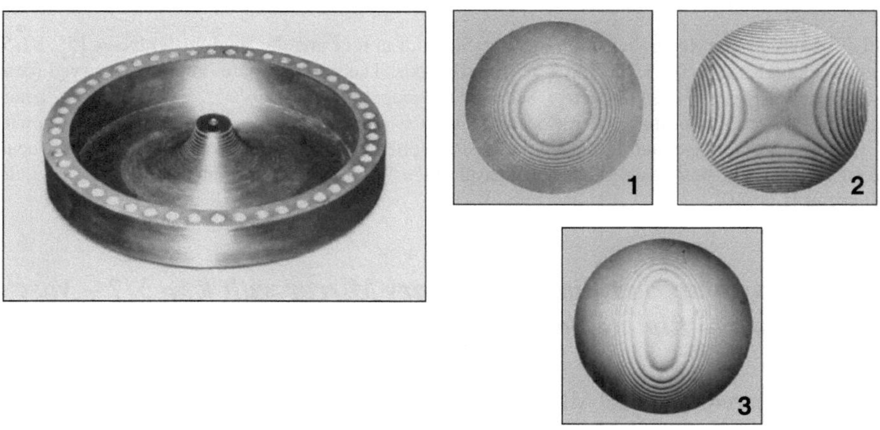

Fig. 5.13 (*Left*) View of a circular primary mirror designed in the VTD class with $k = 0$. The stainless steel mirror for the FAUST experiment is shown without its deforming cell. The dimensionless thickness $\mathcal{T}_{40} = (1/\rho^{8/(3+\nu)} - 1/\rho^2)^{1/3}$ allows one to compensate for aberrations at f/1.5 with a beam deviation of 30°. (*Right*) He-Ne interferograms of the elastic aspherization with respect to a plane. (1) *Sphe* 3 mode achieved by a central force F_{40} in reaction with the outer ring. (2) *Astm* 5 mode achieved by two equal axial forces F_{42} applied on the perimeter ring at $\theta = 0, \pi$ in reaction with two forces $-F_{42}$ at $\theta = \pm\pi/2$. (3) Superposition of the two modes [16] (LOOM)

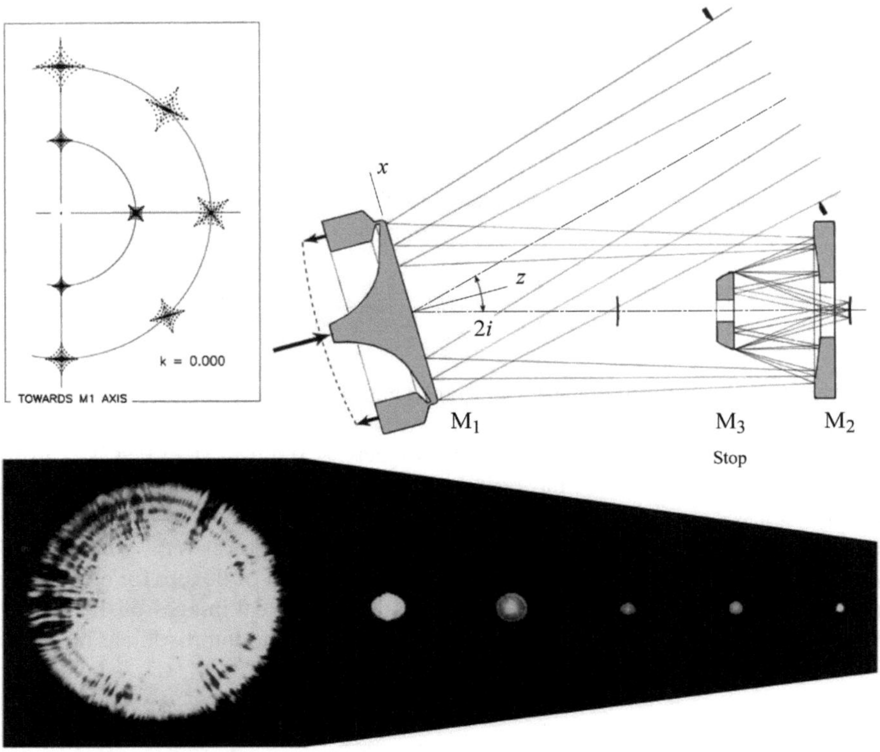

Fig. 5.14 (*Up-left*) Ray traces of a reflective Schmidt in four angular sections with $i = 15°$, f/1.5, and field angles $\varphi = 0°$, 1.25° and 2.5° from the axis. The ray traces are on the best curved focal surface. (*Up-right*) FAUST reflective Schmidt telescope at f/1.5, FOV=5°, $F = 270$ mm, beam deviation $2i = 30°$, primary mirror designed with $k = 0$ and 180 mm circular aperture. (*Down*) Evolution of the image quality during the *in situ* aspherization. The first image at the Gauss focus corresponds to the unstressed state. The measured resolution was 50 line pairs/mm over a 2.5° field diameter (LOOM)

5.3.5 *Bisymmetric Elliptical Primary Mirror with k = 3/2 – Vase Form – Biplate Form*

• **Non-centered systems**: All-reflective Schmidts generally provide better images with a primary of bi-axial symmetry (Sect. 4.3). Considering incident beams of circular cross section, a constant thickness plate with an elliptical contour allows simultaneously to compensate for *Cv* 1, *Sphe* 3 and the astigmatism modes *Astm* 3 and *Astm* 5 determined by (7.49) in Chap. 7. The elliptical contour of the mirror - telescope pupil - is defined by the incidence angle i of the principal beam. It is of interest to aspherize such primary mirrors by active optics methods. This simple process can also be used for other systems than Schmidts requiring aberration corrections on elliptical mirrors preferably used as *pupil mirrors*.

In order to directly obtain affine curvilinear flexures with meridian profiles defined by $k = 3/2$, we will see that a vase form must be modified into a "*biplate form*" made of two identical elliptical vases that are sealed together at their outer ring.

• **Flexure of an elliptical plate in a vase form**: Let us consider the system coordinates of an elliptical plate as shown in Fig. 5.15 and denote n the normal to the contour $\mathcal{C}$ of the optical clear aperture; the equation of $\mathcal{C}$ is represented by

$$\frac{x^2}{a_x^2} + \frac{y^2}{a_y^2} - 1 = 0\,. \tag{5.25}$$

The differential equation of a flexure generated by a uniform load q, is

$$\nabla^4 z \equiv \partial^4 z/\partial x^4 + 2\,\partial^4 z/\partial x^2 \partial y^2 + \partial^4 z/\partial y^4 = q/D\,. \tag{5.26}$$

With boundary conditions defined by a *clamped edge*, i.e. the origin $z|_{\mathcal{C}} = 0$ and curvilinear slope $\partial z/\partial n|_{\mathcal{C}} = 0$ of the flexure along the contour $\mathcal{C}$ of the ellipse (a_x, a_y), the bilaplacian equation is satisfied if the flexure is represented by (Bryan [2], Love [26], Timoshenko and Woinovsky-Krieger [30a])

$$z = z_0 \left(1 - \frac{x^2}{a_x^2} - \frac{y^2}{a_y^2}\right)^2\,. \tag{5.27}$$

The flexural sag z_0 is obtained from substitution in the biharmonic equation

$$z_0 = \frac{q}{8D}\,\frac{a_x^4 a_y^4}{3a_x^4 + 2a_x^2 a_y^2 + 3a_y^4}\,. \tag{5.28}$$

The bending moments M_x, M_y and twisting moment M_{xy} are defined by

$$\begin{cases} M_x = D\left(\dfrac{\partial^2 z}{\partial x^2} + \nu \dfrac{\partial^2 z}{\partial y^2}\right), \quad M_y = D\left(\dfrac{\partial^2 z}{\partial y^2} + \nu \dfrac{\partial^2 z}{\partial x^2}\right), & (5.29)\\[2ex] M_{xy} = -(1-\nu)D\,\dfrac{\partial^2 z}{\partial x \partial y} = -8(1-\nu)\,z_0 D\,\dfrac{xy}{a_x^2 a_y^2}\,. & (5.30) \end{cases}$$

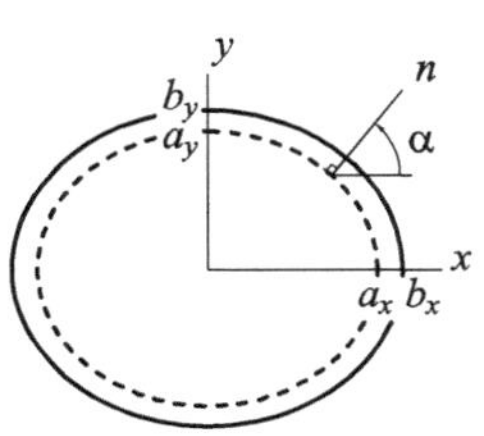

Fig. 5.15 Top view of an *elliptical vase form*. The clear aperture is delimited by the elliptical contour $\mathcal{C}$ defined by semi-axe radii (a_x, a_y). The vase of constant rigidity is clamped into a perimeter ring at $\mathcal{C}$ (*dotted line*) where the n-direction is normal to $\mathcal{C}$. The outer edge of the ring is the elliptical contour $\mathcal{C}'$ determined from (b_x, b_y)

From the flexure z, as represented by (5.26), we obtain

$$M_x = 4z_0 D\left[\frac{3x^2}{a_x^4} + \frac{y^2}{a_x^2 a_y^2} - \frac{1}{a_x^2} + \nu\left(\frac{x^2}{a_x^2 a_y^2} + \frac{3y^2}{a_y^4} - \frac{1}{a_y^2}\right)\right]. \tag{5.31}$$

At the extremities of x, y axes, these moments are respectively

$$M_x\{\pm a_x, 0\} = 8z_0 D/a_x^2, \qquad M_y\{0, \pm a_y\} = 8z_0 D/a_y^2. \tag{5.32}$$

Then, the largest bending moment is applied to the ends of the shortest principal axis of $\mathcal{C}$.

The curvilinear expressions of the bending moment M_n and net shearing force V_n to apply at the contour $\mathcal{C}$ of the ellipse (a_x, a_y) define the boundaries. At any point of the contour, let α be the angle between the axis x and normal n. Since $ds^2 = dx^2 + dy^2$,

$$\sin\alpha = -\frac{dx}{ds} = \frac{a_x^2 y}{\sqrt{a_y^4 x^2 + a_x^4 y^2}}, \qquad \cos\alpha = \frac{dy}{ds} = \frac{a_y^2 x}{\sqrt{a_y^4 x^2 + a_x^4 y^2}}. \tag{5.33}$$

The curvilinear moments M_n and M_{nt} are represented by

$$M_n = M_x \cos^2\alpha + M_y \sin^2\alpha - 2M_{xy}\sin\alpha\cos\alpha\,, \tag{5.34}$$

$$M_{nt} = M_{xy}(\cos^2\alpha - \sin^2\alpha) + (M_x - M_y)\sin\alpha\cos\alpha\,. \tag{5.35}$$

After substitutions, we obtain

$$M_n = 4z_0 D\frac{a_x^4 a_y^4}{a_y^4 x^2 + a_x^4 y^2}\left[\frac{3x^4}{a_x^8} + \frac{3y^4}{a_y^8} - \frac{x^2}{a_x^6} - \frac{y^2}{a_y^6} + \left(\frac{1}{a_x^4} + \frac{1}{a_y^4} - 4\frac{1-2\nu}{a_x^2 a_y^2}\right)\frac{x^2 y^2}{a_x^2 a_y^2}\right], \tag{5.36}$$

$$M_{nt} = 4(1-\nu)z_0 D\frac{a_x^2 a_y^2\, x\, y}{a_y^4 x^2 + a_x^4 y^2}\left(\frac{x^2}{a_x^4} - \frac{y^2}{a_y^4} - \frac{x^2 - y^2}{a_x^2 a_y^2}\right). \tag{5.37}$$

The shearing forces Q_x and Q_y are

$$Q_x = -D\frac{\partial}{\partial x}\nabla^2 z = -8z_0 D\left(\frac{3x}{a_x^4} + \frac{x}{a_x^2 a_y^2}\right), \qquad Q_y = -D\frac{\partial}{\partial y}\nabla^2 z, \tag{5.38}$$

the curvilinear shearing force is

$$Q_n = Q_x\cos\alpha + Q_y\sin\alpha\,, \tag{5.39}$$

and after substitution,

$$Q_n = -8z_0 D \frac{a_x^2 a_y^2}{\sqrt{a_y^4 x^2 + a_x^4 y^2}} \left(\frac{1}{a_x^2 a_y^2} + \frac{3x^2}{a_x^6} + \frac{3y^2}{a_y^6} \right). \tag{5.40}$$

The net shearing force V_n which is the axial reaction at the edge, is given by

$$V_n = Q_n - \frac{\partial M_{nt}}{\partial s}. \tag{5.41}$$

The previous equations representing M_n and V_n at the contour $\mathcal{C}$ completely determine the boundary conditions for the elliptical plate. The net shearing forces can be readily determined in the x and y directions. Considering V_n on y-axis ($x = 0$) for $y = a_y$ at the contour, the second term in (5.40) vanishes and $V_y\{0, \pm a_y\} = Q_n\{0, \pm a_y\}$ at the ends of the minor axis a_y, that is

$$V_y\{0, \pm a_y\} = -8z_0 D \frac{3a_x^2 + a_y^2}{a_x^2 a_y^3} \quad \text{for} \quad a_y < a_x. \tag{5.42a}$$

Similarly, on the x-axis and at the contour $\mathcal{C}$, the reaction at the ends of the major axis a_x is $V_x\{\pm a_x, 0\} = Q_n\{\pm a_x, 0\}$, and

$$V_x\{\pm a_x, 0\} = -8z_0 D \frac{a_x^2 + 3a_y^2}{a_x^3 a_y^2} \quad \text{for} \quad a_y < a_x. \tag{5.42b}$$

Its absolute value gives a slightly lower intensity than for $|V_y\{0, \pm a_y\}|$. Degenerating the ellipse into a circle with $a_x = a_y = a$, the two latter relations allow recovering the well-known result for circular plates, $V_r\{a\} = Q_r\{a\} = -qa/2$.

While the above analysis applies to a *built-in contour*, let us consider now a flexure z composed of a co-added quadratic mode such as

$$z = z_0 \left[\left(1 - \frac{x^2}{a_x^2} - \frac{y^2}{a_y^2} \right)^2 + \kappa \left(1 - \frac{x^2}{a_x^2} - \frac{y^2}{a_y^2} \right) \right], \tag{5.43}$$

where κ is a free parameter. The new sag is $z\{0, 0\} = (1 + \kappa) z_0$ and this flexure also satisfies $\nabla^4 z = q/D$ with z_0 as given by (5.28). $\kappa = 0$ is corresponding to the previous case of null slope at edge. So, with $\kappa \neq 0$, the principal curvatures are changed which implies modification of the bending moments along the plate edge.

At the extremities of the x, y axes, the bending moments are respectively

$$M_x\{\pm a_x, 0\} = 2z_0 D \left(\frac{4 - \kappa}{a_x^2} - \nu \frac{\kappa}{a_y^2} \right), \tag{5.44a}$$

$$M_y\{0, \pm a_y\} = 2z_0 D \left(\frac{4 - \kappa}{a_y^2} - \nu \frac{\kappa}{a_x^2} \right), \tag{5.44b}$$

while the net shearing forces $V_x\{\pm a_x, 0\}$ and $V_y\{0, \pm a_y\}$ remain unchanged since the $\partial M_{nt}/\partial s$ component of the co-added mode in κ of (5.43) is zero.

Considering a *vase form*, and denoting z_1 the previous flexure for the inner plate, the outer ring of finite geometry could provide a rotation around its neutral axis. This would allow us to set the null powered zone at the location defined by the optics rule $k = 3/2$ in the axisymmetric case $z \propto 2k\rho^2 - \rho^4$ leading to $2k = 2 + \kappa$, which corresponds here to $\kappa = 1$. We can represent this elliptic null power zone by

$$U^2 - \frac{3}{2} = 0 \quad \text{with} \quad U^2 \equiv \frac{x^2}{a_x^2} + \frac{y^2}{a_y^2}. \tag{5.45}$$

Assuming an elliptical clear aperture (a_x, a_y) linked to the inner side of the ring, could the ring generate *independently* a part of the quadratic deflexure? Its flexure z_2 cannot be represented by functions of $1 - U^2$ for the two logarithm terms – that would correspond to $r^2 \ln r^2$ and $\ln r^2$ of a circular ring – because the bilaplacian equation $\nabla^4 z_2 = 0$ would not be satisfied, i.e. $\nabla^4 \ln U \neq 0$. So, the flexure of an elliptical ring cannot provide affine curvilinear ellipses without requiring special moments and forces at the contour in addition to the uniform load reacting along it. However, an alternative consists of the use of two vase forms linked together with an outer elliptic cylinder in a closed shape (see hereafter). This avoids requiring $F_{a,k}$ and $F_{c,k}$ forces as treated in Sect. 7.6.2.

The case of an elliptical plate with a *simply supported edge* can be envisaged from the case of a circular plate. It would correspond to $\kappa = 4/(1 + \nu)$ and

$$z = z_0\left(1 - U^2\right)\left(\frac{5 + \nu}{1 + \nu} - U^2\right), \tag{5.46}$$

which would provide a null powered zone along the elliptical line defined by

$$U^2 - \frac{3 + \nu}{1 + \nu} = 0. \tag{5.47}$$

The boundary conditions of a simply supported edge should correspond to a curvilinear bending moment $M_n = 0$ and a net shearing force $V_n = 0$ at the contour. This is not exactly satisfied since

$$M_x\{\pm a_x, 0\} = \frac{8\,\nu z_0 D}{1 + \nu}\left(\frac{1}{a_x^2} - \frac{1}{a_y^2}\right), \quad M_y\{0, \pm a_y\} = \frac{8\,\nu z_0 D}{1 + \nu}\left(\frac{1}{a_y^2} - \frac{1}{a_x^2}\right). \tag{5.48}$$

Compared to the *clamped edge* case and with the incidence angle i defining the ellipticity ratio by $a_y/a_x = \cos i$, we see that these moments are small. Nevertheless, except for the clamped edge case, for a *simply supported edge* or *quasi-simply supported edge* obtaining a flexure whose contour in a plane strictly requires *additional* bending moments M_n and net shearing forces V_n applied to the outer cylinder by an appropriate external force distribution.

Various flexures of an elliptical plate of interest in optics can be represented by

$$z = \frac{q}{8D} \frac{a_x^4 a_y^4}{3a_x^4 + 2a_x^2 a_y^2 + 3a_y^4} \left(1 - U^2\right)\left(\kappa + 1 - U^2\right), \quad \kappa = 2\,(k-1)\,, \quad (5.49)$$

where the associated parameters k and κ determining the location of the null power zone are listed as follows.

Flexure geometry	k	Null power zone U	κ
Clamped edge	1	1	0
Reflective Schmidt	3/2	$\sqrt{3/2} = 1.225$	1
Quasi-simply supp. edge	U^2	$\sqrt{(3+\nu)/(1+\nu)}$	$4/(1+\nu)$

Since Poisson's ratio $\nu \in [0, 1/2]$, the zero power ellipse would be at $1.527 \leq U \leq 1.732$ for a quasi-simply supported edge; with $\nu = 1/4$, this gives $U = x/a_x = 1.612$ if $y = 0$. For reflective Schmidt systems, we see that the optical position of the zero power ellipse is intermediate between clamped and quasi-simply supported edge.

• **Flexure of an elliptical plate in a closed biplate form**: For obtaining $k = 3/2$ profiles, i.e. $\kappa = 1$, let us consider a *closed biplate form* (Fig. 5.16) as the primary mirror of a reflective Schmidt system. This form is constructed with two identical plates of equal thickness linked together at their elliptical contour C with a cylinder. For continuity reasons, the ends of the elliptic cylinder are rounded along C in order to transmit the bending moment to the plates. Taking into account the effect of stress concentration due to a right angle junction, it can be shown that the exact geometry of a plate clamped into an infinite size corner is provided by a *quarter-circle junction* whose radius of curvature is equal to the thickness of the plate.

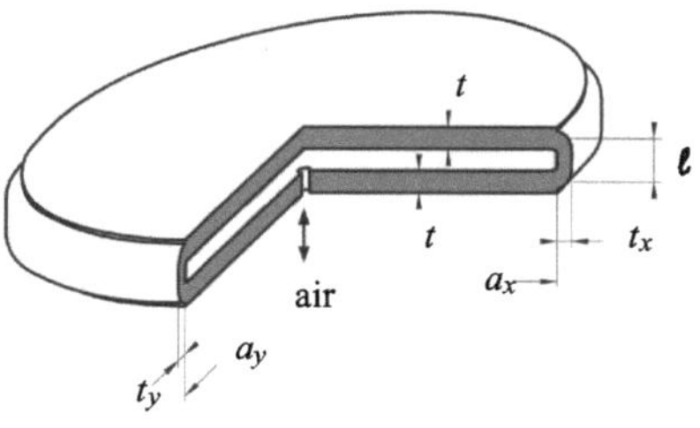

Fig. 5.16 *Closed biplate form* generating affine elliptical flexures by uniform loading only. The elliptical contour C of the plates is linked with a cylinder. The radial thickness of the cylinder realizes a *semi-built-in condition* and is varying along C to satisfy the best optical profile $\kappa = 1$ (i.e. $k = 3/2$). In the (z, x) and (z, y) sections, the principal radial thicknesses of the outer cylinder are $t_x = b_x - a_x$ and $t_y = b_y - a_y$, respectively

In the aspherization process, starting from a plane mirror, the flexure directly provides the optical figure if

$$R_{\text{Tool}} = \infty, \quad Z_{\text{Opt}} + \wp\, Z_{\text{Elas}} = 0 \quad \text{and} \quad k = 3/2 \;\text{ i.e. }\; \kappa = 1\,, \tag{5.50}$$

where $\wp = \pm 1$ depends on considering an *in situ stressing* or a *stress figuring*. Assuming an *in situ stressing*, thus taking the positive value, the flexure is represented by [$\kappa = 1$ in (5.49)]

$$Z_{\text{Elas}} \equiv z = z_0 \left[1 - \frac{x^2}{a_x^2} - \frac{y^2}{a_y^2} + \left(1 - \frac{x^2}{a_x^2} - \frac{y^2}{a_y^2} \right)^2 \right], \tag{5.51}$$

where z_0 is given by (5.28). From the mirror optical shape Z_{Opt} in (5.20d), we deduce the equivalences of the semi-axe radii at clear aperture

$$a_x = x_{\text{m}} = y_{\text{m}} / \cos i \quad \text{and} \quad a_y = y_{\text{m}} = r_{\text{m}}\,.$$

From the identification of U^4 coefficients, we obtain

$$\wp\, z_0 = \frac{y_{\text{m}}}{2^9 \Omega^3 \cos i}\,.$$

After substitution of the rigidity and simplifications, the thickness of each plate of the closed form is

$$t = 4\Omega \left[\frac{12\, \wp\, (1-\nu^2) \cos i}{3 + 2\cos^2 i + 3\cos^4 i}\, \frac{q}{E} \right]^{1/3} y_{\text{m}} \tag{5.52}$$

This defines the first part of execution conditions, the second is the design of the perimeter elliptic cylinder.

• **Closed biplate form - Design of elliptic cylinder contour**: At the contour $\mathcal{C}$, and in the x- and y-direction, the bending moment M_n and net shearing forces V_n are

$$M_x\{\pm a_x, 0\} = 2 z_0 D \left(\frac{3}{a_x^2} - \frac{\nu}{a_y^2} \right), \quad V_x\{\pm a_x, 0\} = -8 z_0 D \frac{a_x^2 + 3a_y^2}{a_x^3 a_y^2} \tag{5.53a}$$

$$M_y\{0, \pm a_y\} = 2 z_0 D \left(\frac{3}{a_y^2} - \frac{\nu}{a_x^2} \right), \quad V_y\{0, \pm a_y\} = -8 z_0 D \frac{3a_x^2 + a_y^2}{a_x^2 a_y^3} \tag{5.53b}$$

For a circular plate, denoting $a_x = a_y = a$ and s_{pl} its slope at the edge, the bending moment is

$$M\{r = a\} = 2(3-\nu) z_0 D \,/\, a^2 = -(3-\nu) D s_{\text{pl}} \,/\, a\,. \tag{5.54}$$

In (z, x) and (z, y) sections, the radial thicknesses t_x, t_y of the elliptic cylinder to link with the plates, can be defined by dimensionless coefficients $\gamma_x,\ \gamma_y$ such as

$$\gamma_x = \frac{t_x}{t} = \frac{b_x - a_x}{t}, \qquad \gamma_y = \frac{t_y}{t} = \frac{b_y - a_y}{t}, \tag{5.55}$$

where t is the thickness of the two plates.

In order to close the two separated plates by a cylinder linked at their edge, let us consider at first the axisymmetric case of a circular cylinder in setting $\gamma = \gamma_x = \gamma_y = t_{\rm cyl}/t$, where $t_{\rm cyl}$ is the *radial* thickness of the cylinder. Let $b = b_x = b_y$ be its external radius and $a_{\rm cyl} = (a+b)/2$ its mean radius. Without changing the frame notations, let us move the (z, r) axes of a plate frame in parallel directions so the origin of the coordinates are now at the median point of the neutral surface of the cylinder. The flexure is a solution of the differential equation [30b]

$$\frac{d^4 r}{dz^4} + 4\beta^4 r = \frac{q}{D_{\rm cyl}} \tag{5.56}$$

where the rigidity of the cylinder $D_{\rm cyl}$ and the coefficient β are defined by

$$D_{\rm cyl} = \frac{E\, t_{\rm cyl}^3}{12(1-\nu^2)}, \qquad \beta^4 = \frac{3(1-\nu^2)}{a_{\rm cyl}^2 t_{\rm cyl}^2}. \tag{5.57}$$

In order to determine the thickness $t_{\rm cyl}$ and the length ℓ of the cylinder, first we notice that the cylinder is bent by opposite moments M at its edges $z = \pm\ell/2$, so the flexure $r(z)$ must be even. Then, the solution of (5.56) is

$$r = C_1 \sinh\beta z \sin\beta z + C_2 \cosh\beta z \cos\beta z + f(z) \tag{5.58}$$

where the odd terms, i.e. C_3 and C_4 coefficients, have been cancelled, and

$$f(z) = q/4\beta^4 D = constant$$

is the particular solution due to loading q. Assuming hereafter the case of very short cylinders, the extensional deformations due to loading q can be neglected in comparison to the effect of the bending moments M and $-M$ applied to the extremities. We have seen with (5.53) that the reactions $V_x\{\pm a_x, 0\}$, $V_y\{0, \pm a_y\}$ only depend on the loading intensity q whatever the boundaries at the edge (clamped, semi-clamped, or simply supported). In a first-order approximation, we will also neglect the deformation of the cylinder in the z direction.

Since linked to the plates, the extremities of the cylinder cannot appreciably move in the radial direction r, but can rotate. The bending moment is defined by $M = D\, d^2r/dz^2$. Therefore, the boundary conditions for the outer cylinder are as follows

$$(r)_{z=\ell/2} = 0, \qquad \left(\frac{d^2 r}{dz^2}\right)_{z=\ell/2} = \frac{M}{D_{\rm cyl}}. \tag{5.59}$$

Denoting

$$\omega = \beta\ell/2,$$

we find after substitution,

$$C_1 \sinh\omega \sin\omega + C_2 \cosh\omega \cos\omega = 0\,, \tag{5.60a}$$

$$C_1 \cosh\omega \cos\omega - C_2 \sinh\omega \sin\omega = \frac{M}{2\beta^2 D}\,. \tag{5.60b}$$

Using $\cosh^2\omega \cos^2\omega + \sinh^2\omega \sin^2\omega = \frac{1}{2}(\cosh 2\omega + \cos 2\omega)$, the coefficients are

$$C_1 = \frac{M}{\beta^2 D} \frac{\cosh\omega \cos\omega}{\cosh 2\omega + \cos 2\omega}\,, \qquad C_2 = -\frac{M}{\beta^2 D} \frac{\sinh\omega \sin\omega}{\cosh 2\omega + \cos 2\omega}\,. \tag{5.61}$$

With these coefficients, the slope of the flexure at the edge of the cylinder is

$$s_{\rm cyl} = \left(\frac{dr}{dz}\right)_{z=\ell/2} = \frac{M}{2\beta D_{\rm cyl}} \frac{\sinh 2\omega + \sin 2\omega}{\cosh 2\omega + \cos 2\omega} = \frac{M\ell}{2D_{\rm cyl}}\left(1 - \frac{8}{15}\,\omega^4 + \cdots\right). \tag{5.62}$$

In order to obtain a basic formulation, we can restrict the term in parentheses to unity by choosing enough small values of ω. Taking $\frac{8}{15}\,\omega^4 \leq 10^{-2}$, we obtain the inequality

$$16\,\omega^4 = \beta^4\ell^4 = \frac{3(1-\nu^2)\ell^4}{a_{\rm cyl}^2 t_{\rm cyl}^2} \leq \frac{3}{10}\,. \tag{5.63}$$

From (5.62), we see that the first-order term of the slope $s_{\rm cyl}$ does not depend on the cylinder mean radius $a_{\rm cyl} = a + \frac{1}{2}\,t_{\rm cyl}$. Making $a_{\rm cyl}$ equal to a in the above inequality, the condition on length for the cylinder is

$$\ell^2 \leq \frac{a\,t_{\rm cyl}}{\sqrt{10(1-\nu^2)}}\,. \tag{5.64}$$

Since the rigidities are related by $D_{\rm cyl} = \gamma^3 D$, if using the same material for the plates and cylinder, (5.54) and (5.62) give respectively

$$\left|\frac{M}{s_{\rm pl}}\right| = \frac{(3-\nu)D}{a}\,, \qquad \text{and} \qquad \left|\frac{M}{s_{\rm cyl}}\right| = \frac{2\gamma^3 D}{\ell}\,. \tag{5.65}$$

The equality of the plate and cylinder slopes being de facto realized by the *closed form*, the semi-clamping condition is satisfied if

$$\gamma = \frac{t_{\rm cyl}}{t} = \sqrt[3]{\frac{3-\nu}{2}\,\frac{\ell}{a}}\,. \tag{5.66}$$

In addition, it would be difficult to achieve in practice the boundary condition if $\ell \leq t$ because the cylinder should be preferably linked to the base of the plates in a quarter-circle shape. With $\ell \gtrsim t$, the link of the two plate edges will provide the respect of Saint-Venant's principle and a compact transmission of the bending moment. Hence, the z-length ℓ of the cylinder should be somewhat larger than the thickness t of each of the two plates.

1 → *In the axisymmetric case, a* closed biplate form *provides a full aperture mirror, of diameter* $2a$, *as the primary mirror of a reflective Schmidt system if the thickness* $t_{\rm cyl}$ *of the outer cylinder, of length* ℓ *between the two plates of thickness* t, *satisfies the conditions* $t_{\rm cyl} = [(3-\nu)\ell/2a]^{1/3}\, t$ *and* $t < \ell \leq [10(1-\nu^2)]^{-1/4}\sqrt{a\, t_{\rm cyl}}$.

Taking Poisson's ratio $\nu = 1/4$ and a thickness ratio $\gamma = t_{\rm cyl}/t = 1/2$, these conditions give $a/\ell = 11$, and thus $t < \ell \leq 1.8\, t$. For example, this allows the choice of $\ell = 3t/2$. For $\nu = 1/4$ and $\gamma = 1$, the inequality is not satisfied. For $\nu = 1/4$ and $\gamma = 2/3$, the inequality is just satisfied. For ν varying in $\{0,\, 1/2\}$, the results are quite
similar.

2 → *In the axisymmetric case, a* closed biplate form *provides solutions for the primary mirror of a reflective Schmidt system with a cylinder-plate thickness ratio* $t_{\rm cyl}/t \leq 2/3$, *where* $t_{\rm cyl}$ *is in the radial direction and* t *for each of the two plates is axial.*

From (5.66), the outer radius of the axisymmetric cylinder is

$$b = a + t_{\rm cyl} = a + [(3-\nu)\ell\,/2a]^{1/3}\, t\,. \tag{5.67}$$

Returning to the case of the elliptical flexure of plates with $\kappa = 1$, and taking into account the bending moments M_x and M_y at the principal axes and at the contour $\mathcal{C}$ as given by (5.53), the absolute values of theses moments with respect to the corresponding local slope are

$$\left|\frac{M_x}{s_x}\right|_{\rm pl} = D\left(\frac{3}{a_x} - \nu\frac{a_x}{a_y^2}\right), \qquad \left|\frac{M_y}{s_y}\right|_{\rm pl} = D\left(\frac{3}{a_y} - \nu\frac{a_y}{a_x^2}\right). \tag{5.68}$$

The corresponding M/s ratios for the elliptic cylinder are respectively

$$\left|\frac{M_x}{s_x}\right|_{\rm cyl} = D\,\frac{2\gamma_x^3}{\ell}\,, \qquad \left|\frac{M_y}{s_y}\right|_{\rm cyl} = D\,\frac{2\gamma_y^3}{\ell}\,. \tag{5.69}$$

By identification of these ratios, the conditions of semi-clamping of the closed form at the principal axes of the ellipse are

$$\gamma_x = \frac{t_x}{t} = \sqrt[3]{\left(3 - \nu\frac{a_x^2}{a_y^2}\right)\frac{\ell}{2a_x}}\,, \qquad \gamma_y = \frac{t_y}{t} = \sqrt[3]{\left(3 - \nu\frac{a_y^2}{a_x^2}\right)\frac{\ell}{2a_y}}\,. \tag{5.70}$$

Similarly as with (5.64), we obtain the inequalities

$$\ell^2 \leq \frac{a_x\, t_x}{\sqrt{10(1-\nu^2)}}\,, \qquad \ell^2 \leq \frac{a_y\, t_y}{\sqrt{10(1-\nu^2)}}\,. \tag{5.71}$$

3 → *In the elliptical case, a* closed biplate form *provides a full aperture mirror, of dimensions* $2a_x \times 2a_y$, *as primary mirror of a reflective Schmidt system* *if the thicknesses t_x, t_y of the outer cylinder, of length ℓ between the two plates of thickness t, satisfy the conditions $t_{x,y} = [(3-\nu a_{x,y}^2/a_{y,x}^2)\ell/2a_{x,y}]^{1/3}\,t$ and $t < \ell \le [10(1-\nu^2]^{-1/4}\sqrt{a_{x,y}\,t_{x,y}}$.*

From (5.70), the thicknesses t_y and t_x of the elliptical cylinder are related by

$$\left(\frac{t_y}{t_x}\right)^3 = \frac{a_x}{a_y}\,\frac{3-\nu\dfrac{a_y^2}{a_x^2}}{3-\nu\dfrac{a_x^2}{a_y^2}} \tag{5.72a}$$

The principal incident beam of circular cross-section reflects at the mirror – telescope input pupil – of elliptical contour defined by $a_x/a_y = 1/\cos i$, where i is the incident angle. Thus, for the principal axes of the cylinder, the thickness ratio expands as

$$\frac{t_y}{t_x} = 1 + \frac{1}{2}\,\frac{1+\nu}{3-\nu}\,i^2 + \cdots, \tag{5.72b}$$

which, for a Poisson's ratio $\nu = 1/4$, a deviation angle $2i \le 60°$, and the first two terms of this expansion, gives an error that deviates of $\le 6\times 10^{-3}$ from unity compared to (5.72a).

4 → *In the elliptical case, the ellipticity of the outer contour (b_x, b_y) of the cylinder differs from that of its inner contour (a_x, a_y). The principal radial thicknesses t_y, t_x of the cylinder must satisfy*

$$t_y^3 a_y(3-\nu a_x^2/a_y^2) = t_x^3 a_x(3-\nu a_y^2/a_x^2) \quad \textit{with} \quad a_x/a_y = 1/\cos i\,, \tag{5.73}$$

where i is the incidence angle of the circular cross-section beam from an object at infinity. Therefore, the radial thickness of the cylinder is the largest at the small axis of the close biplate form.

Table 5.4 displays the ratios t_y/t_x of the radial thicknesses in the principal directions as a function of Poisson's ratios and beam deviations $2i$ at the mirror of elliptical geometry $a_x/a_y = 1/\cos i$, i.e. $a_x \ge a_y$, corresponding to a principal incident beam of circular cross-section from an object at infinity.

Table 5.4 Closed biplate form. Radial thickness ratios t_y/t_x of the peripheral cylinder as a function of Poisson's ratio and beam deviation $2i$. The inner elliptical contour is defined by $a_x/a_y = 1/\cos i$ i.e. $a_x \ge a_y$

Beam dev. $2i$	15°	30°	45°	60°	75°	90°	105°
$\nu = 0$	1.003	1.011	1.027	1.049	1.080	1.122	1.180
$\nu = 1/6$	1.003	1.014	1.033	1.061	1.101	1.156	1.237
$\nu = 1/5$	1.004	1.015	1.034	1.064	1.105	1.164	1.250
$\nu = 1/4$	1.004	1.016	1.037	1.068	1.112	1.176	1.271
$\nu = 1/3$	1.004	1.017	1.040	1.075	1.125	1.197	1.310
$\nu = 1/2$	1.005	1.021	1.049	1.091	1.153	1.248	1.410

The principal outer radii of the elliptical cylinder closing the biplate are alternatively for x and y,

$$b_{x,y} = a_{x,y} + t_{x,y} = a_{x,y} + \left[\left(3 - \nu \frac{a_{x,y}^2}{a_{y,x}^2} \right) \frac{\ell}{2a_{x,y}} \right]^{1/3} t \, . \qquad (5.74)$$

Given an elastic material, the eight geometrical quantities $t,\ a_x,\ a_y,\ \ell,\ b_x,\ b_y,\ t_x,\ t_y$ of the closed vase form and the uniform load q completely determine the aspherization conditions of an elliptical primary mirror used as input *pupil* of a non-centered system reflective Schmidt telescope.

• **Other applications of an elliptic closed biplate form**: The above results for the primary mirror of all-reflective non-centered Schmidts can also apply to mirrors of other optical systems where a parallel beam must be folded by a plane mirror. The aspherization of such elliptical contour mirrors will provide the best results in terms of aberration corrections when a pupil is located near or at this mirror.

5.3.6 LAMOST: A Segmented Bisymmetric Elliptical Primary

LAMOST is a giant all-reflective Schmidt, 4-meter clear aperture and 5° field of view, fully dedicated to spectroscopic studies. Its aspherical primary mirror is a bisymmetric elliptical shape segmented into 24 hexagons of 1.1 m in the diagonal. Each segment being flat when at rest, an active optics *closed-loop control system* generates, through actuators, the required asphericity which is a function of the sky position of observed objects.

Because the LAMOST segments are *multimode deformable mirrors* (MDM) and variably aspherized in-situ at the telescope accounts of these developements are given in Sect. 7.6.4 and Sect. 8.7.3.

5.4 Aspherized Reflective Diffraction Gratings

5.4.1 Active Optics Replication for Grating Aspherization

All diffraction grating manufacturers are well aware of replication techniques. In the basic technique, the optical surface to replicate – or "master" surface – is first coated with a gold layer; then a layer of convenient epoxy resin is pressed onto this surface while slightly heated. After polymerization, the gold layer facilitates the separation of the two surfaces. Low to medium dispersion gratings for astronomical purposes are usually *replicas*; this allows a substantial decreases of the cost. Furthermore, it has long been known from ancient diamond *ruled* gratings that a single or even a double replication increases the grating efficiency by removing the sharp edges caused by the diamond.

The *active optics replication technique* is a straightforward process for obtaining aspherics. Applied to the making of an aspherized grating, this technique uses two replication stages and requires an *active optics submaster*. Starting for instance from a plane master grating, a first stage provides a plane replica on the submaster; after changing the stress state of the submaster, the second replication stage provides the final aspherized grating replica on a classical rigid substrate.

5.4.2 Optical Profile of Aspherical Reflective Gratings

Similarly to all-reflective Schmidts, the shape of an aspherical reflective grating that minimizes the field residual aberrations is obtained by Lemaitre's condition $k = 3/2$ (Sect. 4.3.1). Thus, the local radial curvatures of the grating are extremals at its center and at its clear aperture edge, and of opposite sign for these radii (see Fig. 5.10).

Because of the *beam anamorphosis* of the diffracted beams, and since the incident beams are usually of circular cross section – diameter $2r_{\rm m}$ –, one defines the basic camera f-ratio as $\Omega = {\rm f/d} = R/4r_{\rm m} = R/4y_{\rm m}$ in the $(y,\, z)$ plane containing the grating central line, where $y_{\rm m} = r_{\rm m}$ is the semi-length of this line. Since the grating lines are parallel to the spectrograph slit, Ω is called *nebular f-ratio*.

Summarizing the results in Sect. 4.4 for all reflective grating spectrographs, such as for all-reflective telescopes, the best images are obtained with design parameters

$$k = 3/2\,, \quad \text{i.e.}\;\; M = 3/64\,\Omega^2 \quad \text{with} \quad \Omega = R/4y_{\rm m} = R/4r_{\rm m}\,. \tag{5.75}$$

With recent developments of large size spectrographs designed with f-ratios as fast as f/1.5, it is useful to summarize hereafter both high- and third-order representations of a grating optical shape. The following representations are for an incident principal beam of circular cross section.

• **Axisymmetric gratings** ($\beta_0 = 0$): For a normal diffraction mounting [see (4.23)], $\beta_0 = 0$, the grating shape is *axisymmetric* and the contour of its clear aperture is an ellipse. If the spectrograph only uses a spherical camera mirror – curved field –, then the best under-correction factor is $s = \cos^2\varphi_{\rm m}$. From (4.26) and denoting $\rho = r/r_{\rm m}$ a dimensionless radius, the grating shape is

$$z_{\rm Opt} = \frac{\cos^2\varphi_{\rm m}}{1+\cos\alpha} \sum_{2,4,6,\ldots} \frac{\mathcal{A}_n\, r^n}{R^{n-1}} \simeq \frac{\cos^2\varphi_{\rm m}}{2^8\,\Omega^3\,(1+\cos\alpha)} \left(3\,\rho^2 - \rho^4\right) r_{\rm m}\,, \tag{5.76a}$$

$$\text{Elliptic clear aperture} \begin{cases} 0 \le \rho \le 1 & \text{in nebular direction } y, \\ 0 \le \rho \le 1/\cos\alpha & \text{in dispersion direction } x, \end{cases}$$

where $\mathcal{A}_n$ are given by Table 4.1.

The location of the circular null power zone r_o is *outside* the clear aperture radius r_m and defined in the third-order by $\rho_o = r_o/r_m = \sqrt{k} = \sqrt{3/2}$. For higher-order correction profiles, the determination of ρ_o can be accurately derived from $d^2z/dr^2|_0 = -d^2z/dr^2|_{r_m}$.

• **Bisymmetric elliptical gratings** ($\beta_0 \neq 0$): For off-normal diffraction mountings, $\beta_0 \neq 0$, the grating shape must be with biaxial symmetry. Without a field-lens flattener and for a spherical camera mirror, then $s = \cos^2\varphi_m$. From (4.28), the grating shape is

$$Z_{Opt} = \frac{\cos^2\varphi_m}{\cos\beta_0 + \cos\alpha} \sum_{2,4,6,\ldots} \frac{\mathcal{A}_n}{R^{n-1}} \left(\cos^2\beta_0\, x^2 + y^2\right)^{n/2}. \tag{5.76b}$$

$$\text{Elliptic clear aperture} \begin{cases} 0 \leq y \leq y_m & \text{in nebular direction } y, \\ 0 \leq x \leq x_m & \text{in dispersion direction } x, \end{cases}$$

where $\mathcal{A}_n$ coefficients are given in Table 4.1. Assuming a collimated incident beam of circular cross-section and aperture $2r_m$, the iso-level lines of the surface are ellipses represented by

$$\cos^2\beta_0\, x^2 + y^2 = constant\,. \tag{5.76c}$$

In the third-order approximation, the elliptical null powered zone located outside the clear aperture is the ellipse

$$\cos^2\beta_0\, x^2 + y^2 = r_o^2 = \tfrac{3}{2}\, r_m^2\,. \tag{5.76d}$$

For a collimated beam of circular cross-section, $2r_m$ in diameter, the contour line of the grating clear aperture – as well as in a Littrow mounting –, is the ellipse

$$\cos^2\alpha\, x^2 + y^2 = r_m^2, \quad \text{i.e.} \quad x_m = r_m/\cos\alpha\,, \quad y_m = r_m\,. \tag{5.76e}$$

These relations are useful for the design of basic spectrographs with a curved focal surface. The undercorrection ratio $s = \cos^2\varphi_m$ provides the best blur images along the *central* spectrum but would be slightly different for a balance with sideways spectra or *multi-object spectroscopy*. A slight modification of s allows maintaining the $k = 3/2$ condition either for the latter aberration balance or for the design of flat fielded systems with a positive corrector lens often used as the detector cryostat window.

• **Aspherization replication process**: In all the following Sections, the aspherization of a reflective grating is achieved via two replication stages by use of an intermediate *active optics submaster* as follows. Starting from a classically plane diffraction grating – "master grating" –, the first stage is performed on the plane surface of the submaster when in a null stress state. Then, the deformable submaster is aspherized in an *opposite* shape to that of the final grating replica on a rigid substrate which is obtained in the second stage.

5.4.3 Axisymmetric Gratings with k = 3/2 and Circular Built-in Submasters

Aspherized reflective gratings with a geometrical profile $k = 3/2$ are easily generated from the replication of built-in deformable submasters. In the axisymmetric grating case – corresponding to the normal diffraction angle $\beta_0 = 0$ for the central wavelength – the best design is a circular active submaster with a built-in radius at the null powered radius $a = r_o$. The aspherized grating is used with an elliptic clear aperture included into the built-in circle; this condition, $r_m/\cos\alpha \leq r_o$, is always satisfied for usual incidence angles.

Let Z_{Sub} be the shape of the active submaster when in a stressed stage and assume that it was figured in a flat shape when unstressed ($Z_{Sphe}=0$). The aspherization of a grating replica Z_{Opt} is realized if the components of the *active optics co-addition law* write

$$Z_{Elas} \equiv Z_{Sub}\,, \qquad Z_{Sphe} = 0 \quad \text{and} \quad Z_{Sub} + Z_{Opt} = 0\,, \tag{5.77}$$

where Z_{Elas} is the elastic flexure.

• **Constant thickness submasters for slow focal ratio cameras**: The design of *active optics submasters* for making gratings of low f-ratio spectrographs can be easily derived in the third-order optic theory, i.e. limited to the *Cv* 1 and *Sphe* 3 terms. In this case the elasticity theory provides a constant thickness solution which is a *built-in plate* bent by a uniform load. Hence, the submaster for grating replications is a *vase form* made of a constant thickness plate built-in to a thick outer ring.

From the identification of r^4 terms of Z_{Elas} and Z_{Sub} in (5.5) and (5.76a) respectively, we obtain, with a uniform load q,

$$\frac{q}{64D} = -\frac{\cos^2\varphi_m}{1+\cos\alpha}\,\frac{\mathcal{A}_4}{R^3}\,, \tag{5.78a}$$

which, after substitution of the coefficient $\mathcal{A}_4$ and rigidity D, writes

$$\frac{q}{E} = \frac{4\cos^2\varphi_m}{3(1-\nu^2)(1+\cos\alpha)}\left[1 - \frac{4M(1-M^3)}{(1-M)(1+M)^3}\right]\frac{t^3}{R^3}\,. \tag{5.78b}$$

Using the definitions of M and Ω in (5.75), the constant thickness of the plate, built-in at r_o, is

$$\begin{cases} t = 4\,\Omega\left[\dfrac{3(1-\nu^2)(1+\cos\alpha)}{4\cos^2\varphi_m}\,\dfrac{q}{E}\right]^{1/3}\left(1 + \dfrac{1}{16\,\Omega^2} + \cdots\right) r_m\,, \\ r_o = \sqrt{3/2}\, r_m\,. \end{cases} \tag{5.78c}$$

This determines the execution conditions of the deformable submaster for focal ratios slower, say, than f/2.5. At the built-in edge r_o, the junction to a thick outer ring is achieved by a quarter-circle of curvature $1/t$. Metal submasters such as chromium stainless steel are preferred for linearity and safety when applying several atmospheres inner air pressure. The *shrinkage effect* of epoxy resins during the polymerization process is corrected at the final aspherization process of replication by a slight overload Δq that usually is $\Delta q/q \simeq 4\%$. This also requires use of a minimal resin quantity; hence the rigid Zerodur substrate of the final grating replica must be given a *slightly* concave spherical shape before the replication where the thickness of the resin is minimum even at center and edge.

• **Variable thickness submasters for fast focal ratio cameras**: The design of active submasters for making gratings of fast f-ratio spectrographs requires use of the power series (5.76a) for a higher-order correction of spherical aberration modes. With an *active optics submaster* made of a plate built-in to a thick outer ring, we show hereafter that these corrections lead to an active plate thickness slightly decreasing from center to edge.

The static equilibrium of the plate is derived from the radial bending moment, tangential bending moment, and shearing force. These are defined by the equation set (Sect. 2.1.2)

$$\begin{cases} M_r + \dfrac{dM_r}{dr}\, r \;-\; M_t \;+\; Q_r\, r = 0\,, \\ M_r = D\left(\frac{d^2z}{dr^2} + \frac{\nu}{r}\frac{dz}{dr}\right), \\ M_t = D\left(\nu\frac{d^2z}{dr^2} + \frac{1}{r}\frac{dz}{dr}\right), \\ Q_r = -\frac{1}{2\pi r}\int_0^r q\, 2\pi r\, dr = -\frac{1}{2}\, q\, r\,. \end{cases}$$

After substitution of M_r, M_t, and Q_r, the equilibrium equation becomes

$$\frac{dD}{dr}\left(\frac{d^2z}{dr^2} + \frac{\nu}{r}\frac{dz}{dr}\right) + D\frac{d}{dr}\left(\frac{d^2z}{dr^2} + \frac{1}{r}\frac{dz}{dr}\right) = \frac{1}{2}\, q\, r\,, \tag{5.79}$$

where the flexure of the submaster plate is, from (5.76a) and (5.77),

$$z \equiv Z_{\mathrm{Sub}} = -\frac{\cos^2\varphi_{\mathrm{m}}}{1+\cos\alpha}\sum_{2,4,6,\ldots}\frac{\mathcal{A}_n\, r^n}{R^{n-1}}\,. \tag{5.80}$$

For solving the rigidity $D(r)$ with given number p of expansion terms $\mathcal{A}_n(M)$, we first determine the boundary radius r_{o} of the clamped edge. This radius is the root of

$$\sum_{2,4,\ldots,p} n\,\mathcal{A}_n\left(\frac{r}{R}\right)^{n-1} = 0 \quad \text{with} \quad M = 3/64\Omega^2\,. \tag{5.81}$$

Defining dimensionless quantities

$$\zeta = \frac{z}{R}\,, \quad \rho = \frac{r}{R}\,, \quad g = -\frac{\cos^2\varphi_{\mathrm{m}}}{1+\cos\alpha}\,, \quad \mathcal{D} = \frac{2\,g}{q\,R^3}\, D\,, \tag{5.82}$$

the substitutions and simplifications lead to

$$\begin{cases} \frac{d\mathcal{D}}{\rho\, d\rho}\left(\frac{d^2\zeta}{d\rho^2}+\frac{\nu}{\rho}\frac{d\zeta}{d\rho}\right)+\mathcal{D}\frac{d}{\rho\, d\rho}\left(\frac{d^2\zeta}{d\rho^2}+\frac{1}{\rho}\frac{d\zeta}{d\rho}\right)=g\,, & (5.83)\\ \zeta = g \sum\limits_{2,4,\ldots,p} \mathcal{A}_n\,\rho^n \quad \text{with } p \text{ even}\,, & (5.84)\end{cases}$$

from which the dimensionless rigidity is a solution of

$$\frac{d\mathcal{D}}{\rho\, d\rho}\sum_{2,4,\ldots,p} n(n-1+\nu)\mathcal{A}_n\,\rho^{n-2} + \mathcal{D}\sum_{4,6,\ldots,p}(n-2)n^2\mathcal{A}_n\,\rho^{n-4} = 1\,. \quad (5.85)$$

Let us represent the rigidity $\mathcal{D}(\rho)$ up to the null power zone $\rho_o = r_o/R$ which will be the dimensionless clamping radius of the active plate at the outer thick ring. From the even expansion

$$\mathcal{D} = \sum_{0,2,4,\ldots p'} X_n\,\rho^n \quad \text{with} \quad 0 \le \rho \le \rho_o\,, \quad (5.86)$$

the substitution in (5.85) and identifications of the *constant*-, ρ^2-, ρ^4-terms, etc, lead to a linear system,

$$32\mathcal{A}_4\,X_0 + 4(1+\nu)\mathcal{A}_2\,X_2 = 1\,,$$
$$18\mathcal{A}_6\,X_0 + (7+\nu)\mathcal{A}_4\,X_2 + (1+\nu)\mathcal{A}_2\,X_4 = 0\,,$$

were the next null sums successively include the unknown X_6, X_8, etc.

If $\mathcal{A}_6 = 0$ and the higher-order coefficients are null, then $X_n = 0$ for $n \ge 2$ and $\mathcal{D} = X_0 = 1/(32\mathcal{A}_4)$ which, from (5.82) and (5.78a), provides the well-known constant thickness plate solution correcting *Sphe* 3. If $\mathcal{A}_6 \ne 0$ and the higher-order coefficients are null, then the third identification equation for ρ^4 terms, in (5.85), is needed to determine the rigidity that generates the additional *Sphe* 5 mode correction. In this latter three-equation set, $X_6 = 0$; therefore the dimensionless rigidity is of the form $\mathcal{D} = X_0 + X_2\rho^2 + X_4\rho^4$. A general result is that, for higher-order spherical aberration corrections than *Sphe* 3, a polynomial flexure represented by $n/2 = p/2$ terms is generated by a rigidity represented by $p/2 - 1$ even terms.

For the design of active optics submasters generating corrected gratings from f/3 to f/1.22 in the nebular direction, computational resolutions of the equation set (5.85) were carried out up to the order including $n = 8$ in (5.86). These rigidities are for normal- or quasi-normal-diffraction reflective grating spectrographs that have been built (Fig. 5.17 and Table 5.5). For instance since $\rho_m = r_m/R$, for the clear aperture edge $\Omega = 1/4\rho_m = 1.22$ – i.e. an f-ratio of f/1.22 in the nebular direction, we obtain $\mathcal{D}(\rho_o)/\mathcal{D}(0) = 0.785$; since the thickness and rigidity are related by $t \propto \mathcal{D}^{1/3}$, this corresponds to a *thinner* plate edge than at the center in a ratio $t(\rho_o)/t(0) = 0.922$.

The final design of the submaster must take into account the stress level at the clamping boundary by appropriate set up of the plate thickness. The maximum stresses caused by the inner uniform pressure load q (>0) come from the radial bending moment and stand at the built-in radius r_o. At the outer surfaces of the plate these stresses are

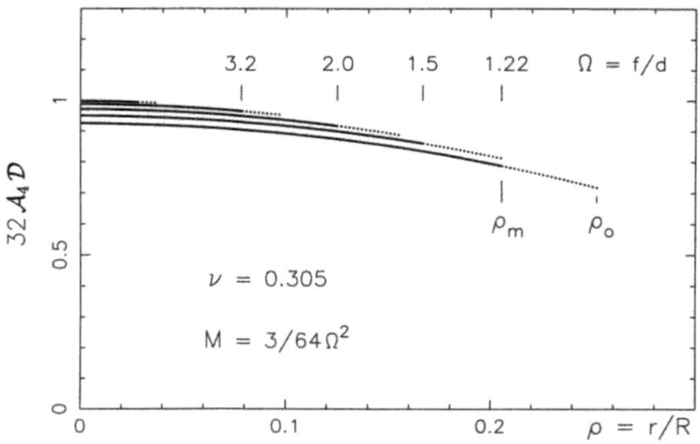

Fig. 5.17 Five dimensionless rigidities of *active optics submasters* from integration of (5.85) with $\nu = 0.305$. Coefficients $\mathcal{A}_n$ in Table 4.1 are set with $M = 3/64\Omega^2$ providing $k \simeq 3/2$ (see Table 5.5). The diagram plots $32\mathcal{A}_4\,\mathcal{D}$ vs $\rho = r/R$. (*Full lines*): Rigidities corresponding to grating clear aperture ρ_m in slit direction and associated focal ratios $\Omega = 1/4\rho_m$. (*Dotted line*): Extensions from ρ_m to the built-in edge ρ_o partly used by the elliptic pupil in dispersion direction

Table 5.5 Dimensionless rigidities $\mathcal{D}(\Omega, \rho)$ of *vase-form submasters* for various spectrogrpah f-ratios $\Omega = 1/4\rho_m$ in nebular direction (y). Position's ratio $\sqrt{k} = \rho_o/\rho_m$. Built-in radius ρ_o root of (5.81). Values $32\mathcal{A}_4 X_n$ from integration of (5.85) with $\mathcal{D} = \sum X_n \rho^n$ and $\nu = 0.305$

f/ Ω	$\sqrt{k}$	ρ_o	$32\mathcal{A}_4X_0$	$32\mathcal{A}_4X_2$	$32\mathcal{A}_4X_4$	$32\mathcal{A}_4X_6$
f/9.0	1.2247	0.0340	0.9986	−3.6881	0.8293	0.2109
f/3.2	1.2209	0.0953	0.9890	−3.6312	0.7590	−0.8093
f/2.0	1.2126	0.1517	0.9720	−3.5330	0.6485	−2.6210
f/1.5	1.2064	0.2010	0.9506	−3.4143	0.5312	−4.9023
f/1.22	1.1979	0.2459	0.9256	−3.2807	0.4199	−7.5311

$$\sigma_r\Big|_{max} = \frac{6\,M_r}{t}\Big|_{r=r_o} \simeq \pm\,\frac{3}{4}\,\frac{r_o^2}{t^2(r_o)}\,q\,. \tag{5.87}$$

Submasters for grating aspherization are generally made of high-strength metal alloys. The maximum stress must be lower than the elasticity limit $\sigma_{E\,0.2\%}$. Up to f/1, a convenient determination of the thickness is easily obtained with stainless steel Fe87 Cr13 alloy. After a quenching and prestressing cycle for linear stress-strain increase, the submaster is polished flat. Although uniform loads up to 15 bar have been used, more typical values are in the range 4–5 bar. A typical design of an active submaster requires a thick boundary ring strengthened by the enclosing plate (Fig. 5.18).

For instance, a submaster design generating corrected gratings for LAMOST f/1.5 second generation low-resolution spectrographs, by Zhu and Lemaitre [31] (see Sect. 4.4.11), also includes the spherical aberration corrections of a field lens flattener and spherical collimator. The 246 mm built-in plate is a quenched stainless steel Fe87 Cr13 ($E = 201$ GPa, $\nu = 0.305$). For continuity reasons, the junction

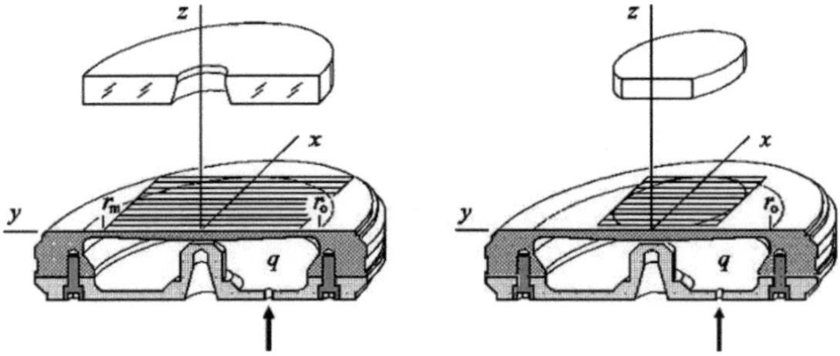

Fig. 5.18 Typical design of a metal *active optics submaster* generating aspherized gratings by inner uniform load q and replication. The thickness of the built-in plate decreases from center to edge. (*Left*): On-axis replications. (*Right*): Off-axis replications for telescope beams without central obstruction

at the built-in radius r_o is realized by a quarter torus of radius 8 mm tangent to the ring inner side. The asphenization load – air pressure – is $q = 5$ bar and the maximum radial stress at the surface of the submaster does not exceed $\sigma_r = 92$ MPa (Table 5.6).

Aspherized gratings are generally obtained on fused silica substrates or vitro ceram substrates such as Schott-Zerodur. Because of extremely low tensions existing after resin polymerization, the replication process can be carried out equally with circular, square, rectangular, or other perimeter shape substrates. Also, the process can use holed substrates for on-axis input of the telescope beams. This was the case of the UVPF spectrograph at f/1.22 of CFHT [1, 19–21], (cf. Table 4.3). The moderate focal-ratio spectrographs MARLY 1, MARLY 2, and CARELEC have been built with aspherized gratings [22, 23] for Haute-Provence and Purple Mountain observatories (see Table 4.4).

Table 5.6 Design of *vase-form submasters* generating the aspherized gratings of LAMOST low-resolution spectrographs, $\mathcal{R} = 2,000$, red and blue channels at f/1.5. From Table 4.5 in Sect. 4.4.11, the flexure is $Z_{\mathrm{Sub}} = -2.0150\,10^{-5}\,r^2 + 6.068\,10^{-10}\,r^4 + 2.34\,10^{-15}\,r^6 + 1.2\,10^{-20}\,r^8$

ρ^2	$\mathcal{D}$	r [mm]	t [mm]	σ_r [MPa]
0.000	0.95093	0.00	8.734	50.6
0.005	0.93461	42.42	8.684	36.3
0.010	0.91828	60.00	8.633	21.6
0.015	0.90192	73.48	8.581	6.6
0.020	0.88555	84.85	8.529	−8.8
0.025	0.86915	94.87	8.476	−24.5
0.030	0.85273	103.92	8.422	−40.6
0.035	0.83428	112.25	8.368	−57.2
0.040	0.81982	120.00	8.313	−74.2
0.04205	0.81320	123.03^-	8.256	−91.6
		123.03^+	60.000	–
		170.00	60.000	–

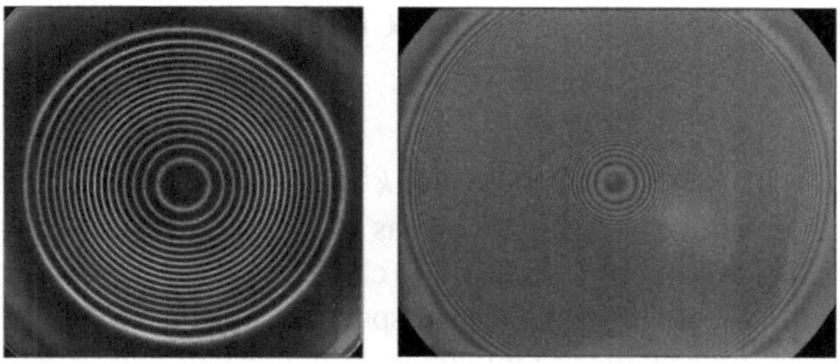

Fig. 5.19 He-Ne interferograms of aspherized reflective gratings in zero-order with respect to a plane. Zerodur substrates. (*Left*): One of the gratings of f/3.18 MARLY spectrographs $\delta = 30°$, 1,200 ℓ/mm, 84×84 mm area, 66.04×76.25 mm elliptical clear aperture. (*Right*): One of the gratings of f/2.5 CARELEC spectrograph $\delta = 30°$, 1,200 ℓ/mm, 102×128 mm area, 100×115 mm elliptical clear aperture (Coll. LOOM–Horiba Jobin Yvon Corp.)

The optical tests of the final aspherized grating replicas were carried out from He-Ne interferograms with respect to a plane (Fig. 5.19). All the submasters were made in quenched Fe87 Cr13 alloy as follows.

- MARLY spectrographs: $r_o = 40.44$ mm built-in radius, $t(0) = 5.45$ mm, $q = 4.22$ bar.
- CARELEC spectrographs: $r_o = 65.70$ mm built-in radius, $t(0) = 6.83$ mm, $q = 4.00$ bar.

Off-axis designed spectrographs avoid the central obstruction of usual collimated telescope beams towards the grating and make the detector access to the focal plane easier. These designs result from a sideways section of the previous designs. Thus, grating replicas can be aspherized by off-axis print of an axisymmetric submaster. In order to obtain the same focal ratios than for on-axis designs, the amount of the submaster flexure must be at least 2–2.5-times larger which could lead to difficulties for the construction of *fast* off-axis spectrographs. Off-axis reflective imager-spectrographs have been developed and built with ISARD for faint object studies at the 2-m Bernard Lyot Telescope at Pic-du Midi, and with OSIRIS/ODIN for a survey in space (Lemaitre and Richardson [24]). *Off-axis aspherics* – mirror and gratings – were obtained by a double replication process with quenched Fe87 Cr13 alloy submasters, first with a flat optical surface when unstressed, and then stressed (Fig. 5.20).

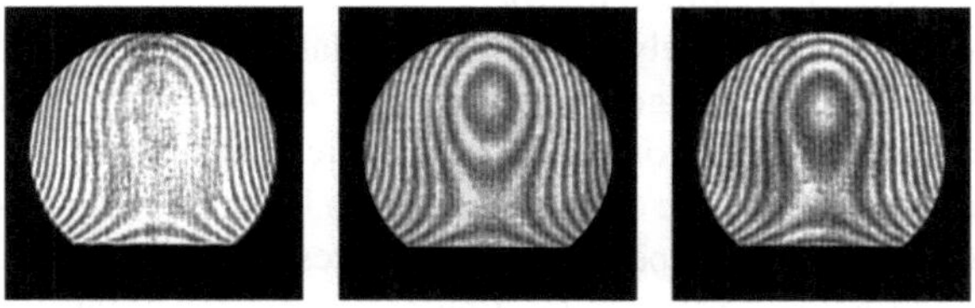

Fig. 5.20 He-Ne interferograms with respect to a plane of *off-axis aspherics* generated from double replication of an axisymmetric submaster for ISARD imager and spectrograph at Pic du Midi 2-m telescope. (*Left*): Aspherized mirror. (*Right*): Aspherized gratings 75 and 150 ℓ/mm in zero-order (Coll. LOOM – Horiba Jobin Yvon Corp. [24])

5.4.4 Axisymmetric Gratings with $k = 0$ and Circular Simply Supported Submasters

Aspherized reflective gratings with a geometrical profile $k = 0$ – corresponding to $M = 0$ – are less performing than with $k = 3/2$ because, as we have seen, *Astm* 5 is not balanced in the field with a small amount of *Astm* 3 (cf. Sect. 4.4.5). However, these pure r^4 profiles can be of interest for single object spectrographs using single array detectors where long rectangular pixels are much larger in the y direction than in the x direction of dispersion.

Starting from a plane master grating, the replication conditions are the same as in (5.75) except that $M = 0$. The aspherization of the grating is generated from replication of a *tulip-form* deformable submaster. With spectrograph mountings working in normal diffraction $\beta_0 = 0$ at the center of the spectral range, axisymmetric submasters provide straightforward solutions. The aspherization is achieved by a central force under perimeter reaction. The thickness distribution corresponds to the case VTD 2 in Sect. 3.3.2. For this configuration, the thickness t of the simply supported submasters is, from (3.30a),

$$t = \mathcal{T}_{40}\, t_0 = \left(\frac{4}{1-\nu}\right)^{1/3} \left[\left(\frac{a^2}{r^2}\right)^{4/(3+\nu)} - \frac{a^2}{r^2}\right]^{1/3} t_0\,. \tag{5.88}$$

Assuming that the submaster is flat when unstressed [see conditions (5.77)], the identification of the A_{40} coefficient of the flexure $z_{40} = A_{40} r^4$ with the coefficient in r^4 of (5.76a) leads to

$$A_{40} = -\frac{\cos^2\varphi_{\rm m}}{1+\cos\alpha}\,\frac{\mathcal{A}_4}{R^3} = \frac{\cos^2\varphi_{\rm m}}{2^8\,\Omega^3(1+\cos\alpha)}\,\frac{1}{r_{\rm m}^3}\,.$$

The incident circular beam defines an elliptic pupil at the grating whose semi-axis lengths are $r_{\rm m}$ and $r_{\rm m}/\cos\alpha$, so the submaster edge radius must be such as $a \geq r_{\rm m}/\cos\alpha$. From (3.30b), the t_0 thickness parameter of the submaster is derived as a function of the central force F and of the material $(E,\ \nu)$ by

$$t_0 = \left[\frac{3(1-\nu^2)(1+\cos\alpha)}{4\pi\,\cos^2\varphi_{\rm m}}\,\frac{F}{a^2 E}\right]^{1/3} R \quad \text{with} \quad a \geq \frac{r_{\rm m}}{\cos\alpha}\,, \tag{5.89}$$

where R is the radius of curvature of the camera mirror.

Representations (5.88) and (5.89) completely determine the elasticity parameters of the submaster design for generating $k = 0$ geometry gratings. A classical way of obtaining such gratings is the double replication process by replicating first a plane grating on an unstressed submaster. The active optics co-addition law (5.77) applies and the final grating replica is of opposite shape to that of the stressed submaster.

The first developments of the *active optics replication technique* for obtaining aspheric gratings were carried out by Lemaitre and Flamand [17] with a metal submaster generating an r^4 flexure (Fig. 5.21).

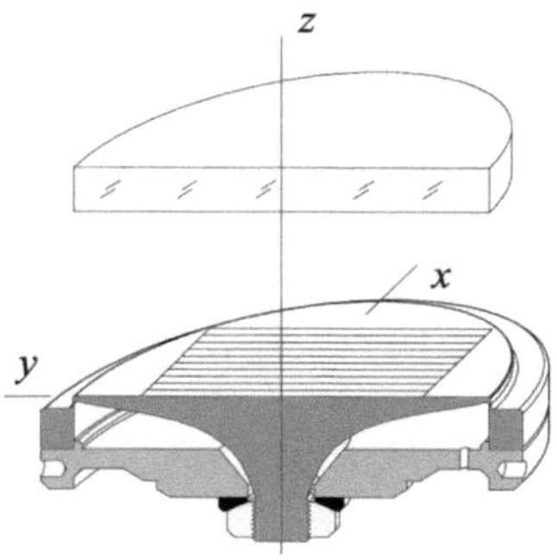

Fig. 5.21 Design of a *tulip-form submaster* generating $z = A_{40}r^4$ aspherized gratings by central force and double replications. The active VTD is with $t(a/2) = 8$ mm and simply supported at the contour by a thin collar of diameter $2a = 170$ mm. The VTD, collar and ring form an holosteric piece of quenched Fe87 Cr13 alloy [17]

In fact, for the *first* aspherized grating built by the active optics replication method, the small *Sphe* 5 correction of the camera mirror was also taken into account in the design of the submaster thickness distribution. Although not really necessary for a 90 mm circular collimated beam whose central wavelength diffracted beam is merging from the camera mirror with an f/2×f/1.8 anamorphosis, this correction is easily obtained. Since $Cv\,1 = 0$ and from the first terms in (5.76a), the active surface of the submaster is represented by *Sphe* 3 and *Sphe* 5 correction modes as

$$Z_{\text{Sub}} = -Z_{\text{Opt}} = -\frac{\cos^2\varphi_{\text{m}}}{1+\cos\alpha}\left[\frac{\mathcal{A}_4}{R^3}\,r^4 + \frac{\mathcal{A}_6}{R^5}\,r^6 + \cdots\right].$$

From Table 4.1, $\mathcal{A}_4 = -1/4$ and $\mathcal{A}_6 = -3/8$, then

$$Z_{\text{Sub}} = \frac{\cos^2\varphi_{\text{m}}}{1+\cos\alpha}\left[\frac{1}{4R^3}\,r^4 + \frac{3}{8R^5}\,r^6 + \cdots\right]. \qquad (5.90)$$

Using the same dimensionless quantities as in (5.82), where q is substituted by $F/\pi a^2$, a numerical integration of (5.83) provides the dimensionless rigidity $\mathcal{D}(\rho)$. Since theoretically infinite at the origin, the integration of the rigidity is carried out from a small radius, e.g. $\rho = a/(10R)$, by starting with the corresponding $\mathcal{D}$-value derived from the previously obtained VTD for A_{40} coefficient alone (5.88). After successive increments of the radius, next iterations are made by adjusting the starting $\mathcal{D}$-value up to $\mathcal{D} = 0$ for the radius $\rho = a/R$.

This elasticity design was followed up by construction of a submaster in quenched Fe87 Cr13 stainless steel. The rear side and the outer collar and ring were machined in a single piece – forming a holosteric substrate – by a computer-controlled lathe. This submaster allowed obtaining the first aspherized grating by an active optics replication process. The substrate of the reflective grating replica was Zerodur from Schott (Fig. 5.22).

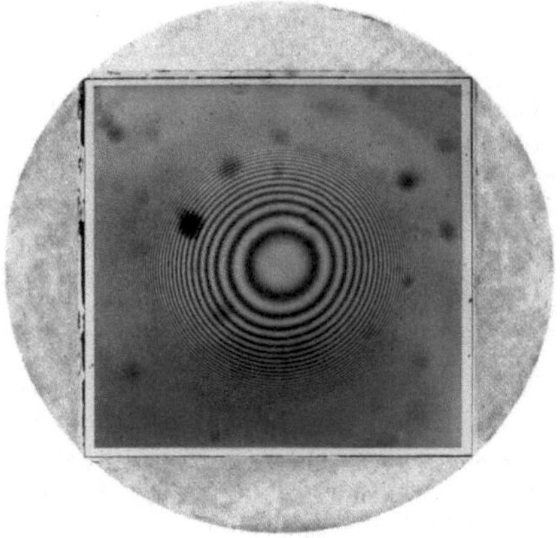

Fig. 5.22 He-Ne interferograms of a plane replica grating, 102×102 mm, 1,200 ℓ/mm, when stressed on the *tulip-form submaster*. The next replication provided the aspherized grating on a rigid Zerodur substrate for an f/2×f/1.8 spectrograph, 14° curved field, λλ [300–500 nm], $\alpha =$ 28°7, $\beta_0 = 0°$ at 400 nm, R =360 mm, collimator beam diameter $2r_m$ =90 mm. Submaster: alloy Fe87 Cr13, active diameter $2a$ =170 mm, $A_{40} = 2.812\ 10^{-9}\,\mathrm{mm}^{-3}$, $A_{60} = 3.2\ 10^{-14}\,\mathrm{mm}^{-5}$, thickness $t(a/2)$ = 8 mm (Coll. LOOM - HORIBA-Jobin-Yvon [17])

A laboratory spectrograph in the ultraviolet spectral range λλ [300–500 nm] was built for optical evaluation of the grating spherical aberration correction ($Sphe\,3$ and $Sphe\,5$). A circular cross-section collimated beam, 90 mm in diameter, illuminated the 1,200 ℓ/mm grating onto an elliptic area 90×100 mm in size. At the wavelength 400 nm ($\beta_0 = 0$), the output beam from the camera mirror was with an f/2×f/1.8 anamorphosis. Both Fe arc and low-pressure Cd lamp provided spectral line widths sharper than 5 μm over the 15° curved field of view.

The spectra obtained with this grating showed that the resulting spectral resolution were in accordance with the theoretical blur images for $k = 0$ in Sect. 4.4.5 (see Fig. 4.14). This also demonstrates that the *active optics replication technique* perfectly applies to the *asphérization of diffraction gratings*.

5.4.5 Bisymmetric Gratings with $k = 3/2$ and Elliptic Built-in Submasters

Considering now the case of a diffraction angle $\beta_0 \neq 0$ for the center of the output spectrograph field, the best design is still provided by the geometry $k = 3/2$. The shape for these gratings, now defined by (5.76b), is made of iso-level lines that are homothetical ellipses. From (5.76c), their semi-axis ratio is $a_y/a_x = \cos\beta_0$. Assuming hereafter a circular incident beam of aperture $2r_m$, the following properties are useful to the design of the deformable submaster.

→ *The flexure is generated from a built-in elliptical contour determined from* β_0 *in* (5.76d),

→ *The final grating clear aperture is determined from α by the elliptical contour* (5.76e).

Let us denote $Z_{\rm Sub}$ the submaster flexure. If the grating aspherization process is a double replication, then the *active optics co-addition law* writes

$$Z_{\rm Sub} + Z_{\rm Opt} = 0\,, \tag{5.91}$$

where $Z_{\rm Opt}$ is the figure of the final aspherized grating replica. If the grating is directly used on the submaster, then the sign of its flexure – and then the uniform load q considered hereafter – must be opposite.

• **Constant thickness submaster for slow focal ratio cameras**: For camera focal ratios slower than f/3 or f/2.5, a constant thickness submaster with an elliptic built-in edge $\mathcal{C}$ defined by

$$\cos^2\beta_0\; x^2 + y^2 = \frac{3}{2}\, r_{\rm m}^2 \equiv r_{\rm o}^2 \tag{5.92}$$

can generate an elliptic symmetry grating. The flexure shape of the inner built-in plate of the submaster is given by (5.27) in Sect. 5.3.5. The elliptic machining of such a submaster only consists of the inner elliptical cylinder of the contour ring, a groove for the O-ring seal and holes distributed over an ellipse for tightening the air pressure closing plate; all other surfaces can be axisymmetrical. The O-ring length is selected from the determination of a *Legendre elliptic integral* of second kind.

Representing the built-in contour $\mathcal{C}$ by (5.25), i.e. $x^2/a_x^2 + y^2/a_y^2 = 1$, leads to

$$a_x = r_{\rm o}/\cos\beta_0\,, \qquad a_y = r_{\rm o}\,. \tag{5.93}$$

The flexural sag from the center to the elliptic contour $\mathcal{C}$ is, from (5.28),

$$Z_{\rm Elas}\big|_{\mathcal{C}} \equiv -z_0 = -\frac{q}{8D}\,\frac{r_{\rm o}^4}{3+2\cos^2\beta_0+3\cos^4\beta_0}\,. \tag{5.94a}$$

Since $Z_{\rm Sub}$ is opposite to $Z_{\rm Opt}$, the submaster optical sag from center to contour $\mathcal{C}$ is easily expressed as a function of the optics parameters by substitution of (5.92) in (5.76b). This gives

$$Z_{\rm Sub}\big|_{\mathcal{C}} = -\frac{\cos^2\varphi_{\rm m}}{\cos\beta_0+\cos\alpha}\sum_{2,4}\frac{\mathcal{A}_n}{R^{n-1}}\, r_{\rm o}^n\,,$$

where, due to the constant thickness of the built-in inner plate, only the two coefficients $\mathcal{A}_2$, $\mathcal{A}_4$ are possible which, then, provides a spherical aberration correction *limited* to the *Sphe* 3 term. For a meridian flexure with $k = 3/2$ in accordance to (5.75), the approximations $\mathcal{A}_2 \simeq M = 3/64\Omega^2 = r_{\rm o}^2/2R^2$ and $\mathcal{A}_4 = -1/4$ leads to

$$Z_{\rm Sub}\big|_{\mathcal{C}} = -\frac{\cos^2\varphi_{\rm m}}{\cos\beta_0+\cos\alpha}\,\frac{r_{\rm o}^4}{4R^3}\,. \tag{5.94b}$$

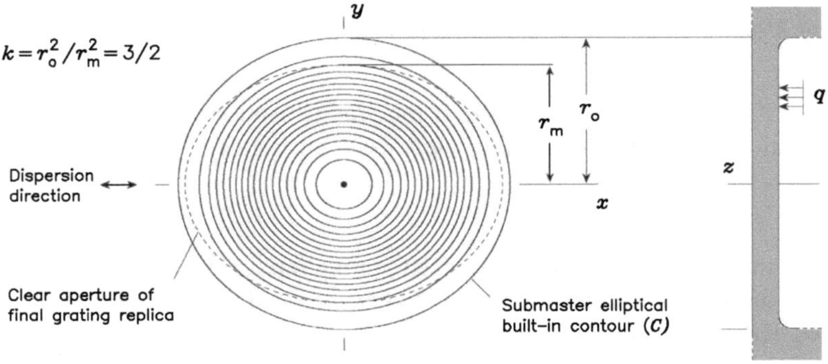

Fig. 5.23 Theoretical isolevel lines of the flexure for *Sphe* 3 correction only. The constant thickness elliptical plate is built-in to an infinitely large outer ring. *Full line*: built-in contour ellipse of semi-axes $(r_o/\cos\beta_0,\ r_o)$ of the submaster. *Dashed line*: clear aperture ellipse of semi-axes $(r_m/\cos\alpha,\ r_m)$ of the final grating replica

Equalizing the elastic and optical sags of the substrate, we obtain

$$\frac{q}{E} = \frac{\cos^2\varphi_m\,(3 + 2\cos^2\beta_0 + 3\cos^4\beta_0)}{6\,(1-\nu^2)\,(\cos\beta_0 + \cos\alpha)}\ \frac{t^3}{R^3}\,, \tag{5.95a}$$

which provides the following thickness of the built-in plate

$$\begin{cases} t = \left[\dfrac{6\,(1-\nu^2)\,(\cos\beta_0 + \cos\alpha)}{\cos^2\varphi_m\,(3 + 2\cos^2\beta_0 + 3\cos^4\beta_0)}\,\dfrac{q}{E}\right]^{1/3} R\,, \\ r_o = \sqrt{3/2}\,r_m \quad \text{and} \quad R = 4\,\Omega\,r_m\,. \end{cases} \tag{5.95b}$$

This determines the execution conditions for a grating with *Sphe* 3 correction mode *only* and illuminated by a circular collimated beam of semi-aperture r_m. The two ellipses representing the built-in edge of the submaster and the clear aperture contour of the final grating replica are displayed by Fig. 5.23.

• **Variable thickness submasters for fast focal ratio cameras**: Diffraction gratings for fast camera mirrors require higher-order corrections. Similarly as axisymmetric submasters for $\beta_0 = 0$ (cf. Sect. 5.4.3), this can be achieved by a built-in plate whose thickness is decreasing from center to edge. The thickness function must follow an homothetic ellipse distribution.

Let us denote variables v, ϑ, and ζ respectively defined by

$$v^2 = \cos^2\beta_0\,x^2 + y^2\,, \qquad \vartheta = \frac{v}{R} \quad \text{and} \quad \zeta = \frac{Z_{\mathrm{Sub}}}{R}\,. \tag{5.96}$$

The dimensionless flexure ζ of the submaster plate must be, from (5.76b),

$$\zeta = -\frac{\cos^2\varphi_m}{\cos\beta_0 + \cos\alpha}\sum_{2,4,6,\ldots} \mathcal{A}_n\,\vartheta^{\,n}\,, \tag{5.97}$$

where $\mathcal{A}_n(M)$ coefficients in Table 4.1 can be considered up to a large n-value. The boundary ellipse of semi axes $r_o/\cos\beta_0$, r_o, which determines the clamped edge location, is derived from the positive root of

$$\sum_{2,4,6,\ldots,p} n\,\mathcal{A}_n\,\vartheta^{n-1} = 0 \quad \text{and} \quad M = 3/64\Omega^2\,. \tag{5.98}$$

With the exact representation of the $\mathcal{A}_n(M)$ coefficients in Table 4.1, this root gives the accurate ellipse of the null powered zone. This leads to a k-ratio, $k = r_o^2/r_m^2$, which is slightly smaller than the $k = 3/2$-value of the third-order condition. Since ϑ in (5.98) would correspond to ρ in (5.81) for the axisymmmetric case, the exact k-ratios as a function of Ω are identical to those of an axisymmetric submaster (see Table 5.5, 2nd-column).

Comparing the required q/E ratios in (5.78b) and (5.95a), which corresponds to constant thickness built-in plates of circular and elliptical contour respectively, one notes that the expression of the load transforms as

$$q \;\rightarrow\; \frac{8}{3+2\cos^2\beta_0+3\cos^4\beta_0}\,q\,. \tag{5.99}$$

Then, from (5.82), let us define dimensionless geometry parameter g and rigidity $\mathcal{D}$ as

$$g = -\frac{\cos^2\varphi_m}{\cos\alpha+\cos\beta_0} \quad \text{and} \quad \mathcal{D} = \frac{3+2\cos^2\beta_0+3\cos^4\beta_0}{4\,q\,R^3}\,g\,D\,. \tag{5.100}$$

In the transformation from circular to elliptic symmetry, $\rho^2 \rightarrow \vartheta^2$, hence from (5.85), the rigidity is a solution of

$$\frac{d\,\mathcal{D}}{\vartheta\,d\vartheta}\sum_{2,4,\ldots,p} n(n-1+\nu)\mathcal{A}_n\,\vartheta^{n-2} + \mathcal{D}\sum_{4,6,\ldots,p}(n-2)n^2\mathcal{A}_n\,\vartheta^{n-4} = 1\,. \tag{5.101}$$

Representing it, up to the ellipse of built-in radius parameter $\vartheta_o = \upsilon_o/R$, by

$$\mathcal{D} = \sum_{0,2,4,\ldots p'} X_n\,\vartheta^n \quad \text{with} \quad 0 \le \vartheta \le \vartheta_o\,, \tag{5.102}$$

a dimensionless solution $\mathcal{D}(\Omega,\vartheta)$ shows the same variation of the rigidity as $\mathcal{D}(\Omega,\rho)$ in (5.86) for a circular plate. The derived thickness $t(\Omega,x,y)$ and the built-in boundary completely determine the execution conditions of a deformable submaster generating bisymmetric gratings for the aberration correction of a fast focal-ratio camera mirror and field flattener lens. A typical design of an elliptical vase from the submaster uses a similar vase enclosure either welded at the ring rear surface or linked via discrete points on an homothetic elliptic line. The ring outer edge of both submaster and enclosure can be made circular (Fig. 5.24).

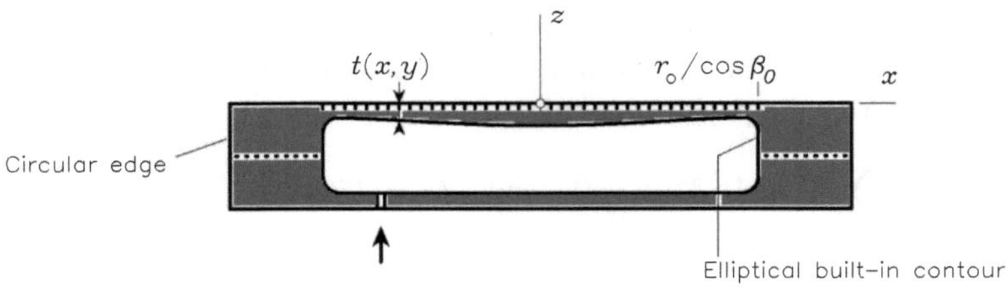

Fig. 5.24 Typical design of an *elliptical vase from the submaster* generating bisymmetric gratings for high-order correction by uniform load q and replication. The active plate is built-in at its elliptical contour. Its thickness $t(x, y)$ – made of homothetical ellipses $(r_o/\cos\beta_0,\ r_o)$ – decreases from center to edge

• **Submasters of elliptic geometry**: Whether a slow or fast camera mirror, the following results can be summarized:

The elasticity design of a bisymmetric submaster generating an aspherized reflective grating for high-order aberration corrections, where the grating is illuminated by a collimated beam of clear aperture diameter $2r_m$*, mounted at diffraction angle* $\beta_0 \neq 0$ *and correcting a fast focal ratio* $\Omega = R/4r_m$ *of a camera mirror of curvature* $1/R$*, can be achieved by the following features.*

1 → *A plate of slightly decreasing center-to-edge thickness whose rear side is made of homothetic isolevel ellipses and edge is built-in to the elliptic contour*

$$\cos^2\beta_0\, x^2 + y^2 = v_0^2 = k\, r_m^2\,, \tag{5.103}$$

where the y*-direction is the grating line direction.*
2 → *The exact position ratio* $\sqrt{k} = v_o/r_m$ *which determines the location of the null powered zone – which is also the built-in zone – is such as given in* Table 5.5 *for axisymmetric submasters.*
3 → *In the* y*-direction, the dimensionless rigidity variation* $\mathcal{D}(\Omega, 0, y)$ *of the plate is identical to that of the axisymmetric case* (see $\mathcal{D}(\Omega, \rho)$ in Table 5.5).
4 → *In the* x*-direction, the dimensionless rigidity* $\mathcal{D}(\Omega, x, 0)$ *is similar to* $\mathcal{D}(\Omega, 0, y)$ *but distributed along the anamorphosed scale* $1/\cos\beta_0$*. Therefore, the isolevel lines of the plate rear side are homothetic ellipses to that of the contour.*
5 → *Compared to an axisymmetric submaster corresponding to* $\beta_0 = 0$*, the dimensioned rigidity at the center,* $D(\Omega, 0, 0)$*, is*

$$\frac{\cos\alpha + \cos\beta_0}{\cos\alpha + 1} \times \frac{8}{3 + 2\cos^2\beta_0 + 3\cos^4\beta_0} \tag{5.104}$$

times larger.
6 → *The grating clear aperture at the spectrograph is determined by the grating incidence angle* α*. Its aperture contour is the ellipse*

$$\cos^2\alpha\, x^2 + y^2 = r_m^2\,, \tag{5.105}$$

where the spectrograph collimated beam is assumed of circular cross section.

5.4.6 Constructional Replication Condition for Active Optics Process

The exact k-value condition for an optimal grating geometry is given in Table 5.5 as a function of the camera f-ratio Ω. A *constructional replication condition* for making possible the replication of reflective grating concerns its clear aperture length in the dispersion direction (here the x-direction): the x-semi-length of the grating must *not* exceed the built-in radius of the submaster in this direction. Setting $y = 0$ in (5.103) and (5.105), this condition is satisfied if $1/\cos^2\alpha \leq k/\cos^2\beta_0$, i.e.

$$\frac{\cos^2\beta_0}{\cos^2\alpha} \leq k \quad \text{with} \quad k = \frac{v_\text{o}^2}{r_\text{m}^2}\,. \tag{5.106}$$

We have seen that for a low-order aberration correction, as for a camera mirror slower than ~f/3 in the nebular direction, then the grating geometry is optimal for $k = 3/2$. For a high-order correction, as for a fast f/1.25 camera mirror in the nebular direction, then $k = 1.44$ from Table 5.5. Thus, for any spectrograph aperture slower than f/1, the best aberration correction is a value

$$1.4 \leq k \leq 1.5 \tag{5.107}$$

depending on the nebular f-ratio $\Omega = R/4r_\text{m}$ of the camera mirror.

From the *grating law* [see (4.23)], one can deduce whether the incident and diffraction angles, α and β_0, are compatible or not with the constructional replication condition.

• **All-reflective spectrographs of the Schmidt type**: The simplest case is when the diffraction angle is normal to the grating for the central wavelength ($\beta_0 = 0$). Both submaster and replicated grating have a rotational symmetry, $v_\text{o} \equiv r_\text{o}$, and condition (5.106) reduces to $1/\cos\alpha \leq r_\text{o}/r_\text{m}$. Then, for a slow focal-ratio spectrograph where $r_\text{o}/r_\text{m} = \sqrt{3/2}$, the circular cross-section collimated beam – which determines the elliptical clear aperture of the grating – must not exceed the *maximum incidence angle* $\alpha_\text{Lim} = 35.26°$. In all other diffraction cases ($\beta_0 \neq 0$), constructional condition (5.106) must be checked with the $\sqrt{k}$-values in Table 5.5.

• **Spectrographs with grating in Littrow mounting**: Other spectrograph types, such as *Littrow mounting spectrographs* for high spectral resolution, are usually designed with magnification $M = -1$, double-pass mirror (or singlet lens) and strongly elliptical clear aperture grating (see, for instance, Fig. 9.9 in Sect. 9.3). The Littrow mounting condition is $\alpha = -\beta_0$ in (4.23). The substitution in (5.106) entails $1 \leq k$ where $k = 3/2$ for a usual slow focal-ratio. Therefore, the constructional replication condition for their aspherization by an active optics submaster is always satisfied.

References

1. O. Boulade, G.R. Lemaitre, L. Vigroux, UV prime focus spectrograph for the CFHT, Astron. Astrophys. **163**, 301–306 (1986)
2. G.H. Bryan, see Love: *Theory of Elasticity*, Dover edt. and Timoshenko & Woinowsky-Krieger: *Theory of Plates and Shells*, McGraw-Hill edt., New York 310 (1959)
3. H. Chrétien, *Le Calcul des Combinaisons Optiques*, Sennac Edit., Paris, 346–350 (1958)
4. B.A.J. Clark, Journ. Astron. Soc. Victoria, Australia, **17**, 65–68 (1964)
5. M. Cohendet, Laboratoire d'Astronomie Spatiale, LAS report Opt-FAUST-011 (1972)
6. A. Couder, Sur l'exécution [par élasticité] des surfaces optiques non sphériques: application au télescope de Schmidt, C. R. Acad. Sc. Paris, **210**, 327–329 (1940)
7. G. Courtès, P. Cruvellier, M. Detaille, M. Saïsse, *Progress in Optics*, **XX**, E. Wolf edt., North-Holland, 1–61 (1983)
8. E. Everhart, Making corrector plates by Schmidt's vacuum method, Appl. Opt., **5**(5), 713–715 (1966)
9. C. Fehrenbach, *Des hommes, des télescopes, des étoiles*, CNRS edt., 299 (1990) and 2nd issue with complements, Vuibert edt. (2007)
10. R.N. Haward, *The Strength of Plastics and Glass*, Interscience Publ., New York, 51 (1965)
11. G.R. Lemaitre, Sur les dioptres asphériques de révolution en optique astronomique, C. R. Acad. Sc. Paris, **270** Série A, 266–269 (1970)
12. G.R. Lemaitre, French patent No 2097216 (1972), US patent No 3693301 (1972)
13. G.R. Lemaitre, New procedure for making Schmidt corrector plates, Appl. Opt. **11**(7), 1630–1636 (1972), and **11**(10), 2264 (1972)
14. G.R. Lemaitre, Sur les diotres aphériques en optique astronomique, C. R. Acad. Sc., Paris, **276** Série B, 145–148 (1973)
15. G.R. Lemaitre, Asphérisation par élasticité d'une lame de 50 cm pour le télescope de Schmidt de l'Observatoire de Lyon, Astron. Astrophys. **44**(2), 305–313 (1975)
16. G.R. Lemaitre, Reflective Schmidt anastigmat telescopes and pseudo-flat made by elasticity, Journ. Opt. Soc. Am., **66**(12), 1334–1340 (1976)
17. G.R. Lemaitre, J. Flamand, Spectrographic development of diffraction gratings aspherized by elastic relaxation, Astron. Astrophys., **59**(2), 249–253 (1977)
18. G.R. Lemaitre, Asphérisation par relaxation élastique de miroirs astronomiques dont le contour circulaire ou elliptique est encastré ou semi-encastré, C. R. Acad. Sc., Paris, **290** Série B, 171–174 (1981)
19. G.R. Lemaitre, Combinaisons optiques à réseaux asphériques: Le spectrographe UV-Prime CFHT, Astron. Astrophys. Letters, **103**(2), L14–L16 (1981)
20. G.R. Lemaitre, Un spectrographe à réseau asphérique pour télescope f/4, in *Instrumentation for Astronomy with Large Optical telescopes*, C.M. Humphries edt., Reidel Publ. Co., 137–141 (1982)
21. G.R. Lemaitre, L. Vigroux, All-reflective aspherized grating spectrographs at the prime focus of the CFHT, *Instrumentation for Ground-based Astronomy*, L.B. Robinson ed., Springer-Verlag, New York, 275–295 (1987)
22. G.R. Lemaitre, D. Kohler, All-reflective aspherized grating spectrographs: MARLY spectrographs of the Haute-Provence and Nanjing observatories, C. R. Acad. Sc. Paris, **308 II**, 381–387 (1989)
23. G.R. Lemaitre, D. Kohler, D. Lacroix, J.-P. Meunier, A. Vin, All reflective aspherized grating spectrographs for Haute-Provence and Nanjing observatories: MARLYS and CARELEC, Astron. Astrophys., **228**, 546–558 (1990)
24. G.R. Lemaitre, E.H. Richardson, Ground-based and orbital off-axis aspherized grating imager-spectrographs: ISARD/Pic-du-Midi and OSIRIS/ODIN, in *Optical Astronomical Instrumentation*, Kona, SPIE Proc., **3355**, 682–695 (1998).

25. G.R. Lemaitre, Active optics and aberration correction with multimode deformable mirrors (MDMs) - Vase form and meniscus form, in *Laser Optics 2003: Wavefront Transformation and Laser Beam Control*, St. Petersburg, SPIE Proc. **5481**, 70–81 (2004)
26. A.E.H. Love, *Theory of Elasticity*, Dover edt., Dover 484 (1927)
27. G. Monnet, R. Zaharia, G.R. Lemaitre, Programme Spatial FAUST du Centre National d'Etudes Spatiales, LAS/CNES report FAUST-PJ-02-70 (1970)
28. S.D. Poisson, Mémoires Acad. Sci., Paris, **VIII**, 357–627 (1829)
29. The Springfield science museum team, Appl. Opt., **11**, 222–225 (1972)
30. S.P. Timoshenko, S. Woinowsky-Krieger, *Theory of Plates and Shells*, McGraw-Hill edt.,(a: p. 310), (b: p. 468), (1959)
31. Y.-t. Zhu, G.R. Lemaitre, LAMOST multi-object spectrographs with aspherized gratings, in *Instrument Design for Ground-based Telescopes*, SPIE Proc. **4841**, 1127–1133 (2002)

Chapter 6
Theory of Shells and Aspherization of Axisymmetric Mirrors – Meniscus, Vase and Closed Forms

6.1 Active Optics Aspherization of Fast f-Ratio Mirrors

We consider in this Chapter axisymmetric mirror substrates characterized by highly curved surfaces i.e. surfaces with a significant *cambrure*. The elasticity theory of thin plates basically assumes a plane middle surface. Thus, its validity field remains limited to mirrors of moderate f-ratio.

For mirrors faster than f/4 or f/3, the *theory of shallow spherical shells*, in French "coques surbaissées," allows taking into account the significant curvature of the middle surface and the "in-plane" radial and tangential tensions appearing on it during the flexure. This theory elaborated by E. Reissner [25, 26] in the 1940s brings a remarkable accuracy in the elasticity analysis and is of utmost importance for the active aspherization of telescope mirrors in astronomy.

In its general form, this theory also concerns shallow shells loaded in a non-axisymmetric manner. We only consider hereafter the rotational symmetry case where a uniform load is applied all over the surface. In this Chapter, a free parameter is the determination of a slightly variable thickness distribution – VDT – of a shell made of successive and continuous shell rings in order to generate from a sphere – the mirror surface – and a flexure the resulting shape required by the optics. The results presented hereafter show that the shallow shell theory has proved accurate for the flexure calculation of mirror f-ratios up to f/1.7 and is probably still valid for faster mirrors.

Various boundary conditions occur at the shell contour for the radial displacement and tangential rotation. With respect to these conditions, we consider several geometrical configurations: (i) a *meniscus form* with a simply supported and radially free edge, (ii) a *vase form* semi-built-in with an outer cylinder, and (iii) a *closed form* made of two shells built-in together via an outer cylinder.

6.2 Theory of Shallow Spherical Shells

Let us consider a *shallow spherical shell* and denote $<R>$ the radius of curvature of its spherical middle surface. In a cylindrical coordinate system (z, r, θ), the shape $z(r)$ of the middle surface can be represented by

G.R. Lemaitre, *Astronomical Optics and Elasticity Theory,* Astronomy and Astrophysics Library, DOI 10.1007/978-3-540-68905-8_6,

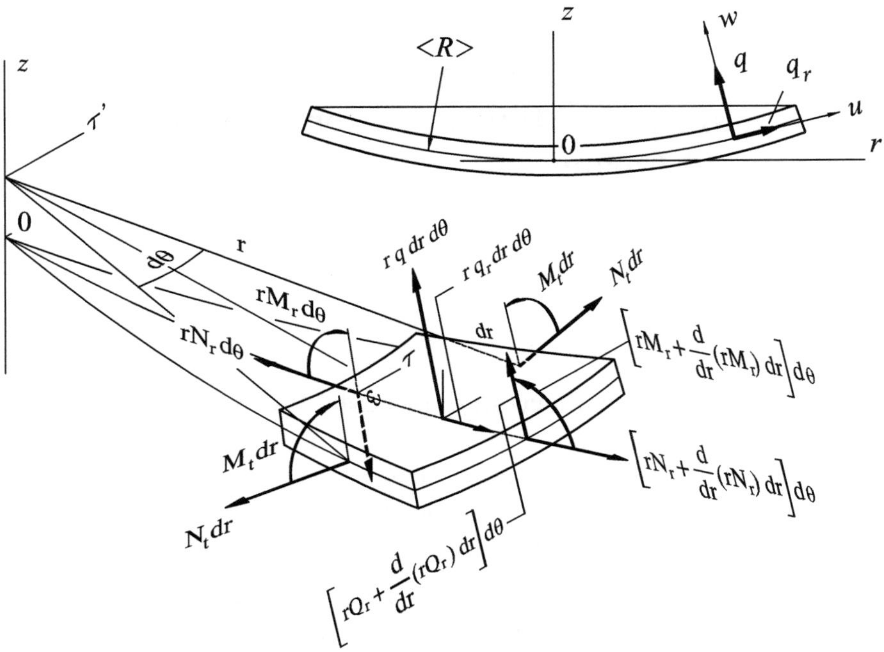

Fig. 6.1 Shallow spherical shell and equilibrium force components acting on a shell element

$$\frac{z}{<R>} = 1 - \sqrt{1 - \frac{r^2}{<R>^2}} \equiv \sum_{n=1}^{\infty} \frac{(2n-2)!\, n}{2^{2n-1}\, (n!)^2} \frac{r^{2n}}{<R>^{2n}}, \tag{6.1}$$

where r is the distance from the origin of the surface to the x, y projection of a current point of the surface (Fig. 6.1).

Denoting r_{m} the maximum radius of this surface – corresponding to an optics clear aperture $2r_{\mathrm{m}}$ in diameter – the assumption of shallowness is expressed by the limitation of the slope maximum value

$$\left[dz/dr\right]_{\max} \simeq r_{\mathrm{m}}/<R> \ll 1. \tag{6.2}$$

6.2.1 Equilibrium Equations for Axisymmetric Loadings

Restraining hereafter to axisymmetric loading cases, let q be a uniform load applied in the normal direction to the midsurface and q_r a load applied in its meridian in-plane direction (see Fig. 6.1), the differential equations of the equilibrium of bending moments M_r, M_t, shearing force Q_r, and tensions N_r, N_t are (1, and 3a) [1]

[1] In all generality, Reissner (1) originally elaborated the elasticity theory for a non-axisymmetric loading – involving q, q_r, and q_t – and derived the complete equilibrium equation set corresponding to this case.

$$\frac{d(rN_r)}{dr} - N_t - \frac{r}{<R>} Q_r + rq_r = 0\,, \tag{6.3a}$$

$$\frac{d(rQ_r)}{dr} + \frac{r}{<R>} (N_r + N_t) + rq = 0\,, \tag{6.3b}$$

$$\frac{d(rM_r)}{dr} - M_t + rQ_r = 0\,. \tag{6.3c}$$

At the midsurface of curvature $1/<R>$ the radial and tangential strains ε_{rr} and ε_{tt} are functions of the radial and tangential tensions and of the displacements w and u in the normal and meridian line directions, respectively. These quantities are related by

$$\varepsilon_{rr} = \frac{1}{Et} (N_r - \nu N_t) = \frac{du}{dr} - \frac{w}{<R>}\,, \tag{6.3d}$$

$$\varepsilon_{tt} = \frac{1}{Et} (N_t - \nu N_r) = \frac{u}{r} - \frac{w}{<R>}\,. \tag{6.3e}$$

The bending moments are

$$M_r = D\left(\frac{d^2w}{dr^2} + \frac{\nu}{r}\frac{dw}{dr}\right), \qquad M_t = D\left(\frac{1}{r}\frac{dw}{dr} + \nu\frac{d^2w}{dr^2}\right), \tag{6.3f}$$

where the rigidity is

$$D = E\,t^3 / \left[12\,(1-\nu^2)\right]. \tag{6.3g}$$

The membrane type flexure due to shearing force Q_r can be neglected in (6.3a) if the shell thickness is thin, i.e. $t/r_{\mathrm{m}} \ll 1$. Thus, all concentric levels of the shell have the same flexure in the normal direction as that of the midsurface. Assuming that the load q_r acting along the meridian lines are derivable from a load potential Ω, we set

$$q_r = -d\Omega/dr\,, \tag{6.4}$$

where (6.3a) can be satisfied by using a *stress function* F for representing the tensions as follows

$$N_r = \frac{1}{r}\frac{dF}{dr} + \Omega\,, \qquad N_t = \frac{d^2F}{dr^2} + \Omega\,. \tag{6.5a}$$

From the Laplacian $\nabla^2\cdot = d^2\cdot/dr^2 + (1/r)\,d\cdot/dr$, we obtain

$$N_r + N_t = \nabla^2 F + 2\,\Omega\,. \tag{6.5b}$$

6.2.2 General Equation of Shallow Spherical Shells

Combining (6.3d) and (6.3e) with the above equations, a first fundamental relation for the stress function F and normal flexure w is

$$\nabla^2\nabla^2 F + \frac{E\,t}{<R>}\,\nabla^2 w = -(1-\nu)\,\nabla^2\Omega\,. \tag{6.6}$$

The shearing force Q_r in (6.3c) is derived from the bending moments in Eqs. (6.3f),

$$Q_r = -D \frac{d}{dr} (\nabla^2 w) . \tag{6.7}$$

A second fundamental relation between the F and w functions is obtained by substitution of the shearing force Q_r in (6.3b). Use of (6.5b) leads to [2]

$$\nabla^2\nabla^2 w - \frac{1}{D<R>} \nabla^2 F = \frac{q}{D} + \frac{2\,\Omega}{D<R>} . \tag{6.8}$$

Let us now assume that the meridian loads are always null, $q_r = 0$, for all loading cases investigated hereafter. From (6.4), this leads to only considering the two following cases for the load potential,

$$q = \Omega = 0 \quad \text{or} \quad q = -2\,\Omega/<R> = constant . \tag{6.9}$$

Hence, the left-hand terms of the two fundamental relations both vanish, so these relations can be combined into a single equation. Let us define quantities λ and ℓ by the following set, where $i = \sqrt{-1}$,

$$\lambda Et\,/<R> = i/\ell^2 , \qquad 1\,/\,(\lambda D<R>) = -i/\ell^2 , \tag{6.10a}$$

which gives

$$\lambda = i\sqrt{12\,(1-\nu^2)}\;/\,E\,t^2 , \qquad \ell = \sqrt{<R>t\;/\sqrt{12\,(1-\nu^2)}} . \tag{6.10b}$$

If $q_r = 0$ and if one of conditions (6.9) on q is satisfied, then, multiplying (6.6) by $-\lambda$ and summing with (6.8) leads to

$$\nabla^2\nabla^2(w - \lambda F) - \frac{i}{\ell^2} \nabla^2(w - \lambda F) = 0 , \tag{6.11}$$

This is the general *equation of shallow spherical shells* for a load $q = 0$ or $q = constant$. It was first derived by Eric Reissner [2] in 1947. This equation is integrable by writing

$$w - \lambda F = \Phi + \Psi , \tag{6.12}$$

where the Φ and Ψ functions are general solutions of

$$\nabla^2\Phi = 0 \quad \text{and} \quad \nabla^2\Psi - \frac{i}{\ell^2}\Psi = 0 . \tag{6.13}$$

Assuming complex constants A_1 to A_4, the respective solutions are

[2] For a non-axisymmetric load where the tangential component q_t is derivable from a load potential, as $q_t = -\frac{1}{r}\frac{\partial\Omega}{\partial\theta}$, Reissner [1] demonstrated that the two fundamental equations (6.6) and (6.8) still apply.

$$\Phi = A_1 + A_2 \ln(r/\ell), \tag{6.14a}$$

$$\Psi = A_3 [\psi_1(r/\ell) - i\,\psi_2(r/\ell)] + A_4 [\psi_3(r/\ell) - i\,\psi_4(r/\ell)], \tag{6.14b}$$

where the ψ_n functions are zero-order Kelvin functions as follows [3]

$$\psi_1 = \mathrm{ber}\,(r/\ell), \quad \psi_2 = \mathrm{bei}\,(r/\ell), \quad \psi_3 = \mathrm{ker}\,(r/\ell), \quad \psi_4 = \mathrm{kei}\,(r/\ell), \tag{6.15}$$

which are explicitly given in Sect. 6.2.3.

The general solutions for w and F are derived by identification of the real and imaginary parts. Introducing real coefficients C_1 to C_8 and the *characteristic length* ℓ, we set

$$\begin{aligned} A_1 &= \ell\,(C_5 + i\,C_8), & A_2 &= \ell\,(C_7 + i\,C_6), \\ A_3 &= \ell\,(C_1 + i\,C_2), & A_4 &= \ell\,(C_3 + i\,C_4), \end{aligned} \tag{6.16}$$

and representing the $\psi_i(r/\ell)$ functions by the concise writing ψ_i,

$$\begin{aligned} \Psi/\ell &= C_1\psi_1 + C_2\psi_2 + C_3\psi_3 + C_4\psi_4 \\ &\quad + i\,(C_2\psi_1 - C_1\psi_2 + C_4\psi_3 - C_3\psi_4), \\ \Phi/\ell &= C_5 + i\,C_8 + (C_7 + i\,C_6)\ln(r/\ell). \end{aligned}$$

where, from (6.3g) and the second of equations (6.10a), the fourth power of the *characteristic length* also writes

$$\ell^4 = \frac{D <R>^2}{E\,t} = \frac{<R>^2\, t^2}{12\,(1-\nu^2)}. \tag{6.17}$$

Returning to the above expressions of Ψ/ℓ and Φ/ℓ, the identifications of the real and imaginary parts in (6.12) provide the determination of the normal displacement and stress function respectively as

$$w = \ell\,[C_1\psi_1 + C_2\psi_2 + C_3\psi_3 + C_4\psi_4 + C_5 + C_7 \ln(r/\ell)], \tag{6.18a}$$

$$F = \frac{E\,t\,\ell^3}{<R>}\,[-C_2\psi_1 + C_1\psi_2 - C_4\psi_3 + C_3\psi_4 - C_8 - C_6 \ln(r/\ell)]. \tag{6.18b}$$

Thus, in the general form corresponding to an axisymmetric load $q = 0$ or $q = constant$, each function w and F is represented by eight unknown coefficients where C_1 to C_4 are common coefficients.

---

[3] In Timoshenko W-K [29], these authors use the following representation for the ψ_i Kelvin functions: $\psi_1 = \mathrm{ber}$, $\psi_2 = -\mathrm{bei}$, $\psi_3 = -(2/\pi)\,\mathrm{kei}$, $\psi_4 = -(2/\pi)\,\mathrm{ker}$ on p. 560 and in the Tables on pp. 491–494. We adopt hereafter the Reissner representation [25, 26].

6.2.3 Kelvin Functions

Denoting $x = r/\ell$, the four $\psi_n(x)$ functions in (6.15) are Kelvin functions. These are real and imaginary parts of zero-order Bessel functions $I_0(\sqrt{i}\,x)$ and $K_0(\sqrt{i}\,x)$, defined from Dwight [5] and Abramovitz & Stegun [1] by

$$I_0(\sqrt{i}\,x) = \operatorname{ber} x + i \operatorname{bei} x\,, \tag{6.19a}$$

$$K_0(\sqrt{i}\,x) = \operatorname{ker} x + i \operatorname{kei} x\,. \tag{6.19b}$$

The Kelvin functions are represented by the following series

$$\operatorname{ber} x = 1 - \frac{(x/2)^4}{(2!)^2} + \frac{(x/2)^8}{(4!)^2} - \cdots \tag{6.20a}$$

$$\operatorname{bei} x = \frac{(x/2)^2}{(1!)^2} - \frac{(x/2)^6}{(3!)^2} + \frac{(x/2)^{10}}{(5!)^2} - \cdots \tag{6.20b}$$

$$\begin{aligned}\operatorname{ker} x = &-\left[\ln\left(\tfrac{x}{2}\right) + \gamma\right] \operatorname{ber} x + \tfrac{\pi}{4} \operatorname{bei} x \\ &- \left(1+\tfrac{1}{2}\right)\frac{(x/2)^4}{(2!)^2} + \left(1+\tfrac{1}{2}+\tfrac{1}{3}+\tfrac{1}{4}\right)\frac{(x/2)^8}{(4!)^2} - \cdots\end{aligned} \tag{6.20c}$$

$$\begin{aligned}\operatorname{kei} x = &-\left[\ln\left(\tfrac{x}{2}\right) + \gamma\right] \operatorname{bei} x - \tfrac{\pi}{4} \operatorname{ber} x \\ &+ \frac{(x/2)^2}{(1!)^2} - \left(1+\tfrac{1}{2}+\tfrac{1}{3}\right)\frac{(x/2)^6}{(3!)^2} + \cdots\end{aligned} \tag{6.20d}$$

where γ is the Euler constant $\gamma = 0.57721\ 56649\ 01532\ ...$

Denoting Δ^2 a Laplacian operator with respect to x, i.e. $\Delta^2 = \ell\,\nabla^2$, these functions have the following properties

$$\Delta^2 \operatorname{ber} x = -\operatorname{bei} x\,,\quad \Delta^2 \operatorname{bei} x = \operatorname{ber} x\,, \tag{6.21a}$$

$$\Delta^2 \operatorname{ker} x = -\operatorname{kei} x\,,\quad \Delta^2 \operatorname{kei} x = \operatorname{ker} x\,, \tag{6.21b}$$

and satisfy independently

$$\Delta^2\Delta^2\psi + \psi = 0\,. \tag{6.22}$$

For $x \leq 10$, the four Kelvin functions are plotted in Fig. 6.2.

For large values of the argument, the determination of the Kelvin functions require use of asymptotic relations as follows

$$\operatorname{ber} x \approx \frac{1}{\sqrt{2\pi x}}\, e^{x/\sqrt{2}}\left[L_0(x)\cos\left(\tfrac{x}{\sqrt{2}}-\tfrac{\pi}{8}\right) - M_0(x)\sin\left(\tfrac{x}{\sqrt{2}}-\tfrac{\pi}{8}\right)\right], \tag{6.23a}$$

$$\operatorname{bei} x \approx \frac{1}{\sqrt{2\pi x}}\, e^{x/\sqrt{2}}\left[L_0(x)\sin\left(\tfrac{x}{\sqrt{2}}-\tfrac{\pi}{8}\right) + M_0(x)\cos\left(\tfrac{x}{\sqrt{2}}-\tfrac{\pi}{8}\right)\right], \tag{6.23b}$$

$$\operatorname{ker} x \approx \sqrt{\frac{\pi}{2x}}\, e^{-x/\sqrt{2}}\left[L_0(-x)\cos\left(\tfrac{x}{\sqrt{2}}+\tfrac{\pi}{8}\right) + M_0(-x)\sin\left(\tfrac{x}{\sqrt{2}}+\tfrac{\pi}{8}\right)\right], \tag{6.23c}$$

$$\operatorname{kei} x \approx \sqrt{\frac{\pi}{2x}}\, e^{-x/\sqrt{2}}\left[-L_0(-x)\sin\left(\tfrac{x}{\sqrt{2}}+\tfrac{\pi}{8}\right) + M_0(-x)\cos\left(\tfrac{x}{\sqrt{2}}+\tfrac{\pi}{8}\right)\right], \tag{6.23d}$$

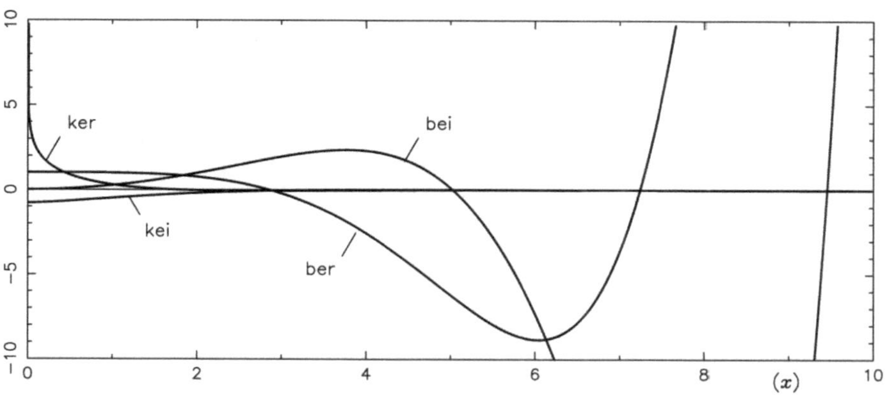

Fig. 6.2 Kelvin functions for $x \leq 10$

where L_0 and M_0 are the expansions

$$L_0(x) = 1 + \frac{1^2}{1!\,8x}\cos\frac{\pi}{4} + \frac{1^2.3^2}{2!\,(8x)^2}\cos\frac{2\pi}{4} + \frac{1^2.3^2.5^2}{3!\,(8x)^3}\cos\frac{3\pi}{4} + \cdots \quad (6.24a)$$

$$M_0(x) = -\frac{1^2}{1!\,8x}\sin\frac{\pi}{4} - \frac{1^2.3^2}{2!\,(8x)^2}\sin\frac{2\pi}{4} - \frac{1^2.3^2.5^2}{3!\,(8x)^3}\sin\frac{3\pi}{4} + \cdots \quad (6.24b)$$

For instance, we use either equation set (6.20) for $x \leq 10$ or sets (6.23) and (6.24) for $x \geq 12$. For argument values such as $10 \leq x \leq 12$, linear cross interpolations are obtained from the two forms. Some computed values of the Kelvin functions are given in Table 6.1.

The asymptotic relations for the derivatives are

$$\text{ber}'x \approx \frac{1}{\sqrt{2\pi x}}\, e^{x/\sqrt{2}}\left[S_0(x)\cos\left(\frac{x}{\sqrt{2}}+\frac{\pi}{8}\right) - T_0(x)\sin\left(\frac{x}{\sqrt{2}}+\frac{\pi}{8}\right)\right], \quad (6.25a)$$

$$\text{bei}'x \approx \frac{1}{\sqrt{2\pi x}}\, e^{x/\sqrt{2}}\left[S_0(x)\sin\left(\frac{x}{\sqrt{2}}+\frac{\pi}{8}\right) + T_0(x)\cos\left(\frac{x}{\sqrt{2}}+\frac{\pi}{8}\right)\right], \quad (6.25b)$$

$$\text{ker}'x \approx \sqrt{\frac{\pi}{2x}}\, e^{-x/\sqrt{2}}\left[-S_0(-x)\cos\left(\frac{x}{\sqrt{2}}-\frac{\pi}{8}\right) - T_0(-x)\sin\left(\frac{x}{\sqrt{2}}-\frac{\pi}{8}\right)\right] \quad (6.25c)$$

$$\text{kei}'x \approx \sqrt{\frac{\pi}{2x}}\, e^{-x/\sqrt{2}}\left[S_0(-x)\sin\left(\frac{x}{\sqrt{2}}-\frac{\pi}{8}\right) - T_0(-x)\cos\left(\frac{x}{\sqrt{2}}-\frac{\pi}{8}\right)\right] \quad (6.25d)$$

where S_0 and T_0 are the following series

$$S_0(x) = 1 - \frac{1{\cdot}3}{1!\,8x}\cos\frac{\pi}{4} - \frac{1^2{\cdot}3{\cdot}5}{2!\,(8x)^2}\cos\frac{2\pi}{4} - \frac{1^2{\cdot}3^2{\cdot}5{\cdot}7}{3!\,(8x)^3}\cos\frac{3\pi}{4} + \cdots \quad (6.26a)$$

$$T_0(x) = \frac{1{\cdot}3}{1!\,8x}\sin\frac{\pi}{4} + \frac{1^2{\cdot}3{\cdot}5}{2!\,(8x)^2}\sin\frac{2\pi}{4} + \frac{1^2{\cdot}3^2{\cdot}5{\cdot}7}{3!\,(8x)^3}\sin\frac{3\pi}{4} + \cdots \quad (6.26b)$$

Some values of the Kelvin function derivatives are listed in Table 6.2.

Table 6.1 Kelvin functions – order 0

x	$\psi_1 \equiv \mathrm{ber}(x)$	$\psi_2 \equiv \mathrm{bei}(x)$	$\psi_3 \equiv \mathrm{ker}(x)$	$\psi_4 \equiv \mathrm{kei}(x)$
0.0	1	0	∞	−0.78539 81634
0.1	0.99999 84375	0.00249 99996	2.42047 39810	−0.77685 06465
0.2	0.99997 50000	0.00999 99722	1.73314 27537	−0.75812 49330
0.3	0.99987 34379	0.02249 96836	1.33721 86375	−0.73310 19123
0.4	0.99960 00044	0.03999 82222	1.06262 39028	−0.70380 02120
0.5	0.99902 34640	0.06249 32184	0.85590 58721	−0.67158 16951
0.6	0.99797 51139	0.08997 97504	0.69312 06956	−0.63744 94947
0.7	0.99624 88284	0.12244 89390	0.56137 82742	−0.60217 54516
0.8	0.99360 11377	0.15988 62295	0.45288 20936	−0.56636 76507
0.9	0.98975 13567	0.20226 93635	0.36251 48126	−0.53051 11224
1.0	0.98438 17812	0.24956 60400	0.28670 62087	−0.49499 46365
1.2	0.96762 91558	0.35870 44199	0.16894 55928	−0.42616 36043
1.4	0.94007 50567	0.48673 39336	0.08512 60483	−0.36166 47815
1.6	0.89789 11386	0.63272 56770	+0.02602 98616	−0.30256 54736
1.8	0.83672 17942	0.79526 19548	−0.01469 60868	−0.24941 70690
2.0	0.75173 41827	0.97229 16273	−0.04166 45140	−0.20240 00678
2.2	0.63769 04571	1.16096 99438	−0.05833 88341	−0.16143 07014
2.4	0.48904 77721	1.35748 54765	−0.06737 34934	−0.12624 14877
2.6	0.30009 20903	1.55687 77737	−0.07082 57000	−0.09644 28914
2.8	+0.06511 21084	1.75285 05638	−0.07029 63212	−0.07157 06486
3.0	−0.22138 02496	1.93758 67853	−0.06702 92333	−0.05112 18840
3.2	−0.56437 64305	2.10157 33881	−0.06198 48330	−0.03458 23127
3.4	−0.96803 89953	2.23344 57503	−0.05589 65499	−0.02144 62871
3.6	−1.43530 53217	2.31986 36548	−0.04931 55564	−0.01123 10962
3.8	−1.96742 32727	2.34543 30614	−0.04264 68639	−0.00348 66647
4.0	−2.56341 65573	2.29269 03227	−0.03617 88479	+0.00219 83993
4.2	−3.21947 98323	2.14216 79867	−0.03010 75737	0.00619 36127
4.4	−3.92830 66215	1.87256 37958	−0.02455 68923	0.00882 56237
4.6	−4.67835 69372	1.46103 68359	−0.01959 50241	0.01037 88649
4.8	−5.45307 61749	0.88365 68537	−0.01524 81878	0.01109 73989
5.0	−6.23008 24787	+0.11603 43816	−0.01151 17272	0.01118 75865
6.0	−8.85831 59660	−7.33474 65408	−0.00065 30375	0.00721 64915
7.0	−3.63293 02425	−21.23940 25796	+0.00192 20216	0.00270 03651
8.0	+20.97395 56107	−35.01672 51649	0.00148 58341	+0.00036 95840
9.0	73.93572 98576	−24.71278 31687	0.00063 71642	−0.00031 91529
10.0	138.84046 59416	+56.37045 85538	0.00012 94663	−0.00030 75247
15.0	$-0.29672\times10^{+04}$	$-0.29527\times10^{+04}$	$+0.34678\times10^{-07}$	$+0.79627\times10^{-05}$
20.0	$+0.47489\times10^{+05}$	$+0.11477\times10^{+06}$	-0.78010×10^{-07}	-0.18553×10^{-06}
50.0	$-0.11762\times10^{+15}$	$-0.50102\times10^{+14}$	-0.29020×10^{-16}	$+0.72610\times10^{-16}$
100.	$+0.73689\times10^{+29}$	$+0.19069\times10^{+30}$	-0.99183×10^{-32}	-0.22356×10^{-31}

6.2.4 Flexure and Stress Function of Shallow Spherical Shells

We will show that from the eight coefficients C_i in the equation set (6.18) only *five* coefficients must be considered.

First, the coefficient C_8 is immaterial since only the derivatives of F occur for the tensions in the equilibrium equations, and C_7 only occurs in non-axisymmetric cases to solve problems on meridian opened ring shells [26]; then

Table 6.2 Kelvin function derivatives – order 0

x	ber$'(x)$	bei$'(x)$	ker$'(x)$	kei$'(x)$
0.0	0	0	$-\infty$	0
0.1	−0.00006 25000	0.04999 99740	−9.96095 93945	0.14597 48114
0.2	−0.00049 99993	0.09999 91667	−4.92294 85203	0.22292 68147
0.3	−0.00168 74881	0.14999 36719	−3.21986 52536	0.27429 20987
0.4	−0.00399 99111	0.19997 33335	−2.35206 99343	0.30951 40025
0.5	−0.00781 20761	0.24991 86211	−1.81979 97533	0.33320 37916
0.6	−0.01349 84813	0.29979 75068	−1.45653 85507	0.34816 44251
0.7	−0.02143 30321	0.34956 23451	−1.19094 33196	0.35630 94756
0.8	−0.03198 86227	0.39914 67577	−0.98733 51336	0.35904 24956
0.9	−0.04553 65525	0.44846 25284	−0.82586 86531	0.35744 31916
1.0	−0.06244 57522	0.49739 65115	−0.69460 38911	0.35236 99133
1.2	−0.10780 56420	0.59352 34988	−0.49464 32458	0.33447 39427
1.4	−0.17092 83240	0.68600 81757	−0.35105 50585	0.30964 15839
1.6	−0.25454 46385	0.77273 99216	−0.24511 46495	0.28090 38357
1.8	−0.36118 21248	0.85092 69510	−0.16594 24225	0.25043 85416
2.0	−0.49306 71247	0.91701 36134	−0.10660 09659	0.21980 79099
2.2	−0.65200 02440	0.96660 86142	−0.06233 72570	0.19011 37417
2.4	−0.83920 27205	0.99442 86435	−0.02971 22571	0.16210 69153
2.6	−1.05513 18152	0.99426 29440	−0.00613 57611	0.13626 89174
2.8	−1.29926 41124	0.95896 54561	+0.01039 90120	0.11287 48199
3.0	−1.56984 66322	0.88048 23241	0.02147 61869	0.09204 30505
3.2	−1.86361 69538	0.74992 36905	0.02836 03141	0.07377 52137
3.4	−2.17549 51752	0.55768 98006	0.03206 61916	0.05798 80989
3.6	−2.49825 25273	+0.29366 24209	0.03340 86788	0.04453 93672
3.8	−2.82216 38502	−0.05252 66206	0.03303 99488	0.03324 80329
4.0	−3.13465 39628	−0.49113 74406	0.03147 84898	0.02391 06138
4.2	−3.41995 12244	−1.03186 21695	0.02913 24188	0.01631 36670
4.4	−3.65876 53064	−1.68325 09472	0.02631 86752	0.01024 33125
4.6	−3.82801 03480	−2.45201 26972	0.02327 90805	0.00549 22592
4.8	−3.90059 92163	−3.34218 12989	0.02019 39082	+0.00186 47791
5.0	−3.84533 94733	−4.35414 05148	0.01719 34038	−0.00081 99865
6.0	−0.29307 99665	−10.84622 33287	0.00563 17093	−0.00522 39209
7.0	+12.76452 25603	−16.04148 88882	+0.00042 05095	−0.00345 95086
8.0	38.31132 57009	−7.66031 84136	−0.00087 97241	−0.00133 63129
9.0	65.60077 09994	+36.29938 44231	−0.00071 12309	−0.00020 80794
10.0	51.19525 83936	135.30930 17156	−0.00031 55972	+0.00014 09133
15.0	$+0.91055\times10^{+02}$	$-0.40877\times10^{+04}$	$+0.55578\times10^{-05}$	-0.59171×10^{-05}
20.0	$-0.48805\times10^{+05}$	$+0.11185\times10^{+06}$	-0.73278×10^{-07}	$+0.19060\times10^{-06}$
50.0	$-0.46498\times10^{+14}$	$-0.11816\times10^{+15}$	$+0.72024\times10^{-16}$	-0.31597×10^{-16}
100.	$-0.83105\times10^{+29}$	$+0.18598\times10^{+30}$	-0.87258×10^{-32}	$+0.22924\times10^{-31}$

$$C_8 = C_7 = 0\,.$$

For the determination of C_6, let us consider the total shearing force Q_z issued from the transverse shearing force Q_r and the axial component of radial tension N_r. For a shallow shell, the equilibrium with a uniform load is given by

$$Q_z = Q_r - \frac{r}{\langle R\rangle} N_r = \frac{1}{2} q r\,. \tag{6.27}$$

Taking into account (6.5a) and (6.7), this becomes

$$D\frac{d}{dr}(\nabla^2 w) + \frac{1}{<R>}\frac{dF}{dr} + \frac{r}{<R>}\Omega + \frac{pr}{2} = 0\,,$$

and since, from (6.9), $q = -2\Omega/<R> = constant$ provides a uniform normal load, we obtain after integration

$$D\,\nabla^2 w + \frac{1}{<R>}F = C_9\,.$$

Noticing in (6.15) that the Kelvin functions have the following properties,

$$\nabla^2\psi_1 = -\psi_2/\ell^2,\quad \nabla^2\psi_2 = \psi_1/\ell^2,\quad \nabla^2\psi_3 = -\psi_4/\ell^2,\quad \nabla^2\psi_4 = \psi_3/\ell^2, \tag{6.28}$$

and using equation set (6.18), the substitutions lead to the residual expression $C_9 = \frac{Et\ell^3}{<R>^2}C_6 \ln(r/\ell)$ which can be only satisfied if

$$C_6 = C_9 = 0\,.$$

Hence, the normal displacement and stress function are related by

$$\nabla^2 w + \frac{1}{D<R>}F = 0\,. \tag{6.29}$$

Therefore, if a uniform normal load $q = -2\Omega/<R> = constant$ is applied to a shallow shell, or a uniform bending moment is only applied to its perimeter ($q = 0$), then the general forms of the normal displacement, stress function, tensions, tangential elongation, and meridian line displacement are represented by

$$\begin{cases}
w = \ell\,(C_1\psi_1 + C_2\psi_2 + C_3\psi_3 + C_4\psi_4 + C_5)\,, & (6.30a)\\
F = \dfrac{Et\ell^3}{<R>}(-C_2\psi_1 + C_1\psi_2 - C_4\psi_3 + C_3\psi_4), & (6.30b)\\
N_r = \dfrac{1}{r}\dfrac{dF}{dr} - \dfrac{q}{2}<R>\,, & (6.30c)\\
N_t = \dfrac{d^2F}{dr^2} - \dfrac{q}{2}<R>\,, & (6.30d)\\
\varepsilon_{tt} = \dfrac{1}{Et}\left[\dfrac{d^2F}{dr^2} - \dfrac{\nu}{r}\dfrac{dF}{dr} - \dfrac{1-\nu}{2}q<R>\right], & (6.30e)\\
u = \varepsilon_{tt}\,r + w\dfrac{r}{<R>}\,, & (6.30f)
\end{cases}$$

Equation set (6.30) applies to constant thickness plain shells but also to constant thickness *ring shells*.

6.3 Variable Thickness Shell and Continuity Conditions

Equation set (6.30) applies to a plain or holed shallow shell provided the only condition for the shearing force is $Q_z = qr/2$ [cf. (6.27)] where the load is either $q = 0$

or $q = constant$. In order to determine the flexure of a variable thickness shell, we will consider a shell made of several concentric ring elements of constant thickness that are continuously linked together and where the normal thickness of each of them is varied by small increments.

6.3.1 Shell Relations for a Constant Thickness Ring Element

The five unknowns C_1 to C_5 in set (6.30) require use of five relations which determine the boundary conditions at center and edge of a constant thickness shell either in a plain or ring form. These five relations are hereafter chosen as the normal displacement, slope, radial bending moment, radial tension, and tangential elongation. From sets (6.30) and (6.3), we obtain respectively

$$\text{displac.} \quad w = \ell\Big(\sum_{i=1}^{4} C_i\,\psi_i + C_5\Big), \tag{6.31a}$$

$$\text{slope} \quad \frac{dw}{dr} = \ell\sum_{i=1}^{4} C_i\,\frac{d\psi_i}{dr}, \tag{6.31b}$$

$$\text{moment} \quad M_r = D\,\ell\sum_{i=1}^{4} C_i\Big(\frac{d^2\psi_i}{dr^2} + \frac{\nu}{r}\frac{d\psi_i}{dr}\Big), \tag{6.31c}$$

$$\text{tension} \quad N_r = \frac{E\,t\,\ell^3}{<R>}\Big(-\sum_{1,3} C_{i+1}\frac{d\psi_i}{r\,dr} + \sum_{2,4} C_{i-1}\frac{d\psi_i}{r\,dr}\Big). \tag{6.31d}$$

$$\text{strain} \quad \varepsilon_{tt} = \frac{\ell^3}{<R>}\Big\{-\sum_{1,3} C_{i+1}\Big[\frac{d^2\psi_i}{dr^2} - \frac{\nu}{r}\frac{d\psi_i}{dr}\Big]$$

$$+\sum_{2,4} C_{i-1}\Big[\frac{d^2\psi_i}{dr^2} - \frac{\nu}{r}\frac{d\psi_i}{dr}\Big]\Big\} - (1-\nu)\,\frac{q<R>}{2\,E\,t}. \tag{6.31e}$$

Since $w(0)$ can be set to zero or any arbitrary value for the first element, we notice that the constant C_5 for the displacement w in (6.31a) can be derived separately after solving $C_{1,\dots,4}$ from the four latter equations.

6.3.2 Various Boundaries and Constant Thickness Plain Shells

Given a plain shell – or cup-like shell – of mean curvature $<R>$, normal thickness $t = constant$ and elastic constants E, ν, its rigidity D and characteristic length ℓ are determined by (6.3g) and (6.10b). The functions $\psi_i(r/\ell)$ in equation set (6.31) allow deriving the normal displacement $w(r)$ as a function of five boundary conditions for solving the C_i unknowns.

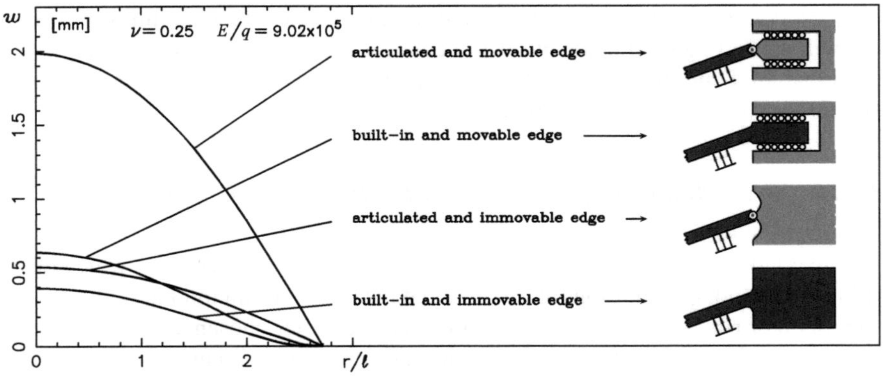

Fig. 6.3 Compared normal flexures w of a constant thickness f/2 menicus under four differing boundary conditions. The meniscus is in Zerodur with $<R> = 6$ m, $2r_m = 1.5$ m, $t = 42.5$ mm. Uniform load $q = 10^5$ Pa

At the center of the shell, we set

$$w\{0\} = 0\,, \quad dw/dr|_{r=0} = 0\,, \quad \varepsilon_{tt}\{0\} = 0\,,$$

and at the edge radius $r = r_{\max}$, we must introduce two other conditions among the following

$$dw/dr = 0\,, \quad M_r = 0\,, \quad N_r = 0\,, \quad \varepsilon_{tt} = 0\,,$$

where, from (6.30 f), we also have $\varepsilon_{tt} = u/r - w/<R>$ whatever r.

Typical edge boundary conditions for a shell can be defined at $r_{\max}$ by one of the following sets.

- *Articulated and movable edge*: $M_r = 0$ and $N_r = 0$, (6.32a)
- *Built-in and movable edge*: $dw/dr = 0$ and $N_r = 0$, (6.32b)
- *Articulated and immovable edge*: $M_r = 0$ and $\varepsilon_{tt} = 0$, (6.32c)
- *Built-in and immovable edge* : $dw/dr = 0$ and $\varepsilon_{tt} = 0$. (6.32d)

The results from calculation of a 1.5-m aperture Zerodur meniscus at f/2, with thickness $t = constant = 42.5$ mm and uniform load $q = 10^5$ Pa, show that for the above four cases the normal displacements $w(r)$ substantially differ (Fig. 6.3).

6.3.3 Some Quantities Involved in a Variable Thickness Shell

A basic geometry for defining a variable thickness shell is a central cup element surrounded by successive ring elements linked together.

Let us consider a shell made of N ring elements with a thickness distribution $\{\, t_1, \ldots t_n \ldots, t_N\}$ slowly varying from center to edge (Fig. 6.4).

To element number n, expanding in the region $r_{n-1} \le r \le r_n$, we associate a thickness t_n, a rigidity D_n and a characteristic length ℓ_n; the two latter quantities are

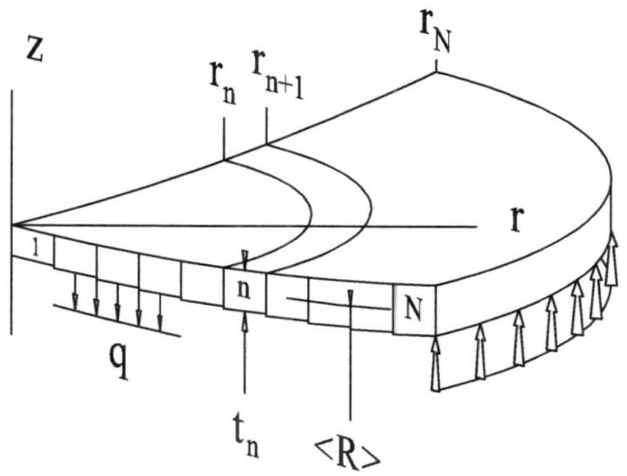

Fig. 6.4 Thickness distribution of a shallow shell – or fast meniscus mirror – made of N continuously linked elements

defined by (6.3g) and (6.17). Some proportionality relations between these quantities are

$$D_n \propto t_n^3\,, \quad t_n \propto \ell_n^2\,, \quad D_n \propto \ell_n^6\,. \tag{6.33}$$

All elements $n \in N$ are with same quantities E, ν, $<R>$ and submitted to the same load q.

6.3.4 *Continuity Conditions of a Shell Element Ring*

From equation set (6.14), we have seen that the dimensionless radial variable $x = r/\ell_n$, as defined by the characteristic length ℓ_n of ring number n, must be the argument of the four functions ψ_i. Let us denote $C_{i,n}$ the five unknown coefficients of the element ring n whose associated thickness and rigidity are t_n and D_n. From (6.31a), the coefficient set $C_{5,n}$ is straightforwardly derived from $C_{1,n}$ to $C_{4,n}$ once known. By use of (6.3g), (6.17), and (6.33), the four remaining continuity conditions – corresponding to equation set (6.31b–e) – may be written at the junction of radius r_n between rings n and $n+1$ as [10, 15]

$$\frac{dw}{dr} \;\rightarrow\; \sum_{i=1}^{4} C_{i,n}\frac{d\psi_i}{dx} = \sum_{i=1}^{4} C_{i,n+1}\frac{d\psi_i}{dx} = \text{invariant for } n \rightarrow n+1\,, \tag{6.34a}$$

$$M_r \;\rightarrow\; \ell_n^5 \sum_{i=1}^{4} C_{i,n}\Big(\frac{d^2\psi_i}{dx^2} + \frac{\nu}{x}\frac{d\psi_i}{dx}\Big) = \text{inv.}, \tag{6.34b}$$

$$N_r \;\rightarrow\; \ell_n^3 \Big(-\sum_{i=1,3} C_{i+1,n}\frac{d\psi_i}{dx} + \sum_{i=2,4} C_{i-1,n}\frac{d\psi_i}{dx}\Big) = \text{inv.}, \tag{6.34c}$$

$$\varepsilon_{tt} \;\rightarrow\; \ell_n \Big[-\sum_{i=1,3} C_{i+1,n}\Big(\frac{d^2\psi_i}{dx^2} - \frac{\nu}{x}\frac{d\psi_i}{dx}\Big) + \sum_{i=2,4} C_{i-1,n}\Big(\frac{d^2\psi_i}{dx^2} - \frac{\nu}{x}\frac{d\psi_i}{dx}\Big) - (1-\nu)\,\frac{q\,\ell_n^3}{2D_n} \Big] = \text{inv.} \tag{6.34d}$$

After solving the above N-set coefficients $C_{i,n}$ for $i = 1, \ldots 4$, the N-set for $C_{5,n}$ is derived from continuity of the normal displacements by use of (6.31a) where the origin of the normal displacement w must be fixed; for instance, one may set $w_1(0) = 0$. Given two boundary conditions at the outer contour of the outmost element, solving all $C_{i,n}$ determines all $\{w_n,\ u_n\}$ displacements.

If the central element, $n = 1$, is not a ring but a plain meniscus ($r_0 = 0$), then

$$C_{3,1} = C_{4,1} = 0\,. \tag{6.35}$$

Since $\psi_1(0) \equiv \text{ber}\,(0) = 1$ and $\psi_2(0) \equiv \text{bei}\,(0) = 0$, the set up of the center of the first element as origin of the displacements is achieved if $C_{5,1} = -\,C_{1,1}$. Thus, only the two unknowns $C_{1,1}$ and $C_{2,1}$ must be considered.

The outermost ring element optically used is numbered $n = N$ and extends from r_{N-1} to r_N. This element may either receive the mirror external reaction at its edge or be *linked* to a special outer ring or cylinder.

For instance, for a shell whose thickness variation is made of $N = 3$ elements, and where the central element $n = 1$ is a cup, the continuity conditions lead to $2+4+4$ unknowns. Taking also into account the boundary conditions for the external contour of the outmost element $n = 3$ leads to introducing 2 supplementary unknowns. These latter unknowns, hereafter chosen as the radial bending moment $M_{r,4}$ and radial tension $N_{r,4}$, require two additional columns in the matrix to solve. Therefore, denoting $\dot{w} \equiv dw/dr$, the square matrix and associated column of unknowns are

$$\begin{array}{ll}
\text{radius } r_1 & \left\{\begin{array}{l} \dot{w} \Rightarrow \\ M_r \Rightarrow \\ N_r \Rightarrow \\ \varepsilon_{tt} \Rightarrow \end{array}\right. \\
\text{radius } r_2 & \left\{\begin{array}{l} \dot{w} \Rightarrow \\ M_r \Rightarrow \\ N_r \Rightarrow \\ \varepsilon_{tt} \Rightarrow \end{array}\right. \\
\text{radius } r_3 & \left\{\begin{array}{l} \dot{w} \Rightarrow \\ M_r \Rightarrow \\ N_r \Rightarrow \\ \varepsilon_{tt} \Rightarrow \end{array}\right.
\end{array}
\quad
\begin{pmatrix}
\times & \times & \times & \times & \times & \times & 0 & 0 & 0 & 0 & 0 & 0 \\
\times & \times & \times & \times & \times & \times & & & & & & \\
\times & \times & \times & \times & \times & \times & & & & & & \\
\times & \times & \times & \times & \times & \times & & & & & & \\
0 & & \times & \times & \times & \times & \times & \times & \times & \times & & \\
0 & & \times & \times & \times & \times & \times & \times & \times & \times & & \\
0 & & \times & \times & \times & \times & \times & \times & \times & \times & & \\
0 & & \times & \times & \times & \times & \times & \times & \times & \times & & \\
0 & & & & & & \times & \times & \times & \times & \alpha & \beta \\
0 & & & & & & \times & \times & \times & \times & 1 & 0 \\
0 & & & & & & \times & \times & \times & \times & 0 & 1 \\
0 & & & & & & \times & \times & \times & \times & \gamma & \delta
\end{pmatrix}
\begin{pmatrix}
C_{1,1} \\ C_{2,1} \\ C_{1,2} \\ C_{2,2} \\ C_{3,2} \\ C_{4,2} \\ C_{1,3} \\ C_{2,3} \\ C_{3,3} \\ C_{4,3} \\ M_{r,4} \\ N_{r,4}
\end{pmatrix} \tag{6.36}$$

The two boundary conditions at the shell contour can be as well considered as link conditions to a special outer cylinder; the geometry resulting from a shell clamped in an outer cylinder is called a *vase shell*. We assume hereafter that the cross-section dimensions of this outer cylinder are small compared to the radius r_N so the effects of the load q on it can be neglected. Furthermore, the displacement w due to the axial compression or extension Q_r of the cylinder does *not* intervene in the shape of the optical surface for which $r \leq r_N$. Hence, from (6.30f), the tangential strain of the cylinder – numbered $N+1$ – can be simply represented by $\varepsilon_{tt} = u/r_N$ instead of $u/r_N - w/{<}R{>}$. From these two assumptions we may characterize the cylinder elastic link to the outermost meniscus shell element N at $r = r_N$ by the two relations

$$\frac{dw}{dr} = \alpha\, M_{r,N+1} + \beta\, N_{r,N+1}\,, \qquad \frac{u}{r_N} = \gamma\, M_{r,N+1} + \delta\, N_{r,N+1}\,, \tag{6.37a}$$

where

$$\alpha = \frac{\partial}{\partial M_{r,N+1}}\left(\frac{dw}{dr}\right),\ \ \beta = \frac{\partial}{\partial N_{r,N+1}}\left(\frac{dw}{dr}\right),\ \ \gamma = \frac{1}{r_N}\frac{\partial u}{\partial M_{r,N+1}}\,,\ \ \delta = \frac{1}{r_N}\frac{\partial u}{\partial N_{r,N+1}}\,. \tag{6.37b}$$

Particular values of $\alpha,\ \beta,\ \gamma,\ \delta$ coefficients allow analysis with various boundary conditions [cf. (6.32)]. Some of these coefficient sets are determined below.

6.4 Edge Cylinder Link and Boundary Conditions

Let us join the outer contour of the meniscus shell element N to one end of a *cylinder* $N{+}1$ by a continuous link. We then obtain a vase shell configuration where the set up of the cylinder dimensions and boundary conditions at its opposite end shall allow analysis of three manifolds.

6.4.1 Three Geometrical Configurations and Boundaries

The shallow shell analysis will be investigated for the three following geometries: a *meniscus shell* (Fig. 6.4), a *vase shell*, and a *closed shell* (Fig. 6.5).

Each geometry can be characterized by the axial and radial thicknesses of the outer cylinder and a specific couple of boundary conditions. For instance, a vase shell with an outer cylinder of extremely thin radial thickness is an equivalent geometry for a meniscus shell. In practice, only the boundary conditions following below can be easily achieved. We call "external end" of the cylinder the opposite end to that which is linked to the meniscus shell.

• **Meniscus shell boundaries and thin cylinder external end boundaries**: The axial thickness of the outer cylinder is, for instance, set to $t_{z,N+1} = t_{z,N}$ and is given a *negligible* radial thickness compared to t_N. Its external end is simply supported without restraint on the radial displacement. In other terms, the boundary conditions (6.32a) for an articulated and movable external end are

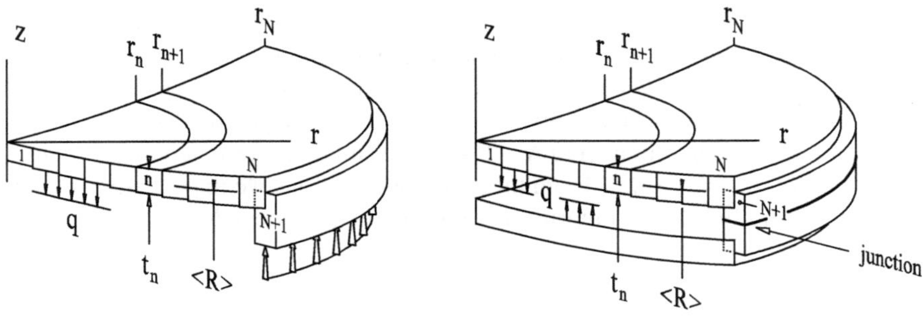

Fig. 6.5 Geometries for a *vase shell* (*left*) and a *closed shell* (*right*)

$$M_{r,E} = 0 \,, \quad N_{r,E} = 0 \,, \tag{6.38a}$$

where the subscript E holds for "external end." These conditions correspond to (6.32a) (see also Fig. 6.3).

• **Vase shell and cylinder external end boundaries**: In this design we always set $t_{z,N+1} \geq t_{z,N}$ which increases the self-rigidity for a perimeter supporting of the mirror [12, 13]. The boundaries for the external end of the outer cylinder are similar to the above case, i.e. (6.38a).

• **Closed shell and cylinder external end boundaries**: This design is made of two vase shells oppositely linked together at their cylinder rear surfaces [11]. The boundaries at this junction are a built-in and movable surface,

$$(dw/dr)_E = 0 \,, \quad N_{r,E} = 0 \,. \tag{6.38b}$$

The junction allows analysis by only considering the first vase shell – with variable thickness – whilst the second complementary vase shell must have an appropriate mean thickness if this latter is a constant. The above conditions correspond to (6.32b) (see also Fig. 6.3).

6.4.2 Outer Cylinder Linked to a Meniscus Shell

The theory of cylindrical shell allows determining the flexure of the outer cylinder of a vase shell. We assume that the effect of the pressure load q on the cylindric element $N+1$ can be neglected because its axial length or thickness is such as

$$t_{z,N+1} \,/\, r_N \ll 1 \,.$$

Let x, z be a local frame in a normal section of the cylinder and passing through the common z-axis of the shell meniscus elements (Fig. 6.6).

A simple case is when the radial thickness t_x is constant and $q = 0$. Referring to Timoshenko and Woinowsky-Krieger [29], the radial displacement $\bar{u}$ in the x-direction is given by

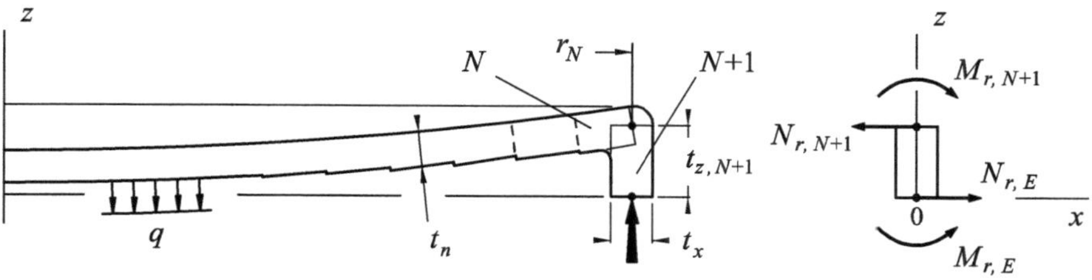

Fig. 6.6 Vase shell geometry and link of the outer cylinder to the meniscus shell

$$\bar{D}\,\frac{d^4\bar{u}}{dz^4} + \frac{Et_x}{r_N^2}\,\bar{u} = 0 \quad \text{with} \quad \bar{D} = \frac{Et_x^3}{12(1-\nu^2)}\,, \tag{6.39}$$

where the bar superscript holds for the cylinder. Using the notation

$$\kappa^4 = \frac{Et_x}{4r_N^2\bar{D}} = \frac{3(1-\nu^2)}{r_N^2 t_x^2}\,, \tag{6.40}$$

we obtain the simplified form

$$\frac{d^4\bar{u}}{dz^4} + 4\kappa^4\,\bar{u} = 0\,. \tag{6.41}$$

The general solution of this equation is

$$\bar{u} = e^{\kappa z}\big(c_1\cos\kappa z + c_2\sin\kappa z\big) + e^{-\kappa z}\big(c_3\cos\kappa z + c_4\sin\kappa z\big)\,, \tag{6.42}$$

in which constants c_1, c_2, c_3, and c_4 are determined by particular boundary cases at the cylinder ends.

The transverse shearing forces, usually denoted Q_r for a cylinder, correspond in fact to z-plane radial tensions N_r in the meniscus shell frame z, r. Similarly the bending moments M_r about a tangential direction of a z-cut of the cylinder correspond to M_r in the meniscus frame. Hence, we use hereafter the notation N_r and M_r for referring to those quantities. They are determined by

$$M_r = \bar{D}\,\frac{d^2\bar{u}}{dz^2}\,, \qquad N_r = \frac{dM_r}{dz} = \bar{D}\,\frac{d^3\bar{u}}{dz^3}\,, \tag{6.43}$$

where M_r is always given the positive sign convention (cf. Sects. 1.13.8 and 3.2). Since $dw/dr \equiv d\bar{u}/dz$ at the meniscus-cylinder junction, local equations (6.37) write for the cylinder variables as

$$\frac{d\bar{u}}{dz} = \alpha\,M_{r,N+1} + \beta\,N_{r,N+1}\,, \qquad \frac{\bar{u}}{r_N} = \gamma\,M_{r,N+1} + \delta\,N_{r,N+1}\,, \tag{6.44a}$$

with

$$\alpha = \frac{\partial}{\partial M_{r,N+1}}\left(\frac{d\bar{u}}{dz}\right), \quad \beta = \frac{\partial}{\partial N_{r,N+1}}\left(\frac{d\bar{u}}{dz}\right), \quad \gamma = \frac{1}{r_N}\frac{\partial \bar{u}}{\partial M_{r,N+1}}, \quad \delta = \frac{1}{r_N}\frac{\partial \bar{u}}{\partial N_{r,N+1}}. \tag{6.44b}$$

Taking the origin $z = 0$ at the external end of the cylinder, each case (6.38a) and (6.38b) for the boundary conditions at this end allows us to determine the c_i constants as follows.

• **Articulated and movable cylinder external end**: In the local frame x, z of the cylinder, the boundary conditions (6.38a) for an articulated and movable external end write

$$M_{r,E} = \bar{D}\left.\frac{d^2\bar{u}}{dz^2}\right|_{z=0} = 0\,, \qquad N_{r,E} = \bar{D}\left.\frac{d^3\bar{u}}{dz^3}\right|_{z=0} = 0\,. \tag{6.45}$$

After substitutions of the first and third $\bar{u}$ derivatives, solution of the two relations leads to

$$c_3 = c_1 \quad \text{and} \quad c_4 = c_2\,. \tag{6.46}$$

At the cylinder end linked to the meniscus the continuity conditions are determined by the bending moment $M_{r,N+1}$ and the radial tension $N_{r,N+1}$. In the meniscus frame, equations (6.44) ensure the continuity of the slope $d\bar{u}/dr$ and of the relative radial displacement $\bar{u}/r_N$.

Let us introduce a second subscript M to denote a c_i coefficient set relative to a *unitary* bending moment $M_{r,N+1} = 1$ while $N_{r,N+1} = 0$, so the c_i write $c_{i,M}$. From the axial ordinate $z = t_{z,N+1}$ of the junction we also introduce the quantity

$$\tau = \kappa\; t_{z,N+1}\,. \tag{6.47}$$

From equations (6.43), we obtain respectively

$$-\,c_{1,M}\,(e^{\tau} - e^{-\tau})\sin\tau \;+\; c_{2,M}\,(e^{\tau} - e^{-\tau})\cos\tau = \frac{1}{2\bar{D}\kappa^2}\,, \tag{6.48a}$$

$$-\,c_{1,M}\left[e^{\tau}(\sin\tau + \cos\tau) \;+\; e^{-\tau}(\sin\tau - \cos\tau)\right] \;-$$
$$c_{2,M}\left[e^{\tau}(\sin\tau - \cos\tau) \;-\; e^{-\tau}(\sin\tau + \cos\tau)\right] = 0\,, \tag{6.48b}$$

so that all coefficients $c_{1,M} = c_{3,M}$ and $c_{2,M} = c_{4,M}$ are known by solving the above system. This allows us to determine

$$\frac{\partial\bar{u}}{\partial M_{r,N+1}} = e^{\tau}\big(c_{1,M}\cos\tau + c_{2,M}\sin\tau\big) \;+\; e^{-\tau}\big(c_{3,M}\cos\tau + c_{4,M}\sin\tau\big) \tag{6.49a}$$

$$\frac{\partial}{\partial M_{r,N+1}}\left(\frac{d\bar{u}}{dz}\right) = \kappa\left[\,c_{1,M}\,e^{\tau}\big(\cos\tau - \sin\tau\big) \;+\; c_{2,M}\,e^{\tau}\big(\cos\tau + \sin\tau\big) \;-\right.$$
$$\left. c_{3,M}\,e^{-\tau}\big(\cos\tau + \sin\tau\big) \;+\; c_{4,M}\,e^{-\tau}\big(\cos\tau - \sin\tau\big)\,\right] \tag{6.49b}$$

from where the γ and α coefficients in matrix (6.36) are straightforwardly obtained from (6.44b).

A second resolution for $M_{r,N+1} = 0$ and *unitary* radial tension $N_{r,N+1} = 1$, now by use of $c_{i,N}$ coefficients, provide two equations similar to (6.48a and b) except

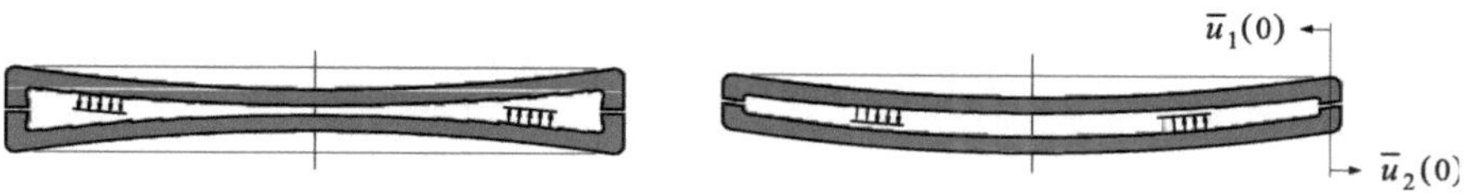

Fig. 6.7 *Left*: Flat closed shell made of identical vase shells. *Right*: Curved closed shell

that the righthand terms are 0 and $1/2\bar{D}\kappa^3$, respectively. Solving them provides the values of the two partial derivatives with respect to $N_{r,N+1}$ in (6.44b) which therefore determine δ and β, respectively.

• **Built-in and movable cylinder external end**: As seen above, a built-in and movable external end of the cylinder will allow analysis of the half part of a closed shell. In the local frame x, z of the cylinder, the boundary conditions (6.38b) for an external end write

$$\left.\frac{dw}{dr}\right|_E = \left.\frac{d\bar{u}}{dz}\right|_{z=0} = 0\,, \qquad N_{r,E} = \bar{D}\left.\frac{d^3\bar{u}}{dz^3}\right|_{z=0} = 0\,. \tag{6.50}$$

After substitutions of the first and third $\bar{u}$ derivatives, solution of the two relations leads to

$$c_3 = c_1 \quad \text{and} \quad c_4 = -\,c_2\,. \tag{6.51}$$

Hence, from (6.42), the radial displacement of the cylinder can be represented by

$$\bar{u} = 2c_1 \cosh\kappa z \cos\kappa z \,+\, 2c_2 \sinh\kappa z \sin\kappa z\,, \tag{6.52}$$

from where we now note that the flexure is *even*, thus in accordance to a closed form made of two strictly *identical* vase shells. Similar determinations to the latter boundary case, with unitary bending moments and radial tensions at the meniscus-cylinder junction for the determination of $c_{i,M}$ and $c_{i,N}$ set values, thus provide the α, β, γ, and δ coefficients.

Note: The above conditions strictly apply to the analysis of a closed shell made of two identical vase shells linked together at the plane base of their outer ring because the linked bases remain plane during flexure (Fig. 6.7-*Left*). This form may be called a *flat closed shell.*

If the mean curvatures $<R_1>, <R_2>$ of two vase shells are similar but with opposite sign, then the radial displacements $\bar{u}_1\{0\}$ and $\bar{u}_2\{0\}$ may have opposite signs. Now when two such vase shells are linked together to form a closed shell, this effect introduces a perimeter bending moment during loading (Fig. 6.7-*Right*). We have seen in Sect. 2.1.1 that a uniform bending moment acting on the contour of a plain plate only generates a pure $Cv\,1$ mode (parabolic flexure); this result also applies to a moderately curved plain shell. Hence, for a *curved closed shell*, final calculation by finite element analysis is appropriate.

• **Other external end boundary cases**: It must be underlined that the two above latter cases accurately apply to the simplest implementations for practical execution conditions. Other boundary conditions such as, for instance, a built-in and immovable external end of the cylinder are set by

$$\bar{u}\big|_{z=0} = 0\,, \qquad \frac{d\bar{u}}{dz}\bigg|_{z=0} = 0\,, \tag{6.53}$$

would lead to

$$c_3 = -\,c_1 \quad \text{and} \quad c_4 = -\,2c_1 - c_2\,. \tag{6.54}$$

However, the radial and axial thicknesses of the cylinder should be infinite which then is rather an academic case.

6.5 Determination of a Variable Thickness Vase Shell

6.5.1 Flexure Representation in the Shell z, r Main Frame

In the previous sections the local orthogonal displacements w_n, u_n at any ring element links $r = r_n$ are curvilinear quantities. Hence, the axial and radial displacements in the main coordinate frame (z, r) of the shell are

$$\delta z_n = \; w_n\,\cos\tan^{-1}\!\left(\tfrac{r}{<R>}\right) + u_n\,\sin\tan^{-1}\!\left(\tfrac{r}{<R>}\right), \tag{6.55a}$$

$$\delta r_n = -\,w_n\,\sin\tan^{-1}\!\left(\tfrac{r}{<R>}\right) + u_n\,\cos\tan^{-1}\!\left(\tfrac{r}{<R>}\right). \tag{6.55b}$$

In this frame, the flexure z_{Flex} of the middle surface is the locus of points of coordinates $\delta z_n,\ r_n + \delta r_n$ resulting from the displacements after deformation. Given an n-value, the corresponding flexure at this point is

$$z_{\text{Flex}}\{r_n + \delta r_n\} \;\equiv\; \delta z_n\,. \tag{6.56}$$

Since the Kelvin functions are polynomial expansions for small or moderate argument values r/ℓ, a polynomial smoothing can be obtained for the axial flexure in the form of the following even series

$$z_{\text{Flex}} = \sum_{i=1,2,3,\cdots}^{\infty} a_{2i}\, r^{2i}, \tag{6.57}$$

where, for a shell made of N linked elements, the determination of the a_{2i} coefficients requires taking into account the flexure of each junction point, therefore solving an N-unknown system.

The shallow shell theory assumes that the largest of thicknesses $\{t_n\}$ is negligible compared to the mean radius of curvature $<R>$ of the shell. In accordance with this condition, we will consider in the next sections $|\, t_n/<R>\,|$ ratios equal to or lower than 1/100. Furthermore, we assume that the curvilinear thicknesses $\{t_n\}$ and the associated thickness $t_z(r)$ measured along the z-axis remain unchanged during stressing so the flexure of the shell outer surfaces are the *same* as that of the middle surface.

6.5.2 Inverse Problem and Thickness Distribution

Given a vase shell with a discrete thickness distribution $\{t_n\}$ made of N meniscus rings and an outer cylinder $N+1$, a square $4N$-matrix similar to (6.36) allows us to solve the $C_{i,n}$, $M_{r,N+1}$, and $N_{r,N+1}$ unknowns. Then the associated displacements $\{w_n, u_n\}$ are known which determines the flexure $z_{\rm Flex}$ in the form (6.57). To summarize, the successive calculations are

$$\{t_n\} \;\rightarrow\; C_{i,n} \;\rightarrow\; \{w_n, u_n\} \;\rightarrow\; z_{\rm Flex}(r) \tag{6.58}$$

Now given a flexure $z_{\rm Flex}(r)$ to be achieved in an even expansion form the inverse problem is to determine the associated thickness distribution $\{t_n\}$. In general, starting from a constant thickness shell, such a distribution can be determined by an iterative code.

A dedicated code for this purpose was elaborated on the basis of variational parameters where the coefficient ratios a_6/a_4, a_8/a_4, a_{10}/a_4, and a_{12}/a_4 in flexure (6.57) form the target. After preliminary calculations for determining a correction vector, iterations toward the required coefficient ratios are carried out by linear algorithms which modify the $\{t_n\}$-set for $n > 1$ up to obtaining the convergence. The iterations also provide a correction factor for the load; then the whole process is repeated with a new t_1-value up to obtaining the required load q. In a final stage, the curvilinear solution $\{t_n\}$ is transformed into a $t_z(r)$ axial function for practical realization of the thickness distribution on the shell meniscus rear side.

6.6 Active Optics Aspherization of Telescope Mirrors

6.6.1 Active Optics Co-addition Law

We consider here the case of an optical surface which can be aspherized by active optics as a result from the co-addition of a flexure and a spherical surface. The position of the problem is to determine the curvature $1/R_{\rm Sphe}$ of this sphere and also the associated parameters for the flexure. These parameters are the curvature $1/R_{\rm Flex}$, the mirror thickness distribution $t(r)$, the uniform load q, and the boundary conditions. This concerns the above inverse problem.

The aspherization process that generates the optical figure $z_{\rm Opt}$ by flexure $z_{\rm Flex}$ of a spherical surface $z_{\rm Sphe}$ results from the *active optics co-addition law*

$$z_{\rm Opt} = z_{\rm Sphe} + z_{\rm Flex}\,, \tag{6.59}$$

from where a basic theorem can be stated as follows.

→ *Whatever the aspherization process, either*

- in situ stressing *after spherical figuring without stress, or*
- elastic relaxation *after spherical figuring while stressed,*

the sphere and also the flexure are algebraically the same for obtaining the same optical figure. The uniform loads q have opposite signs and equal absolute value.

In the second process only the sign of the flexure *during* stressing is opposite to the above $z_{\rm Flex}$. However, the sign of the flexure $z_{\rm Flex}$ is the same for in situ stressing or elastic relaxation.

6.6.2 Parabolization of Concave Mirrors

Active optics parabolization of concave mirrors can be theoretically carried out by one of the two above processes. Unfortunately, because of practical difficulties to realize in a single piece a closed shell aspherized by inner air pressure and stress figuring, use of a vase or a meniscus shell aspherized by *in situ stressing* is preferred. However, from the above theorem, the following results are valid for both processes [6, 20].

The coefficient of the first term in flexure (6.57) is $a_2 \equiv 1/2R_{\rm Flex}$. From the co-addition law (6.59), the first quadratic terms – i.e. the curvature terms – of the expansions must satisfy

$$i = 1 \;\rightarrow\; 1/R_{\rm Opt} = 1/R_{\rm Sphe} + 1/R_{\rm Flex} \tag{6.60a}$$

whilst the sums $z_{\rm Sphe} + z_{\rm Flex}$ at any higher order exactly cancel. Hence, from (6.57) and the expansion of a sphere (see (1.38c) in Sect. 1.7.1), next coefficients of the flexure write

$$i > 1 \;\rightarrow\; a_{2i} = -\,\frac{(2i-2)!\,i}{2^{2i-1}\,(i\,!)^2}\,\frac{1}{R_{\rm Sphe}^{2i-1}}\,. \tag{6.60b}$$

There are infinitely many a_{2i} sets satisfying the active optics co-addition law. Since the sign of a_2 is opposite to that of any higher-order coefficient a_{2i} we can always find a flexure coefficient set with a *balanced* shape. From (6.57) and a third-order approximation, one shows that an *acceptable* range for obtaining a balanced flexure $z_{\rm Flex}$ is when

$$\frac{1}{2} \le -\,\frac{a_2}{a_4\, r_N^2} \le 8\,, \tag{6.61a}$$

where the left and right limit values correspond to $dz_{\rm FLex}/dr = 0$ for $r/r_N = 1/2$ and 2, respectively. An optimal ratio is when the volume to remove is *minimal*; it can be shown that this is achieved for a null slope of the flexure located at $r/r_N = 1/\sqrt[4]{3} = 0.7598...$ which corresponds to $-a_2/a_4 r_N^2 = 2/\sqrt{3} = 1.1547...$ Because $z_{\rm Flex}$ is derived from the displacements w_n, u_n in (6.55) which are dependent on the shell boundary conditions, a strict condition for the above null slope radius-ratio would be excessive. For this reason, we assume as acceptable any flexure solution where the null slope radius r_0 is such as

$$1/2 \le r_0/r_N \le 2\,. \tag{6.61b}$$

We consider hereafter the three shallow shell geometries defined in Sect. 6.4.1 and use Schott Zerodur substrates (cf. the elasticity constants in Table 1.10). For instance, we can solve the inverse problem for 40-cm or 2-m clear aperture mirrors made of $N = 10$ meniscus elements.

- **Parabolization of 40-cm clear aperture mirrors**: The calculations of 40-cm optical diameter mirrors with $N = 10$ elements use a 2-cm radius increment; the radius of the circle where the load reaction applies is $r_N=r_{10}=20$ cm. The uniform load is $q=\pm 80$ kPa $\simeq \pm 0.8$ Atm where, as stated in Sect. 6.6.1, the negative sign is for a depressure (in-situ stressing) and the positive for a pressure (stress figuring). The iteration process allows us to determine the associated radii of curvatures and normal thicknesses $\{t_n\}$ (Table 6.3).

- **Parabolization of 2-m clear aperture mirrors**: Similar calculations of 2-m optical diameter mirrors with $N = 10$ elements use a 10-cm radius increment; the radius of the circle where the load reaction applies is $r_N=r_{10}=1$ m (Table 6.4).

- **Concluding remarks**: Whatever the above shell geometries and associated boundaries, we obtain the following results.

→ *The thickness distributions* $\{t_n\}$ *always increase from center to edge. In all cases, the faster the* f-*ratio, the larger the relative thickness increase.*

In the design shown by Fig. 6.6, an optical clear aperture of $2r_N$ is not exactly achieved because the inner part of the cylinder $N+1$ locally increases the rigidity. In order to effectively obtain a clear aperture up to $2r_N$ in diameter, slight modifications of the shape and position of the outer cylinder must be made. Since the radial thickness t_x of the cylinder always satisfies $t_x/r_N \ll 1$, the inner and outer radii of the cylinder can be modified such as respectively

$$r_N - t_x/2 \rightarrow r_N \quad \text{and} \quad r_N + t_x/2 \rightarrow r_N + t_x \tag{6.62}$$

without significant change of the results in Tables 6.3 and 6.4. From the axial thickness $t_{z,N+1}$ used for the calculation, the geometry of the outer cylinder can be somewhat modified for the final mirror design; equivalences can be obtained by a widened local base and a shortened axial thickness.

To prevent from local stress concentrations at the link of the meniscus with the outer cylinder, a toric inner junction of radius R_J must be made such as

$$1/2 \le R_J/t_N \le 1\,. \tag{6.63}$$

An example of such a modified vase shell is shown in Fig. 6.8.

One shows that if a plane plate of constant thickness t is clamped at $r = a$ into a solid of semi-infinite axial and radial thickness, then the radius R_J of the inner junction – at the rear side of the plate – must approximately satisfy $R_J/t \simeq 1 - \nu$. This value provides the deformation sag corresponding to that given by the thin plate theory for a built-in radius a realized by a square-corner junction. This result – which also includes the flexure effect of the shearing forces – was derived from the results of deformation tests of samples having various two-dimensional shapes (Fig. 6.9).

Table 6.3 Parameters for the parabolization of various shell geometries and f-ratios. Mirror clear aperture $2r_N = 40$ cm. Zerodur. Load $q = \pm 80$ kPa

1 – Meniscus Shell. Simply supported and movable edge.									[Units: mm]
	t_1	t_4	t_6	t_8	t_N	R_{Opt}	$<R>$	R_{Sphe}	R_{Flex}
f/3.0	25.36	25.43	25.54	25.71	25.91	2,400	2,529	2,421.5	270,010
f/2.4	19.63	19.77	20.00	20.33	20.71	1,920	2,075	1,946.8	139,706
f/2.0	15.35	15.61	16.00	16.58	17.21	1,600	1,782	1,631.9	81,819
2 – Vase Shell. Simply supported and movable base. $t_x = 20,\ t_{z,N+1} = 25$.									[Units: mm]
	t_1	t_4	t_6	t_8	t_N	R_{Opt}	$<R>$	R_{Sphe}	R_{Flex}
f/3.0	25.64	25.70	25.80	25.95	26.13	2,400	2,519	2,419.6	296,382
f/2.4	20.06	20.18	20.37	20.65	20.98	1,920	2,054	1,943.1	161,750
f/2.0	15.96	16.17	16.49	16.94	17.46	1,600	1,747	1,626.0	100,113
f/1.7	12.37	12.70	13.20	13.90	14.67	1,360	1,522	1,388.6	66,077
3 – Vase of a Closed Shell. Built-in and movable base. $t_x = 20,\ t_{z,N+1} = 25$.									[Units: mm]
	t_1	t_4	t_6	t_8	t_N	R_{Opt}	$<R>$	R_{Sphe}	R_{Flex}
f/3.0	25.52	25.56	25.61	25.69	26.77	2,400	2,474	2,411.6	497,806
f/2.4	20.00	20.07	20.17	20.30	20.45	1,920	1,996	1,932.8	290,372
f/2.0	16.04	16.15	16.32	16.58	16.90	1,600	1,680	1,614.2	182,012
f/1.7	12.68	12.87	13.14	13.50	13.88	1,360	1,488	1,375.9	117,810

Table 6.4 Parameters for the parabolization of various shell geometries and f-ratios. Mirror clear aperture $2r_N = 2$ m. Zerodur. Load $q = \pm 20$ kPa

1 – Meniscus Shell. Simply supported and movable edge.									[Units: mm]
	t_1	t_4	t_6	t_8	t_N	R_{Opt}	$<R>$	R_{Sphe}	R_{Flex}
f/3.0	76.76	77.37	78.32	79.72	81.33	12,000	12,892	12,107.5	1,354.5 10^3
f/2.4	55.91	57.15	59.05	61.76	64.72	9,600	10,705	9,733.5	699.9 10^3
f/2.0	38.28	40.77	44.18	48.72	54.21	8,000	9,427	8,158.6	411.7 10^3
2 – Vase Shell. Simply supported and movable base. $t_x = 50,\ t_{z,N+1} = 100$.									[Units: mm]
	t_1	t_4	t_6	t_8	t_N	R_{Opt}	$<R>$	R_{Sphe}	R_{Flex}
f/3.0	77.72	78.27	79.14	80.42	81.91	12,000	12,821	12,098.7	1,470.3 10^3
f/2.4	57.54	58.62	60.26	62.62	65.27	9,600	10,565	9,717.0	797.2 10^3
f/2.0	40.57	42.88	45.60	49.36	53.77	8,000	9,158	8,132.4	491.4 10^3
f/1.7	24.24	26.97	32.25	37.98	44.53	6,800	8,120	6,945.3	325.0 10^3
3 – Vase of a Closed Shell. Built-in and movable base. $t_x = 50,\ t_{z,N+1} = 100$.									[Units: mm]
	t_1	t_4	t_6	t_8	t_N	R_{Opt}	$<R>$	R_{Sphe}	R_{Flex}
f/3.0	77.94	78.30	78.82	79.66	80.58	12,000	12,536	12,064.2	2,252.4 10^3
f/2.4	58.80	59.45	60.40	61.73	63.22	9,600	10,158	9,669.0	1,346.7 10^3
f/2.0	44.01	45.11	46.68	48.77	51.00	8,000	8,602	8,074.6	865.7 10^3
f/1.7	29.96	31.98	34.57	37.84	41.09	6,800	7,486	6,881.8	571.8 10^3

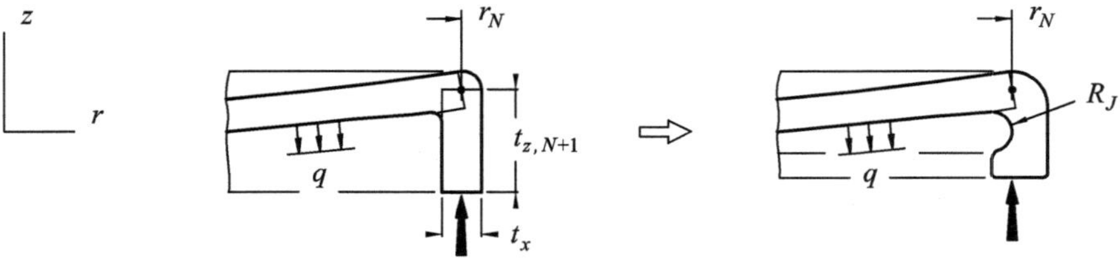

Fig. 6.8 *Left*: Calculated vase shell. *Right*: Edge-modified vase shell

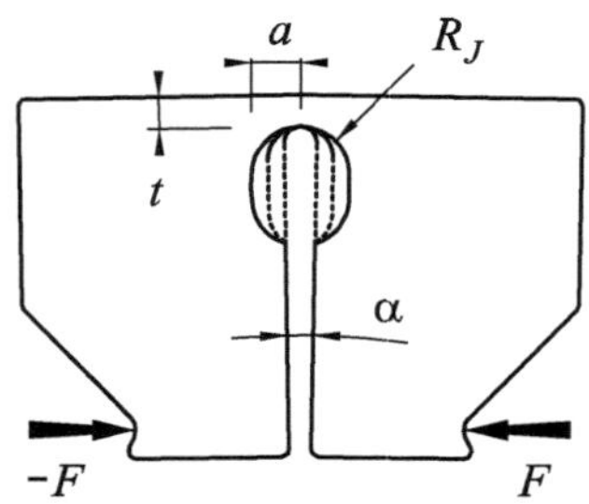

Fig. 6.9 Flexure tests of plane samples bent by moments

The samples were made of quenched Fe87 Cr13 steel alloy well known for its stress-strain linearity. Their geometries were given the same central thickness t but various curvatures R_J at the junctions to rigid ends. The bending moments generated by forces F and $-F$ acting at large distances from the junctions allowed determining the deviation angle α. From these measures, one shows that there exists only one R_J/t-value for which parameters t and a are in accordance with the results of the beam theory: bending moments applied at the beam ends, via square-corner junctions, provide a *constant* curvature flexure. For larger R_J/t-ratios, the a-value which gives the same flexure as that of the beam theory becomes smaller than the semi-separation a of the semi-infinite ends and conversely.

• **Approached values of the maximal stresses**: The maximum stress at the surface of the mirror can be accurately derived from finite element codes. However, a basic estimate of the maximum radial and tangential stresses, σ_{rr} and σ_{tt}, can be easily obtained at the center of the shell from Love-Kirchhoff hypotheses of the plane plate theory (cf. Sect. 1.13.6). In this simplified case, whatever the boundaries, we have

$$\sigma_{rr}(0) = \sigma_{tt}(0) = \pm\frac{6M_r(0)}{t_1^2}\,, \qquad M_r(0) = \frac{3+\nu}{16}\, q\, r_N^2\,,$$

and the coefficient a_2 of the flexure, in (6.57), is

$$a_2 \equiv \frac{1}{2R_{\text{Flex}}} = \frac{3+\nu}{1+\nu}\,\frac{q\, r_N^2}{32\, D}\,,$$

which then leads to

$$\sigma_{rr}(0) = \sigma_{tt}(0) = \pm \frac{1}{2(1-\nu)} \frac{t_1}{R_{\rm Flex}} E \,. \tag{6.64}$$

This maximum stress at the center of a plane plate provides approximate stress values for cases 1 and 2 in the above tables. Compared to the tensile rupture stress $\sigma_{10^3\,\rm sec} = 90\,\rm MPa$ for Zerodur in Table 5.2 (cf. Sect. 5.2.5), the maximum stress obtained from Table 6.3 for cases 1 and 2 lead to a safety factor of at least ~9. The similar two cases in Table 6.4 lead to a safety factor of at least ~18.

For the third case of a built-in and movable edge, used to form a closed shell, the maximal surface stresses are at the edge; From the plane plate theory, one easily shows that these stresses are derived from $\sigma_{rr}(r_N)/\sigma_{rr}(0) = 2/(1+\nu)$ which gives 1.61 for Schott Zerodur. For the edge one also substitutes thickness t_N to t_0 in (6.64). Hence, from case 3 in Table 6.3, we find a safety factor of at least ~ 8. For the same case in Table 6.4, this factor is at least ~13.

Although the bending stresses always dominate in the deformation of these shells, their exact values can be determined from finite element analysis.

6.6.3 Concave Paraboloid Mirrors with a Central Hole

Concave mirrors with a central hole require the introduction of some conditions for the active optics aspherization. We will see that the lack of stiffness caused by the removed material at the hole can be compensated by a local extra-thickness around the hole. In addition, boundary conditions at the hole edge must be found to allow the aspherization by a uniform load.

• **Conditions for a holed mirror aspherizable by active optics**: In most astronomical primary mirrors, the size of the hole is smaller than 1/6 or 1/5 of its clear aperture diameter. The high-order flexure terms, such as the r^6 and r^8 terms which provide the optical shape, have extremely small values in the region near the hole. Hence, the study of a holed geometry can be accurately done within the third-order theory of optics. For a mirror at f/2, the edge region is little larger than f/12 or f/10. From these features, use of the plane plate theory of constant thickness plate is appropriate.

Let a (instead of r_N in the latter sections) be the outer clear aperture radius of the mirror. Considering a built-in edge at $r = a$, we want to obtain the same flexure as that of a plain mirror (see Case 3 in Sect. 1.13.11). Introducing a radius $r = b$ somewhat larger than that of the hole radius $r = c$, we state that the flexure in the region corresponding to the optical clear aperture must be

$$z_1 = \frac{q}{64D_1} \left[r^4 - 2a^2r^2 \right], \quad b \le r \le a \,, \tag{6.65}$$

where the load q is negative for a partial vacuum (in situ aspherization) and positive for an air pressure (stress figuring).

Now if the region $c \le r \le b$ around the hole is with the increased rigidity

$$D_h = Et_h^3 \,/\, [12(1-\nu^2)] \;>\; D_1 \tag{6.66}$$

and submitted to the *same* load q, the general form of a flexure satisfying Poisson's equation is

$$z_2 = \frac{q}{64D_h}\left[\, r^4 + C_1a^2r^2 + C_2a^2r^2\ln r + C_3a^4\ln r + C_4 \,\right], \quad c \le r \le b\,, \tag{6.67}$$

where the five constants C_i and D_h are unknowns determined by the continuity and boundary conditions.

The radial shearing forces and radial bending moments are respectively

$$Q_r = -\,D\,\frac{d}{dr}\left(\nabla^2 z\right) \quad\text{and}\quad M_r = D\left(\frac{d^2z}{dr^2} + \frac{\nu}{r}\frac{dz}{dr}\right). \tag{6.68}$$

After substitutions of z_1 and z_2 and their associated derivatives, the results for the shearing forces are

$$Q_{1r} = -\,\frac{1}{2}\,q\,r\,, \qquad Q_{2r} = -\,\frac{q}{64}\left(32\,r + 4C_2\,\frac{a^2}{r}\right). \tag{6.69}$$

At the junction of the two zones, the continuity condition for the load implies that $Q_{2r}(b) = Q_{1r}(b)$. Therefore, this can be only achieved if

$$C_2 = 0\,,$$

and if a circular force

$$f = \frac{\pi q\,c^2}{2\pi c} = \frac{1}{2}\,q\,c \tag{6.70}$$

per unit length is exerted along the hole edge. The superposition of force f applied at $r = c$ with the load q applied in the region $c \le r \le a$ can be considered as providing the *equivalence* of a uniform load q applied *all over* the surface of a plain mirror (Fig. 6.10).

The unknown constant C_4, which holds for the condition $z_1(b) = z_2(b)$, is not involved in the determination of the three remaining unknowns C_1, C_3, and D_h/D_1. These are solutions of the following conditions,

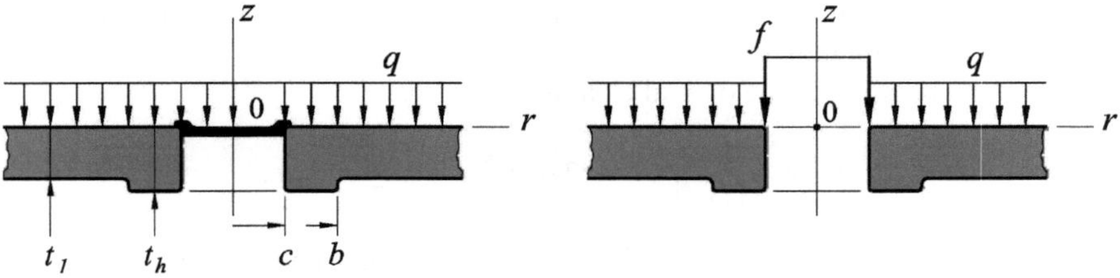

Fig. 6.10 Geometry and equivalent load configurations for a holed mirror aspherizable by active optics

$$\left.\frac{dz_2}{dr}\right|_{r=b} = \left.\frac{dz_1}{dr}\right|_{r=b}, \qquad M_{2r}(b) = M_{1r}(b)\,, \qquad M_{2r}(c) = 0\,. \tag{6.71}$$

After solving this system, and since $t_h/t_1 = (D_h/D_1)^{1/3}$, we obtain

$$\frac{t_h}{t_1} = \left\{1 + 2\,\frac{\left[(1+\nu)a^2 - (3+\nu)c^2\right]c^2}{(1-\nu^2)(a^2-b^2)(b^2-c^2)}\right\}^{1/3} \tag{6.72}$$

for a *built-in holed mirror*. We note here that there exists a solution $t_h = t_1$ for $\nu = 1/4$ and $c/a = 0.620$ which is completely useless in practice.

Referring to Case 2 in Sect. 1.13.11, where the mirror edge is simply supported, the associated flexure z_1 is straightforwardly obtained by the substitution

$$a^2 \;\rightarrow\; \frac{3+\nu}{1+\nu}\,a^2$$

in (6.65). Hence, the same substitution in the above equation leads to the thickness ratio

$$\frac{t_h}{t_1} = \left\{1 + 2\,\frac{(3+\nu)(a^2-c^2)c^2}{(1-\nu)\left[(3+\nu)a^2 - ((1+\nu)b^2)\right](b^2-c^2)}\right\}^{1/3} \tag{6.73}$$

for a *simply supported holed mirror*. For instance, for $a = 1$, $b = 0.2$, $c = 0.1$, and $\nu = 1/4$, (6.72) gives $t_h/t_1 = 1.238$ whilst (6.73) gives $t_h/t_1 = 1.231$, thus a quite similar result.

The small difference between these two results for the usual sizes of a central hole, say, $c/a \lesssim 0.2$ and $b/a \lesssim 0.25$, allows neglecting some of the above b^2 and c^2 terms compared to the a^2 terms. Hence, (6.72) and (6.73) join together into the simplified form [17]

$$\frac{t_h}{t_1} = \left[1 + 2\,\frac{c^2}{(1-\nu)(b^2-c^2)}\right]^{1/3}, \qquad \frac{c}{a} < \frac{b}{a} \lesssim \frac{1}{4}\,. \tag{6.74}$$

Whatever the supporting conditions at the mirror edge, this general *thickness-ratio relation* provides the complete geometry in the region of the hole for active optics aspherization. This relation directly applies for implementing a central hole in any of the three cases previously treated of a *meniscus shell*, a *vase shell*, and a *closed shell*. Therefore, a concluding result for the active optics aspherization of holed mirrors is as follows.

→ *When implementing a small central hole of radius c in a meniscus, a vase or a closed shell, except for a local increase of the thickness in a narrow region $c < r < b$ and use of a small circular force f at the hole edge for the continuity of the load q, the mirror geometry for $r > b$ remains unchanged.*

• **Parabolization of a holed concave mirror by in situ stressing**: The mirror geometry for the active optics aspherization of a paraboloid with a central hole is first determined for a plain mirror. In a second stage, an increased constant thickness in the region $c < r < b$ surrounding the hole of radius c is determined from (6.74) provided the two conditions attached to it are satisfied. The geometry in the region $r > b$ remains unchanged after the geometric modification for the hole implementation. This allows the determination of holed meniscus, vase or closed shell geometries without change of their associated boundaries at the edge $r = a \equiv r_N$ or at the base of the outer cylinder $N+1$.

For instance, we consider hereafter the case of an f/1.75 *holed vase shell mirror* parabolized by in situ stressing. A prototype mirror of 186-mm clear aperture diameter semi-built-in at $a \equiv r_N = 95$ mm was made of Schott Zerodur vitroceram (cf. E and ν in Table 1.10). The analytical shell theory in Sect. 6.3.4 with $N = 10$ shell elements, including a simply supported movable base of the outer cylinder $N+1 = 11$, allows the iterative determination of the normal thickness distribution $\{t_n\}$ of a plain vase shell. All successive ring-shell elements are 9.5 mm width. Relations (6.74) provide the thickness-ratio t_h/t_1 in the normal direction for the design of the final holed vase shell (Table 6.5).

In the final design, the outer cylinder of axial length $t_{z,N+1}$ is modified into a more compact L-shaped ring providing an equivalent rigidity. From the third condition in (6.71), we have shown that *no* bending moment is necessary at the hole edge. After spherical figuring and polishing at the radius of curvature R_{Sphe} while the mirror was unstressed, the paraboloid is obtained by permanent loads q and f applied in situ. The uniform load q is obtained by partial vacuum inside the mirror whilst, from (6.70), the small ring force $f = qc/2$ at the hole edge provides the load continuity (Figs. 6.11 and 6.12).

A helium-neon point source located at the mirror focus allowed optical testing by double pass of a collimated beam reflected on a flat mirror. Interferometric analysis of this prototype mirror showed that the double-pass reflected wavefront errors from a perfect sphere were smaller than $\lambda/5$ ptv.

• **Parabolization of a holed concave mirror by stress figuring**: From the theorem stated in Sect. 6.6.1, if the mirror is aspherized by *stress relaxation* after spherical figuring under stress, then the geometry is the same as that of in situ stressing of a spherical surface. Only the uniform load q and circle force f change to opposite signs but remain with same absolute values (Fig. 6.13).

During the figuring, this requires use of an outer reaction ring to absorb the pressure reaction of the load. Because this ring must be located near the edge of the optical surface this alternative presents some practical difficulties. These can be partly avoided by use of a closed shell made of two sealed vase shells in which air or liquid pressure applies the load q during the stress figuring.

6.6.4 Aspherization of Concave Spheroid Mirrors

The active optics aspherization of a concave spheroid mirror requires some conditions relative to the sign and amplitude of the consecutive polynomial terms which represent the mirror figure. In certain cases, it is clear that those conditions entail that the *active optics co-addition law* (6.59) cannot provide a spherical surface for which an elastic flexure exists.

Table 6.5 Normal thickness distribution $\{t_n\}$ for in situ parabolization of an f/1.75 holed vase shell. Mirror clear aperture 186 mm. Schott Zerodur. Load $q = -73.6$ kPa, $a = 95$ mm, $b = 25$ mm, $c = 15$ mm, $t_h/t_1 = 1.3547$, R_{Opt}=650 mm, R_{Sphe}=658.37 mm, R_{Flex}=51,126 mm, $<R>= 696.9$ mm

1. Plain Vase Shell. $t_x = 8$ mm, cylinder outer edge radius $r_{OE} = 103$ mm.										[Units: mm]
t_1	t_2	t_3	t_4	t_5	t_6	t_7	t_8	t_9	t_N	$t_{z,N+1}$
6.057	6.071	6.104	6.152	6.216	6.295	6.387	6.488	6.594	6.696	36.000
2. Holed Vase Shell. $t_x = 8$ mm, cylinder outer edge radius $r_{OE} = 103$ mm.										[Units: mm]
–	t_h	t_3	t_4	t_5	t_6	t_7	t_8	t_9	t_N	$t_{z,N+1}$
–	8.205	6.104	6.152	6.216	6.295	6.387	6.488	6.594	6.696	36.000

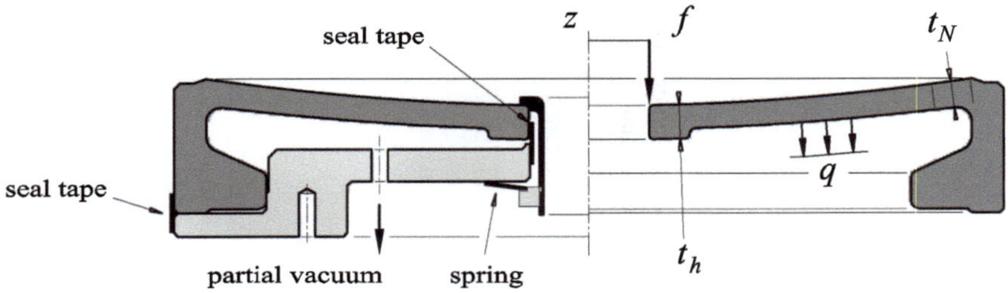

Fig. 6.11 Geometry from Table 6.5 and load configuration for the parabolization of an f/1.75 holed vase shell mirror by *in situ stressing* of a spherical surface

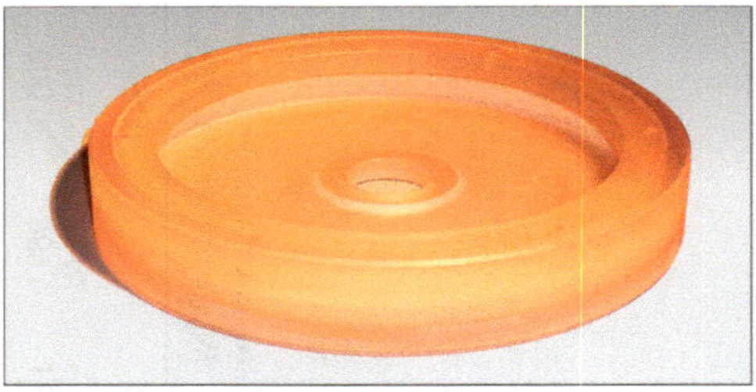

Fig. 6.12 Rear view of the Zerodur substrate mirror after diamond boring by numerical command (LOOM)

Fortunately, most – if not all – concave mirrors for telescope optics are represented by series expansions for which solution pairs ($z_{\rm Sphe}$, $z_{\rm Flex}$) of the co-addition law can be found.

Primary mirrors of two-mirror telescopes which are not of paraboloid shape are mainly spheroids or hyperboloids belonging to the Ritchey-Chrétien form (see Sect. 1.9.2). In general, the shape of these mirrors can be accurately approximated by a hyperboloid and their representation expansion limited to the three first terms.

Except RC telescopes especially designed for extended-field sky surveys – where $R_1 \simeq R_2$ and equipped with additional two-lens correctors (see Fig. 6.18 in

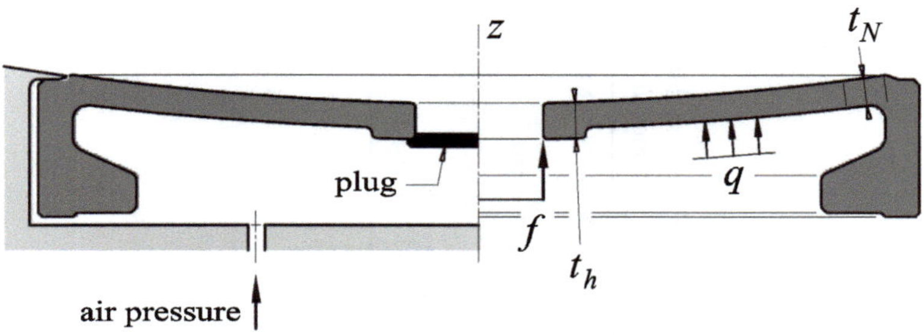

Fig. 6.13 Geometry from Table 6.5 and load configuration for the parabolization of an f/1.75 holed vase shell mirror by *stress relaxation* after spherical figuring under stress

Sect. 6.6.6) –, most primary mirrors in RC telescopes have a hyperboloid figure which departs slightly from that of a paraboloid, i.e. their conic constants are generally included in the range $\kappa_1 \in [-1.20; -1.02]$. Hence the thickness and load configurations, as developed in the previous section, do not differ widely from the case of a paraboloid. For this reason, we do not treat this case here.

Another special case is the wide-field three-mirror Rumsey telescope where all mirrors are hyperboloids with large negative conic constants, particularly for the tertiary mirror. Developments and results for this remarkable telescope form – a flat field anastigmat – are given in Sect. 6.6.7.

6.6.5 Aspherization of Cassegrain Mirrors

Most usual telescope forms are two-mirror systems where the primary and secondary mirrors are a paraboloid-hyperboloid pair (PH) or a Ritchey-Chrétien with a hyperboloid-hyperboloid pair (RC). The afocal telescope form is a paraboloid-paraboloid pair (PP). In all possible forms with a convex secondary, this mirror is usually called a Cassegrain mirror. The PH, RC, and PP forms are with hyperboloid or paraboloid Cassegrain mirrors. Active optics aspherization of these mirrors by *elastic relaxation* after spherical figuring while stressed is always preferred.

However, some exceptions are non-conventional telescope forms such as, for instance, two-mirror systems with a spherical primary where the aspherization of the *oblate ellipsoid* Cassegrain mirror is more easily obtained by *in situ stressing* after spherical figuring without stress. This form was investigated to simplify the segmentation of large primary mirrors [2–4]. A 1.4-m prototype telescope with an in-situ aspherized vase shell Cassegrain mirror was built for imaging evaluations on the sky at Haute Provence observatory [14].

Returning to the case of paraboloid or hyperboloid Cassegrain mirrors, we develop hereafter some examples for these two cases.

• **Paraboloid Cassegrain mirrors**: In a two-mirror system, a paraboloid Cassegrain mirror is the second optical component of an afocal Mersenne form. This remarkable form is free from $Sphe\,3$, $Coma\,3$, and $Astm\,3$, therefore anastigmatic (see Sects. 1.9.3 and 2.3). Although sometimes used in the past for a coudé telescope arrangement, afocal forms are the basic systems of high-resolution telescope arrays.

Using Schott Zerodur vitroceram, a purely meniscus shell, an enlarged meniscus shell, and three varied vase shells have been computed with $N = 10$ equal width element rings (Fig. 6.14 and Table 6.6).

• **Geometric scale factor**: Similarly as for all the previous data given in Sects. 6.6.2 and 6.6.3 for concave mirrors, each parameter set defining the geometry of convex mirrors in Tables 6.6 can be scaled up or down provided the material elasticity constants E, ν, and the load q are the same.

This general linear property also applies to the design parameters of paraboloid and hyperboloid convex mirrors in the examples below.

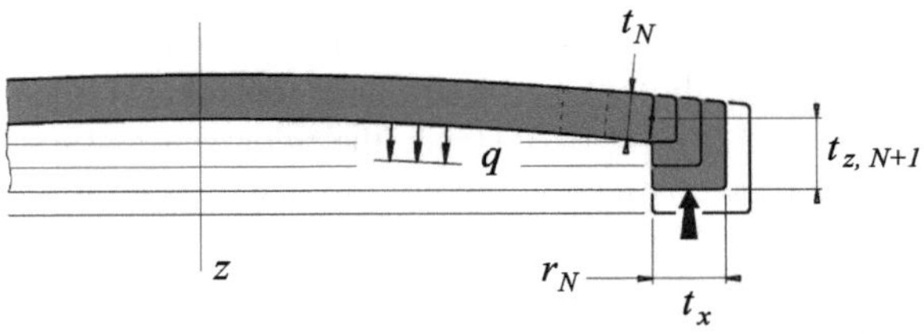

Fig. 6.14 Five varied shapes of Cassegrain mirrors for active optics aspherization by spherical figuring while stressed

• **Maximum stress**: For a meniscus shell and vase shells where the outer cylinder is such that $t_x \leq t_N$, a basic estimate of the maximum stress at the center of the shell $\sigma_{rr}(0)$ and $\sigma_{tt}(0)$ is given by relation (6.64) of the constant thickness plate theory. For the five convex paraboloids at f/2.5 in Tables 6.6, this relation leads to a maximum stress $\sigma_{rr}(0) = \sigma_{tt}(0) \lesssim 7.7$ MPa. This value is well below the Zerodur tensile maximum stress for safety, $\sigma_{T\,\max} = 22$ MPa (cf. Table 1.10) and $\sigma_{10^3\,\text{sec}} = 90$ MPa (cf. Table 5.2) and thus provides a safety factor of ~ 11.7.

For a vase shell where the dimensions of the outer cylinder are $t_{z,N+1} = t_x = 2t_N$ – the last cases in Tables 6.6 –, the maximum stress at the junction radius $r = r_N$ are dominating with $\sigma(r_N) \simeq 1.3\,\sigma(0)$. Although the bending stresses always dominate in the deformation of these shells, their exact values can be determined from finite element analysis.

• **Execution of a paraboloid Cassegrain mirror**: Two identical paraboloid Cassegrain mirrors at f/2.33 were parabolized by stress figuring for the 1.5 m telescopes of the GI2T high-resolution interferometer near Nice. These two Mersenne afocal telescopes were designed for a *beam compression* $k = R_1/R_2 = 20$. The Cassegrain mirrors were made of Corning fused silica in low expansion ULE grade for which the Young modulus was found to be abnormally large, showing $E = 85$ GPa, instead of $E = 68.8$ GPa as given by Corning (cf. Table 1.10).

The analytical shell theory in Sect. 6.3.4 with $N = 10$ shell elements, including a simply supported movable base of the outer cylinder $N+1$, allowed the iterative determination of the normal thickness distribution $\{t_n\}$ of the vase shell where all successive ring-shell elements have a r_N/N radial width (Fig. 6.15 and Table 6.7)

From the result of iterations with the shell theory, the active optics superposition law was accurately satisfied. The resulting flexure writes, in [mm],

$$z_{\text{Flex}} = 0.722932\ 10^{-5}\,r^2 - 0.16567\ 10^{-8}\,r^4 - 0.4639\ 10^{-14}\,r^6 - 0.162\ 10^{-19}\,r^8 \tag{6.75}$$

The dimensionless radius $\rho = r/r_N$ for which $dz_{\text{Flex}}/dr = 0$ is ρ_0=1.038. The total flexure sag in the range $\rho \in [\,0;\,1\,]$ is Δz_{Flex}=7.81 μm. The vase geometry of two sample Zerodur secondary mirrors were obtained by computer numerical control diamond turning. After spherical figuring under stress and elastic relaxation, the results from He-Ne interferometric tests showed that the paraboloid surfaces were obtained within $\lambda/12$ ptv for each sample.

Table 6.6 Parameters for the parabolization of various shell geometries and f-ratios. Mirror clear aperture $2r_N = 20$ cm. Zerodur. Load q=80 kPa. Outer cylinder geometries with $t_x = t_{z,N+1}$, $t_{z,N+1} = k\, t_N$ and k=0,$\frac{1}{2}$,1,$\frac{3}{2}$,2

1 – Convex Paraboloids at f/3.0. Simply supported movable edge. R = 1200										[Units: mm]
t_1	t_4	t_6	t_7	t_8	t_9	t_N	$t_{z,N+1}$	$<R>$	R_{Sphe}	R_{Flex}
12.61	12.66	12.73	12.77	12.83	12.89	12.95	0	1,151	1,210.8	134,932
12.62	12.66	12.73	12.78	12.84	12.90	12.96	$t_N/2$	1,151	1,210.7	135,713
12.81	12.85	12.91	12.95	13.00	13.05	13.10	t_N	1,157	1,209.4	153,643
13.04	13.07	13.10	13.13	13.16	13.19	13.23	$3t_N/2$	1,168	1,207.0	205,680
13.09	13.10	13.13	13.15	13.17	13.20	13.22	$2t_N$	1,175	1,205.6	259,807
2 – Convex Paraboloids at f/2.5. Simply supported movable edge. R = 1000										[Units: mm]
t_1	t_4	t_6	t_7	t_8	t_9	t_N	$t_{z,N+1}$	$<R>$	R_{Sphe}	R_{Flex}
10.14	10.22	10.35	10.44	10.55	10.66	10.77	0	942	1,012.9	78,704
10.13	10.21	10.34	10.44	10.54	10.66	10.77	$t_N/2$	942	1,012.8	78,811
10.31	10.38	10.50	10.58	10.67	10.77	10.87	t_N	947	1,011.7	86,170
10.64	10.69	10.77	10.83	10.89	10.96	11.03	$3t_N/2$	961	1,009.0	112,250
10.77	10.82	10.87	10.91	10.95	11.00	11.04	$2t_N$	970	1,007.0	144,000
3 – Convex Paraboloids at f/2.0. Simply supported movable edge. R = 800										[Units: mm]
t_1	t_4	t_6	t_7	t_8	t_9	t_N	$t_{z,N+1}$	$<R>$	R_{Sphe}	R_{Flex}
7.041	7.267	7.598	7.816	8.062	8.317	8.550	0	720	815.99	40,845
6.973	7.203	7.539	7.761	8.010	8.268	8.504	$t_N/2$	719	816.05	40,678
7.097	7.311	7.623	7.830	8.063	8.307	8.533	t_N	723	815.38	42,417
7.652	7.804	8.021	8.166	8.332	8.509	8.681	$3t_N/2$	742	812.43	52,274
8.062	8.170	8.337	8.411	8.521	8.638	8.753	$2t_N$	759	809.55	67,757

Fig. 6.15 Rear view of an f/2.33 vase shell Cassegrain mirror, in Corning fused silica ULE, after active optics parabolization by spherical figuring under stress (LOOM)

Table 6.7 Normal thickness distribution $\{t_n\}$ for in situ parabolization of an f/2.33 vase shell Cassegrain mirror. Mirror clear aperture 90 mm. Fused silica ULE $E = 85$ GPa. Load $q = 80$ kPa. $r_N = 45$ mm, $t_x = 15$ mm. $R_{\rm Opt} = 420$ mm, $R_{\rm Sphe} = 422.56$ mm, $R_{\rm Flex} = 69162$ mm, $<R>=$ 409.2 mm

VASE SHELL MIRROR. Cylinder outer edge radius r_{OE}=60.										[Units: mm]
t_1	t_2	t_3	t_4	t_5	t_6	t_7	t_8	t_9	t_N	$t_{z,N+1}$
4.550	4.552	4.557	4.565	4.576	4.589	4.605	4.621	4.639	4.655	20.000

• **Execution of a hyperboloid Cassegrain mirror**: Hyperboloid Cassegrain mirrors are classical secondary mirrors for PH or RC telescopes. The active optics apherization was carried out for two sample secondary mirrors of the 90-cm solar RC telescope THEMIS, f/3.5–f/17, installed in Tenerife [24] (Fig. 6.16). The optical parameters of the telescope main mirrors M_1 and M_2 are R_1=6,270 mm, R_2=1,683 mm, $\kappa_1 = -1.036$, $\kappa_2 = -2.62$ and their axial separation is $M_1M_2 =$ 2,465 mm. The telescope tube is equipped with a 1-m enclosure plate for a vacuum. A tip-tilt mirror allows atmospheric corrections for both imaging and spectroscopic modes. This latter mode provides high spectral resolution capabilities.

Fig. 6.16 THEMIS France-Italia-Spain solar observatory (courtesy CNRS-CNR-IAC)

Beside the active optics advantage of obtaining a smooth surface secondary mirror, the thin clear aperture area – aspect-ratio $2r_N/t_0 \simeq 19$ – of a vase shell geometry allows a fast *thermal exchange* by cooling system located inside the vase shell.

Two sample secondaries were hyperbolized by elastic relaxation after spherical figuring under stress. The analytical shell theory in Sect. 6.3.4 with N=10 shell elements, including a simply supported movable base of the outer cylinder $N+1$, allowed the iterative determination of the normal thickness distribution $\{t_n\}$ of the vase shell where all successive ring-shell elements have a r_N/N radial width (Fig. 6.17 and Table 6.8).

From the result of iterations with the shell theory, the active optics superposition law of the surfaces writes, in [mm],

$$\begin{aligned}
z_{\mathrm{Sphe}} &= 0.294684\ 10^{-3}\, r^2 + 0.25590\ 10^{-10}\, r^4 + 0.4444\ 10^{-17}\, r^6 + 0.964\ 10^{-24}\, r^8\\
z_{\mathrm{Flex}} &= 0.002405\ 10^{-3}\, r^2 - 0.68070\ 10^{-10}\, r^4 - 0.4441\ 10^{-17}\, r^6 - 0.964\ 10^{-24}\, r^8\\
z_{\mathrm{Sum}} &= 0.297089\ 10^{-3}\, r^2 - 0.42480\ 10^{-10}\, r^4 - 0.3179\ 10^{-46}\, r^6 + 0.481\ 10^{-50}\, r^8\\
z_{\mathrm{Opt}} &= 0.297089\ 10^{-3}\, r^2 - 0.42480\ 10^{-10}\, r^4 + 0.0000\ 10^{+00}\, r^6 + 0.000\ 10^{+00}\, r^8
\end{aligned} \tag{6.76}$$

from where we satisfy $z_{\mathrm{Sphe}} + z_{\mathrm{Flex}} = z_{\mathrm{Opt}}$. The dimensionless radius $\rho = r/r_N$ for which $dz_{\mathrm{Flex}}/dr = 0$ is ρ_0=1.055. The total flexure sag in the range $\rho \in [\,0;\,1\,]$ is Δz_{Flex}=20.98 μm. The vase geometry of two Zerodur secondary mirror samples were obtained by computer numerical control diamond turning. After spherical figuring under stress and elastic relaxation, the results from He-Ne interferometric tests showed that the hyperboloid surfaces were obtained within $\lambda/12$ ptv for each sample.

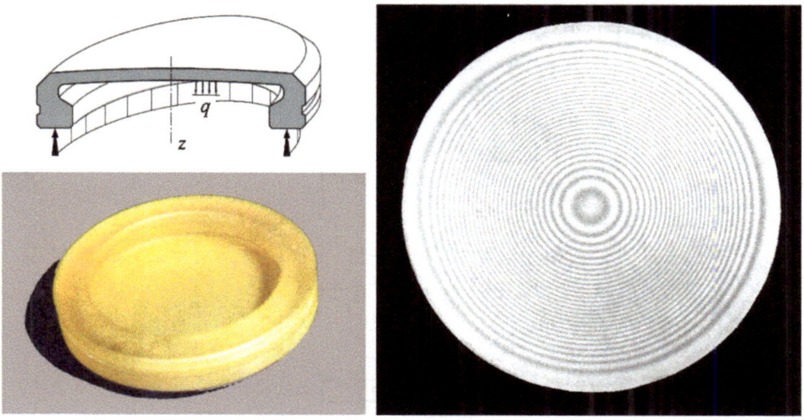

Fig. 6.17 *Left*: Design and rear view of the f/3.33 vase shell THEMIS secondary mirror in Schott Zerodur. *Right*: Fizeau He-Ne interferogram with respect to a sphere after stress polishing and elastic relaxation (LOOM)

Table 6.8 Normal thickness distribution $\{t_n\}$ for the hyperbolization of THEMIS f/3.33 vase secondary mirrors. Mirror clear aperture 252 mm. Schott Zerodur E =90.2 GPa. Load q =80 kPa. r_N =126 mm, t_x =30 mm. R_{Opt} =1,683 mm, R_{Sphe} =1,696.7 mm, R_{Flex} =208,029 mm, $<R>$=1,657 mm

VASE SHELL MIRROR. Cylinder outer edge radius r_{OE}=150.										[Units: mm]
t_1	t_2	t_3	t_4	t_5	t_6	t_7	t_8	t_9	t_N	$t_{z,N+1}$
13.16	13.16	13.17	13.18	13.20	13.22	13.24	13.26	13.29	13.31	50.000

6.6.6 *Comparison of Various Wide-Field Telescope Designs*

Although Schmidt telescopes with refractive corrector plates provide an accurate imaging correction for a 5° sky field, this design suffers from aperture size limitation – due to the difficulty of obtaining homogeneous glass plates larger than 1 or 1.2 m –, spherochromatism residuals and field curvature. Larger Schmidt systems of the all-reflection type similarly suffer from field curvature which is a difficulty for the implementation of a flat and large CCD detector array.

For the purpose of wide-field astronomical sky surveys, typically 1–2° fields with large telescope diameters, various designs in the 2–4 m class have been investigated and some of them have been built. A comparison of possible various *wide-field telescope designs* show important variations in the overall length and chromatism residual correction (Fig. 6.18).

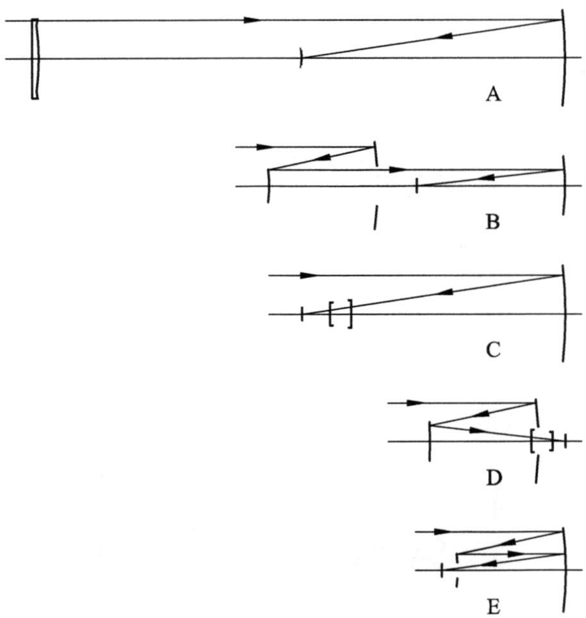

Fig. 6.18 Comparison of five wide-field telescope designs having the same input beam diameter, focal length and field of view

- **(A) Schmidt telescope with refractive corrector plate**: The overall length is two times the focal length, $L = 2f$. Convex field curvature, $c_P = 1/f$.
$\rightarrow$ **Three** optical surfaces including **one** aspheric.

- **(B) Mersenne-Schmidt telescope by Willstrop [30]**: The overall length slightly exceeds the focal length, $L \simeq f$, when the focal surface is near M_1. Flat field curvature, $c_P = 0$. The *Sphe 3* correction of the spherical tertiary mirror is achieved by deformation of the paraboloid secondary mirror, so the center of curvature of M_3 mirror must be located at the M_2 vertex.
$\rightarrow$ **Three** optical surfaces including **two** aspherics.

- **(C) Paraboloid and Wynne triplet lens corrector**: The overall length is equal to the focal length, $L = f$. Flat field curvature, $c_P = 0$. For a wide spectral range as $\lambda\lambda$ [350 nm–1 μm] and field of view larger than 1°, this system suffers from chromatism residuals larger than 1 arcsec. With a moderate spectral range, this system operates at Canada France Hawaii telescope (CFHT) [23].
$\rightarrow$ **Seven** optical surfaces including **one** aspheric.

- **(D) Flat-field Ritchey-Chrétien and doublet lens corrector**: The overall length is about half the focal length, $L = f/2$. The flat field curvature is achieved by equalizing the curvature of the two mirrors, $c_P = 0$. The null power doublet lens provides the correction of *Astm* 3. For a wide spectral range and field larger than 1°, curvature, $c_P = 0$. For a wide spectral range as $\lambda\lambda$ [350 nm–1 μm] and field of view larger than 1°, this system also suffers from chromatism residuals larger than 1 arcsec. For limited spectral ranges, this design was built as a 2.5-m aperture telescope at f/2.5/5 for the Slaon Digital Sky Survey (SDSS) operating at the Apache Point observatory, New-Mexico [28]. Similar form survey telescopes with the VST and VISTA should star observations in 2008.
$\rightarrow$ **Six** optical surfaces including **two** aspherics.

- **(E) Three-reflection Rumsey telescope**: The compact three-reflection telescope proposed in 1969 by Norman Rumsey [27] shows an overall length about two-times smaller than the focal length, $L \simeq f/2$, thus is very similar length to design (D). This anastigmatic system provides a null Petzval sum, $c_P = 0$. The design does not suffer from the chromatism residuals inherent to the above lens corrector systems. For good seeing sites, the spherochromatism due to the thickness of the pass-band filters and cryostat window can be made negligible because the image sampling requires a telescope design of low f-ratio, such as f/5. For a large spectral range, this allows obtaining a flat field up to two degrees, i.e. $1.5 \times 1.5°$ whose aberration residuals can be made smaller than 0.25 arcsec and where the beam baffling and vignetting is similar to design D.
$\rightarrow$ **Three** aspherical surfaces.

Other optical designs based on three-mirror concepts with additional corrector lenses are under investigation such as, for instance, the Large Synoptic Survey Telescope (LSST) project; however these designs are more complex and obtaining both high throughput and large spectral range capabilities from UV to IR is a major difficulty.

From the above five systems, the Rumsey design is a compact system – about 4-times shorter than a Schmidt – which provides the best optical performances for a 2° FOV and large spectral range capability from UV to IR. The central obstruction of the Rumsey is *similar* to that of the flat-field Ritchey-Chrétien (D). We shall see in the next sections that, with a slightly modified Rumsey form, the active optics aspherization of the mirrors provides a high accuracy and important simplifications by figuring only *two* spherical surfaces for *three* mirrors.

6.6.7 Modified-Rumsey Three-Reflection Telescope Mirrors

Astronomical sky surveys require the development of dedicated telescopes in large size. As shown above, by comparing several telescope designs, the optical properties of a three-mirror telescope in the Rumsey form [27] provide many significant advantages: (i) the anastigmatic flat field design does not require use of any corrector lens and thus is free from these chromatic residual errors over an extended spectral range, (ii) the system is as compact – and with the same central obstruction – as a flat-field RC with a two-lens corrector but uses *three* optical surfaces instead of *six*, and (iii) the primary and tertiary mirrors are on a single disk of glass which provides the advantages of a permanent alignment of these mirrors and less diffraction light by avoiding a spider support.

A *modified-Rumsey* form was proposed by Lemaitre [16] with active optics apherization of the mirrors. The active process only requires the figuring of *only two* spherical surfaces for the whole optics. The first one elastically generates the primary and tertiary mirrors whilst the second generates the secondary mirror.

Starting from the Rumsey design where all mirrors are spheroids of the hyperboloid type, and given an effective focal length and output f-ratio, one shows that the eight available free parameters – three curvatures, three conic constants, and two axial separations – can be slightly modified for obtaining the simultaneous aspherization of the primary and tertiary mirrors, M_1 and M_3, as belonging to a *single* deformable substrate. This fully retains the anastigmatism properties $Sphe\,3{=}Coma\,3{=}Astm\,3{=}0$ with a flat-field $Petz\,3{=}0$ and determines the modified-Rumsey design. Given a focal length, f-ratio and back focal distance there exists one and only one design that satisfies these four conditions.

The basic features of a modified-Rumsey design whose mirrors are generated by active optics aspherizations are the following:
(i) the M_1–M_3 mirror substrate is made of four concentric zones called a *double vase form*. Each zone corresponding to the mirror clear apertures have an increasing thickness from center to edge. The link zone between M_1 and M_3 is a narrow ring not used optically – of larger axial thickness than those of the reflective zones – for which the flexural continuity conditions also apply. The outer zone surrounding M_1 is a thick cylinder as for usual vase forms,

(ii) the elastic aspherization of M_1 and M_2 mirrors are simultaneously generated by the same uniform load. The axial reaction to the load is only exerted at the basis of the outer cylinder of the double vase form.
(iii) the optical aspherization alternatives for M_1 and M_2 mirrors are *stress figuring* at the laboratory or *in situ stressing* at the telescope.

• **Telescope optical design**: The modified-Rumsey design is of interest for developing large space- and ground-based *survey telescopes* in ultraviolet, visible, or infrared. In order to validate the active optics aspherization methods the design and construction of two identical telescopes MINITRUST-1 and -2, f/5, 1.5 × 1.5° FoV, were carried out by Lemaitre et al. [18, 19]. The hyperbolizations of M_1 and M_3 mirrors were simultaneously obtained by *in-situ stressing* of a *double vase form*. The hyperbolization of M_2 was obtained by *stress figuring* of a *tulip form* (see Sect. 3.3.5 for this latter mirror).

The elasticity continuity conditions for the slope and sag variations in the area of the intermediate ring linking M_1 to M_3 lead to cross optimizations between optical and elasticity designs. From these optimizations, and setting the MINITRUST input *pupil* on M_2, we obtain an optical design which leads to an optimal balance of the M_1 and M_3 clear aperture areas (Table 6.9 and Fig. 6.19).

• **Double vase shell primary and tertiary mirrors – In situ stressing**: The elasticity design of a *double vase shell* as the common substrate of M_1 and M_3 mirrors requires that the aspherization process – hereafter carried out by in situ stressing –

Table 6.9 Modified-Rumsey telescope optical design – MINITRUST f/5–1.5×1.5° FoV – λλ [380–1,000 nm] – Efl=2,265.6 mm. [Units: mm]

i	Surface	R_i	AS_i	$A_{4,i}$	$A_{6,i}$	Aperture	$[\kappa_i]^\star$
1	Primary	$-2,208.0$	-631.000	$6.3530\ 10^{-12}$	$5.217\ 10^{-19}$	220–440	[−1.547]
Sto	Second.	$-1,091.5$	-631.005	$2.8043\ 10^{-10}$	$-1.950\ 10^{-16}$	100–200	[−3.917]
3	Tertiary	$-2,197.2$	-766.224	$7.5716\ 10^{-11}$	$-1.897\ 10^{-17}$	d_{ext} 180	[−7.425]
4	Fused	∞	-10.000			59×59	
5	silica	∞	-25.000			58×58	
6	Focus	∞				56×56	

Equation of mirrors: $z_i = (1/2R_i)r^2 + A_{2,i}\,r^4 + A_{4,i}\,r^6$. Axial separations : AS_i.
* Conic constant from third order : $\kappa_i = 8R_i^3 A_{4,i} - 1$ and $\kappa = -1$ for a paraboloid.

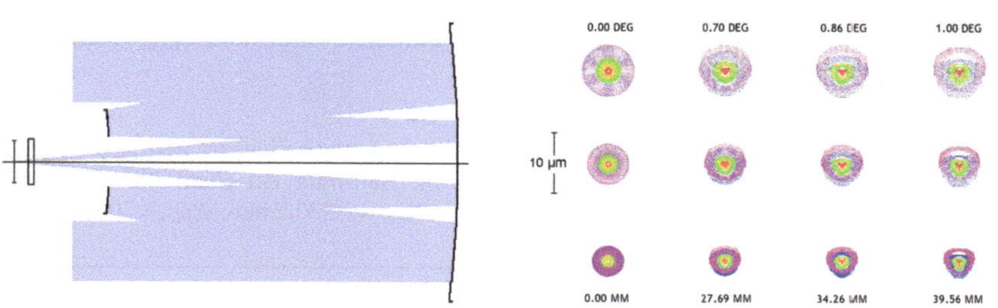

Fig. 6.19 *Left*: MINITRUST optical design from Table 6.9. Entrance pupil at M_2. *Right*: Residual spherochromatism blur images for 10, 8, and 6-mm thickness plate as equivalent thickness for filter and cryostat window. Corresponding blur rms diameters 0.42, 0.33, and 0.25 arcsec, respectively

is achieved from the flexure of the same polished spherical surface for both mirrors. The optical design in Table 6.9, which results from cross optimizations between optics and elasticity, allows one to obtain this common sphere for a Schott Zerodur substrate ($E = 90.2\,\mathrm{GPa}$, $\nu = 0.243$).

First, the elasticity design of $\mathrm{M_1}$ is determined as a full aperture vase shell mirror. The analytical shell theory in Sect. 6.3.4 with N=10 shell elements, including a simply supported movable base of the outer cylinder N+1, allows the iterative determination of the normal thickness distribution $\{t_n\}$ of the vase shell with successive ring-shell elements of r_N/N radial width (Table 6.10).
Associated to this $\{t_n\}$ distribution is the active optics co-addition law of the surfaces which writes, in [mm],

$$
\begin{aligned}
z_{1,\mathrm{Sphe}} &= 0.223988\ 10^{-3}\,r^2 + 0.11238\ 10^{-10}\,r^4 + 0.1127\ 10^{-17}\,r^6 + 0.141\ 10^{-24}\,r^8\\
z_{1,\mathrm{Flex}} &= 0.002461\ 10^{-3}\,r^2 - 0.17591\ 10^{-10}\,r^4 - 0.1649\ 10^{-17}\,r^6 - 0.141\ 10^{-24}\,r^8\\
z_{1,\mathrm{Sum}} &= 0.226449\ 10^{-3}\,r^2 - 0.63530\ 10^{-11}\,r^4 - 0.5217\ 10^{-18}\,r^6 + 0.123\ 10^{-50}\,r^8\\
z_{1,\mathrm{Opt}} &= 0.226449\ 10^{-3}\,r^2 - 0.63530\ 10^{-11}\,r^4 - 0.5217\ 10^{-18}\,r^6 + 0.000\ 10^{+00}\,r^8
\end{aligned}
\tag{6.77}
$$

from where the dimensionless radius $\rho = r/r_N$ for which $dz_{\mathrm{Flex}}/dr = 0$ is $\rho_0 = 1.202$. The total flexure sag in the range $\rho \in [\,0;\,1\,]$ is $\Delta z_{1,\mathrm{Flex}} = 77.73\,\mu\mathrm{m}$.

In a second stage, we similarly calculate for $\mathrm{M_3}$ an isolated built-in vase shell aspherized by the same load q. Because the conic constants are in the ratio $\kappa_3/\kappa_1 = 4.7994$, this shows that the mean value of $\{t_n\}$ for $\mathrm{M_3}$ is smaller than that of $\mathrm{M_1}$. Since the $\mathrm{M_3}$ mirror is designed with clear aperture radius $r_{3,\max} = 90\,\mathrm{mm}$, thus at f/6.1, the thin plate theory applies. Hence, the mean thickness ratio is $<\{t_n\}_{\mathrm{M3}}/\{t_n\}_{\mathrm{M1}}> = (\kappa_1/\kappa_3)^{1/3} = 0.5928$, which determines the mean thickness of $\mathrm{M_3}$ in accordance to the result from the shell analysis.

Now we depart from an $\mathrm{M_3}$ vase shell perfectly built-in by introducing an intermediate ring of thickness t_{ir} which links $\mathrm{M_3}$ to $\mathrm{M_1}$ and, thus, realizes a double vase shell. This ring extends in the narrow region from $r_{3,\max}$ to $r_{1,\min}$ i.e. 90–110 mm. Its constant thickness to the normal must be determined such as the above flexural sag of $\mathrm{M_1}$ remains unchanged in this transformation; this value was set to $t_{\mathrm{ir}} = 30.220\,\mathrm{mm}$. The transformation of a built-in condition into a continuity condition also implies a rotation of the $\mathrm{M_3}$ shell at $r = r_{3,\max}$. Hence, the co-addition law for this mirror must include a rotation component z_{Rota} which can

Table 6.10 Normal thickness distribution $\{t_n\}$ for the hyperbolization of MINITRUST vase shell $\mathrm{M_1}$ mirror f/2.5. Mirror clear aperture 440 mm. Substrate ext diam 480 mm. Load q=−80 kPa. r_N=220 mm, t_x=30 mm. R_{Opt}=2,208 mm, R_{Sphe}=2,232.2 mm, R_{Flex}=203,140 mm, $<R>$=2,318 mm. [Units: mm]

t_1	t_3	t_4	t_5	t_6	t_7	t_8	t_9	t_N	$t_{z,N+1}$
20.163	20.205	20.248	20.307	20.382	20.470	20.570	20.678	20.785	60.000

be accurately assumed as that of M_1 at $r_{3,\max}$. Therefore, the optical surface results from the sum of the following three terms

$$z_{3,\mathrm{Opt}} = z_{3,\mathrm{Sphe}} + z_{3,\mathrm{Flex}} + z_{3,\mathrm{Rota}} \,. \tag{6.78}$$

We know that a bending moment applied at the perimeter of a constant thickness plate (cf. Chap. 2) generates a purely quadratic flexure. Since we will see that the variation of the $\{t_n\}$ distribution for M_3 is less than 1%, the flexural rotation caused by its link to M_1 can be accurately taken into account by the single term representation

$$z_{3,\mathrm{Rota}} = \frac{1}{2r_{3,\max}} \left(\frac{dz_{1,\mathrm{Flex}}}{dr}\right)_{r_{3,\max}} r^2 \,. \tag{6.79}$$

With $N = 10$ successive shell rings all included inside the continuous link at $r_N = r_{3,\max}$, the analytical determination of the normal thickness distribution $\{t_n\}$ of the M_3 meniscus made of r_N/N radial width elements leads to a slightly increasing distribution from center to edge (Table 6.11).
From (6.78) and associated to this $\{t_n\}$ distribution, the co-addition law of the surfaces writes, in [mm],

$$\begin{aligned}
z_{3,\mathrm{Sphe}} &= 0.223988\ 10^{-3}\,r^2 + 0.11238\ 10^{-10}\,r^4 + 0.0113\ 10^{-16}\,r^6 + 0.141\ 10^{-24}\,r^8\\
z_{3,\mathrm{Flex}} &= 0.001396\ 10^{-3}\,r^2 - 0.86954\ 10^{-10}\,r^4 + 0.1784\ 10^{-16}\,r^6 - 0.141\ 10^{-24}\,r^8\\
z_{3,\mathrm{Rota}} &= 0.002178\ 10^{-3}\,r^2\\
z_{3,\mathrm{Sum}} &= 0.227562\ 10^{-3}\,r^2 - 0.75716\ 10^{-10}\,r^4 + 0.1897\ 10^{-16}\,r^6 + 0.106\ 10^{-48}\,r^8\\
z_{3,\mathrm{Opt}} &= 0.227562\ 10^{-3}\,r^2 - 0.75716\ 10^{-10}\,r^4 + 0.1897\ 10^{-16}\,r^6 + 0.000\ 10^{+00}\,r^8
\end{aligned} \tag{6.80}$$

where the total radius of curvature of the flexure R_{Flex} is deduced from the A_2 terms of the sum $z_{\mathrm{Flex}} + z_{\mathrm{Rota}}$ of a perfectly built-in edge plus the tangential edge rotation of the intermediate ring assumed equal to that of the M_1 mirror at $r_N = r_{3,\max}$. The final design of the double vase shell is derived from each thickness distribution in the two latter tables (Table 6.12 and Fig. 6.20-Left).

The spherical figuring without stress was carried out at the same radius of curvature for both mirrors by means of a single full size tool. The finite element analysis of the shell when stressed at $q = -0.80 \times 10^5$ Pa gives maximal stresses

Table 6.11 Normal thickness distribution $\{t_n\}$ for the hyperbolization of MINITRUST vase shell M_3 mirror f/6.1. Mirror clear aperture 180 mm. Strictly built-in at the contour. Load q=−80 kPa. r_N = 90 mm. R_{Opt} = 2,197.2 mm, R_{Sphe} = 2,232.2 mm, R_{Flex} = 139,900 mm, $<R>$=2,267 mm. [Units: mm]

t_1	t_3	t_4	t_5	t_6	t_7	t_8	t_9	t_N	$t_{z,N+1}$
12.032	12.035	12.037	12.040	12.045	12.050	12.056	12.062	12.069	30.220

Table 6.12 Geometry of the double vase shell of the M_1–M_3 mirrors. Normal thickness distribution $\{t_n\}$. Axial thickness distribution $\{t_z\}$. [Units: mm]

r	0	22.5	40.5	58.5	67.5	76.5	85.5	90.0^-	90.0^+
t_n	12.032	12.035	12.040	12.050	12.056	12.062	12.069	12.072	30.220
t_z	12.032	12.035	12.042	12.054	12.061	12.069	12.078	12.082	30.245
r	110^-	110^+	121	165	187	209	220^-	226^+	240^-
t_n	30.220	20.344	20.382	20.570	20.678	20.785	20.839	–	–
t_z	30.257	12.369	20.409	20.622	20.745	20.870	20.933	68.00	68.000

$\sigma_{\max} = \pm 102 \times 10^5$ Pa at the substrate surfaces near the junction with the outer cylinder.

The Fizeau He-Ne interferometric tests of the mirrors under stress showed accurate hyperboloid figures. For each mirror M_1 and M_2, the autocollimation tests with respect to a sphere were carried out at each Kerber zone, i.e. at $\sqrt{3}/2$ of the clear aperture radii $r_{1,\max}$ and $r_{3,\max}$ respectively; hence the source is moved ~13.3 mm towards the shell to pass from the M_1- to the M_3-interferogram (Fig. 6.20-Right).

• **Tulip-form secondary mirror – Stress figuring**: We have seen, in Sect. 6.6.5, that the active optics parabolization and hyperbolization of a convex mirror can be readily realized by use of a vase shell and stress figuring under uniform load. However, in the case of a convex mirror with a noticeable central hole – as for the present design – another alternative with a uniform load consists of using a variable thickness distribution (VTD) surrounding a rigid ring. Because this VTD is infinite

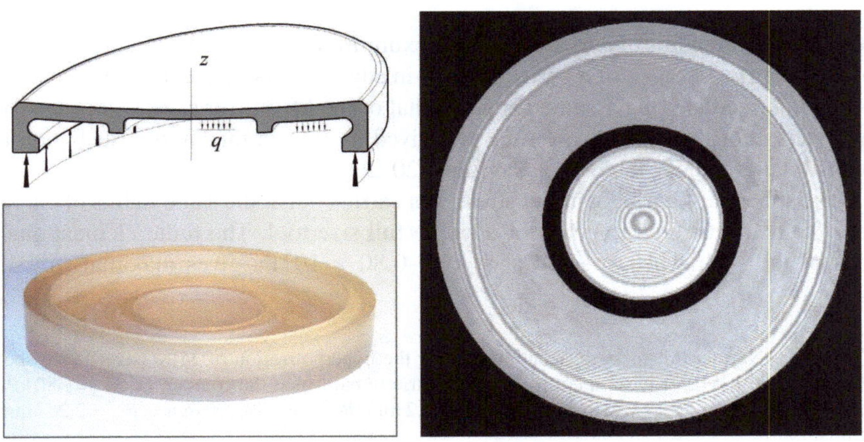

Fig. 6.20 *Left*: Elasticity design and rear view of the double vase shell as MINITRUST M_1–M_3 mirror substrate. *Right*: Composite Fizeau He-Ne interferograms with respect to sphere references after spherical polishing without stress and in situ stressing (LOOM)

at the center and becomes zero at the mirror edge, we have called it a *"tulip form"* (cf. Sect. 2.1.2).

A *holed tulip form* secondary mirror was investigated and developed for the hyperbolization of the MINITRUST secondary mirror. The mirror clear aperture area, defined by the radii $r_{\min}$ and $r_{\max}$, is built-in at $r_{\min}$ into a thick ring whilst the free outer edge of radius r_{ext} is slightly larger than $r_{\max}$. A waterproof paste at the free external edge allows application of the bending load q by partial vacuum on the rear area of the VTD (see Fig. 3.11 in Chap. 3). The hyperbolization of the M_2 mirror is achieved by elastic relaxation after spherical and stress figuring where the axial reaction is absorbed by the thick ring at $r \simeq r_{\min}$. The radial shearing force corresponding to the uniform load q is

$$Q_r = \frac{q}{2}\left(1 - \frac{r^2}{r_{\text{ext}}^2}\right) r \,. \tag{6.81}$$

Assuming that the inner thick ring is strictly undeformable, the built-in condition $dz_{\text{Flex}}/dr = 0$ at the inner clear aperture $r_{\min}$ entails, from the co-addition law $z_{\text{Opt}} = z_{\text{Sphe}} + z_{\text{Flex}}$, that the slopes for the figuring sphere z_{Sphe} and optical surface z_{Opt} are identical at this radius,

$$dz_{\text{Sphe}}/dr\,|_{r\,\min} = dz_{\text{Opt}}/dr\,|_{r\,\min} \,. \tag{6.82}$$

This fully determines the radius of curvature R_{Sphe} of the figuring tool and the elastic deformation to generate z_{Flex}.

Since the mirror is convex and the VTD decreases from center to edge, the middle surface of the substrate can be accurately assumed as slightly departing from a plane. Hence the elasticity theory of thin plates applies to the determination of the M_2 VTD. The unknown rigidity $D(r) = Et^3(r)/[12(1-\nu^2)]$ is determined by integration of the derivative equation

$$D\frac{d}{dr}\left(\frac{d^2z}{dr^2} + \frac{1}{r}\frac{dz}{dr}\right) + \frac{dD}{dr}\left(\frac{d^2z}{dr^2} + \frac{\nu}{r}\frac{dz}{dr}\right) = Q_r \,, \tag{6.83}$$

where $z \equiv z_{\text{Flex}}$. Using a Schott Zerodur vitroceram (see E, ν in Table 1.10) and a load $q = -0.8 \times 10^5$ Pa, the numerical integration is carried out from $r_{\min}$ toward increasing radii with a small increment dr, starting with a provisional thickness value $t(r_{\min})$. The starting thickness is modified and the integration process repeated up to obtaining $t(r_{\max}) = 0$. The final VTD is determined by use of a small enough increment dr providing non-significant change in the integration result (Table 6.13).

From the result of iterations with the thin plate theory, the sphere and the flexure write, in [mm],

Table 6.13 Axial thickness distribution $\{t_n\}$ for the hyperbolization of MINITRUST tulip form M_2 mirror f/5.4. Mirror clear aperture radii $r_{min} = 50$ mm and $r_{max} = 100$ mm. $r_{ext} = 103$ mm. Load $q = -80$ kPa. $R_{Opt} = 1{,}091.5$ mm, $R_{Sphe} = 1{,}096.0$ mm, $R_{Flex} = 266{,}812$ mm, $\langle R \rangle \simeq \infty$. [Units: mm]

r	30^+	50^-	50^+	60	70	80	90	95	100	103*
t_z	32.00	32.27	14.300	9.962	7.083	4.878	2.988	2.062	1.042	0.350

*Avoiding zero for a practical realization, this thickness was actually set from the tangent at $t(r\ \max)$.

$$\begin{aligned}
z_{Sphe} &= 0.456211\ 10^{-3}\ r^2 + 0.09495\ 10^{-9}\ r^4 + 0.0395\ 10^{-15}\ r^6 + 0.205\ 10^{-22}\ r^8 \\
z_{Flex} &= 0.001874\ 10^{-3}\ r^2 - 0.37538\ 10^{-9}\ r^4 + 0.1555\ 10^{-15}\ r^6 - 0.205\ 10^{-22}\ r^8 \\
z_{Sum} &= 0.458085\ 10^{-3}\ r^2 - 0.28043\ 10^{-9}\ r^4 + 0.1950\ 10^{-15}\ r^6 + 0.000\ 10^{-22}\ r^8 \\
z_{Opt} &= 0.458085\ 10^{-3}\ r^2 - 0.28043\ 10^{-9}\ r^4 + 0.1950\ 10^{-15}\ r^6 + 0.000\ 10^{+00}\ r^8
\end{aligned} \tag{6.84}$$

from where we satisfy the active optics co-addition law $z_{Sphe} + z_{Flex} = z_{Opt}$. During stress figuring the maximum radial stresses arise at $r = 70$ mm with values $\sigma_r = \pm 64 \times 10^5$ Pa at the substrate surfaces. Three substrate samples were built whose rear profiles were obtained by computer control diamond turning (Fig. 6.21).

The tulip-form design of M_2 provides a very lightweight mirror. Although more difficult to elaborate accurately, this design avoids the outer cylinder weight of a vase-form geometry. The quasi-conical shape in the perimeter region shows high resistance to vibrations. *Holed tulip-form mirrors* supported in their inner region are interesting lightweight configurations for space telescopes [21].

• **Telescope optical tests in the laboratory**: Two samples of the telescope optics were built for MINITRUST-1 and -2. The optics schematic including the shape of the two substrates for the three mirrors, and the baffles, is shown by Fig. 6.22.

The tube of the first prototype telescope was designed as a classical Serrurier truss with a central square frame which allows maintaining the mirrors coaxially whatever the gravity orientation. Since the M_1 and M_3 mirrors are *de facto* perfectly coaxial as belonging to the double shell substrate, the only necessary alignments

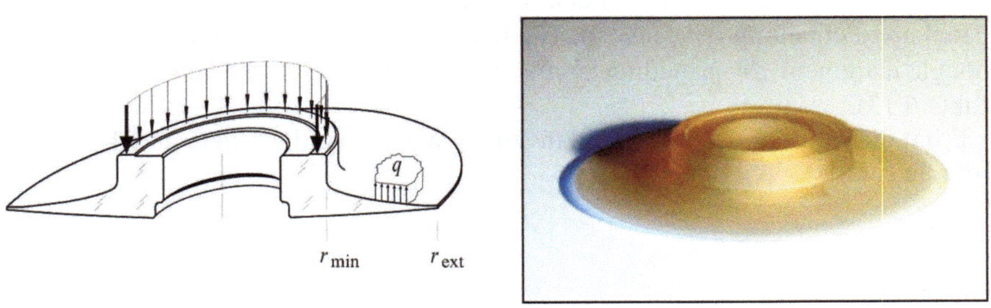

Fig. 6.21 Elasticity design and rear view of the tulip-form M_2 mirror (LOOM)

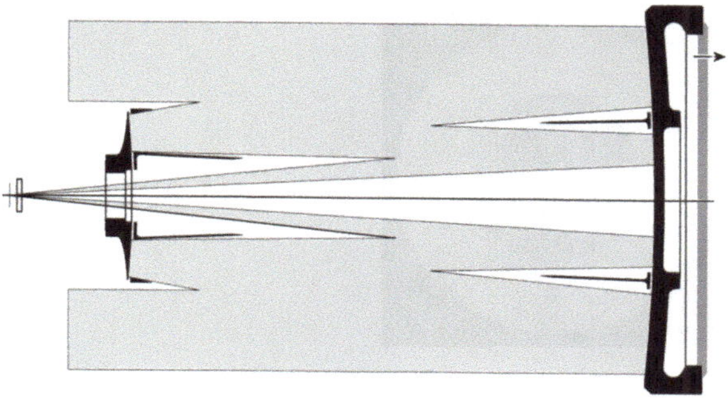

Fig. 6.22 MINITRUST on-axis beams, mirror substrate geometries, and baffles. The entrance pupil is on the secondary mirror

were (i) the orientation set up of M_1–M_3 with the telescope tube axis, and (ii) the centering and orientation set up of M_2 with M_1–M_3. These alignments were carried out by cross-wire reticles and retroreflection of a He-Ne laser beam at the M_3 vertex.

The telescope optical tests were realized on-axis by autocollimation with a Fizeau interferometer – MINIFIZ from Phase Shift Technology – imaging a point source at the telescope focus. After passing a first time in the telescope, the output beam was reflected by a plane mirror and was passed a second time in the telescope (Fig. 6.23).

From data reductions of the wavefronts issued from a double pass through the telescope, the optimal *in situ* load for M_1–M_3 was $q = 0.794 \times 10^5$ Pa; the theoretical value was 0.8×10^5 Pa. After a preliminary alignment sequence, the first double-pass interferogram displayed a dominating *Coma* 3 due to a residual decentering error of M_2. In the final alignment phase, this aberration was reduced to a negligible value (Fig. 6.24).

The results from final alignments and data reductions gave the following peak-to-valley (ptv) main residuals onto a He-Ne wavefront coming from infinity, i.e. in a single-pass or direct star wavefront,

Fig. 6.23 View of the modified-Rumsey telescope MINITRUST-1 under optical tests (LOOM)

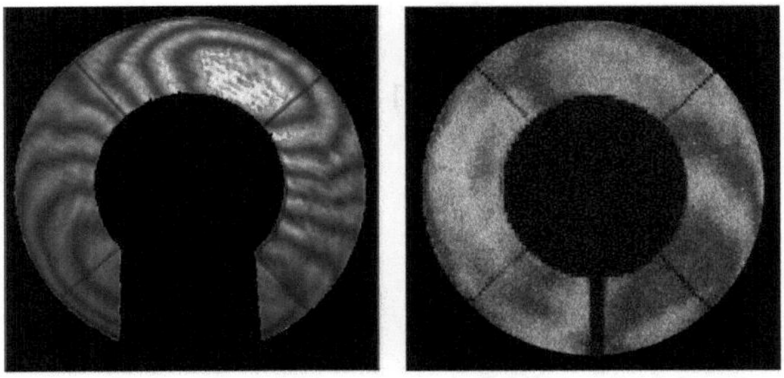

Fig. 6.24 MINITRUST-1 optical tests : He-Ne wavefronts from telescope double-pass (LOOM)

$$Sphe\,3 = 0.032\,\lambda\,, \qquad Coma\,3 = 0.034\,\lambda\,, \qquad Astm\,3 = 0.210\,\lambda\,.$$

The overall sum, including all order aberrations, is $0.280\,\lambda_{\text{He-Ne}}$ ptv corresponding to $0.048\,\lambda_{\text{He-Ne}}$ rms for MINITRUST-1. This telescope presently stays at LOOM. The second prototype telescope used the second mirror set and a mechanical mounting designed and built by IAS-Frascati [28].

The excellent imaging performances obtained with the active optics process of aspherization opens up the way to potential applications in 2–3 m class mirrors, for instance, for large wide-field sky-survey telescopes.

6.6.8 Mirror Aspherizations of a Large Modified-Rumsey Telescope

From the optical and elasticity designs of telescopes and mirror substrates in the previous sections, all the geometries can be scaled up or down, thus straightforwardly obtaining the optical and elasticity design parameters of any larger or smaller instrument. Doing so, all the mirror individual and resulting telescope focal ratios remain unchanged, as are the maximum stresses $\sigma_{r,\,\max} \propto (r^2/t^2)\,q$ provided the load q is unchanged and the elastic constants $(E,\ \nu)$ of the material are the same. However, a significant scale up in the size of a mirror requires introducing a somewhat larger safety factor to avoid a glass rupture from the active optics aspherization.

Passing from a 0.44-m to a 2.2-m clear aperture telescope (scale factor of 5) is done hereafter by decreasing the load to $q = -0.5 \times 10^5$ Pa instead of $q = -0.8 \times 10^5$ Pa in Sect. 6.6.7. Hence, because of the load change, the results in the latter section cannot be scaled up for the *double vase shell* as the M_1–M_3 mirror substrate. The following optical design has a smaller back focal distance. Next, the M_1–M_3 substrate design provides decreased maximal stresses. A 2.2-m modified-Rumsey telescope is proposed hereafter by Lemaitre, Ferrari, Viotti and La Padula as a "three-reflection sky-survey"(TRSS) telescope [7–9, 22].

• **Optical design of a 2.2-m modified-Rumsey telescope – TRSS proposal**: A Three-Reflection Sky Survey (TRSS) ground-based telescope has been proposed for wide spectral range observations extending from the UV band to IR bands. As in the previous section, its design is a modified-Rumsey form. The elasticity continuity conditions for the slope and sag variations in the area of the intermediate ring linking M_1 to M_3 lead to cross optimizations between optical and elasticity designs. Setting the TRSS input *pupil* at M_2 provides an optimal balance of the M_1 and M_3 clear aperture areas (Table 6.14 and Fig. 6.25).

• **Double vase shell primary and tertiary mirrors – In situ stressing**: The elasticity design of a *double vase shell* as the common substrate of M_1 and M_3 mirrors is carried out by similar iteration processes to those in the previous section. This determines the common spherical surface for simultaneously hyperbolizing both mirrors. In order to reduce the stress level for large mirrors, the in situ stressing of a Schott Zerodur substrate is hereafter achieved with the reduced load $q = -0.50 \times 10^5$ Pa ($\simeq -0.5$ atm). Hence, the substrate thickness distributions are thinner than those in Sect. 6.6.7 when scaled up by a factor of five, and the maximum stress level does not exceed $\sigma_{\max} = \pm 8$ MPa at the surface of the double vase shell (Tables 6.15 and 6.16).
The active optics co-addition law associated to the above M_1 normal thicknesses $\{t_n\}$ for a simply supported movable edge writes, in [mm],

$$\begin{aligned}
z_{1,\mathrm{Sphe}} &= 0.464455\ 10^{-4}\ r^2 + 0.100192\ 10^{-12}\ r^4 + 0.4322\ 10^{-21}\ r^6 - 0.141\ 10^{-29}\ r^8 \\
z_{1,\mathrm{Flex}} &= 0.006355\ 10^{-4}\ r^2 - 0.149236\ 10^{-12}\ r^4 - 0.6528\ 10^{-21}\ r^6 - 0.141\ 10^{-29}\ r^8 \\
z_{1,\mathrm{Opt}} &= 0.470810\ 10^{-4}\ r^2 - 0.049044\ 10^{-12}\ r^4 - 0.2206\ 10^{-21}\ r^6 + 0.000\ 10^{+00}\ r^8
\end{aligned} \tag{6.85}$$

For mirror M_3, the total radius of curvature R_{Flex} of the flexure is deduced from the A_2 terms of the sum $z_{\mathrm{Flex}} + z_{\mathrm{Rota}}$ of a perfectly built-in edge plus the tangential edge rotation of the intermediate ring assumed equal to that of mirror M_1 at $r = r_{3,\max}$. We have seen that this latter flexure can be assimilated as a single term curvature mode [cf. (7.79)]. Hence, the co-addition law for the optical surface associated to the above thickness distribution $\{t_n\}$ is the sum of three series which write, in [mm],

Table 6.14 Modified-Rumsey telescope design – TRSS Proposal f/5–2° diagonal FOV – λλ [380–1,200 nm] – Efl=11 m. [Units: mm]

i	Surf.	R_i	AS_i	$A_{4,i}$	$A_{6,i}$	Aperture	$[\kappa_i]^\star$
1	M_1	−10, 620.0	−3, 125.000	4.90447 10^{-14}	2.206 10^{-22}	2,200–920	[−1.470]
Sto	M_2	−5, 256.6	−3, 125.021	2.71885 10^{-12}	−8.686 10^{-20}	830–400	[−4.159]
3	M_3	−10, 560.0	−3, 403.070	8.52957 10^{-13}	−1.065 10^{-20}	d_{ext} 780	[−9.035]
4	Fused	∞	−10.000			277×277	
5	silica	∞	−25.000			275×275	
6	Focus	∞				270×270	

Equation of mirrors : $z_i = (1/2R_i)r^2 + A_{2,i}\, r^4 + A_{4,i}\, r^6$. Axial separations : AS_i.
* Conic constant from third order : $\kappa_i = 8R_i^3 A_{4,i} - 1$ and $\kappa = -1$ for a paraboloid.

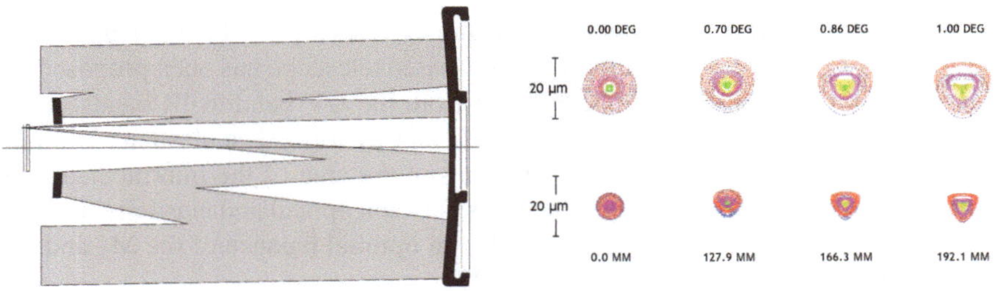

Fig. 6.25 *Left*: TRSS telescope optical design from Table 6.14. Entrance pupil at M_2. *Right*: Residual spherochromatism blur images for 20 and 10-mm thickness plates as equivalent thickness of the filter and cryostat window. The corresponding blur rms diameters are 0.20 and 0.10 arcsec, respectively

$$\begin{aligned}
z_{3,\text{Sphe}} &= 0.464455\ 10^{-4}\ r^2 + 0.100192\ 10^{-12}\ r^4 + 0.0043\ 10^{-19}\ r^6 + 0.233\ 10^{-29}\ r^8\\
z_{3,\text{Flex}} &= 0.003016\ 10^{-4}\ r^2 - 0.953149\ 10^{-12}\ r^4 + 0.1022\ 10^{-19}\ r^6 - 0.233\ 10^{-29}\ r^8\\
z_{3,\text{Rota}} &= 0.006014\ 10^{-4}\ r^2\\
z_{3,\text{Opt}} &= 0.473485\ 10^{-4}\ r^2 - 0.852957\ 10^{-12}\ r^4 + 0.1065\ 10^{-19}\ r^6 + 0.000\ 10^{+00}\ r^8
\end{aligned} \tag{6.86}$$

The final geometry of the double vase shell is derived from the thickness distributions in the two latter tables. The normal thickness of the M_1–M_3 intermediate ring – which links the mirrors in the region $r_{3,\max} \leq r \leq r_{1,\min}$ – must be set such as the flexural sag of M_1 at $r = r_{3,\max}$ remains unchanged when passing from the single vase to the double vase (Table 6.17 and Fig. 6.26).

• **Case of a wide-field space telescope – Stress figuring**: As stated in Sect. 6.6.1 from the active optics co-addition law, the above double vase shell geometry for obtaining M_1 and M_3 hyperboloid mirrors is valid whatever the asphérization process, either in situ stressing or stress figuring.

For a wide-field space telescope, the conditions in space entail that the aspherization of M_1 and M_3 must be carried out by elastic relaxation after spherical figuring

Table 6.15 Normal thickness distribution $\{t_n\}$ for the hyperbolization of TRSS telescope vase shell M_1 mirror. Mirror clear aperture 2,200 mm. Substrate d_{ext} =2,320 mm. Load q =−50 kPa. r_N =1,100 mm, t_x =70 mm. R_{Opt} =10,620 mm, R_{Sphe} =10,765.3 mm, R_{Flex} =786,760 mm, $<R>$=11,362. [Units: mm]

t_1	t_3	t_4	t_5	t_6	t_7	t_8	t_9	t_N	$t_{z,N+1}$
82.125	82.458	82.810	83.286	83.890	84.615	85.442	86.337	87.236	200.0

Table 6.16 Normal thickness distribution $\{t_n\}$ for the hyperbolization of TRSS telescope vase shell M_3 mirror. Mirror clear aperture 780 mm. Strictly built-in at the contour. Load q =−50 kPa. r_N = $r_{3,\max}$ =390 mm. R_{Opt} =10,560 mm, R_{Sphe} =10,765.3 mm, R_{Flex} =553,700 mm, $<R>$= 10,913. [Units: mm]

t_1	t_3	t_4	t_5	t_6	t_7	t_8	t_9	t_N	$t_{z,N+1}$
46.701	46.710	46.720	46.733	46.750	46.770	46.793	46.818	46.845	∞

Table 6.17 Geometry of the double vase shell of the M_1–M_3 mirrors. Normal thickness distribution $\{t_n\}$. Axial thickness distribution $\{t_z\}$. [Units: mm]

r	0	91.25	164.25	237.25	273.75	310.25	346.75	390.0^-	390.0^+
t_n	46.701	46.710	46.733	46.770	46.793	46.818	46.845	46.858	126.40
t_z	46.701	46.712	46.739	46.782	46.810	46.840	46.872	46.887	126.48
r	460^-	460^+	605	825	935	1,045	$1,100^-$	$1,100^+$	$1,160^-$
t_n	126.40	82.905	83.890	85.442	86.337	87.236	87.685	205.0	205.0
t_z	126.50	82.972	84.009	85.668	86.631	87.607	88.093	205.9	206.0

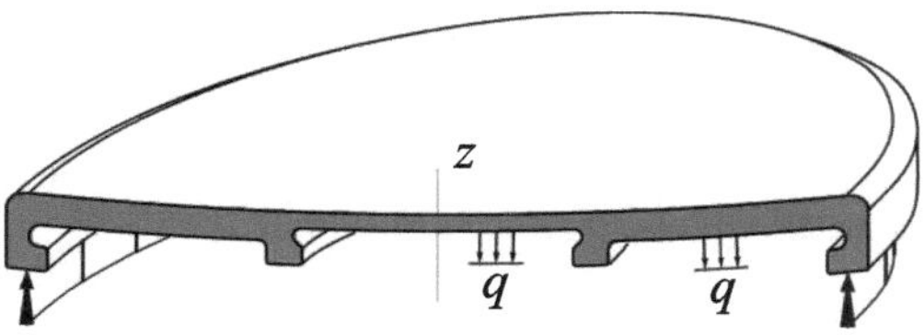

Fig. 6.26 Elasticity design from Table 6.17 of a double vase shell as M_1–M_3 mirror substrate of a 2.2-m TRSS telescope. The intermediate ring – not usable optically – is designed as a narrow L-shaped section. After spherical figuring of a *single* surface without stress, both hyperbolizations are achieved by in situ stressing

under stress. This can be achieved by air or liquid pressure inside the double substrate and perimeter axial reaction. A sufficiently large number of perimeter reaction forces must be distributed all along the shell outer cylinder. This requires some local modifications of this outer cylinder surface in order for these forces to act on a large enough area. Another similar alternative consists of the use of a closed shell – made of two opposite sealed vase shells (cf. Sect. 6.4.1) – and a spherical stress figuring with inner pressure.

References

1. M. Abramovitz, I.A. Stegun, *Handbook of Mathematical Functions*, Dover Publ., Issue 10, New York, 374–384 and 430–431 (2003)
2. A. Baranne, G.R. Lemaitre, Combinaisons optiques pour très grands télescopes spécialisés, C.R. Acad. Sc. Paris, **291** Série B, 39–41 (1980)
3. A. Baranne, G.R. Lemaitre, Combinaisons optiques pour très grands télescopes spécialisés: le concept TEMOS, C.R. Acad, Sc. Paris, **305** Série II, 445–450 (1987)
4. A. Baranne, G.R. Lemaitre, TEMOS, Opt. Acta, **29** 847–860 (1982)
5. H.B. Dwight, *Table of Integrals and Other Mathematical Data*, Macmillan Co. edit., 3rd edition, New York, 184–188 and 276–279 (1957)
6. M. Ferrari, G.R. Lemaitre, Active optics methods for highly aspheric mirrors, in *Optical Design, Materials, Fabrication, and Maintenance*, SPIE Proc., **4003**, 34–42 (2000)
7. M. Ferrari, G.R. Lemaitre, R.F. Viotti, C. La Padula, G. Comte, M. Blanc, M. Boer, Three reflection telescope proposal at Dome C, in *Astronomie et Astrophysique au Dome C*, EAS Pulications Series, EDP Sciences, **14**, 325–330 (2005)

8. C.D. La Padula, A.Carusi, R.F. Viotti, A.Vignato, G.R. Lemaitre, Proposal for a mini-satellite with a wide-field TRT, Mem. Soc. Astron. Ital., **74**, 63–67 (2003)
9. G.R. Lemaitre, A 2.2 m three reflection telescope – Optics and elasticity design study, LOOM internal report (2006)
10. G.R. Lemaitre, A. Baranne, The LOOM project III. A variable asphericity secondary mirror, Workshop on Large Telescopes, K.J. Fricke edt., *Mitteilungen der Astronmischen Gesellschaft*, **67**, 236–242 (1986)
11. G.R. Lemaitre, Actively aspherized and active pupil mirrors for ELTs, in *Extremely Large Telescopes*, ESO Conf. and Proc., **57**, 121–128 (1999)
12. G.R. Lemaitre, Asphérisation par relaxation élastique des miroirs astronomiques dont le contour circulaire ou elliptique est encastré ou semi-encastré, C. R. Acad. Sc. Paris, **290** Série B, 171–174 (1980)
13. G.R. Lemaitre, French patent No 2472198 (1981), US Patent No 4382657 (1983)
14. G.R. Lemaitre, M. Wang, Optical results with TEMOS 4: a 1.4 meter telescope with a primary mirror of spherical segments and a secondary mirror actively aspherized, in *Metal Mirrors Conference*, SPIE Proc., **1931**, 43–52 (1992)
15. G.R. Lemaitre, Modified Rumsey telescope and elasticity design for active aspherization, in *Optical Telescope of Today and Tomorrow*, SPIE Proc., **2871**, 326–340 (1996)
16. G.R. Lemaitre, Modified-Rumsey telescope and associated elasticity for active aspherization, in *Optical Telescopes of Today and Tomorrow*, SPIE Proc., **2871**, 326–340 (1996)
17. G.R. Lemaitre, Optical figuring with elastic relaxation methods, in *Current Trends in Optics*, Taylor & Francis Publ., London, 135–149 (1981)
18. G.R. Lemaitre, P. Montiel, P. Joulié, K. Dohlen, P. Lanzoni, Active optics and the axisymmetric case: MINITRUST wide-field three-reflection telescopes with mirrors aspherized from tulip and vase form, SPIE Proc., **5494**, 426–433 (2004)
19. G.R. Lemaitre, P. Montiel, P. Joulié, K. Dohlen, P. Lanzoni, Active optics and modified-Rumsey wide-field telescopes: MINITRUST demonstrators with vase- and tulip-form mirrors, Appl. Opt., **44**(34), 7322–7332 (2005)
20. G.R. Lemaitre, R.N. Wilson, S. Mazzanti, Proposal for a 1.8 m metal meniscus mirror aspherized at f/1.8 by active optics, in *Metal Mirror Conference*, University College London, SPIE Proc., **1931**, 67–75 (1992)
21. G.R. Lemaitre, Sur la flexion des miroirs secondaires des télescopes, Nouv. Rev. Opt., **74**, 389–397 (1976)
22. G.R. Lemaitre, TRSS: A three reflection sky survey at Dome C with an active optics modified-Rumsey telescope, in *Wide Field Survey telescopes on Dome C and A*, Beijing conf. proc. of the Chinese Acad. Sc., 62–71 (2005)
23. MEGAPRIME-MEGACAM. http://cfht.hawaii.edu/Instruments/Imaging/MegaPrime
24. G. Molodij, J. Rayrole, P.Y. Madec, F. Colson, THEMIS: Télescope héliographique pour l'étude du magnétisme et des instabilités solaires, Astron. Astrophys. Suppl. Ser., **118**, 169–179 (1996), and http://webast.ast.obs-mip.fr/people/paletou/Themis
25. E. Reissner, Stresses and small displacements of shallow spherical shells – I, J. Math. Phys., **25**, 80–85 (1946)
26. E. Reissner, Stresses and small displacements of shallow spherical shells – II, J. Math. Phys., **25**, 279–300 (1947)
27. N.J. Rumsey, A compact three-reflection camera, in *Optical Instruments and Techniques*, ICO 8 Symposium London, ICO 8 Proc., H. Dickson edt., Oriel publ. – Newcastle, 514–520 (1969)
28. SDSS: A.P. Sloan Foundation, NSF and joint institutions. http://www.sdss.org
29. S.P. Timoshenko, S. Woinowsky-Krieger, *Theory of Plates and Shells*, McGraw Hill edt., New York, a: 558–560, b: 466–468 (1959)
30. R.V. Willstrop, The Mersenne-Schmidt: a three reflection survey telescope, Mont. Not. Roy. Astron. Soc., **210**, 597–609 (1984)

Chapter 7
Active Optics with Multimode Deformable Mirrors (MDM) Vase and Meniscus Forms

7.1 Introduction – Clebsch-Seidel Deformation Modes

This chapter treats the general case of obtaining aspherical mirrors such as off-axis mirrors locally defined from an axisymmetric shape, or on-axis mirrors for aberration correction of a non-centered system.

This requires the simultaneous correction of several wavefront modes such as 1st-order curvature, 3rd-order coma, astigmatism and spherical aberration, 5th-order triangle, etc. Although a mirror belonging to the *variable thickness distribution* (VTD) class may be bent within a diffraction limited tolerance for generating some superposed modes – such as, for instance, the $Cv\,1$ and $Astm\,3$ modes with the cycloid-type VTD $\mathcal{T}_{20} \equiv \mathcal{T}_{22} = (1-\rho^2)^{1/3}$ (see Sects. 2.1.2 and 3.5.2) –, in general, it is clear that a mirror belonging to the *constant thickness distribution* (CTD) class provides a higher superposition capability than a mirror of the VTD class.

From the analysis of an optical system showing aberrations, the optical wavefront function or the normalized Zernike form allows determining the various aberration modes to be corrected by a mirror. On the other hand, the elastic deformation modes of a CTD plate are represented by Clebsch's modes. These latter modes are perfectly similar to Seidel's optics modes, thus in the Fourier form $\Sigma\, A_{nm} r^n \cos m\theta$.

For some high-order Clebsch's modes, generating the elastic deformation implies that all the mirror clear aperture must be submitted to a *non-uniform* loading as a prismatic load, quadratic load, etc. Such loads are extremely difficult to generate for practical applications. However, restraining to the two cases where *no* load and *uniform* load are applied to the mirror clear aperture, we will see hereafter that a large variety of Clebsch's modes are easily practicable. The elastic modes generated from loads $q = 0$ or $q = constant$ belong to a subclass of the optics triangle matrix modes; we propose to call them *Clebsch-Seidel modes.*

Multimode deformable mirrors (hereafter MDMs) have been developed for the purpose of generating and superposing Clebsch-Seidel modes.

G.R. Lemaitre, *Astronomical Optics and Elasticity Theory,* Astronomy and Astrophysics Library, DOI 10.1007/978-3-540-68905-8_7,

7.2 Elasticity and Vase-Form MDMs

Given a continuous deformation mode, the elasticity theory of thin plates applied to a constant thickness disk – flat or with a small curvature – allows deriving the forces and moments acting on it. Some of the optical aberration modes in the form $z = A_{nm} r^n \cos m\theta$ can be easily obtained by a convenient distribution of axial forces and bending moments acting along the disk perimeter. Generating a radial bending moment distribution requires use of two axial force sets. When discrete forces are applied close to the mirror edge – and equally distributed in azimuth angles –, the shear components of the elastic deformation due to the point forces provide slope discontinuities at these regions. In order to generate a *smooth* and *continuous* surface, it is much preferable to avoid point forces applied near the clear aperture edge of the mirror. Hence, the best continuity is obtained with point forces applied at *maximal* distances from the optical surface.

Following Saint Venant's principle [26, 60] (see Sect. 1.13.12), leads to a *vase-form* design, i.e. a mirror having two concentric zones of *constant* rigidity as proposed by Lemaitre [41, 44]. Similar *drum-shaped mirrors* were suggested in the past by Couder [12] for obtaining low-weight axisymmetric mirrors. In a vase form, the outer zone is thicker than that of clear aperture which is the inner zone. The two zones are clamped together in a single – or *holosteric* – piece. Whatever the symmetry of the modes to generate, the radial bending moments to distribute along the outer ring may reach important values. Therefore, the vase form is completed with external radial arms built-in to the ring on which axial forces are applied to the outer ends. Apart from minimizing the shear flexure effect to negligible value, another advantage of the ring is to provide a *smooth* angular modulation of the perimeter flexure from the discrete forces. The number of radial arms is strongly dependent on the highest order of the flexural mode $z(r, \theta)$ to be generated and must also be optimized for all the required modes. A design with a vase form and radial arms is called *multimode deformable mirror* (Fig. 7.1).

Let us consider a plane MDM with a clear aperture zone defined by $0 \leq r < a$, a built-in ring zone defined by $a < r \leq b$, where t_1, t_2 and D_1, D_2 are the thicknesses and associated rigidities of the inner and outer zones, respectively. The axial forces applied to the ring inner radius, $r = a$, are denoted $F_{a,k}$; those applied to the arm outer-end, at $r = c$, are $F_{c,k}$. With a total number of k_m arms, each arm is numbered by $k \in [1, 2, \ldots, k_m]$ and $k = 1 \Leftrightarrow \theta = 0$. In addition, positive or negative uniform loads q can be superposed into the vase inner zone by mean of air pressure or depressure.

Denoting E and ν the Young modulus and Poisson's ratio respectively, in a cylindrical coordinate system z, r, θ, the flexure surface Z is given by Poisson's equation (7.6)

$$\nabla^2\nabla^2 Z(r, \theta) = q/D, \tag{7.1}$$

where q and D are the load per unit surface area and the flexural rigidity,

$$D = Et^3/[12(1 - \nu^2)] = constant, \tag{7.2a}$$

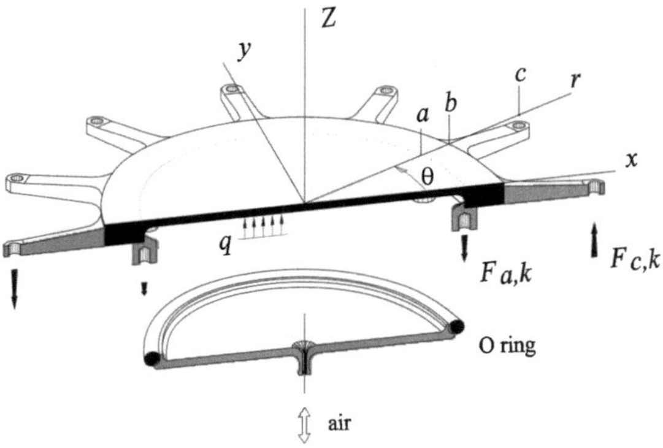

Fig. 7.1 Elasticity design of a *vase-form* MDM based on two concentric rigidities and radial arms. The clear aperture zone is built-in at $r = a$ into a thicker ring. This *holosteric* design allows generating and co-adding the Clebsch-Seidel deformation modes, $Cv\,1$, $Sphe\,3$, $Coma\,3$, $Astm\,3$, etc, by axial forces $F_{a,k}$ and $F_{c,k}$ applied at the ring inner radius $r = a$ and outer end $r = c$ of k_m arms

and $\nabla^2 \cdot = \partial^2 \cdot /\partial r^2 + \partial \cdot /r\partial r + \partial^2 \cdot /r^2\partial\theta^2$ is the Laplacian. Inner and outer constant rigidities are defined by

$$\begin{cases} D = D_1 & \text{if} \quad 0 \leq r \leq a, \\ D = D_2 & \text{if} \quad a \leq r \leq b. \end{cases} \tag{7.2b}$$

• **Inner zone** ($0 \leq r \leq a$): At the inner zone of the vase form, we assume hereafter that a flexural mode to generate is with a polynomial representation having a symmetry with respect to z, x plane ($\theta = 0$), i.e. we only consider $\cos m\theta$ terms; $\sin m\theta$ terms would give a similar representation. Thus, in a cylindrical coordinate system, a representation of the flexure is

$$Z = \sum z_{nm} = \sum A_{nm} r^n \cos m\theta, \quad n + m \text{ even}, \quad m \leq n, \tag{7.3}$$

where n and m are positive integers and A_{nm} coefficients are identical to those belonging to the *optics triangular matrix* (see Sect. 1.8.2) expressing the shape of a wavefront or a mirror, i.e. using the same composition rules for n and m. Given a mode z_{nm}, the substitution in (7.1) leads to

$$A_{nm}(n^2 - m^2)[(n-2)^2 - m^2]r^{n-4} \cos m\theta = q/D, \qquad \text{with } n \geq 2. \tag{7.4a}$$

The only combinations of n and m for which the equation can be solved for practicable applications are:

$$\begin{cases} \text{case } q = 0 \rightarrow m = n, & \text{i.e. } z_{22}, z_{33}, z_{44}, \ldots \text{ terms}, \\ \qquad\qquad\;\; \rightarrow m = n-2, & \text{i.e. } z_{20}, z_{31}, z_{42}, \ldots \text{ terms}, \\ \text{case } q = constant \rightarrow n = 4, & m = 0, \text{ i.e.the } z_{40} \text{ term}. \end{cases} \tag{7.4b}$$

These cases define a *subclass* proposed to be called *Clebsch-Seidel modes*. Except for the z_{40} mode, these modes belong to the two lower diagonals of the optics triangle matrix. The generation of $z_{20} \equiv Cv\,1$, $z_{40} \equiv Sphe\,3$, $z_{31} \equiv Coma\,3$, $z_{22} \equiv Astm\,3$, $z_{42} \equiv Astm\,5$, $z_{33} \equiv Tri\,5$, $z_{53} \equiv Tri\,7$, $z_{44} \equiv Squa\,7, \ldots$ modes is obtained, while it is found not possible to generate the two other 5th-order modes $z_{51} \equiv Coma\,5$ or $z_{60} \equiv Sphe\,5$ by only using $q = 0$ or a uniform loading $q = constant$. Generating z_{51} would require a prismatic loading while for z_{60} a parabolic loading. Because of the extreme difficulties to achieve them in practice, such non-uniform loading distributions will not be considered hereafter.

• **Outer zone** ($a \leq r \leq b$): On the outer zone of the vase mirror, a uniform load is never applied, so that the equation to be solved for $a \leq r \leq b$ is (7.1) with $q = 0$. The solutions are

$$Z = \sum z_{nm} = R_{n0} + \sum_{m=1}^{\infty} R_{nm} \cos m\theta + \sum_{m=1}^{\infty} R'_{nm} \sin m\theta, \tag{7.5}$$

in which $R_{n0}, R_{n1}, \ldots, R'_{n1}, \ldots$ are a function of the radial distance only. In our case, one considers the same azimuth of the deformation as for (7.3), so that the third functions R'_{nm} vanish.

The functions R_{nm} are Clebsch's solutions [11] of

$$\left(\frac{d^2}{dr^2} + \frac{1}{r}\frac{d}{dr} - \frac{m^2}{r^2}\right)\left(\frac{d^2 R_{nm}}{dr^2} + \frac{1}{r}\frac{d R_{nm}}{dr} - \frac{m^2}{r^2} R_{nm}\right) = 0. \tag{7.6}$$

For $m = 0$, $m = 1$ and $m > 1$, the functions R_{nm} – or *Clebsch's polynomials* – have the following forms

$$\begin{aligned} R_{n0} &= B_{n0} + C_{n0} \ln r + D_{n0} r^2 + E_{n0} r^2 \ln r\,, \\ R_{n1} &= B_{n1} r + C_{n1} r^{-1} + D_{n1} r^3 + E_{n1} r \ln r\,, \\ R_{nm} &= B_{nm} r^m + C_{nm} r^{-m} + D_{nm} r^{m+2} + E_{nm} r^{-m+2}. \end{aligned} \tag{7.7}$$

• **Bending moments and shearing Forces:** The boundaries between the two zones at $r = a$ must provide a continuity of the flexure z_{nm}, slope dz_{nm}/dr, bending moment M_r and net shearing force V_r. Denoting $z \equiv z_{nm}$ for a single mode, the bending moments M_r, M_t, and twisting moment M_{rt} are respectively defined by

$$M_r = D\left[\frac{\partial^2 z}{\partial r^2} + \nu\left(\frac{1}{r}\frac{\partial z}{\partial r} + \frac{1}{r^2}\frac{\partial^2 z}{\partial \theta^2}\right)\right], \tag{7.8a}$$

$$M_t = D\left[\frac{1}{r}\frac{\partial z}{\partial r} + \frac{1}{r^2}\frac{\partial^2 z}{\partial \theta^2} + \nu\frac{\partial^2 z}{\partial r^2}\right], \tag{7.8b}$$

$$M_{rt} = (1-\nu)\, D\left[\frac{1}{r^2}\frac{\partial z}{\partial \theta} - \frac{1}{r}\frac{\partial^2 z}{\partial r \partial \theta}\right]. \tag{7.8c}$$

The radial and tangential shearing force Q_r and Q_t derived from the static equilibrium [cf. Sect. 3.2, eqns. (3.3) and (3.4)], are

$$Q_r = -\frac{\partial M_r}{\partial r} - \frac{1}{r}\left(M_r - M_t - \frac{\partial M_{rt}}{\partial \theta}\right) = -D\frac{\partial}{\partial r}(\nabla^2 z), \tag{7.9a}$$

$$Q_t = -\frac{1}{r}\left(\frac{\partial M_t}{\partial \theta} - 2\,M_{rt}\right) + \frac{\partial M_{rt}}{\partial r} = -D\frac{1}{r}\frac{\partial}{\partial \theta}(\nabla^2 z), \tag{7.9b}$$

and their equilibrium with an external load q [cf. Sect. 3.2, (3.5)] demonstrates Poisson's biharmonic equation

$$q = -\frac{1}{r}\left[\frac{\partial}{\partial r}(r\,Q_r) + \frac{\partial Q_t}{\partial \theta}\right] = D\,\nabla^2\nabla^2 z. \tag{7.9c}$$

The net shearing force V_r, first derived by Kirchhoff [38, 39, 65], takes into account the variation of the twisting moment M_{rt}. Accordingly to the sign convention for the three flexural moments, the net shearing force is[1]

$$V_r = Q_r - \frac{1}{r}\frac{\partial M_{rt}}{\partial \theta}. \tag{7.9d}$$

The force V_r, which represents the axial resultant acting at a plate radius r, is useful to define a boundary condition – known as *Kirchhoff's condition* – at the edge: if a plate is with *free edge*, then $V_r = 0$ at its contour. After substitution of Q_r and M_{rt}, and since here $D = constant$, we obtain

$$V_r = -D\frac{\partial}{\partial r}(\nabla^2 z) + (1-\nu)D\frac{1}{r}\frac{\partial}{\partial r}\left(\frac{1}{r}\frac{\partial^2 z}{\partial \theta^2}\right). \tag{7.9e}$$

• **Rigidity ratio:** Let γ be the *rigidity ratio* between the two zones as

$$\gamma = D_1/D_2 = t_1^3/t_2^3 \qquad \text{where } \gamma < 1 \text{ for a vase form.} \tag{7.10}$$

• **Continuity conditions:** The continuity conditions on z, $\partial z/\partial r$, M_r, and V_r at the junction $r = a\forall\theta$ are respectively, after simplifications,

$$A_{nm}\,a^n = R_{nm}(a), \tag{7.11a}$$

$$A_{nm}\,n\,a^{n-1} = \left[\frac{dR_{nm}}{dr}\right]_{r=a}, \tag{7.11b}$$

$$A_{nm}[n(n-1) + \nu(n-m^2)]\,a^{n-2} = \frac{1}{\gamma}\left[\frac{d^2R_{nm}}{dr^2} + \frac{\nu}{r}\frac{dR_{nm}}{dr} - \frac{\nu m^2}{r^2}R_{nm}\right]_{r=a}, \tag{7.11c}$$

[1] As in Sect. 1.13.10 and the other chapters, the positive sign convention for the bending moments M_r and M_t provides a more logical representation of the flexure than the negative sign convention currently used by other authors: as shown in Fig. 7.4, a positive curvature mode z_{20} is generated by a positive bending moment M_r for a positive x-value. N.B.: For authors using the negative sign convention, an erroneous definition of M_{rt} is often encountered (see footnote in Sect. 3.2).

$$A_{nm}[(n-2)(n^2-m^2)+(1-\nu)(n-1)m^2]\,a^{n-3}$$
$$=\frac{1}{\gamma}\left[\frac{d^3R_{nm}}{dr^3}+\frac{1}{r}\frac{d^2R_{nm}}{dr^2}-\frac{1+\nu\,m^2}{r^2}\frac{dR_{nm}}{dr}+\frac{(1+\nu)m^2}{r^3}R_{nm}\right]_{r=a}. \quad (7.11d)$$

The above set allows us to determine B_{nm}, C_{nm}, D_{nm}, and E_{nm} as functions of A_{nm} and then the bending moment $M_r(b,\theta)$ and net shearing force $V_r(b,\theta)$ distributions to apply at the ring edge $r=b$.

• **First Clebsch-Seidel modes:** For the first Clebsch-Seidel modes, the substitution of each z_{nm} mode into eqs. (7.11) and solving the associate system set leads to the following relationships.

Curvature 1st-order mode – $Cv\,1$, $n=2$, $m=0$,

$$\begin{aligned}
B_{20}&=(1-\gamma)(1+\nu)(1-\ln a^2)a^2A_{20}/2\\
C_{20}&=(1-\gamma)(1+\nu)a^2\,A_{20}\\
D_{20}&=[2-(1-\gamma)(1+\nu)]A_{20}/2\\
E_{20}&=0\\
M_r(b,0)&=D_2[-(1-\nu)C_{20}/b^2+2(1+\nu)D_{20}+(3+\nu)E_{20}+(1+\nu)E_{20}\ln b^2]\\
Q_r(b,0)&=-4\,D_2E_{20}/b\\
V_r(b,0)&=Q_r(b,0)
\end{aligned} \quad (7.12a)$$

Spherical aberration 3rd-order mode – $Sphe\,3$, $n=4$, $m=0$, with $q=64D_1A_{40}$,

$$\begin{aligned}
B_{40}&=\{\nu+\gamma(5-\nu)-[(1+\nu)+\gamma(1-\nu)]\ln a^2\}a^4A_{40}\\
C_{40}&=2\,[(1+\nu)+\gamma(1-\nu)]a^4A_{40}\\
D_{40}&=[1-\nu-\gamma(5-\nu+4\ln a^2)]a^2A_{40}\\
E_{40}&=8\,\gamma a^2A_{40}\\
M_r(b,0)&=D_2[-(1-\nu)C_{40}/b^2+2(1+\nu)D_{40}+(3+\nu)E_{40}+(1+\nu)E_{40}\ln b^2]\\
Q_r(b,0)&=-4\,D_2E_{40}/\,b\\
V_r(b,0)&=Q_r(b,0)
\end{aligned} \quad (7.12b)$$

Coma 3rd-order mode – $Coma\,3$, $n=3$, $m=1$,

$$\begin{aligned}
B_{31}&=(1-\gamma)[3+\nu-(1-\nu)\ln a^2]a^2A_{31}/2\\
C_{31}&=-(1-\gamma)(1+\nu)a^4A_{31}/2\\
D_{31}&=\gamma A_{31}\\
E_{31}&=(1-\gamma)(1-\nu)a^2A_{31}\\
M_r(b,0)&=D_2[2(1-\nu)C_{31}/b^3+2(3+\nu)D_{31}b+(1+\nu)E_{31}/\,b]\\
Q_r(b,0)&=-2\,D_2[4D_{31}-E_{31}/\,b^2]\\
V_r(b,0)&=-D_2[-2(1-\nu)C_{31}/\,b^4+2(5-\nu)D_{31}-(1+\nu)E_{31}/\,b^2]
\end{aligned} \quad (7.12c)$$

Astigmatism 3rd-order mode – $Astm\,3$, $n = 2$, $m = 2$,

$$
\begin{aligned}
B_{22} &= [4 + (1-\gamma)(1-\nu)]A_{22}/4\\
C_{22} &= -(1-\gamma)(1-\nu)a^4 A_{22}/12\\
D_{22} &= -(1-\gamma)(1-\nu)a^{-2} A_{22}/6\\
E_{22} &= 0\\
M_r(b,0) &= 2\,D_2\,[(1-\nu)B_{22} + 3(1-\nu)C_{22}/\,b^4 + 6D_{22}b^2 - 2\nu E_{22}/\,b^2]\\
Q_r(b,0) &= -8\,D_2[3D_{22}b + E_{22}/b^3]\\
V_r(b,0) &= -4\,D_2[(1-\nu)B_{22}/\,b - 3(1-\nu)C_{22}/\,b^5 + 3(3-\nu)D_{22}b\\
&\quad + (1+\nu)E_{22}/\,b^3]
\end{aligned}
\tag{7.12d}
$$

Astigmatism 5th-order mode – $Astm\,5$, $n = 4$, $m = 2$,

$$
\begin{aligned}
B_{42} &= 3(1-\gamma)(3-\nu)a^2 A_{42}/4\\
C_{42} &= -(1-\gamma)(1+\nu)a^6 A_{42}/\,4\\
D_{42} &= [\gamma - (1-\gamma)(1-\nu)]A_{42}/\,4\\
E_{42} &= -3(1-\gamma)(1-\nu)a^4 A_{42}/4\\
M_r(b,0) &= 2\,D_2\,[(1-\nu)B_{42} + 3(1-\nu)C_{42}/b^4 + 6D_{42}b^2 - 2\nu E_{42}/b^2]\\
Q_r(b,0) &= -8\,D_2[3D_{42}b + E_{42}/b^3]\\
V_r(b,0) &= -4\,D_2[(1-\nu)B_{42}/b - 3(1-\nu)C_{42}/b^5 + 3(3-\nu)D_{42}b\\
&\quad + (1+\nu)E_{42}/\,b^3]
\end{aligned}
\tag{7.12e}
$$

Triangle 5th-order mode – $Tri\,5$, $n = 3$, $m = 3$,

$$
\begin{aligned}
B_{33} &= [2 + (1-\gamma)(1-\nu)]A_{33}/2\\
C_{33} &= -(1-\gamma)(1-\nu)a^6 A_{33}/8\\
D_{33} &= -3\,(1-\gamma)(1-\nu)a^{-2} A_{33}/8\\
E_{33} &= 0\\
M_r(b,0) &= 2\,D_2[3(1-\nu)B_{33}b + 6(1-\nu)C_{33}/b^5 + 2(5-\nu)D_{33}b^3\\
&\quad + (1-5\nu)E_{33}/\,b^3]\\
Q_r(b,0) &= -24\,D_2[2D_{33}b^2 + E_{33}/\,b^4]\\
V_r(b,0) &= -6\,D_2[3(1-\nu)B_{33} - 6(1-\nu)C_{33}/b^6 + 2(7-3\nu)D_{33}b^2\\
&\quad + (1+3\nu)E_{33}/b^4]
\end{aligned}
\tag{7.12f}
$$

• **Monomode forces $F_{a,k}$ and $F_{c,k}$:** In order to generate the bending moments M_r and net shearing forces V_r at $r = b$, we may remark that the MDM design gains in *compactness* by applying axial forces at $r = a$ and $r = c$ instead of at $r = b$ and $r = c$. With this choice, the axial forces denoted $F_{a,k}$ and $F_{c,k}$ are determined from the statics equilibrium equations (cf. Fig. 7.1)

$$F_{a,k} + F_{c,k} = b \int_{\pi(2k-3)/k_m}^{\pi(2k-1)/k_m} V_r(b,\theta)\, d\theta, \tag{7.13a}$$

$$(a-b)F_{a,k} + (c-b)F_{c,k} = b \int_{\pi(2k-3)/k_m}^{\pi(2k-1)/k_m} M_r(b,\theta) d\theta, \tag{7.13b}$$

with $k = 1, 2, \ldots, k_m$ for a MDM having k_m arms.

• **Resultant multimode forces $\mathcal{F}_{a,k}$ and $\mathcal{F}_{c,k}$**: The forces $F_{a,k}$ and $F_{c,k}$ are determined for each mode by solving this system. The co-addition of various modes is obtained by summing the corresponding forces. The resultant forces $\mathcal{F}_{a,k}$ and $\mathcal{F}_{c,k}$ to apply to the MDM are

$$\mathcal{F}_{a,k} = \sum_{nm \text{ modes}} F_{a,k} \quad \text{and} \quad \mathcal{F}_{c,k} = \sum_{nm \text{ modes}} F_{c,k} \tag{7.14}$$

With $k_m = 12$ arms, Table 7.1 displays the geometrical parameters of a metal MDM and the associated intensities of $F_{a,k}$ and $F_{c,k}$ forces for some Clebsch-Seidel modes.

A first 12-arm prototype MDM was built in a Fe87 Cr13 stainless steel alloy (Fig. 7.2).

A diagram showing the distribution of Clebsch-Seidel modes of the optics triangle matrix and some He-Ne interferograms obtained with this MDM are displayed in Fig. 7.3.

Fig. 7.2 View of the 12-arm *vase form* and plane MDM. Geometrical parameters are $a = 80$ mm, $b/a = 1.25$, $c/a = 1.8125$, $t_1 = 4$ mm, $t_2/t_1 = 1/\gamma^{1/3} = 3$. Elasticity constants of quenched stainless steel Fe87 Cr13 are $E = 2.05 \times 10^4$ daN.mm^{-2} and $\nu = 0.305$. Deformation modes generated by rotation of differential screws at $r = a$ and $r = c$. Air pressure or depressure can be applied onto rear side of clear aperture $r \leq a$ for generating the *Sphe* 3 mode [Loom]

Table 7.1 Axial distribution of forces $F_{a,k}$ and $F_{c,k}$ of a 12-arm plane MDM ($k_m = 12$). Fe87 Cr13 stainless steel, $E = 205\,10^9$ Pa, $\nu = 0.305$, designed with $t_1 = 8$ mm, $\gamma = (t_1/t_2)^3 = 1/27$, $a = 100$ mm, $b/a = 1.24$ and $c/a = 1.6$. The PtV amplitude of each Clebsch-Seidel mode is $w = 0.1$ mm i.e. with $A_{20} = w/a^2$, $A_{40} = w/a^4$, $A_{31} = w/2a^3$, $A_{22} = A_{20}/2$, $A_{42} = A_{40}/2$, and $A_{33} = A_{31}$ [Units: daN]

Angle	Arm	*Cv* 1		*Sphe* 3*		*Coma* 3		*Astm* 3		*Astm* 5		*Tri* 5	
	nb.	$n=2$	$m=0$	$n=4$	$m=0$	$n=3$	$m=1$	$n=2$	$m=2$	$n=4$	$m=2$	$n=3$	$m=3$
θ	k	$F_{a,k}$	$F_{c,k}$	$F_{a,k}$	$F_{c,k}$	$F_{a,k}$	$F_{c,k}$	$F_{a,k}$	$F_{c,k}$	$F_{a,k}$	$F_{c,k}$	$F_{a,k}$	$F_{c,k}$
0	1	−113.3	113.3	−464.0	302.4	−84.0	71.6	154.0	−17.0	168.2	29.9	792	37.3
$\pi/6$	2	−113.3	113.3	−464.0	302.4	−72.7	62.0	77.0	−8.5	84.1	14.9	0	0
$\pi/3$	3	−113.3	113.3	−464.0	302.4	−42.0	35.8	−77.0	8.5	−84.1	−14.9	−792	−37.3
$\pi/2$	4	−113.3	113.3	−464.0	302.4	0.0	0.0	−154.0	17.0	−168.2	−29.9	0	0
$2\pi/3$	5	−113.3	113.3	−464.0	302.4	42.0	−35.8	−77.0	8.5	−84.1	−14.9	792	37.3
$5\pi/6$	6	−113.3	113.3	−464.0	302.4	72.7	−62.0	77.0	−8.5	84.1	14.9	0	0
π	7	−113.3	113.3	−464.0	302.4	84.0	−71.6	154.0	−17.0	168.2	29.9	−792	−37.3
$7\pi/6$	8	−113.3	113.3	−464.0	302.4	72.7	−62.0	77.0	−8.5	84.1	14.9	0	0
$4\pi/3$	9	−113.3	113.3	−464.0	302.4	42.0	−35.8	−77.0	8.5	−84.1	−14.9	792	37.3
$3\pi/2$	10	−113.3	113.3	−464.0	302.4	0.0	0.0	−154.0	17.0	−168.2	−29.9	0	0
$5\pi/3$	11	−113.3	113.3	−464.0	302.4	−42.0	35.8	−77.0	8.5	−84.1	−14.9	−792	−37.3
$11\pi/6$	12	−113.3	113.3	−464.0	302.4	−72.7	62.0	77.0	−8.51	84.1	14.9	0	0

* The uniform load to generate the *Sphe* 3 mode is $q = 64D_1 A_{40} = 0{,}06172$ MPa

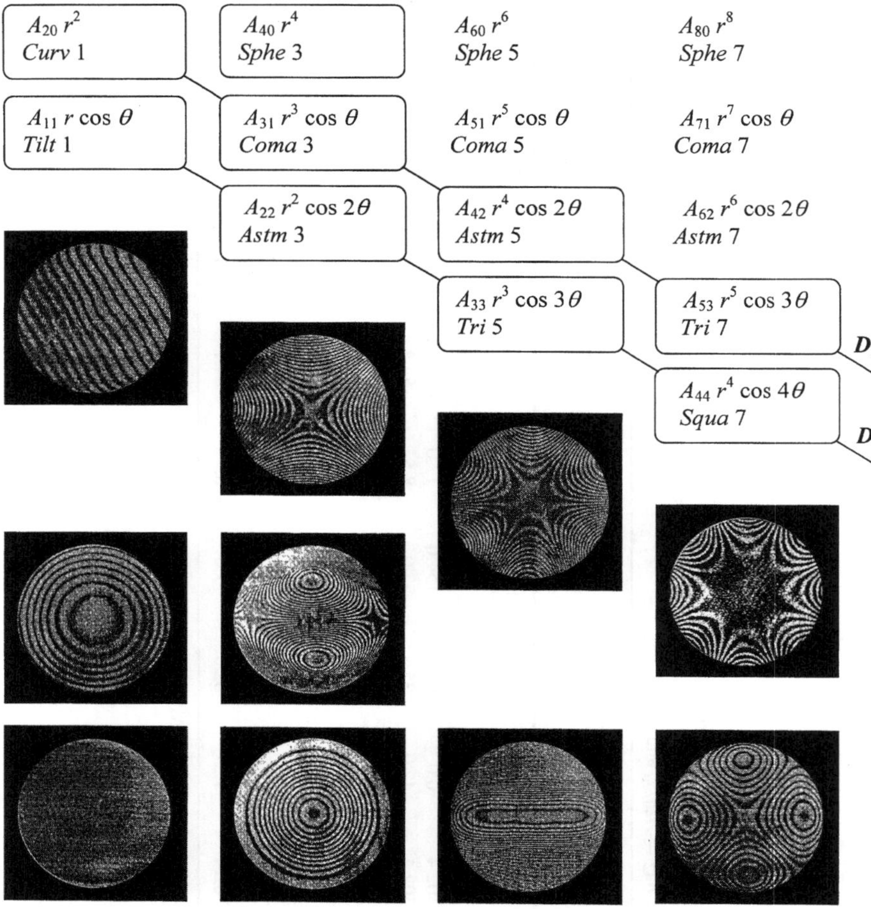

Fig. 7.3 (*Up*) Distribution of Clebsch-Seidel modes into the optics triangle matrix (piston z_{00} not shown). $\rightarrow q = 0$ provides $m = n$ modes, and $m = n - 2$ modes (D_1 and D_2 diagonals). $\rightarrow q = constant$ provides $m = 0, n = 4$ mode (z_{40} coupled with z_{20}). (*Down*) Deformation interferograms obtained with the 12-arm vase MDM described and displayed by Fig. 7.2. From left to right, the interferograms are as follows: (*Up diagonal boxes*) *Tilt* 1, *Astm* 3, *Tri* 5, and *Squa* 7 modes, (*Middle line boxes*) *Cv* 1 and *Coma* 3 modes, (*Bottom line boxes*) Mirror at rest with respect to a plane, co-addition of *Cv* 1 and *Sphe* 3 modes, co-addition of *Cv* 1, *Sphe* 3, and *Astm* 3 modes, and co-addition of *Astm* 3 and *Squa* 7 modes [LOOM]

7.3 Elasticity and Meniscus-Form MDMs

In order to reduce the flexure singularities due to the shear component of the flexure, the number of discrete forces and moments applied along the mirror perimeter has to be optimized. This arm number depends on the order of the considered modes and on the stress level in the material. Low curvature meniscuses and flat plates can generate Clebsch-Seidel modes (Fig. 7.4).

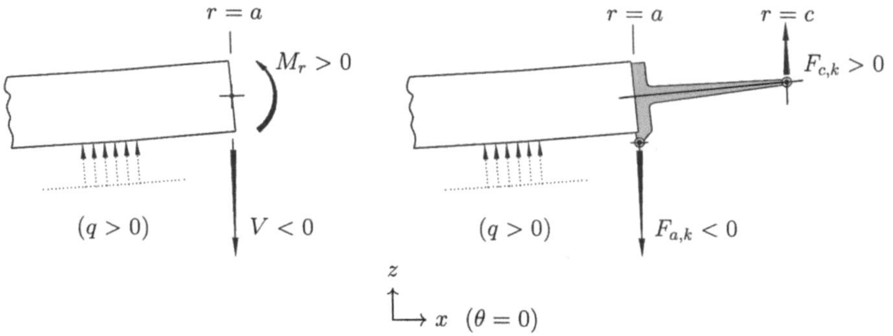

Fig. 7.4 *Meniscus-form multimode deformable mirror* deflected by clamped radial arms. The constant rigidity allows one to achieve the superposition of Clebsch-Seidel modes by action of discrete axial forces $F_{a,k}$ and $F_{c,k}$ that are equivalent to the bending moment M_r and net shearing force V_r per unit length on $r = a$. For the single curvature mode $z = A_{20}r^2$, if $A_{20} > 0$, then the sign convention gives $M_r > 0$

Denoting D the mirror rigidity and representing the elastic flexure to generate by

$$Z = \sum A_{nm} r^n \cos m\theta,$$

from (7.8) and (7.9), the bending moments M_r, shearing forces Q_r and net shearing forces V_r are

$$M_r = D \sum [n(n-1) + \nu(n - m^2)]\, A_{nm}\, r^{n-2} \cos m\theta \tag{7.15}$$

$$Q_r = -D \sum (n-2)(n^2 - m^2)\, A_{nm}\, r^{n-3} \cos m\theta \tag{7.16}$$

$$V_r = -D \sum [(n-2)(n^2 - m^2) + (1-\nu)(n-1)m^2]\, A_{nm}\, r^{n-3} \cos m\theta \tag{7.17}$$

For each of first modes these moments and forces at the mirror perimeter $r =$ and at $\theta = 0$ are the following.

Mode	n	m	$M_r(a,0)$	$Q_r(a,0)$	$V_r\,(a,0)$	
Cv 1	2	0	$2(1+\nu)DA_{20}$	0	0	(7.18a)
Sphe 3	4	0	$4(3+\nu)Da^2A_{40}$	$-32DaA_{40}$	$-32DaA_{40}$	(7.18b)
Coma 3	3	1	$2(3+\nu)DaA_{31}$	$-8DA_{31}$	$-2(5-\nu)DA_{31}$	(7.18c)
Astm 3	2	2	$2(1-\nu)DA_{22}$	0	$-4(1-\nu)Da^{-1}A_{22}$	(7.18d)
Astm 5	4	2	$12Da^2A_{42}$	$-24DaA_{42}$	$-12(3-\nu)DaA_{42}$	(7.18e)
Tri 5	3	3	$6(1-\nu)DaA_{33}$	0	$-18(1-\nu)DA_{33}$	(7.18f)

With *Sphe* 3 mode, the uniform loading is $q = 64DA_{40}$. The axial forces $F_{a,k}$ and $F_{c,k}$ to apply at $r = a$ and at the end $r = c$ of radial arms clamped onto the edge, are obtained by the static equilibrium relationships

$$F_{a,k} + F_{c,k} = a \int_{\pi(2k-3)/k_m}^{\pi(2k-1)/k_m} V_r(a, \theta)\, d\theta, \qquad (7.19a)$$

$$(c - a) F_{c,k} = a \int_{\pi(2k-3)/k_m}^{\pi(2k-1)/k_m} M_r(a, \theta)\, d\theta, \qquad (7.19b)$$

with $k = 1, 2, \ldots, k_m$ for a MDM having k_m arms. The co-addition of various modes is obtained by summing the corresponding forces. The resulting forces to apply $\mathcal{F}_{a,k}$ and $\mathcal{F}_{c,k}$ are

$$\mathcal{F}_{a,k} = \sum_{n,m \text{ modes}} F_{a,k} \quad \text{and} \quad \mathcal{F}_{c,k} = \sum_{n,m \text{ modes}} F_{c,k} \qquad (7.20)$$

7.4 Degenerated Configurations and Astigmatism Mode

7.4.1 Special Geometry for the Astigmatism Mode

Special configurations for monomode flexures with $m = n \geq 2$ can be derived where forces $F_{a,k} = 0$, thus proving important practical simplifications in the deformable design by reducing the number of action points to a *minimum* (see the case $m = n = 3$ of triangle mode in Sect. 7.9). Because only requiring use of $F_{c,k}$ forces, we call them *degenerated configurations.*

One restrains hereafter to the case of a flexure purely corresponding to third-order astigmatism $z_{22} = A_{22} r^2 \cos 2\theta$, i.e. *Astm* 3, and determines the associated arm geometry of degenerated vase or meniscus forms (fifth-order triangle mode – or *Tri* 5 mode – treated in Sect. 7.9). From (7.13a) and (7.13b), the condition $F_{a,k} = 0$ is satisfied if

$$(c - b) \int_{\pi(2k-3)/k_m}^{\pi(2k-1)/k_m} V_r(b, \theta)\, d\theta = \int_{\pi(2k-3)/k_m}^{\pi(2k-1)/k_m} M_r(b, \theta)\, d\theta. \qquad (7.21)$$

Using the expressions of $M_r(b, 0)$ and $V(b, 0)$ given by (7.12d) for *Astm* 3 mode, the results from calculation show that such degenerated configurations can be obtained if the radial arms have the special geometry [41]

$$\frac{c}{b} = 1 - \frac{1 + (1-\gamma)(1-\nu)\left[\dfrac{1}{4} - \dfrac{1}{1-\nu}\dfrac{b^2}{a^2} - \dfrac{a^4}{4b^4}\right]}{2 + (1-\gamma)(1-\nu)\left[\dfrac{1}{2} - \dfrac{3-\nu}{1-\nu}\dfrac{b^2}{a^2} + \dfrac{a^4}{2b^4}\right]}, \qquad (7.22a)$$

which is valid for flat to moderately curved vase or meniscus mirrors.

7.4.2 Single Astm 3 Mode and Degenerated Meniscus Form

For a meniscus form, then $\gamma = (t_1/t_2)^3 = 1$, so the c/b ratio does not depend on Poisson's ratio. In this case one must set $b = a$ since the outer ring vanishes. From (7.22a), the result is as follows.

→ *A meniscus form of aperture* $2a$ *provides a pure third-order astigmatic flexure* $Astm\,3$ *by requiring only four alternatively opposite forces* $F_{c,k}$ *applied to the ends of folded arms at the radial distance c such as*

$$\frac{c}{a} \equiv \frac{c}{b} = \frac{1}{2}. \tag{7.22b}$$

With $k_m = 4$, the general geometry of a vase or meniscus form satisfying the condition $F_{a,k} = 0$, only uses $F_{c,k}$ forces applied to the extremities of four folded arms (Fig. 7.5).

From (7.12d) and (7.13a), these forces are

$$F_{c,k} = (-1)^k(1-\gamma)(1-\nu)^2\left[\frac{4}{(1-\gamma)(1-\nu)} + 1 - 2\frac{3-\nu}{1-\nu}\frac{b^2}{a^2} + \frac{a^4}{b^4}\right] D_2\,A_{22}. \tag{7.23a}$$

→ *For a meniscus mirror with four arms* ($k_m = 4$)*, the intensity and direction of these forces are represented by*

$$F_{c,k} = (-1)^k \frac{E\,t^3}{3(1+\nu)} A_{22} \quad \text{with} \quad k = 1, 2, 3, 4. \tag{7.23b}$$

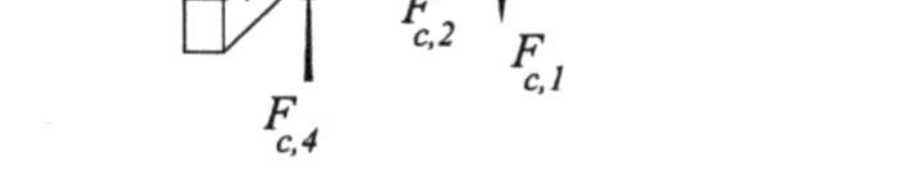

Fig. 7.5 Degenerated configuration of a *four-arm meniscus-form* mirror generating an $Astm\,3$ mode. Only two pairs of opposite forces $F_{c,k}$ act on folded arm ends in the directions $\theta = 0$ and $\pi/2$. With $b = a$ and $\gamma = 1$, condition (7.22a) gives $c/a = 1/2$. $\forall\ F_{a,k} = 0$ and $F_{c,1} = F_{c,3} = -F_{c,2} = -F_{c,4}$. The built-in contour condition is realized by arches providing a best azimuth modulation of $M_r(a,\theta)$ and $V_r(a,\theta)$ for large deformations

7.4.3 Single Astm 3 Mode and Degenerated Vase Form

Comparing the geometries between vase form and meniscus form, from (7.22a), we obtain the following result.

→ *A vase-form configuration provides much shorter arms than for a meniscus form, hence improves the stability of the flexure mounting by compact designs.*

A degenerated configuration of a *four-arm vase-form* mirror was designed from condition (7.22a) for $F_{a,k} = 0$. Several vase forms were built in Hooke's linear stainless steel alloy for large flexural sags. With respect to higher-order modes of the astigmatism family, the reduction of the interferograms showed that the purity of this mode was characterized by low amplitude harmonics, $|A_{42}/A_{22}| \leq 0.023$ and $|A_{62}/A_{22}| \leq 0.005$ (Fig. 7.6).

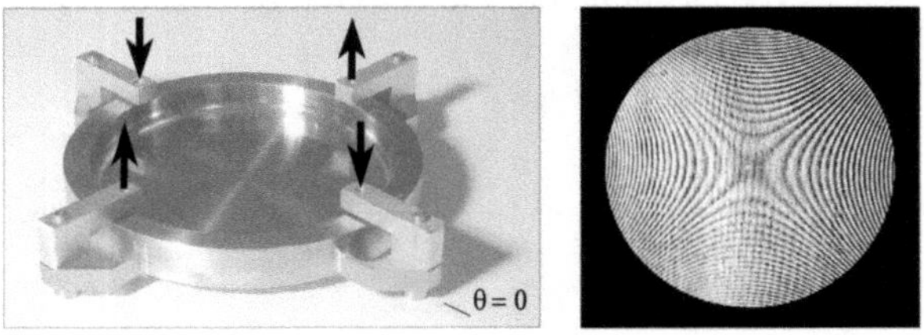

Fig. 7.6 Degenerated configuration of a *four-arm vase-form* mirror providing an $Astm\,3$ flexure mode. This solution satisfies condition (7.22a) for $F_{a,k} = 0$. Hence only four $F_{c,k}$ forces act on the ends of folded arms clamped at $r = b$. Substrate: quenched Fe87 Cr13 stainless steel, $\nu = 0.305$. Geometry: clear aperture $2a = 100$ mm, $2a/t_1 = 20$, $\gamma = 1/27$, $b/a = 6/5$, $c/b = 0.7582$, then $c/a = 0.9098$. (*Left*) Rear view of the mirror. (*Right*) He-Ne interferogram of the flexure [Loom]

- **Note on single** $Astm$ 3 **mode**: The above CTD configurations can be compared to the VTD obtained by use of a *cycloid-like distribution* simply bent at their edge by an outer ring modulation (see Fig. 3.25, *Right*).

7.5 Meniscus Form and Segments for Large Telescopes

The segmentation of telescope mirrors is an inevitable step in the development of large telescopes. Aspherical mirrors such as off-axis segments of an optical surface can be obtained with vase or meniscus-form MDMs. Using meniscus segments, the stress figuring method was applied for the construction of the primary mirror of the 10-m Keck Telescope (see Lubliner & Nelson [45], Nelson et al. [53]). With this method, the aspherization of an off-axis segment, whose optical shape is denoted z_{Opt}, is the result of co-addition of an elastic relaxation z_{Elas} with a spherical shape z_{Sph} obtained by a tool of full lap aperture. Thus, the *active optics co-addition law* writes

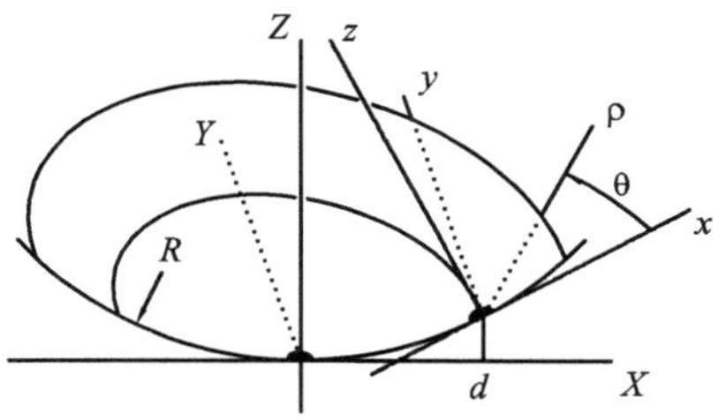

Fig. 7.7 Diagram of the global (X, Y, Z) and local (x, y, z) or (r, θ, z) coordinates of a mirror

$$z_{\mathrm{Opt}} = z_{\mathrm{Sph}} + z_{\mathrm{Elas}}. \tag{7.24}$$

In a coordinate frame (X, Y, Z), the global shape $Z_{\mathrm{Opt}}(X, Y)$ of a conicoid mirror is represented by (cf. (1.38) in Sect. 1.7.1)

$$Z_{\mathrm{Opt}} = \frac{X^2 + Y^2}{R\left[1 + \sqrt{1 - (1+\kappa)(X^2+Y^2)/R^2}\right]}, \tag{7.25a}$$

which expands as

$$Z_{\mathrm{Opt}} = \frac{1}{2R}(X^2+Y^2) + \frac{1+\kappa}{8R^3}(X^2+Y^2)^2 + \frac{(1+\kappa)^2}{16R^5}(X^2+Y^2)^3 + \cdots, \tag{7.25b}$$

where $1/R$ and κ are the curvature and the conic constant.

In order to represent the shape of a segment mirror with respect to its vertex, we consider a local coordinate frame (x, y, z) whose origin is a point at the surface of the mirror, and at distance d from the Z-axis (Fig. 7.7).

These coordinates are set such as

$$X = d, \quad Y = 0, \quad Z = Z_{\mathrm{Opt}}\{X^2 + Y^2 = d^2\},$$

the (x, y) plane is tangent to the global surface at this origin, and the y- and Y-axis are parallel. Let us define the dimensionless quantities

$$u = d/R, \quad s = \sin(\arctan u), \quad c = \cos(\arctan u), \tag{7.26}$$

where $\arctan u$ is the inclination angle of the (x, y) plane with respect to the (X, Y) plane. The coordinates of the two frames are linked together by

$$\begin{cases} X = d + cx - sz, \\ Y = y, \\ Z = Z_{\mathrm{Opt}}\{d^2\} + sx + cz. \end{cases} \tag{7.27}$$

7.5.1 *Off-Axis Segments of a Paraboloid Mirror*

Now considering the case where the mirror is a paraboloid, i.e. $\kappa = -1$ in eqs. (7.25), the system writes

$$\begin{cases} X = d + cx - sz, \\ Y = y, \\ Z = \dfrac{d^2}{2R} + sx + cz. \end{cases} \tag{7.28}$$

Since $Z = (X^2 + Y^2)/2R$ at the optical surface, after substitution we obtain

$$\frac{d^2}{2R} + sx + cz = \frac{1}{2R}\left[(d + cx - sz)^2 + y^2\right], \tag{7.29}$$

and since $s/c = d/R \equiv u$, this becomes a quadratic equation in z,

$$s^2c\,z^2 - 2(R + sc^2x)z + c(c^2x^2 + y^2) = 0. \tag{7.30}$$

The root satisfying the condition $z\{0, 0\} = 0$ provides the optical shape z_{Opt} of a segment in the local system. This is

$$z_{\mathrm{Opt}} = \frac{1}{s^2c}\left[R + sc^2x - (R^2 + 2Rsc^2x - s^2c^2y^2)^{1/2}\right], \tag{7.31a}$$

which may be written

$$z_{\mathrm{Opt}} = \frac{R + sc^2x}{s^2c}\left[1 - \left[1 - \frac{s^2c^2(c^2x^2 + y^2)}{(R + sc^2x)^2}\right]^{1/2}\right], \tag{7.31b}$$

so the expansion of the square root gives

$$z_{\mathrm{Opt}} = \frac{c}{2R}\left(c^2x^2 + y^2\right)\left[\frac{1}{1 + sc^2x/R} + \frac{s^2c^2(c^2x^2 + y^2)}{4R^2(1 + sc^2x/R)^3} + \cdots\right]. \tag{7.31c}$$

By expanding the $1 + sc^2x/R$ terms, the representation of the shape of a segment becomes

$$\begin{aligned} z_{\mathrm{Opt}} = \left(c^2x^2 + y^2\right)\bigg[&\frac{c}{2R} - \frac{sc^3}{2R^2}x + \frac{s^2c^3}{8R^3}\left(5c^2x^2 + y^2\right) \\ &- \frac{s^3c^5}{8R^4}\left(7c^2x^2 + 3y^2\right)x + O\{x^4, y^4\} + \cdots\bigg]. \end{aligned} \tag{7.31d}$$

Let us consider the corresponding cylindrical coordinates (ρ, θ, z) where $\rho = r/a$ is the normalized radius, $r = (x^2 + y^2)^{1/2}$ and a is the segment semi-aperture. We set $\theta = 0$ in the (x, z) plane. The local shape of a paraboloid segment can be expanded in the form

$$z_{\mathrm{Opt}} \equiv \sum O_{nm}\,\rho^n \cos m\theta \quad \text{with} \quad 0 \le \rho \le 1, \tag{7.32}$$

where the optics coefficients O_{nm} are determined from identifications. Also using $c = (1 + u^2)^{-1/2}$ and $s = u(1 + u^2)^{-1/2}$ leads to the sags

$$\begin{aligned}
O_{20} &= \frac{a^2 c}{4R}(c^2+1) &&= \frac{a^2}{4R}\frac{2+u^2}{(1+u^2)^{3/2}} && Cv\,1\\
O_{22} &= \frac{a^2 c}{4R}(c^2-1) &&= \frac{a^2}{4R}\frac{-u^2}{(1+u^2)^{3/2}} && Astm\,3\\
O_{31} &= -\frac{a^3 s c^3}{8R^2}(3c^2+1) &&= \frac{a^3}{8R^2}\frac{-u(4+u^2)}{(1+u^2)^3} && Coma\,3\\
O_{40} &= \frac{3a^4 s^2 c^3}{64R^3}(5c^4+2c^2+1) &&= \frac{3a^4}{64R^3}\frac{u^2(8+4u^2+u^4)}{(1+u^2)^{9/2}} && Sphe\,3\\
O_{33} &= -\frac{a^3 s c^3}{8R^2}(c^2-1) &&= \frac{a^3}{8R^2}\frac{u^3}{(1+u^2)^3} && Tri\,5\\
O_{42} &= \frac{a^4 s^2 c^3}{16R^3}(5c^2-1) &&= \frac{a^4}{16R^3}\frac{u^2(4-u^2)}{(1+u^2)^{7/2}} && Astm\,5\\
O_{51} &= -\frac{a^5 s^3 c^5}{64R^4}(35c^4+10c^2+3) &&= \frac{3a^5}{4R^4}\left(-u^3+\frac{23}{4}u^5+\cdots\right) && Coma\,5
\end{aligned} \tag{7.33}$$

From the curvature sag, O_{20}, we may notice that:

→ *The curvature of a segment decreases when the off-axis distance increases.*

For large telescope mirrors, the *linear segment number* (N) and the *global mirror f-ratio* (Ω) are fundamental design parameters. These two numbers must be set such as the optical modes O_{nm} which would introduce too much complexity to generate are reduced to negligible quantities for the marginal segment. Let us consider for instance a 42-m aperture paraboloid at f/1, $\Omega = 1$, with segments of 2-m in diameter, i.e. $a = 1$ m. For a marginal segment, the off-axis distance is $d_{\mathrm{m}} = 20$ m so that $u_{\mathrm{m}} = d_{\mathrm{m}}/R = 1/4.2$. From (7.33), the optical ptv sags are respectively

O_{20}	O_{22}	O_{31}	O_{40}	O_{33}	O_{42}
5.635 mm,	−155 μm,	−14 μm,	29 nm,	203 nm,	19 nm.

The choice of the (N, Ω) numbers always allows to minimize the *Sphe* 3 and *Coma* 5 modes, say reducing them to sags lower than $\lambda_{\mathrm{Vis}}/20$ or $\lambda_{\mathrm{Vis}}/30$ ptv, i.e.

$$O_{40} \text{ and } O_{51} \le \lambda_{\mathrm{Vis}}/30 = \text{negligible sags}, \tag{7.34}$$

which entails that the higher-order modes, not included in the above list (such as O_{60}), become straightforwardly negligible.

• **Determination of the elastic relaxation – Active optics co-addition law:** The sphere generated by the figuring, with a tool of curvature $1/R_{\mathrm{S}}$, is represented by

$$z_{\mathrm{Sph}} = \sum_{2,4,6,\ldots} S_{n0}\,\rho^n \tag{7.35a}$$

with

$$S_{20} = a^2/2R_S \quad \text{and} \quad S_{n0} \simeq 0 \text{ if } n > 2, \tag{7.35b}$$

since this single term representation comes from the fact that if O_{40} is negligible as stated in (7.34), then $S_{40}, S_{60}, \ldots$ are also negligible; within criterion (7.34), we cannot distinguish whether the surface $z_{\rm Sph}$ is a sphere or a paraboloid. The sag of the total flexure will be, of course, much smaller than that of any segment curvature.

The elastic relaxation and its α_{nm} coefficients are obtained from the *active optics co-ddition law* expressed by (7.24). This flexure is represented by

$$z_{\rm Elas} = z_{\rm Opt} - z_{\rm Sph} \equiv \sum_{n,m} \alpha_{nm}\, \rho^n \cos m\theta, \tag{7.36}$$

from where we derive

$$\alpha_{n0} = O_{n0} - S_{n0} \quad \text{for } m = 0, \tag{7.37a}$$

$$\alpha_{nm} = O_{nm} \quad \text{for } m \neq 0. \tag{7.37b}$$

Taking also into account the modes that have been given negligible values, the elastic relaxation is represented by the set

$$\begin{cases} \alpha_{20} = -\frac{a^2}{2R}\left[\frac{R}{R_S} - \frac{2+u^2}{2(1+u^2)^{3/2}}\right], & \alpha_{n0} \simeq 0 \text{ if } n \geq 4, \\ \alpha_{nm} = O_{nm} \text{ if } 2 \leq m \leq 4, & \alpha_{nm} \simeq 0 \text{ if } n \geq 5 \text{ and } m \geq 1, \end{cases} \tag{7.38}$$

thus only concerning the five Clebsch-Seidel modes $\alpha_{20}, \alpha_{22}, \alpha_{31}, \alpha_{33}, \alpha_{42}$, the three first are negative for the outermost segment of a paraboloid. The flexure to generate during the stress figuring is represented by the *opposite* coefficients.

• **Determination of the figuring curvature $1/R_S$ – Minimal stress**: For simplification of the execution process we may assume that the spherical curvature of the figuring tool, $1/R_S$, is the same for all the segments. Next, the determination of R_S must minimize the stress level arising at the surface of the meniscus. Because all the four modes $\alpha_{22}, \alpha_{31}, \alpha_{33}, \alpha_{42}$ have a null mean curvature, the stresses introduced for generating them cannot be minimized. Hence the practical way for generating the remaining mode α_{20} is only concerned with a stress minimization.

Denoting $u_{\rm m} = d_{\rm m}/R$ the slope of the marginal segment, the minimization of the stress is achieved when the sag is the mean value of the sags for $u = 0$ and $u = u_{\rm m}$. From (7.33a)

$$\frac{a^2}{2R_S} = \frac{1}{2}\left[\frac{a^2}{2R} + \frac{2+u_{\rm m}^2}{(1+u_{\rm m}^2)^{3/2}}\frac{a^2}{4R}\right], \tag{7.39a}$$

$$R_S = 2\left[1 + \frac{2+u_{\rm m}^2}{2(1+u_{\rm m}^2)^{3/2}}\right]^{-1} R\,, \tag{7.39b}$$

which, with (7.38), fully determines the execution conditions. Therefore, the stressing is as follows.

→ *The paraxial segments are bent convex whilst the marginal ones are bent concave.*
→ *A spherical tool of curvature $1/R_S$ allows to aspherize all the segments with minimal bending stresses.*

If a paraboloid mirror is at f/1, then $u_m = 1/4.2$, and from (7.39), $R_S \simeq 1.02737\,R$.

• **Determination of the forces and moments at the contour**: The conversion of the α_{nm} coefficients into the A_{nm} coefficients which are basically used in this chapter is simply achieved by

$$A_{nm} = \alpha_{nm}/a^n. \tag{7.40}$$

The associated axial forces $\mathcal{F}_{a,k}$ and $\mathcal{F}_{c,k}$ generating the net shearing force and bending moment at the contour of the meniscus can be readily determined from the Clebsch-Seidel modes such as derived in Sect. 7.3.

• **Linear segment number and global mirror f-ratio:** The reasons for making some optical modes in a segment negligible come from elasticity: The *Sphe* 3 elastic mode requires a uniform load applied all over the segment which would introduce some complexity in practical applications of the stressing process. Next elasticity modes such as *Coma* 5 and *Sphe* 5 are not Clebsch-Seidel modes – in our definition of these modes (cf. the two diagonal lines of the matrix in Fig. 7.3) – since requiring prismatic and quadratic loadings respectively.

Hence, the reduction of *Sphe* 3 and *Coma* 5 modes to negligible values must be achieved by a convenient design of the segmentation. This involves an optimal setting of the *linear* segment number (N) and *global* mirror f-ratio (Ω) which are fundamental parameters for the primary mirror of extremely large telescopes (ELTs).

7.5.2 *Off-Axis Segments of a Conicoid Mirror*

The shape of large primary mirrors of presently large segmented telescopes are either a sphere (HET, SALT), a hyperboloid slightly deviating from a paraboloid (KECK, GTC) or an actively controlled pseudo-plane-aspheric surface (LAMOST). For the next telescope generation – which will have aperture diameters in the class 25–50 m –, it is useful to determine the shape of a segment for a conicoid primary mirror which may appreciably deviate from a paraboloid.

Still representing the optical shape of a segment by $z_{\mathrm{Opt}} = \sum O_{nm}\,\rho^n \cos m\theta$, we give hereafter the result of the analysis by Nelson and Temple-Raston [54], also including a small correction in one of them. The O_{nm} coefficients are derived by substituting the conicoid shape into the equation set (7.27) and expanding them by modes similarly as in the previous subsection. These optical sags are

$$O_{20} = \frac{a^2}{4R}\frac{2-\kappa u^2}{(1-\kappa u^2)^{3/2}} \qquad Cv1 \tag{7.41}$$

$$O_{22} = \frac{a^2}{4R} \frac{\kappa u^2}{(1-\kappa u^2)^{3/2}} \qquad Astm3$$

$$O_{31} = \frac{a^3}{8R^2} \frac{\kappa u[1-(\kappa+1)\,u^2]^{1/2}(4-\kappa u^2)}{(1-\kappa u^2)^3} \qquad Coma3$$

$$O_{40} = \frac{a^4}{64R^3} \frac{8(1+\kappa)-24\kappa u^2+3\kappa^2(1-3\kappa)u^4-\kappa^3(2-\kappa)u^6}{(1-\kappa u^2)^{9/2}} \qquad Sphe3$$

$$O_{33} = \frac{a^3}{8R^2} \frac{\kappa^2 u^3[1-(\kappa+1)u^2]^{1/2}}{(1-\kappa u^2)^3} \qquad Tri5$$

$$O_{42} = \frac{a^4}{16R^3} \frac{-\kappa^2 u^2[1+5\kappa-\kappa(6+5\kappa)u^2]}{(1-\kappa u^2)^{7/2}} \qquad Astm5 \qquad (7.42)$$

These coefficients allow one to determine the convenient linear segment number and global mirror f-ratio, (N, Ω), by making several modes of the marginal segments negligible such as, for instance, the sag O_{40} of the *Sphe* 3 mode. Hence this provides important simplifications in the stress figuring process and allows the execution of a *hyperboloid* or an *elongated ellipsoid* whose global sag may substantially deviate from that of a paraboloid.

7.5.3 Segments of the Keck Telescope

With its two units of 10 m aperture, the Keck Telescope is historically the first built segmented telescope which surpassed in size the large monolithic telescopes. Its optical design is a Ritchey-Chrétien form; the conic constant $\kappa = -1.00379$ of the primary mirror slightly deviates from a paraboloid. A set of 36 segments of aperture 1.8 m and thickness 75 mm, are sub-mirrors of the f/1.75 primary mirror. Lubliner and Nelson [45] carried out the analysis for determining the off-axis shape and elastic deformation of the circular meniscus segments.

The stress polishing – preferably called stress figuring , since operated first at a fine grinding stage – was applied by Nelson et al. [53] and Mast and Nelson [46] for the aspherization of the segments. A set of 24 Invar blocks bonded along the perimeter were used for the clamping of radial arms which generated the stressing. Axial forces $\mathcal{F}_{a,k}$ and $\mathcal{F}_{c,k}$ were applied on each arm end via accurate lever systems. These forces are derived from the bending moments M_r and net shearing forces V (cf. Sect. 7.3).

The active optics figuring method provides extremely smooth surfaces as shown by the results of the interferomeric tests (Fig. 7.8).

Because of slight deviations from the theoretical shape, due to figuring residuals and to the cutting of the meniscuses into hexagons, final figuring retouches were

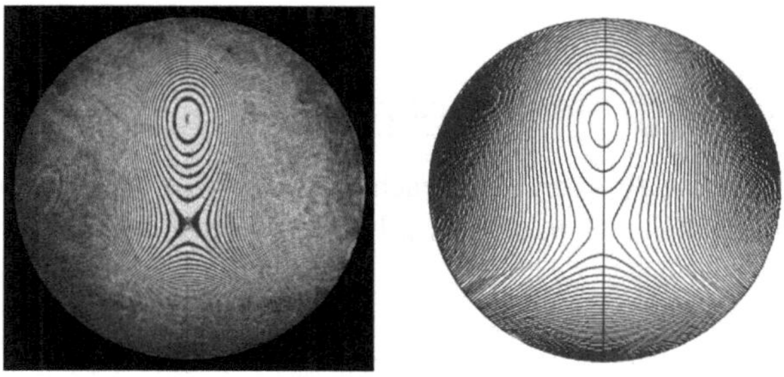

Fig. 7.8 He-Ne interferogram from *stress figuring* of a marginal segment of the Keck telescope with respect to a sphere and reference theoretical interferogram [53]

performed by developing the ion beam technique [1, 2, 71]. Another alternative by active supporting of the segments has been investigated [9] but not used. Finally, the manufacturing steps were:

→ *Convex Side Polish* → *Stress Figuring* → *Cutting Circular to Hexagonal* → *Boring for Lateral Support* → *Passive Support Mounting* → *Ion Beam Finishing*

A detailed summary on the construction of the Keck segments is given by Wilson [72].

7.6 Vase and Meniscus MDMs for Reflective Schmidts

7.6.1 Centered Systems with a Circular Vase-Form Primary

For all-reflective Schmidts, the study of MDMs allows determining the geometry of vase mirrors for obtaining the shape of the primary mirror by active deformation. In the case of a centered system used off-axis, we will demonstrate that the use of radial arms can be avoided by a convenient choice of the rigidity ratio $\gamma = D_1/D_2 = (t_1/t_2)^3$ between clear aperture and outer ring. In the 3rd-order approximation, using the dimensionless radius $\rho = r/r_{\mathrm{m}}$ with r_{m} as clear aperture radius, the shape of the primary mirror is represented by [see (5.20b)]

$$Z_{\mathrm{Opt}} = \frac{r_{\mathrm{m}}}{2^9 \Omega^3 \cos i} (3\,\rho^2 - \rho^4), \tag{7.43}$$

where $\Omega = f/d = R/4r_{\mathrm{m}}$, R is the radius of curvature of the concave secondary mirror and i the incidence angle of the principal ray at the primary mirror (the null powered zone is outside r_{m} at $r = \sqrt{3/2}\, r_{\mathrm{m}}$). With the denotation of Sect. 7.2 in this chapter, let us define A_{nm} mirror coefficients from

$$Z_{\mathrm{Opt}} = A_{20}\, r^2 + A_{40}\, r^4.$$

This leads to

$$A_{20} = \frac{3}{2^9\,\Omega^3\,r_{\rm m}\,\cos i} \quad \text{and } A_{40} = \frac{-1}{2^9\,\Omega^3\,r_{\rm m}^3\,\cos i}. \tag{7.44}$$

From Sect. 7.2, we can derive the $F_{c,k}$ forces for each of these two modes. After substitution of corresponding coefficients B_{nm} to E_{nm} by their respective value in function of A_{nm}, we obtain

$$F_{c,k}|_{Cv\,1} = \frac{2\pi D_2 b}{k_m(c-a)} A_{20}(1+\nu)(1-\gamma)\left[\frac{2}{1-\gamma} - (1+\nu) - (1-\nu)\frac{a^2}{b^2}\right], \tag{7.45a}$$

$$F_{c,k}|_{Sphe\,3} = \frac{4\pi D_2 ba^2}{k_m(c-a)} A_{40}\times$$
$$\left[(1-\nu^2)\left(1-\gamma-\frac{a^2}{b^2}\right) - \gamma(1-\nu)^2\frac{a^2}{b^2} + 16\gamma\frac{a}{b} - 8\gamma - 8\gamma(1+\nu)\ln\frac{a}{b}\right]. \tag{7.45b}$$

The co-addition of $Cv\,1$ and $Sphe\,3$ modes allows avoiding the radial arms if verifying the condition

$$F_{c,k}|_{Cv\,1} + F_{c,k}|_{Sphe\,3} = 0. \tag{7.46}$$

For the primary mirror of a reflective Schmidt, (7.44) provides the relationship between A_{20} and A_{40}, that is

$$A_{20} = -3a^2 A_{40}, \tag{7.47}$$

which does not depend on the f-ratio, and where $a = r_{\rm m}$ i.e. a clear aperture corresponding to the inner radius of the perimeter ring. After substitution of (7.47) in (7.46), the resulting condition for the reciprocal rigidity ratio is

$$\frac{1}{\gamma} = \frac{16\left[2\frac{a}{b} - 1 - (1+\nu)\ln\frac{a}{b}\right] - (5+\nu)\left[1+\nu+(1-\nu)\frac{a^2}{b^2}\right]}{(1-\nu^2)\left(1-\frac{a^2}{b^2}\right)} \tag{7.48}$$

This condition fully defines the geometry of the vase form as an implicit relationship $(t_2/t_1, a/b, \nu)$ and provides effective solutions *without* radial arms. An example of a vase-form primary mirror in Zerodur vitroceram is displayed by Fig. 7.9-*Up*.

For this *centered system used off-axis*, the co-addition of the $Cv\,1$ and $Sphe\,3$ flexure modes are simultaneously achieved by the appropriate vase-form geometry and uniform loading $q = 64D_1A_{40}$. An inner enclosure plate is sealed into the vase. The asphérization can be obtained either by air pressure and stress polishing or by air depressure inside the mirror during observations i.e. in situ stressing.

7.6.2 Non-Centered Systems and Circular Vase-Form Primary

With a *non-centered system*, the shape of the primary mirror is represented by (see 5.20c)

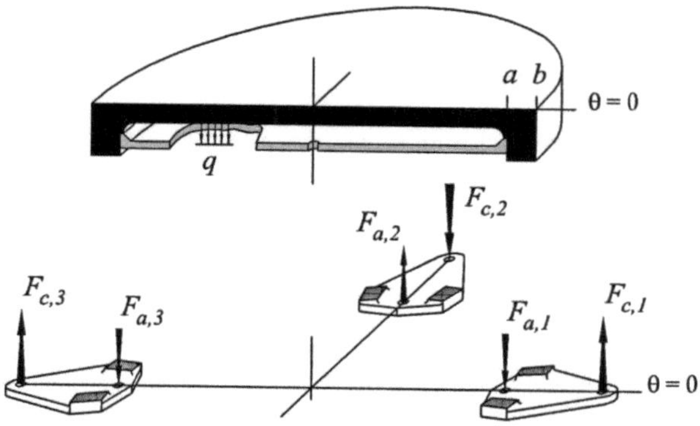

Fig. 7.9 True proportion *vase form* as the primary mirror of a reflective Schmidt. MDM in Zerodur vitroceram. Poisson's ratio $\nu = 0.240$. Geometry $b/a = 1.150$, $t_2/t_1 = (1/\gamma)^{1/3} = 2.808$ [from (7.48)]. (*Up*) Co-addition of $Cv\,1$ and $Sphe3$ modes for a *centered system used off-axis*. (*Down*) Co-addition of $Astm\,3$ and $Astm\,5$ modes for a *non-centered system*

$$Z_{\mathrm{Opt}} = \frac{r_{\mathrm{m}}}{2^9 \Omega^3 \cos i} \left[3(1-t)\rho^2 - 3t\rho^2 \cos 2\theta - (1-2t)\rho^4 + 2t\rho^4 \cos 2\theta \right], \tag{7.49}$$

where $t = \frac{1}{2}\sin^2 i$. For a non-folded Schmidt, the incident angle of principal ray is $i = \varphi_{\mathrm{m}} + 1/4\Omega$ (cf. Chap. 4). With the denotation of Sect. 7.2 in this chapter, let us define the mirror coefficients A_{nm} as

$$Z_{\mathrm{Opt}} = \sum A_{nm}\, r^n \cos m\theta,$$

with

$$A_{20} = \frac{3(1-t)}{2^9\,\Omega^3\, r_{\mathrm{m}} \cos i}, \qquad A_{22} = \frac{-3t}{2^9\,\Omega^3\, r_{\mathrm{m}} \cos i},$$
$$A_{40} = \frac{-(1-2t)}{2^9 \Omega^3\, r_{\mathrm{m}}^3 \cos i}, \qquad A_{42} = \frac{2t}{2^9\,\Omega^3\, r_{\mathrm{m}}^3 \cos i}. \tag{7.50}$$

Setting the radius aperture at the inner ring radius, $r_{\mathrm{m}} = a$, we can derive from Sect. 7.2, the $F_{c,k}$ for each of these modes. As for previous *centered systems* it is also possible to find a co-addition of $Cv\,1$ and $Sphe\,3$ that does not require radial arms i.e. $\Sigma F_{c,k} = 0$ for these modes. This is obtained by substitution of

$$A_{20} = -3\frac{1-t}{1-2t}a^2 A_{40},$$

in (7.45a), so obtaining with (7.46) an implicit relationship $(t_2/t_1, a/b, \nu, i)$ which is valid for small to moderate values of the incidence angle i. As for centered systems, this fully defines the geometry of the circular vase, i.e. the dimensions of the ring with respect to the thickness of the clear aperture area.

For this *non-centered system*, because A_{20} and A_{40} are not exactly the same as in Sect. 7.6.1, the rigidity-ratio γ must be slightly modified. In addition to the uniform load, the *Astm* 3 and *Astm* 5 flexure modes are also required by use of $F_{a,k}$ and $F_{c,k}$ forces. These forces act on the MDM via bridge-shaped Invar arms sealed on the ring rear side (Fig. 7.9-*Down*).

As a primary mirror of an f/5 reflective Schmidt telescope, a *four-arm vase MDM* ($k_m = 4$) has been designed with Zerodur vitroceram. Its geometry, the in situ partial vacuum q for *Cv* 1 and *Sphe* 3 modes, and the bending forces $F_{a,k}$, $F_{c,k}$ for *Astm* 3 and *Astm* 5 modes are given in Table 7.2.

Table 7.2 *Four-arm vase MDM* as the primary mirror of an f/5 reflective Schmidt. Force distributions $F_{a,k}$ and $F_{c,k}$. Aperture $2r_m = 2a = 400$ mm. Field of view $2\varphi_m = 5°$. Incidence angle of principal ray at $M_1 i = \varphi_m + 1/4\Omega$. Coefficients $A_{20} = 2.34410^{-5}$, $A_{40} = -1.94510^{-10}$, $A_{22} = -1.02910^{-7}$ and $A_{42} = 1.71510^{-12}$ in mm^{1-n} from (7.50). Zerodur $E = 90.6$ GPa, $\nu = 0.240$. MDM $t_1 = 20$ mm, $t_2/t_1 = 2.791$, $b/a = 1.150$, $c/a = 1.5$

[Units: daN]

Angle	Arm	*Cv* 1		*Sphe* 3*		*Astm* 3		*Astm* 5	
	nb.	$n = 2,\ m = 0$		$n = 4,\ m = 0$		$n = 2,\ m = 2$		$n = 4,\ m = 2$	
θ	k	$F_{a,k}$	$F_{c,k}$	$F_{a,k}$	$F_{c,k}$	$F_{a,k}$	$F_{c,k}$	$F_{a,k}$	$F_{c,k}$
0	1	−39.318	39.318	64.376	−39.318	−0.231	0.033	0.144	0.040
$\pi/2$	2	−39.318	39.318	64.376	−39.318	0.231	−0.033	−0.144	−0.040
π	3	−39.318	39.318	64.376	−39.318	−0.231	0.033	0.144	0.040
$3\pi/4$	4	−39.318	39.318	64.376	−39.318	0.231	−0.033	−0.144	−0.040

* Uniform loading $q = 64\, D_1 A_{40} = -0.0798$ MPa.

7.6.3 Non-Centered Systems and Elliptical Vase-Form Primary

Does an elliptical contour allow us to eliminate the arms of an MDM to simultaneously generate the astigmatism *Astm* 3, *Astm* 5 modes as represented by A_{22}, A_{42} in (7.50). In fact, we have seen in Chapter 5 that a single vase MDM form cannot provide a solution because the flexure of an elliptical ring having an homothetic geometry with the contour line $\mathcal{C}$ of the clear aperture would not satisfy the biharmonic equation $\nabla^4 z = constant$. However, a solution to this problem can be obtained by "*closed biplate forms*" that are made of two identical elliptical vases sealed at their outer ring (cf. Sect. 5.3.5).

7.6.4 In-situ Aspherized Meniscus Segments of LAMOST

With a 4-m aperture, LAMOST is by far the largest of all Schmidt telescopes. Compared to other wide-field telescopes also dedicated to spectroscopy, its 5° field of view confers the instrument an outstanding *optical étendue* (cf. Sect. 1.6.3) which allows obtaining simultaneously spectra of 4,000 stellar or quasi-stellar objects (cf. Sect. 4.3.4 and Sect. 8.7.3).

LAMOST is a *non-centered system* with a segmented primary mirror M_1 made of 24 hexagonal meniscus submirrors arranged in a surface of 4.4×5.7 meters [67]. The spherical secondary mirror M_2 is with 37 hexagonal segments inscribed in a 6.7-m circular aperture.

The M_1 segments are polished flat when in a non-stressed state. A large dedicated machine allows the simultaneous flat polishing of four segments under immersion. The surface reference for optical testing is a *plane silicon oil mirror*. Each M_1 segment is stressed by 37 force-actuators including 3 displacement-actuators for a pre-calibrated asphеrization. Some of the basic features of LAMOST are described by Wang, Su et al. [67] and Su & Cui [63]. At the telescope, starlight sensors and an *active optics closed-loop system* control the *in situ aspherization* and ensure the asphericity variation as a function of the alt-az angles of the observed region (Cui et al. [13–15]). Observations for a given declination angle $\delta \in [-10°, 90°]$ and integration time of 1.5 hour also require the active optics to generate a variation of the x, y ellipticity of M_1. The *alignment control* maintains the cofocusing of the M_1 segments whilst their cophasing is not necessary for spectroscopy. Whatever the declination δ of the observed field, the central incident beam at M_1 is with a circular cross-section of 4 m in diameter. In the y-direction this determines the clear aperture width, $2y_{\max} = 4$ m, of the primary mirror (Fig. 7.10).

If $1/R$ is the curvature of M_2, (x, z) the telescope symmetry plane, and Ω the telescope f-ratio, a first approximation shape for M_1 is, from Sect. 4.3.2,

$$z = \frac{1}{\cos i}\left[\frac{k}{64\Omega^2 R}\left(x^2\cos^2 i + y^2\right) - \frac{1}{8R^3}\left(x^2\cos^2 i + y^2\right)^2\right], \qquad (7.51)$$

where $k = 3/2$, $\Omega = R/4y_{max} = 5$, and i is the incidence angle at M_1 mirror of the beam principal ray for the center of the field of view.

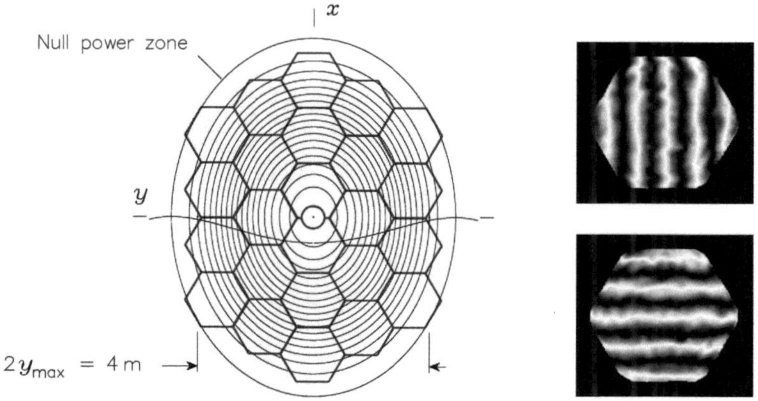

Fig. 7.10 (*Left*) LAMOST segmented primary mirror and iso-level lines for the maximum ellipticity – corresponding to the sky region declination $\delta = 90°$ – with respect to a plane surface. (*Right*) Double-pass He-Ne wavefronts in two orthogonal directions of a *stressed* most marginal segment pair under autocollimation test (courtesy X. Cui, NIAOT/CAS, Nanjing)

7.7 Vase MDMs for Liquid Mirror Telescopes

7.7.1 Zenithal Observations with LMTs

Liquid mirror telescopes (LMTs) have been developed by Borra [8] who obtained a diffraction limited 1.5 meter mirror followed by a 2.5 m mirror of the same optical quality [7]. Next, Hickson et al. [30] built a 2.7-m–f/1.9 LMT for astronomical sky surveys at zenithal fields. During the diurnal motion, the drift scanning technique of the CCD pixel columns allows integrating astronomical images over the field of view, thus obtaining star limiting magnitudes of 21 in red. Several other LMTs are used as a receiver for LIDAR applications in atmospheric sciences. In 2005, P. Hickson started operating a 6-meter LMT at f/1.5 for astronomy [31] (Fig. 7.11).

The mirror paraboloid shape is naturally achieved by the rotation of a liquid in its cell. The radius of curvature $R = g/\omega^2$ results from the hydrostatic equilibrium where g is the local gravity and ω the rotation rate. With the dimensionless radius $\rho = r/r_{\rm m}$, the paraboloid shape and the f-ratio $\Omega = f/2r_{\rm m}$ are represented by

$$Z_{\rm Opt} = \frac{r_{\rm m}}{8\Omega}\rho^2 \quad \text{and} \quad \Omega = \frac{g}{4\omega^2 r_{\rm m}}. \tag{7.52}$$

For a 20-meter aperture mirror at f/1 on the Earth, the rotation rate is $\omega \simeq 0.5$ round$\times$sec^{-1} and decreases with the mirror size $2r_{\rm m}$. The rotation is obtained by using hydrostatic air pads and thin drive belts that compensate for the low friction. The constancy of the rotation is achieved by an optical tacheometer in a closed-loop system. From the first developments of liquid mirror technologies, significant milestones have been reached by Borra in the construction of much larger mirrors [6].

For very large liquid mirrors, the Earth is rotation causes significant deformation modes such as *Cv* 1, *Sphe* 3, *Coma* 3, and *Astm* 3 arising at the liquid surface. It has been shown by Gibson [27] and Mulrooney [52] that the astigmatism due to the

Fig. 7.11 The 6-m aperture liquid zenith telescope (LZT) at The University of British Columbia. Galaxy image from this telescope [31]

Coriolis force can be eliminated by a fixed compensated tilt of the mirror rotation axis. The tilt angle is in the terrestrial meridian plane. Considering an LMT located at latitude ℓ, and denoting $R_\oplus$ and $\omega_\oplus$ the radius and rotation rate of the Earth – assumed as a spherical geoid –, the tilt angle is given by

$$\tau = \frac{R_\oplus\,\omega_\oplus^2}{2g}\left(1 + \frac{R_\oplus \omega_\oplus^2}{g}\cos^2\ell + \cdots\right)\sin 2\ell - \frac{\omega_\oplus}{\omega}(1 + \cdots)\cos\ell. \qquad (7.53)$$

This angle is zero at the poles and maximum near $\ell = \pm\pi/4$. In practice, liquid mirror bearings are levelled using a precise bubble level that is sensitive to the vertical direction defined by the acceleration vector $\mathbf{g} - \mathbf{a}_c$ where $\mathbf{a}_c$ is the centripetal acceleration. The deviation of $\mathbf{g} - \mathbf{a}_c$ from the true vertical is corresponding to the $\sin 2\ell$ term of (7.53). In this case, the required adjusting angle is just the $\cos\ell$ term; this term defines the direction of vector $\boldsymbol{\omega} + \boldsymbol{\omega}_\oplus$ to coalign with $\mathbf{g} - \mathbf{a}_c$.

Under this condition of cancelling the Coriolis effect, *Astm* 3 is removed, and, in a cylindrical reference (Z, r, θ) tangent to the mirror vertex with $\theta = 0$ in the terrestrial meridian, the mirror shape is

$$Z_{\mathrm{Opt}} = \left[\frac{r_{\mathrm{m}}}{8\Omega}\left(1 + \frac{\omega_\oplus}{\omega}\sin\ell\right) + \frac{r_{\mathrm{m}}^2\cos\ell}{2R_\oplus\Omega^2}\right]\rho^2 - \frac{r_{\mathrm{m}}^2\cos\ell}{64\,R_\oplus\,\Omega^2}\rho^4 - \frac{\omega_\oplus^2\,r_{\mathrm{m}}^2\,\sin 2\ell}{8\,g\,\Omega}\rho^3\cos\theta, \qquad (7.54)$$

where $\rho = 1$ at the mirror edge. Applying the quarter-wave rule to the *Sphe* 3 term provides the limit in which a dioptric corrector does not need to compensate the spherical aberration; this gives a mirror size $2r_{\mathrm{m}} = \Omega\sqrt{8\,R_\oplus\,\lambda}$. At the wavelength $\lambda_{\mathrm{He-Ne}}$, this would correspond to a 10 meter mirror at f/1.76. The *Coma* 3 term is fully negligible even for a 100 meter mirror.

These results remove the last fundamental obstacle to achieve diffraction-limited performance with extremely large LMTs (Hickson [33, 34]). Thus, ELMTs may be foreseen for future astronomical observations from the ground or at a lunar pole.

7.7.2 Field Distortions and Four-Lens Correctors for LMTs

For LMTs, which are transit telescopes, the time exposure of a moving image on the detector is achieved by time delay integration, also called drift scanning. Since the image moves at a constant velocity because of the sidereal rotation, the electronic drive of the detector is operated at a constant rate for each pixel column by readout of the column reaching the detector edge.

Of all telescope systems, LMTs are an unprecedented case requiring control of the third order distortion aberration – *Dist* 3 – (which is usually removed *a posteriori* by image processing) and also of the distortion effect due to non-rectilinear sky projection of the field. The correction of these distortion effects was first solved by E.H. Richardson.

Considering celestial equatorial coordinates (α, δ) and an LMT located at the terrestrial latitude, say $0 < \ell < 60°$, the meridian lines – $\alpha =$ constant – are field imaged as straight lines all converging in a common point whilst declination lines – $\delta =$ constant – are projected as circles all centered at this point. The optimum drift scan integration of the image by the detector requires that the curvilinear field frames (α, δ) shall be optically transformed into strictly cartesian frames (x, y).

If an LMT is at the Earth's equator, $\ell = 0$, then only $Dist3 = 0$ is needed.

If an LMT is at $0 < \ell < 60°$, in addition to $Dist\,3 = 0$, the corrector lenses shall provide an asymmetrical distortion which is opposite to the above *sidereal distortion*. The correction is achieved by (i) setting the sag $s_x = -s_\alpha$ due to the north-south differential sidereal rate in the field, and (ii) setting the sag $s_y = -s_\delta$ due to the star trail curvature. These conditions can be simultaneously satisfied. Denoting the semi-field angles $\varphi_{x\mathrm{m}}$, $\varphi_{y\mathrm{m}}$ in α and δ directions respectively, these sags are, from Hickson [33],

$$s_x = -2\Omega r_\mathrm{m}\, \varphi_{x\mathrm{m}}\, \varphi_{y\mathrm{m}} \tan \ell, \quad s_y = \Omega\, r_\mathrm{m}\, \varphi_{x\mathrm{m}}^2 \tan \ell, \tag{7.55}$$

where s_x, representing the de-rotation sag of the meridian straight lines, is maximum at the diagonal field edges for which if the integration surface of the detector is square $\varphi_{y\mathrm{m}} = \pm\varphi_{x\mathrm{m}}$.

Richardson [59] has shown that fortunately both $Dist\,3$ and asymmetrical sidereal distortion can be removed: the best result is achieved by a quadruplet lens that includes one of the lenses wedged. The design starts from a *three-lens Wynne corrector* [74] which annuls $Sphe\,3$, $Coma\,3$, $Astm\,3$, and $Petz\,3$, and consists of a positive meniscus followed by negative and positive lenses, all of the same glass. The *four-lens Richardson corrector* [32] introduces an off-axis wedged lens as fourth element near the focus. This corrector is presently used for typical fields of 20×20 arcmin but may be used for larger fields. This is a key component of all LMTs.

7.7.3 LMT Concepts with MDMs for Off-Zenith Observations

A paraboloid mirror observing a 10 arcmin typical field of view, with a field center located at several degrees from its axis, would present high amplitude $Coma\,3$, $Astm\,3$ and several higher-order aberrations. The design of an off-axis dioptric corrector would be extremely difficult to achromatize. In order to observe with a liquid mirror at a substantial off-zenith angle, a preliminary study by Wang [68], Moretto [50] investigated the case of off-axis corrections with a three-mirror system. The telescope consist of a parabolic primary, and two additional tilted mirrors – secondary and tertiary – having biaxial symmetries.

The starting design is derived from a Paul-Baker telescope [4, 57]. In the original Paul form [62, 73], the concept starts off with a primary-secondary afocal pair, that is concave-convex confocal paraboloids which provide an *anastigmatic* Mersenne beam compressor. Let us define the beam reduction ratio by $k = R_2/R_1$. Paul added a tertiary whose center of curvature is placed at the vertex of the primary.

In collimated beams the tertiary can be used spherical is similarly as a Schmidt spherical mirror. The spherical aberration of the tertiary can be compensated by the secondary. If $R_2 = R_3$, we can modify the parabolic secondary into a sphere since the $Sphe\,3$ contributions are just opposite. The system is again free from $Sphe\,3$. It was first noted by Paul that the system is also free of $Coma\,3$ and $Astm\,3$. The focal surface curvature is $2/R_1$ since the Petzval sum is $2(1/R_1 - 1/R_2 + 1/R_3)$. The Paul system can be generalized to give also a zero Petzval curvature with $R_2 \neq R_3$. It was first noted by Baker that the 3rd-order aberrations of this modified system remains at zero if the spherical secondary is replaced by an ellipsoid whose conic constant is $c_{\kappa 2} = -1 + (R_2/R_3)^3 = -1 + (1-k)^3$ so that the tertiary still receives collimated beams and remains a sphere ($c_{\kappa 3} = 0$). The first large Paul-Baker telescope

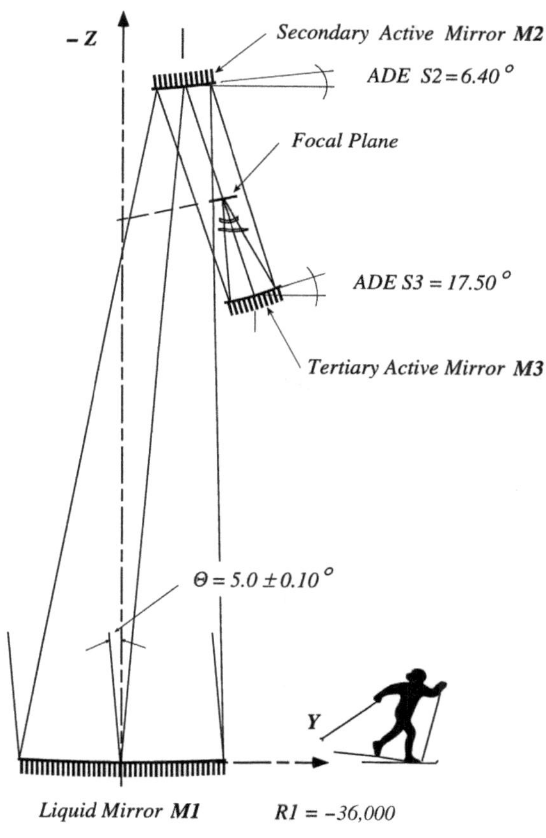

Fig. 7.12 Optical design of a three-mirror telescope with doublet-lens corrector for observaions at 5° from zenith with 15 arcmin FOV. The liquid primary is 4 m in diameter. The secondary and tertiary mirrors are 1.15 m in diameter; their design with vase-form MDM allows *in-situ* aspherizations. The rms size of residual images is smaller than 0.5" over the FOV [48]

was built at the instigation of Angel et al. [3] as a transit automatized instrument for CCD sky surveys. This telescope [47] has a 1.8 meter aperture, f/2.2 and is flat fielded over 1°. With its beam reduction ratio of $k \simeq 1/3$, one has $R_2 = R_1/3$, $R_3 = R_1/2$, $c_{\kappa 1} = -1$, $c_{\kappa 2} = -0.704$ and $c_{\kappa 3} = 0$.

The study by Moretto [48, 49] of a telescope concept for off-zenith observations with a liquid mirror is derived from Paul-Baker flat-fielded anastigmats. The concave secondary and tertiary are vase-form MDMs. The x, y field compensations for the distortion and sky projection has been carried out by including a doublet-lens corrector (spherical lenses) to allow the CCD drift scan imaging. This design has a 4 meter primary and a beam reduction of 3.5 at the secondary for observations up to 5 and 7.5° from the zenith (Fig. 7.12).

The first MDM was designed with the parameters of Fig. 7.2 for the development of such secondary and tertiary mirrors. This 12-arm MDM was built and various obtained Clebsch-Seidel modes were controlled by interferometry. The resulting co-addition of $Tilt$ 1 mode and four flexural modes is shown by the interferogram of Fig. 7.13.

The *stress-strain linearity* – Hooke's law – of the Fe87 Cr13 alloy for the 12-arm MDM has been verified by experiment in generating a maximal stress $\sigma = 50\,\mathrm{daN.mm^{-1}}$ into the outer ring. This test was performed for the $Astm$ 3 mode $z_{22} = A_{22} r^2 \cos 2\theta$. If the rigidity of the inner disk is neglected, then the stress-strain relation derived from the thicker ring is

$$A_{22} \approx (1+\nu)\,\sigma_{\max}/E\,t_2, \tag{7.56}$$

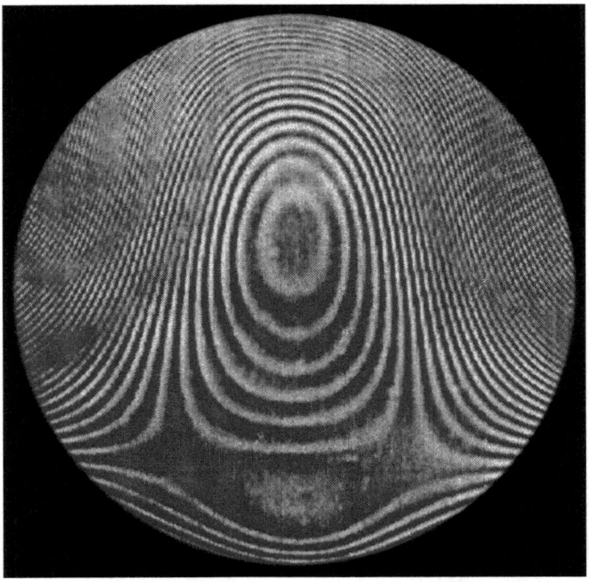

Fig. 7.13 He-Ne interferogram of superposed modes $Z = z_{11} + z_{20} + z_{22} + z_{33} + z_{40}$ with the 12-arm MDM described in Fig. 7.2 [49] [Loom]

leading to a ptv deformation sag of $z_{22} = \pm 1.7$ mm at clear aperture radius $a = 80$ mm. This large deformation was found to lie within the elasticity domain, then showing the large amplitude capability of active methods.

7.8 MDMs as Recording Compensators for Holographic Gratings

7.8.1 *Holographic Gratings Correcting Aberrations*

The recording process for obtaining holographic diffraction gratings requires the formation of an interference pattern which is frozen into the photosensitive layer of the grating substrate. Straight interference lines allow recording of plane gratings. The use of a Rowland mounting allows recording of concave gratings free from $Astm\,3$ by curved and variable spaced interference lines. For correcting several aberrations, at least one of the two recording wavefronts must be aspheric. Up to now, the formation of an aspherical wavefront requires the design and construction of special optical systems providing exactly the opposite shape of the wavefront to be corrected. Classically, such a compensating system is complex, expensive and furthermore, only usable for making a particular grating. Also, various types of aberrations cannot be simultaneously achieved with such optical systems, therefore leading to great difficulties for correcting high-order aberrations with holographic gratings. However, an exception is given in "Third generation Rowland holographic mounting" by Duban [19, 23, 24], which uses two spherical auxiliary holographic gratings to produce the two aberrated recording wavefronts. This allows aberration correction of the recorded spherical grating up to and including some fifth-order modes. The construction of the auxiliary gratings for the recording of a corrected grating increases the cost and the auxiliary gratings remain dedicated to this.

Recording methods for making *aberration corrected holographic gratings* are greatly simplified by using a plane MDM upon one of the two recording beams. MDM compensators provide easily the superposition of many aberration modes by active deformations. The available Clebsch-Seidel modes allow a higher degree of correction than the Rowland mounting. Aberration compensators based on *plane* MDMs provide a general method for recording corrected diffraction gratings without requiring the above sophisticated optical systems.

As an example of the method, we have considered the recording of the holographic gratings of the HST Cosmic Origins Spectrograph (COS) [28, 51]. Very substantial improvements in the image quality have been found by Duban [18, 20, 22] by use of a MDM as a *recording compensator*. The result is that with such holo-corrected gratings:

→ *Many higher-order aberrations can be simultaneously corrected,*

→ *The residual monochromatic images of the spectra have much smaller areas.*

7.8.2 *Design Example for the COS Gratings of HST–Recording Parameters*

The gratings of the Cosmic Origins Spectrograph must correct the original spherical aberration of the HST. Before correction, the maximum amount of this aberration is up to 200-times larger than that of the spectrograph alone. No holographic correction can reduce such an amount over the whole spectral range. Therefore, it is not possible to keep the grating substrates purely spherical; we have introduced fourth and sixth degree deformations on the grating substrates, i.e. z_{40} and z_{60} terms.

Since the COS incident beam is located 5.40 arcmin off the HST optical axis, we also have been led to correct the HST field aberrations, mainly $Astm\,3$ since the HST produces an astigmatism length of 1.20 mm. This correction is made holographically, by using a MDM to record the gratings onto the Optimized Rowland Mounting [24]. The mounting cancels $Astm\,3$ at two points $P1$ and $P2$ of the spectrum. In a very general result which is also valid for the COS gratings, this mounting is the only one really suitable for obtaining the $Astm\,3$ compensation.

For three COS gratings, Table 7.3 shows the spectral data of the spectrograph. Table 7.4 gives the grating parameters, where N is the groove density in $l.\,\mathrm{mm}^{-1}$, R the radius of curvature of the grating substrates in mm, λ_0 the laser recording wavelength, i the incidence angle at the HST, α and β the recording angles in degree. Table 7.5 gives the deformation coefficients of the grating substrates in mm^{-n+1}. Substrates of gratings #1 and #2 are identical. Table 7.6 gives the deformation coefficients in mm^{-n+1} and the incidence angle i_{MDM} upon the MDM.

The optical design of each corrected grating minimizes the blur images for five wavelengths. For grating #1, the wavelengths are those listed in Table 7.3 and correspond – from left to right – to $\lambda_{\min}$, P_1, the middle of the spectrum λ_{med}, P_2, and $\lambda_{\max}$, plus two other intermediate wavelengths around the central one. The correction of astigmatism at points P_1 and P_2 is evident (Fig. 7.14).

Table 7.3 Grating spectral data at the spectrograph in Å [28]

Grating	$\lambda_{\min}$	$P1$	λ_{med}	$P2$	$\lambda_{\max}$
#1	1,150	1,185	1,295.5	1,382	1,449
#2	1,405	1,456	1,589.5	1,684	1,774
#3	1,230	1,320	1,615.0	1,810	2,000

Table 7.4 Gratings and geometrical recording parameters [21]

Grating	N	R	λ_0	i	α	β
#1	3,800	1,652.0	3,511	19.886	−36.089	48.171
#2	3,052.6	1,652.0	3,511	19.538	−25.750	39.592
#3	380	1,613.4	4,880	2.106	−4.025	6.618

Table 7.5 Grating substrate coefficients (maximum flexures in μm) [21]

Grating	A_{40}	A_{60}
#1	1.913E-9	9.14E-14
	[2.68]	[0.15]
#2	1.913E-9	9.14E-14
	[2.68]	[0.15]
#3	1.822E-9	1.03E-13
	[2.33]	[0.15]

Table 7.6 MDM coefficients and incidence angle [21]

Grating	A_{31}	A_{33}	A_{42}	i_{MDM}
#1	4.821E-8	−5.582E-8	−2.172E-9	29.96°
#2	1.880E-8	−2.671E-8	−2.360E-9	16.92°
#3	0.512E-8	−0.003E-8	−0.180E-9	10.00°

→ *In x, the* Cos *spectral resolution* $\lambda/\delta\lambda$ *is increased by a factor 10,*
→ *In y, the limiting magnitude is increased by 1–1.2 magnitude* [21]

The FWHM of the blur image at 1,300 Å is $2.5 \times 88\,\mu\text{m}^2$. The diffraction will increase the width in the direction of dispersion up to $3.8 \times 88\,\mu\text{m}^2$. Compared to the results of imaging and resolution tests by the Cos team [56], with a 3,800 l/mm classically *Astm* 3 corrected grating at 1,284 Å showing the FWHM image of $38 \times 264\,\mu\text{m}^2$, the high-order corrected grating provides images 10-times better in dispersion and 2.8-times better in cross direction.

The resolving power of the f/24 HST images at the input of Cos is $1.22\,\lambda\, f/d = 3.8\,\mu\text{m}$ at 1,300 Å and the concave gratings provide a magnification $\simeq -1$. The images given by gratings #1 and #2 are diffraction limited with regard to the resolution over the main part of the spectral range, and are nearly diffraction limited at the lower wavelength. With grating #3, of low dispersion, both the widths and heights of the images are diffraction limited.

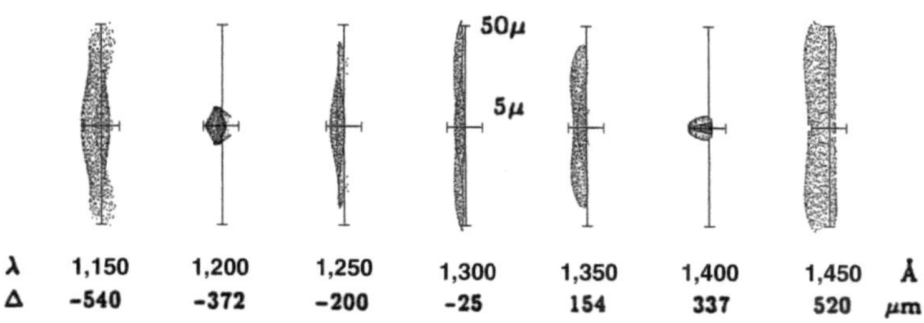

Fig. 7.14 Spot-diagram of Cos with grating#1, 3,800 l.mm^{-1}. For each color, Δ is defocussing onto the principal ray with respect to the Rowland circle. A positive Δ means an increase of the image distance to the grating vertex. The pixel size of the Cos detector is $2.4 \times 33\,\mu\text{m}^2$. Compared to the design by Osterman et al. [56], the gain is 28 in light concentration (after Duban [21])

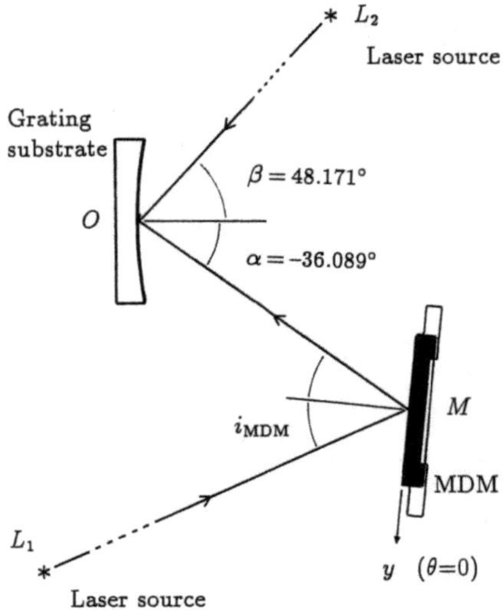

Fig. 7.15 Basic recording mounting. The recording angles α and β of the principal rays at the vertex O of the grating from laser source points L_1 and L_2 respectively are given by Table 7.4. For Cos grating #1, the α and β value are shown, the incidence angle at the vertex M of the MDM is $i_{\mathrm{MDM}} = 29.96°$; the Rowland circle optical paths are $L_1O = R\cos\alpha$ and $L_2O = R\cos\beta$ with $R = 1,652$ mm and $L_1M = 1,100$ mm (after Duban [21])

The recording distance from the laser source 1 to the MDM is 1,100 mm for gratings #1 and #2, and 1,000 mm for grating #3 (Fig. 7.15).

The f/24 HST projects central beam areas for all three gratings that are contained in a 73.2 mm circle since $i \leq 20°$ (see Table 7.4); for recording 80-mm circular aperture Cos gratings, the corresponding size of the recording projected beam at the MDM is $42.9 \times 53.3\,\mathrm{mm}^2$ for #1 and #2, and a little smaller for #3.

7.8.3 Elasticity Design of a Six-Arm MDM as Recording Compensator

Plane six-arm MDMs have been developed and experimented by Lemaitre [40, 42]. These mirrors can provide the co-addition of six Clebsch-Seidel modes but only three modes have to be generated to achieve the recording of the Cos-HST gratings. Aberration corrections achieved by an active MDM are straightforwardly obtained though the other alternative based on the use of a digital computer-generated interferogram is also of interest [17, 29].

Compared to glass or vitroceram materials, metal mirrors present several features that are of interest for the achievement of high curvature variations and high asphericity variations as well. The gain in *flexibility-ratio* σ_{lim}/E, is larger than 100. This is basically due to the much higher yield strength σ_{lim} of metal alloys.

A second selective criterion for metal substrates is a perfect *stress-strain linearity* in the sense of Hooke's law. With respect to these criteria, the Fe87 Cr13 alloy is well known as having a large linear range and no polishing problems. Otherwise metal alloys such as Cu62 Ni18 Zn20 or Ti90 Al6 V4 have not been characterized but might show good stress-strain linearities on large linear ranges while more flexible Al predominant alloys show more restricted linear ranges and need a Ni overcoat for polishing. The selected metal alloy was Fe87 Cr13 – referenced in the series AISI 420 – brought in a *quenched* state to Brinell hardness BH = 300. This material used for a long time at LOOM provides very smooth polished surfaces. The optimization for a convenient flexibility has been done by determining the rigidities D_1 and D_2 i.e. the thicknesses t_1 and t_2, with respect to the maximum stress given by a fifth-order triangle aberration $Tri\,5$ as defined by the A_{33} coefficient of the COS grating #1 (Table 7.6). The maximum stress has been kept lower than the yield strength of the Fe87 Cr13 alloy which is $1,200\text{N/mm}^2$. In order to ensure the best 3D-homogeneity of the substrate, the vase form and radial arms were machined out from the mass of a disk. Nine differential screws linked between the MDM and its support allow generating the flexure while the three remaining screws located at $\theta = 0, \pm 2\pi/3$ and $r = a$, define the reference plane of the deformations. A six-arm MDM was appropriate as a COS grating recording compensator (Fig. 7.16).

The axial distribution of forces $F_{a,k}$ and $F_{c,k}$ applied to the MDM has been determined for each of six Clebsch-Seidel modes having a ptv deformation of $1\,\mu$m at $r = a = 40$ mm over $\theta \in [0, 2\pi]$. These forces and the associated geometrical parameters of the built MDMs are given in Table 7.7.

Five flexure modes were evaluated from He-Ne interferograms with respect to a plane (Fig. 7.17).

With the above optical designs of COS, the grating substrates must also be axisymmetric aspherics – caused by the *Sphe* 3 residual of the HST primary mirror – and the holographic recording requires the co-addition onto the MDM of the three A_{31}, A_{33}, and A_{42} Clebsch-Seidel modes such as given in Table 7.5. For COS grating #1, the MDM geometry of the *holographic recording compensator* and $F_{a,k}$ and $F_{c,k}$ forces that generate those flexures are listed in Table 7.8.

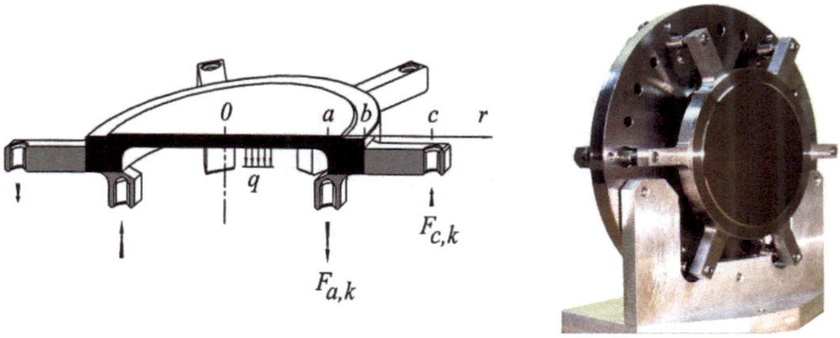

Fig. 7.16 Six-arm vase-form MDM as *recording compensator* for aberration corrected holographic gratings. Clear aperture diameter $2a = 80$ mm. *(Left) True proportion design. (Right) View in its stressing cell* [LOOM]

Table 7.7 Six-arm plane MDM. Force distributions $F_{a,k}$ and $F_{c,k}$ derived from $M_r(b)$ and $V(b)$. MDM geometry $t_1 = 5$ mm, $\gamma = (t_1/t_2)^3 = 1/27$, $a = 40$ mm, $b/a = 1.35$, $c/a = 2$. Fe87 Cr13 alloy $E = 205 \times 10^9$ Pa and $\nu = 0.305$. Forces generating $z_{nm} = A_{nm}\, r^n \cos m\theta$ modes with a ptv flexure of 1μm at $a = 40$ mm corresponding to Clebsch-Seidel coefficients $A_{20} = 6.250 \times 10^{-7}$, $A_{40} = 3.906 \times 10^{-10}$, $A_{22} = A_{20}/2$, $A_{31} = 7.812 \times 10^{-9}$, $A_{33} = A_{31}$ and $A_{42} = A_{40}/2$ in mm^{1-n}. Clear aperture $2a$. [Units: daN]

Angle	Arm nb.	$Cv\,1$ $n=2, m=0$		$Sphe\,3$* $n=4, m=0$		$Astm\,3$ $n=2, m=2$		$Coma\,3$ $n=3, m=1$		$Tri\,5$ $n=3, m=3$		$Astm\,5$ $n=4, m=2$	
θ	k	$F_{a,k}$	$F_{c,k}$	$F_{a,k}$	$F_{c,k}$	$F_{a,k}$	$F_{c,k}$	$F_{a,k}$	$F_{c,k}$	$F_{a,k}$	$F_{c,k}$	$F_{a,k}$	$F_{c,k}$
0	1	−2.76	2.76	−11.95	7.02	5.62	−0.14	−1.90	1.54	26.58	2.66	6.98	0.86
$\pi/3$	2	−2.76	2.76	−11.95	7.02	−2.81	0.07	−0.95	0.77	−26.58	−2.66	−3.99	−0.43
$2\pi/3$	3	−2.76	2.76	−11.95	7.02	−2.81	0.07	0.95	−0.77	26.58	2.66	−3.99	−0.43
π	4	−2.76	2.76	−11.95	7.02	5.62	−0.14	1.90	−1.54	−26.58	−2.66	6.98	0.86
$4\pi/3$	5	−2.76	2.76	−11.95	7.02	−2.81	0.07	0.95	−0.77	26.58	2.66	−3.99	−0.43
$5\pi/3$	6	−2.76	2.76	−11.95	7.02	−2.81	0.07	−0.95	0.77	−26.58	−2.66	−3.99	−0.43

* Uniform load $q = 64 D_1 A_{40} = 0.00589$ MPa.

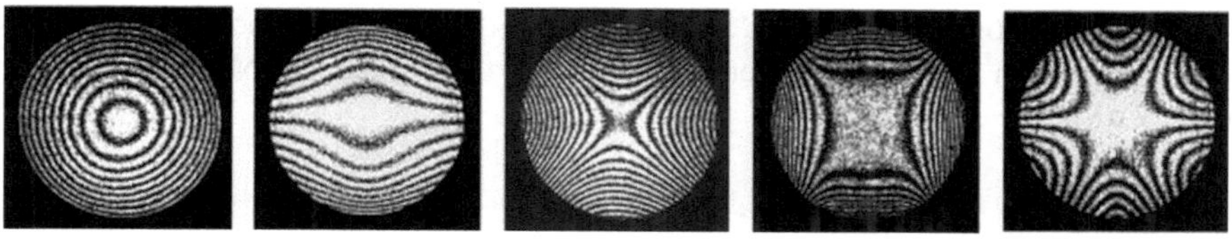

Fig. 7.17 He-Ne interferograms with Six-arm MDM at full aperture 80 mm showing single modes *Cv* 1, *Coma* 3, *Astm* 3, *Astm* 5, and *Tri* 5. The *Sphe* 3 mode – not shown – could be obtained by air pressure or partial vacuum inside the mirror [Loom]

Table 7.8 Six-arm plane MDM as *holographic recording compensator* for the Cos gratings. Force distributions $F_{a,k}$ and $F_{c,k}$ for the recording of Cos concave grating #1. MDM geometry $t_1 = 5\,\mathrm{mm}$, $\gamma = (t_1/t_2)^3 = 1/27$, $a = 40\,\mathrm{mm}$, $b/a = 1.35$, $c/a = 2$. Elasticity constants $E = 205\,\mathrm{GPa}$, $\nu = 0.305$. Coefficients of $z_{nm} = A_{nm} r^n \cos m\theta$ modes $A_{31} = 4.821 \times 10^{-8}$, $A_{33} = -5.582 \times 10^{-8}$, $A_{42} = -2.172 \times 10^{-9}$ in mm^{1-n}. [Units: daN]

Angle	Arm nb.	*Coma* 3 $n = 3, m = 1$		*Tri* 5 $n = 3, m = 3$		*Astm* 5 $n = 4, m = 2$		Co-addition *	
θ	k	$F_{a,k}$	$F_{c,k}$	$F_{a,k}$	$F_{c,k}$	$F_{a,k}$	$F_{c,k}$	$\Sigma\mathcal{F}_{a,k}$	$\Sigma\mathcal{F}_{c,k}$
0	1	−11.72	9.54	−190.00	−19.98	−77.54	−9.58	−279.2	−20.0
$\pi/3$	2	−5.86	4.77	190.00	19.98	38.72	4.79	222.8	29.5
$2\pi/3$	3	5.86	−4.77	−190.00	−19.98	38.72	4.79	−145.5	−19.9
π	4	11.72	−9.54	190.00	19.98	−77.54	−9.58	124.2	0.8
$4\pi/3$	5	5.86	−4.77	−190.00	−19.98	38.72	4.79	−145.5	−19.9
$5\pi/3$	6	−5.86	4.77	190.00	19.98	38.72	4.79	222.8	29.5

* Since $m \neq 0$ for all three modes, $\Sigma_k F_{a,k} = \Sigma_k F_{c,k} = \Sigma_k \mathcal{F}_{a,k} = \Sigma_k \mathcal{F}_{c,k} = 0$.

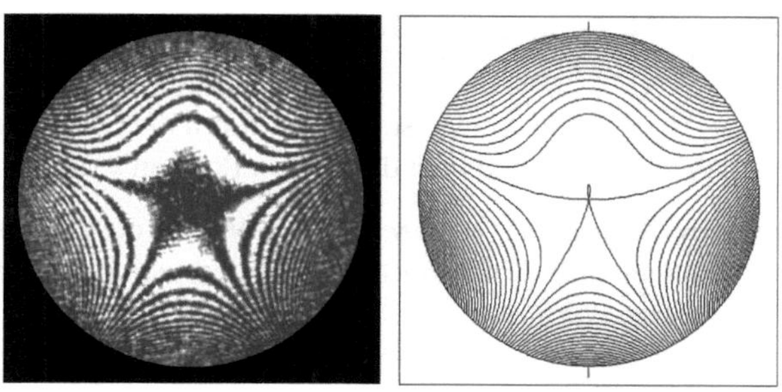

Fig. 7.18 Full aperture He-Ne interferogram obtained with the six-arm MDM tuned as optical path compensator for the holographic recording of Cos grating #1 (3, 800l/mm). The three modes *Coma*3, *Tri*5, and *Astm* 5 are co-added by generating the resulting forces $\mathcal{F}_{a,k}$ and $\mathcal{F}_{c,k}$ such as given in the two last columns of Table 7.8. (*Left*) Obtained shape. (*Right*) Synthetic theoretical shape. The θ-origin is downwards [Loom]

The interferogram of flexure resulting from the three superposed modes was found to be in excellent agreement with the theoretical He-Ne interferogram (Fig. 7.18).

7.9 Degenerated Configurations and Triangle Mode

7.9.1 Special Geometry for the Triangle Mode

In the case of the 5th-order triangle mode, denoted $Tri5$ for $z_{33} = A_{33}r^3 \cos 3\theta$, *degenerated configurations* can be obtained; this means that the external forces $F_{a,k}$ applied to the inner end of the arms can be cancelled. Similarly to the case for $Astm3$, if the condition $F_{a,k} = 0$ is introduced into the equation set (7.13), then

$$(c - b)V_r(b) = M_r(b), \quad \forall\theta.$$

The expressions of $V_r(b)$ and $M_r(b)$ are given by the equation set (7.12f). After substitution into the above equation we obtain a MDM whose arms have the special geometry [41]

$$\frac{c}{b} = 1 - \frac{1 + \frac{1}{4}(1-\gamma)(1-\nu)\left[2 - \frac{5-\nu}{1-\nu}\frac{b^2}{a^2} - \frac{a^6}{b^6}\right]}{3 + \frac{3}{4}(1-\gamma)(1-\nu)\left[2 - \frac{7-3\nu}{1-\nu}\frac{b^2}{a^2} + \frac{a^6}{b^6}\right]}, \tag{7.57a}$$

which is valid for flat to moderately curved vase or meniscus mirrors.

7.9.2 Single Tri 3 Mode and Degenerated Meniscus Form

With a meniscus form, $\gamma = (t_1/t_2)^3 = 1$, and we can set $b = a$. From (7.9.2a), and similarly to $Astm$ 3 in Sect. 7.4, the b/c ratio does not depend on Poisson's ratio.

→ *A meniscus form of aperture* $2a$ *provides a pure fifth-order triangle flexure* Tri 5 *by requiring only six alternatively opposite forces* $F_{c,k}$ *applied to the ends of folded arms at the radial distance* c *such as*

$$\frac{c}{a} \equiv \frac{c}{b} = \frac{2}{3}. \tag{7.57b}$$

In the general case of a vase form or a meniscus form, the $F_{c,k}$ forces applied to the arm ends can be derived from (7.13a)

$$F_{c,k} = b\int_{\pi(2k-3)/k_m}^{\pi(2k-1)/k_m} V_r(b,\theta)d\theta. \tag{7.58}$$

With $k_m = 6$ and since $E_{33} = 0$ in (7.12f), we obtain after integration

$$F_{c,k} = (-1)^k 4(1-\nu)\left(3\,B_{33} - 6\frac{C_{33}}{b^6} + 2\frac{7-3\nu}{1-\nu}b^2 D_{33}\right) D_2 b, \qquad (7.59)$$

and after substitution of the coefficients

$$F_{c,k} = (-1)^k 3(1-\gamma)(1-\nu)^2 \left[\frac{4}{(1-\gamma)(1-\nu)} + 2 - \frac{7-3\nu}{1-\nu}\frac{b^2}{a^2} + \frac{a^6}{b^6}\right] D_2 b A_{33}. \qquad (7.60a)$$

→ *For a meniscus mirror with six arms ($k_m = 6$), the intensity and direction of the six axial forces are represented by*

$$F_{c,k} = (-1)^k \frac{E\,t^3 a}{1+\nu} A_{33} \quad \text{with} \quad k = 1,\ 2, \cdots, 6. \qquad (7.60b)$$

7.9.3 Single Tri 3 Mode and Degenerated Vase Form

Two identical *six-arm vase form* mirrors of a degenerated configuration for a $Tri\,5$ mode flexure have been built. Their design satisfied analytical condition (7.9.2a) for $F_{a,k} = 0$. The choice of a chromium stainless steel material allowed both a large stress-strain range and much larger flexure sags than with a vitroceram substrate. Interferometric optical tests showed that this mode was obtained with high accuracy. Within the triangle mode family, the results from this analysis were $|A_{53}/A_{33}| \le 0.033$ and $|A_{73}/A_{33}| \le 0.0013$ (Fig. 7.19).

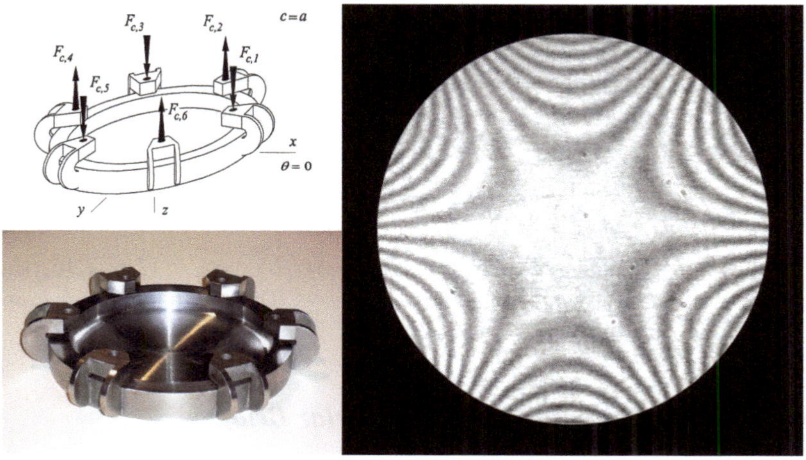

Fig. 7.19 Degenerated configuration of a *six-arm vase form* providing a $Tri\,5$ flexure mode. This solution satisfies condition (7.9.2a) for $F_{a,k} = 0$. Hence only the six forces $F_{c,k}$ act to the ends of folded arms clamped at $r = b$. Substrate : Fe87 Cr13 stainless steel, $\nu = 0.305$. Geometry : clear aperture $2a = 100$ mm, $2a/t_1 = 20$, $t_2/t_1 = 3$, $b/a = 1.2$, $c/b = 0.8382$, then $c/a = 1.0058 \simeq 1$. (*Left*) Design and view of the holosteric mirror. (*Right*) He-Ne interferogram of the flexure [LOOM]

7.10 Single Mode and Deformable Outer Ring

7.10.1 Outer Ring Designs for High Accuracy Correction

For MDMS generating a single mode $m = n$ the rigidity of the outer ring is generally set much larger than that of the meniscus, say, the reciprocal rigidity-ratio is $1/\gamma = D_2/D_1 = (t_2/t_1)^3 \geq 3$. Hence, the ring is preeminent in the *overall* rigidity of a vase form. The forces that are applied to the ring must provide an accurate flexure in $\cos m\theta$. For practical reasons those forces are *not* generally applied in the form of continuous cosine distributions because this would require an extremely large number of force points. It is always preferred to reduce this number of forces to the strict *minimum* for generating the ring flexure.

For instance, if $m = n = 2$, then the minimum number of forces is four in a vase form degenerated configuration, i.e. two equal and opposite pairs in orthogonal directions. If these forces are all applied to the *rear* side of the ring at discrete azimuth positions $0,\ \pi/2,\ \pi,\ 3\pi/2$, then the ring will basically bend in $\cos 2\theta$. However, finite element analysis with four axial forces applied to the rear side of a *constant* thickness ring shows that there exist small deviations from this modulation.

In order to obtain a $\cos 2\theta$ modulation within, say, 5–10 nm accuracy, one of the two following alternatives can be used.

7.10.2 Ring with Axial Thickness Variation

For the above reason, when directly applying the axial forces at the extremals of $\cos m\theta$, the constant axial thickness t_2 of the ring must be transformed into a slightly variable one $t_2(\theta)$ which shows m wedges at these extremals. The m-wedged shape is realized at the ring rear side. For $m = n = 2$ – *Astm* 3 mode – and a metal ring with $\nu = 0.315$, its axial thickness in a quadrant is given by Hugot [35, 36] as

$$t_2(\theta) \simeq 1.11 \left[\frac{(b+a)\,F}{(b-a)\,EA_{22}} \frac{(1/2 - 2\theta/\pi)}{\cos 2\theta} \right]^{1/3}, \qquad 0 \leq \theta \leq \pi/2, \qquad (7.61)$$

which satisfies $t_2(\theta) = t_2(\frac{\pi}{2} - \theta)$ and leads to $t_2(0)/t_2(\frac{\pi}{4}) = \sqrt[3]{\pi/2} \simeq 1.16$. The thickness distribution is the same in the other quadrants. The external forces $\pm F$ are applied to the rear surface wedges (Fig. 7.20-Left).

7.10.3 Ring with Forces Acting on Angular Bridges

Instead of directly applying an external force in a direction where the angular modulation is an extremum, one subdivides this force in to two equal components and applies them to a ring of axial thickness $t_2 = constant$ via a bridge whose angular separation of the two acting ends is conveniently optimized (Fig. 7.20-Right).

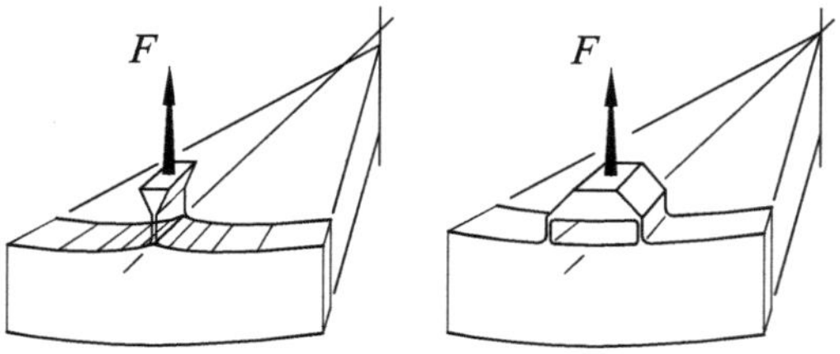

Fig. 7.20 Alternative solutions for a pure $\cos m\theta$ flexure of a ring by discrete axial forces. ($Left$) Wedged ring with axial thickness variation. ($Right$) Angular bridges with ring of constant axial thickness

7.11 Future Giant Telescopes and Segment Aspherization

7.11.1 Current Trends in Giant Telescope Concepts

Many investigations on the design of future *giant telescopes* with three, four, or five consecutive mirrors were done with a *spherical* primary mirror. For instance, Baranne and Lemaitre [5], Sasian [61], Wilson and Delabre [70], Dierickx, Delabre and Noethe [16] discussed solutions with a spherical primary mirror which also include a pupil transfer for upgrading the image quality by use of active and adaptive optics. Rakich [58] described a method that gave the complete solution set for four-mirror anastigmats in which all the mirrors were spherical. Even though a large variety of such systems were shown to exist, most of these were not practical due to large central obstructions. In a modified approach for setting up of the partial first-order characteristics, surveys of useful four-mirror anastigmats with one, two or three mirrors kept strictly spherical can be carried out.

However, several severe requirements such as diffraction-limited imaging in the visible, largest possible field of view and minimum number of mirrors are imperative features. At $\lambda = 550$ nm and with a 30-m aperture mirror, the spatial resolution to be achieved in the field of view by the optical design is $1.2\lambda/D = 4 \times 10^{-3}$ arcsec. Hence, it turns out that all the giant telescopes of the next generation are planned to use an *aspherical* primary mirror close to a paraboloid.

At this moment, in the 30–40 m telescope class, the Nasmyth TMT project [66] and the Cassegrain GSMT project [55] are classical Ritchey-Chretien or aplanatic Gregory designs, thus would have diffraction-limited fields of view of $\sim$2 arcsec in the visible. The segment shape of the primary mirror may be hexagonal or sector-shaped. The 25-m GMT project [10] is with a similar field of view and a primary mirror made of seven circular segments. In all these projects, the adaptive optics developments are assumed operating for a 5–6 m mirror diameter used as the *secondary* mirror which then is the telescope input *pupil*.

Other 30–40 m telescope projects have a seeing-limited field of view of 10 arcmin and include a diffraction-limited field of at least 5 arcmin in the visible. The following projects involve an *adaptive pupil mirror* of 2.5–3 m in diameter.

The four-mirror CFGT project of China, by Su, Wang and Cui [64], is a Nasmyth mounting with three aspheric mirrors. The telescope input pupil is the adaptive secondary mirror. The telescope horizontal beams are obtained by a slightly tilted aspheric tertiary mirror followed by a 120° deviation plane quaternary mirror. The tertiary mirror is a quasi-plane non-axisymmetric sphere (Fig. 7.21-Left).

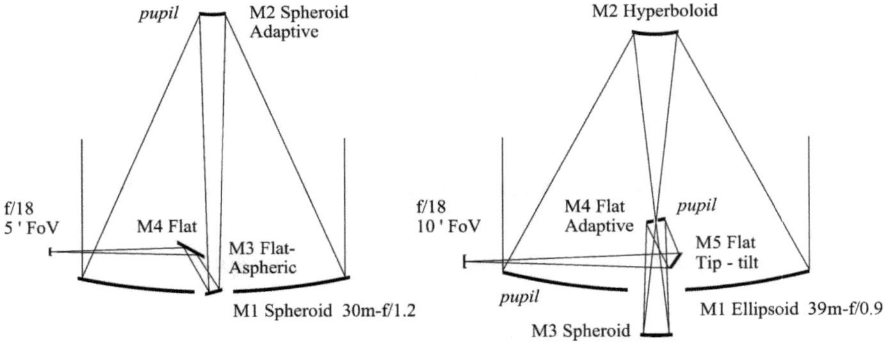

Fig. 7.21 Optical concepts for future 30–40 m telescopes with extended diffraction-limited field of view. (Left) The four-mirror CFGT project has a quasi-plane aspherical tertiary mirror (after CAS). (Right) The five-mirror E-ELT project has a pupil transfer on the quaternary mirror (after ESO)

The six-mirror JELT project [37] is a Nasmyth with three powered aspheric mirrors and three flat mirrors. A pupil transfer of the primary mirror on the sixth mirror is realized by the fifth mirror which is located sideways from the Nasmyth platform. The adaptive sixth mirror, located at the Nasmyth platform, works with a small beam deviation.

The five-mirror E-ELT project by Delabre, Cayrel et al. [25] is a Nasmyth with a tertiary mirror downwards of the primary mirror cell which provides a pupil transfer of the primary on the holed quaternary mirror. The flat quaternary mirror is dedicated to adaptive optics corrections. It has been shown [43] that, with a slight alteration of the image quality, if the imaged pupil quaternary mirror is given a non-axisymmetric asphericity – smaller than 1 μm ptv –, then the primary mirror becomes a paraboloid. The flat fifth mirror ensures the fast tip-tilt mode (Fig. 7.21-Right).

7.11.2 Active Optics Asphcerization of Mirror Segments

For future 30–40 m aperture telescopes the segmentation of the primary mirror will be an inescapable feature. A segment perimeter could be either circular, hexagonal or sector-shaped with a largest aperture size in the range 1–2 m. Given the large number of segments to be made (1,000–1,600), the realization processes must be carefully optimized. The replication technique from a deformable matrix – such as developed for the grating asphcerization (cf. Sect. 5.4) – is not clearly valid in these sizes and would probably give rise to difficulties for the mirror re-coating operations. A higher performing technique is stress figuring.

In order to obtain very smooth meniscus aspheric segments with a maximum surface error of, say, 20 nm rms, the recovering of a MDM geometry during stress figuring seems a promising alternative. Such a process is based on the newly developed "releasable property" of special epoxy resins by local heating or strong lighting. Hence a meniscus segment should be temporarily bonded on a smoothly deformable *ring harness* during active optics controlled figuring. The meniscus and ring harness with arms reconstruct the complete geometry of a MDM. The release property of the resin should allow a harness to generate many aspherized meniscuses. (Fig. 7.22).

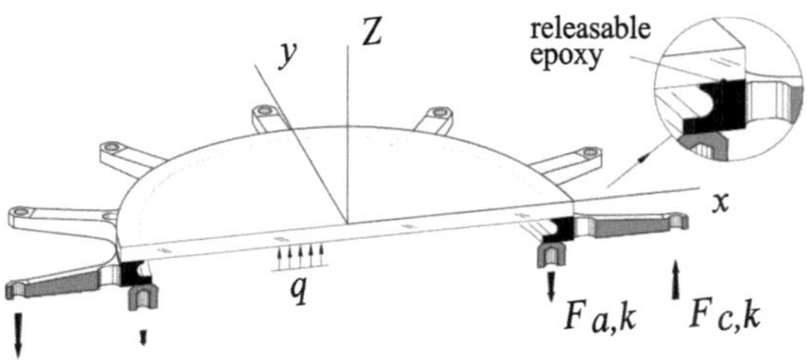

Fig. 7.22 The Meniscus segment bonded at its rear edge on a *ring harness* for stress figuring (here with $k_{\mathrm{m}} = 12$ arms). The releasable property of the resin should allow using the same harness to aspherize many meniscuses

At controlled room temperature, a vitroceram meniscus segment bonded on a metal ring following Hooke's law should not provide important thermoelastic effects. In any case, the possible thermoelastic effects – as $Cv\,1$ and $Sphe\,3$ modes – are *a fortiori* controlled by the harness loading system because the curvature correction mode must be also controlled for each segment as a function of its height from the mirror axis.

7.12 Vase Form and Middle Surface

With vase-form mirrors, the extra-thickness $t_2 - t_1$ due to the outer ring is only on one side: the rear side of the optical surface. Distributing symmetrically thicknesses $(t_2 - t_1)/2$ on each side of the built-in plate would not allow a polishing access for the mirror surface. So the middle surface of a vase form is not a plane or a sphere – surfaces of constant curvature – but is *curved* at its edge towards the rear side.

Now, from the point of view of the above flexures as derived from the thin plate elasticity theory, the whole middle surface is assumed to be a plane or a sphere up to $r = b$. Thus, the ring thickness t_2 is theoretically considered as equally distributed on each side of the middle surface, i.e. in a "T-shaped" mirror edge. This difference in the shape of the middle surface is traduced in practice by a small curvature mode – let it be $Cv\,1'$ – which arises in addition to the generated modes.

Since a MDM uses two forces per arm, the $Cv\,1'$ effect can be exactly cancelled by generating anti-moments at the contour with those forces. With a degenerated configuration, since there is only one force per arm, the exact balance of the azimuthal amplitudes can be recovered either by refocussing or, if not possible, by polishing the mirror at a new spherical shape changed by the small opposite amount $-Cv\,1'$. When generating z_{22} or z_{33} mode, the $Cv\,1'$ effect is positive, i.e. corresponding to a flexure sag towards z positive. The mirror displayed by Fig. 7.19 was designed with a plane surface and polished slightly convex to a clear aperture sag of one He-Ne wavelength for obtaining a balanced interferogram.

Finite element analysis codes accurately allow one to determine the small amount of this curvature mode correction.

7.13 Vase Form and Saint-Venant's Principle

With respect to *Saint-Venant's principle* (cf. Sect. 1.13.13), the deformation interferograms show that a *vase-form* MDM is useful to avoid the slope discontinuities of the shear component at the optical surface. Such discontinuities are caused by the application of concentrated point forces near the optical surface, as can locally occur at the contour of a bent meniscus form.

In contrast to this point-force proximity, one can find configurations like that of an optical meniscus whose perimeter is clamped to the end of a deformable tube – as a *drum-like form* generating the $Astm\,3$ mode [69] – where point forces are applied at large distance from the optical surface, i.e. act at the other end of the long drum-tube. However, such an elasticity design is not compact enough to be applied to large mirrors.

With either a single mode degenerated configuration only requiring a single $F_{c,k}$ force set – such as derived in Sects. 7.4 and 7.9 for the $Astm\,3$ and $Tri\,5$ modes – or a multimode configuration requiring $\mathcal{F}_{a,k}$ and $\mathcal{F}_{c,k}$ sum force sets, a conclusion is the following.

→ *A vase form with radial arms, or MDM, is a compromise for an optimal fulfilment of Saint-Venant's principle.*

References

1. L.N. Allen, H.W. Romig, in *Advanced Optical Manufacturing*, SPIE Proc., **1333**, 22 (1990)
2. L.N. Allen, J.J. Hannon, R.W. Rambach, in *Active and Adaptative Components*, SPIE Proc., **1543**, 190 (1991)
3. J.R.P. Angel, N.J. Woolf, H.W. Epps, Good images with very fast paraboloidal primaries: an optical solution and applications, in *International Conference on Advanced Technology Telescopes*, SPIE Proc., **332**, 134 (1982)

4. J.G. Baker, On improving the effectiveness of large telescopes, IEEE Trans. Aerosp. Electron. Syst., **AES-5**, 261 (1969)
5. A. Baranne, G.R. Lemaitre, Combinaisons optiques pour très grands télescopes ; le concept TEMOS, C.R. Acad. Sc. Paris, **305** Série II, 445–450 (1987)
6. E.F. Borra, Liquid mirrors, Can. J. Phys., **73**, 109–125 (1995)
7. E.F. Borra, R. Content, L. Girard, Optical shop tests of a f/1.2 – 2.5 meter diameter liquid mirror, Ap. J., **418**, 943–946 (1993)
8. E.F. Borra, R. Content, L. Girard, S. Szapiel, L.M. Tremblay, E. Boily, Liquid mirrors: Optical shop tests and contributions to the technology, Ap. J., **393**, 829–847 (1992)
9. M.P. Budiansky, Ten Meter Telescope Technical Note No. 95, University of California, Berkeley
10. Carnegie Observatories report, The 24.5 m Giant Magellan Telescope project (2007)
11. A. R. F. Clebsch, in *Theorie der Elasticität fester Körper*, Teubner edit. Leipzig (1862), [French translation: *Théorie de l'Élasticité des Corps Solides* with annotations and complements by Saint-Venant and Flamant, Dunod edit., Paris (1881)]
12. A. Couder, Sur les miroirs de télescopes, Bulletin Astronomique, Paris, 2ème Série, Tome VII, Fasc. VI, 219 et seq. (1931)
13. X. Cui, Y. Li, X. Ni, D. Yang, Z.-q. Yao, G. Li, Z. Yao, Y. Wang, Active support system of LAMOST reflective Schmidt plate, SPIE Proc., **4837**, 628–636 (2003)
14. X. Cui, D.-q. Su, G. Li, Z. Yao, Z. Zhang, Y. Li, Z. Yong, W. You, Z. Xu, W. Hai, Experiment system of LAMOST active optics, SPIE Proc., **5489**, 974–985 (2004)
15. X. Cui, D.-q. Su, Y.-n. Wang, G. Li, G. Li, Y. Zhang, Y. Li, The optical performance of LAMOST telescope, SPIE Proc., **7733** (2010)
16. P. Dierickx, B. Delabre, L. Noethe, OWL 100-m telescope optical design, active optics and error budget, in *Optical Design, Materials, Fabrication, and Maintenance*, SPIE Proc., **4004**, 203–209 (2000)
17. S.A. Dimakov, Comparison properties of analog and digital correction of a primary mirror aberration in observing telescopes, in *Laser Optics 2003 : Wavefront Transformation and Laser Beam Control*, ICLO XI, SPIE Proc., **5481**, 59–70 (2004)
18. M. Duban, G.R. Lemaitre, R. Malina, A new recording way to obtain high resolution holographic gratings through use of MDMs, Appl. Opt. **37**(16), 3438–3439 (1998)
19. M. Duban, Holographic aspheric gratings printed with aberration waves, Appl. Opt. **26**, 4263–4273 (1987)
20. M. Duban, K. Dohlen, G.R. Lemaitre, Illustration of the use of MDMs to record high resolution concave gratings: Results for the COS gratings of HST, Appl. Opt. **37**(31), 7214–7217 (1998)
21. M. Duban, Theory and computation of three Cosmic Origins Spectrograph aspheric gratings recorded with a MDM, Appl. Opt. **38**(7), 1096–1102 (1999)
22. M. Duban, Theory of spherical holographic gratings recorded by use of a MDM, Appl. Opt. **37**, 7209–7213 (1998)
23. M. Duban, Third-generation holographic Rowland mounting: Third order theory, Appl. Opt. **38**(16), 3443–3449 (1999)
24. M. Duban, Third-generation Rowland holographic mounting, Appl. Opt. **30**, 4019–4025 (1991)
25. E-ELT – Summary of telescope optical designs, ESO conf. report, Marseille, Doc. ESO E-ELT 5 mirrors (2006) and E-ELT Programme, Doc. ESO (2011)
26. P. Germain, P. Muller, in *Introduction à la Mécanique des Mileux Continus*, Masson edit., Paris (1994)
27. B.K. Gibson, P. Hickson, Ap. J., **391**, 391 (1992)
28. J.C. Green, The Cosmic origins spectrograph: A Hubble replacement instrument, in *Space Telescope and Instruments*, SPIE Proc. **3356**, 265–270 (1998)
29. M. Gruneisen, Computer-generated diffractive optics for large aberration correction, in *Laser Optics 2003 : Wavefront Transformation and Laser Beam Control*, ICLO XI, SPIE Proc., **5481**, 82–93 (2004)

30. P. Hickson, E.F. Borra, R. Cabanac, R. Content, B.K. Gibson, G.A.H. Walker, Ap. J. Lett., **436**, 201 (1994)
31. P. Hickson, E.F. Borra, R. Cabanac, S.C. Chapman, V. de Lapparent, M. Mulrooney, G.A. Walker, Large Zenith Telescope project: A 6-m mercury-mirror telescope, in *Avanced Technology Optical/IR Telescopes VI*, SPIE Proc., **3352**, 226–232 (1998)
32. P. Hickson, E.H. Richardson, A curvature-compensated corrector for drift-scan observations, PASP **110**, 1081–1086 (1998)
33. P. Hickson, Eliminating the Coriolis effect in liquid mirrors, PASP, **113**, 1511–1514 (2001)
34. P. Hickson, Private communication to E.H. Richardson (1995)
35. E. Hugot, G.R. Lemaitre, M. Ferrari, Active optics : single actuator principle and angular thickness distribution for astigmatism compensation by elasticity, Appl. Opt., **47**, 1401–1409 (2008)
36. E. Hugot, *Optique Astronomique et Elasticité*, Ph. D. dissertation, Universite de Provence, Aix Marseille I (2007)
37. M. Iye and JELT Working group, Concept study of Japan Extremely Large Telescope, in *Ground-based Telescopes*, SPIE Proc., **5489**, 417–428 (2004)
38. G.R. Kirchhoff, Uber das gleichgewicht und die bewegung einer elastischen scheibe, Journ. Crelle **40**, 51 (1850). See also the comment on this paper by A.E.H. Love, *A Treatise on the Mathematical Theory of Elasticity*, Dover publ., 458 (1927) and I. Todhunter, K. Pearson, *The Theory of Elasticity*, Dover Pub., **2**-Part 2, 44 and 406 (1960)
39. G.R. Kirchhoff, Vorlesungen über Mathematische Physik, *Mechanik*, 450 (1877)
40. G.R. Lemaitre, Active optics and aberration correction with multimode deformable mirrors (MDMs) – Vase form and meniscus form, in *Laser Optics 2003 : Wavefront Transformation and Laser Beam Control*, ICLO XI, SPIE Proc., **5481**, 70–81 (2004)
41. G. R. Lemaitre, Active Optics: Vase or meniscus multimode mirrors and degenerated monomode configurations, Meccanica, Springer, **40**, vol. 3, 233–249 (2005)
42. G.R. Lemaitre, M. Duban, Universal method for holographic grating recording: MDMs generating Clebsch-Zernike polynomials, Appl. Opt. **40**(4), 461–471 (2001)
43. G.R. Lemaitre, Note 1 and Note 2 to ESO E-ELT telescope design working group (2006). Note 1 includes an option with a paraboloid primary mirror design, instead of a slightly elongated eliipsoid primary, that would use an asphericity sag smaller than 600-nm PtV at the pupil conjugate adaptive mirror M4
44. G.R. Lemaitre, Various Aspects of Active Optics, in *Telescopes and Active Systems*, Orlando, FA, SPIE Proc., **1114**, 328–341 (1989)
45. J. Lubliner, J.E. Nelson, KECK Telescope : Stressed mirror polishing, Appl. Opt., **19**, 2332–2340 (1980)
46. T.S. Mast, Nelson J.E., in *Advanced Technology Telescopes IV*, SPIE Proc., **1236**, 1236 (1990)
47. J. McGraw, H. Stockman, R. Angel, H. Epps, SPIE Proc. **331**, 137 (1982)
48. G. Moretto, A corrector design using vase mirrors that allows a fixed telescope to access a large region of sky, Appl. Opt. **36**(10), 2114–2122 (1997)
49. G. Moretto, G.R. Lemaitre, T. Bactivelane, M. Wang, M. Ferrari, S. Mazzanti, E.F. Borra, Active mirrors warped using Zernike polynomials for correcting off-axis aberrations of fixed primary mirrors: Optical testing and performance evaluation, Astron. Astrophys. Suppl. Ser. **114**, 379–386 (1995)
50. G. Moretto, in *Optical Designs for Fixed Primary Mirrors Observing Off-Axis*, Ph. D. thesis, Université Laval, Chap. 5 (1996)
51. J.A. Morse, J.C. Green, D. Ebbets et al., Performance and science goals of the Cosmic Origins Spectrograph for the HST, in *Space Telescope and Instruments*, SPIE Proc. **3356**, 365–368 (1998)
52. M.K. Mulrooney, Ph. D. thesis, Rice Univerity (2002)
53. J.E. Nelson, G. Gabor, J. Lubliner, T. Mast, KECK Telescope : Stressed mirror polishing, Appl. Opt., **19**, 2341–2350 (1980)
54. J.E. Nelson, M. Temple-Raston, Off-axis expansions of conic surfaces, KECK Observatory Report 91 (1982)

55. NOAO report, Developing the future Giant Segmented Mirror Telescope (GSMT) (2006)
56. S. Osterman, E. Wilkinson, J.C. Green, K. Redman, FUV grating performance for the Cosmic Origins Spectrograph, in *UV, Optical and IR Space Telescopes and Instruments*, SPIE Proc. **4013**, 360–366 (2000)
57. M. Paul, Systèmes correcteurs pour réflecteurs astronomiques, Rev. Opt. **14**(5), 169–202 (1935)
58. A. Rakich, Four-mirror anastigmats with useful first order layouts and minimum complexity, in *Novel Optical Systems Design and Optimization* VII, SPIE Proc., **5524**, 101–114 (2004)
59. E.H. Richardson, Corrector lens design for the UBC 5-Meter Liquid Mirror Telescope, private communication to P. Hickson (1995)
60. A. Saint-Venant (Barré de), in *Résumé des Leçons de Navier sur l'Application à la Mécanique*, Dunod edit., Paris (1881)
61. J.M. Sasian, Four-mirror optical system for large telescopes, Opt. Eng., **29**(10), 1181–1185 (1990)
62. D.J. Schroeder, in *Astronomical Optics*, Academic Press ed., San Diego 115 (1987)
63. D.-q. Su, X. Cui, Active optics in LAMOST, Chin. J. Astron. Astrophys., **4**(1), 1–9 (2004)
64. D.-q. Su, Y.-n. Wang, X. Cui, A configuration for [Chinese] future giant telescope, Chin. J. Astron. Astrophys., **28**, 356–366 (2004)
65. S. P. Timoshenko, S. Woinowsky-Krieger, in *Theory of Plates and Shells*, McGraw-Hill edit., New York 282 (1959)
66. TMT observatory corp., Thirty Meter Telescope construction proposal (TMT) (2007)
67. S.-g. Wang, D.-q. Su, Y.-q. Chu, X. Cui, Y.-n. Wang, Special configuration of a very large Schmidt telescope for extensive astronomical spectroscopic observation, Appl. Opt., **35**(25), 5155–5161 (1996)
68. M. Wang, G. Moretto, E.F. Borra, G.R. Lemaitre, A single active corrector for liquid mirror telescopes observing off zenith, Astron. Astrophys. **285**, 344–353 (1994)
69. M. Wang, G.R. Lemaitre, Diffraction-limited toroid mirrors aspherized by active optics and drum-like forms, Astron. Astrophys., **240**, 551–555 (1990)
70. R.N. Wilson, B. Delabre, New optical solutions for very large telescopes using a spherical primary, Astron. Astrophys., **294**, 322–338 (1995)
71. S.R. Wilson, D.W. Reicher, J.R. McNeill, in *Advances in Fabrication and Metrology for Optics and Large Optics*, SPIE Proc., **966**, 74 (1988)
72. R.N. Wilson, in *Reflecting Telescope Optics II*, Springer edit. Berlin, Chap. 1, 28–37 (1999)
73. R.N. Wilson, in *Reflecting Telescope Optics I*, Springer edit. Berlin, Chap. 3, 219–220 (1996)
74. C.G. Wynne, Field correctors for telescopes at better observing sites, Mon. Not. Astr. Soc. **189**, 279 (1979)
75. Y. Zhang, X. Cui, Calculations for the pre-calibration of LAMOST active optics, Chin. J. Astron. Astrophys., **5**–3, 302–314 (2005)

Chapter 8
Own Weight Flexure and Figure Control of Telescope Mirrors

8.1 Primary Mirror Support Systems Against Gravity

8.1.1 Introduction

Foucault solved the problem of maintaining a mirror with a high reflective coating by introducing *glass* mirrors, which then can be easily re-silvered after chemical removal of the tarnished coating. This ended the era of speculum metal mirrors which required, when tarnished, a repolishing within a seeing limited – or diffraction limited – criterion. The chemical process was later replaced by the vacuum deposition process by J. Strong (cf. Sect. 1.1.5).

The next problems for the primary mirror of large telescopes were to minimize their elastic deflection under *gravity* – or *own weight deformation* – and also that resulting from *thermal* gradients. This latter problem was appropriately solved by the inventions and developments of low expansion materials such as Pyrex, Sitall, fused silica, vitrocerams, and silicon carbide.

8.1.2 Axial and Lateral Support System Concepts

The design of *axial* and *lateral* supports for large telescope mirrors necessarily act onto several mirror subareas which support a small fraction of the mirror weight. In addition to some reference subareas, an appropriate number of additional subareas are distributed all over the back surface and possibly on the edge surface of the mirror. Whatever the geometrical distribution where these forces act, the associated system of forces must be non-hyperstatic. In other terms, the set of the supporting forces must be in *astatic* equilibrium so the 3D-orientation of the mirror remains unchanged with respect to the reference zones.

Three small reference areas, generally located at 120° and near the mirror edge, allow us to determine its axial position whilst two or three small reference areas are required to define the lateral position. The number of these latter areas depends on whether the telescope is an equatorial mount or alt-azimuthal mount.

G.R. Lemaitre, *Astronomical Optics and Elasticity Theory,* Astronomy and Astrophysics Library, DOI 10.1007/978-3-540-68905-8_8,

Let P denote the weight of the mirror, N_a the number of axial supports and N_l the number of lateral supports. We shall hereafter refer to *passive* axial and lateral support systems as astatic systems that deliver reaction sets of equal forces f_a and mean forces $< f_l >$ in the *axial* and *lateral* directions respectively; thus $N_a^2 f_a^2 + N_l^2 < f_l >^2 = P^2$ and these forces are

$$f_a = (P/N_a)\cos z, \qquad < f_l >= (P/N_l)\sin z, \tag{8.1}$$

where z is the zenith distance angle of the telescope axis.

In such *passive* support systems, the assumption of invariance of the mirror orientation is strictly valid only if the flexure of the cell structure remains axisymmetric in all orientations of the telescope. We note here that the telescope star-pointing error resulting from deviation from this hypothesis is easier to compensate for an altazimuth mount than with an equatorial mount. These corrections are achieved by the telescope control system.

Several types of passive astatic systems have been invented for axial support (Fig. 8.1) and lateral support (Fig. 8.2).

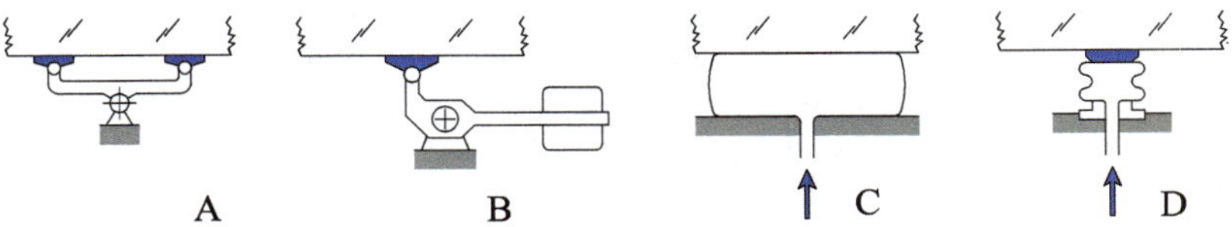

Fig. 8.1 Passive astatic systems for the *axial support* of large mirrors. (**A**) Articulated stacked segments – ridges, triangles, or both – invented by T. Grubb [32] and known as *whiffletree assembly*. (**B**) *Folded astatic levers* invented by Lassel [40]. (**C**) Open-loop *air pressure cushions* introduced by Foucault [28]. (**D**) Open-loop *hydraulic pads*

The mirror support system with articulated stacked segments – or *whiffletree assembly* – was introduced by T. Grubb [32] for the axial support of the 1.8-m Lord Rosse telescope. Because of the difficulty to obtain such stiff and lightweight systems, these devices are limited to thin mirrors whose aperture diameter does not exceed 2 m, hence preventing resonance instability problems that are generated from wind buffeting. For instance each 1.8-m hexagon segment mirror of the KECK telescopes ($t/d = 1/24$) is supported by three whiffletrees acting on $3 \times 12 = 36$ pads,

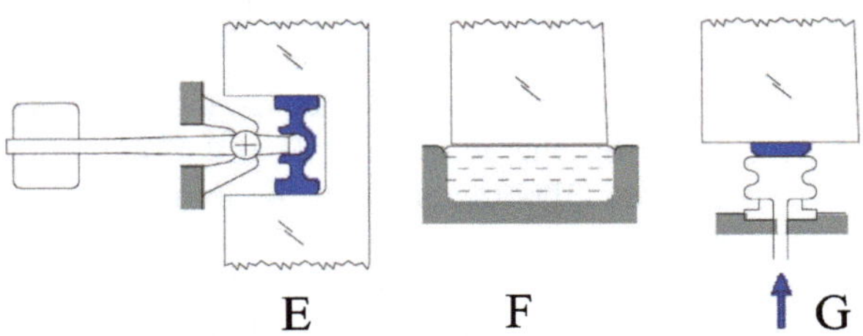

Fig. 8.2 Passive astatic systems for the *lateral support* of large mirrors. (**E**) *Direct astatic levers* acting on rear apertures or on edge. (**F**) *Perimeter bag* filled with liquid of the same mirror density or narrow mercury bag. (**G**) Open-loop radial *hydraulic pads* acting on edge

which for the 36 segments of a telescope gives $N_a = 1,296$ pads [56]. Within this mirror size, the other classic alternative uses *folded astatic levers*.

For monolithic mirror sizes in the range 4–8 m, although an air pressure system is the "perfect" system, the most common passive axial support systems are *astatic levers* or controlled *hydraulic pads* linked together in each of three sectors. An option combining folded and direct astatic levers in a single device was adopted for the 5 m Hale telescope mirror (see the description by Bowen [9]) and 6-m SAO telescope mirror as emphasized by Balega [6]. The option with an *air pressure cushion* was developed by Stepp et al. for the 8-m primary mirrors of the GEMINI telescopes [81, 82] in addition to an active control system that incorporates $N_a = 120$ hydraulic pads distributed on five concentric rings. With its other alternative of partial vacuum, it turns out that this latter option is sometimes preferred for the supporting of smaller mirrors such as Cassegrain mirrors (for instance, see Bely [8] for CFHT). The lateral support system of 8 m mirrors alternatively employs direct astatic levers or hydraulic pads.

8.1.3 Some Examples of Primary Mirror Geometries

Various schools of thought progressively led to noticeably differing solutions for the mirror geometry and the mirror weight per unit surface area. For instance, let us consider some existing large telescope mirrors (cf. focal-ratios in Table 1.1) with their aspect-ratios t/d and the N_a, N_l pad-numbers of their associated support systems.

• **Moderately lightweight ribbed mirrors**: A typical case of a moderately lightweight mirror cast with a ribbed structure is the 5-m aperture mirror of the Hale telescope at Palomar [9, 10]. The plane-concave blank in Pyrex (aspect-ratio $t/d = 1/8.33$, $N_a \equiv N_l = 36$ pads) includes cylindrical and equally spaced cavities where the axial and lateral astatic support acts. These open cavities are 36 in number and distributed in a 3-fold symmetry where the ribs converge. All ribs and the mirror surface are 10–12 cm thick. Very elaborated and precise astatic support systems, 33 in number, designed in a double astatic form, provide in a single device both axial and lateral supporting forces within an accuracy of 0.2%.

The primary mirror of the 6 m telescope of the Special Astrophysical Observatory in Caucasus is a meniscus of 65 cm thickness ($t/d = 1/9.3$) with $N_a \equiv N_l = 60$ combined axial-lateral astatic levers installed in 60 holes of 31 cm diameter and 43 cm depth that are distributed along four concentric rings. From chief designer B.K. Ioannisiani, the mirror is made from the Pyrex type glass known in USSR as "glass No. 316" whose coefficient of thermal expansion is $\alpha = 3 \pm 0.3 \times 10^{-6}$ [6, 33].[1]

• **Thick solid mirrors**: A classical case of a thick solid mirror is the 4 m mirror of the Mayall-KPNO telescope at Kitt Peak ($t/d = 1/8$, $N_a = 36$) with large support

[1] The 6-m telescope of the Special Astrophysical Observatory was the first large telescope erected with an alt-az mount, a concept which was followed up for all subsequent large telescopes.

Fig. 8.3 Views of two geometrical distributions of axial mirror support pads. (*Left*) Support pads of the 8.2-m VLT meniscus mirrors (courtesy ESO, Garching). (*Right*) Support pads of the 8.4-m LBT honeycomb mirror (after J.M. Hill, courtesy LBT)

pads along two concentric rings. This geometry with flat rear side was adopted for many telescope primaries in the 4 m class.

• **Sandwich honeycomb mirrors**: The spin casting technique applied to sandwich honeycomb mirrors provided the two 8.4 m mirror blanks of the LBT at Mount Graham. The two Pyrex borosilicate mirrors are plano-convex with 0.9 m thickness at the edge (aspect-ratio $t/d = 1/9.3$, $N_a \simeq 418$) and the lateral support system, described by Parodi et al., mainly acts on the rear faceplate [64] (Fig. 8.3).

• **Thin solid meniscus mirrors**: The spin casting technique allowed production of 8.2 m menicus blanks for the main four primary mirrors of the VLT at Cerro Paranal [26]. The Zerodur vitoceram blanks were cast by Schott and figured into a thin shape ($t/d = 1/47$). The axial support system allows an active control of the mirror shape by use of 147 hydraulic systems. These and the three reference positions act on the mirror via 150 articulated tripods ($N_a = 450$ pads) distributed over six concentric rings. The lateral support system is achieved by radial hydraulic pads surrounding the mirror [25] (Fig. 8.3). Similar 8-m thin meniscus mirrors were cast in fused silica by Corning for the primaries of GEMINI-1, -2 [54] and SUBARU [37].

8.2 Density and Thermal Constants of Mirror Substrates

Several important features enter into the choice of a mirror substrate. Beside the ability to obtain a fine surface polish and easily renew the reflective coating, we may briefly list some of them: large mechanical rigidity, linear strain-response for active optics shaping, low total mass, high resonant frequency of the fundamental vibration mode against wind buffeting, low deformation to environmental thermal change.

Some important constants in the choice of a mirror substrate are the following:

• **Stiffness**: The flexure of a mirror supported into gravity and its resonance frequency of the fundamental mode from wind buffeting are functions of the Young modulus E to the *density* μ of the material. Given a material, several merit-ratios in the form E/μ^m where $m \in [0, 1, 3/2, 2]$ may characterize a *stiffness* (see Sect. 8.4.1) depending on whether comparisons are with equal volume, equal mass, external or internal bending forces. The higher the stiffness, the smaller the flexure and the higher the natural resonance frequency of the fundamental mode.

• **Coefficient of thermal expansion** α (CTE): The coefficient of thermal expansion is the dimensional response of a material to temperature change.

• **Thermal diffusivity** d_t. This constant enters in all differential equations on heat propagation through a solid; it is defined as[2]

$$d_t = \frac{k_t}{\mu C_p}, \tag{8.2}$$

and characterizes how quickly a thermal equilibrium is achieved.

• **Specific heat** C_p: The specific heat is the heat energy required to raise the temperature of a given amount of a substance by one degree. For a solid substance it is typically measured under constant pressure and expressed in [J Kg^{-1} K^{-1}].

• **Thermal conductivity** k_t: the thermal conductivity is the intensive property of a material that indicates its ability to conduct heat. It is defined as the quantity of heat, Q, transmitted in time t through a thickness L, in a direction normal to a surface of area $\mathcal{A}$, due to a temperature difference ΔT, under steady state conditions and when the heat transfer depends only on the temperature gradient. The thermal conductivity = heat flow rate × distance / (area × temperature difference): $k_t = \frac{Q}{t} \times \frac{L}{\mathcal{A} \times \Delta T}$, expressed in [W m^{-1} K^{-1}].

Quantities k_t and d_t are both time dependent. Quantities μ, α, and d_t are given in Table 8.1.

[2] A. Couder introduced the reciprocal of the thermal diffusivity d_t for astronomical mirrors, which he also checked with several experiments, and listed the numerical values α/d_t for various glass, Pyrex, and metal substrates ([17], p. 308). Since, after a thermal disturbance, metal mirrors recover equilibrium much more quickly than glass or Pyrex mirrors, Couder pointed out that a thin metal mirror would be almost completely insensitive to thermal shocks. Thus, he proposed a *drum-shaped mirror* design – or *vase form* where a thin meniscus is surrounded by a ring – and experimented with some of them in cast iron. Couder further developed such mirrors with a 20 cm aperture and enamel deposits, and obtained interesting results.

D. Maksutov, apparently without knowledge of Couder's table, quite similarly established a thermo-mechanical table for the merit features of mirror materials. Also arriving at the conclusion that metal mirrors are almost insensitive to thermal shocks, he developed and built several stainless steel mirrors the largest of them was ~80 cm in diameter for his 70-cm aperture wide-field telescope at Pulkovo Observatory [98].

Table 8.1 Density μ, coefficient of thermal expansion α, and thermal diffusivity d_t of some linear strain-stress materials at 20°C

Material	μ [10^3 Kg/m^3]	α [10^{-6}/K]	d_t [10^{-6}m^2/s]
Beryllium pure VHP	1.85	11.3	57.2
Fused silica – Suprasil	2.20	0.41	1.0
Fused silica ULE	2.20	±0.03	0.8
Borofloat, Pyrex, Duran 50	2.23	3.25	0.7
Soda-lime glass	2.44	8.60	1.8
BK7 glass, UBK7 glass	2.51	7.10	0.5
Zerodur vitroceram	2.53	±0.05	0.7
ZPF Ceratec vitroceram	2.54	±0.03	2.6
Silicon carbide polycr. CVD	3.21	3.30	88.4
Sapphire polycrystal. Al_2O_3	3.98	10.0	12.6
Titanium alloy Ti 90 Al6 V4	4.45	8.80	2.9
Stainless steel Fe87 Cr13	7.72	10.2	6.8

* Elasticity constants E, ν, $\sigma_{\max}$ are included in Sect. 1.13.2 (Table 1.10).

8.3 Substrates for Large Mirrors

Despite metal mirrors all having a high thermal diffusivity, and thus being the best potential materials for telescopes having a metal tube, most large telescope mirrors are made from brittle materials. One of the dominating criteria in this choice is that mirror substrates with low or zero CTE have been deliberately preferred over other features. With the recent introduction of closed-loop active optics control of large telescope mirrors, and since we now know how to remove a reflective coating on corrosion-resistant substrates such as nickel or stainless steel [29, 88], the criteria of choice of a mirror substrate could be revised.

• **Borosilicate glass**: Low expansion borosilicate glass such as Pyrex from Corning – or the equivalent Duran 50 from Schott, or glass No 316 from Russia – have been of current use for large mirrors, e.g. the 5 m blank of Hale's telescope and 6 m of the Russian telescope. This material can be easily recast to obtain thin-walled sandwiched honeycomb mirrors. This option, proposed by Angel [1] and developed by Hill et al. [30], allowed the construction of honeycomb mirrors for several large telescopes such as the two 8 m blanks of the LBT [30].

• **Fused silica –** ULE: Fused silica are difficult to elaborate in large size because the fusion temperature of pure quartz is ~1, 700°C, i.e. a high value for refractory materials used in a furnace. The fused silica in "ultra low expansion" grade (ULE) was developed by Corning by appropriate addition of titanium oxide TiO_2. 8-m mirror blanks for the GEMINI and SUBARU telescopes [37, 54] were obtained by the classical assembly process of partial remelting at ~1, 300°C of several sub-blanks.

• **Zero-expansion vitrocerams**: Vitrocerams are zero expansion materials that first appeared as Cer-Vit from Owens-Illinois in the 1960s. Later vitrocerams of quasi-equivalent properties were known as Zerodur and Astro-Sitall. The latest developed vitroceram, ZPF Ceratex from Japan, has a higher Young modulus and ultimate tensile strength than the previous ones but blanks are only available in ~1.5-m maximum diameter. After a first casting or spin casting, the elaboration process called ceramization requires an accurate reheat cycle which gives rise to a controlled growth of tiny monocrystal grains of negative CTE whilst the remaining vitreous phase is of positive CTE. The zero CTE, at given temperature, is achieved if the proportion of the two phases is carefully adjusted (∼74% weight of crystalline phase). Seven 8.2-m Zerodur blanks have been produced by Schott, of which six of them were for the VLT and GEMINI telescopes.

• **Silicon carbide**: Silicon carbide or SiC is a brittle substance whose hardness on *Mohs' hardness scale* – extending from 1 to 10 – comes after diamond (10), boron carbide (9.5), and is the same as aluminum oxide (sapphire or corundum) and tungsten carbide with hardness (9) (see Table 8.2). SiC is known by the trademark carborundum and is often used as an abrasive. Its high Young modulus makes it of interest for thin meniscus or lightweight substrate of secondary, tertiary, or chopping telescope mirrors. The elaboration process of mirror blanks in SiC is the "chemical vapor deposition" (CVD) currently used to produce high-purity solid materials, thus providing blanks able to be super-polished within ∼10 Å rms micro-roughness. The mirror surfacing requires use of diamond grains and takes much more time than other substrates. The long elaboration process presently limits the size of the blanks to ∼1 m in diameter.

• **Beryllium**: Beryllium is a very light metal with high Young modulus, thus showing high stiffness merit factors (cf. Sect. 8.4.1). Its noticeable CTE is compensated by a high thermal diffusivity. These features make it a first rank substrate for fast chopping mirrors for ground-based observations in the thermal infrared. The elaboration process requires use of a fine grain powder. The grains are fused together by "vacuum hot pressing" (VHP): a mold containing the powder is placed under vacuum while heated (∼900°C) and compressed (∼10^3 atm.). The present technical limitations of the process restrain making blank sizes larger than ∼2 m in diameter. Beryllium can be classically machined by lathes and mills; however the powder and small Be particles, easily becoming airborn, must not be breathed in because

Table 8.2 Mean stiffness $E/\mu^{3/2}$, elastic deformability ratio $\sigma_{T\,\max}/E$, and Mohs' hardness of some linear strain-stress materials at 20°C *

Material	$E/\mu^{3/2}$ [10^5 SI]	$\sigma_{T\,\max}/E$ [10^{-4}]	Mohs
Borofloat glass, Pyrex, Duran	6.1	1.1	5
Fused silica ULE $SiO_2 + TiO_2$	6.6	2.9	(6)
Fused silica SiO_2 Suprasil	7.5	2.5	6
BK7 optical glass	6.5	1.2	6
U-BK7 optical glass	7.2	1.3	6
Zerodur vitroceram	7.1	2.4	6.2
Titanium alloy Ti90 Al6 V4	4.1	73.8	4.5
ZPF Ceratech vitroceram	11.7	4.0	(6.2)
Stainless steel Fe87 Cr13	2.9	49.8	6
Stainless steel Fe87 Cr13 quenched	3.0	69.7	7
Beryllium pure VHP	36.1	13.9	5.5
Silicon carbide polycrystal. CVD	23.6	34.9	9
Sapphire polycrystalline Al_2O_3	17.5	22.7	9

* E and $\sigma_{T\,\max}$ are given in Table 1.10; μ in Table 8.1.
Mohs' hardness scale is maximum at 10 for diamond.

this may entail a lung desease similar to silicosis. For a surface polish with low micro-roughness, beryllium is electroless nickel plated by a *Kanigen process*, a thermochemical process which uses metallic electrochemical potentials with a liquid catalyst; this avoids the difficulty of edge-sharpening encountered in the electrolytic process. For field stabilization at 15 Hz and thermal infrared studies, the ribbed and nickel-plated substrates of the four Cassegrain mirrors of the 8-m VLT units are beryllium blanks of 1.1 m diameter from Brush-Wellman in USA; more details are given by Stanghellini et al. [86]. The 18 segments of the primary mirror of the JWST project are being built in beryllium with this technology.

• **Stainless steel**: Stainless steel is an extremely low cost material compared to the above substrates. Its elastic deformability $\sigma_{T\,\max}/E$ is the highest of mirror materials (cf. Sect. 8.4.2 and Table 8.2). Its thermal diffusivity makes it weakly sensitive to thermal shocks. This latter advantage was pointed out by Couder [17], and also Maksutov [98] whose best and final work is generally regarded to be the Pulkovo 70-cm aperture double-meniscus telescope with a 80-cm diameter *stainless steel mirror*. The best alloy for a linear stress-strain relationship – Hooke's law – is the martensitic steel Fe87 Cr13 (in % weight, also known as AISI 420) with less than 0.15% carbon. Current applications are turbine blades and springs. Recent developments of variable curvature mirrors, operating since 2006 at the focal plane of the delay lines of the VLT Interferometer (zoom range [f/∞–f/2.5], aspect ratio $t/d = 1/60$), show that the Fe87 Cr13 alloy is a stable material. As for Ti90 Al4 V6 alloy, the casting of Fe87 Cr13 delivers a fine structural material that can be smoothly polished to high reflectivity without the need of a Kanigen process. Its large *elastic deformability ratio* (cf. Sect. 2.6.1) would allow the complete aspherization of large mirrors by active optics. For instance, Lemaitre, Wilson et al. [78] proposed an f/1.75 stainless steel primary mirror of diameter 1.8 m and thickness 30 mm (aspect-ratio $t/d = 1/60$) for the VLT auxiliary interferometric telescopes to be at once actively aspherized and actively supported. The cost of the casting and machining of the Fe87 Cr13 meniscus blank – and also in a vase-shaped option – from Ferry-Capitain in France was an order of magnitude lower than that of a Zerodur blank.

• **Liquid materials**: In contrast to solid materials, it has been known for centuries that liquids at rest naturally provide diffraction limited plane surfaces (in fact of radius that of the Earth). Of course, such surfaces are totally free from ripple surface errors since no figuring is needed. High density liquids, such as glycerine or mercury, are of classical use in the laboratory as high-quality reference surfaces – currently at $\lambda/100$ – for optical testing such as large Fabry-Perot plates by Marioge [49]; furthermore these are low-cost references. At the surface boundary, the capillarity is an extremely local effect which only requires increasing the liquid surface area by a few percent. In 1909, Wood [97] introduced and built a 50-cm diameter liquid telescope, and performed astronomical observations with it. The mercury mirror was rotated and stabilized to a constant angular velocity for which the hydrodynamic equilibrium equation under gravity shows that the mirror surface is a perfect *parabola*. The main difficulty is to obtain a high accuracy in the constancy of the angular velocity because ripples arise from accelerations. Solving this requires the use of an extremely low-friction rotating pad and minimum quantity of mercury. Pneumatic pads rotated by belts of ancient magnetic tapes more recently allow zenith observations with 2.5 and 2.7-m liquid mirror telescopes. The last development of liquid mirrors is the 6 m Telescope of the University of Vancouver (cf. Sect. 7.7).

8.4 Stiffness and Elastic Deformability Criteria

The duality between (1) low flexural deformation of a mirror to gravity or to wind buffeting and (2) high elastic deformability for stress aspherization of a mirror, though partly incompatible, must be interpreted as two different active optics purposes. In the first one various stiffness criteria of a mirror substrate may be determined from the quantities (E, μ), the Young modulus and the density. In the second one, because of the ability of a mirror to be actively bent – and possibly completely aspherized – by external forces, an elastic deformability criterion may be determined from the quantities $(\sigma_{T\,\max}, E)$, the tensile maximum stress and the Young modulus.

8.4.1 *Mirror Materials and Stiffness Criteria*

In order to determine merit factors relative to the *stiffness* – "raideur" in French –, let us consider a circular plate of diameter d and thickness t in a material of density μ and Young modulus E. First, one may determine two merit factor types:

- a merit factor for equal volume plates (S_V), where all dimensions are identical,
- a merit factor for equal mass plates (S_M), where only diameters are the same.

The volume and mass of these plates are

$$V \propto t^2 d \qquad \text{and} \qquad M \propto \mu\, t^2 d.$$

Whatever the equilibrium configuration of these plates and provided those configurations are identical for plates in various materials, it is well known from the thin plate theory that the flexure z_F due to an external force and the flexure z_{ow} due to the own weight under gravity (cf. Sect. 1.13.12) are of the form

$$z_F = \frac{1}{E}\frac{Fd^2}{t^3}\,\zeta_F(\rho, \nu), \qquad z_{ow} = \frac{\mu}{E}\frac{d^4}{t^2}\,\zeta_{ow}(\rho, \nu), \tag{8.3}$$

where dimensionless flexures $\zeta_F(\rho, \nu)$ and $\zeta_{ow}(\rho, \nu)$ are only dependent on the reduced radial variable $\rho = 2r/d$ and Poisson's ratio ν of the material. Generally Poisson's ratio contribution in the amplitude of the flexure is quite small so we shall neglect the effect of this dimensionless quantity in a first approximation. Hence, with this restriction, the flexure functions ζ_F and ζ_{ow} are not dependent on the material.

• **Merit factor for equal volume stiffness**: Introducing the volume of the plate into (8.3), we obtain after substitution

$$z_{\mathrm{F,\,V}} \propto \frac{1}{E}\frac{F\,d^{7/2}}{V^{3/2}}, \qquad z_{\mathrm{ow,\,V}} \propto \frac{\mu}{E}\frac{d^5}{V}, \tag{8.4a}$$

where the first right-hand fraction is only material dependent. The reciprocal fraction allows defining merit-factors for *equal volume stiffness* as

$$\mathcal{S}_{\mathrm{F,\,V}} = E, \qquad \mathcal{S}_{\mathrm{ow,\,V}} = E/\mu, \tag{8.4b}$$

where the first one, E, is a stiffness classically introduced in the definition of the rigidity since determining the slope of Hooke's law.

• **Merit factor for equal mass stiffness**: Introducing the mass of the plate into (8.3), we obtain after substitution

$$z_{\mathrm{F,\,M}} \propto \frac{\mu^{3/2}}{E}\frac{F\,d^{7/2}}{M^{3/2}}, \qquad z_{\mathrm{ow,\,M}} \propto \frac{\mu^2}{E}\frac{d^5}{M}, \tag{8.4c}$$

where the first right-hand fraction is only material dependent. The reciprocal fraction allows defining merit-factors for *equal mass stiffness* as

$$\mathcal{S}_{\mathrm{F,\,M}} = E/\mu^{3/2}, \qquad \mathcal{S}_{\mathrm{ow,\,M}} = E/\mu^2. \tag{8.4d}$$

• **The four stiffness criteria**: From (8.4b) and (8.4d), the four stiffness merit-ratios may be written in the form

$$\mathcal{S} = E/\mu^m, \qquad \text{where} \qquad m \in [0,\ 1,\ 3/2,\ 2], \tag{8.5}$$

each ratio depending on whether comparisons are with equal volume, equal mass, external or internal gravity bending forces. The higher these ratios, the less the flexure and the higher the resonance frequency of the fundamental mode. The stiffness merit factor for equal mass substrates, $\mathcal{S}_{\mathrm{F,\,M}} = E/\mu^{3/2}$, is an average value of the three latest merit factors. Furthermore, this mean stiffness is convenient for comparisons of equal mass mirrors to wind buffeting vibrations (Table 8.2).

8.4.2 Mirror Materials and Elastic Deformability Criterion

In the application of active optics methods, such as stress surfacing aspherization or in-situ deformation, important features are the choice of a material substrate that shows a linear stress-strain law and the largest linear range possible. The combination of those two features is the so-called *elastic deformability ratio*. Glass, vitro ceramics, and silicon carbide possess this linear characteristic up to rupture; however the linear range of all these brittle materials is quite moderate partly because their maximum tensile stress is time dependent (cf. Sect. 5.2.5 and Table 5.2).

In contrast several metal materials exhibit stress-strain linearity over extended ranges.

It is well known from unidirectional rupture tests that for any material the *tensile* ultimate stress is much lower than the *compression* maximum stress. Let $\sigma_{T\,\max}$ be the *tensile maximum stress* of a material as an acceptable limit that must not be exceeded to avoid rupture or plastic strain (see data in Table 1.10).

In all generality, a local curvature may be elastically generated by stress distributions σ_{rr}, σ_{tt} arising through the thickness of a plate. In the thin plate theory, these distributions are linear over the thickness of the plate and have null values at its middle surface. In Chap. 2 on curvature mode flexures, where $\sigma_{rr} = \sigma_{tt} = \sigma$ in the thin plate theory [cf. (2.10) or (2.17b)], we have shown that the ratio $|\sigma|/E$ governs the deformation sag of a plate. If $\sigma_{rr} \neq \sigma_{tt}$, so the local principal curvatures differ, the maximum stress of them will give a limitation in the flexure when the loading intensity is increased. Hence the corresponding maximum acceptable ratio for obtaining a maximum flexure – or *elastic deformability ratio* of a material (cf. (2.48d)) – is

$$\mathcal{D}_{F,\,V} = \frac{\sigma_{T\,\max}}{E}, \tag{8.6}$$

where the significance of suffix F, V is that the bending is generated by a load F which is *external* to the plate and that the sag is for a given volume V determined by the diameter d and thickness t; these two latter dimensions must be the same for comparing materials elastic deformability in all generality (Table 8.2). The *Mohs' hardness* – as classically defined in the range 1 to 10 – of some linear stress-strain materials is included in the last column of Table 8.2.

8.5 Axial Flexure of Large Mirrors Under Gravity

Important features for large telescope mirrors are the concept and design of its axial support system. In general the quasi-perfect concept of a mirror support by air pressure is avoided for technical reasons, so the remaining support system options all use a *discrete* pad distribution. An optimal design of the axial support system consists in the determination of the minimal number of pads for which the flexure under gravity will satisfy a convenient optics tolerance such as, for instance, a diffraction limited criterion.

8.5.1 Density Distribution of Mirror Support Pads

A convenient investigation for the determination of the flexure of a large mirror under gravity is to consider an infinitely large mirror where the supporting pads are equally spaced in a 3-fold symmetry of step $2a$. As a preliminary parameter, one requires defining the pad density at the back surface of the mirror (Fig. 8.4).

The parallelogram in the figure contains four pads, thus the pad density per unit surface area is

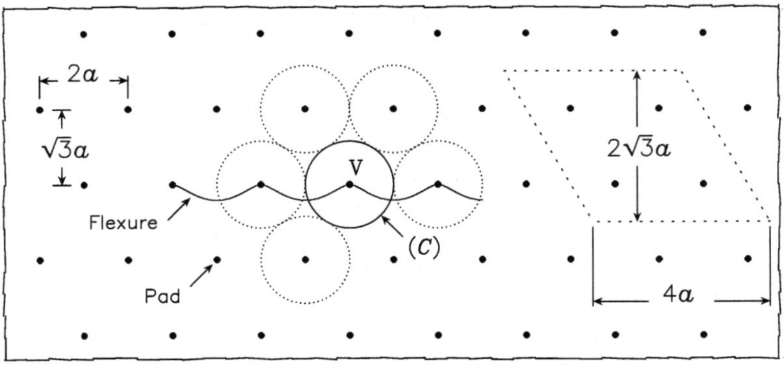

Fig. 8.4 Infinitely large mirror axially supported by equal spaced pads in a 3-fold symmetry

$$n_p = 1/2\sqrt{3}a^2. \tag{8.7}$$

From this definition, if the mirror is of finite diameter d, then the total number of axial support pads is

$$N_p = \frac{\pi}{4} n_p d^2 = \frac{\pi}{8\sqrt{3}} \frac{d^2}{a^2} \simeq 0.2267 \frac{d^2}{a^2}. \tag{8.8}$$

Returning to the case of an infinitely large mirror, one shall calculate hereafter the flexure of a mirror sub-element that surrounds a pad.

8.5.2 Flexure of a Mirror Sub-Element Supported by a Ring Pad

In the above 2-D distribution, the infinitely large mirror may be seen as an assembly of successive hexagon elements, each of them supported by a ring pad. To a sufficient approximation for determining the flexure of this hexagon element, one may assume that its contour is a circle (C) which is tangent to six proximate circles (Figs. 8.4 and 8.5).

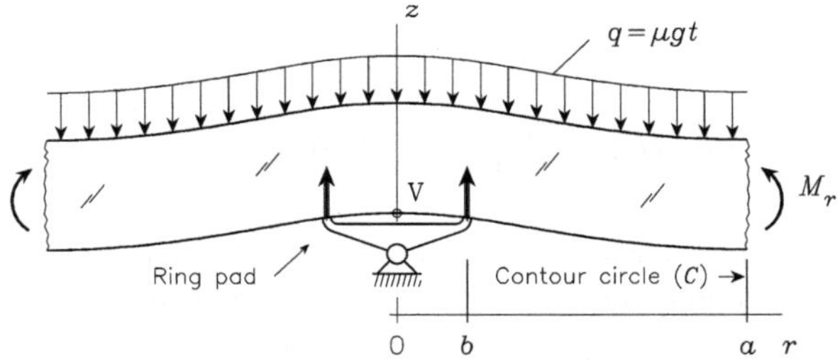

Fig. 8.5 Circular mirror sub-element of an infinitely large mirror axially supported in a 3-fold symmetry by ring pads of radius b. The outer radius a of the mirror sub-element is the radius of circle (C) in Fig. 8.4

Hence the problem of the flexure of a large mirror under gravity reduces to that of a circular mirror element, of outer radius a and thickness t, supported by a ring pad of radius b where the gravitational volume load is equivalent to a surface load $q = \mu g t$, $(g < 0)$. If the thickness of the mirror is constant and if its curvature relatively large, then the flexure $z(r)$ of the mirror "sub-element" is derived from Poisson's bilaplacian equation (see Sect. 1.13.10)

$$\nabla^2\nabla^2 z = \frac{q}{D}, \qquad q = \mu\, g\, t, \qquad D = \frac{E\, t^3}{12(1-\nu^2)}, \tag{8.9}$$

where D is the rigidity. The general solution is

$$z = \frac{q}{64D}\left(r^4 + c_1\, r^2 + c_2\, r^2 \ln r + c_3 \ln r + c_4\right), \tag{8.10}$$

where c_i are constants and term in r^4 is the particular solution.

Let us assume a circular pad acting at its contour. The determination of the flexure must be dissociated into inner and outer zones with respect to radius $r = b$ of the supporting pad. It is useful to introduce dimensionless current radius ρ and flexure function $\zeta(\rho)$ as

$$\rho = r/a, \qquad z(r) = \frac{q\, a^4}{64D} \times \begin{cases} \zeta_1(\rho), & 0 \le \rho \le b/a, \\ \zeta_2(\rho), & b/a \le \rho \le 1. \end{cases} \tag{8.11}$$

Since there is no hole in the inner zone ($\rho < b/a$), and no ring force is applied to it, the two logarithm terms vanish. Setting the origin of the flexure at $\rho = 0$, the two components of the dimensionless flexure may be represented as

$$\begin{cases} \zeta_1 = \rho^4 + 2\, C_{1,1}\, \rho^2, \\ \zeta_2 = \rho^4 + 2\, C_{2,1}\, \rho^2 + 8\, C_{2,2}\, \rho^2 \ln \rho + 4\, C_{2,3} \ln \rho + 4\, C_{2,4}, \end{cases} \tag{8.12}$$

where the numbers in front of the five coefficients avoid use of fractions in further calculations.

The boundary conditions at the outer circle (C) of the sub-element – where $\rho = 1$ and the bending moment M_r contributes to the flexures – is a null radial shearing force, $Q_r = 0$, and a slope perpendicular to the gravity vector **g**, thus $d^2\zeta/d\rho^2 = 0$. From (1.182), the radial shearing force writes

$$Q_r = -D\frac{d}{dr}\nabla^2 z = -\frac{D}{a^2}\frac{d}{d\rho}\Delta^2\zeta, \tag{8.13}$$

where Δ^2 is a Laplacian operator with respect to ρ. The continuity conditions at $\rho = b/a$ are equal sags, equal slopes and equal radial bending moments M_r. From the expression (1.181) of M_r, the latter condition may be readily substituted by equal radial curvatures $d^2\zeta/d\rho^2$. Hence, the five conditions are the following,

$$\begin{cases} \text{for } \rho = a/b: & \zeta_1 = \zeta_2, \quad \dfrac{d\zeta_1}{d\rho} = \dfrac{d\zeta_2}{d\rho}, \quad \dfrac{d^2\zeta_1}{d\rho^2} = \dfrac{d^2\zeta_2}{d\rho^2}, \\ \text{for } \rho = 1: & \dfrac{d}{d\rho}\Delta^2\zeta_2 = 0, \quad \dfrac{d^2\zeta_2}{d\rho^2} = 0, \end{cases} \tag{8.14}$$

from which one notes that they all are independent of Poisson is ratio. After solving this equation set, we find

$$C_{1,1} = -3 + 2\frac{b^2}{a^2} - 2\ln\frac{b^2}{a^2}, \qquad C_{2,1} = 1 + 2\frac{b^2}{a^2},$$
$$C_{2,2} = -1, \quad C_{2,3} = -2\frac{b^2}{a^2}, \qquad C_{2,4} = -2\frac{b^2}{a^2} + \frac{b^2}{a^2}\ln\frac{b^2}{a^2}. \tag{8.15}$$

After substitution into functions ζ_1, ζ_2, one beholds that the total sag of the flexure is strongly dependent on the pad radius ratio b/a (Fig. 8.6).

Coefficients $C_{1,1}$ and $C_{2,i}$ allow determining the maximum sag as well as the maximum slope of the flexure. For instance, from the flexures in Fig. 8.6, we see that the maximum sag is

$$\Delta\zeta_{\max} = \zeta_1\{b/a\} - \zeta_1\{0\} + \zeta_2\{1\} - \zeta_2\{b/a\}, \quad b/a \in [0, 0.549].$$

Since we have set $\zeta_1\{0\} = 0$ and $\zeta_1\{b/a\} = \zeta_2\{b/a\}$, the second equation in (8.12) entails

$$\Delta\zeta_{\max} = \zeta_2\{1\} = 1 + 2\,C_{2,1} + 4\,C_{2,4}.$$

After substitution of the coefficients, the dimensionless maximum flexure sag is

$$\Delta\zeta_{\max} = 3 - 4\frac{b^2}{a^2}\left(1 - \ln\frac{b^2}{a^2}\right), \quad \frac{b}{a} \in [0, 0.549]. \tag{8.16a}$$

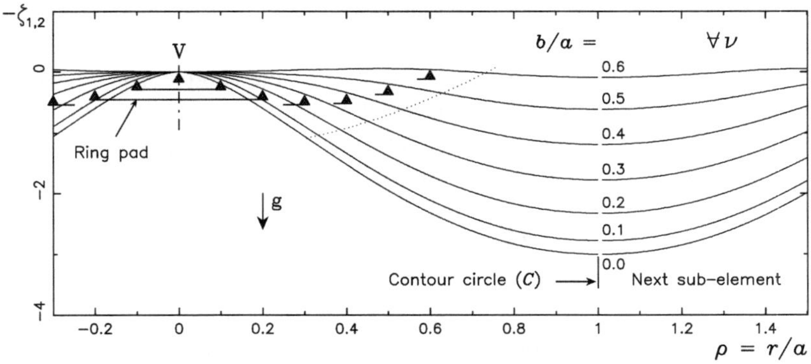

Fig. 8.6 Flexure dependence of a mirror sub-element as a function of the pad radius ratio b/a in a constant spaced-pad distribution of step $2a$ with three-fold symmetry. For $b/a = 0$, the support pad reduces to a single point. The locus of some inflexion points is shown by the *dotted line*

When $0.549 < b/a < 0.70$, the flexure $\zeta_{1,2}$ shows a second inflexion point, so that (8.16a) does not apply. We obtain $\zeta_2(1) = \zeta_1(0) = 0$ for $b/a \simeq 0.6184$, where a ripple arises at $\rho \simeq 0.56$. The corresponding smallest possible maximum flexure sag, which occurs here, is

$$\Delta\zeta_{\max} = 0.0978 \quad \text{for} \quad \frac{b}{a} = 0.6184. \tag{8.16b}$$

To summarize, from equations (8.16) and further calculations, the dimensionless maximum flexure sag as a function of the pad radius ratio is as follows (Fig. 8.7),

b/a	0	0.10	0.20	0.30	0.40	0.50	0.549	0.60	0.6184,
$\Delta\zeta_{\max}$	3	2.775	2.324	1.773	1.187	0.613	0.348	0.145	0.0978.

(8.16c)

Therefore, the analysis from thin plate theory provides the following general conclusions.

1 → *If the minimization of the axial flexure of a mirror under gravity must be achieved by use of a minimum pad number N_p, of step $2a$, then the pad radius ratio must be $b/a \simeq 0.62$ whatever Poisson's ratio.*

2 → *For a narrow support pad ($b/a \simeq 0$) the flexure curve of a mirror sub-element shows that the absolute value of the curvature at $\rho = 0$ is 7.75 times as large as that of the radial curvature at $\rho = 1$.*

The local flexure is much more marked near the supporting pad when it reduces to a point force. Thus, axial support systems that are, for instance, with a pad radius ratio $b/a \simeq 0.05$ require a larger pad number N_p than with a larger b/a-value. Current values for various large telescope mirrors that have been built are varying in the range $b/a \in [0.05, 0.6]$. For instance, the 4 m mirror of the Mayall KPNO telescope is supported by ring pads that are distributed along two concentric circles; in a radial direction of this mirror, the pad radius ratio is $b/a \simeq 0.6$ (Fig. 8.8).

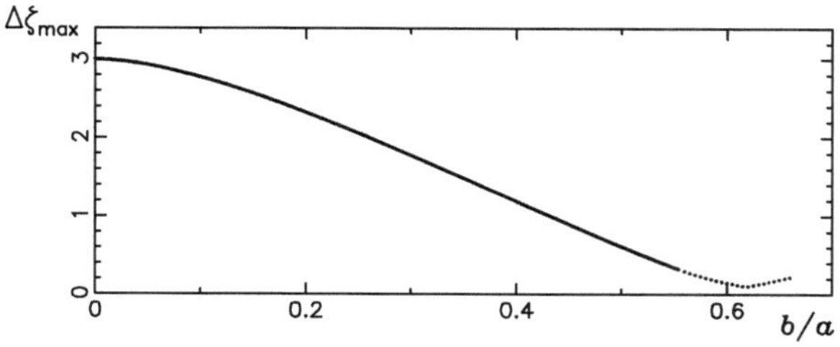

Fig. 8.7 Dimensionless maximum flexure sag $\Delta\zeta_{\max}$ as a function of the pad radius ratio b/a. *The dotted line* corresponds to the region $0.549 < b/a < 0.70$ where the mirror sub-element shows a flexure with two inflexion zones

Fig. 8.8 Mirror support system of the KPNO 4-m telescope (courtesy Kitt Peak National Observatory)

8.5.3 Density Criterion for Pad Distribution – Couder's Law

From the latter study of an infinitely large mirror supported by a uniform pad density, we now have all the elements for determining the dimensioned maximum flexure sag $\Delta z_{\max}$ of a mirror sub-element in its complete region $r \in [0,\ a]$. In addition, we also know the dimensionless maximum sag $\Delta\zeta_{\max}(b/a)$ as a function of the pad radius ratio b/a. From the flexure (8.11) of a sub-element, the maximum flexure sag writes

$$\Delta z_{\max} = \frac{q\,a^4}{64D}\,\Delta\zeta_{\max}(b/a), \tag{8.17}$$

where $\Delta\zeta_{\max}$ is given by the distribution (8.16c) for $b/a \in [0,\ 0.62]$. From equations (8.9), the ratio q/D is represented by

$$\frac{q}{D} = 12(1-\nu^2)\frac{\mu\,g}{Et^2}. \tag{8.18}$$

After substitution, the maximum flexure sag for the mirror sub-element of circular contour (C) is

$$\Delta z_{\max} = (1-\nu^2)\frac{3\,\mu g\,a^4}{16E\,t^2}\,\Delta\zeta_{\max}(b/a). \tag{8.19}$$

In a 3-fold symmetry, the slab filling is with *hexagon* contour sub-elements. Compared to the radius a of the circle (C), the radial distance of a hexagon corner is $2a/\sqrt{3} \simeq 1.154a$. On the other hand, the contour line (H) of the hexagon does remain in a plane, so the flexure is somewhat larger than that at circle (C). For continuity and minimum strain energy reasons, the maximum flexure sag $\Delta Z_{\max}$ at a corner of the hexagon sub-element – i.e. for all the mirror – can be accurately related to that of the circular sub-element by use of the scaling factor $2/\sqrt{3}$. Hence, the maximum flexure sag for an infinitely large mirror generated by hexagon slabs is

$$\Delta Z_{\max} = \frac{2}{\sqrt{3}}\Delta z_{\max} = (1-\nu^2)\frac{\sqrt{3}\,\mu g\,a^4}{8\,E\,t^2}\,\Delta\zeta_{\max}(b/a). \tag{8.20}$$

• **Extended Couder's law**: Now for a large mirror of finite diameter d, from (8.8), $a^2 = \pi d^2/8\sqrt{3}N_p$ so that a^2 can be substituted into (8.20). If we restrain the pad radius ratio such as $b/a \leq 0.549$, then equation (8.16a) for $\Delta\zeta_{\max}(b/a)$ applies. Therefore, the substitutions entail the following general result.

→ *For a large telescope mirror, of finite diameter d and thickness t, supported by N_p axial ring pads of radius b that are distributed in a 3-fold symmetry of step 2a, the maximum flexure sag is*

$$\begin{cases} \Delta Z_{\max} = \dfrac{\pi^2(1-\nu^2)}{512\sqrt{3}\,N_p^2}\left[3 - 4\dfrac{b^2}{a^2}\left(1 - \ln\dfrac{b^2}{a^2}\right)\right]\dfrac{\mu\, g\, d^4}{E\, t^2}, \\ \textit{with ring pad radius ratio condition} \quad 0 \leq b/a \leq 0.549. \end{cases} \tag{8.21}$$

Although in a somewhat different formulation, a first similar result was enounced by Couder which would correspond here to point-pads i.e. $b = 0$; this is known as the d^4/t^2 *flexure scaling law* or *Couder's law* [18].[3] Equation set (8.21) includes the *extension* for current values of the pad radius ratio. A somewhat larger extension may be used from the distribution (8.16c).

This result readily allows us to introduce an optics tolerance criterion. After reflection of a wavefront at the mirror, let w_{ptv} be the maximum wavefront deflection error introduced by the flexure of the mirror, so we have $w_{\text{ptv}} = 2\Delta Z_{\max}$. Now if we state that an optics tolerance error is defined by a *wavefront ptv criterion* such as, say, $w_{\text{ptv}} \leq \lambda/5$, then we obtain the inequality,

$$w_{\text{ptv}} = 2\Delta Z_{\max} \leq \lambda/5, \tag{8.22}$$

where we assume that the flexure of the mirror can be potentially corrected by an adaptive optics system. Hence, from (8.21), a *wavefront tolerance criterion* for the choice of the mirror material (E, ν, μ), the mirror thickness (t) and its pad support system $(b/a, N_p)$, is given by the inequalities

$$\begin{cases} w_{\text{ptv}} = \dfrac{\pi^2(1-\nu^2)}{256\sqrt{3}N_p^2}\left[3 - 4\dfrac{b^2}{a^2}\left(1 - \ln\dfrac{b^2}{a^2}\right)\right]\dfrac{\mu\, g\, d^4}{E\, t^2} \leq \dfrac{\lambda}{5}, \\ \text{ring pad radius ratio condition } 0 \leq b/a \leq 0.549. \end{cases} \tag{8.23}$$

Current designs of large telescope mirrors and associated support systems satisfy these inequalities for the shortest wavelength they are optimized for. If one prefers using the pad density per unit surface area n_p instead of the total number of pads, then, from (8.8), $N_p = \pi n_p d^2/4$ must be substituted in (8.23).

Let us consider, for instance, the case of the primary mirror meniscuses of the ESO-VLT in Zerodur from Schott. The axial support pads are tripods that have a

[3] From the thin plate theory, André Couder analytically derived the flexure of a constant thickness plane plate supported by a continuous concentric ring. He applied the superposition principle of the flexure for passing from one to several support rings and minimized the resulting flexure sag from a best fit balance of the radius of these rings and of the reacting forces per unit length acting on them.

relatively small radius b with respect to a (Fig. 8.3-Left); thus we can accurately assume that their effect is equivalent to a ring pad of same ratio b/a. The quantities concerned in (8.22) are as follows,

$$E = 90.2 \times 10^9\text{Pa}, \quad \nu = 0.243, \quad \mu g = 24.82 \times 10^3\text{Pa/m},$$
$$d = 8.2\text{m}, \quad t = 0.175\text{m}, \quad b/a = 0.2, \quad N_p = N_a/3 = 150,$$

where, from (8.16a), $\Delta\zeta_{\max}(0.2) = 2.324$. After substitution in (8.23), we find for the wavefront deflection $w_{\text{ptv}} = 0.439\,\mu$m. Thus, at the wavelength $\lambda = 0.5\,\mu$m, the wavefront error due to the flexure of the mirror under gravity is $w_{\text{ptv}} = \lambda/5.7$.

• **Wavefront variance and rms criterion**: The wavefront deflection w caused by the flexure of a mirror under gravity may be minimized by use of a *wavefront rms criterion* from which the best wavefront deflection is denoted w_{rms}. The variance w^2_{rms} of the wavefront deflection w is defined by

$$w^2_{\text{rms}} = \frac{1}{\mathcal{A}} \iint_{\mathcal{A}} (w + A_{00})^2 d\mathcal{A}, \tag{8.24}$$

where the integration is taken over the surface area $\mathcal{A}$ of the wavefront, and A_{00} is the unknown piston term. If a small defocussing is tolerated, then the determination of the variance is obtained from the substitution $A_{00} \rightarrow A_{00} + A_{20}r^2$.

Note that the rms values of the slopes $(\nabla w)_{\text{rms}}$ are also relevant for the image degradation. In this case, the variance $(\nabla w)^2_{\text{rms}}$ should be calculated by simply minimizing the resulting image size.

• **Pad density, mirror thickness, and ptv criterion**: Whatever the diameter of a large telescope mirror, the pad density n_p per unit surface area, the thickness t of the mirror and the ring pad radius ratio b/a are the three fundamental parameters. Once the ratio b/a is fixed – which determines either a ring pad or a point pad –, from (8.8), the substitution of $N_p = \pi n_p d^2/4$ into (8.23) provides a tolerance wavefront ptv criterion for the pad density n_p relative to the mirror thickness t. For a wavefront tolerance, say $w_{\text{ptv}} \le \lambda/5$, the general result in dimension [L^{-2}] is

$$\begin{cases} n_p^2\, t^2 \ge \dfrac{1-\nu^2}{16\sqrt{3}} \left[3 - 4\dfrac{b^2}{a^2}\left(1 - \ln\dfrac{b^2}{a^2}\right)\right] \dfrac{\mu g}{E} \times \dfrac{5}{\lambda}, \\ \text{ring pad radius ratio condition} \quad 0 \le b/a \le 0.549. \end{cases} \tag{8.25}$$

Given a material (E, ν, μ), a pad radius ratio (b/a) and an optical tolerance (above $\lambda/5$ on the wavefront), the latter relations are of more convenient use in practice since they are *only* dependent on the thickness of the mirror.

• **Pad density and mirror thickness geometry**: We have seen that the above criterion is derived from the *local* flexure of a hexagon slab element in a 3-fold symmetry of an infinitely large mirror. On the other hand, large telescope mirrors may have either a constant thickness t (meniscus) or a flat rear side (plano-concave). For these various thickness geometries, the local property of criterion (8.25) must be interpreted as follows.

1 → *Whatever the thickness distribution $t(r)$ of a large telescope mirror, the tolerance criterion for its flexure under gravity determines the pad support density n_p per unit surface area from the law "$n_p t = \text{constant}$."*

2 → *Hence, for a mirror geometry with a* flat *rear side, if the thickness distribution is represented by $t = t_0(1 + \kappa r^2)$, where $\kappa > 0$, then the pad density is $n_p(r) \propto [t_0(1 + \kappa r^2)]^{-1}$.*

A general theoretical conclusion is that a concave mirror with flat rear side requires a smaller pad density at the edge than at the center.

8.5.4 Other Axial Flexure Features

The calculation for minimizing the flexure under gravity of mirrors having *finite* size originated with the case of constant thickness plates simply supported by a concentric and continuous ring. The first theoretical analysis for this problem was investigated by Couder [18] in a classic study based on the thin plate theory. Couder also carried out interferometric experiments with glass plates allowing him to determine the maximum elastic deflection when the plate is supported by three points in a 3-fold symmetry.[4] For two concentric and continuous support rings, the optimizing parameters are the radii of the rings and the reaction distribution per unit length along each ring. In a first stage, separated flexures $z_i(r)$ are calculated for each support ring of radius r_i by use of continuity conditions at the ring and free edge boundary conditions. Then the superposition principle allows summing the flexures z_i such that the sum of each reaction R_i along the corresponding ring is equal to the mirror weight. Variation of parameters r_i and R_i allows determining the minimum flexure. For mirrors with flat rear side, further flexure and reaction distribution corrections allow taking into account the thickness variation, thus providing the final flexure. Later investigations introduced the free inner edge boundary conditions for the case of central holed mirrors; for mirror apertures in the 3–4 m class, Schwesinger [74] optimized a 2-ring support and Lemaitre [44] a 3-ring support. However for thick mirrors, say with aspect ratios $t/d \gtrsim 1/8$, other considerations can also be taken into account as follows.

• **Shear stresses and thick plate theory**: From *Love-Kirchhoff hypotheses* of the thin plate theory (cf. Sect. 1.13.6), and in the axisymmetric bending of circular plates, the stress component σ_{zz} arising along normal lines to the mid-surface of the plate is *not* taken into account.

For instance, in a thick cantilever beam loaded at its free end, Saint Venant showed that a normal plane to the beam section becomes an *S-shaped* surface. This means that the stresses σ_{xx} acting through the thickness of the beam are not linear but vary accordingly to an odd *cubic* law from one face to the other. Around 1910 or

[4] At that time it was unclear whether or not a glass material may show a slight *viscosity*. In a dedicated interferometric experiment, where a glass plate was bent by weights for more than a year, Couder [18] showed that a variation of the fringe pattern could not be detected.

somewhat earlier, taking into account the shear deformations in circular and elliptical plates, Augustus Love elaborated the shear theory usually referred to as *thick plate theory*.[5]

When a uniform load is applied to a face of a plain constant thickness plate, Love [45] derived the bending and shear flexure of the middle surface for a circular plate simply supported at the edge, and for circular or elliptic plates built-in at the edge. In these cases, he also derived the complete equation set expressing the radial and axial displacements for surfaces located at distance $z \in [-t/2,\ t/2]$ from the middle surface. Woinowsky-Krieger [96] showed that in the simple case of a plain built-in plate with an aspect ratio $t/d = 1/10$, the shear flexure contributes $\sim$17% of the total flexure. Comparisons of the flexures between thin and thick plate theories are also given in Timoshenko and Woinowsky-Krieger [90]. As shown hereafter, this effect is less pronounced for a simply supported plate.

For instance, if we denote z the bending flexure of a constant thickness plate simply supported at its edge $r = a$, we have seen in Sect. 1.13.10 [case 2, (1.184b)], that Poisson's equation $\nabla^2\nabla^2 z = q/D$ of the thin plate theory leads to the flexure

$$z = \frac{qa^4}{64D}\left(\frac{r^2}{a^2} - 2\frac{3+\nu}{1+\nu}\right)\frac{r^2}{a^2}, \tag{8.26}$$

where $q = \mu\, g\, t$ is negative. Referring to Love [45] or Woinowsky-Krieger in [90], the shear flexure of the *middle* surface of a thick plate simply supported at the edge is

$$z_S = -\frac{qa^4}{40D}\,\frac{8+\nu+\nu^2}{1-\nu^2}\,\frac{t^2r^2}{a^4} \quad \text{[Love]}, \tag{8.27a}$$

$$z_S = -\frac{qa^4}{48D}\,\frac{3+\nu}{1-\nu^2}\,\frac{t^2r^2}{a^4} \quad \text{[W.-K.]}, \tag{8.27b}$$

but, from comparison with the finite element analysis result, it is more accurate to adopt the expression

$$z_S = -\frac{qa^4}{16D}\,\frac{1-\nu}{1+\nu}\,\frac{t^2r^2}{a^4}. \tag{8.27c}$$

From (8.27c), a more exact expression for the flexure including the shear component is

$$Z = z + z_S = \frac{qa^4}{64D}\left(\frac{r^2}{a^2} - 2\frac{3+\nu}{1+\nu} - 4\frac{1-\nu}{1+\nu}\frac{t^2}{a^2}\right)\frac{r^2}{a^2}, \tag{8.28}$$

[5] Analytical investigations of the shear deformations of beams were originated by Saint Venant [69] in the case of a cantilever beam. He also pointed out that the shear deformations can be considered as additive components to those of the bending deformations. Further development of the shear theory was briefly generalized to Michell [51], in 1900, who obtained the equilibrium equations including the shear stress components. This led to the elaboration of the *thick plate theory* by Love [45].

so the maximum flexure sag, obtained for $r = a$, is

$$Z_{max} = z(a) + z_S = -\frac{qa^4}{64D}\left(\frac{5+\nu}{1+\nu} + 4\frac{1-\nu}{1+\nu}\frac{t^2}{a^2}\right). \tag{8.29}$$

The two quantities inside the parentheses allow evaluation of the effect of the shear component relative to the bending component of the flexure. At the edge, where this effect is maximum, this can be deduced from the ratio

$$\frac{z_S}{z(a)} = 4\,\frac{1-\nu}{5+\nu}\,\frac{t^2}{a^2}. \tag{8.30}$$

For instance, with a Poisson's ratio $\nu = 1/5$ and aspect ratios $t/d \equiv t/2a = 1/12,\ 1/8,\ 1/6$, we obtain for the shear effect at the middle surface $z_S/z(a) = 1.7\%,\ 3.8\%,\ 6.8\%$ respectively (Fig. 8.9).

The above flexure component z_S of the shear deformation only applies to the middle surface of the plate. The contribution of the shear deformation in the vertical direction of the plate is not a constant function over the thickness. The shear strains entail that the volume elements where the supporting forces act are much more deformed than those of the middle surface whilst the elements near the opposite outer surface of the plate are less deformed. In the local regions where the support acts, the shear flexure shows a slope variation somewhat similar to that of a footprint.

The shear effects are implicitly taken into account in *three-dimensional* finite element analysis.

• **Meniscus mirrors and shallow shell theory**: An important advance in elasticity theory was realized by Eric Reissner [66, 67] who introduced the so-called *shallow shell theory*. In this theory, the tensions and compressions arising in the two tangential directions of the middle surface of a meniscus plate are taken into account.

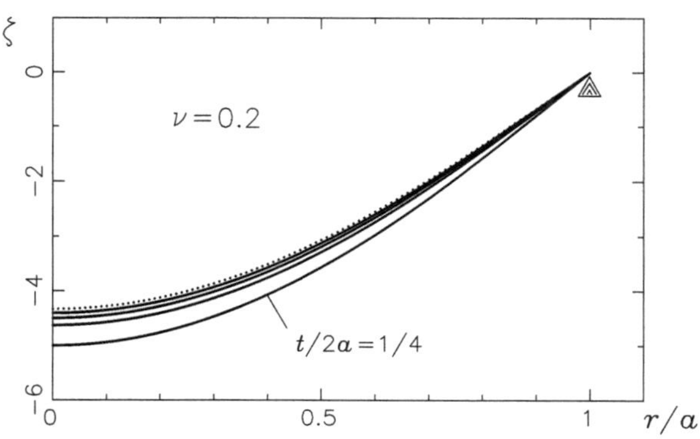

Fig. 8.9 Normalized flexures $\zeta = [Z(a) - Z(r)]/(qa^4/64D)$ of the middle surface of constant thickness plates simply and continuously supported at the edge $r = a$. Aspect ratios $t/2a = 1/12,\ 1/8,\ 1/6,\ 1/4$. The flexures represented by *full lines* include the shear component in t^2r^2/a^4. *The dotted line* is from the thin plate theory

Including this shell effect, Reissner derived a complete equation set allowing the determination of the bending and shear deformations. For astronomical mirror meniscuses faster than, say, f/2.5, and in the axisymmetric case, this equation set allows introducing continuity conditions at the ring support radius; however solving such problems requires the complex use of four Kelvin functions (cf. Sect. 8.6.3).

From the shallow shell theory, Selke [79] derived some flexures – although not optimized – of a plain mirror meniscus with 1-ring and 2-ring supports. For an aspect ratio $t/d = 1/12$ and a single ring support, he showed that the shear component of the flexure does not exceed ~6%. Schwesinger [73] commented on the non-linear stress distribution over the thickness of a plate.

The thin shallow shell theory was also used by Arnold [2] to derive the flexure of a meniscus mirror from a discrete support pad distribution operating with active optics correction modes. This required expressing the load function $q(r, \theta)$ in a Fourier series form and solving both the stress function F and normal flexure z from the following equations (cf. (6.6) and (6.8) in Chap. 6),

$$\nabla^2\nabla^2 F + \frac{E\,t}{<R>}\nabla^2 z = 0, \tag{8.31}$$

$$\nabla^2\nabla^2 w - \frac{1}{D<R>}\nabla^2 F = \frac{1}{D}\, q(r,\theta), \tag{8.32}$$

where $1/ < R >$ is the mean curvature of the mirror meniscus. Several explicit forms of the load function $q(r, \theta)$ are given hereafter.

• **Variable thickness mirrors**: For large mirrors which are not meniscuses, such as mirrors with a flat rear side, Poisson's equation does not apply since the flexural rigidity $D \neq constant$. A classic process for determining the flexure z under gravity is to distribute the mirror thickness variation into N constant thickness segments. For instance, the thin plate theory leads to solving the equation set

$$\nabla^2\nabla^2 z_n(r,\theta) = q/D_n, \qquad r \in [r_n,\, r_{n+1}], \quad \forall\, n \in [1,\, N], \tag{8.33}$$

where the flexural rigidities D_n are constants. Four continuity conditions at each segment junction and two boundary conditions at the center and at the edge must be solved.

Wan et al. [91] make the assumption that the shear component of the flexure can be related to that of the bending flexure by use of $z_S \propto t^2\nabla^2 z$. Although giving a demonstration, this assumption is not clearly proved; from (8.26) and one of the equations in (8.27), it appears to be *not* verified for axisymmetric flexures. These authors derived the Clebsch polynomials expressing the total flexure $z + z_S$ of a large plano-concave mirror supported by discrete point forces by use of the two-dimensional Dirac function (see hereafter).

• **Infinitely small support pad areas**: Schwesinger [78] proposed a method to extend the flexure analysis of a mirror under gravity for k discrete point pads distributed along a concentric ring in a k-fold symmetry. This method was subsequently investigated by Nelson et al. [55] and Arnold [3, 4] by use of Dirac functions.

Let W and a be the total weight (here negative because of the negative acceleration) and outer radius of the mirror. A two-dimensional *Dirac δ function* must satisfy

$$\iint \frac{1}{b_j}\,\delta(r-b_j)\,\delta(\theta-\theta_j)\,r\,dr\,d\theta = 1, \tag{8.34}$$

which allows expressing the load q per unit area as

$$q(r,\theta) = \frac{W}{\pi a^2} + \sum_{j=0}^{k} \frac{f_j}{b_j}\delta(r-b_j)\,\delta(\theta-\theta_j), \tag{8.35}$$

where the support system has k-fold symmetry and each support point j is characterized by its force f_j located on a circle of radius $r = b_j$ at the azimuth angle $\theta_j = \theta_0 + 2j\pi/k$. Replacing the Dirac function $\delta(\theta-\theta_j)$ by its complex Fourier series leads to

$$q(r,\theta) = \frac{W}{\pi a^2} + \sum_{j=0}^{k} \frac{f_j}{2\pi b_j}\delta(r-b_j) \sum_{m=-\infty}^{\infty} \exp\left[i\,m(\theta-\theta_j)\right]. \tag{8.36}$$

The condition of the static equilibrium for the forces and moments are

$$\sum_{j=1}^{k} f_j + W = 0, \qquad \sum_{j=1}^{k} f_j b_j \exp(i\theta_j) = 0. \tag{8.37}$$

From the thin plate theory, the flexure is expressed by

$$z = \sum_{m=0}^{\infty} R_{nm}(r)\cos\left[km(\theta-\theta_0)\right], \tag{8.38}$$

where the Clebsch solutions $R_{nm}(r)$ are polynomial forms expressed by the set (7.7) in Sect. 7.2. However, the Dirac functions restrain the problem to infinitely small support points, which is a quite pessimistic option.

• **Finite support pad areas**: Monolithic mirrors of presently large telescopes are often designed in a meniscus shape with aspect ratios $t/d \leq 1/20$. Given the small value of such ratios, the shear flexure from the support pad is fully negligible. Although the f-ratios of these primary mirrors often reach values such as f/1.8 or f/1.6 that would require use of Reissner's shallow shell theory, the low level of stress involved in their flexure under gravity renders it unnecessary. It is also reasonable to assume that the maximum flexure sag Δz_{max} for generating any active optics correction modes $z_{nm} \propto r^n \cos m\theta$ does not give rise to significant stresses in the "in plane" middle surface of the meniscus. Hence, for most of large telescope mirror meniscuses, the basic thin plate theory can be assumed as *accurate* enough for the determination of the mirror thickness and its passive and active optics support system.

In establishing a support pad density criterion for an infinitely large mirror (Sect. 8.5.3), we have seen from the thin plate theory that when a mirror is supported by *ring pads*, the pad radius ratio is a useful free parameter [cf. criterion (8.25)].

Also using the thin plate theory, another approach by Arnold [5] consists of determining the flexure under gravity and the influence function from *plain* actuated pads having a finite acting area. Compared to infinitely small pads represented by Dirac functions, pads with finite surface area spread the load under the mirror and smooth the optical surface. For mathematical convenience, and without changing significantly the final result, the pads considered by Arnold are almost squares of dimension d_p and analytically defined by the product of two *top-hat* Π *functions* as

$$\Pi\left[\frac{(r-b_j)}{d_p}\right]\Pi\left[\frac{b_j(\theta-\theta_j)}{d_p}\right], \tag{8.39}$$

where b_j and $\theta_j = \theta_0 + 2j\pi/k$ are the polar coordinates of the k pads in the mirror frame r, θ, z, and

$$\Pi[x] = 1 \quad \text{for} \quad x < 0.5, \qquad \Pi[x] = 0 \quad \text{for} \quad x \geq 0.5. \tag{8.40}$$

For a mirror of weight W with a hole of radius $r = c$, the general expression of the load q can be written as

$$q(r,\theta) = \frac{W}{\pi(a^2-c^2)} + \sum_{j=1}^{k} \frac{f_j}{d_p^2} \Pi\left[\frac{(r-b_j)}{d_p}\right]\Pi\left[\frac{b_j(\theta-\theta_j)}{d_p}\right], \tag{8.41}$$

or, denoting the *sine cardinal function* $\text{sinc}(x) = \sin(\pi x)/\pi x$ and using a Fourier expansion,

$$\begin{aligned} q(r,\theta) = {} & \frac{W}{\pi(a^2-c^2)} + \sum_{j=1}^{k} \frac{f_j}{2\pi b_j d_p} \Pi\left[\frac{(r-b_j)}{d_p}\right] \\ & \times \sum_{m=-\infty}^{\infty} \text{sinc}\left(\frac{m d_p}{2\pi b_j}\right) \exp\left[im(\theta-\theta_j)\right], \end{aligned} \tag{8.42}$$

where the k acting forces f_j and their moments about any axis in an x, y plane satisfy the static equilibrium equations similarly as set (8.37). Given a k-set of pads on a circle of radius b_j, the substitution of $q(r,\theta)$ into Poisson's equation leads to the determination of the coefficients of the Clebsch polynomials $R_{nm}(r)$ (cf. Sect. 7.2) so the flexure can be represented by $z = \Sigma R_{nm} \cos[km(\theta-\theta_0)]$. Further similar determinations of the flexure z are calculated for pads distributed on circles of various radius b_j. Then, redistributing partial loads for each flexure, a linear superposition of them provides the resulting flexure for all the mirror. Variations of the partial loads and of the design parameters d_p, b_j and θ_j associated k-values allow obtaining a support topology which minimize the resulting flexure. In the case of pads distributed on four concentric circles, Arnold obtained somewhat different topologies when optimizing only the flexure under gravity or when also including active optics correction of a large *Astm* 3 mode.

8.5.5 *Finite Element Analysis*

For large primary mirrors, *finite element analysis* has become the classic computational modeling to accurately calculate the flexure owing to gravitational loading on discrete supporting areas. The above approximation conditions do not need to be considered. Finite element analysis allows solving the three-dimensional problem with almost no limiting assumptions (cf. Sect. 1.13.14). From the boundary conditions, the strains ε_{ik} are determined for all finite elements which, then, provides the components

$$u(r,\theta,z), \quad v(r,\theta,z), \quad w(r,\theta,z) \tag{8.43}$$

of the displacement vector for each element. Iteration algorithms allow repeating the solving process until no variation occurs in the displacement vectors, which thus corresponds to the static equilibrium. A convenient accuracy is reached when increasing the number of finite elements entails quasi-equivalent displacements. Current codes can take into account all needed features such as:

- mirror geometry, thickness distribution, and central hole,
- support pad geometry and associated pad force distribution,
- mirror shell effects and middle surface tension function,
- mirror shear effects and their three-dimensional variations.

Furthermore, in the case of gravitational elastic deformations where volume loads act, finite element analysis can easily treat this case which, for a constant thickness mirror, is not exactly equivalent to that of a uniform load applied to the upper or middle surface of a mirror.

In 1968, Malwick and Pearson [48] pioneered finite element analysis in the case of astronomical mirrors with the 4-m primary mirror of the Ritchey-Chrétien Mayall-KPNO telescope. This primary mirror at f/2.6 is a plain fused silica blank with a flat rear side, large central hole and aspect ratio $t/d = 1/8$. The support mirror cell includes $N_a = 36$ axial pads distributed along two concentric rings (see Fig. 8.8). In reference [77], the authors consider a preliminary f/2.75 mirror with aspect ratio $t/d = 1/7$ and obtain several *axial* flexures for supports with two continuous rings and for three-fold symmetry pads; they also obtain *lateral* flexures for various support cases.

8.6 Lateral Flexure of Large Mirrors Under Gravity

8.6.1 *Various Supporting Force Distributions*

The passive lateral supporting of a mirror requires the use of one or several system elements that are shown in Fig. 8.2. The support concept and the number of supporting elements depends on the geometry and weight of the mirror. For a 1.8-m aperture mirror, as for a mirror segment of the KECK telescope, it has been found

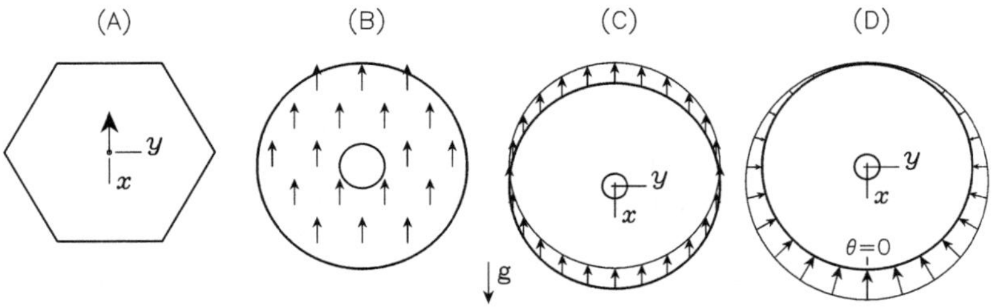

Fig. 8.10 Various force distributions mirror lateral support. (**A**) Single direct astatic lever at the center of a mirror segment. (**B**) Direct astatic levers uniformly distributed in a 3-fold symmetry at the mirror middle surface. (**C**) *Parallel and equal push-pull forces* uniformly distributed along the mirror contour. (**D**) *Radial push forces* distributed in $1 + \cos\theta$ along the mirror contour

that a single direct astatic lever is convenient (Fig. 8.10A). For large and moderately lightweight primary mirrors, direct astatic levers uniformly distributed at the mirror middle surface were designed, for instance, for the 5-m Hale telescope ($N_l = 36$ lateral levers) and 6-m SAO telescope ($N_l = 60$); in these latter telescopes, the axial and lateral supports – of the same number ($N_a = N_l$) – are combined into double systems (Fig. 8.10B).

For large monolithic plano-concave or meniscus mirrors, the best result is obtained with *parallel push-pull forces* uniformly distributed along the mirror contour (Fig. 8.10C). If N_l lateral forces support the mirror weight W, then the components of each force numbered $n \in [1,\ N_l]$ are

$$F_{x,n} = -W/N_l, \qquad F_{y,n} = 0, \tag{8.44}$$

however the parallel push-pull force system is somewhat difficult to implement in practice. Another alternative easily practicable is with only *radial push forces* f_r per unit length,

$$f_r = -\frac{W}{\pi a}(1 + \cos\theta), \tag{8.45a}$$

distributed by hydraulic pads along the mirror contour (Fig. 8.10D). The distribution of N_l *even* lateral forces acting at angles $\theta_n = (2n - 1)\pi/N_l$ along arches corresponding to $\Delta\theta = 2\pi/N_l = constant$, determines the radial and tangential components as

$$F_{r,n} = \int_{\theta_n-\Delta\theta/2}^{\theta_n+\Delta\theta/2} f_r \cos(\theta - \theta_n) a\, d\theta, \qquad F_{t,n} = 0. \tag{8.45b}$$

These forces are implicitly generated by pads pushing normally to the edge surface.

Compared to other possible edge force distributions such as, for instance, with two concentrated forces at $\theta = \pm 45°$ located each side of the lowest part of the edge, the two latter distributions provide the main advantage of a smooth flexure of the mirror by avoiding large m-values of spacial frequency modes in $\cos m\theta$. The radial push force distribution, which corresponds to forces generated by a *narrow* mercury bag, only requires use of normal *pushing* forces.

8.6.2 *Flexure of a Mirror Supported at its Lateral Edge*

Elasticity analysis using Fourier expansions were developed early by Schwesinger, [72, 74, 77], who derived mirror surface flexures caused by various lateral support systems. In order to avoid solving the complex three-dimensional problem of the determination of the three displacement components $u,\ v,\ w$ in (8.43), it was assumed that the radial and tangential displacements $u,\ v$, and axial displacement w – here denoted z for active optics reasons – at the middle surface of the mirror surface can be accurately expressed from Love-Kirchhoff hypotheses (cf. Sect. 1.13.6), i.e. reduce to

$$u|_{z=0} = 0, \qquad v|_{z=0} = 0, \qquad w = z(r,\theta). \tag{8.46}$$

In this approximation corresponding to the thin plate theory, the small axial displacements – hereafter called the *flexure* – can also be assumed as the same for all mirror points of coordinate r, θ, so the problem reduces to the determination of the axial flexure of the middle surface when the mirror axis is horizontal.

Using the radius variable $\rho = r/a$ normalized with respect to the edge radius, a Fourier expansion of the z-flexure of the middle surface due to the lateral support can be represented by

$$z = \frac{\mu g a^4}{E t^2} \sum_{m=0}^{\infty} c_m P_m(\rho) \cos m\theta, \tag{8.47}$$

where c_m and P_m are coefficients and polynomials.

If the middle surface of the mirror is without curvature (e.g. a mirror with equiconcave faces), then all $P_m = 0$, which entails $z \equiv u_z = 0$ from the same above approximation that $u_r = u_t = 0$.

In addition to the force distribution applied to the contour, the two parameters entering into the determination of the flexure of the mirror are the curvature $1/<R>$ of the middle surface and the thickness distribution $t(\rho)$. These latter parameters can be characterized by dimensionless quantities α and κ as

$$\alpha = a/<R>, \qquad t/t_0 = 1 + \kappa\rho^2. \tag{8.48}$$

Let $1/R$ be the curvature of the optical surface, the radius of curvature of the middle surface is

$$<R> = 2R \qquad \text{for a plane-concave mirror,}$$
$$<R> = \frac{1}{2}t_0 + R \qquad \text{for a meniscus mirror,}$$

and $\kappa = 0$ for a meniscus mirror.

Assuming hereafter that the mirror is without central hole, the bending and twisting moments $M_x,\ M_y,\ M_{xy}$ are determined from the components $F_{x,k}$, $F_{y,k}$ of the perimeter force distribution and from the α and κ parameters. For each mode of order m, the Fourier expansions for $F_{x,k}$, $F_{y,k}$ allow expressing the three equilibrium equations of the moments as functions of the stresses σ_{xx}, σ_{yy}, σ_{xy} (see equations (1.165) in Sect. 1.13.8), and then provide the flexure.

For instance, considering the radial push force distribution $f_r \propto 1 + \cos\theta$ and a plano-concave mirror for which, from (8.48), $\kappa = \alpha\, a/t_0 = a^2/2Rt_0$, Schwesinger [77] derived the rms flexure (cf. (8.24) in the form

$$z_{\mathrm{rms}} = \frac{\mu g a^2}{E}\left[\sum_m a_m^2 Y_m(\kappa)\right]^{1/2}, \tag{8.49}$$

where the harmonic m of the lateral force distribution leads to the associated flexural mode m represented by the product of the square of a dimensionless amplitude a_m – depending on the force distribution type – with the dimensionless function $Y_m(\kappa)$ (Table 8.3).

For the mode $m = 0$, it appears from the Y_0-value that the piston, $Cv\,1$, $Sphe\,3$ and higher-order spherical aberration modes have a negligible resulting effect. For $m = 1$, which represent a lateral displacement of the field of view, the effect is extremely small. The dominant effect is the all-order astigmatism mode given by $m = 2$. The next aberration mode $m = 3$, representing coma, is much smaller than astigmatism, whilst the influence of the higher modes decreases continuously.

Whether a plano-concave or a meniscus shape of the mirror, use of parallel and equal push-pull forces $F_{x,n} = -W/N_l$ uniformly distributed at the contour provide a smaller residual flexure than that of the radial push force distribution $f_r = -(W/\pi a)(1+\cos\theta)$. In practice, direct axial astatic levers can easily generate such a push-pull distribution either with equatorial or altazimuth telescope mount, the latter mount requiring the levels to pivot only around one direction. The lateral support of the meniscus mirror of the 3.5-m ESO-NTT is with $N_l = 24$ parallel and equal push-pull forces uniformly distributed at the contour in the plane of the center of gravity [71].

Table 8.3 Value of coefficients $Y_m(\kappa)$ for the various flexure modes in $\cos m\theta$ of a vertical plano-concave mirror supported at its edge. Geometrical ratio $t_0/2a = 1/8$. Poisson's ratio $\nu = 0.25$. Radial push force distribution $f_r \propto 1 + \cos\theta$ [77]

Flexure mode	$\kappa = 0.0$(*)	$\kappa = 0.1$	$\kappa = 0.2$	$\kappa = 0.3$
$m = 0$	0	1.01	4.04	9.1
1	0	0.03	0.13	0.3
2	4.6	140.5	465	978
3	3.4	20.7	52.5	99
4	2.7	8.9	18.5	32
5	2.3	5.5	10.1	16
.	.	.	.	.
10	1.3	1.9	2.7	3.7

(*) values derived from the thick plate theory.

8.6.3 Other Force Distributions and Skew Surface of Forces

Telescopes with altazimuth mounts allow use of more complex distributions for the lateral support forces of the mirror. Further, these forces may not necessarily be distributed in a plane. Hereafter distributions are considered for large monolithic mirrors in alt-az mount telescopes.

• **Mirror slices and parallel push-pull force distribution**: Mack [47] found it appropriate to distribute parallel push-pull forces $F_{x,n}$ such that the total force of each bottom-top pair, acting in the same direction for a given y-value, is equal to the weight of the vertical mirror *slice* element, dy in width. Let W be the weight of a mirror without central hole. Assuming that each force of a pair is equal, the element components of each force of this pair are

$$dF_x = -\frac{W}{\pi}\cos^2\theta d\theta, \qquad F_y = 0. \tag{8.50}$$

The distribution of N_l *even* lateral forces numbered $n \in [1, N_l]$, acting at angles $\theta_n = (2n-1)\pi/N_l$ along arches corresponding to $\Delta\theta = 2\pi/N_l = constant$, determines the location and size of each $N_l/2$ vertical slice. From integration, the nth vertical force $F_{x,n}$ is

$$F_{x,n} = -\frac{W}{N_l}\left[1 + \text{sinc}\frac{2}{N_l}\cos\frac{(2n-1)2\pi}{N_l}\right], \qquad F_{y,n} = 0, \tag{8.51}$$

where $\text{sinc}\, x = \sin\pi x/\pi x$. Compared to the classical parallel push-pull force distribution defined by equations (8.44), the above distribution acts with larger forces at the vertical mid-plane of the mirror whilst the forces in the regions $y \simeq \pm a$ are much smaller.

A somewhat different alternative leads to redefining the width of the slices such that all the forces $F_{x,n}$ are set equal. Such a *variable* y-spacing parallel push-pull force distribution was proposed by Mack (Fig.8.11-A) and adopted, with $N_l = 24$ parallel push-pull forces generated by direct astatic levers, for the plano-concave mirror of the 4.2 m UK-WHT.

• **Skew surface of forces**: A second proposal by Mack [47] takes under consideration the fact that, for any plano-concave or meniscus mirror, the individual centers of gravity of the vertical slice elements are not at a same Z-value. Thus, constraining

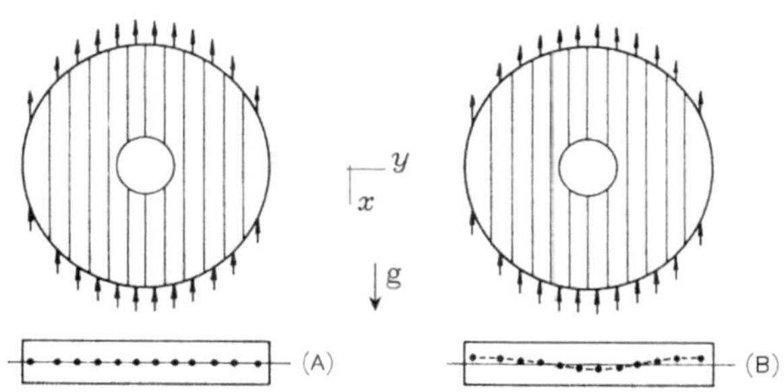

Fig. 8.11 Variable y-spacing parallel and equal push-pull forces. (**A**) Top view of the plane surface of forces passing through the center of gravity of the mirror. (**B**) Top view of the skew surface of forces in which the individual centers of gravity lie [71]

the equal and parallel forces to lie in the skew surface which is defined from vertical lines passing through the individual centers of gravity (Fig. 8.11-B), it is shown from finite element analysis that, compared to the case where the forces act into a plane surface passing through the mirror center of gravity, the flexure can be reduced by about a third.

Use of a skew surface of forces allows us to introduce an interesting free parameter for the optimization of lateral mirror supports.

• **Push-pull force distribution with increased shear**: Schwesinger [75, 76] investigated force distributions where the tangential component is a free parameter. Denoting $F_{r,n}$ and $F_{r,t}$ the radial and tangential components of the applied lateral forces, where from Pythagoras $F_{r,n}^2 + F_{t,n}^2 = F_{x,n}^2 + F_{y,n}^2$, various distributions can be compared together by use of the *fraction* β of the weight W supported by the tangential forces $F_{t,n}$. Thus, β can be determined from

$$\sum_{n=1}^{N_l} F_{t,n} \cos\theta = -\beta W. \tag{8.52}$$

With the radial push force function $f_r \propto 1 + \cos\theta$ shown in Fig. 8.10-D, we have $\beta = 0$.

With the equi-spaced push-pull forces $F_{x,n} = -W/N_l,\ F_{y,n} = 0$ in Fig. 8.10-C, $\beta = 0.5$.

For the 8.2-m VLT meniscus mirror at f/1.8 with aspect ratio $t/2a = 1/47$, compared to the two latter distributions by Mack, it was found that if the tangential forces $F_{t,n}$ are given a certain geometry so that $\beta \simeq 0.75$, then the flexure is substantially reduced [76]. The lateral forces act along a circle located somewhere between the rear and midpoint edge of the mirror. This circle is in a plane which does *not* pass through the center of gravity. Therefore, the final acting forces were given small $F_{z,n}$ force components to recover the static moment equilibrium which, then, avoids use of lateral forces at the inner contour of the hole. The primary mirror lateral support of the ESO-VLT is with $N_l = 48$ forces equally spaced at the outer contour (Fig. 8.12). The force distribution $F_{r,n},\ F_{t,n}$ that generates the shear function is obtained by hydraulic levers acting on extra-peripheral pivots. Each pivot

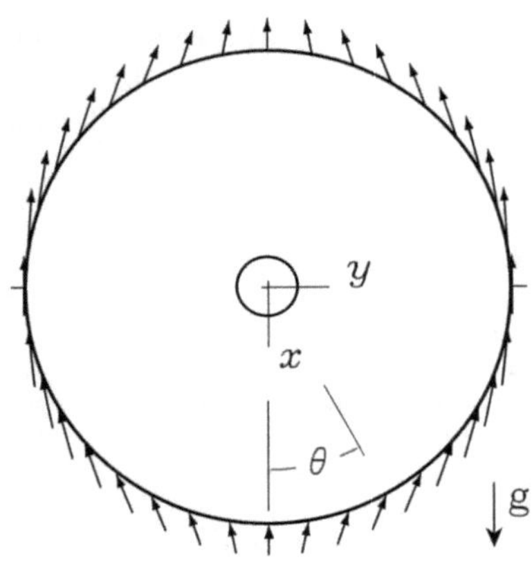

Fig. 8.12 Equal-spacing push-pull force distribution for the lateral support of the 8.2 m primary mirrors of the ESO-VLT. $F_{r,n},\ F_{t,n}$ forces are with increased shear components $\beta = 0.7529$ generated via extra-peripheral pad pivots lying in a circle. Because this circle is off-plane to the center of gravity, the final forces are inclined by addition of $F_{z,n}$ components [76]

pad is sealed to the mirror edge in the two regions $\theta_n \pm \pi/N_l$; thus, the effect of the successive moments generated by the forces acting on the pivots is almost cancelled and gives rise to the required increased shear push-pull distribution.

8.6.4 *Finite Element Analysis*

In addition to their first computational modeling of large astronomical mirrors axially supported by concentric continuous rings or discrete pads in three-fold symmetry, Malvick and Pearson [48] derived the flexure in various cases of lateral support force distributions. Some of these early results from finite element analysis are shown in Fig. 8.13.

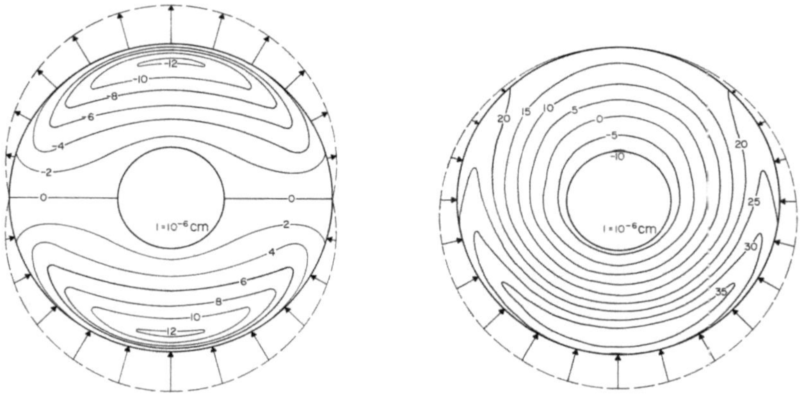

Fig. 8.13 Finite element analysis flexures from lateral supports of the 4-m mirror Mayall-KPNO telescope. *Left*: Special push-pull force distribution. *Right*: Radial push force distribution [48]

Finite element analysis has now become an indispensable method for the accurate determination of the elastic deflections of a mirror. When finite element analysis is conjugated with Seidel wavefront mode analysis, the residual low spacial frequency modes resulting from the lateral support flexure can also be corrected by closed-loop *in situ active optics* applied to the axial support system.

8.7 Active Optics and Active Alignment Controls

8.7.1 *Introduction and Definitions*

The main optics of a large telescope can benefit from active corrections allowing recovery of the image quality for which the telescope was designed. For a ground-based telescope, numerous reasons cause the starlight image to deteriorate: optics manufacturing residuals, theoretical errors of mirror passive support systems – such as described in the latter sections –, focussing and centering errors from the mechanical structure, maintenance errors of the structure and mirror supports, thermal distortions of the mirrors, telescope structure and ambient air, mirror deformation from wind buffeting, atmospheric degradation from turbulence, and tracking errors. All these sources of error were systematically quantified in terms of time-dependent frequency bandpass by Wilson et al. [92]. We hereafter mainly follow and summarize the account by Wilson [95].

• **The HST early study**: For a space-based telescope, all these image degradations vanish except the thermal gradient effects on the optics and structure, and tracking errors; the effect of the gravity change from $0\,g$ to $1\,g$ may also alter the mirror shape. Although a space telescope is the best choice for high-resolution imaging up to the diffraction limit, early investigations (1960s) on the active control of a primary mirror figure were undertaken at NASA by Creedon and Lindgren [19] for the HST. Howell and Creedon [31] developed a sensing wavefront and mirror reshaping system that was applied on the ground to a 30-inch diameter model mirror with 0.5-inch thickness. Using 58 actuators, the initial error of $\lambda/2$ rms at $\lambda = 0.663$ nm was reduced to $\lambda/50$ rms. The control system used the *natural lateral vibration modes* – i.e. *minimum energy modes* – of the mirror [19].

With 24 actuators at the 2.4-m primary mirror of the HST, the system would have compensated for half a micron of astigmatism sag. This early study must be considered as a pioneering work. However, because of its complexity, its application to the HST was finally rejected. In can be noted that the correction range would have been far too small to compensate for the spherical aberration due to a calibration error in the optical testing. This problem was then solved by the introduction of an aspherical correcting plate in a post-focal pupil. At that time, such an active optics system would not have been very efficient since CCD detectors appeared later.

• **Active optics control and active alignment control**: From definitions in Sect. 1.14.3, one may define the active controls of a telescope as follows.

An *active optics control system* is a low time frequency bandpass process involving corrections of wavefront errors by changing the shape of a telescope mirror, generally the primary mirror; these corrections may include *partial* wavefront errors caused by misalignments with other mirrors. Concerned actuators are included in the highest level of the mirror axial support system.

An *active alignment control system* is a low time frequency bandpass process involving corrections of wavefront errors caused by misalignments between the telescope mirrors. Concerned actuators are dedicated to the relative positioning in space of, generally, the secondary mirror. The associated displacements – or "*optical despaces*" – are the piston (z translation), lateral centering (x and y translations) and tip-tilt (x and y rotations). These five despaces can be obtained by use of an *active hexapod truss* which links the secondary mirror to its reference support plane.[6]

[6] A *hexapod truss* is a three-dimensional structure made of three sets of adjacent bar pairs in a "VVV" arrangement in a way that the upper and lower parts of the V's form two distinct triangles located in parallel planes. The system links three hinge points of the mirror to three hinge points

The interdependence conditions between the two above control systems must be strictly set.

For both the above systems, the "open-loop" control can be restrained to passive or quasi-passive corrections whilst "closed-loop" control is relative to active.

A higher time frequency closed-loop level for adaptive optics corrections can be introduced on a telescope mirror if this mirror is extremely thin and possesses its own actuators.

• **Aberration evaluation algorithm for the alignment process**: Misalignments of the primary and secondary mirrors provide on-axis coma and an astigmatism variation in the field and, in general, the axes of the two mirrors do not intersect. From wavefront analyses of several star beams distributed over the field of view, say at a Cassegrain focus, a tilt, a lateral shift or both of the secondary mirror allow setting the on-axis coma to zero. The result is that now the two axes intersect at the so-called *coma neutral point* details are given by Wilson et al. [92–95] and Rakich [65], but because of the residual angle formed by the two axes there exist two points in the field where the astigmatism is minimal, usually called *field binodal astigmatism.* The final alignment consists of dispaces of the secondary mirror about the coma neutral point up to obtaining the superposition of the mirror axes. This is achieved when the binodal astigmatism merges into an axisymmetric distribution all over the field.

8.7.2 Monolithic Mirror Telescopes

• **The ESO NTT and VLT**: The first ground-based monolithic mirror telescopes equipped with a complete active optics and alignment control systems were the 3.5 m NTT whose first light was in 1989, and the 8.2 m VLT of ESO whose first telescope unit started observations in 1998. The principle of the closed-loop control is based on the analysis at Nasmyth or Cassegrain focus of starlight wavefronts by Shack-Hartmann (S-H) systems using a lenslet and CCD. Several guide star beams can either be selected from an on-axis beamsplitter or from offset regions to the field edge (Fig. 8.14).

---

of its reference support, thus forming a statically rigid 6-hinge truss. Independent system controls of each bar length of the V's provide the above five required displacements.

The conditions for *rigid trusses* were first discussed by August Ferdinand Möbius [46], in 1837, who was the professor of astronomy at the University of Leipzig and is best known for his work in analytic geometry and in topology. Denoting the number of the hinges by N and that of the bars by B, he found that *necessary* conditions for two- and three-dimensional trusses are respectively

$$2N - 3 = B, \qquad \text{and} \qquad 3N - 6 = B.$$

In addition to these fundamental results, Möbius indicated that even if these conditions are satisfied there are exceptional cases where the truss is not absolutely rigid because the triangulation is improperly done. In studying such exceptional cases, he found that they occur when the determinant of the static equilibrium equation system at the truss joints vanishes (see *History of Strength of Materials* by Timoshenko [89]).

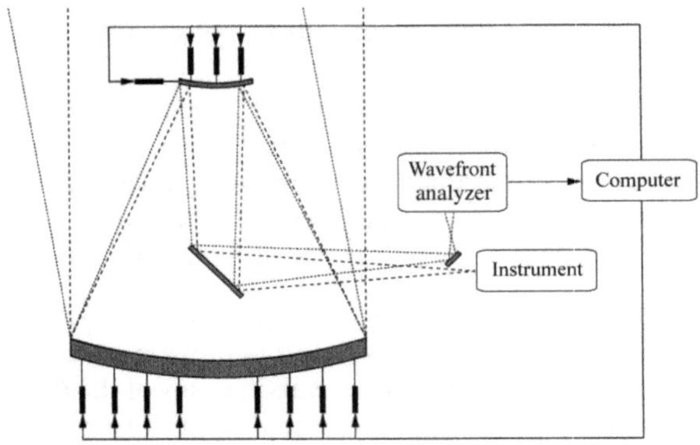

Fig. 8.14 Principle of active optics and alignment controls for a monolithic mirror telescope [59]

Taking into account the azimuth ϕ_{nm} for a non-axisymmetric wavefront, each Seidel mode of a wavefront may be written as

$$w_{nm} = A_{nm}\rho^n \cos(m\theta + \phi_{nm}), \tag{8.53}$$

where A_{nm} coefficients characterize the sag of the wavefront error and $\rho = 1$ is the edge normalized radius of the wavefront pupil.

The principle of the Shack-Hartmann test is to image the telescope mirror input pupil into subpupils by use of a two-dimensional microlens array; thus, the CCD image analysis provides a measure of the local slope errors of the wavefront $\partial w/\partial x$, $\partial w/\partial y$, where $w = \Sigma w_{nm}$. While the telescope is tracking a star of magnitude $V = 12$–13, closed-loop cycles – typically 10 minutes and 60 seconds of time for the NTT and VLT respectively – deliver resulting correction signals that are transduced either into a new force configuration of the axial support system of the primary mirror or into relative displacement and rotation of the secondary mirror.

A preliminary experiment by Wilson et al. [94] and Noethe et al. [60] on a thin meniscus mirror of one meter diameter was carried out for checking some basic features. With the NTT, a primary mirror aspect ratio of $t/d = 1/15$ and $N_a = 75$ axial supports allows corrections of the seven Seidel terms w_{11}, w_{20}, w_{22}, w_{31}, w_{40}, w_{33} and w_{44}. Because of a calibration error in the optical testing, the primary mirror showed a *Sphe* 3 error of ~3.50 μm in ρ^4. Instead of refiguring the mirror, the problem was solved by the axial support system in introducing a first level of passive correction with springs; the second level for passive diurnal corrections is with astatic levers, and the third active level uses actuators modifying the axial forces. Located on the same site as the conventional – i.e. passive – 3.6-m ESO telescope (La Silla, Chile), the results from the rather *thick* primary mirror of the NTT clearly proved the efficiency of active optics and alignment controls. More details are given in Wilson [93].

Much more optimized for active optics control is the *thin* primary meniscus mirror of the VLT (Fig. 8.15) with an aspect ratio $t/d = 1/47$ and $N_a = 150$ axial supports. The *flexibility* $z/F \propto d^2/Et^3$ to an external force F is ~13-times larger than that of the NTT. The axial support system consists of a first level with hydraulic

Fig. 8.15 View of a VLT Unit at Cerro Paranal (courtesy ESO)

tripod pads and a second active level with spring-hydraulic actuators [70]. Similarly to the NTT, all the axial forces can be only generated in the push mode, which then does not allow maintenance operations of the telescope in the horizontal direction. Noethe [59] investigated the modal form of the correction to be generated by the closed-loop control command and determined, from the shallow shell theory, the orthogonal polynomials representing the *natural lateral vibration modes* of a meniscus; these minimum energy modes, which differ from Zernike's polynomials, provide an optimal process for generating the force distribution corrections. Within the correction range of the VLT primary mirror, it was shown by Noethe that the active component of the axial forces can be considered as linear with the deformation (Hooke's law). Given a total piston sag Δz of a minimum energy mode (n, m), representing a mirror shape error composed of Seidel modes $m = constant$ in the order $K_O = n + m - 1$ (with conditions (1.47) in Sect. 1.8.2), one can derive this linear relation. Denoting K the number of the S-H subpupils, the relation between the column vector of the force change by the N_a actuators and the corresponding z-displacement vector of the mirror measured at the sensor writes

$$\Delta F_l = a_{k\,l}^{-1} \Delta z_k, \qquad l \in [1,\ N_a], \qquad k \in [1,\ K], \tag{8.54}$$

where $a_{k\,l}^{-1}$ is the *stiffness* matrix. For each minimum energy mode n, m, the associated $a_{k\,l}^{-1}$ coefficients can be determined from experiment but were more accurately determined from theoretical analysis. The largest active optics effect is the configuration conversion from Nasmyth to Cassegrain which, because of an appreciable axial image shift, introduces for $m = 0$ an error polynomial of about 12 μm in ρ^4;

this leads to generation of maximum active force components of about +413 and −156 N within an actuator z-step increment (absolute accuracy) corresponding to a typical force of 0.05 N [58]. A review on active optics control aspects for large optical telescopes is given by Noethe [62].

• **Other Active Optics Telescopes**: The next ground-based monolithic mirror telescopes with active optics control for visible and infrared observations were designed from preliminary active optics experiments on small mirrors by Itoh et al. [34] and Iye et al. [36], which led to the construction of the 8.2-m SUBARU telescope; a combined axial-lateral support system uses 264 cylindrical cavities in the rear side of the meniscus mirror where direct astatic levers act as lateral passive forces [35]. The two 8-m GEMINI telescopes are equipped with an active support system [82]; the passive level is by air pressure supporting 75% of the weight whilst the active level is with hydraulic pressure actuators.

8.7.3 Segmented Mirror Telescopes

• **Early Approaches**: Some preliminary aspects on the recent development of large segmented telescopes may be briefly recalled. For instance, a 4 m, f/1.4/13, segmented infrared collector was built in the 1970s by Chevillard, Connes et al. [15]. The initial goal was to obtain high-resolution infrared spectra of planets and astronomical objects by use of a Michelson-type Fourier transform interferometer installed at coudé focus. A detailed review is given by Connes and Michel [16]. The alt-az collector was built with a spherical primary mirror of 36 square segments actively controlled and a highly elliptic secondary provided a stigmatic narrow field; although the active alignment control of the segments showed tracking star images with FWHM of ∼5 arcsec, the project was stopped by lack of support and the interferometer never installed.

Also with a spherical segmented primary mirror concept, Baranne and Lemaitre [7] proposed a five-mirror two-axes multifocus telescope (TEMOS) where, apart from the M_1 mirror and flat 45°M_3, one of the other mirrors M_i can be actively in situ aspherized after initial spherical polishing without stress. Let q be an active uniform load applied to the clear aperture of a *vase-form* mirror M_i of curvature $1/R$. Its shape from *in situ stressing* – at the telescope – can be represented by

$$z = \left[R + \sqrt{R^2 - (1+\kappa)r^2}\right]^{-1} r^2 + q(c_2 r^2 + c_4 r^4 + c_6 r^6 + \cdots), \qquad (8.55)$$

where $\kappa = 0$ and $q = 0$ when the mirror M_i is unstressed (sphere). The total *in situ aspherization* of a mirror correcting the important *Sphe* 3 amount of a spherical primary mirror is an interesting challenge. The simplest way for an experimental validation is to choose a two-mirror system and mirror M_i as the Cassegrain mirror, thus M_2. This leads to qc_2, qc_4, ... values that provide a quasi-elliptic secondary corresponding to a partial vacuum inside the vase-form secondary. An experimental telescope was built and tested on the sky with 1.4 m TEMOS 4, f/2/6.6, in a configuration using four circular spherical segment replicas of 0.5 m diameter, a stainless steel vase-form secondary of 0.35 m diameter and a doublet lens corrector. With a mean aspect ratio $t/d = 1/53$, mirror M_2 was in situ aspherized by a load $q = 0.789 \times 10^5$ Pa corresponding to a center-edge sag of 204.7 μm. Seeing limited images of stars with FWHM $< 0.8 - 1$ arcsec were obtained, but the seeing conditions at Haute Provence did not allow full evaluation of the angular resolution [41].

• **Alignment and cophasing of the KECK Telescope**: The 36 hexagon segments of each primary mirror of the Keck telescope (Fig. 8.16) were mainly aspherized by active optics stress polishing to form a hyperboloid primary mirror of 10 m aperture at f/1.7 (cf. Sect. 7.5.1). A segment is axially supported by three 12-point whiffletrees and laterally supported by a passive single metal diaphragm mounted in a central cavity. Each of the four subpivots of a whiffletree are equipped with straight springs that can generate moments for partial passive correction of the segment figure residuals. Each of the three whiffletrees is mounted on an axial translation actuator ensuring the piston and tilt control of a segment. Each actuator uses a ball screw pushing a two-stage demultiplying hydraulic bellows that acts on the central pivot of the whiffletree with a translation increment accuracy of 30 nm. Details are given by Mast and Nelson [80]. Pairs of capacitive sensors located at intersegment edges provide accurate measures of surface boundary errors between adjacent segments. When the segments are aligned (see below), the reading of 168 sensors is stored in rms values as a reference set. Then, the readout of the sensors is done every tenth of a second and the piston and tilt corrections by the actuators twice a second.

The alignment of the mirrors and the cophasing of the passive primary segments are determined from a wavefront sensor, the phasing camera system (PCS) – based

Fig. 8.16 View of the Keck telescopes at Mauna Kea (courtesy W. M. Keck Observatory)

on the Shack-Hartmann principle with starlight –, which can operate in *four* modes [12]. The "passive tilt mode" collects the light of each segment into one spot per segment and measures the tilt error of the segments. The "fine screen mode," where the light of each segment is sampled in 13 subareas, can measure also the defocus and decentering coma aberrations of the telescope optics, generated by despace (tilt, lateral, and axial shifts) of the secondary mirror. A telescope global defocus and coma introduce, for each segment subaperture, local defocus and astigmatism respectively. After evaluations, these errors are corrected by movements of the secondary mirror. Instead of conventional lenslets, the first mode uses 36 prisms and a lens and the second uses additional multi-aperture masks [14]. The "ultra fine screen mode" is a classical Shack-Hartmann test with ~200 lenslets for precise checking of a segment shape. The "segment phase mode", developed by Chanan, Troy et al. [13], is a physical optics generalization of the Shack-Hartmann technique. The phasing procedure uses the starlight reflected by 78 circular subapertures, 120 mm in diameter, which are located at the middle of the intersegment edges. For an atmospheric coherence diameter of $r_0(0.5\,\mu\mathrm{m}) = 20\,\mathrm{cm}$ and rms piston errors varying in the range $\Delta z_{\mathrm{rms}} \in [0,\ 200\,\mu\mathrm{m}]$, theoretical results show that the FWHM of the images is of 0.5 arcsec at $\lambda = 0.5\,\mu\mathrm{m}$ whatever the Δz_{rms}-value and that the phasing is effectively irrelevant. However, since r_0 scales as $\lambda^{6/5}$, the results also show that the phase errors significantly reduce the image central intensity for $\lambda \geq 1\,\mu\mathrm{m}$; for $\lambda = 5\,\mu\mathrm{m}$ and $\Delta z_{\mathrm{rms}} = 500\,\mathrm{nm}$, the central intensity is reduced by ~60%. For this reason, and also for efficient use of adaptive optics, the phasing tolerance goal of the Keck telescopes is $\Delta z_{\mathrm{rms}} \leq 30\,\mathrm{nm}$, i.e. corresponding to an actuator incremental step. The Strehl ratio due to piston errors Δz_{rms} for a telescope made of n_s segments is [11]

$$S = \frac{1}{n_s}\left[1 + (n_s - 1)\exp\left(-16\pi^2 \Delta z_{\mathrm{rms}}^2/\lambda^2\right)\right] \tag{8.56}$$

which, for $\Delta z_{\mathrm{rms}} = 30\,\mathrm{nm}$ and $\lambda = 0.7\,\mu\mathrm{m}$, leads to $S = 0.76$ for $n_s = 36$ segments.

For the first three modes, the integration time of the PCS is of the order of 30 seconds. The passive mode tilt uses stars of magnitude $V = 9$ whilst the other three modes require $V = 45$. The segment phase mode can take 20 or 90 minutes with the narrow- or broad-band algorithms respectively. In a routine situation, a full alignment takes approximately one hour.

• **LAMOST and in-situ aspherized segments**: The large sky area multi-object spectroscopic telescope (LAMOST) (Fig. 8.17) is a 4-m all-reflective Schmidt telescope of the siderostat mount type with a 5° field of view (cf. Sects. 1.6.3, 4.3.4, and 5.3.6). It was designed by Su, Cui et al. [86]. Both plane in situ aspherized primary mirror M_1 and concave spherical secondary mirror M_2 are made of hexagon segments, 1.1 m in diameter. Input pupil mirror M_1 is mounted at a nodal plane of an alt-az hemispherical truss. Mirror M_2 is in a fixed position. For typical integration times of 1.5 hour, the effect on the images of the *atmospheric differential refraction* over the large field of view was analyzed by Su and Wang [87].

Three purposes are required for the active optics control: (i) in situ aspherization of the plane mirror M_1 into a shape determined by homothetic ellipse level lines,

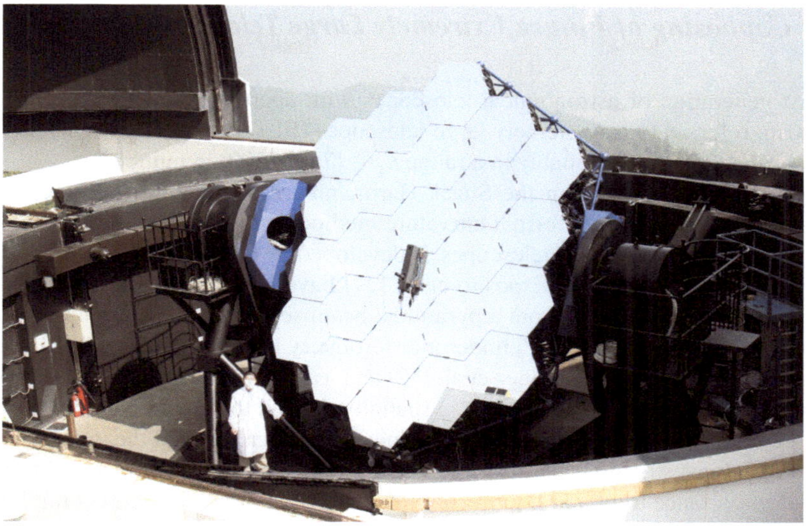

Fig. 8.17 View of LAMOST primary mirror M_1 in its alt-az hemispheric mount. This mirror is the telescope input pupil and reflects 4-m circular beams whatever the declination angle $\delta \in [-10°, 90°]$ of the observed region in the sky (courtesy NIAOT/CAS)

whose sag is a function of the declination angle $\delta \in [-10°, 90°]$, (ii) alignment and co-focusing of 24 M_1 segments, and (iii) alignment and co-focusing of 37 M_2 segments.

Preliminary active optics experiments for in situ aspherization were carried out by Su, Jiang et al. [84] with an M_1-type outer-most segment of 500 mm diameter and 6 mm thickness (aspect-ratio $t/d = 1/83$); a set of 58 actuators and three fixed points distributed over concentric circles allowed correcting the spherical aberration of a spherical M_2. Shack-Hartmann analysis of up to ten Seidel modes in the form (8.53) showed resulting wavefront errors $\leq$ 30 nm rms. Using an on-scale LAMOST segment pair $M_1 - M_2$ in the outer-most configuration, active optics tests on the sky by Cui et al. [22, 23] with M_1 aspherized by 37 actuators including 3 definer points provided S-H wavefront errors $\leq$ 80 nm rms. Experiments by Su, Zou et al. [85] concerned the alignment and co-focusing of three subsegments by six actuators and capacitive sensors; Shack-Hartmann tests showed tilt errors $\leq$ 0.035 arcsec and diffraction limited images. The final design for the active optics aspherization of all 24 M_1 segments, 1.1 m long in the diagonal, uses 37 actuators including 3 definer points and 18 actuators near the hexagonal edge (Cui et al. [24])

• **Other segmented telescopes**: Other built large telescopes are the 11 m HET [38] at McDonald Observatory and 10 m SALT in South Africa [83]. With a segmented spherical primary mirror, both designs are a tilted-Arecibo concept where the telescope platform can rotate 360° in azimuth. A four-mirror spherical aberration corrector (cf. for instance O'Donoghue [63]) mounted a the prime focus enables a celestial object to be tracked across 12° sky. These telescopes are mainly used for spectroscopy with fibre feed options as by D. Buckley for SALT.

8.7.4 Cophasing of Future Extremely Large Telescopes

The next generation of astronomical telescopes with aperture diameters reaching 30 or 40 m, referred to as extremely large telescopes (ELTs), will have highly segmented primary mirrors, probably in a range $n_s \in [750, 1500]$ in number.

Other sensing systems than the Shack-Hartmann are applicable to segmented mirrors. For instance, the wavefront curvature method, originated by Roddier [68] for the testing of ground-based telescopes, is developed by Cuevas et al. [20, 21] and Montoya-Martinez et al. [53]. Esposito et al. [27] have shown that piston and tip-tilt errors can also be derived from a pyramidal beam separator. Wavefront sensing based on the principle of Mach-Zehnder interferometry have been investigated and experimented by Montoya-Martinez et al. [52] and Yaitskova et al. [101]; the wavefront in one of the interferometer arms is spatially filtered to provide the reference wavefront so the two recombined beams produce an interferogram of the wavefront.

Theoretical investigations on the effects of segment cophasing errors are of fundamental importance to obtain high resolution imaging. For an n_s segment mirror, the effect of piston errors (z translation) on the Strehl ratio is given by (8.56). The effect of tip-tilt errors ($x,\ y$ rotations) on the point-spread-function have been analyzed by Yaitskova and Dohlen [96, 100] showing that, for n_s hexagonal segments, the Strehl ratio can be approximated by

$$S \simeq 1 - \sigma_{\rm rms}^2 + \frac{1}{2}\left(1.17 + \frac{1}{n_s}\right)\sigma_{\rm rms}^4, \qquad \sigma_{\rm rms}^2 = 16\pi^2\gamma^2\frac{\Delta z_{\rm rms}^2}{\lambda^2}, \tag{8.57}$$

where $\Delta z_{\rm rms}$ is the sag of the tilt from flat to flat edges, and γ a shape contour factor whose value is $\gamma = \sqrt{5}/6$ for a hexagon. To the second order, this expression coincides with Maréchal's approximation. Unlike the case of piston errors [cf. (8.56)], this equation has no strong dependence on the segment number n_s. For a tip-tilt error of $\sigma_{\rm rms} = 2\pi/30$, corresponding to $\Delta z_{\rm rms} = 30$ nm at $\lambda = 0.7\ \mu$m, we obtain $S = 0.96$.

In these two latter papers, other effects such as the diffraction caused by the intersegment gaps and segment edge misfigure have also been evaluated.

Active optics corrections of the figure residual errors of each segment is under investigation by Noethe [61]; such systems may use an appropriate number of controlled three-directional stressing strips distributed on the rear side of the segments.

8.8 Special Cases of Highly Variable Thickness Mirrors

8.8.1 Introduction – Mirror Flexure in Fast Tip-Tilt Mode

Secondary or tertiary telescope mirrors may be required to generate either tip-tilt *functions* for the adaptive optics correction of first-order atmospheric degradation modes – *field stabilization mirror* – or a single axis rotation *function* for the thermal noise substraction of infrared observations – *wobbling mirror*.

Because of the necessarily high frequency of the tip-tilt mode (at least 100 Hz), a special mirror design with a variable thickness distribution (VTD) allows minimizing the elastic deformation. The problem of the determination of the mirror geometry can be enounced as follows: *Given a material and a finite volume – or mass –, what are the mirror VTD and associated support geometry that provide a minimal flexure during tip-tilt motions?*

A preliminary approach for solving this problem consists of determining a VTD that minimizes the flexure due to gravity **g**. For a mirror with rotational symmetry, VTD solutions for several supporting cases are investigated. In the **g** case, one will see that the resulting thickness geometries lead to "*linear prismatic edge mirrors*." Such thickness shapes are useful distributions for minimizing the k**g** flexure from fast tip-tilt motions.

8.8.2 Minimum Flexure in Gravity of a Plate Supported at its Center

Within the Love-Kirchhoff hypotheses of elasticity theory, one preliminarily investigates the basic theoretical case of minimizing the flexure of an axisymmetric plate in the field of gravity **g** when in a horizontal plane and supported at its center.

The static equilibrium of an elementary segment of size dr, $r\,d\theta$, t, is derived about its local tangential axis. This involves the components of the radial and tangential bending moments M_r, M_t and the shearing force Q_r arising at the element. The equilibrium is achieved if (cf. for instance Sect. 2.1.2)

$$M_r + r\frac{dM_r}{dr} - M_t + r\,Q_r = 0, \tag{8.58}$$

where the bending moments are defined from the rigidity $D(r) = Et^3(r)/[12(1-\nu^2)]$ and the flexure $z(r)$ by

$$M_r = D\left(\frac{d^2z}{dr^2} + \frac{\nu}{r}\frac{dz}{dr}\right), \qquad M_t = D\left(\nu\frac{d^2z}{dr^2} + \frac{1}{r}\frac{dz}{dr}\right). \tag{8.59}$$

For a plate supported at its center, one can easily define the radial shearing force Q_r at radius $r \in [0, a]$ as the *partial* weight ΔW of the plate corresponding to the region $r \in [r, a]$ taken per length unit, i.e. $Q_r = \Delta W/2\pi r$. At the outer edge $r = a$ of the plate, this entails $Q_r = 0$. Hence, the radial shearing force is expressed by

$$Q_r = \frac{\mu\,g}{r}\int_r^a r\,t(r)\,dr, \tag{8.60}$$

where μ and g are the density of the plate and the acceleration of gravity. From the latter equations, the substitutions into (8.58) give

$$\frac{d^3z}{dr^3} + \left(\frac{1}{D}\frac{dD}{dr} + \frac{1}{r}\right)\frac{d^2z}{dr^2} + \left(\frac{\nu}{rD}\frac{dD}{dr} - \frac{1}{r^2}\right)\frac{dz}{dr} + \frac{\mu\,g}{rD}\int_r^a r\,t\,dr = 0. \tag{8.61}$$

Only using the flexure $z(r)$ and thickness $t(r)$ as unknowns, we obtain

$$\frac{d^3z}{dr^3} + \left(\frac{1}{t^3}\frac{dt^3}{dr} + \frac{1}{r}\right)\frac{d^2z}{dr^2} + \left(\frac{\nu}{rt^3}\frac{dt^3}{dr} - \frac{1}{r^2}\right)\frac{dz}{dr}$$
$$+ 12(1-\nu^2)\frac{\mu g}{E}\frac{1}{rt^3}\int_r^a r\,t\,dr = 0\,, \tag{8.62}$$

which is the general equation of the flexure of an axisymmetric VTD plate under its own weight and supported at its center.

For comparisons of various VTD mirrors, it is useful to define a *mean* thickness $\bar{t}$ for which the mass of the mirrors $M = \pi\mu a^2\bar{t}$ is the same, so the mean thickness is

$$\bar{t} = \frac{2}{a^2}\int_0^a r\,t(r)\,dr. \tag{8.63}$$

This leads to definition of the associated dimensionless thickness $\mathcal{T}(r)$ such as

$$\mathcal{T}(r) = t(r)/\bar{t}, \qquad \int_0^1 \rho\mathcal{T}(\rho)d\rho = 1/2. \tag{8.64}$$

where $\rho = r/a$ is the normalized radius.

Two flexure cases are investigated and compared hereafter: a constant thickness plate and a plate which shows a parabolic flexure.

• **Constant thickness plate**: The basic case of a constant thickness plate provides a reference center-edge flexure for further comparisons with VTDs. In this case [39],

$$t(r) = \bar{t} = constant, \qquad \mathcal{T}(r) \equiv 1,$$

and since $d\nabla^2\cdot/dr = d/dr[d^2\cdot/dr^2 + (1/r)d\cdot/dr]$, (8.62) reduces, after a first integration, to

$$\nabla^2 z = -\frac{\beta}{4}\left(2a^2\ln r - r^2 + 2C_1\right), \qquad \beta = 12(1-\nu^2)\frac{\mu g}{E\bar{t}^2}, \tag{8.65}$$

where C_1 is a constant. Since the Laplacian also writes $\nabla^2\cdot = (1/r)d/dr(r\,d\cdot/dr)$, the second integration leads to

$$\frac{dz}{dr} = -\frac{\beta}{16}\left[4a^2r\ln r - 2\,(1+2C_1)\,a^2r - r^3 + 8C_2a^4\frac{1}{r}\right].$$

The thickness of the plate is finite and also is the flexure. Since $dz/dr|_{r=0} \to \infty$, one must set $C_2 = 0$. The bending moment M_r must be zero for the edge, thus, from (8.59),

$$\left[\frac{d^2z}{dr^2} + \frac{\nu}{r}\frac{dz}{dr}\right]_{r=a} = 0.$$

This boundary condition entails, after calculation, that

$$C_1 = \ln a - \frac{1}{4}\frac{1+3\nu}{1+\nu},$$

which leads to the slope

$$\frac{dz}{dr} = -\frac{\beta}{16}\left[4a^2 r \ln\frac{r}{a} - \frac{1-\nu}{1+\nu}a^2 r - r^3\right]. \tag{8.66}$$

Setting as usual the origin of the flexure at $r = 0$, one obtains after a last integration

$$z = (1-\nu^2)\frac{3\mu\, g}{16E\bar{t}^2}r^2\left[r^2 + 2\frac{3+\nu}{1+\nu}a^2 - 4a^2\ln\frac{r^2}{a^2}\right]. \tag{8.67}$$

This allows derivation of the total sag $\Delta z = z\{a\}$ of the flexure of a constant thickness plate supported at it's center as

$$\Delta z_{\mathrm{CT}} = (1-\nu)(7+3\nu)\frac{3\,\mu\, g\, a^4}{16\, E\bar{t}^2}, \tag{8.68}$$

which is negative for $g < 0$.

• **Plate with parabolic flexure**: Because of the discontinuity arising in the above case at $r = 0$, where the slope of the flexure becomes *infinite* – as does the shearing force –, an efficient means of minimizing the flexure of a plate in gravity is to determine the VTD for which this effect is cancelled. As shown by Lemaitre [43] in comparing four distributions, a VTD that provides a purely *parabolic flexure* allows minimizing the center-edge sag Δz.

Assuming a parabolic flexure $z \propto r^2$, the first left-hand term in (8.62) vanishes; after substitution of $t(r)$ by $\bar{t}\mathcal{T}(r)$ the remaining terms are

$$\frac{d\mathcal{T}^3}{dr}\frac{d^2 z}{dr^2} + \frac{\nu}{r}\frac{d\mathcal{T}^3}{dr}\frac{dz}{dr} + \frac{\beta}{r}\int_r^a r\,\mathcal{T}\,dr = 0.$$

For further simplifications, the parabolic flexure may be introduced as

$$z = \frac{1}{2(1+\nu)}\alpha\beta a^2 r^2, \tag{8.69}$$

where α is an unknown. Thus, after substitution and using $\rho = r/a$, we obtain

$$\rho\mathcal{T}^2\frac{d\mathcal{T}}{d\rho} + \frac{1}{3\alpha}\int_\rho^1 \rho\,\mathcal{T}\,d\rho = 0. \tag{8.70}$$

After derivation and division by ρ, the researched thickness is a solution of the general equation

$$\nabla^2\mathcal{T}^3 + \frac{1}{\alpha}\mathcal{T} = 0, \tag{8.71}$$

where ∇^2 is a Laplacian with respect to the radial variable ρ. However, it is preferable to solve the thickness $\mathcal{T}(\rho)$ from the integro-differential form (8.70). For this, one remarks that for $\rho \to 0$, the second term is with a negligible variation which entails $\mathcal{T} \propto (-\ln\rho)^{1/3}$. In the region of the edge, from first and second-order variations of $1-\rho$, we find the asymptotic form

$$\mathcal{T}|_{\rho\to 1} \to (1-\rho)/\sqrt{6\alpha}. \tag{8.72}$$

Therefore, at some distance from its center, the solution for the plate VTD is a *quasi-conical thickness distribution*.

The asymptotic representation (8.72) allows the numerical integration from the edge of the plate which also must satisfy the comparison condition (8.64) of the normalized thickness for equal mass plates, that is $\int_0^1 \rho\mathcal{T}(\rho)\,d\rho = 1/2$, and then determines $\bar{t}$ from (8.63). The results from integration of (8.70), which is *not* Poisson's ratio dependent, are

$$\alpha = 0.022316... \simeq 1/45 \tag{8.73}$$

and a quasi-conical $\mathcal{T}(\rho)$ distribution given in Table 8.4 and shown by Fig. 8.18.

From (8.73) and (8.65), the substitutions of α and β into (8.69) provide the dimensioned form of the *parabolic flexure* as

$$z = (1-\nu)\frac{2\,\mu\,g\,a^2}{15\,E\,\bar{t}^2}r^2, \tag{8.74}$$

from which the center-edge sag is

$$\Delta z_{\mathrm{PF}} = (1-\nu)\frac{2\,\mu\,g\,a^4}{15\,E\bar{t}^2}. \tag{8.75}$$

Table 8.4 $\mathcal{T}(\rho)$ values of the normalized thickness distribution from (8.70) of a plate with *parabolic flexure* in the gravity field when supported at its center (Lemaitre [43])

ρ	0.00	0.01	0.02	0.04	0.06	0.08	0.10	0.20
$\mathcal{T}$	∞	4.372	4.082	3.746	3.519	3.339	3.187	2.617
ρ	0.30	0.40	0.50	0.60	0.70	0.80	0.90	1.00
$\mathcal{T}$	2.184	1.811	1.471	1.153	0.850	0.559	0.276	0.000

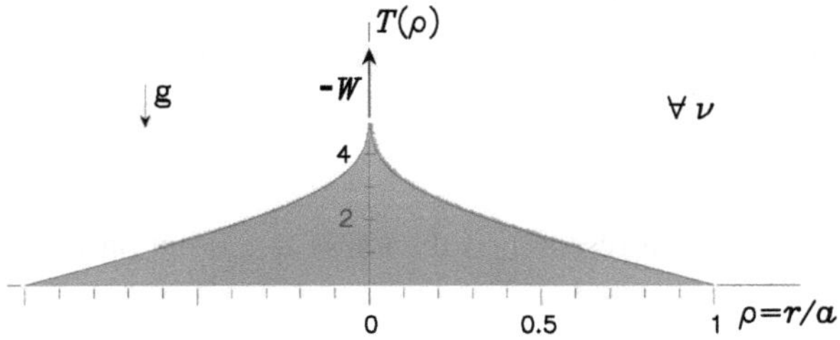

Fig. 8.18 Quasi-conical thickness distribution $\mathcal{T}(\rho)$ providing a parabolic flexure from Table 8.4. The infinite branch in $(-\ln\rho)^{1/3}$ at the central region is very narrow. The branch can be given a finite length with a small section area for the substitution of a central load to the central force, thus avoiding infinite local stresses

A flexure sag ratio can be derived for comparing the sag of the quasi-conical plate to that of the constant thickness plate. Since the normalized thickness $\bar{t}$ is the same in (8.68) and (8.75) – same mass normalization –, for identical materials we obtain

$$\frac{\Delta z_{CT}}{\Delta z_{PF}} = 45\,(7+3\nu)/32. \tag{8.76}$$

For $\nu \in [0,\ 1/2]$, the corresponding flexure sag ratio is in the range [9.8, 11.9]. Therefore, a conclusion is as follows [43].

→ *In the field of gravity, whatever the Poisson's ratio, the flexural sag of a plate with a quasi-conical thickness supported at its center is at least 10-times smaller than that with a constant thickness, for the same material and mass.*

8.8.3 Field Stabilization Mirrors and Infrared Wobbling Mirrors

Because of the fast time frequency of the tip-tilt movements of atmospheric *field stabilization mirrors* and *wobbling mirrors* used in astronomy, these mirrors must be designed with the best compromise to obtain both high rigidity and lightweight. This constraint requires the choice of a material with low density and high Young modulus such as Be or SiC.

In addition, if a plate is supported in gravity $\mathbf{g}$ by an intermediate ring $r = b$, it can be shown that, similarly to the result in Sect. 8.8.2, a quasi-conical thickness distribution also applies in the outer region $b < r < a$. This substantially contributes to the light weighting of the mirror.

Now considering field stabilization mirrors actuated by three 120° forces located at $r = b$, the inertial flexure from tip-tilt motions must be a given negligible value when the acceleration becomes $k\mathbf{g}$.

• **Case of a mirror without central hole**: Assuming that a field stabilization mirror optically requires a full aperture radius $0 < r < a$ and is with a constant thickness in its central region, $r < b$, it can be shown that, given a total mass, a conical or quasi-conical distribution at the outer region $b < r < a$ provides a minimal flexural sag by an appropriate determination of the b/a ratio. An appropriate choice of the total mass allows reducing the inertial flexure within a negligible value (Fig. 8.19A).

For instance, a 0.4-m flat tertiary mirror with an axisymmetric conical-edge geometry was developed by Neufeld, Bennet et al. [57] as a 50-Hz atmospheric field stabilization mirror of the SOAR telescope. However, these authors use a mirror design with deep radial ribs which probably is not fully satisfactory.

Because the tip-tilt motions are actuated by forces located at 120° subareas of the circle $r = b$, a three-fold symmetry of the thickness distribution allows a higher weight reduction than with an axisymmetric thickness. The redistribution of the thickness is done by adding mass near the actuated zones and removing it in the most distant regions. If the VTD is of the form $t(r, \cos 3\theta)$ and the actuators are distributed at $\theta = 0$ and $\pm 2\pi/3$, then the thickness can be *increased* in the regions

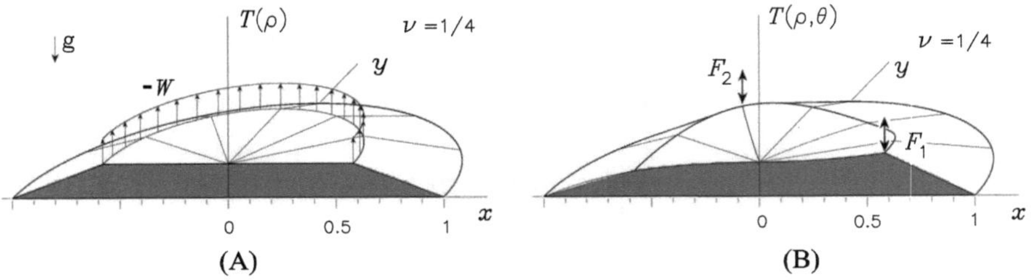

Fig. 8.19 Mirror thickness geometries $\mathcal{T}(\rho)$ or $\mathcal{T}(\rho, \theta)$ for atmospheric field stabilization mirrors. (**A**): Moderately lightweight axisymmetric mirror with a *conical edge* optimized for a minimal axisymmetric flexure with a ring support. (**B**): Increased lightweight three-fold symmetry mirror with *linear prismatic edge*

of the actuators and *decreased* in the most distant regions from them. For instance, the dimensionless thickness distribution

$$\mathcal{T}(\rho, \theta) = \mathcal{T}_0 \left(1 + \frac{\rho^2}{4\beta^2} \cos 3\theta \right), \qquad 0 < \rho < \beta, \tag{8.77a}$$

$$\mathcal{T}(\rho, \theta) = \mathcal{T}_0 \frac{1-\rho}{1-\beta} \left(1 + \frac{1}{4} \cos 3\theta \right), \qquad \beta < \rho < 1, \tag{8.77b}$$

with a *linear prismatic outer edge* is shown in Fig. 8.19B for $\beta = b/a = 0.582$ and $\nu = 1/4$.

• **Case of a mirror with central hole**: Some future large ELTs will be designed with the primary mirror pupil imaged at or near one or two mirrors of the main optical train. For instance, the pupil of the 42-m primary mirror of the Eso E-ELT project – a five-mirror design – is imaged by the concave mirror M_3 on the flat mirror M_4 which should be an adaptive optics mirror. Close to mirror M_4 is the flat mirror M_5, $2.3 \times 2.75\ \mathrm{m}^2$ in size, planned as an atmospheric field stabilization mirror. Because of the central obstruction of the secondary mirror and of its proximity to M_4, the central zone of M_5 is of no use; hence a holed mirror design will be appropriate for M_5.

Let $r = c$ be the radius of the hole where $c < b < a$. The mirror geometry must be determined such as the flexure due to the inertial tip-tilt motions is reduced to a negligible value (diffraction limited criterion). A first approach to solve this problem is to consider the gravity case. In this case, we showed in Sect. 8.8.2 that the flexure of a plain mirror supported at its center is minimal when its shape is a *paraboloid*. Now, for a holed mirror supported on a ring of radius $r = b$, an efficient minimization of the flexure in gravity will be obtained when the flexure shape in the region $c \leq r \leq a$ is purely *quadratic* and shows identical deflections at edges, $z(a) = z(c)$. One may assume that, for extremely small deflections, its quadratic flexure shape $z(r)$ does differ from that of an arc of a circle taken in a radial section.

→ *The minimal flexure $z(r)$ of a holed axisymmetric mirror supported in gravity* **g** *along a circle of radius b is obtained when the flexure shape is a part of a* torus. *This surface is generated by revolving a horizontal circle arc, of ends $r = c$ and $r = a$, about the z-axis, along a circle of radius $(a + c)/2 = b'$, where generally $b' \neq b$.*

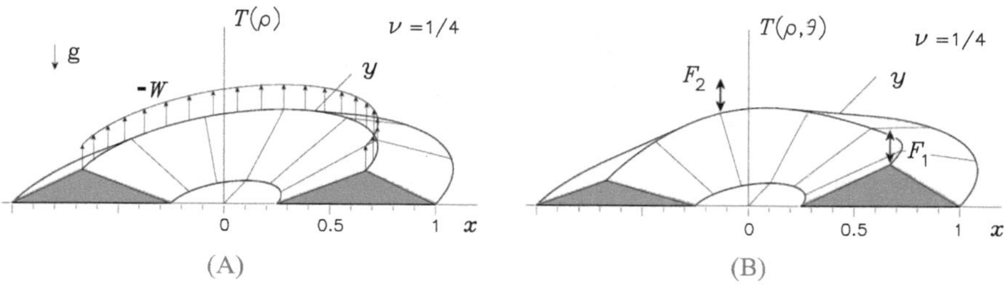

Fig. 8.20 Mirror thickness geometries $\mathcal{T}(\rho)$ or $\mathcal{T}(\rho, \theta)$ for atmospheric field stabilization *holed* mirrors. (**A**): Moderately lightweight axisymmetric mirror with *conical edges* optimized for a minimal axisymmetric flexure – as part of a *torus surface* – with a ring support. (**B**): Increased lightweight three-fold symmetry mirror with *linear prismatic edges*

From this statement, one can derive thickness distributions for minimal flexures in $k\mathbf{g}$ gravity cases such as for *holed* field stabilization mirrors.

One may distinguish two possible alternative thickness families: (i) an axisymmetrical distribution with inner and outer *conical edges*, or (ii) a three-fold symmetry distribution with inner and outer *linear prismatic edges* (Fig. 8.20).

8.8.4 Design of Low Weight Wobbling Mirrors

Infrared or far-infrared observations generally require use of a wobbling mirror. When this mirror is the telescope secondary mirror, it is appropriate to locate the telescope input *pupil* on it. At the detector, the rotation generated by the wobbling mirror allows a single pixel to alternatively receive a target position beam and a reference position beam. Because the reference beam must be free from any emitting astronomical object – black sky –, the rotation angle of the wobbling mirror must reach a large off-target value, such as 2–5 arcmin. Thus, contrarily to field stabilization mirrors, wobbling mirrors must provide a large rotation angle at low frequency. Typical values are 2 arcmin at 10 Hz.

If, as usual, a wobbling mirror only provides a rotation about a *single* axis, then a best rigid geometry for its substrate should differ from the axi- or three-fold symmetry in Fig. 8.20. If the tip motion is about the x-axis, the rigidity in the x-direction can be somewhat decreased. In both cases, the thickness geometries can be derived from those in Fig. 8.20 by use, for instance, of the transformation $\mathcal{T}(x) \rightarrow (1-k^2)\,\mathcal{T}(x)$ where $k^2 < 1$, while $\mathcal{T}(y)$ is unchanged. Hence, this again decreases the total weight of the mirror.

References

1. J.R.P. Angel, B. Martin, D. Sandler et al., The next generation space telescope: a monolithic mirror candidate, SPIE Proc., **2807**, 354 (1996)
2. L. Arnold, Influence functions of a thin shallow meniscus-shaped mirror, Appl. Opt., **36**(10), 2019–2028 (1997)

3. L. Arnold, Optimized axial support topologies for thin telescope mirrors, Opt. Eng., **34**, 567–574 (1995)
4. L. Arnold, Optimized mirror supports and active primary mirrors, SPIE Proc., **2199**, 239–250 (1994)
5. L. Arnold, Uniform-load and actuator influence functions of a thin annular mirror: application to active mirror support optimization, Appl. Opt., **35**(7), 1095–1106 (1996)
6. Y.Y. Balega, private communication (2007)
7. A. Baranne, G.R. Lemaitre, Combinaison optique pour très grands télescopes: le concept TEMOS, C.R. Acad. Sc. Paris, **305**, Série II, 445–450 (1987)
8. P.Y. Bely, *The Design and Construction of Large Optical Telescopes*, Springer edt., New York, 221 (2002)
9. I.S. Bowen, The 200-inch Hale Telescope, in *Telescopes*, G.P. Kuiper & B.M. Midlehurst edts., The University of Chicago Press, Chicago, 2nd issue, 1–15 (1962)
10. I.S. Bowen, The 200-inch Hale telescope, IAU Proc., vol. VIII, **5**,750–754 (1953)
11. G. Chanan, C. Ohara, M. Troy, Phasing the primary mirror of the Keck telescopes II, Appl. Opt., **39**(25), 4706–4714 (2000)
12. G. Chanan, J. Nelson, T. Mast, P. Wisinowich, B. Schaefer, The Keck telescope phasing camera system, SPIE Proc., **2198**, 1139–1150 (1994)
13. G. Chanan, M. Troy, Strehl ratio and modulation transfer function for segmented mirror telescopes as function of the segment phase error, Appl. Opt., **38**(31), 6642–6647 (2000)
14. G. Chanan, T. Mast, J. Nelson, Phasing the mirror segments of the Keck Telescope, SPIE Proc., **2199**, 622–637 (1994)
15. J.-P. Chevillard, P. Connes et al., Near infrared astronomical light collector, Appl. Opt., **16**(7), 1817–1833 (1977)
16. P. Connes, G. Michel, Astronomical Fourier spectrometer, Appl. Opt., **14**(9), 2067–2084 (1975)
17. A. Couder, Recherches sur les déformations des grands miroirs – Effets thermiques: Déformations des miroirs, agitation de l'air, Bull. Astronomique Obs. Paris, **VII**, Fasc. 7, 19–312 (1932). [Cf. also transl. E.T. Pearson, KPNO Library (1966)]
18. A. Couder, Recherches sur les déformations des grands miroirs employés aux observations astronomiques, Bull. Astronomique Obs. Paris, **VII**, Fasc. 6, 201–281 (1931). [Cf. also transl. E.T. Pearson, KPNO Library (1966)]
19. J.F. Creedon, A.G. Lindgren, Automatica, **6**(5), 643 (1970)
20. S. Cuevas, V.G. Orlov, F. Garfias et al., Curvature equation for a segmented telescope, SPIE Proc., **4003**, 291–302 (2000)
21. S. Cuevas, V.G. Orlov, F. Garfias, V.V. Voitsekhovich, L. Sanchez, Curvature equation for a segmented telescope, SPIE Proc., **4003**, 291–302 (2000)
22. X. Cui, D.-q. Su, G. Li et al., Experiment system of LAMOST active optics, SPIE Proc., **5489**, 974–985 (2004)
23. X. Cui, Y. Li, X. Ni et al., The active support system of LAMOST reflective Schmidt plate, SPIE Proc., **4837**, 628–636 (2003)
24. X. Cui, D.-q. Su, Y.-n. Wang, G. Li, G. Li, Y. Zhang, Y. Li, The optical performance of LAMOST telescope, SPIE Proc., **7733** (2010)
25. P. Dierickx, D. Enard, R. Geyl, J. Paseri, M. Cayrel, P. Béraud, The VLT primary mirrors, SPIE Proc., **2871**, 385–392 (1996) (see also www.eso.org/projects/vlt/unit-tel/m1unit.html)
26. D. Enard, ESO-VLT: status of the main 8-m telescopes, SPIE Proc., **2199**, 394–403 (1994)
27. S. Esposito, E. Pinna, A. Puglisi, A. Tozzi, P. Stefanini, Pyramid sensor for segmented mirror alignment, Opt. Lett., **30**(19), 2572–2574 (2005)
28. L. Foucault, Annales de l'Observatoire de Paris, **5**, 197–237 (1859)
29. P. Giordano, Internal ESO report, Paranal (2002)
30. J.M. Hill, J.R.P. Angel, R.D. Lutz et al., Casting the first 8.4 m borosilicate honeycomb mirror for the Large Binocular Telescope, SPIE Proc., **3352**, 172 (1998)
31. W.E. Howell, J.F. Creedon, NASA Technical Note, NASA TN D-7090 (Jan. 1973)
32. http://grus.berkeley.edu/~jrg/MIDDLE/

33. http://w0.sao.ru/Doc-en/Telescopes/bta/descrip.html
34. N. Itoh, I. Mikami et al., Active optics experiments II, Publ. Natl. Astron. Obs. Japan., **1**, 57–61 (1989)
35. M. Iye, K. Kodaira, Primary support system for the SUBARU telescope, SPIE Proc., **2199**, 762–772 (1994)
36. M. Iye, R. Noguchi, Y. Torii et al., Active optics experiments with a 62 cm thin mirror, SPIE Proc., **1236**, 929–939 (1990)
37. N. Kaifu, SUBARU project: current status, SPIE Proc., **2199**, 56–63 (1994)
38. V.L. Krabbendam, T.A. Sebring, F.B. Ray, S.R. Fowler, Development and performance of Hooby-Herberly Telescope 11 meter segmented mirror, SPIE Proc., **3352**, 436–445 (1998)
39. L.D. Landau, E.M. Lifshitz, *Theory of Elasticity* in *Course of Theoretical Physics* – Vol. 7, USSR Acad. of Sc., Butterworth & Heinemann eds, 3rd edition, Pergaman press, Oxford, 67 (1986)
40. W. Lassell, Mem. Roy. Astron. Soc., **XII**, 265 (1842)
41. G.R. Lemaitre, M. Wang, Optical results with TEMOS 4 and a metal secondary mirror actively aspherized, SPIE Proc., **1931**, 43–52 (1992)
42. G.R. Lemaitre, R.N. Wilson, S. Mazzanti, Proposal for a 1.8 m meter metal meniscus mirror at once actively aspherized and actively supported, SPIE Proc., **1931**, 67–75 (1992)
43. G.R. Lemaitre, Sur la flexion des miroirs secondaires de télescopes, Nouv. Rev. Optique, **7**(6), 389–397 (1976)
44. G.R. Lemaitre, Sur la flexion du grand miroir de 3.6 m ESO, ESO Bull., **8**, 21–31 (1971)
45. A.E.H. Love, *Mathematical Theory of Elasticity*, first and second edition. See also fourth revised enlarged issue (1927). Reissued by Dover publ., New York, Sects. 299, 309 and 312c (1944)
46. A.F. Möbius, in *Lehrbuch der Statik*, 2 vols., Leipzig, Vol. 2, Chaps. 4 and 5 (1837)
47. B. Mack, Deflection and stress analysis of a 4.2 m primary mirror of an alt-azimuth mounted telescope, Appl. Opt., **19**(6), 1000–1010 (1980)
48. A.J. Malvick, E.T. Pearson, Theoretical elastic deformation of a 4-m diameter optical mirror using dynamic relaxation, Appl. Opt., **7**(6), 1207–1212 (1968)
49. J-P. Marioge, C. Mahé, Contribution au contrôle des surfaces planes non traitées, Journ. Mod. Optics, Taylor & Francis publ., **20**(6), 413–433 (1973)
50. T. Mast, G. Nelson, SPIE Proc., **1236**, 670 (1990)
51. J.H. Michell, London Math. Soc. Proc., **31**, 100 (1900)
52. L. Montoya-Martinez, N. Yaitskova, P. Dierickx, K. Dohlen, Mach-Zender wavefront sensor for phasing of segmented telescopes, SPIE Proc., **4840**, 564–573 (2003)
53. L. Montoya-Martinez, M. Reyes, A. Schumacher, E. Hernández, DIPSI: the diffraction image phase sensing instrument for APE, SPIE Proc., **6267**, (2006)
54. C.M. Mountain, R. Kurz, J. Oschmann, GEMINI 8-m telescope project, SPIE Proc., **2199**, 41–55 (1994)
55. J.E. Nelson, J. Lubliner, T.S. Mast, Telescope mirror supports: Mirror deflection on point supports, SPIE Proc., **332**, 212 (1982)
56. J.E. Nelson, J. Lubliner, T.S. Mast, Telescope mirror supports, SPIE Proc., **332**, 212–228 (1982)
57. C. Neufeld, V. Bennet, T. Sebring, V. Krabbendam et al., Development of an active optics system for the SOAR telescope, SPIE Proc., **5489**, 1052–1060 (2004)
58. L. Noethe et al., Proc. ESO Conf. on *Progress in Telescope and Instrumentation technologies*, ESO, Garching, 195 (1992)
59. L. Noethe, Active optics in modern large optical telescopes, Progress in Optics, Elsevier publ., **43**, 1–13 (2002)
60. L. Noethe, F. Franza, P. Giorgano, R.N. Wilson, Active Optics II. Results of an experiment with a thin 1 m test mirror, J. Mod. Opt., **35**, 1427–1457 (1988)
61. L. Noethe, private communication (2006)
62. L. Noethe, Use of minimum-energy modes for modal-active optics corrections of thin meniscus mirrors, J. Mod. Opt., **38**(6), 1043–1046 (1991)

63. D. O'Donoghue, The correction of spherical aberration in the Southern African Large Telescope (SALT), SPIE Proc., **4003**, 363–370 (2000)
64. G. Parodi, G.C. Cerra, J.M. Hill, W.B. Davison, P. Salinari, LBT primary mirror: the final design of the supporting system, SPIE Proc., **2871**, 352–359 (1997)
65. A. Rakich, J.M. Hill, C.J. Biddick, D.L. Miller, T. Leibold, Use of field aberrations in the alignment of the Large Binocular Telescope optics, SPIE Proc., **7012** (2008)
66. E. Reissner, J. Appl. Mech., **12**, A-69 (1945)
67. E. Reissner, Quart. Appl. Math., **5**, 55 (1947)
68. C. Roddier, F. Roddier, Wavefront reconstruction from defocused images and the testing of ground-based optical telescopes, J. Opt. Soc. Am., A, **10**(11), 3433–3436 (1988)
69. A. Saint-Venant (Barré de), Flamant, *Théorie de l'Élasticité des Corps Solides de Clebsch*, Dunod edt., Paris, 858–859 (1881). (French transl. of Clebsch's book including important annotations and complements; sometimes referred to as "Clebsch Annoted Version")
70. M. Schneermann, X. Cui et al., SPIE Proc., **1236**, 920 (1990)
71. G. Schwesinger, An analytical determination of the flexure of the 3.5 m primary and 1 m mirror of the ESO New Technology Telescope for passive support and active control, J. Mod. Opt., **35**, 1117–1149 (1988)
72. G. Schwesinger, Comparative assessment of aberrations originating in telescope mirrors from the edge support. Astron. J., **74**, 1243–1254 (1969)
73. G. Schwesinger, E.D. Knol, Comments on a series of articles by L.A. Selke, Appl. Opt., **11**, 200–201 (1972)
74. G. Schwesinger, General characteristics of elastic mirror flexure in theory and applications, Symposium Proc. on *Support and Testing of Large Astronomical Mirrors*, KPNO, Tucson, 10–23 (1966)
75. G. Schwesinger, Lateral support of very large telescope mirrors by edge forces only, J. Mod. Opt., **38**, 1507–1516 (1991)
76. G. Schwesinger, Non-distorting lateral edge support of large telescope mirrors, Appl. Opt., **33**(7), 1198–1202 (1994)
77. G. Schwesinger, Optical effect of flexure in vertically mounted precision mirrors, J. Opt. Soc. Am., **44**, 417 (1954)
78. G. Schwesinger, Support configuration and elastic deformation of the 1.5 m prime mirror of the ESO Coudé Auxiliary Telescope (CAT), European Southern Observatory Tech. Rep. 9, Garching (1972)
79. L.A. Selke, Theoretical elastic deflections of a thick horizontal circular mirror on a double-ring support, Appl. Opt., **9**(6), 1453–1456 (1970)
80. S. Stanghellini, E. Manil, M. Schmid, K.Dost, Design and preliminary tests of the VLT secondary mirror unit, SPIE Proc., **2871**, 105–116 (1996)
81. L. Stepp, Conceptual design of the primary mirror cell assembly, GEMINI Report O-G0025 (1993)
82. L. Stepp, E. Huang, M. Cho, GEMINI primary mirror support system, SPIE Proc., **2199**, 223–238 (1994)
83. B. Stobie, K. Meiring, D.A.H. Buckley, Design of the Southern African Large Telescope, in *Optical Design, Material, Fabrication and Maintenance*, SPIE Proc., **4003**, 355–362 (2000)
84. D.-q. Su, S.-t. Jiang, W.-y. Zou et al., Experiment system of thin-mirror active optics, SPIE Proc., **2199**, 609–621 (1994)
85. D-q. Su, W-j. Zou, Z-c. Zhang et al., Experiment system of segmented-mirror active optics, SPIE Proc., **4003**, 417–425 (2000)
86. D.-q. Su, X. Cui, Y.-n. Wang, Z. Yao, LAMOST and its key technology, SPIE Proc., **3352**, 76–90 (1998)
87. D.-q. Su, Y.-n. Wang, A computational study of the star-image displacement due to differential atmospheric refraction during observations, Acta Astrophysica Sinica, **17**, 202–212 (1997)
88. H. Tafelmaier, Dünnschicht-Tecknik Co., www.tafelmaier.de/eng/default.asp
89. S.P. Timoshenko, in *History of Strength of Materials*, Dover Publ. Inc., New York, 304 (1983)

90. S.P. Timoshenko, S. Woinowsky-Krieger, in *Theory of Plates and Shells*, McGraw-Hill edt., New York, second issue, Sect. 20, 74 (1959)
91. D.S. Wan, J.P.R. Angel, R.E. Parks, Mirror deflection on multiple axial supports, Appl. Opt., **28**, 354–362 (1989)
92. R.N. Wilson, F. Franza, L. Noethe, Active optics I. A system for optimizing the optical quality and reducing the costs of large telescopes, J. Mod. Opt., **34**, 485–509 (1987)
93. R.N. Wilson, F. Franza, L. Noethe, G. Andreoni, Active Optics IV. Set-up and performance of the optics of the ESO New Technology Telescope (NTT) in the observatory, J. Mod. Opt., **38**, 219–243 (1991)
94. R.N. Wilson, F. Franza, P. Giordano, L. Noethe, M. Tarenghi, Active Optics III. Final results with the 1 m test mirror and NTT 3.56 m primary in the workshop, J. Mod. Opt., **36**, 1415–1425 (1989)
95. R.N. Wilson, *Reflecting Telescope Optics* II, Springer-Verlag edt., New York, (1999)
96. S. Woinowsky-Krieger, Ingr. Arch., **4**, 305 (1933)
97. R.W. Wood, Astrophys. J., **29**, 164 (1909)
98. www.telescopengineering.com/company/DmitriMaksutov.html
99. N. Yaitskova, K. Dohlen, P. Dierickx, Analytical study of diffraction effects in extremely large segmented telescopes, JOSA A, **20**(8), 1563–1575 (2003)
100. N. Yaitskova, K. Dohlen, Tip-tilt error for extremely large segmented telescopes: detailed theoretical point-spread-function analysis and numerical simulation results, JOSA A, **19**(7), 1274–1285 (2002)
101. N. Yaitskova, L.-M. Montoya-Martinez, K. Dohlen, P. Dierickx, A Mack-Zender phasing sensor for extremely large segmented telescopes, SPIE Proc., **5489**, 1139–1151 (2004)

Chapter 9
Singlet Lenses and Elasticity Theory of Thin Plates

9.1 Singlet Lenses

Because of the axial chromatism inherent to the term $n-1$ in the power of a singlet lens, a combination of two or several lenses is preferred to achieve this correction simultaneously with those of other aberrations. It is well known that even at low aperture the axial chromatism of a lens is strongly dominant – this is a first-order aberration – whilst *Sphe*3 may be fully negligible. The efforts of Descartes aimed at developing stigmatic singlet lenses were not really convincing since in fact the developments in the early period of optical astronomy were based on aerials with long focal length which were later improved by stigmatic achromats (cf. Chap. 1).

Hence rather then aspherizing a lens, it has been often preferred to deform a refractive element when it has a null power, as for instance with aspherical plates. Of course, the situation is different if a lens is used in monochromatic light such as with a laser.

9.1.1 *Aberrations of a Thin Lens with Spherical Surfaces*

We consider here the case where the entrance pupil *coincides* with the lens whose surfaces are spherical. In the determination of the primary aberrations of a thin lens, we follow the analysis by Welford [10] who points out the advantage of using Coddinton's notation. These determinations can also be found, for instance, in Chrétien [3] and in Born and Wolf [2]. Let us denote c_1 and c_2 the surface curvatures of the lens. For a thin lens in air, from (1.28b) and (1.24), the power and transverse magnification are respectively

$$K=\frac{1}{f'}=-\frac{1}{f}=\frac{1}{z'}-\frac{1}{z}=(n-1)(c_1-c_2),\qquad M=\frac{u_1}{u_2'},\tag{9.1}$$

where u_1 and u_2' are the conjugate *aperture angles* of the input and output rays. For symmetry reasons, it is convenient to use the Coddington variables [4]. These two *dimensionless* variables characterize the bending or *cambrure* of the lens

G.R. Lemaitre, *Astronomical Optics and Elasticity Theory,* Astronomy and Astrophysics Library, DOI 10.1007/978-3-540-68905-8_9,

$$B = \frac{(n-1)(c_1+c_2)}{K} = \frac{c_1+c_2}{c_1-c_2}, \tag{9.2a}$$

and the configuration of the conjugates

$$C = \frac{u_1+u'_2}{x\,K} = \frac{u_1+u'_2}{u_1-u'_2}, \tag{9.2b}$$

where x is the height of the ray at the lens.

If $B = 0$, the lens is equiconvex; if $B = -1$, the lens is plano-convex and, if $B = 1$, the lens is convexo-plane (Fig. 9.1). Similarly, if $C = 0$, the conjugates are in symmetry with the lens, i.e.$z' = -z$; if $C = -1$, the object is at infinity; if $C = 1$, the object is at the first focal plane. With Coddington's variables, the surface curvatures of the lens and the convergence angles are expressed by

$$\begin{cases} c_1 = \frac{K}{2(n-1)}\,(B+1), & u_1 = \frac{xK}{2}\,(C+1), \\ c_2 = \frac{K}{2(n-1)}\,(B-1), & u'_2 = \frac{xK}{2}\,(C-1), \end{cases} \tag{9.3}$$

hence the curvature ratio c_1/c_2 and the transverse magnification M are

$$\frac{c_1}{c_2} = \frac{B+1}{B-1}, \qquad M = \frac{u_1}{u'_2} = \frac{C+1}{C-1}. \tag{9.4}$$

Except for some special cases, a spherical incident wavefront does not remain spherical after refraction through the spherical surfaces of a thin lens. The resulting shape of the emerging wavefront can be represented by Seidel's primary aberration coefficients S_I to S_V such as shown for the general case of the aberration wavefront function (cf. eq. (1.41) in Sect. 1.80). The calculation of Seidel coefficients representing the sum of the aberration contribution of the two surfaces is lengthy but not particularly difficult. From Welford [10], the resulting Seidel sums for a lens in air and a pupil at the lens are

$$S_I = \frac{x^4}{4f'^3}\left[\left(\frac{n}{n-1}\right)^2 + \frac{n+2}{n(n-1)^2}\left(B + \frac{2(n^2-1)}{n+2}\,C\right)^2 - \frac{n}{n+2}\,C^2\right], \tag{9.5a}$$

$$S_{II} = -\frac{x^2 H}{2f'^2}\left[\frac{n+1}{n(n-1)}\,B + \frac{2n+1}{n}\,C\right], \tag{9.5b}$$

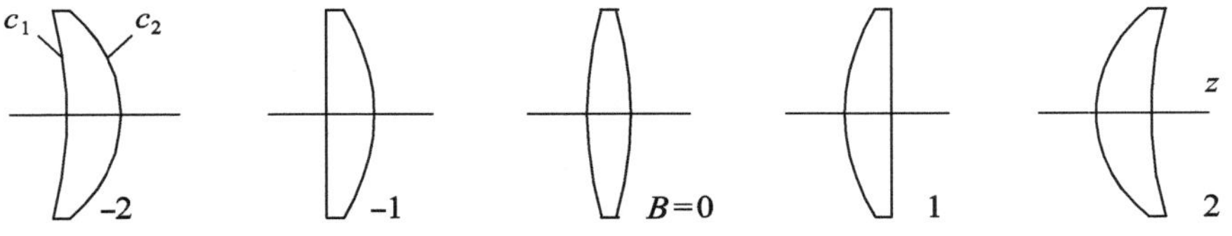

Fig. 9.1 B values defining the bending or *cambrure* of a lens: meniscus, plano-convex, equiconvex, convexo-plane, and meniscus

$$S_{III} = \frac{H^2}{f'}, \quad S_{IV} = \frac{H^2}{nf'}, \quad S_V = 0. \tag{9.5c}$$

where H is the Lagrange invariant (cf. Sect. 1.6.1). Since we have assumed that the entrance pupil coincides with the lens, the distortion S_V is null because the principal ray which passes through the vertex of the lens is undeviated.

If the power and conjugates are fixed, the variation of the primary spherical aberration, *Sphe* 3, is quadratic with the shape. For opposite conjugates ($C = 0$, i.e. $z' = -z$), the minimum value is for the equiconvex lens i.e. $B = 0$. The quantity $4S_I f'^3/x^4$ in (9.5a) is plotted in Fig. 9.2 with respect to B and n for opposite conjugates $C = 0$ i.e. $M = -1$.

- **Lenses with minimal spherical aberration**: If both conjugates are real, it can be shown that, whatever the bending B, the amount of spherical aberration of a thin lens cannot be cancelled. Setting minimal the above expression of the S_I coefficient in writing $dS_I/dB = 0$ entails

$$B + 2\frac{n^2-1}{n+2}C = 0, \tag{9.6}$$

and the minimal value for *Sphe* 3 is

$$S_I\,|_{\min} = \frac{x^4}{4f'^3}\left[\left(\frac{n}{n-1}\right)^2 - \frac{n}{n+2}C^2\right], \tag{9.7a}$$

$$= \frac{x^4}{4f'^3}\left[\left(\frac{n}{n-1}\right)^2 - \frac{n(n+2)}{4(n^2-1)^2}B^2\right]. \tag{9.7b}$$

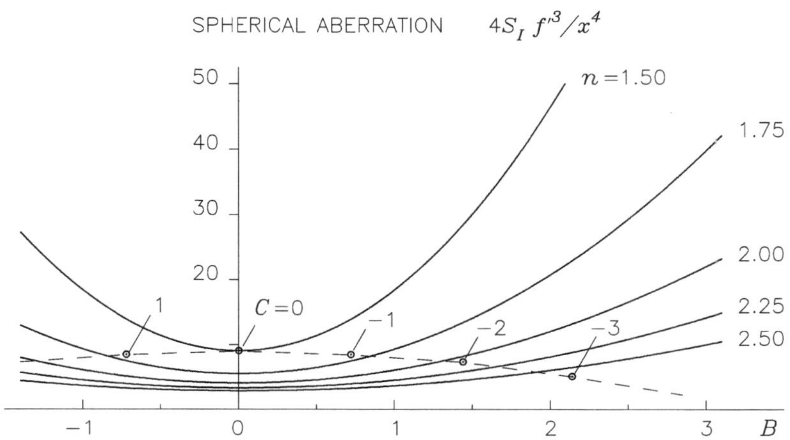

Fig. 9.2 Variation of *Sphe* 3 of a thin lens with the shape variable B. The full line parabolas are for opposite conjugates, $C = 0$ i.e. a transverse magnification $M = -1$, and for the refractive indices indicated. For other values of C the curves are translated; for example, the curve for $n = 1.5$ is moved so that its vertex stays on the *dashed line*, the appropriate value of C being labelled [10]

- **Lenses with a conjugate at infinity**: If a conjugate is at infinity, $C = \pm 1$, the lens with *Sphe* 3 minimal must have the shape

$$B = \mp 2\frac{n^2-1}{n+2}, \tag{9.8a}$$

which leads to $c_1 c_2 \le 0$ up to a refractive index $n \le (\sqrt{33}-1)/2 = 2.372$. For this index, $B^2 = 1$, so the minimum aberration lens would be with a flat surface. Returning to (9.8) and substituting B into (9.4), the curvature ratio of such lenses for an object at infinity, $C = -1$, is

$$\frac{c_2}{c_1} = -\frac{4+n-2n^2}{n(2+n)}, \tag{9.8b}$$

and is the reciprocal for an image at infinity, $C = 1$. If $n = 3/2$, we obtain the curvature ratios $c_2/c_1 = -4/21$ and $-21/4$, respectively. In addition, the following rule must be applied:

→ *If a lens is with spherical surfaces, then the* Sphe 3 *aberration is minimal when the surface of higher curvature is towards the conjugate at infinity.*

- **Stigmatic lenses in a convergent beam**: For conjugates where one of them, is virtual, equations (9.7) give the values of B^2 and C^2 for $S_I = 0$. Since, from (9.6), these variables must have opposite signs, the conditions are

$$B = \pm 2(n+1)\sqrt{\frac{n}{n+2}} \quad \text{and} \quad C = \mp\frac{\sqrt{n(n+1)}}{n-1}, \tag{9.9}$$

which corresponds to meniscus lenses since $|B| \ge 5\sqrt{3/7} = 3.273$ for usual materials $n \ge 3/2$.

9.1.2 Stigmatic Lens with Descartes Ovoid and Spherical Surface

If one of the surfaces of a lens is a sphere centered on one of the Gaussian conjugate points, Descartes [5] demonstrated that the sigmatism at *all* orders is achieved if the surface of the other diopter is an *ovoid*, known as *Descartes ovoid*. Creating analytic geometry for this purpose, Descartes also gave a classical construction of the ovals with the marked straight edge and a string (cf. Sect. 1.1.4 and Fig. 1.4).

Let us consider a source point P in a medium of refractive index unity, and its conjugate P' in a next medium n. The locus of points M determining the surface separation of the two media can be derived from the sum of light propagation times $\Delta t_{\mathrm{PM}} + \Delta t_{\mathrm{MP'}}$ which is a constant. For isotropic media, these terms are proportional to PM/c and $n\mathrm{MP'}/c$ respectively, where c is the velocity of light. Hence the stigmatism condition is achieved if the optical path is stationary,

$$\mathrm{PM} + n\,\mathrm{MP}' = 2a, \tag{9.10a}$$

where a is the optical path constant. Considering a polar coordinate system r, θ whose origin is P with $\theta = \mathrm{PP}', \mathrm{PM}$ and denoting the axial distance $\mathrm{PP}' = 2b$, the above relation writes

$$(n^2 - 1)r^2 - 4(b\,n^2\cos\theta - a)\,r + 4\,(b^2n^2 - a^2) = 0. \tag{9.10b}$$

With dimensionless quantities $\rho = r/2a$ and $k = b/a$, the *Descartes ovoids* are represented by

$$\rho^2 - 2\frac{1 - k\,n^2\cos\theta}{1 - n^2}\,\rho + \frac{1 - k^2n^2}{1 - n^2} = 0. \tag{9.10c}$$

These are fourth degree curves which includes the *conicoid* class. The roots $\rho_1(\theta)$ and $\rho_2(\theta)$ allow plotting the ovoids (Fig. 9.3).

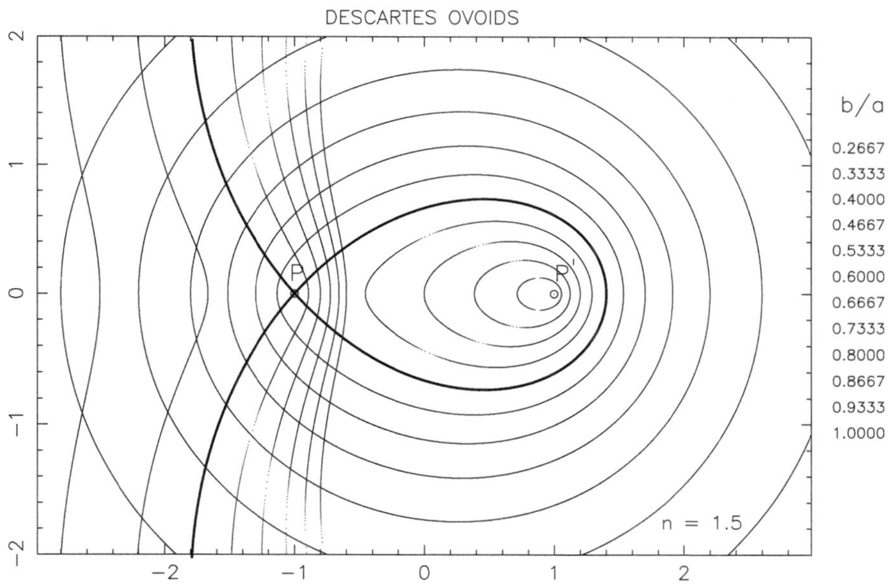

Fig. 9.3 *Descartes ovoids* for various values of $k = b/a$. These curves provide stigmatic lenses at *all* orders. The *thick curve* is the Pascal limaçon $k\,n = 1$ associated with the refractive index n

9.1.3 Aplanatic and Anastigmatic Singlet Lenses

A special case of a singlet lens with spherical surfaces is when one of the conjugates is virtual. Then the emergent beam has a convergence angle u_2' of the same sign as u_1 of the incident beam, and it is possible to obtain thin and thick lenses which, for one wavelength, satisfy the sine condition and even are anastigmatic at all orders. This corresponds to a limit case of the ovoids (Fig. 9.4).

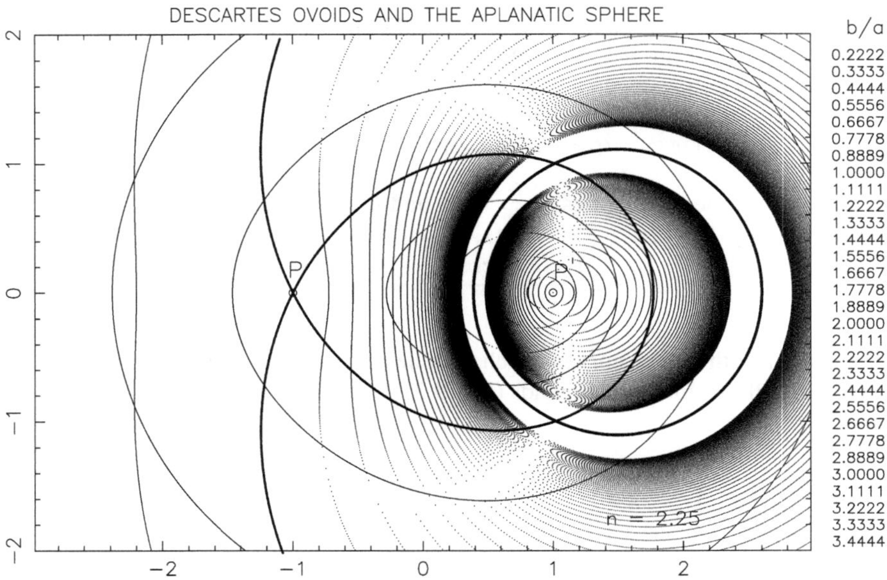

Fig. 9.4 A limit case of ovoids is a sphere (*thick line*) providing anastigmatic conjugates. This sphere, obtained for $a = 0$ and which center C is located at PC/P′C $= n^2$, satisfies the *all*-order correction of spherical aberration and coma, i.e. the Abbe sine condition

An all-order *anastigmatic thick lens* for one wavelength is a *monocentric thick lens* of index n with front and back spherical radii R/n and R centered in C. If the emerging beams virtually converge behind C on a sphere of radius nR also centered at C, then all the incident beams converge on the lens front-surface of radius R/n (Fig. 9.5). The Abbe sine condition (cf. Sect. 1.9.2) is satisfied and due to the monocentric symmetry, the astigmatism is also removed at *all* orders.

An *aplanatic thick lens* with spherical surfaces can be obtained with conjugates at similar locations R'/n, nR' from the center of a sphere of radius R', where the beam is refracted, whilst the other surface is in normal incidence. Setting the third-order aplanatism condition for a thin lens as $S_I = S_{II} = 0$ in (9.5a) and (9.5b) allows solving for the B and C variables. After simplifications, we find

$$B = \pm(2n+1) \quad \text{and} \quad C = \mp\frac{n+1}{n-1}. \tag{9.11}$$

If $n = 3/2$, this gives $B = \pm 4$ and $C = \mp 5$, i.e. a meniscus lens and a virtual conjugate for both cases.

Substituting the C variable into (9.4), the transverse magnification is

$$M = \frac{n+1 \pm (n-1)}{n+1 \mp (n-1)}, \tag{9.12}$$

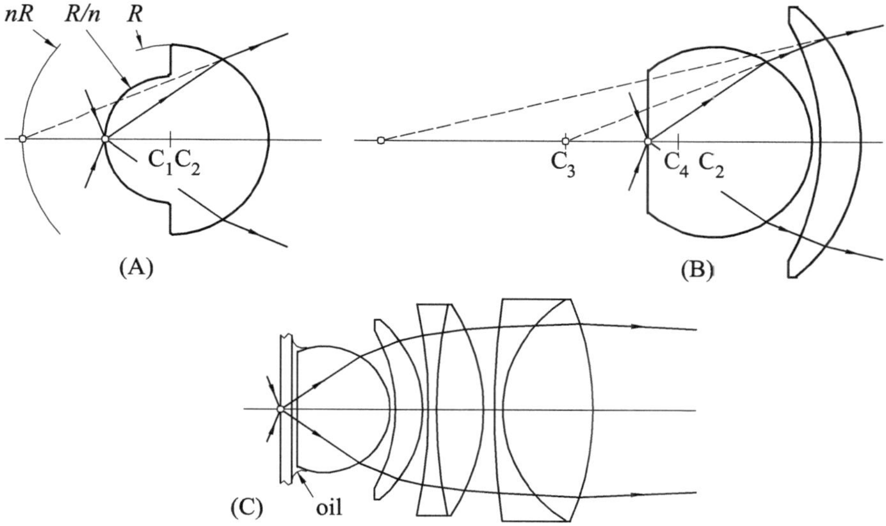

Fig. 9.5 Anastigmatic monocentric magnifier (**A**) corrected in all orders for the three first aberrations. Lens components (**B**) individually satisfying the Abbe sine condition. In a high power microscope, these are classically used for reducing the beam divergence in the front part of the objective (**C**) where the final imaging convergence must be achieved by cemented lenses

which for the two cases with a virtual conjugate entails $M = n$ or $M = 1/n$. In this cases one shows that these meniscus lenses also satisfy the Abbe sine condition.

Such thick and thin lenses have many applications for the design of high power *microscope objectives*. Given the extremely narrow field of view for *numerical aperture* (cf. Sect. 1.9.2) close to N.A. $= n \sin |U_{\max}| = 1$ – or 1.25 with oil immersion –, the first face of the monocentric thick lens can be designed flat. One or several meniscus lenses with their first face in normal incidence allow reduction of the divergence of the beam (Fig. 9.5). However, such lenses assembled together cannot provide a real image by fulfilling the aplanatism condition and correcting the chromatic aberrations, so the final imaging stage of a microscope objective must be formed by additional doublet lenses.

9.1.4 Isoplanatic Singlet Lenses and Remote Pupil

Since *Sphe* 3 of a thin lens with spherical surfaces cannot be corrected for both real conjugates, we may consider lenses that compensate for *Coma* 3 only, i.e. where $S_I \neq 0$. Such a *coma-free lens* is called an *isoplanatic lens* because it is *not* stigmatic. For real conjugates, the convergence angles are such as $u_1 u_2' \leq 1$ which, from (9.2b), entails $C \leq 1$.

- **Pupil at the lens**: Setting the pupil at the lens, the conditions for an isoplanatic lens, $S_{II} = 0$ from (9.5b), and for real conjugates are

$$\frac{n+1}{n(n-1)} B + \frac{2n+1}{n} C = 0 \quad \text{and} \quad C \leq 1. \tag{9.13}$$

If the conjugates are opposites with $M = -1$, i.e. $C = 0$, we obtain $B = 0$:

→ *At transverse magnification $M = -1$, a Coma 3 corrected lens is an equiconvex lens.*

If one of the conjugates is at infinity, $u_1 = 0$ or $u'_2 = 0$, and

$$B = -\frac{(n-1)(2n+1)}{n+1} C \quad \text{with} \quad C = \pm 1, \tag{9.14}$$

The substitution of B into (9.4) gives

$$\frac{c_2}{c_1} = -\left(\frac{1+n}{n^2} - 1\right)^{-C} \quad \text{with} \quad C = \pm 1. \tag{9.15}$$

For $n = 3/2$, we obtain $c_2/c_1 = -(1/9)^{\pm 1}$. If the conjugate at infinity is the object, $C = -1$, and from (9.3) the curvatures are

$$c_1 = \frac{9K}{5} = \frac{9}{5f'}, \quad c_2 = -\frac{K}{5} = -\frac{1}{5f'}. \tag{9.16}$$

→ *A Coma 3 corrected lens is obtained when the surface of higher curvature is towards the conjugate at infinity.*

→ *From (9.15), if $n = (1+\sqrt{5})/2$, then a Coma 3 corrected lens becomes convexo-plane or plano-convex for objects at $\mp\infty$, and the curvature of the convex surface is $c = \pm n\,K$.*

These latter cases, $c_1 c_2 = 0$, are the limit cases between biconvex and meniscus isoplanatic lenses, i.e. $c_1 c_2 < 0$ or > 0 respectively. Since $n^2 = n + 1$, if the object is at infinity, then $c_1 = -n\,K$ and $c_2 = 0$. From (1.78), (1.45), and (1.46), the radii of curvature of the Petzval, sagittal, mean and tangential surfaces are $R_P = -(n+1)R_1$, $R_s = -R_1$, $R_m = -R_1/n$, and $R_t = -R_1/(2n-1)$, respectively (Fig. 9.6).

- **Lens and remote pupil**: The amount of *Coma* 3, *Astm* 3, and *Dist* 3 of a lens are also functions of the location of the entrance pupil or stop, whilst the *Sphe* 3 and *Petz* 3 amounts remain unchanged. The stop-shift effect can be introduced into the analytic representation of the Seidel sums S_{II}, S_{III}, and S_V (cf. for instance Welford [10]).
As a basic example of an isoplanatic arrangement with a *remote pupil*, let us consider an extended object at infinity whose optical beams first pass through an aperture stop (Fig. 9.6). A plano-convex thin lens, $c_1 = 0$, is then located at a distance d from the stop so a ray passing at the center of the stop is perpendicular to the second surface of the lens, i.e. $dc_2 = 1/n$ if we neglect the thickness of

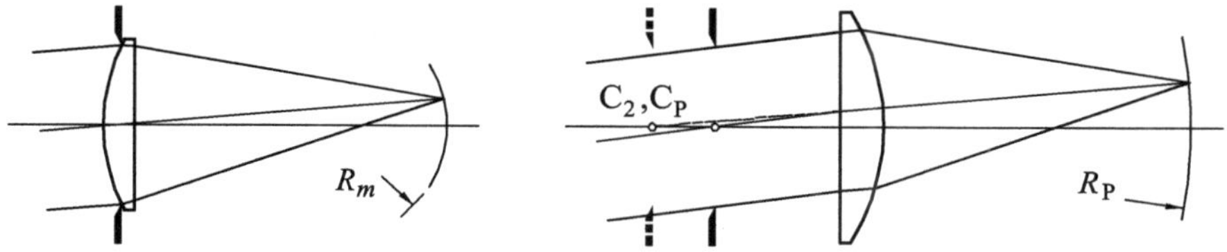

Fig. 9.6 Isoplanatic lenses ($S_I \neq 0$, $S_{II} = 0$) with spherical surfaces for an object at infinity. *Left*: Convexo-plane lens of refractive index $n = (1+\sqrt{5})/2 = 1.618$ with entrance pupil at the lens ($S_{III} \neq 0$). *Right*: Entrance pupil and remote plano-convex lens; the center of the entrance pupil – or aperture stop – is located at the conjugate of the center of curvature of the lens second surface with respect to its plane first surface (hence $S_{III} = 0$ also)

the lens. The result is that *Coma* 3 is corrected, $S_{II} = 0$. In addition, this system is also free from *Astm* 3, ($S_{III} = 0$), which provides interesting properties for very wide field systems if the chromatism and *Sphe* 3 are corrected by additional lenses.

- **Photographic objectives**: The *remote pupil* principle is a key feature in the development of photographic objectives. This was implicitly used in 1840 by Joseph Max Petzval in the first calculation of a *portrait objective* which consisted of a cemented lens plus a pair of air-spaced elements. Some important accounts of the design and evolution of the various forms of *photographic objectives* are given, for instance, by Zeiss [11], Chrétien [3], Kingslake [6] and Laikin [7].

9.1.5 Aspheric Lenses in the Third-Order Theory

Instead of considering a lens where one of the surfaces is a Descartes ovoid, we may fall back to the approximation of the optics third-order theory by representing one of its surfaces or both by the two first terms of a conicoid (cf. Sect. 1.7.1). The equation of first surface is then

$$z_1 = \frac{1}{2} c_1 r^2 + \frac{1}{8}(1+\kappa_1) c_1^3 r^4, \tag{9.17}$$

whilst the second surface is with the subscript 2.

It can be shown (cf. for instance Born and Wolf [2]) that, for an object at infinity and the entrance pupil on the lens, the condition which cancels *Sphe* 3 of the lens, i.e. $S_I = 0$, is satisfied if

$$\kappa_1 c_1^3 - \kappa_2 c_2^3 = -\frac{n^3}{(n-1)^3 (n+1)^2} K^3. \tag{9.18a}$$

where the quantity

$$z_{\text{asp}} = \frac{1}{8} (\kappa_1 c_1^3 - \kappa_2 c_2^3) r^4 \tag{9.18b}$$

represents the algebraic sum of the asphericity sags of the two surfaces.

Since the power of a lens is $K = (n-1)\,(c_1 - c_2)$, after substitution the condition is

$$\kappa_1\, c_1^3 - \kappa_2\, c_2^3 = -\frac{n^3}{(n+1)^2}\,(c_1 - c_2)^3\,. \tag{9.18c}$$

For $n = 3/2$ this gives

$$\kappa_1\, c_1^3 - \kappa_2\, c_2^3 = -\frac{27}{50}(c_1 - c_2)^3, \tag{9.18d}$$

from where, given the power of the lens, the three available parameters are the *cambrure* (or bending) c_2/c_1 and the two conic constants. For instance, this allows removing $Sphe\,3$ and $Coma\,3$ simultaneously.

- **Stigmatic aspherical lens**: If the power is positive and $Sphe\,3$ only removed, and if $c_2 = \kappa_2 = 0$ and $n = 3/2$, then $c_1 > 0$ and the amount of asphericity $z_{asp} = \frac{1}{8}\,\kappa_1 c_1^3 r^4$ must be set by $\kappa_1 = -n^3/(n+1)^2 = -27/50 = -0.54$. The other solution with a plane surface, $c_1 = \kappa_1 = 0$, is with $c_2 < 0$ and $\kappa_2 = -27/50$ which, of course, is the same in the Gaussian theory of thin lenses.
- **Aplanatic aspherical lens**: If $Sphe$ and $Coma\,3$ are removed, then $S_I = S_{II} = 0$ where the second condition entails c_2/c_1 as given by (9.15). Hence, for $n = 3/2$, we obtain $c_2/c_1 = -1/9$ from (9.16), and (9.18c) writes

$$9^3\,\kappa_1 + \kappa_2 = -20 \times 3^3, \tag{9.18a}$$

showing, of course, that there is an infinity of aplanatic lenses with two aspherical surfaces. When only one surface is aspherical, the two solutions are $\kappa_1 = 0$, $\kappa_2 = -540$ with c_2 negative, and $\kappa_1 = -20/27$, $\kappa_2 = 0$ with c_1 positive.
For an object at infinity and a positive lens with the entrance pupil on it, the conic constants in the above results are all negative. This can be easily generalized whatever the refractive index.

→ *In the third-order theory, whether a stigmatic or an aplanatic lens, and whether one or both surfaces are aspherized, a single lens must be thickened towards its edge.*

9.1.6 Power of a Two-Lens System

From formula (9.1) expressing the power of a thin lens, one shows that the resulting power K of a system with two thin lenses which are axially air-spaced by a distance d is

$$K = K_1 + K_2 - d\,K_1 K_2. \tag{9.19}$$

For two lenses which are in contact, the resulting power simply reduces to the sum of the individual powers.

It is clear that the thin lens concept gives rigorous analytical results for a lens whose central thickness is zero which is manifestly impossible to achieve in practice. However, in a first stage, the concept allows accurate predictions and studies of the basic properties of any lens system in the first- and third-order theories of dioptrics and aberrations.

9.2 Thin Lens Elastically Bent by Uniform Load

9.2.1 Equilibrium Equation of the Thin Plate Theory

The flexure of an axisymmetric lens by a uniform load applied over all its surface with a reaction at the circular edge involves taking into account its non-uniform thickness $t(r)$. The rigidity of the lens is defined by

$$D(r) = E\,t^3(r)/[12(1-\nu^2)] \tag{9.20}$$

where E and ν are the Young modulus and Poisson's is ratio of the material.

The equilibrium equation of the thin plate theory, (3.3) in Sect. 3.2, applies to the case of a variable thickness plate. For a lens with an axisymmetric thickness this equation reduces to

$$M_r + r\,\frac{dM_r}{dr} - M_t + r\,Q_r = 0\,, \tag{9.21}$$

where the radial and tangential bending moments are, from (3.1a) and (3.1b),

$$M_r = D(r)\left(\frac{d^2z}{dr^2} + \frac{\nu}{r}\frac{dz}{dr}\right) = D(r)\left(\frac{d\varphi}{dr} + \nu\frac{\varphi}{r}\right), \tag{9.22a}$$

$$M_t = D(r)\left(\frac{1}{r}\frac{dz}{dr} + \nu\,\frac{d^2z}{dr^2}\right) = D(r)\left(\frac{\varphi}{r} + \nu\,\frac{d\varphi}{dr}\right), \tag{9.22b}$$

and the slope of the flexure is

$$\varphi = \frac{dz}{dr}. \tag{9.22c}$$

At a circle of radius r, the shearing force per unit length is

$$Q_r = -\frac{1}{2\pi r}\int_0^r q\,2\pi r\,dr = -\frac{1}{2}\,q\,r\,, \tag{9.23}$$

where q is the intensity of the uniform load. As for all other chapters,

→ *the load q is positive when directed towards the positive z direction.*

Substituting the expressions of the bending moments and shearing force into (9.21), the equilibrium equation may be written

$$\frac{d^2\varphi}{dr^2} + \left(\frac{1}{r} + \frac{3}{t}\frac{dt}{dr}\right)\frac{d\varphi}{dr} - \left(\frac{1}{r^2} - \frac{3\nu}{t\,r}\frac{dt}{dr}\right)\varphi = 6(1-\nu^2)\frac{q}{E}\frac{r}{t^3} \tag{9.24}$$

where the unknown is the slope φ of the flexure. The two coefficients in parentheses on the left side are determined as input data from the geometry of the lens.

9.2.2 Lens Deformation and Parabolic Thickness Distribution

References on studies of the flexure of variable thickness plates by various authors can be found in Timoshenko's book [9]. However, none of them directly represent the geometry of a lens, thus we developed the following analysis for this purpose [8].

Approximating the spherical surfaces of a lens by the curvature term of their power expansion, the thickness distribution is then represented by the addition of a constant thickness plate and a paraboloid,

$$t(r) = t_0 - \tfrac{1}{2}(c_1 - c_2)\, r^2 \,, \tag{9.25}$$

where $t_0 = t(0)$ is the thickness at its center.

The thin plate theory of elasticity assumes that the middle surface of the plate is flat, and then does not take into account the *cambrure* of the lens. Hence for a noticeable cambrure, say $|B| \geq 2$ as defined in Sect. 9.1.1, a more accurate analysis would require using the theory of shells. We adopt for the lens curvatures the same sign convention as in the previous sections. The quantity $c_1 - c_2$ is proportional to the optical power, and thus is positive for a lens with positive power K [cf. (9.1)]. A simple configuration for the aspherization of a positive lens by stress polishing consists of bonding its edge into a thin L-square ring absorbing the reaction and allowing it to tangentially rotate when the load is applied (Fig. 9.7).

Let us introduce dimensionless quantities with respect to the outer radius $r = a$ of the lens, as

$$\rho = \frac{r}{a}, \qquad \mathcal{T} = \frac{t}{a}, \qquad \mathcal{T}_0 = \frac{t_0}{a} \qquad \text{and} \qquad \varpi = \tfrac{1}{2}\, a(c_1 - c_2) \tag{9.26}$$

where the sign of the pseudo-power ϖ is

$$\varpi > 0 \quad \text{for a positive lens } (K > 0),$$
$$\varpi < 0 \quad \text{for a negative lens } (K < 0).$$

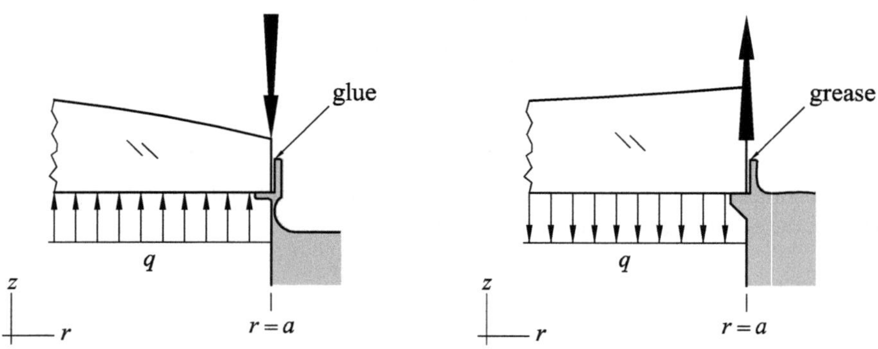

Fig. 9.7 Lens bent by a uniform load q in reaction with simply supported edge. *Left*: Positive lens and positive load. *Right*: Negative lens and negative load

The dimensionless thickness is

$$\mathcal{T}(\rho) = \mathcal{T}_0 - \varpi\,\rho^2 \qquad \text{with} \quad 0 \le \rho \le 1, \tag{9.27}$$

and substituting these quantities into (9.24), we obtain

$$\frac{d^2\varphi}{d\rho^2} + \left(\frac{1}{\rho} - \frac{6\varpi\rho}{\mathcal{T}}\right)\frac{d\varphi}{d\rho} - \left(\frac{1}{\rho^2} + \frac{6\nu\varpi}{\mathcal{T}}\right)\varphi = 6(1 - \nu^2)\frac{q}{E}\frac{\rho}{\mathcal{T}^3}. \tag{9.28}$$

This can be simplified by defining the quantities

$$\delta = \frac{\mathcal{T}_0}{\varpi} \qquad \text{and} \qquad p = 6(1 - \nu^2)\frac{q}{E\varpi^3}. \tag{9.29}$$

Hence the differential equation of the slope-flexure of a lens, bent by a uniform load in reaction at its edge of diameter $2a$, and its associated reduced thickness $\mathcal{T} = t/a$ assumed of quadratic shape, take the form

$$\left\{\begin{array}{l} \left(\delta - \rho^2\right)\dfrac{d^2\varphi}{d\rho^2} + \left(\dfrac{\delta}{\rho} - 7\rho\right)\dfrac{d\varphi}{d\rho} - \left(\dfrac{\delta}{\rho^2} - 1 + 6\nu\right)\varphi = \dfrac{p\,\rho}{(\delta - \rho^2)^2}, \quad (9.30\text{a}) \\[2ex] \quad \mathcal{T} = \varpi\left(\delta - \rho^2\right) \quad \text{with} \quad \mathcal{T}_0 = \varpi\,\delta > 0 \quad \text{and} \quad 0 \le \rho \le 1. \qquad (9.30\text{b}) \end{array}\right.$$

One can easily show that the particular solution φ_0 of (9.30a) is of the same form as the left side term but only differs by from a multiplicative factor. After identification of this factor, the particular solution is

$$\varphi_0 = \frac{p}{2(5 - 3\nu)}\frac{\rho}{(\delta - \rho^2)^2}. \tag{9.31}$$

For deriving the solution φ_1 without left side, let us denote μ such as $2(5-3\nu) = 4(4 - \mu)$ i.e.

$$\mu = \frac{3}{2}(1 + \nu), \tag{9.32}$$

which gives

$$\left(\delta - \rho^2\right)\frac{d^2\varphi_1}{d\rho^2} + \left(\frac{\delta}{\rho} - 7\rho\right)\frac{d\varphi_1}{d\rho} - \left(\frac{\delta}{\rho^2} - 7 + 4\mu\right)\varphi_1 = 0. \tag{9.33}$$

Considering the function φ_1 in the form of an odd expansion

$$\varphi_1 = \sum_{n=0}^{\infty} a_{2n+1}\rho^{2n+1}, \tag{9.34}$$

we implicitly satisfy, as for φ_0, a first boundary condition $\varphi_1|_{\rho=0} = 0$ of a null slope of the deformation at the origin, a condition which would not be achieved by a holed lens (in this latter case a $\ln \rho$ term must be included in the odd expansion). For $n \geq 1$ we obtain a recurrent relation between the coefficients,

$$\frac{a_{2n+1}}{a_{2n-1}} = \frac{\mu + (n+3)(n-1)}{n\,(n+1)\,\delta} \tag{9.35}$$

and with respect to the first one,

$$a_{2n+1} = \frac{\mu\,(\mu+5)\cdots[\mu+(n+3)(n-1)]}{n!\,(n+1)!\delta^n} a_1. \tag{9.36}$$

Setting $a_1 = 1$, the basic solution of the equation without left side is

$$\varphi_1 = \rho + \frac{\mu}{2\delta}\,\rho^3 + \cdots + \frac{\mu\,(\mu+5)\cdots[\mu+(n+3)(n-1)]}{n!(n+1)!\,\delta^n}\rho^{2n+1} + \cdots. \tag{9.37}$$

The general solution of (9.30a) is a linear combination of φ_0 and φ_1 which both already satisfy the boundary condition of null slope at the origin. Hence, it can be represented by

$$\varphi(\rho) = \frac{p}{2(5-3\nu)}\left[\Lambda\,\varphi_1(\rho) + \frac{\rho}{(\delta-\rho^2)^2}\right] \tag{9.38}$$

where Λ is a constant.

We here assume that the lens has a simply supported edge during its aspherization; hence the second boundary condition must express that the radial bending moment M_r vanishes at the edge. From (9.22a), this entails

$$\left[\frac{d\varphi}{d\rho} + \nu\,\frac{\varphi}{\rho}\right]_{\rho=1} = 0, \tag{9.39a}$$

which, from (9.38), leads to

$$\begin{aligned}\Lambda\left[\frac{d\varphi_1}{d\rho} + \nu\,\frac{\varphi_1}{\rho}\right]_{\rho=1} &= -\left\{\frac{d}{d\rho}\left[\frac{\rho}{(\delta-\rho^2)^2}\right] + \frac{\nu}{(\delta-\rho^2)^2}\right\}_{\rho=1}\\ &= -\frac{2}{3}\,\frac{\mu(\delta-1)+6}{(\delta-1)^3}.\end{aligned} \tag{9.39b}$$

After substitution of φ_1, and using the notation with parameter μ, we obtain

$$\begin{aligned}\left[\frac{d\varphi_1}{d\rho} + \nu\,\frac{\varphi_1}{\rho}\right]_{\rho=1} &= \frac{2}{3}\left[\mu + \frac{(\mu+3)\mu}{2\delta} + \cdots\right.\\ &\left.\cdots + \sum_{n=2}^{\infty}\frac{\mu+3n}{n+1}\prod_{m=1}^{n}\frac{\mu+(m+3)(m-1)}{m^2\delta}\right],\end{aligned} \tag{9.39c}$$

and, from (9.39c), the integration constant is

$$\Lambda = -\frac{\mu(\delta-1)+6}{(\delta-1)^3}\left[\mu + \frac{(\mu+3)\mu}{2\,\delta} + \cdots\right.$$

$$\left.\cdots + \sum_{n=2}^{\infty}\frac{\mu+3n}{n+1}\prod_{m=1}^{n}\frac{\mu+(m+3)(m-1)}{m^2\,\delta}\right]^{-1}. \qquad (9.40)$$

Finally from (9.22c), we determine the flexure $z = a\int \varphi\, d\rho$. Setting the integration constant equal to zero, i.e. choosing the origin of the flexure at the center of the lens, the flexure is represented by (Lemaitre [8])

$$z = \frac{a\,p}{8(4-\mu)}\left\{\Lambda\left[\rho^2 + \frac{\mu}{4\delta}\rho^4 + \cdots\right.\right.$$

$$\left.\left.\cdots + \sum_{n=2}^{\infty}\left(\prod_{m=1}^{n}\frac{\mu+(m+3)(m-1)}{(m+1)^2\delta}\right)\rho^{2n+2}\right] + \frac{\rho^2}{\delta(\delta-\rho^2)}\right\}. \qquad (9.41)$$

For expanding the flexure z, the absolute value of the coefficient δ must be larger than unity, i.e. $|\delta| = t_0/a|\varpi| > 1$. From definition (9.26), the pseudo-power ϖ is related to the power K by

$$\varpi = \frac{1}{2}a(c_1 - c_2) = aK/2(n-1), \qquad (9.42)$$

where n is the refractive index. This entails $|\delta| = 2(n-1)t_0/a^2|K|$. Considering the f-number of the lens, $\Omega = \mathrm{f/D} = \mathrm{f}/2a = 1/2a|K|$, the condition for δ writes

$$|\delta| = 4(n-1)\frac{t_0}{a}\Omega \equiv 4(n-1)\mathcal{T}_0\Omega > 1. \qquad (9.43)$$

For instance, for a lens at f/10 with $\mathcal{T}_0 \equiv t_0/a = 1/10$ and $n = 3/2$, we obtain $\varpi = 1/20$ and the condition gives $|\delta| = 2$. If the lens is positive, then $\delta = 2$.

9.2.3 *Expansion Representation of the Flexure*

From the previous analysis, we consider now a representation of the flexure of a lens by an expansion. Assuming that the inequality (9.43) is satisfied, the last term of the flexure in (9.41) can be expanded as

$$\frac{\rho^2}{\delta(\delta-\rho^2)} = \frac{\rho^2}{\delta^2} + \frac{\rho^4}{\delta^3} + \frac{\rho^6}{\delta^4} + \cdots,$$

so the first three terms in the expansion of the flexure are

$$z = \frac{ap}{8(4-\mu)}\left[\left(\Lambda + \frac{1}{\delta^2}\right)\rho^2 + \left(\frac{\mu\Lambda}{4\delta} + \frac{1}{\delta^3}\right)\rho^4 + \left(\frac{\mu(\mu+5)\Lambda}{36\,\delta^2} + \frac{1}{\delta^4}\right)\rho^6 + \cdots\right]. \tag{9.44}$$

To summarize, the first three terms of the flexure of a single lens can be represented by the following equation set

$$\begin{cases} z = aq(A_2\rho^2 + A_4\rho^4 + A_6\rho^6 + \cdots), \quad 0 \le \rho \le 1 \\ A_2,\ A_4,\ A_6 \text{ obtained from (9.44) with } |\delta| > 1 \\ \Lambda \text{ given by (9.40)} \\ \mu = 3(1+\nu)/2 \\ p = 6(1-\nu^2)q/E\varpi^3 \\ \varpi = a(c_1 - c_2)/2 \equiv aK/2(n-1) \\ \mathcal{T} = \varpi(\delta - \rho^2) \quad \text{with} \quad \mathcal{T}_0 = \varpi\,\delta > 0. \end{cases} \tag{9.45}$$

Given a central reduced thickness $\mathcal{T}_0 = t_0/a$ and elasticity constants E, ν of the glass, the variation of A_2, A_4, and A_6 coefficients of the reduced flexure per load unit, z/aq, with respect to the pseudo-power ϖ are displayed by Fig. 9.8 and Table 9.1.

9.2.4 Maximum Stresses at the Lens Surfaces

In order to avoid the risk of breakage of the glass during the figuring, one must select a convenient maximal value for the maximum stresses which appear at the surface of the lens. Given a material and a loading delay, this value can be established from Table 5.2 by introducing a reduction factor, say $\sim$ 1/2 or 1/3, from the *ultimate tensile strength* (cf. Sect. 5.2.5). The other parameters available to avoid the rupture are the central thickness t_0 of the lens and the intensity q of the load.

The radial and tangential maximum stresses at the surface are

$$\sigma_r = \pm 6\,M_r/t^2, \quad \sigma_t = \pm 6\,M_t/t^2, \tag{9.46}$$

where the sign depends on the considered surface of the lens.

From (9.22a), we obtain for the radial component

$$\sigma_r = \pm\frac{6D}{t^2}\left(\frac{d\varphi}{dr} + \nu\frac{\varphi}{r}\right) = \pm\frac{Et}{2(1-\nu^2)a}\left(\frac{d\varphi}{d\rho} + \nu\frac{\varphi}{\rho}\right). \tag{9.47}$$

After substitution of the slope φ as given by (9.38), the term in parentheses becomes

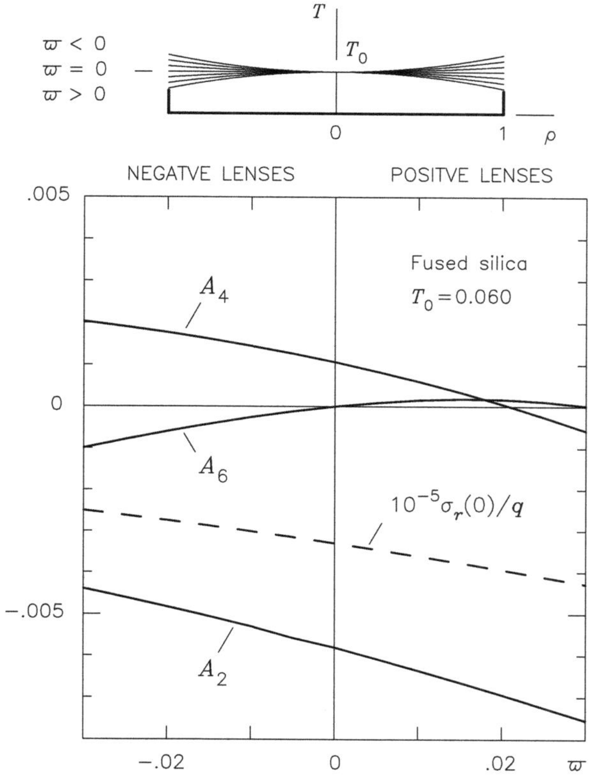

Fig. 9.8 Variation of the coefficients of the reduced flexure $z/aq = A_2\rho^2 + A_4\rho^4 + A_6\rho^6$, from (9.45), of single lenses of same central thickness $\mathcal{T}_0 = t_0/a = 0.06$ in function of the pseudo-power ϖ. Material: Fused Silica $E = 7.90 \times 10^5$ kgf/cm^2, $\nu = 0.165$. Maximum radial stress at center: $\sigma_r(0)$. Load q in kgf/cm^2

Table 9.1 Deformation coefficients A_{2n} and stress of fused-silica lenses of various thickness and power bent by uniform load. Simply supported edge. Quantities defined by set (9.45) with $E = 7.90 \times 10^5$ kgf/cm^2 and $\nu = 0.165$

$\mathcal{T}_0$	ϖ	$\delta = \frac{\mathcal{T}_0}{\varpi}$	Λ	$A_2 \times 10^3$	$A_4 \times 10^3$	$A_6 \times 10^3$	$\sigma(0)/q$
0.120	−0.065	−1.846	−0.06609	−0.53671	0.26062	−0.13800	60.9
	−0.060	−2.000	−0.03942	−0.54936	0.25361	−0.12476	62.4
	−0.055	−2.182	−0.01809	−0.56227	0.24620	−0.11182	63.8
	−0.050	−2.400	−0.00186	−0.57553	0.23838	−0.09921	65.3
	−0.045	−2.667	0.00969	−0.58913	0.23013	−0.08697	66.9
	−0.040	−3.000	0.01698	−0.60301	0.22142	−0.07513	68.5
	−0.035	−3.429	0.02052	−0.61722	0.21226	−0.06373	70.1
	−0.030	−4.000	0.02089	−0.63189	0.20263	−0.05282	71.7
	−0.025	−4.800	0.01876	−0.64666	0.19247	−0.04243	73.4

Table 9.1 (continued)

T_0	ϖ	$\delta = \frac{T_0}{\varpi}$	Λ	$A_2 \times 10^3$	$A_4 \times 10^3$	$A_6 \times 10^3$	$\sigma(0)/q$
	−0.020	−6.000	0.01486	−0.66202	0.18181	−0.03262	75.2
	−0.015	−8.000	0.01005	−0.67739	0.17060	−0.02341	76.9
	−0.010	−12.000	0.00525	−0.69347	0.15886	−0.01487	78.7
	−0.005	−24.000	0.00152	−0.70934	0.14652	−0.00705	80.5
	0.000	$\mp\infty$	0.00000	−0.72597	0.13361	0.00000	82.4
	0.005	24.000	0.00196	−0.74292	0.12009	0.00623	84.3
	0.010	12.000	0.00880	−0.75997	0.10594	0.01157	86.3
	0.015	8.000	0.02202	−0.77727	0.09117	0.01597	88.2
	0.020	6.000	0.04329	−0.79483	0.07574	0.01936	90.2
	0.025	4.800	0.07439	−0.81309	0.05961	0.02168	92.3
	0.030	4.000	0.11723	−0.83110	0.04284	0.02288	94.4
	0.035	3.429	0.17389	−0.84938	0.02538	0.02287	96.4
	0.040	3.000	0.24659	−0.86790	0.00722	0.02160	98.5
	0.045	2.667	0.33769	−0.88666	−0.01165	0.01899	100.7
	0.050	2.400	0.44973	−0.90567	−0.03125	0.01498	102.8
	0.055	2.182	0.58512	−0.92423	−0.05145	0.00954	104.9
	0.060	2.000	0.74717	−0.94369	−0.07253	0.00250	107.1
	0.065	1.846	0.93825	−0.96271	−0.09421	−0.00609	109.3
0.110	−0.060	−1.833	−0.06879	−0.69532	0.33915	−0.18075	72.4
	−0.055	−2.000	−0.03942	−0.71322	0.32926	−0.16197	74.2
	−0.050	−2.200	−0.01642	−0.73152	0.31873	−0.14366	76.1
	−0.045	−2.444	0.00058	−0.75038	0.30757	−0.12587	78.1
	−0.040	−2.750	0.01208	−0.76973	0.29575	−0.10866	80.1
	−0.035	−3.143	0.01867	−0.78961	0.28322	−0.09209	82.2
	−0.030	−3.667	0.02105	−0.80988	0.26996	−0.07622	84.3
	−0.025	−4.400	0.01999	−0.83082	0.25596	−0.06113	86.5
	−0.020	−5.500	0.01643	−0.85230	0.24116	−0.04689	88.7
	−0.015	−7.333	0.01140	−0.87383	0.22552	−0.03357	90.9
	−0.010	−11.000	0.00608	−0.89660	0.20907	−0.02126	93.3
	−0.005	−22.000	0.00179	−0.91894	0.19171	−0.01004	95.6
	0.000	$\mp\infty$	0.00000	−0.94251	0.17346	0.00000	98.0
	0.005	22.000	0.00236	−0.96659	0.15427	0.00876	100.6
	0.010	11.000	0.01068	−0.99081	0.13411	0.01615	103.1
	0.015	7.333	0.02695	−1.01541	0.11297	0.02206	105.7
	0.020	5.500	0.05336	−1.04043	0.09082	0.02640	108.3
	0.025	4.400	0.09227	−1.06583	0.06764	0.02905	110.9
	0.030	3.667	0.14627	−1.09161	0.04340	0.02990	113.6
	0.035	3.143	0.21813	−1.11775	0.01809	0.02885	116.3
	0.040	2.750	0.31085	−1.14424	−0.00831	0.02577	119.1
	0.045	2.444	0.42765	−1.17112	−0.03584	0.02055	121.9
	0.050	2.200	0.57200	−1.19848	−0.06453	0.01305	124.7
	0.055	2.000	0.74717	−1.22517	−0.09416	0.00325	127.5
	0.060	1.833	0.95728	−1.25231	−0.12496	−0.00905	130.3
0.100	−0.055	−1.818	−0.07210	−0.92312	0.45269	−0.24310	87.3
	−0.050	−2.000	−0.03942	−0.94929	0.43824	−0.21559	89.8
	−0.045	−2.222	−0.01445	−0.97614	0.42278	−0.18881	92.4
	−0.040	−2.500	0.00330	−1.00384	0.40630	−0.16289	95.0

Table 9.1 (continued)

T_0	ϖ	$\delta = \frac{T_0}{\varpi}$	Λ	$A_2 \times 10^3$	$A_4 \times 10^3$	$A_6 \times 10^3$	$\sigma(0)/q$
	−0.035	−2.857	0.01455	−1.03231	0.38873	−0.13792	97.7
	−0.030	−3.333	0.02008	−1.06171	0.37003	−0.11401	100.4
	−0.025	−4.000	0.02089	−1.09190	0.35014	−0.09128	103.3
	−0.020	−5.000	0.01809	−1.12267	0.32897	−0.06985	106.2
	−0.015	−6.667	0.01300	−1.15458	0.30654	−0.04987	109.2
	−0.010	−10.000	0.00710	−1.18700	0.28274	−0.03146	112.3
	−0.005	−20.000	0.00213	−1.21993	0.25753	−0.01478	115.4
	0.000	$\mp\infty$	0.00000	−1.25448	0.23088	0.00000	118.6
	0.005	20.000	0.00289	−1.28987	0.20270	0.01273	122.0
	0.010	10.000	0.01323	−1.32550	0.17297	0.02323	125.4
	0.015	6.667	0.03371	−1.36172	0.14164	0.03132	128.8
	0.020	5.000	0.06729	−1.39861	0.10868	0.03682	132.3
	0.025	4.000	0.11723	−1.43614	0.07403	0.03953	135.9
	0.030	3.333	0.18709	−1.47432	0.03765	0.03925	139.5
	0.035	2.857	0.28074	−1.51317	−0.00049	0.03579	143.2
	0.040	2.500	0.40216	−1.55135	−0.04022	0.02898	146.8
	0.045	2.222	0.55619	−1.59135	−0.08197	0.01852	150.6
	0.050	2.000	0.74717	−1.63070	−0.12533	0.00433	154.3
	0.055	1.818	0.98049	−1.67077	−0.17057	−0.01390	158.1
0.090	−0.050	−1.800	−0.07627	−1.26251	0.62313	−0.33774	107.5
	−0.045	−2.000	−0.03942	−1.30218	0.60115	−0.29573	110.9
	−0.040	−2.250	−0.01214	−1.34322	0.57752	−0.25499	114.4
	−0.035	−2.571	0.00634	−1.38560	0.55212	−0.21571	118.0
	−0.030	−3.000	0.01698	−1.42936	0.52486	−0.17809	121.7
	−0.025	−3.600	0.02097	−1.47449	0.49564	−0.14232	125.6
	−0.020	−4.500	0.01971	−1.52091	0.46436	−0.10865	129.5
	−0.015	−6.000	0.01486	−1.56922	0.43097	−0.07731	133.6
	−0.010	−9.000	0.00840	−1.61819	0.39526	−0.04857	137.8
	−0.005	−18.000	0.00258	−1.66900	0.35722	−0.02270	142.1
	0.000	$\mp\infty$	0.00000	−1.72082	0.31670	0.00000	146.5
	0.005	18.000	0.00363	−1.77404	0.27365	0.01922	151.1
	0.010	9.000	0.01681	−1.82849	0.22795	0.03463	155.7
	0.015	6.000	0.04329	−1.88405	0.17953	0.04589	160.4
	0.020	4.500	0.08725	−1.94071	0.12830	0.05265	165.3
	0.025	3.600	0.15332	−1.99843	0.07419	0.05455	170.2
	0.030	3.000	0.24659	−2.05723	0.01712	0.05120	175.2
	0.035	2.571	0.37265	−2.11730	−0.04301	0.04220	180.3
	0.040	2.250	0.53729	−2.17654	−0.10590	0.02727	185.3
	0.045	2.000	0.74717	−2.23689	−0.17191	0.00594	190.5
	0.050	1.800	1.00942	−2.29855	−0.24117	−0.02225	195.7
0.080	−0.045	−1.778	−0.08159	−1.79069	0.89099	−0.48848	135.5
	−0.040	−2.000	−0.03942	−1.85409	0.85594	−0.42107	140.3
	−0.035	−2.286	−0.00937	−1.91995	0.81790	−0.35596	145.3
	−0.030	−2.667	0.00969	−1.98831	0.77668	−0.29352	150.5
	−0.025	−3.200	0.01918	−2.05920	0.73207	−0.23414	155.9
	−0.020	−4.000	0.02089	−2.13262	0.68386	−0.17828	161.4

Table 9.1 (continued)

T_0	ϖ	$\delta = \frac{T_0}{\varpi}$	Λ	$A_2 \times 10^3$	$A_4 \times 10^3$	$A_6 \times 10^3$	$\sigma(0)/q$
	−0.015	−5.333	0.01698	−2.20813	0.63181	−0.12639	167.1
	−0.010	−8.000	0.01005	−2.28618	0.57578	−0.07901	173.0
	−0.005	−16.000	0.00318	−2.36731	0.51558	−0.03668	179.2
	0.000	$\mp\infty$	0.00000	−2.45015	0.45093	0.00000	185.4
	0.005	16.000	0.00468	−2.53558	0.38170	0.03041	191.9
	0.010	8.000	0.02202	−2.62328	0.30768	0.05389	198.6
	0.015	5.333	0.05749	−2.71307	0.22870	0.06973	205.3
	0.020	4.000	0.11723	−2.80496	0.14458	0.07720	212.3
	0.025	3.200	0.20806	−2.89700	0.05543	0.07562	219.3
	0.030	2.667	0.33769	−2.99246	−0.03931	0.06410	226.5
	0.035	2.286	0.51431	−3.08779	−0.13924	0.04201	233.7
	0.040	2.000	0.74717	−3.18495	−0.24478	0.00845	241.1
	0.045	1.778	1.04577	−3.28163	−0.35550	−0.03719	248.4
0.070	−0.040	−1.750	−0.08860	−2.65940	1.33702	−0.74375	176.1
	−0.035	−2.000	−0.03942	−2.76762	1.27767	−0.62853	183.3
	−0.030	−2.333	−0.00600	−2.88026	1.21240	−0.51777	190.8
	−0.025	−2.800	0.01331	−2.99780	1.14086	−0.41232	198.5
	−0.020	−3.500	0.02076	−3.11990	1.06255	−0.31308	206.6
	−0.015	−4.667	0.01919	−3.24695	0.97709	−0.22108	215.0
	−0.010	−7.000	0.01217	−3.37950	0.88404	−0.13742	223.8
	−0.005	−14.000	0.00403	−3.51644	0.78285	−0.06329	232.9
	0.000	$\mp\infty$	0.00000	−3.65738	0.67311	0.00000	242.2
	0.005	14.000	0.00626	−3.80353	0.55443	0.05106	251.9
	0.010	7.000	0.03005	−3.95439	0.42632	0.08840	261.9
	0.015	4.667	0.07975	−4.10974	0.28838	0.11044	272.2
	0.020	3.500	0.16486	−4.26544	0.14070	0.11561	282.5
	0.025	2.800	0.29619	−4.42506	−0.01731	0.10221	293.1
	0.030	2.333	0.48580	−4.58786	−0.18587	0.06849	303.8
	0.035	2.000	0.74717	−4.75421	−0.36538	0.01262	314.9
	0.040	1.750	1.09473	−4.92125	−0.55544	−0.06700	325.9
0.060	−0.035	−1.714	−0.09849	−4.19574	2.13815	−1.21221	238.2
	−0.030	−2.000	−0.03942	−4.39487	2.02889	−0.99808	249.5
	−0.025	−2.400	−0.00186	−4.60427	1.90701	−0.79366	261.4
	−0.020	−3.000	0.01698	−4.82410	1.77140	−0.60104	273.8
	−0.015	−4.000	0.02089	−5.05509	1.62100	−0.42259	287.0
	−0.010	−6.000	0.01486	−5.29613	1.45451	−0.26092	300.6
	−0.005	−12.000	0.00525	−5.54773	1.27086	−0.11899	314.9
	0.000	$\mp\infty$	0.00000	−5.80778	1.06888	0.00000	329.6
	0.005	12.000	0.00880	−6.07975	0.84755	0.09254	345.1
	0.010	6.000	0.04329	−6.35868	0.60590	0.15489	361.0
	0.015	4.000	0.11723	−6.64880	0.34272	0.18300	377.4
	0.020	3.000	0.24659	−6.94317	0.05779	0.17280	394.1
	0.025	2.400	0.44973	−7.24537	−0.24999	0.11985	411.3
	0.030	2.000	0.74717	−7.54951	−0.58021	0.02003	428.6
	0.035	1.714	1.16184	−7.85632	−0.93325	−0.13102	446.0

Table 9.1 (continued)

$\mathcal{T}_0$	ϖ	$\delta = \frac{\mathcal{T}_0}{\varpi}$	Λ	$A_2 \times 10^3$	$A_4 \times 10^3$	$A_6 \times 10^3$	$\sigma(0)/q$
0.050	−0.035	−1.429	−0.22080	−6.79714	3.92569	−2.63486	321.5
	−0.030	−1.667	−0.11306	−7.18340	3.72999	−2.17042	339.8
	−0.025	−2.000	−0.03942	−7.59434	3.50593	−1.72469	359.3
	−0.020	−2.500	0.00330	−8.03072	3.25041	−1.30312	379.9
	−0.015	−3.333	0.02008	−8.49365	2.96024	−0.91208	401.8
	−0.010	−5.000	0.01809	−8.98138	2.63179	−0.55880	424.9
	−0.005	−10.000	0.00710	−9.49603	2.26190	−0.25167	449.2
	0.000	$\mp\infty$	0.00000	−10.03585	1.84704	0.00000	474.7
	0.005	10.000	0.01323	−10.60398	1.38378	0.18584	501.6
	0.010	5.000	0.06729	−11.18884	0.86942	0.29455	529.3
	0.015	3.333	0.18709	−11.79455	0.30122	0.31402	557.9
	0.020	2.500	0.40216	−12.41079	−0.32174	0.23188	587.1
	0.025	2.000	0.74717	−13.04556	−1.00260	0.03462	617.1
	0.030	1.667	1.26100	−13.68180	−1.73930	−0.28985	647.2
	0.035	1.429	1.98873	−14.33180	−2.53580	−0.75617	678.0

$$\frac{d\varphi}{d\rho} + \nu\frac{\varphi}{\rho} = \frac{p}{4(4-\mu)}\left[\Lambda\left(\frac{d\varphi_1}{d\rho} + \nu\frac{\varphi_1}{\rho}\right) + \frac{(1+\nu)\delta + (3-\nu)\rho^2}{(\delta-\rho^2)^3}\right], \tag{9.48}$$

which allows the general determination of σ_r. The tangential maximum stresses σ_t would be derived similarly.

For instance, if $\rho = 0$, (9.37) entails $\varphi_1/\rho = 1$, hence

$$\left[\frac{d\varphi}{d\rho} + \nu\frac{\varphi}{\rho}\right]_{\rho=0} = \frac{(1+\nu)p}{4(4-\mu)}\left(\Lambda + \frac{1}{\delta^2}\right). \tag{9.49}$$

Using the dimensionless thickness $\mathcal{T} = t/a$, the maximum radial stresses at the center of the lens is

$$\sigma_r(0) = \pm\frac{E\mathcal{T}_0 p}{8(1-\nu)(4-\mu)}\left(\Lambda + \frac{1}{\delta_2}\right), \tag{9.50}$$

and since p is related to the pseudo-power ϖ of the lens and to the load q by (9.29), using also $\mu = \frac{3}{2}(1+\nu)$ from (9.32), this finally writes

$$\sigma_r(0) = \pm\frac{\mu}{2(4-\mu)}\frac{\mathcal{T}_0}{\varpi^3}\left(\Lambda + \frac{1}{\delta_2}\right)q, \tag{9.51}$$

where σ and q are the only dimensioned quantities.

For lenses in fused silica, some values of the ratio $\sigma_r(0)/q$ with respect to $\mathcal{T}_0$ and ϖ are displayed in the last column of Table 9.1.

- **Degenerated case of null powered lenses**: The stress formula for $\sigma_r(0)$ also includes the case of a null-power lens, $\varpi = K = 0$, which is a plate or a meniscus of thickness t = constant = t_0. In this case $\delta = \mathcal{T}_0/\varpi \to \infty$, so the above formula is indeterminate. For $\delta > 1$, the constant Λ can be expanded. Since

$$\frac{\mu(\delta-1)+6}{(\delta-1)^3} = \frac{\mu}{\delta^2} + \frac{2(\mu+3)}{\delta^3} + \cdots,$$

from (9.40), we obtain the expansion of the terms in parentheses in eqs. (9.44) as

$$\Lambda + \frac{1}{\delta^2} = -\frac{(3+\mu)(4-\mu)}{2\mu\delta^3} + o_1(\mu)\frac{1}{\delta^4} + \cdots, \tag{9.52a}$$

$$\frac{\mu\Lambda}{4\delta} + \frac{1}{\delta^3} = \frac{4-\mu}{4\delta^3} + o_2(\mu)\frac{1}{\delta^4} + \cdots, \tag{9.52b}$$

$$\frac{\mu(\mu+5)\Lambda}{36\delta^2} + \frac{1}{\delta^4} = o_3(\mu)\frac{1}{\delta^4} + \cdots. \tag{9.52c}$$

Since $\varpi\delta$ remains finite, the maximum radial stress at the center is

$$\sigma_r(0) = \pm\frac{3+\mu}{4}\frac{1}{\mathcal{T}_0^2}q = \pm\frac{3}{8}(3+\nu)\frac{a^2}{t_0^2}q, \tag{9.53}$$

a well-known formula (cf. for instance [9]) which is verified in Table 9.1 when $\varpi = 0$. For a low power lens, one can show that $\sigma_r(0) > \sigma_t(0)$ and that $\sigma_r(0)$ is maximal over $0 \le r \le a$.

Expansions (9.52) lead to a flexure where the coefficient of the ρ^6 term is zeroed as also are the higher-order coefficients. Substituting these expansions into (9.44), the flexure reduces to

$$z = \frac{1}{2}\mu(1-\nu)\frac{qa}{E\varpi^3}\left[-\frac{3+\mu}{2\mu\delta^3}\rho^2 + \frac{1}{4\delta^3}\rho^4\right]$$

and since $\varpi\delta = \mathcal{T}_0$ = constant when the pseudo-power $\varpi = 0$, this may be written, with the rigidity $D_0 = Et_0^3/[12(1-\nu^2)]$ = constant,

$$z = \frac{qa^4}{64D_0}\left(-2\frac{3+\mu}{1+\nu}\rho^2 + \rho^4\right). \tag{9.54}$$

By identification we obtain for the coefficients in (9.45) representing z/aq,

$$A_2 = -\frac{3(1-\nu)(3+\nu)}{8\,E\mathcal{T}_0^3}, \quad A_4 = \frac{3(1-\nu^2)}{16\,E\mathcal{T}_0^3}, \quad A_6 = 0. \tag{9.55}$$

These coefficients are listed in Table 9.1 for the particular cases where $\varpi = 0$. Equation (9.54) represents the well-known bending of a constant thickness plate – or of a meniscus of moderate cambrure – when the uniform load is simply supported at the edge.

9.2.5 *Lenses with Particular Thickness Distributions*

In the sequences listed in Table 9.1, the coefficient A_2 is, of course, always negative for a positive load since it represents the curvature mode which is the first mode of the flexure. Considering next the A_4 and A_6 coefficients, we obtain the following results.

First, one notices that the A_4 coefficient vanishes for particular lenses,

$$A_4 = 0 \quad \text{if} \quad \delta \simeq 3,$$

which are all *positive* lenses such that the thickness is of the form

$$t(r) = \frac{1}{2}(c_1 - c_2)(3a^2 - r^2), \tag{9.56a}$$

hence with an edge thickness of *two-thirds* the central thickness. This result is in complete accordance with the thickness distribution $t \propto (1-\rho^2)^{1/3} = 1-\frac{1}{3}\rho^2+\cdots$ obtained for a variable curvature mirror designed with the same loading configuration (cf. Sect. 2.1.2).

Second, in addition to the case of a null power lens or meniscus where the A_6 coefficient vanishes because $\delta \to \infty$ for which $t(r) = \text{constant}$ [cf. (9.55)], one finds from Table 9.1 that this coefficient also vanishes for particular lenses,

$$A_6 = 0 \quad \text{if} \quad \delta \simeq 2,$$

which are all *positive* lenses such that the thickness is of the form

$$t(r) = (c_1 - c_2)(a^2 - \frac{1}{2}r^2), \tag{9.56b}$$

hence with a central thickness *twice* the edge thickness.

9.2.6 *Conclusions for Active Optics Asphierization*

From the previous results and Table 9.1 for fused silica, the following conclusions for the aspherization of single lenses by active optics methods – stress figuring with spherical tools – are inferred from parameter $\delta = \mathcal{T}_0/\varpi$.

- **Negative power lens**: For a negative power lens, it appears that if, say, $\delta < -4$, the contribution of the ρ^6-term in the expansion of the flexure becomes non-negligible compared to that of the ρ^4-term. For instance, for a fused-silica lens with $\delta = -2$, the A_4 coefficient reaches $\sim - A_2/2$. This also entails that the next order terms also have significant relative values. Hence, unless such significant terms of higher order than *Sphe* 3 are required by a more complex system in order for the lens to compensate such aberrations, a conclusion is as follows:

→ *The correction of Sphe 3 for a virtual stigmatic conjugate can be achieved for an aspherized singlet lens if* $-\infty < \delta \leq -5$, *and there may be an optimum value of* δ *for which Sphe 5 is also corrected.*

- **Positive power lens**: For a positive power lens, the most promising geometries for correcting *Sphe* 3 of the lens also correspond to the cases where A_6 is at least several times smaller than A_4. From Table 9.1, we find that $A_6 = 0$ if $\delta \simeq 2$, i.e. lenses where the edge thickness is twice smaller than the central thickness. From condition $\delta > 1$ stated in (9.43), the flexure of sharp-edge lenses cannot be expanded – since $\delta = 1$ – and thus must be excluded for an active optics aspherization and also for practicable reasons. On the other hand, positive values of δ such as $\delta > 3$ lead to A_4 coefficients with an unacceptable sign. Hence for obtaining A_6 and higher-order coefficients of moderate values relative to usable values of A_4, a conclusion is as follows:

→ *The correction of Sphe 3 for a real stigmatic conjugate can be achieved for an aspherized singlet lens if* $1.666 \leq \delta \leq 2.250$, *and there exists an optimum value of* δ *for which Sphe 5 is also corrected.*

- **Remark**: Lenses with a higher refractive index show a substantially smaller *Sphe* 3 aberration (cf. Fig. 9.2). If the power K is the same and if $n = 1.75$, from (9.18a) the asphericity is 3.18-times smaller than for a fused silica lens ($n_d = 1.458$). Hence, for the case of positive lenses, the somewhat restrictive latter δ-range for fused silica lenses may be relaxed for lenses with a high refractive index. However, except for exotic materials such as fused sapphire, it seems difficult to find a material which shows both high refractive index and high ultimate strength.

9.3 Spectrograph with Single Lens and Corrector Plate

Astronomical spectrographs are usually designed with a parabolic mirror as a collimator. For a collimated beam of slow f-ratio and a design with a reflective grating, the collimator mirror is sometimes substituted by a convexo-plane lens plus an aspherical corrector plate which are air-separated. The plate is located at the position of the remote pupil of the isoplanatic mounting in Fig. 9.6-*Left*. If the lens is assumed thin, the convex spherical surface of the lens of curvature c_1 is located at the distance $d = 1/nc_1$ from the plate. The first Seidel sums are $S_I = S_{II} = S_{III} = 0$, the system is anastigmatic and we know that in this case the location of the input pupil is *free*. The only remaining aberrations are axial chromatism, field curvature, and the chromatic aberration variations. For moderate spectral ranges, i.e. high spectral resolutions, the aberration variations are small and the others can be compensated by a lens flattener and a tilt of the focal plane at the output of the spectrograph.

A *Littrow mounting* – $\alpha = -\beta_0$ [see grating law (4.23)] – of the grating is always preferable to achieve the highest possible dispersion. Hence the dispersed light is retro-reflected through the system which then is directly used as camera optics if no addition of a focal reducer is necessary (Fig. 9.9).

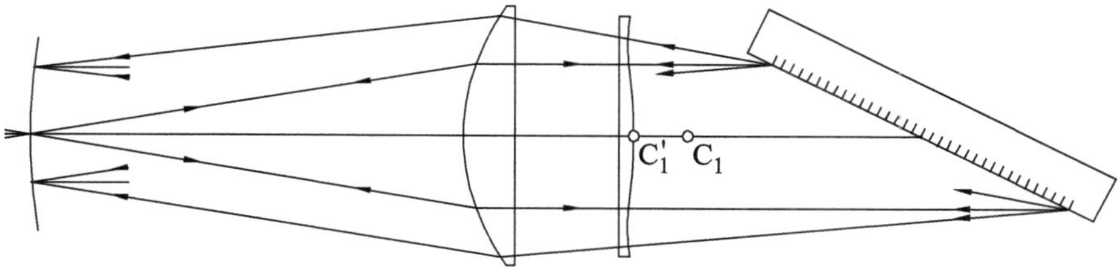

Fig. 9.9 Anastigmatic *Littrow mounting spectrograph* with a double-pass single lens and corrector plate. The mounting is telecentric if the grating is at the distance f'_{L} from the lens

In this design the transverse magnification is $M = -1$ for the central wavelength. The amount of *Sphe* 3 to be corrected by the aspherical plate can be derived from (9.5a). Setting $c_2 = 0$, i.e. $B = 1$ for a convexo-plane lens, and $u'_2 = 0$ i.e. $C = 1$ for an image at infinity, this amount simply reduces to

$$S_I|_{\mathrm{Lens}} = \frac{x^4}{4f'^3_{\mathrm{L}}}\left[\frac{4n^2}{(n-1)^2}\right]. \tag{9.57}$$

From (1.68), *Sphe* 3 of a concave and spherical mirror for an object at infinity is expressed by

$$S_I|_{\mathrm{Mir}} = -\frac{x^4}{4f'^3_{\mathrm{M}}}, \tag{9.58}$$

where its focal length f' is negative.

Let us denote $\Omega_{\mathrm{Lens}} = f'_{\mathrm{L}}/D = f'_{\mathrm{L}}/2x_{\mathrm{max}}$ and $\Omega_{\mathrm{Mir}} = -f'_{\mathrm{M}}/D = -f'_{\mathrm{M}}/2x_{\mathrm{max}}$ the focal ratios of the lens and of the mirror. The above expressions become

$$S_I|_{\mathrm{Lens}} = \frac{D}{64\Omega^3_{\mathrm{Lens}}}\left[\frac{4n^2}{(n-1)^2}\right] \quad \text{and} \quad S_I|_{\mathrm{Mir}} = \frac{D}{64\Omega^3_{\mathrm{Mir}}}, \tag{9.59}$$

If their amount of *Sphe*3 are the same, i.e. $S_I|_{\mathrm{Lens}} = S_I|_{\mathrm{Mir}}$, and also their clear apertures D, we obtain a relation between the focal length of the lens and that of the mirror. Considering their focal ratios, this relation is

$$\Omega_{\mathrm{Lens}} = \left(\frac{2n}{n-1}\right)^{2/3}\Omega_{\mathrm{Mir}}. \tag{9.60}$$

Hence we may compare the f-ratio of a lens-and-plate design with that of a plate-and-mirror Schmidt when they both use the *same* aspherical plate:

→ *If the refractive index of a lens is* $n = 3/2$, *then an aspherical plate correcting a spherical mirror at* f/3.03 *also corrects a convexo-plane lens at* f/$(6^{2/3} \times 3.03) =$ f/10.

For instance, the single lens and corrector plate design was used in cross-dispersion spectrographs with *white pupil* transfer for high spectral resolutions (cf. CORAVEL-OHP and CASHAWEC-CFHT, Baranne et al. [1]).

References

1. A. Baranne, M. Mayor M., J-L. Poncet, CORAVEL – A new tool for radial velocity measurements, Vistas Astron., **23**, 279 (1979)
2. M. Born, E. Wolf, *Principles of Optics*, Cambridge Univ. Press, New York (1999)
3. H. Chrétien, *Calcul des Combinaisons Optiques*, Masson edt., Paris, 5th issue (1980)
4. H. Coddington, *A Treatise on the Reflexion and Refraction of Light*, London (1829)
5. R. Descartes, *La Géometrie Livre II*, and *La Dioptrique*, in *Discours de la Méthode*, Adam & Tannery edt., 389–441 (1637), reissue Vrin edt., Paris (1996)
6. R. Kingslake, *A History of the Photographic Lens*, Academic Press, San Diego 4th issue (1989)
7. M. Laikin, *Lens Design*, Marcel Dekker edt., New York, 2nd issue (1995)
8. G. Lemaitre, *Elasticité et Optique Astronomique*, Doctoral thesis dissertation, Université de Provence, Aix-Marseille I (1974)
9. S.P. Timoshenko, *Theory of Elasticity*, McGraw-Hill edt., New York, 299 (1970)
10. W.T. Welford, *Aberrations of Optical Systems*, Adam Hilger edt., England, 4th edition (2002)
11. C. Zeiss, *Photographic Objectives, Palmos Hand Cameras*, Jena (1902). Facsimile edition of this catalog by N.J. Clifton, The Zeiss Historica Society edt., Toronto (1990)

Chapter 10
X-ray Telescopes and Elasticity Theory of Shells

10.1 X-ray Telescopes

10.1.1 Introduction – The Three Wolter Design Forms

Catoptric systems for X-ray focusing cannot benefit from quasi-normal incidence mirrors, which have extremely low reflectance for this spectral range. They require ray deviation angles limited to a few degrees which avoid absorption by the reflective coating and hence are called *grazing incidence systems*. For *high angular resolution*, the mirror substrates are preferably in vitroceram or glass materials and the reflective coatings are either a single layer Ir, Au, or Pt, or alternate multilayers such as W+Si. Because of atmospheric absorption, X-ray telescopes must be space-based.

X-ray systems can be designed either as quasi-tubular mirrors extremely accurately aligned on a common axis, or as successive segmented sections of them, or as segments that are *not* arranged in a common symmetry plane.

In 1952, for the purpose of X-ray microscope objectives, Hans Wolter [29, 30] described three types of grazing incidence two-mirror systems known as *Wolter two-mirror systems*. All three types are *stigmatic* with a paraboloid primary mirror and a coaxial and confocal conicoid secondary. Type I is with a convergent primary and a convergent hyperboloid secondary. Type II is with a convergent primary and a divergent hyperboloid secondary. Type III is with a divergent primary and a convergent ellipsoid secondary (Fig. 10.1).

Compared to Types II and III, Wolter Type I is the only design where *both* mirrors provide a convergence. Given a maximum graze angle to minimize X-ray absorption, the Wolter Type I is the shortest of these long systems and therefore has been extensively utilized as a *telescope* in X-ray astronomy. We hereafter restrain to the description of this latter form.

10.1.2 Basic Stigmatic Paraboloid-Hyperboloid (PH) Telescopes

In its stigmatic form, the Wolter Type I telescope has paraboloid and hyperboloid mirrors that are both *convergent*, hereafter called the *grazing incidence* PH *telescope*. Its optical design is derived from the reflective properties of a hyperbola.

G.R. Lemaitre, *Astronomical Optics and Elasticity Theory,* Astronomy and Astrophysics Library, DOI 10.1007/978-3-540-68905-8_10,

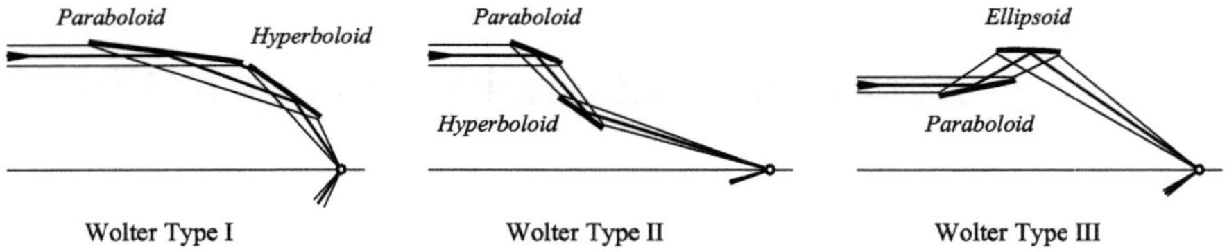

Fig. 10.1 The three stigmatic forms of grazing incidence two-mirror systems by Wolter [29]

Let us represent conic sections having two real foci on a z-axis in a coordinate frame z, r. If the origin is set at the middle of the focii, then these curves are represented by

$$\frac{z^2}{a^2} \pm \frac{r^2}{b^2} - 1 = 0 \tag{10.1}$$

where the positive sign is for the elongated ellipse and the negative for the hyperbola. It is well known that the distances separating their foci are respectively

$$2\sqrt{a^2 - b^2} \quad \text{and} \quad 2\sqrt{a^2 + b^2} \tag{10.2}$$

Setting for both a new origin of the z-axis, now at the vertex V whose concavity of the curve is towards z positive, hence $z \rightarrow z - a$ for the ellipse and $z \rightarrow z + a$ for the hyperbola, their representation becomes

$$\pm \frac{b^2}{a^2} z^2 - 2\frac{b^2}{a} z + r^2 = 0\,, \tag{10.3}$$

where the positive sign is for the elongated ellipse and the negative for the hyperbola.

Reminding that the equation of a conicoid(cf. Sect. 1.7.1) is

$$(1 + \kappa) z^2 - 2R z + r^2 = 0\,, \tag{10.4}$$

from identification, the radius of curvature and conic constant are

$$R = b^2/a, \quad \kappa = (b^2/a^2) - 1 \quad \text{for an ellipsoid } (-1 < \kappa < 0)\,, \tag{10.5a}$$
$$R = b^2/a\,, \quad \kappa = -(b^2/a^2) - 1 \quad \text{for a hyperboloid } \quad (\kappa < -1)\,. \tag{10.5b}$$

From (10.2), the distances separating the two foci are expressed respectively by

$$2\sqrt{-\kappa}\, R/(1+\kappa) \quad \text{and} \quad -2\sqrt{-\kappa} R/(1+\kappa)\,. \tag{10.6}$$

For any conicoid, the distance VF from the vertex to the closest focus and the eccentricity e are respectively

$$\mathrm{VF} = R/(1 + \sqrt{-\kappa}), \quad e = \sqrt{-\kappa}. \tag{10.7}$$

The eccentricity is imaginary for a flattened (or oblate) ellipsoid since $\kappa > 0$.

In a grazing incidence PH telescope of Wolter Type I, the primary and secondary mirrors must be confocal, coaxial and separated by a small gap. Given the maximum value of the ray deviation angle and the aperture diameter, the free parameters are R_1, R_2, and κ_2 since $\kappa_1 = -1$. Denoting F_1 the confocal focus and F_2 the final focus, the above relations and Chasles' relation along the z-axis allow us to obtain the axial separation V_{12} between the mirror vertices. From $V_{12} = V_1F_1 + F_1F_2 + F_2V_2$, we find

$$V_{12} = \frac{R_1}{2} - \frac{R_2}{1 - \sqrt{-\kappa_2}}. \tag{10.8a}$$

This is the *stigmatism* condition for canceling spherical aberration at all orders. If, from convention, the light comes from the left, then R_1 and R_2 must be negative; this entails V_{12} is also negative (Fig. 10.2).

Alternatively, one obtains for the conic constant of the secondary

$$\kappa_2 = -\left(1 + \frac{2R_2}{2V_{12} - R_1}\right)^2. \tag{10.8b}$$

The free parameters R_1, R_2, and V_{12} may be of preferred use in optical design codes. The convergence power of each mirror is not necessarily the same provided the maximum value of the graze angles do not overpass a prescribed value.

Useful quantities defining a telescope design, whose entrance stop is generally located at the input of the primary mirror, are the following

- φ_m maximum semi-angle of the incident field of view,
- r_0 aperture radius at the intersection plane of the mirrors,
- α_1 graze angle at this plane for the primary mirror,
- α_2 graze angle at this plane for the secondary mirror,
- f working focal distance defined from this plane to focus,
- L_1 axial length of the primary mirror,
- L_2 axial length of the secondary mirror.

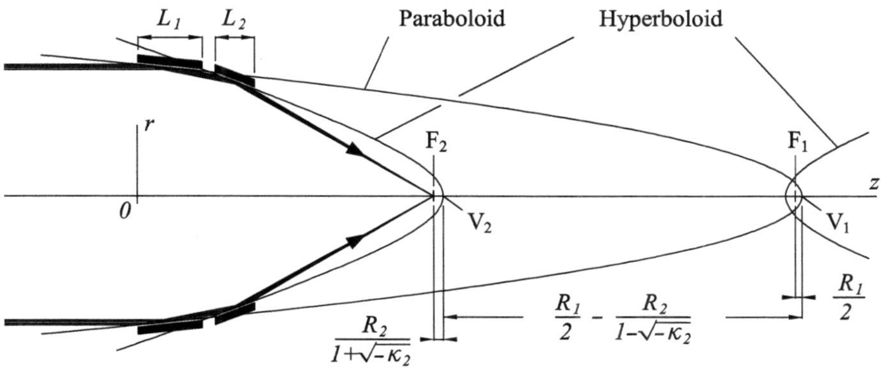

Fig. 10.2 Grazing incidence PH telescope of Wolter Type I

Table 10.1 Optical data of an X-ray grazing incidence PH telescope

	Surf. type	Radius	Thickness	Glass	Semi-dia.	Conic cons.
OBJ	Standard	∞	∞		∞	0.0000000
STO	Standard	∞	2003.049		101.270	0.0000000
2	Standard	−2.56000	−984.000	Mirror	101.270	−1.0000000
3	Alternate	−2.60000	−1.298	Mirror	97.309	−1.0052985
IMA	Alternate	−37.00000	∞	–	4.270	0.0000000

Lengths in [mm]. Associated parameters are $\varphi_m = 0.25° = 0.00436$ rad, $r_0 = 100$, $\alpha_1 = 0.0256$ rad $= 1.467°$, $\alpha_2 = 0.0584$ rad $= 3.346°$, $f = 967$, $L_1 = 50$, $L_2 = 67$.

An example of an optical data set for a telescope that images a field of view corresponding to the angular diameter of the Sun is given in Table 10.1. The best focused blur images from ZEMAX code are shown in Fig. 10.3.

Following the first Wolter study, Mangus and Underwood [26] designed a PH Wolter Type I telescope and evaluated the performance of both laboratory prototype and rocket flight telescope. VanSpeybroeck and Chase [27] elaborated an optical code for the determinations of the field aberrations of these telescopes. Werner [28] attempted to achieve a flat field from raytrace optimizations with spheroid surfaces. Malina, Bowyer et al. [17] designed field optimized instruments by focusing balance. Aschenbach [1] optimized the mirror geometries of the first wide-field X-ray telescope. Nariai [19] analytically investigated the geometrical aberrations and concluded that high-order terms are significant.

In an anastigmatic two-mirror system, the Petzval condition for flat field is $1/R_2 - 1/R_1 = 0$ (cf. Sect. 1.10.1). For a two-mirror system where the astigmatism is small, it is well known that the field curvature of least confusion imaging is significantly reduced when R_1 and R_2 have closed values. For a PH Wolter Type I telescope only R_2 somewhat larger than R_1 can be achieved.

A conclusion is that, whatever the optimization with the free parameters R_1, R_2, and κ_2 in Table 10.1, a grazing incidence PH telescope always shows coma, astigmatism, and low field curvature.

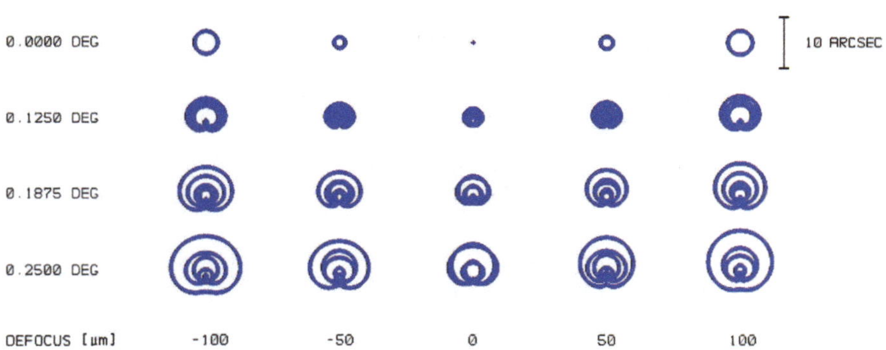

Fig. 10.3 Through focus spot diagram resulting from optical parameters in Table 10.1. The rms angular diameter of the blur image at $\varphi_m = 15$ arcmin semi-field edge is $d_{rms} = 6.3$ arcsec

10.1.3 Sine Condition and Wolter-Schwarzschild (WS) Telescopes

- **Coma aberration**: It is well known that for parallel field rays which lie on a circle $\mathcal{C}$ located in a plane of the input pupil, the gaussian image of this tube-ray is a small circle c. In addition, when each of the rays describes the whole circle $\mathcal{C}$, then circle c is described two times. If the circle $\mathcal{C}$ is the edge of the input pupil, then the diameter of circle c is the tangential size of the coma aberration. Although *not* showing in a meridian direction the typical "V"-pattern of coma because of the annular entrance pupil, the basic grazing incidence PH telescope suffers from coma aberration.

  In a second 1952 paper, Wolter [30] attempted to formulate the equation of the two mirrors for his three system types when strictly satisfying Abbe's sine condition. When free from spherical aberration and coma, the three Wolter grazing incidence systems are known as *Wolter-Schwarzschild (WS) telescopes*. However, only the Wolter Type I form has been of general use in X-ray astronomy. This WS telescope shows a similar geometry to that of a grazing incidence PH telescope (see Fig. 10.2).

- **Implication of sine condition**: For the axial beam, the Abbe sine condition implies that the locus of the intersection points of any ray from infinity with its conjugate passing through the final focus is a sphere centered at the focus, the *Abbe sphere* (cf. Sect. 1.9.2).

  H. Chrétien [5] was the first to derive a *mirror parametric equation set* for solving the general case of the two-mirror telescope family satisfying the sine condition. From these equations, he established the well known and *accurate* conicoid approximations which represent the mirrors in the quasi-normal incidence Cassegrain and Schwarzschild forms. Although Chrétien also derived the fifth-order polynomial coefficients for the two latter forms, it appears that, in practice, these more accurate approximations have negligible effects either in the RC form or aplanatic Gregory form.

  The design of a RC telescope strictly satisfying the sine condition is also that of a WS telescope. In the grazing incidence region, this design corresponds to the Wolter Type II. Whereas, even for all large existing RC telescopes, it is always possible to accurately approximate the shape of the mirrors by *conicoids*, this approximation does *not* apply to a WS telescope because very-high-order terms for linear coma (cf. eq. (1.66) in Sect. 1.9.2) *cannot* be accurately cancelled with conicoid surfaces. In a WS telescope and whatever the Wolter type, the importance of these high-order terms may be glimpsed by considering the grazing incidence regions of the mirrors of an extremely fast Cassegrain telescope satisfying the sine condition, such as determined by Lynden-Bell [16] (Fig. 10.4).

  Since *low*-order polynomial approximations are *inaccurate* in representing grazing incidence mirrors of WS telescopes, each of Wolter Type-I and -III systems must be determined by a similar approach to that of *Chrétien mirror parametric representations*.[1]

[1] Following Chrétien [5], a similar parametric representation of a Cassegrain two-mirror telescope satisfying the sine condition was reformulated by Korsch [12] for near-normal (RC telescope) to

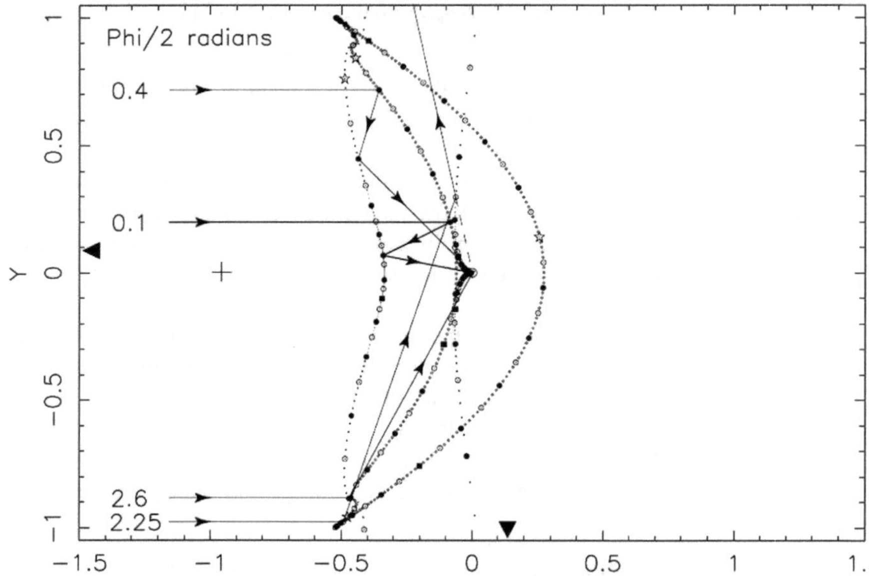

Fig. 10.4 Mirror parametric representations of a Cassegrain telescope strictly satisfying the sine condition up to a very fast f-ratio [16]. Although Wolter Type I uses the *reverse* side of the second optical surface, somewhat similar "S"-shapes exist on the mirrors of grazing incidence WS telescopes

- **Chase and VanSpeybroeck design of WS telescopes**: The first design of a *WS telescope* in the Wolter Type I form – which is the only form of interest for X-ray astronomy – were derived in 1972 by Chase and VanSpeybroeck [4] by use of mirror parametric representations in a study that has become a classic reference. The case of grazing incidence microscopes satisfying the sine condition was subsequently investigated by Chase [3]. A generalized parametric representation for WS telescopes was obtained by Saha [21] and a dedicated raytrace code for this purpose was elaborated by Thompson and Harvey [23].

  Although not entering into detail, Chase and VanSpeybroeck also computed the exact figures of the two mirrors of a WS telescope showing that both mirrors may have *two inflexion* zones. In a systematic study of many WS wide field telescopes, these authors showed that the effect of the field curvature is small and that the astigmatism is dominant.

  Denoting $d_{\rm rms}$ the rms angular diameter of the least confusion astigmatism blur at a semi-field angle φ_m, Chase and VanSpeybroeck derived a relation expressing – here in a somewhat different form – the angular resolution of a WS telescope,

grazing incidence (Wolter Type II). Unfortunately, because of the requirement for much larger ray deviation angles than with Wolter Type I, this system was never adopted for X-ray astronomy. Korsch also investigated the case of grazing incidence three-mirror systems [13].

$$d_{\text{rms}} \simeq 1.08 \left(1 + \frac{\alpha_1}{\alpha_2}\right) \frac{L_1}{r_0}\, \varphi_m^2, \tag{10.9}$$

where the angles are in radians and the quantities α_1, α_2, r_0, L_1, φ_m are defined in Sect. 10.1.2 with grazing angles α_1 and α_2 of the same sign. The validity of this relation is in the region defined by

$$\frac{1}{4} \le \frac{\alpha_1}{\alpha_2} \le 4, \quad \frac{1}{5} \le \frac{L_1}{r_0} \le 6, \quad 0 \le \varphi_m \le 30\,\text{arcmin}. \tag{10.10}$$

In fact the original Chase-VanSpeybroeck relation for the resolution and associated inequalities included three geometrical ratios, but these can be readily reduced to the following two ratios: the grazing angle ratio α_1/α_2, and the primary mirror *aspect ratio* L_1/r_0, so the above relations are totally valid for the *small* grazing angles that are required in X-ray telescopes.

Applying these results to a WS telescope having the same optical parameters as those of the PH telescope in Table 10.1, we find that $\alpha_1/\alpha_2 = 0.438$ and $L_1/r_0 = 1/2$ satisfy the above inequalities. If we also consider the same maximum semi-field angle $\varphi_m = 15\,\text{arcmin}$, the rms diameter of the image at the field-edge, from (10.9), is now $d_{\text{rms}} = 14.8 \times 10^{-6}\,\text{rad} \simeq 3.1$ arcsec. Comparing with the spot-diagram in Fig. 10.3, where $d_{\text{rms}} = 6.3$ arsec for the PH telescope, the gain in rms diameter of field aberrations is here a factor of two in favor of the WS telescope.

To summarize, in a grazing incidence WS telescope the mirror parametric equations allow one to cancel both spherical aberration at all orders and linear coma terms, and to reduce the field curvature to a low amount. However, an important amount of astigmatism remains which, from (10.9), is quadratically dependent on the field angle. Generally the graze angles are set such that $\alpha_2 > \alpha_1$; therefore this aberration can be only decreased if the L_1/r_0 ratio is decreased, i.e. for a short axial length and a large aperture radius of the primary mirror. The primary mirror *aspect ratio* L_1/r_0 appears as the fundamental parameter that governs the dominating residual blur of all grazing incidence Wolter-Schwarzschild telescopes.

→ *In a Wolter-Schwarzschild telescope where the primary and secondary mirrors are joined at the aperture radius* r_0, *if* L_1 *is the axial length of the primary, then the amount of astigmatism is linear with the primary mirror aspect ratio* L_1/r_0.

10.1.4 Aberration Balanced Hyperboloid-Hyperboloid (HH) Telescopes

The complex mirror shapes resulting from the use of parametric equations in grazing incidence WS telescopes, which moreover are not astigmatism-free and also present considerable difficulties in their execution, led to another alternative concept developed for the optical design of Wolter Type I telescopes.

This concept, proposed and developed by Harvey et al. [10, 11], consists of the use of the optimization capability of conventional raytrace codes with mirrors that are pure conicoid surfaces, so the sine condition is no longer applied. Starting from the stigmatic design of a PH telescope, the optimization process is now carried out with the four parameters R_1, κ_1, R_2, κ_2 and also slight variations – or despaces – of the axial separations V_1F_1 and F_2V_2 (cf. Fig 10.1). In addition the axial lengths L_1, L_2 of the mirrors and their gap must be specified. Given an intermediate field angle φ_i, such that $0 < \varphi_i < \varphi_m$, for which the aberrations must be minimized, raytrace optimizations provide aberration balanced fields where both mirrors are hyperboloids. Hence these systems are called *grazing incidence HH telescopes.*

Repeating the iteration process for various values of φ_i, Harvey et al. plotted the rms diameter of the total aberration vs field of view and obtained a grid showing the aberration evolution (Fig. 10.5). They noticed that the locus of the minima φ_i is a straight line and interpreted these raytrace results – shown on the figure as a shaded region – as the uncorrectable linear coma for those field angles. This conclusion is consistent with the results in Sect. 10.3.1 where the sine condition necessarily entails use of mirrors represented by parametric equations which *cannot* be accurately approximated in the conicoid class.

Finding the optimal ratio φ_i/φ_m in this grid corresponds to finding the balanced field for which $d_{\text{rsm}}(0) = d_{\text{rms}}(\varphi_m)$. For a maximum semi-field angle $\varphi_m \leq 20$ arcmin, this ratio is not a constant but a function of φ_m; its value is included in the range [0.75, 0.90]. For a very small semi-field of view, say $\varphi_m = 1$ arcmin, this ratio is close to 0.9.

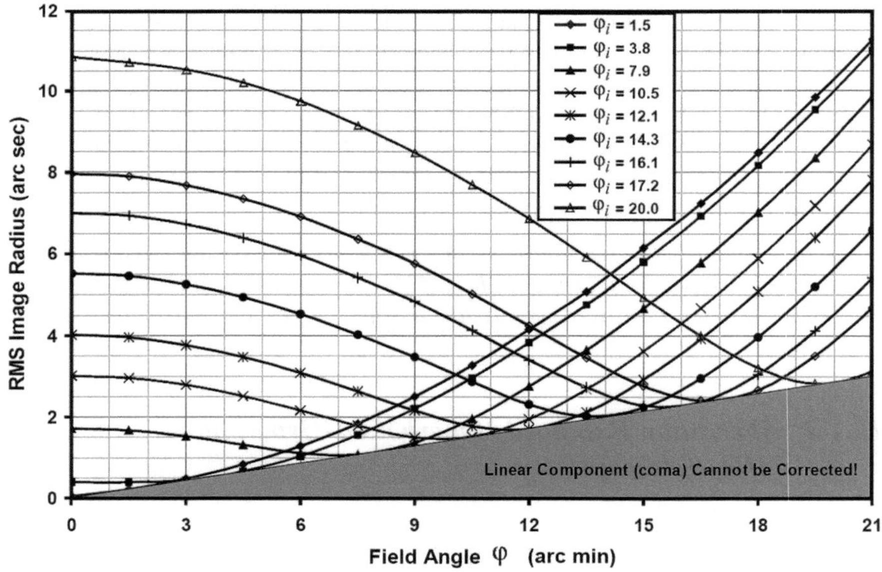

Fig. 10.5 Angular blur-image rms radius vs semi-field angle φ for a grazing incidence HH telescope. Each curve is derived by Harvey et al. from a particular value φ_i of the intermediate design angle for which the aberrations are minimized [11]

In the optical design of grazing incidence HH telescopes with a solar-like field angle and plane detector, the spherical aberration, astigmatism and next-order aberrations – such as cubic coma and linear triangle terms – are substantially reduced whereas the linear coma remains and astigmatism still dominates.

10.1.5 Aberration Balanced Spheroid-Spheroid (SS) Telescopes

As initiated by Werner [28] for wide field telescopes, it is possible to consider mirrors represented by polynomial expansions that roughly approximate the sine condition and reduce astigmatism. As for all previous designs, the field curvature of least confusion imaging is obtained when R_2 is little larger than R_1 (cf. Sect. 10.1.2).

Each mirror equation that generates the polynomial expansion is obtained from a generalization of (10.4). Setting a new origin $z = 0$ at the intersection plane of the two mirrors, if the radius of this circular intersection is r_0, then the mirror surfaces z_1 and z_2 may be represented by

$$a_i\, z_i^2 + 2\, b_i\, R_i\, z_i - r_0^2 + r^2 = 0, \tag{10.11}$$

where R_i is the radius of curvature of the primary or secondary, with the respective suffix $i = 1$ or 2, and $a_i,\ b_i$ four free dimensionless coefficients. The above *implicit* equations lead to expansions $z_i(r)$ which take into account *high order* terms. The shape (10.11) of these mirrors is included in the general class of *spheroids*, hence such systems may be called *grazing incidence SS telescopes*.

This analytic representation, however implicitly expressed from the above trinomial form of $r^2(z_i)$, was proposed and investigated by Burrows et al. [2] with a dedicated raytrace code called OSAC [20]. These latter authors and Concini and Campana [7] introduced a specialized merit function for the image tolerancing.

Subsequently, Saha and Zhang [22] investigated an SS telescope design which was called "equal curvature." This led the authors to define the mirror shapes as a co-addition of simple quadratic sags to the generatrices of a cone-cone telescope. From OSAC's *Legendre polynomials*, a z-shifted quadratic correction can be added for each surface. However, it is unclear whether each mirror is a region (or not) of a torus, such as they state. A torus does belong to the spheroid class.

For a moderate field of view and plane detector, the imaging performance with a grazing incidence SS telescope is somewhat superior to that of the HH telescope design. On the other hand, this latter design seems more successful for a solar-like field. The SS design shows that the residuals for spherical aberration, coma, and dominating astigmatism remain.

10.1.6 Existing and Future Grazing Incidence X-ray Telescopes

Nearly all realized telescopes for X-ray astronomy are in the Wolter Type I form. Because of the short length of the mirrors generally adopted – primary mirror aspect ratio $L_1/r_0 \lesssim 1/2$ – a particular feature is that one or several additional smaller aperture telescopes can be *nested* into the first, thus allowing, in a single volume, simultaneous observation in imaging and spectroscopic modes.

- **PH telescopes**: The early imaging systems were designed with grazing incidence paraboloid and hyperboloid. Such was the case, for instance, for the Einstein-HEAO 2 telescope, ROSAT telescope, and Chandra-AXAF telescope (cf. Table 1.1 in Chap. 1).
- **WS telescopes**: The mirrors defined by the parametric equations for WS telescopes have been found to be extremely difficult to fabricate and test. Each of their mirrors may have two inflexion zones (cf. Sect. 10.1.3 and Fig. 10.13). Compared to the HH or SS designs, their superiority is for a very high angular resolution imaging of small fields. Up to now, astronomical programs for such researches have not been investigated.
- **SS telescopes**: The grazing incidence SS telescope design is attractive for moderate fields of view, say $2\varphi_m \leq 20$ arcmin. The large and high-resolution imaging Chandra-AXAF telescope (see Table 1.1) was originally designed in both PH and SS forms[2] with two nested coaxial mirror pairs (Fig. 10.6). Examples of designs underway are the NASA projects WFXT and XRT.
- **HH telescopes**: The HH design is attractive for a wide-field, say $2\varphi_m \geq 30$ arcmin. It implies mirrors that have simple shapes included in the conicoid class and which thus can be readily optimized with high order corrections by the usual raytrace codes. For instance, this design has been adopted for the Solar X-ray Imager (SXI).
- **Large segmented telescopes**: Similarly as for large ground-based telescopes in the visible and infrared, future large space-based X-ray telescopes will require use of segmented and active mirrors. Examples of such projects under investigation are the Constellation X-ray Mission [9] and XEUS [6].

Fig. 10.6 (*Left*) Chandra X-ray Observatory, launched in 1999, includes a 1 arcsec resolution imaging telescope and a spectrometer using two transmission gratings (credit NASA). (*Right*) X-ray image of the Crab Nebula obtained with Chandra; the dot at the center is the neutron star – or pulsar – spinning 30 times per second (credit NASA/CXC/SAO/J. Hester et al.)

[2] Stephen O'Dell and Martin Weisskopf pointed out that, at the urging of Riccardo Giaconni, the AXAF design was modified from PH to SS polynomials. However, in the end, the PH prescription was selected. Thus, Chandra-AXAF should be included with Einstein and ROAST in the PH telescope family.

10.2 Elasticity Theory of Axisymmetric Cylindrical Shells

10.2.1 X-ray Mirrors and Super-Smoothness Criterion

Whatever the design option of an axisymmetric two-mirror telescope, the fabrication of the mirrors is difficult. The *super-smoothness* of their surface is without doubt one of the most important features because slope errors due to ripples entail absorption and scattering effects which may severely degrade the performances. The X–ray energy domain is [10–0.1 keV] which corresponds in round number to the wavelength range λλ[0.2–20 Å]. Hence, due to the extreme difficulty of surface testing by X-ray in a long tunnel lab, it is presently admitted that the *surface roughness* of the mirrors must *not* exceed a value of 2–3 Å.

For this reason, active optics figuring with rigid lap segments – whose curvature is constant along the mirror axial direction – would greatly improve the performance of tubular mirrors since then avoiding ripple polishing errors. Active optics asphérization can be applied either by mirror stressing or by mandrel stressing which generates a replica mirror.

The stress figuring of long rectangular mirrors – as single focusing element or Kirkpatrick-Baez cross systems – requires a perimeter distribution of the bending moments. For such X-ray mirrors, an optimal surface geometry is further achieved by *in-situ* actuators. This two-stage process is of current use in synchrotron laboratory applications. For instance, some active optics processes applied to long rectangular mirrors were presented by Underwood et al. [26] and Fermé [8].

The stress figuring of grazing incidence mirrors requires use of the elasticity theory of shells. Starting hereafter from investigations with the theory of axisymmetric cylindrical shells, we shall then present an appropriate theory of axisymmetric *weakly conical shells*.

10.2.2 Elasticity Theory of Thin Axisymmetric Cylinders

Because of the very small amount of maximum optical sag of X-ray mirrors, we assume hereafter that the elastic relaxation process provides the complete optical sag distribution with respect to the geometry of a cylinder or a cone, both having *straight* generatrices during the stress figuring.

The elasticity theory of thin axisymmetric cylinders involves radial *extension* and *retraction* of the mid-thickness surface. Considering the static equilibrium of a cylinder element (Fig. 10.7), we hereafter follow the analysis by Timoshenko [25].[3]

[3] Stephen Timoshenko obtained important results in this field. Noticeably, he derived the critical load for the buckling of thin shells [24], results which were applied to improve the strength of large ships. For instance in the case of a cylindrical shell, he established that the critical axial force is $F_{\rm cr} = 2\pi E t^2/\sqrt{3(1-\nu^2)}$. Hence, unlike the length-dependent column buckling (cf. Sect. 1.13.1), the axial shell buckling is independent of the shell length.

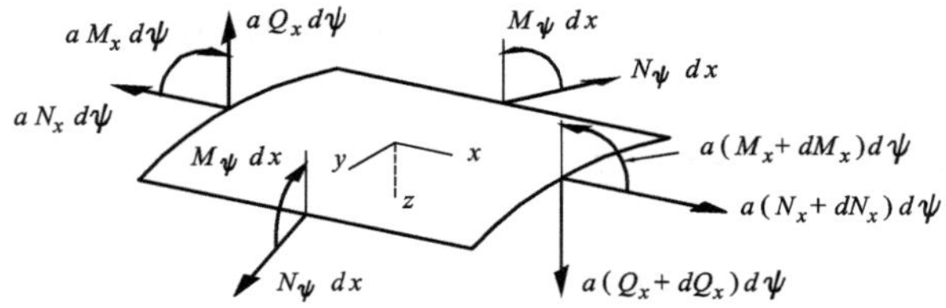

Fig. 10.7 Static equilibrium at the mid-thickness surface of a cylinder element of lengths dx and $a\,d\psi$. The uniform load q – not shown – is normal to the surface

Let us denote a the radius of the *middle* surface of the cylinder and q the intensity of an external uniform load, per unit area, distributed all over the surface. In a small element of this surface, set an x-axis parallel to the cylinder axis and a z-axis passing through this latter axis and positive towards it. The force components involved at this element are the uniform load q, the bending moments M_x and M_ψ, the forces N_x and N_ψ, the shearing force Q_x, and the x-variation of M_x, N_x, and Q_x. From axial symmetry, the force N_ψ and the bending moment M_ψ are constant along the circumference. Writing the equations of equilibrium at the center of the element, the forces in x- and z-directions and the moment about the y-axis, are respectively, after division by the element area $a\,dx\,d\psi$,

$$\frac{dN_x}{dx} = 0, \tag{10.12}$$

$$\frac{dQ_x}{dx} + \frac{1}{a}N_\psi + q = 0, \tag{10.13}$$

$$\frac{dM_x}{dx} - Q_x = 0. \tag{10.14}$$

The first equation implies that the force N_x is a constant. We will assume that no load is applied to the cylinder ends in the x-direction, and thus $N_x = 0$.

Similarly as in Sect. 1.13.3, let us denote $u \equiv u_x$, $v \equiv u_y$ and $w \equiv u_z$ the components of the displacement vector. Since both the shell and load are axisymmetric, the flexure also is, thus $v = 0$. The normal strain components are

$$\varepsilon_{xx} = \frac{du}{dx}, \quad \varepsilon_{yy} = 0, \quad \varepsilon_{\psi\psi} = -\frac{w}{a}, \tag{10.15}$$

and all three shear strains are $\varepsilon_{y\psi} = \varepsilon_{\psi x} = \varepsilon_{xy} = 0$. From Hooke's law, the formulation of the axial and tangential forces leads to

$$N_x = \frac{E\,t}{1-\nu^2}\left(\varepsilon_{xx} + \nu\varepsilon_{\psi\psi}\right) = \frac{E\,t}{1-\nu^2}\left(\frac{du}{dx} - \nu\frac{w}{a}\right) = 0, \tag{10.16}$$

$$N_\psi = \frac{E\,t}{1-\nu^2}\left(\varepsilon_{\psi\psi} + \nu\varepsilon_{xx}\right) = \frac{E\,t}{1-\nu^2}\left(-\frac{w}{a} + \nu\frac{du}{dx}\right), \tag{10.17}$$

where E is the Young modulus, ν Poisson's ratio, and $t(x)$ the thickness of the cylinder. From the first equation, we obtain$du/dx = \nu w/a$, which entails

$$N_\psi = -\frac{E}{a}\, t\, w. \tag{10.18}$$

Still following Timoshenko and using hereafter his convention of negative sign (which does not hold for all previous chapters) for expressing the two bending moments with respect to w and its derivatives (see footnote in Sect. 3.2), we write

$$M_x = -\, D\frac{d^2 w}{dx^2}, \tag{10.19}$$

$$M_\psi = \nu\, M_x, \tag{10.20}$$

where $D(x) = Et^3(x)/12(1-\nu^2)$ is the flexural rigidity. Returning to (10.14) and eliminating dQ_x/dx by using (10.13), we find

$$\frac{d^2 M_x}{dx^2} + \frac{1}{a} N_\psi = -q. \tag{10.21}$$

After substitution of N_ψ and M_x, the general differential equation of the flexure of a cylinder is

$$\frac{d^2}{dx^2}\left(D\frac{d^2 w}{dx^2}\right) + \frac{E}{a^2}\, t\, w = q. \tag{10.22}$$

This equation was first obtained by S. Timoshenko – probably around 1930 – and was included in the 1940 issue of *Theory of Plates and Shells* [25]. [4]

Introducing the dimensionless variable

$$\chi = \left[12(1-\nu^2)\right]^{1/4} \frac{x}{a} \tag{10.23}$$

into (10.22) and multiplying it by a/E, we obtain

$$\frac{d^2}{d\chi^2}\left(\frac{t^3}{a^3}\frac{d^2 w}{d\chi^2}\right) + \frac{t}{a} w = \frac{q}{E} a. \tag{10.24}$$

For clarity it is also useful to define dimensionless thickness $\mathcal{T}$ and flexure $\mathcal{W}$ as the following

$$\mathcal{T} = \frac{t}{a}, \quad \mathcal{W} = C\frac{E w}{q\, a}, \tag{10.25}$$

where C is an unknown constant. After substitutions, we obtain a general *normalized equation of variable thickness cylinders*

[4] For a cylinder with constant thickness, the general equation becomes

$$\frac{t^3}{12(1-\nu^2)}\frac{d^4 w}{dx^4} + \frac{t}{a^2} w = \frac{q}{E}.$$

Although in a somewhat different form, this equation was obtained by Augustus Love [15] around 1915. In the preface of the fourth issue of *Mathematical Theory of Elasticity*, the author refers to the section where it was published (Sect. 339) as one of the sections with letter-A that were added in the third issue.

$$\frac{d^2}{d\chi^2}\left(\mathcal{T}^3\frac{d^2\mathcal{W}}{d\chi^2}\right)+\mathcal{T}\mathcal{W}=C\equiv constant. \tag{10.26}$$

If the flexure $\mathcal{W}(\chi)$ to generate is in a polynomial form, at least quadratic in χ, the integration allows deriving the radial thickness distribution $\mathcal{T}(\chi)$ from a central thickness $\mathcal{T}(0)$.

If the flexure is a constant or a linear form in χ, then the first left-hand term vanishes and the thickness distribution is either the reciprocal constant or the reciprocal linear form, respectively.

10.2.3 Radial Thickness Distributions and Parabolic Flexure

We consider hereafter various cases of obtaining a parabolic flexure. These are generated by a uniform load applied all over the surface of the cylinder.

- **Uniform load in reaction with simply supported ends**: A representation of a purely parabolic flexure can be simply expressed by a quadratic function of the axial ordinate as

$$\mathcal{W}\equiv C_2\frac{Ew}{q\,a}=\beta^2-\chi^2,\quad \chi^2\le\beta^2, \tag{10.27}$$

where β is the length parameter such that $\chi=\pm\beta$ at the cylinder edges (or ends), and $C_2\equiv C$ is the constant in (10.26) for the parabolic case. The radial reacting forces at the edges ensure that the radial displacement is zero at $\chi=\pm\beta$. The intensity of these two equal forces $F_\beta=F_{-\beta}$, per length unit, is determined from the statics by

$$F_\beta+F_{-\beta}+qL=0, \tag{10.28}$$

where $L=2x_{\max}$ is the length of the cylinder.

With the initial variables, this flexure is represented by

$$w=\frac{1}{C_2}\frac{q}{E}\left[\beta^2-\sqrt{12(1-\nu^2)}\frac{x^2}{a^2}\right]a. \tag{10.29}$$

Let us define a mirror length to radius ratio L/a similarly as quantities L_1/r_0 or L_2/r_0 in Table 10.1. From the above equation where $x_{\max}=L/2$, this *mirror aspect ratio* is

$$\frac{L}{a}=\frac{2\beta}{[12(1-\nu^2)]^{1/4}}. \tag{10.30}$$

After substitution of $\mathcal{W}$ into (10.26), we obtain the second-order equation

$$2\frac{d^2}{d\chi^2}\left(\mathcal{T}^3\right)-(\beta^2-\chi^2)\mathcal{T}=-C_2. \tag{10.31}$$

For the integration, we introduce the cubic power of the reduced thickness,

$$u = \mathcal{T}^3, \tag{10.32}$$

so the differential equation writes

$$2\frac{d^2u}{d\chi^2} - (\beta^2 - \chi^2)u^{1/3} = -C_2. \tag{10.33}$$

Denoting t_0 the thickness at the origin, the associated quantities are $\mathcal{T}(0) = t_0/a = u_0^{1/3}$. The quantity u_0 provides the starting condition in the integration from the origin $\chi = 0$. Assuming an infinitesimally small increment $\Delta\chi$, it is well known that

$$\frac{d^2u}{d\chi^2} = \frac{u_{n-1} - 2u_n + u_{n+1}}{\Delta\chi^2},$$

hence the differential equation becomes

$$u_{n+1} = 2u_n - u_{n-1} + \left[(\beta^2 - \chi^2)\, u_n^{1/3} - C_2\right]\frac{\Delta\chi^2}{2} \tag{10.34}$$

where C_2 is the unknown. Starting the integration from a given value of the reduced thickness $\mathcal{T}(0) = t_0/a = u_0^{1/3}$, the first increment u_1 is obtained by noticing that $u(\chi)$ should be continuous and symmetric with respect the central section plane; this entails that $u_{-1} = u_1$. The iterations are carried out by varying C_2 up to obtaining $u_N(\beta) = 0$.

Results from iterations provide the reduced thickness $\mathcal{T}(\chi)$ and the constant C_2 as functions of the reduced thickness at the origin t_0/a and parameter β that defines the mirror aspect ratio L/a (Table 10.2 and Fig. 10.8).

Whatever $\nu \in [0, 1/2]$ and $\mathcal{T}(0)$, there exists a solution β for which $d^2\mathcal{T}/d\chi^2 = 0$ at the origin, i.e. where the cylinder is of constant section in the central region. The application of the radial forces at the simply supported edges requires use of axially thin collars which thus avoids any bending moment. However, even this simple boundary condition is a practical difficulty for a mirror substrate in glass or vitroceram.

- **Uniform load and free ends – Inverse proportional law**: Compared to the case of plane plates, an important particularity of cylinders is that edge reacting forces are *not* necessarily required to generate a flexure by external loads.

Table 10.2 Some values of the integration constant C_2 as a function of the dimensionless thickness at the origin t_0/a and parameter β defining the mirror aspect ratio L/a. Poisson's ratio $\nu = 1/4$, $L/a \in [0.49; 1.47]$

$\beta =$	0.45	0.70	0.90	1.10	1.35
$t_0/a = 0.08$	0.023 126	0.036 939	0.059 360	0.090 721	0.141 072
$t_0/a = 0.10$	0.035 957	0.048 506	0.074 313	0.111 662	0.173 394
$t_0/a = 0.12$	0.053 537	0.062 068	0.090 406	0.133 006	0.205 040

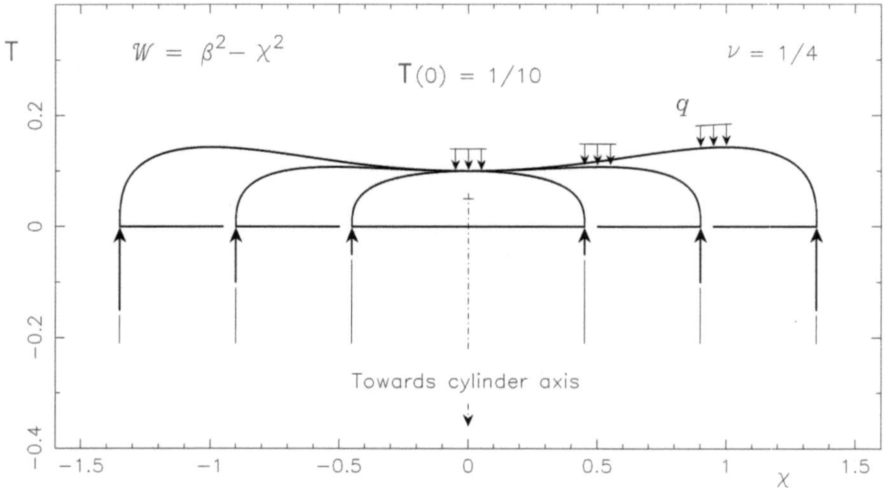

Fig. 10.8 Thickness distributions $\mathcal{T}(\chi)$ of three cylinders generating a χ^2 flexure mode. Radial reactions at edges in equilibrium with uniform load, $F_\beta = F_{-\beta} = qL/2$. Poisson's ratio $\nu = 1/4$. Middle thickness ratio $t(0)/a = 1/10$. Length parameters $\beta = 0.45,\ 0.90,\ 1.35$. Mirror aspect ratios $L/a = 1.092\,\beta$

An axisymmetric *tubular shell* with *free edges* is by itself in a static equilibrium when a uniform load q is applied all over its surface. Returning to (10.19) and using (10.14), we may introduce the expression of the shearing force Q_x. Thus we obtain

$$\frac{dM_x}{dx} = -\frac{d}{dx}\left(D\frac{d^2w}{dx^2}\right) = Q_x, \tag{10.35}$$

and from (10.22), the general differential equation of the flexure of a cylinder may be expressed in a function of the shearing force,

$$-\frac{dQ_x}{dx} + \frac{E}{a^2}\,t\,w = q. \tag{10.36a}$$

Using the dimensionless quantities $\mathcal{T}$ and $\mathcal{W}$ for the thickness and the flexure, equations (10.25) lead to

$$-\frac{dQ_x}{dx} + \frac{q}{C}\mathcal{T}\mathcal{W} = q. \tag{10.36b}$$

If q is the only external force applied to the cylinder, then the shearing force Q_x is null,

$$Q_x = 0, \quad \text{and} \quad M_x = M_\psi = 0, \tag{10.37}$$

so the first left-hand term in the differential equation vanishes. This result is a straightforward *inverse proportional law*,

$$\mathcal{T}\mathcal{W} = C \equiv constant. \tag{10.38}$$

→ *If an axisymmetric cylindrical shell is uniformly loaded all over one of its surfaces, and if no other external forces are applied, then the thickness $\mathcal{T}$ and flexure $\mathcal{W}$ are reciprocal functions.*

This law is of fundamental importance for the aspherization of X-ray mirrors because the boundary conditions at both edges vanish. Therefore only sliding contacts must be set at the end faces to prevent from pressure leak.

1 → *Single-term parabolic flexure*: Let us consider a parabolic flexure where the central section plane of the cylinder cannot extend or retract. This case corresponds to a dimensionless flexure represented by a single quadratic term,

$$\mathcal{W} \equiv C_2 \frac{E\,w}{q\,a} = \chi^2, \quad \chi = x/a. \tag{10.39}$$

From (10.38), the dimensionless thickness is simply

$$\mathcal{T} = \frac{C_2}{\chi^2}, \quad \chi^2 \leq \beta^2. \tag{10.40}$$

Hence the *thin* shell theory provides a valuable solution that is with *infinite* thickness at the plane of central section (Fig. 10.9).

2 → *Two-term parabolic flexure*: In the latter case of a single-term flexure, although the infinite thickness at the symmetry plane may be rendered finite by an optical tolerancing criterion, this difficulty can be easily circumvented by considering that a radial strain may appear all along the cylinder.

Hence if a parabolic flexure is generated from the origin to the edge $\chi = \pm\beta$, then in order to avoid singular poles at the cylinder edges – which, from the inverse proportional law, entails local infinite thicknesses –, it is necessary to

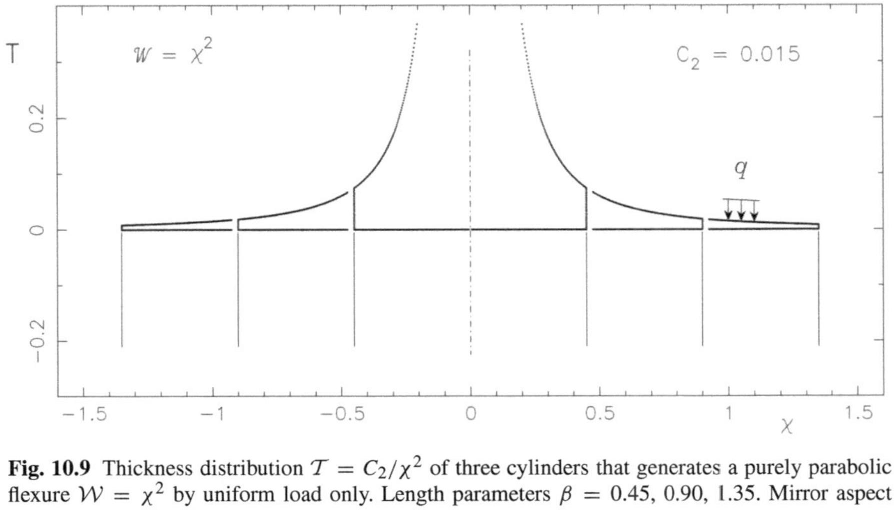

Fig. 10.9 Thickness distribution $\mathcal{T} = C_2/\chi^2$ of three cylinders that generates a purely parabolic flexure $\mathcal{W} = \chi^2$ by uniform load only. Length parameters $\beta = 0.45, 0.90, 1.35$. Mirror aspect ratios $L/a = 1.092\,\beta$

shift these poles outside the edge by introducing an additional constant α^2 in the flexure function. Assuming a quadratic flexure including only a constant and a χ^2 term, this parabolic flexure may be represented by

$$\mathcal{W} \equiv C_2 \frac{E\,w}{q\,a} = \alpha^2 + \beta^2 - \chi^2, \quad \chi = x/a, \tag{10.41}$$

where the mirror aspect ratio is $L/a = 2\beta$ and the total sag of the mirror is $\mathcal{W}(0) - \mathcal{W}(\beta) = \beta^2$. From (10.38), the dimensionless thickness is simply the *reciprocal function*

$$\mathcal{T} = \frac{C_2}{\alpha^2 + \beta^2 - \chi^2}, \quad \chi^2 \leq \beta^2. \tag{10.42}$$

The part $f(\chi) = \beta^2 - \chi^2$ shall be considered as the requested flexure function representing the optical shape to generate, whilst the additional constant α^2 – which represents a *constant* retraction (or extension) along the χ-axis – is a necessary elasticity condition to avoid a large or infinite thickness at any region of the cylinder. An appropriate setting of the free parameters α^2 and C_2 provides a mirror thickness geometry that can be readily fabricated (Fig. 10.10).

- **Parabolic and cone flexure**: A flexure of complete quadratic form can be easily generated. Conserving the additional constant α^2 for rejecting singular poles of the thickness outside the cylinder, we obtain the flexure

$$\mathcal{W} \equiv C_2 \frac{E\,w}{q\,a} = \alpha^2 + \beta^2 + \gamma\,\chi - \chi^2, \quad \chi = x/a, \tag{10.43}$$

where γ is a constant determining the "cone term" of the flexure. From (10.38), the dimensionless thickness distribution is

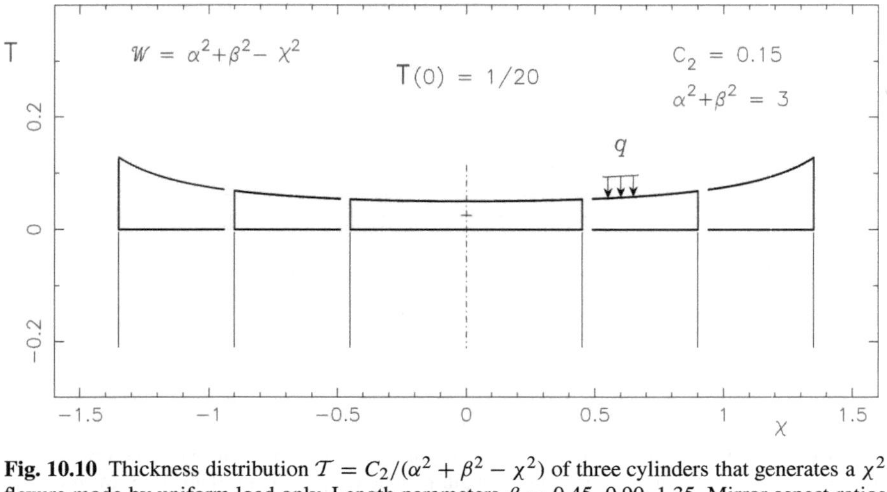

Fig. 10.10 Thickness distribution $\mathcal{T} = C_2/(\alpha^2 + \beta^2 - \chi^2)$ of three cylinders that generates a χ^2 flexure mode by uniform load only. Length parameters $\beta = 0.45, 0.90, 1.35$. Mirror aspect ratios $L/a = 1.092\,\beta$

$$\mathcal{T} = \frac{C_2}{\alpha^2 + \beta^2 + \gamma\chi - \chi^2}, \qquad \chi^2 \le \beta^2. \tag{10.44}$$

One may notice that, in practice, the available cone deformation mode from elasticity remains necessarily small. Telescope grazing incidence mirrors are, at least, always optically surfaced from a passive cone geometry.

10.2.4 Radial Thickness Distributions and 4th-Degree Flexure

We consider various cases of obtaining a fourth-degree flexure where a uniform load is applied all over the tubular surface.

- **Uniform load in reaction with simply supported ends**: Similarly as in the latter Section, we denote the reduced axial ordinate $\chi = [12(1-\nu^2)]^{1/4}x/a$ and the dimensionless flexure $\mathcal{W}$. Let us consider a flexure now expressed by

$$\mathcal{W} \equiv C_4 \frac{Ew}{q\,a} = \beta^4 - \chi^4, \qquad \chi^2 \le \beta^2, \tag{10.45}$$

where β is the reduced length parameter such that $\chi = \pm\beta$ at the cylinder edges and, from (10.26), $C_4 \equiv C$ is the unknown constant. The radial reacting forces F_β and $F_{-\beta}$ at the edges ensure that the radial displacement is zero at $\chi = \pm\beta$. After substitution into (10.26), the dimensionless thickness $\mathcal{T} = t/a$ is a solution of

$$12 \frac{d^2}{d\chi^2}\left(\chi^2 \mathcal{T}^3\right) - (\beta^4 - \chi^4)\mathcal{T} = -C_4, \tag{10.46}$$

which, using similarly the variable $u = \mathcal{T}^3$, becomes

$$12\,\chi^2 \frac{d^2u}{d\chi^2} + 48\,\chi \frac{du}{d\chi} + 24\,u - \left(\beta^4 - \chi^4\right) u^{1/3} = -C_4. \tag{10.47}$$

This equation may be solved by numerical integration from the thickness $\mathcal{T}(0) = t_0/a = u_0^{1/3}$ at the origin. Successive iterations with χ varying from zero to β, up to obtaining $u(\beta) = 0$, provide the unknown C_4. The results would show that, compared to the parabolic flexure case, the thickness of the cylinder in the edge region is relatively smaller for the same flexural sag.
Similarly as the parabolic flexure case with radial forces F_β and $F_{-\beta}$ at the cylinder ends, and because of those forces, this configuration is a difficult practice for glass or vitro-ceram mirrors.

- **Uniform load and free ends – Inverse proportional law**: The *inverse proportional law* (10.38) states that if a uniform load q is applied to the cylinder surface and if no other external forces exist, then the thickness and flexure are reciprocal functions, $\mathcal{T}\,\mathcal{W} = C =$ constant.

1. *Single-term fourth-degree flexure*: Investigating a fourth-degree flexure, where the central section plane of the cylinder cannot extend or retract, leads to the single-term representation $\mathcal{W} = \chi^4$. From (10.38), the dimensionless thickness is $\mathcal{T} = C_4/\chi^4$ with $\chi^2 \le \beta^2$. Similarly as in a parabolic flexure, the *thin* shell theory provides a valuable solution in the form of an *infinite* thickness at the plane of central section. However, this solution is mostly academic.
2. *Two-term fourth-degree flexure*: Assuming that the requested optical shape is of the form $\beta^4 - \chi^4$, the singular poles for the edge thickness, i.e. at $\chi = x/a = \pm\beta$, are avoided if the flexure of the cylinder is set as

$$\mathcal{W} \equiv C_4 \frac{Ew}{q\,a} = \alpha^2 + \beta^4 - \chi^4, \quad \chi = x/a, \tag{10.48}$$

where α^2 is a constant. From (10.38), the dimensionless thickness is, similarly to (10.42),

$$\mathcal{T} = \frac{C_4}{\alpha^2 + \beta^4 - \chi^4}, \quad \chi^2 \le \beta^2. \tag{10.49}$$

If the free constants α^2 and C_4 are appropriately chosen, the thickness geometry can be readily fabricated for glass or vitroceram mirrors (Fig. 10.11).

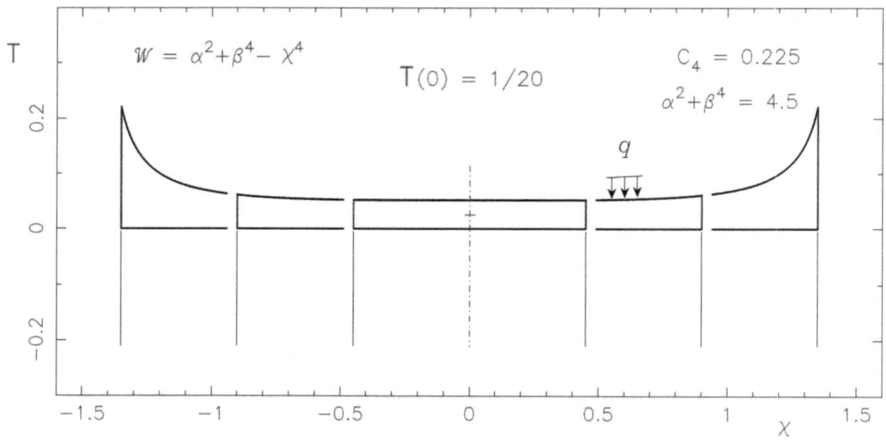

Fig. 10.11 Thickness distribution $\mathcal{T} = C_4/(\alpha^2 + \beta^4 - \chi^4)$ of three cylinders that generates a χ^4 flexure mode by uniform load only. Length parameters $\beta = 0.45, 0.90, 1.35$. Mirror aspect ratios $L/a = 1.092\,\beta$

10.2.5 Thickness Distributions for Tubular Image Transports

Grazing incidence *image transport systems* are preferably made of a single tubular mirror. Let us consider the basic case were the magnification is $M = -1$ and the mirror axisymmetric about a z-axis. The stigmatism of the mirror is achieved by use of the tubular central region of an *elongated ellipsoid*. In a Cartesian $z,\ r$ frame, the equation of the ellipsoid is $z^2/a^2 + r^2/b^2 = 1$ with the z-axis origin at the middle of the foci. Denoting $2\ell = \mathrm{FF'}$ this distance, we may derive the local representation of the surface in the section x, z of the local frame, where z is now a radial axis whose positive values are towards the ellipsoid axis (cf. Fig. 10.7). This calculation gives

$$z = \frac{1}{2R}\,x^2 + \frac{1}{8bR^2}\,x^4 + \frac{1}{16b^2R^3}\,x^6 + \cdots, \tag{10.50a}$$

where

$$R = (\ell^2 + b^2)/\,b, \tag{10.50b}$$

and b is the height of the mirror at the central symmetry plane. In this latter plane, if we return now to the notation a for the distance from the mid-thickness surface of the cylinder to its axis, and use as previously the dimensionless variable χ instead of x, the local shape $z(\chi)$ of the ellipsoid is represented by

$$\frac{z}{a} = \frac{a}{2R}\chi^2 + \frac{a^3}{8bR^2}\,\chi^4 + \cdots, \quad \chi = x/a. \tag{10.51}$$

Depending on the mirror length and on the maximum value of the grazing angle, arctan (b/ℓ), we shall consider hereafter two alternatives for the aspherization by stress figuring and restrain ourself to a third-order expansion of the mirror shape.

- **Stress figuring with straight generatrix segments**: Let us assume that the stress figuring provides both the mirror curvature $1/R$ and the biquadratic term in (10.51). Hence, the inner surface of the mirror is cylindrical when at rest. During the surfacing, a positive load q is applied to its outer surface from the edge $x = -L/2$ to the other at $L/2$ and the surfacing tool is constituted of *straight* segments that are generatrices of the cylinder. For glass or vitroceram materials, we have seen that the easiest case is when the boundary conditions are satisfied with free edges, so the loading system – for instance by air pressure – must avoid any radial force or bending moment at the cylinder ends.

  In the determination of the flexure w with these boundaries, we must set the sag origin at the edges and also introduce a constant α^2 to avoid any pole singularity in the thickness distribution. This latter constant provides the same free retraction all along the cylinder. From (10.51) and with these conditions, the flexure w and its dimensionless associate $\mathcal{W}$ may be represented by

$$\mathcal{W} \equiv C\frac{E\,w}{q\,a} = \alpha^2 + \frac{a}{2R}\left(\beta^2 - \chi^2\right) + \frac{a^3}{8bR^2}\left(\beta^4 - \chi^4\right), \tag{10.52a}$$

$$\text{with} \quad \beta = L/2a, \quad \chi^2 \le \beta^2, \tag{10.52b}$$

  and where C is a constant.

  From the inverse proportional law $\mathcal{TW} = constant$, the dimensionless thickness is

$$\mathcal{T} = C\left[\alpha^2 + \frac{a}{2R}\left(\beta^2 - \chi^2\right) + \frac{a^3}{8bR^2}\left(\beta^4 - \chi^4\right)\right]^{-1}, \tag{10.53}$$

where the constant α^2 is given a value so the expression into the brackets appropriately differs from zero. Introducing the distance 2ℓ between the conjugate foci, from (10.50a), the thickness t/a writes

$$\mathcal{T} = C\left[\alpha^2 + \frac{ab}{2(\ell^2+b^2)}\left(\beta^2 - \chi^2\right) + \frac{a^3 b}{8(\ell^2+b^2)^2}\left(\beta^4 - \chi^4\right)\right]^{-1}. \tag{10.54}$$

Since a is the radius of the mid-thickness surface of the cylinder at its center ($\chi = 0$), the corresponding optical radius of the mirror is $b = a - t(0)/2$. By an appropriate set up of the free parameters $\mathcal{T}(0)$ and α^2, the thickness geometry of the mirror can be readily fabricated either in a *linear* metal alloy, i.e. satisfying Hooke's law, or in glass, or in vitroceram (Fig. 10.12-*Left*).

- **Stress figuring with circle generatrix segments**: Assume now that the stress figuring provides the aspherization by minimizing the quantity of material to remove, thus using circle generatrix segments. For mean graze angles up to 3°, then $\ell/b \geq 19$. Considering the osculating circle defined from the χ^2 coefficient in (10.51), the ratio of the χ^4-term of the osculating circle over that of the ellipse is b/R. From (10.50b), this ratio is $\simeq b^2/\ell^2 \simeq 1/361$ which is small. The χ^4 coefficient of the circle figuring segment will be close to that of the above osculating circle, and then is similarly small.

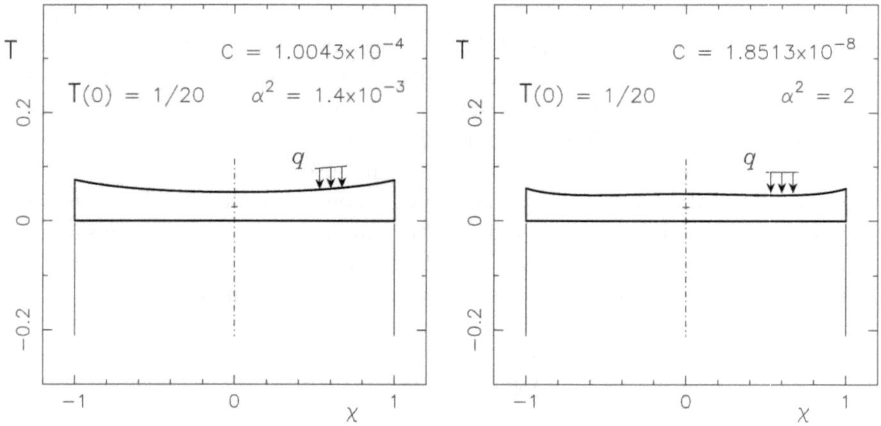

Fig. 10.12 Thickness distribution $\mathcal{T}(\chi)$ for the active optics stigmatism of a tubular image transport $M = -1$. Cylinder length $L = 2a$. Grazing angle $\arctan(b/\ell) = 2°$. (*Left*) Stress figuring by uniform load only and straight generatrix segments. (*Right*) Stress figuring by uniform load only and circle generatrix segments

Thus, we shall assume hereafter that the χ^4-term of the optics is totally generated by the flexure. Minimizing the volume to remove with the circle tool, we obtain for the dimensionless flexure[5]

$$\frac{z}{a}\bigg|_{\text{Flex}} = \frac{a^3}{8bR^2}\left(\tfrac{2}{3}\,\beta^2\chi^2 - \chi^4\right) \tag{10.55}$$

where $\beta = L/2a$ is the dimensionless semi-length of the cylinder. The curvature $1/R_{\text{Cir}}$ of the figuring circle segment is, from (10.51) and $z/a|_{\text{Cir}} = z/a|_{\text{Opt}} + z/a|_{\text{Flex}}$,

$$\frac{1}{R_{\text{Cir}}} = \left(1 + \frac{a^2\beta^2}{6\,bR}\right)\frac{1}{R}. \tag{10.56}$$

Now introducing an additional constant α^2 to avoid the pole singularity, from (10.55), the dimensionless flexure becomes

$$\mathcal{W} \equiv C\,\frac{E\,w}{q\,a} = \frac{a^3}{8\,bR^2}\left(\alpha^2 + \tfrac{2}{3}\,\beta^2\chi^2 - \chi^4\right) > 0, \quad \chi \in [-\beta, \beta]. \tag{10.57}$$

From the inverse proportional law (10.38), $\mathcal{T}\mathcal{W}$ = constant, the dimensionless thickness is

$$\mathcal{T} = \frac{8\,C\,b\,R^2}{a^3(\alpha^2 + \frac{2}{3}\,\beta^2\chi^2 - \chi^4)} \tag{10.58}$$

which, from (10.50b), may be written

$$\mathcal{T} = \frac{8\,C\,(\ell^2 + b^2)^2}{a^3 b(\alpha^2 + \frac{2}{3}\beta^2\chi^2 - \chi^4)}. \tag{10.59}$$

Optimizing the value of the constant C, an appropriate setting of the free parameters $\mathcal{T}(0)$ and α^2 provides a thickness geometry of the mirror which then can be readily fabricated either in a *linear* metal alloy – satisfying Hooke's law –, or in glass, or in vitro ceram (Fig. 10.12-*Right*).

---

[5] Minimal volume with an even biquadratic curve: In an axial section of a cylinder, let a curve $g = p + q\chi^2 - \chi^4$ be tangent to a straight generatrix at $\chi = \pm\sqrt{q/2}$. Setting the curve ordinate such that $g(\sqrt{q/2}) = 0$ entails $p = -q^2/4$, so the curve is $g = -\frac{1}{4}q^2 + q\chi^2 - \chi^4$. In a 2π rotation about the cylinder axis, the volume generated by the curve and its tangent generatrix – the χ-axis –, from the origin to the edge, is

$$v \propto \int_0^\beta \left(-\tfrac{1}{4}\,q^2 + q\,\chi^2 - \chi^4\right) d\chi = -\tfrac{1}{4}\,q^2\beta + \tfrac{1}{3}\,q\,\beta^3 - \tfrac{1}{5}\,\beta^5.$$

This volume is *minimal* when $dv/dq = 0$, which entails $q = \frac{2}{3}\beta^2$. Hence the curve providing a minimal volume is

$$g = -\tfrac{1}{9}\,\beta^4 + \tfrac{2}{3}\,\beta^2\chi^2 - \chi^4.$$

From $dg/d\chi = 0$, the intermediate extremum of this curve is at the abscissa $\chi/\beta = 1/\sqrt{3} = 0.5773\ldots$

10.3 Elasticity Theory of Weakly Conical Tubular Shells

10.3.1 Flexure Condition for Pure Extension of Axisymmetric Shells

In the general case of an axisymmetric two-mirror telescope, let $z(\chi)$ be a section representation of an optical surface where $\chi = x/a_0$ is an axial variable with respect to the radius a_0 of the mid-thickness of a shell at the origin $x = 0$. The function $z(\chi)$ may either be polynomial or parametric as in this latter case for a WS telescope that strictly satisfies Abbe's sine condition (cf. Sect. 10.1.3). For instance, let us assume in the elastic aspherization that the figuring is conical while the mirror is under stress. Then, the radial quantity of material $f(\chi)$ to be removed is opposite to the optical sag i.e. in the form $f(\chi) = -z(\chi) + c_1\chi + c_0$. It is natural to set the constants $c_0,\ c_1$ such that $f(\beta) = f(-\beta) = 0$ at the mirror ends, thus giving the mirror sag with respect to the cone lying to the ends. However, these two conditions lead to radially unmovable ends and we have seen from the cylindrical shell theory that this entails use of radial reactions opposite to the load or infinitely thick ends. Therefore, it is preferable to consider hereafter the other alternative without any radial reaction to the load q.

The inverse proportional law, $\mathcal{T}\mathcal{W} = constant$ (10.38), strictly applies to cylindrical or quasi-cylindrical shells. For a truncated *weakly conical shell*, with a mean slope angle up to a few degrees, another law will be derived in Sect. 10.3.3. Similarly as for a cylindrical shell, an appropriate constant α^2 must be added to the quantity $f(\chi)$ to be removed by stress figuring, in sort that no point of the mid-surface of a conical shell is with a null – or a too small – radial displacement otherwise this would entail a singular pole and thus an infinite – or too large – thickness.

In the sign convention, the flexure function $\mathcal{W}$ is positive when towards the conical shell axis; this corresponds to a retraction of the shell and a positive uniform load q. Hence we have the two following cases of *monotonic sign flexure*

$$\mathcal{W} \equiv C\,\frac{E\,w}{q\,a_0} = \alpha^2 + f(\chi) > 0, \quad q > 0, \tag{10.60a}$$

$$\mathcal{W} \equiv C\,\frac{E\,w}{q\,a_0} = -\alpha^2 + f(\chi) < 0, \quad q < 0, \tag{10.60b}$$

where after stress figuring and elastic relaxation, the constant radial retraction – or extension – of the shell, due to the constant α^2 term, vanishes.

If the function $f(\chi)$ is a polynomial expansion, then it includes odd and even terms. We of course assume hereafter that the main *conical* term, determined by the slope of the mirror ends, is *not* obtained by stress figuring. Hence the flexure may be written in the polynomial form

$$f(\chi) = \sum_{n=0,1,2,3,\ldots}^{N} A_n \, \chi^n \tag{10.61}$$

where the second term A_1 does not represent the total cone angle of the optical surface of the shell but is a much smaller quantity which can be generated by flexure. Avoiding χ-zones with infinite or too large thickness, the general relation expressing the flexure must write

$$\mathcal{W} = \pm\alpha^2 + \sum A_n \, \chi^n \tag{10.62}$$

where the constant α^2 and sign before it are set such as $\mathcal{W}(\chi)$ is with monotonic sign over the range $[-\beta,\ \beta]$ and never approach too closely to zero. This implies a shell in *extension condition only* – or *retraction condition only* – all over its length.

10.3.2 Truncated Conical Shell Geometry and Cylindrical Flexure

The flexure of a *conical shell* by an axisymmetric linearly varying load, whose origin is at the cone vertex, is a homothetic cone. A cylindrical shell uniformly loaded can be considered as a particular case where the cone vertex is at infinity. The elaboration of a linear load function for the extension – or retraction – of a cone is a great practical difficulty; therefore we only consider hereafter the case of a conical shell deformed by a uniform load, $q = constant$. Except for a small axial reaction on its larger ring face, if no other force is applied to the conical shell, then the radial shearing force is null, $Q_x = 0$. Returning to the dimensioned quantities, from (10.36), we have $tw/a^2 = q/E = constant$.

This latter relation allows taking into account the conical geometry. Since with (10.15), the three components of the shear strains are null, a *truncated conical shell* can be assumed as constituted of *separated* element rings that are continuously distributed along the χ-axis. In the χ-variation of the mid-thickness radius a, let us consider that each element ring provides the same amount of flexure w. This is achieved if

$$\frac{t(\chi)}{a^2(\chi)} = \frac{q}{E}\frac{1}{w} = constant. \tag{10.63}$$

Let us denote (L_i), (L_o) the straight segment lines forming the inner and outer surfaces of a conical shell, (L_m) that of the mid-thickness surface, and i, o the associated low-angle slopes (Fig. 10.13). In a frame r, χ where $\chi = x/a_0$, the equation of the segment lines (L_i) and (L_o) are respectively

$$r_i = a_0(1 - i\chi) - \frac{1}{2}t_0, \quad r_o = a_0(1 - o\chi) + \frac{1}{2}t_0, \tag{10.64}$$

which allow defining the mid-thickness line $a(\chi)$ and the thickness $t(\chi)$ as

$$a = a_0\left[1 - \frac{1}{2}(i + o)\chi\right], \quad t = t_0\left[1 + \frac{a_0}{t_0}(i - o)\chi\right]. \tag{10.65}$$

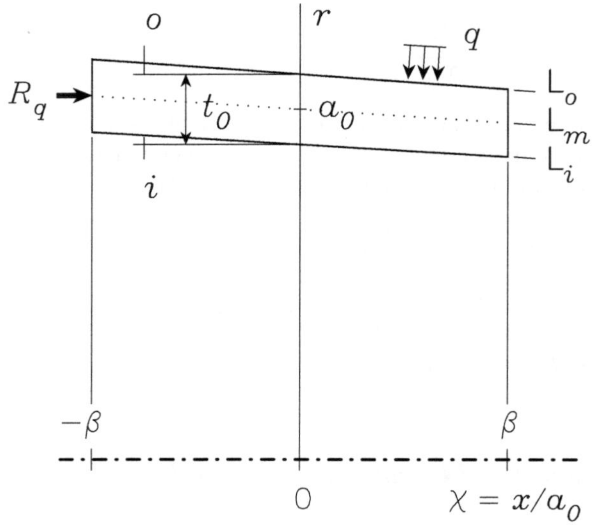

Fig. 10.13 Contour lines of a truncated conical shell

In a grazing incidence telescope, the slope i of a mirror is necessarily small; assuming that o will be small and neglecting the χ^2-term in the expansion of $1/a^2(\chi)$, we obtain

$$\frac{t(\chi)}{a^2(\chi)} \simeq \frac{t_0}{a_0^2}\left\{1 + \left[\frac{a_0}{t_0}(i-o) + i + o\right]\chi\right\}, \tag{10.66}$$

hence the condition of constancy for this ratio is satisfied if the expression into the brackets is equal to zero,

$$o = \frac{1 + t_0/a_0}{1 - t_0/a_0}i, \quad \text{i.e.} \quad i + o = \frac{2i}{1 - t_0/a_0}. \tag{10.67}$$

For instance, if the inner angle slope is with $i = 2°$ and the mirror thickness-ratio is $t_0/a_0 = 1/20$, the outer slope is with $o = 2.210°$; For an outer surface which is homothetic to the inner surface, we would have $o = 2.105°$.

10.3.3 Linear Product Law – Flexure-Thickness Relation

We assume that the *truncated conical shell* is with a small slope angle i, say $i \leq 5°$ typically, and that $t_0/a_0 \leq 1/10$. For practicable reasons, we also assume that the external load q is always perfectly *uniform*. Returning to (10.63) and using the above expression of $a(\chi)$, after substitution of $i + o$, this relation becomes

$$\frac{t(\chi)w(\chi)}{a_0^2} = \frac{q}{E}\left(1 - \frac{2i}{1 - t_0/a_0}\chi\right). \qquad (10.68)$$

Similarly to (10.25), let us introduce a dimensionless thickness and flexure, but now with respect to a_0,

$$\mathcal{T} = \frac{t}{a_0}, \quad \mathcal{W} = C\frac{Ew}{q\,a_0}, \qquad (10.69)$$

where C is a constant. From (10.66), we obtain a relation between the dimensionless thickness and flexure. This is the following *linear product law* (Lemaitre [14])

$$\mathcal{T}\mathcal{W} = C\left(1 - \frac{2i}{1 - t_0/a_0}\chi\right), \quad \chi = \frac{x}{a_0} \in [-\beta, \beta]. \qquad (10.70)$$

→ *If a truncated weakly conical shell is radially submitted to a uniform load q applied all over the outer or inner surface, and if no discrete circle-force is applied except the small axial reaction R_q to the load, then the product thickness-flexure $\mathcal{T}\mathcal{W}$ is a linear function of the axial coordinate χ.*

This law is of fundamental importance for the aspherization of an X-ray mirror because the shell is with both *free edges*. Thus no bending moment or radial force is needed at the boundaries except the small axial reaction R_q which comes from the axial components of load q.

- **Avoiding poles in the linear product law** : For instance, let us assume that the flexure function $\mathcal{W}$ is represented by (10.62) from which the dimensioned flexure is $w = \frac{1}{C}\frac{q}{E}\mathcal{W}a_0$. Then, from the linear product law, the associated thickness is

$$\mathcal{T} = C\left(1 - \frac{2i}{1 - t_0/a_0}\chi\right)\left(\pm\alpha^2 + \sum_{0,1,2,3,\ldots}^{N} A_n\,\chi^n\right)^{-1}. \qquad (10.71)$$

  For rejecting shell solutions with infinite thickness – which would also lead to a sign inversion in the load distribution q – for practicable reasons, it is important to conserve a uniform load $q = constant$ and to avoid pole singularities. Since $|2i\beta/(1 - t_0/a_0)| \ll 1$, this is achieved, $\forall\,\chi$, by setting a convenient value for α^2 and sign before it, in sort that $\pm\alpha^2 + \Sigma A_n\chi^n$ becomes with monotonic sign over $[-\beta, \beta]$.

10.4 Active Optics Aspherization of X-ray Telescope Mirrors

10.4.1 Thickness Distributions for Monolithic Tubular Mirrors

Among the Wolter Type I telescope family reviewed in Sect. 10.1, the Wolter-Schwarzschild (WS) form is the most performing for high angular resolution imaging in X-ray. Although limited to a few arcmin field of view, the fulfilment of the

Abbe sine condition potentially allows an optical imaging towards the diffraction limited tolerance at these very short wavelengths. In counterpart, compared to the PH, SS or HH forms of this family, the shape of WS telescope mirrors is the most difficult to obtain; this fully explains why this telescope design has only been built in one or two laboratory samples – and seems to have never been used in space up to now.

The geometry of a truncated conical shell of small slope angle for any grazing incidence mirror – hereafter simply called *tubular mirror* – can be readily determined from the above linear product law.

- **Linear product law and tubular mirror design** : Whatever the form of a Wolter Type I telescope – PH, WS, SS, or HH forms (cf. Sect. 10.1) –, relation (10.71) allows deriving the associated thickness distributions $\mathcal{T}_1$, $\mathcal{T}_2$ of the primary and secondary mirrors provided the flexure $\mathcal{W}_1$, $\mathcal{W}_2$ for these mirrors have been set such as the radial retraction (or extension) is always of *monotonic* sign all along the mirror surface. This is achieved by convenient choices of the α_1^2 and α_2^2 values and sign before them in (10.71).

- **Application to grazing incidence WS telescopes** : As an application example of active optics aspherization, let us investigate the most difficult case of WS telescope mirrors. We have seen in Sect. 10.1.3 that Chase and VanSpeybroeck [4] derived the mirror parametric representations for the WS telescope (i.e. a telescope form strictly satisfying the sine condition). These authors also gave the resulting shapes for the construction and X-ray tests of a prototype WS telescope. The parameters of the primary mirror are the length $L = 165$ mm, low-angle slope corresponding to $i = 1.5°$, and inner radius $r_j = 152$ mm at the joint of the mirrors which gives $r_0 = 154.16$ mm at the middle of the mirror (Fig. 10.14).

  Using the geometrical parameters corresponding to Fig. 10.14 and introducing, at $\chi = 0$, a thickness $t_0 = 12.33$ mm, we obtain a central radius of the mid-surface $a_0 = r_0 + t_0/2 = 160.32$ mm providing a shell thickness-ratio $L/a_0 = 1/13$. The edge abscissa $\chi = \pm\beta$ is then with $\beta = L/2a_0 = 0.5146$. A polynomial representation of the WS primary mirror in Fig. 10.14 where $\boldsymbol{\Sigma}\, A_n \chi^n$ includes terms up to $n = 8$ allows to obtain, from (10.71), the thickness and complete geometry of this mirror when asphérized with the best circle-segment fit (Fig. 10.15).

  In the previous example, we assumed that the asphérization is generated from the best circle-segment fit. However, both curvature and asphérization can be also

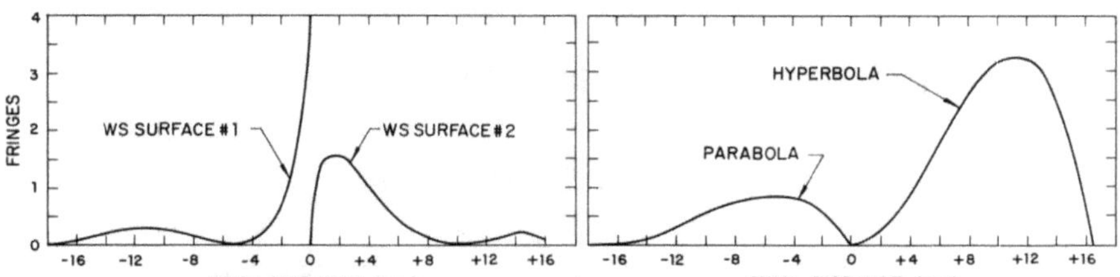

Fig. 10.14 Mirror shapes of a WS telescope and the corresponding PH telescope from the best circles fit to each surface. One fringe is equal to one-half wavelength of 5,461 Å light [4]

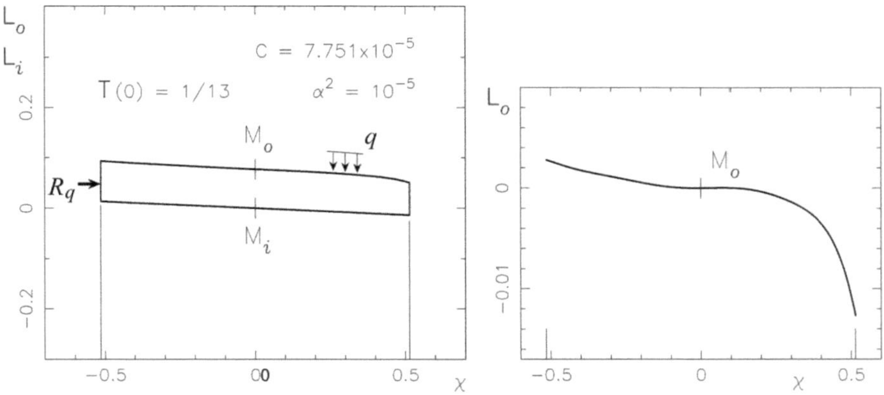

Fig. 10.15 Shell geometry and thickness distribution of the WS primary mirror in Fig. 10.14 aspherized by uniform load and best circle fit. (*Left*) Inner and outer lines L_i, L_o, of the shell meridian section. (*Right*) Enlarged scale of the outer line L_o after origin change and removal of the slope angle component at M_o

achieved from the best straight-segment fit. In any case, a convenient set up of the constant α^2 and thickness $\mathcal{T}(0) = t_0/a_0$ must be done. If these two quantities are increased, then the amplitude of the outer surface with respect to a straight line is decreased, but the intensity of the load is increased.

- **Note** : Slight deviations may affect the accuracy of the flexure from the thickness distribution $\mathcal{T}(\chi)$ as given by (10.71). The axial reaction R_q from the axial component of the uniform load q provides slight displacement effects; the displacements in the axial direction may be assumed negligible whilst those in the radial direction are denoted $\Delta\mathcal{W}_1(\chi)$. Since we assume here that the mid-thickness surface is a perfect conical shape – i.e. $a(\chi)$ is linear –, the thickness variation which is required for the aspherization entails that the middle surface is not exactly a cone; the corresponding deviation is $\Delta\mathcal{W}_2(\chi)$. The linear approximation in the expansion of $a\chi$ also causes a slight deviation $\Delta\mathcal{W}_3(\chi)$. The correction of these three deviations to the requested flexure can be accurately achieved by finite element code analysis in slightly modifying the non-optical surface of the shell.

10.4.2 Boundaries for Segment Mirrors of Large Tubular Telescopes

Beside increasing the angular resolution of X-ray telescopes – presently limited by technological difficulties in the obtention of mirror surfaces having an extremely high smoothness –, astrophysical programs for faint object studies require the development of X-ray telescopes with much larger surface areas. Future space-based

Wolter Type I telescopes will be designed with segmented mirrors allowing construction of large tubular mirrors [6, 9]. The extremely high precision of the tubular surfaces to be figured and aligned will require an *extensive* use of active optics methods.

Active optics shall be mostly elaborated as well for the *stress figuring* of the submirrors as for the *in situ control* of the two successive primary and secondary mirror surfaces resulting from segment assemblies.

Let us consider hereafter the execution by stress figuring of a mirror segment whose contour lines are plane cuts defined by two angular planes passing through the mirror general axis and two parallel planes perpendicular to this axis. Similarly to the second deformation case of a cylinder element (Fig. 10.7), we assume that the static equilibrium of the conical shell segment to a uniform load q is realized by self-compression or self-extension only, i.e. without any radial boundary forces. Then, *no* shearing force is generated through the shell, $Q_x = 0$, and the bending moments along its contour are null, $M_x = M_\psi = 0$ whatever x, ψ or χ, ψ. This means that the load q generates a *pure extension* – or a *pure retraction* – of the shell segment. Therefore, the *linear product law* (10.70) applies and the thickness distribution can be easily derived from this law.

For instance, when the stress figuring is operated on the inner surface of a segment by a pressure load q which acts on the outer surface of low slope angle o – which slightly differs from the inner slope angle i –, the boundaries at a segment reduce to the three following conditions (Fig. 10.16).

- **Boundary condition C1**: The two facets $\pm\psi/2 = constant$ of a segment must be supported by a normal pressure p as

$$q\, d\mathcal{A}_q \cos(\operatorname{atan} o) + p\, d\mathcal{A}_p\, \psi = 0, \tag{10.72}$$

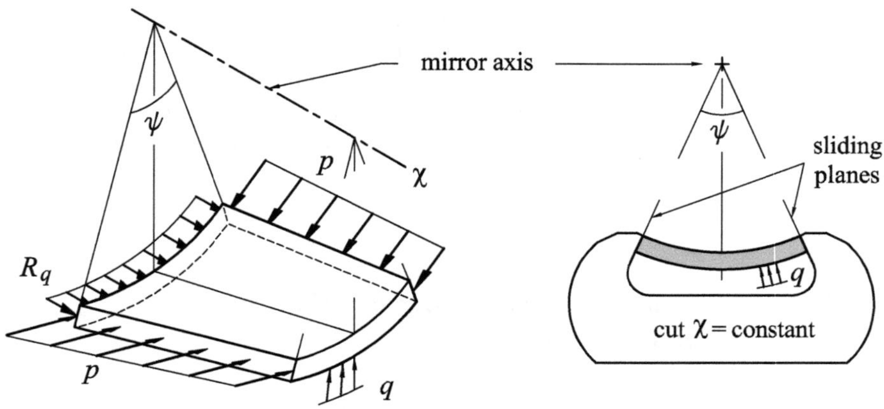

Fig. 10.16 Equilibrium configuration for the aspherization by stress figuring and uniform loading q of a segment mirror of a large X-ray telescope. (*Left*) Force distributions for the boundary conditions **C1** and **C3**. (*Right*) Cut of a harness fulfilling those conditions and also condition **C2**

where $d\mathcal{A}_q$ and $d\mathcal{A}_p$ are infinitesimal areas, of length $d\chi$, of the segment outer surface and of one of these facets. For finite values $\pm\psi/2$, this gives a pressure distribution $p(\chi) = N_\psi / t(\chi)$ which is linear only for a strictly conical thickness. However, from axial symmetry, when passing from the whole tubular cone to the segment cut, the $p(\chi)$ distribution in the harness support is implicitly conserved if the segment is supported by two planes passing through the mirror axis and forming a *dihedral angle* ψ.

- **Boundary condition C2**: In the dihedral planes $\pm\psi/2 = constant$, these two facets must be free to *slide* in the radial directions.
- **Boundary condition C3**: The facet end $\chi = constant$ corresponding to the largest diameter of the loaded area must receive the axial reaction from the uniform load q. The resultant force R_q of this reaction is given by

$$q\,\mathcal{A}_q \,\sin \mathrm{atan} < o > + R_q = 0, \tag{10.73}$$

where $\mathcal{A}_q$ is the segment surface area receiving the load q and $< o >$ a mean-value of this surface slope angle.

10.4.3 Concluding Remarks on the Aspherization Process

Subject to the above conditions of weakly conical shells the active optics aspherization of grazing incidence mirrors can be carried out by a simple uniform load applied to a closed shell with radially sliding ends. Whatever the figuring option – best straight-line fit or best circle-line fit – the rigidity of the surfacing tools must be much *higher* in its axial direction than in the tangential cross-section. This can be also realized by contiguous lapping segments with some freedom to radial motions.

For a large segmented telescope the elasticity design of a mirror segment is simply a cut-out of the elasticity design of the complete tubular mirror. Hence, whether we have a monolithic tubular mirror or simply a segment of it, the thickness distribution which provides the convenient radial displacement distribution can be directly derived from the same *linear product law*.

In a final design, a 3-D finite element code optimization would provide some slight corrections of the mirror substrate geometry such as taking into account the small effect of the load reaction acting at the larger end because of the mirror slope.

Stress figuring of a mirror segment requires use of a support – or "harness" – which must be designed with high stiffness and adequate setups to accurately satisfy the above three boundary conditions, **C1**, **C2**, **C3**, during the process.

Compared to the aspherization of a mirror segment for a large quasi-normal incidence telescope (cf. Chap. 7), the boundary conditions for a weakly conical shell are greatly simplified since the above process avoids the complication of applying bending moments at the mirror contour.

References

1. B. Aschenbach, X-ray telescopes (ROSAT), Rep. Prog. Phys., **48**, 579–629 (1985)
2. C.J. Burrows, R. Burg, R. Giacconi, Optimal grazing incidence optics and its application to wide-field imaging. Astrophys. J., **392**, 760–765 (1992)
3. R.C. Chase, Aplanatic grazing incidence X-ray microscopes: design and performances, Appl. Opt., **15**, 3094–3098 (1976)
4. R.C. Chase, L.P. VanSpeybroeck, Wolter-Schwarzschild telescopes for X-ray astronomy, Appl. Opt., **12**, 1042–1044 (1973)
5. H. Chrétien, Le télescope de Newton et le télescope aplanétique, Rev. d'Optique, **1**, 51–64 (1922)
6. O. Citterio, M. Ghigo et al., Large-size glass segments for the production of the XEUS X-ray mirrors, SPIE Proc., **4815** (2003)
7. P. Concini, S. Campana, Optimization of grazing incidence mirrors and its application to surveying X-ray telescopes, Astron. Astrophys., **372**, 1088–1094 (2001)
8. J-J. Fermé, Improvement in bendable mirrors (Société Européenne de Systèmes Optiques – SESO), SPIE Proc., **3152**, 103–109 (1997)
9. J.H. Hair, J. Stewart et al., Constellation-x soft X-ray telescope segmented optic assembly and alignment implementation, SPIE Proc., **4851** (2002)
10. J.E. Harvey, A. Krywonos, P.L. Thompson, T.T. Saha, Grazing incidence hyperboloid-hyperboloid designs for wide-field X-ray imaging applications, Appl. Opt., **40**, 136–144 (2001)
11. J.E. Harvey, M. Atanassova, A. Krywonos, Balancing detector effects with aberrations in the design of wide-field grazing incidence X-ray telescopes, Opt. Eng., **45**(6) (2006)
12. D. Korsch, Aplanatic two-mirror telescope for near-normal to grazing incidence, Appl. Opt., **19**, 499–503 (1980)
13. D. Korsch, *Reflective optics*, Academic Press edt., Boston, Chap. 11 (1991)
14. G.R. Lemaitre, Active optics and X-ray telescope mirrors, SPIE Proc. on *Space Telescope and Instrumentation II : Ultraviolet to Gamma Ray*, **7011**-37, session 5, p. 1–10 (2008)
15. A.E.H. Love, in *Mathematical Theory of Elasticity*, Dover Publications, New-York, 4th issue, Preface and 569 (1927)
16. D. Lynden-Bell, Exact optics: a unification of optical telescopes, Mont. Not. R. Astron. Soc., **334**, 787–796 (2002)
17. R.F. Malina, S. Bowyer, D. Finley, W. Cash, Wolter-Schwarzschild optics for the extreme uv, Opt. Eng., **19**, 211–217 (1980)
18. J.D. Mangus, J.H. Underwood, Optical design of glancing incidence X-ray telescopes, Appl. Opt., **8**, 95–102 (1969)
19. K. Nariai, Geometrical aberrations of a generalized Wolter Type I. 2. Analytical theory, Appl. Opt., **26**, 4428–4432 (1987)
20. R.J. Noll, P. Glenn, J.F. Osantowski, Optical surface analysis code (OSAC), in *Scattering in Optical Materials II*, S. Mussikant edt., SPIE Proc., **362**, 78–82 (1983)
21. T.T. Saha, General surface equations for glancing incidence telescopes, Appl. Opt., **26**, 658–663 (1987)
22. T.T. Saha, W. Zhang, Equal-curvature grazing incidence X-ray telescopes, Appl. Opt., **42**(22) 4599–4605 (2003)
23. P.L. Thompson, J.E. Harvey, A system engineering analysis of aplanatic Wolter type I X-ray telescopes, Opt. Eng., **39**, 1677–1691 (2000)
24. S.P. Timoshenko, J.M. Gere, in *Theory of Elastic Stability*, McGraw-Hill Book Company, New-York, 2nd issue (1961)
25. S.P. Timoshenko, S. Woinowsky-Kieger, in *Theory of Plates and Shells*, McGraw-Hill Book Company, New-York, 2nd issue, Chap. 15, 466 (1959)
26. J.H. Underwood, P.C. Batson, H.R. Beguiristain, E.M. Gullikson, Elastic bending and water-cooling strategies for producing high-quality synchrotron-radiation mirrors (Laurence Berkeley Nat. Lab.), SPIE Proc., **3152**, 91–98 (1997)

27. L.P. VanSpeybroeck, R.C. Chase, Design parameters of paraboloid-hyperboloid telescopes for X-ray astronomy, Appl. Opt., **11**, 440–445 (1972)
28. W. Werner, Imaging properties of Wolter Type I telescopes, Appl. Opt., **16**, 764–773 (1976)
29. H. Wolter, Generalized Schwarzschild mirror systems with glancing incidence on image producing optics for X-rays, Ann. Phys. (Leipzig), **10**, 286–295 (1952)
30. H. Wolter, Mirror systems with glancing incidence on image producing optics for X-rays, Ann. Phys. (Leipzig), **10**, 94–114 (1952)

Portrait Gallery

The following portraits of some pioneer elasticians, and some opticians and astronomers who used elasticity to improve the performance of astronomical telescopes and instrumentation, are a personal choice of the author.

Fig. 1 Galileo Galilei (1564–1642) was a universal scientist who, from poor "monocular opera glasses," invented the *telescope* in 1609 – just four centuries ago – and discovered with it Jupiter's four *Galilean* satellites. He introduced mathematics in the formulation of the basic laws of dynamics and parabolic trajectories. Galileo's problem of *equal strength cantilevers* and its *scaling law* on the flexure of beams marked the birth of the elasticity theory and strength of materials (*Discorsi e Dimostrazioni Matematiche*, 1638)
PHOTO: (credit: Portrait *Galileo Galilei florentino* by Ottavio Leoni, 1624, Musée du Louvre)

Fig. 2 Robert Hooke (1635–1703) discovered, around 1660, the fundamental law of elasticity which states that the uniaxial stretching of a solid body is proportional to the stress applied in this axial direction. This *linear stress-strain relation*, which allows perfect characterization of elastic materials, is known as *Hooke's law* (*De Potentia Restitutiva*, 1678)
PHOTO: (credit: The MacTutor History of Mathematics archive, University of Saint Andrews)

Fig. 3 Daniel Bernoulli (1700–1782) was a mathematician, physicist, and philosopher particularly remembered for his applications of mathematics to mechanics, especially fluid mechanics, and for his pioneering work in probability and statistics. In 1742, he suggested to Euler that the plane elastic bent curve of a rod without central stretching could be derived from the *principle of minimum total bending energy*, thus by rending minimal the sum of the square curvatures along the rod
PHOTO: (credit: The MacTutor History of Mathematics archive, University of Saint Andrews)

Fig. 4 Leonhard Euler (1707–1783) was a prolific mathematician also renowned for his work in mechanics, optics, and astronomy. From a suggestion by Daniel Bernoulli he initiated the first calculus of variations of a non-linear integral function as a minimal energy principle to determine the bending deformation of unstretchable circular-section objects. Hence Euler's beautiful contributions to elasticity are the *critical load buckling relations* of compressed beams and the first theory of elasticity known as *theory of elasticae* (1744)
PHOTO: (credit: The MacTutor History of Mathematics archive, University of Saint Andrews)

Fig. 5 C. Augustin de Coulomb (1736–1806) was a physicist who worked with mechanical rupture and friction, elasticity and electrostatics. He formulated the inverse square law for the force between two charges, known as Coulomb's law. Inventing the torsion balance and using thin silk and hair threads, Coulomb developed the first *elasticity theory of torsion*. He thus derived the magnitude of the electrostatic forces from the torsion angle of a charge pair when introducing a third charge (1780)
PHOTO: (credit: Archives of the Académie des Sciences and Encyclopedia Wikipedia)

Fig. 6 Thomas Young (1773–1829) was a versatile physicist who brought important contributions on vision, light, elasticity, capillarity, and energy. Illuminating a narrow-slit pair by a distant source point, he proved the wave nature of light. For isotropic materials Young noticed that if a uniaxial stress is applied, then the stress-strain ratio – Young's modulus – is an elasticity coefficient depending only on the material. He also introduced the concept of *shear stresses* in 1807
PHOTO: (credit: The MacTutor History of Mathematics archive, University of Saint Andrews)

Fig. 7 S. Denis Poisson (1781–1840) was a mathematician, geometer, and physicist. Among the derivative equations bearing Poisson's name those of the first form involve the constancy of a single laplacian (gravitational or electrical potentials). In elasticity, those of the second form – at the fourth derivatives – involve the constancy of the bilaplacian of the flexure to a uniform load. Solving this latter form in the axisymmetric case, Poisson derived the general solutions, thus created the *thin plate theory of elasticity* (1828)
PHOTO: (credit: Archives of the Académie des Sciences)

Fig. 8 C.L.M. Henri Navier (1785–1836) was an engineer and physicist who specialized in mechanics. He formulated an equation set for the motion of viscous fluids well known as the Navier-Stokes equations, since they were independently derived by G.G. Stokes. Navier founded the basis of the elasticity theory on the 3-D stress-strain relations which he derived from the equilibrium of a volume element (1826). The *Navier stress-strain relations* were finalized with the shear components by A. Cauchy in 1827–1829
PHOTO: (credit: Ecole des Ponts et Chaussées and The MacTutor History of Mathematics archive)

No portrait of George Green is known

Fig. 9 George Green (1793–1841) was a mathematician and physicist mainly known for Green's divergence theorem which relates the properties of a mathematical function at the surface of a closed volume to other properties inside. Although Poisson, Navier, Cauchy, and Lamé used a bi-constant elasticity theory (E, ν), they implicitly assumed that $\nu = 1/4$ for all materials; this was called the uni-constant theory. Green showed that the number of elastic constants reaches 21 for crystals (1837)

Fig. 10 Gustav R. Kirchhoff (1824–1887) was a mathematician and physicist who made important contributions in the theory of elasticity and, using topology, generalized Ohm's laws to multi-loop circuits. He demonstrated that an electric current flows on a conductor at the velocity of light. In elasticity, after conflicting discussions over decades, Kirchhoff's formulation of the boundary conditions for a free edge when the flexure of a plate is non-axisymmetric were found to be exact; this involves his concept of *net shearing force* (1850)
PHOTO: (credit: The MacTutor History of Mathematics archive, University of Saint Andrews)

Fig. 11 Adhémar J.C. Barré de Saint-Venant (1797–1886) was an engineer and mathematician who significantly contributed to elasticity and fluid mechanics, for which he formulated the equations on open channel flow. In *La Torsion de Prismes*, he derived the shape of various cross-section beams when large torsions occur, thus creating the large torsion theory. In this textbook he also enounced a general principle of equivalence for the application of the boundary conditions, well known as the *Saint-Venant principle* (1855)
PHOTO: (credit: Archives of the Académie des Sciences, Paris)

Fig. 12 R.F. Alfred Clebsch (1833–1872) was a mathematician and physicist who succeeded B. Riemann in the chair of Gauss in Göttingen. His major works are on abelian functions, elasticity, algebraic geometry and invariant theory. In his *Theorie der Elastizität fester Körper*, Clebsch finds the first solutions to Galileo's problem for equal constraint cantilevers, and an infinite set of four-term polynomials – Clebsch polynomials – as radial component solutions to the non-axisymmetric loading of circular plates in the thin plate theory (1862)
PHOTO: (credit: The MacTutor History of Mathematics archive, University of Saint Andrews)

Fig. 13 Augustus Love (1863–1940) was a mathematician and elastician. He formulated the thick plate theory by introducing the shear components of the flexure in addition to the bending ones. Then he completely solved the problem in the case of axisymmetric plates subjected to a uniform load and simply supported at the edge. Love also devised the first *elasticity theory of shells* by establishing the fourth derivative equation of the flexure for constant thickness cylinders around 1915 (*Mathematical Theory of Elasticity*, Sect. 339)
PHOTO: (credit: The MacTutor History of Mathematics archive, University of Saint Andrews)

Fig. 14 Bernhard Schmidt (1879–1935) was an optician and astronomer who in 1928 invented the famous wide field concept with a spherical mirror and an aspherical corrector plate. *Schmidt telescopes* allowed unprecedented complete photographic mappings of the sky in various bandpasses. Recognized for his great skill in the parabolization of mirrors, Schmidt emphasized the idea that the elastic relaxation method, or stress figuring, would provide much smoother surfaces. Therefore he must be considered the *initiator of Active Optics*
PHOTO: (courtesy of Erik Schmidt)

Fig. 15 Stephen P. Timoshenko (1878–1972) was an engineer and mathematician who greatly contributed to the theory of elasticity, elastic stability and buckling, strength of materials, and wrote famous textbooks on these questions. He derived the critical load for the buckling of thin cylindrical shells. These results were applied to improve the strength of very large ships. Timoshenko also established, around 1930, the fourth derivative equation which gave rise to the *theory of axisymmetric cylinders with variable thickness*
PHOTO: (credit: Stanford University and National Academy of Sciences)

Fig. 16 André J.A. Couder (1897–1979) was an optician and elastician who contributed to large telescope optics in writing *Lunettes et Télescopes* with A. Danjon, discovered *Couder's two-mirror anastigmatic telescope*, and invented the *null test* in 1927. He applied the elasticity theory to determine the axial flexure of large mirrors under gravity and improved the image quality by optimized mirror supports. After Schmidt's suggestion that a corrector plate might be aspherized by active optics, Couder solved the theoretical problem (1940)
PHOTO: (courtesy of Charles Fehrenbach)

Fig. 17 Eric Reissner (1913–1996) was a mathematician whose research was dedicated to turbulence and aerodynamic wing theory, and elasticity theory. With W. Martin, he published *Elementary Differential Equations*. Reissner created the *theory of shallow shells* which is one of the most remarkable achievements in elasticity. Compared to the theory of plates, this also takes into account the stresses and strains in the midsurface of the shell. Reissner's theory relates two biharmonic functions in a fourth-order equation pair (1946)
PHOTO: (credit: Scripps Institution of Oceanography Archives, UC San Diego Libraries)

Fig. 18 Gerhard Schwesinger (1913–2001) was an engineer and elastician who developed the elasticity theory for the determination of the lateral flexure of large astronomical mirrors under gravity. Introducing Fourier series for representing the lateral supporting forces (1954), he obtained the first comparison results for various systems, which thus led to *minimal deformation designs*. Advances in finite element codes combined with Schwesinger's expertise led to efficient *lateral support systems* for 8-m monolithic mirrors
PHOTO: (courtesy Raymond N. Wilson, European Southern Observatory)

Fig. 19 Edgar Everhart (1920–1990) was a physicist specializing in atomic collisions, professor at the University of Denver, and director of the associated observatory. He discovered the comets 1964 IX Everhart and 1966 IV Ikeya-Everhart. In elasticity, he independently derived Couder's results for making a Schmidt plate by partial vacuum and spherical figuring tool when the plate is simply supported at its edge. Everhart was the *first to apply Active Optics* to the complete aspherization of a telescope optical surface (1966)
PHOTO: (copyright and courtesy of the University of Denver Penrose Archives Special Collections)

Fig. 20 Raymond N. Wilson's (1928–) inclinations in school were towards the humanities, above all history and Latin, not science and above all not mathematics. Through interest in astronomy and telescope making he finally studied physics and specialized in optics. His invention of *Active Optics* combining optics, mechanics and computer technology has revolutionized modern telescope technology. However, he himself sees his greatest achievement by far as his two books *Reflecting Telescope Optics*, a standard work for specialists
PHOTO: (80th birthday party, courtesy Peter Wilson)

Acronyms

AAT	Anglo-Australian Telescope
ADC	atmospheric dispersion compensator
BIPM	Bureau International des Poids et Mesures
CAS	Chinese Academy of Sciences
CCD	charge-coupled device
CDS	Coronal Diagnostic Spectrometer of SOHO
CFGT	Configuration for Chinese Future Giant Telescope (project)
CFHT	Canada France Hawaii Telescope
COS	Cosmic Origins Spectrograph / HST
CTD	constant thickness distribution
CTE	coefficient of thermal expansion
DM	deformable mirror
efl	effective focal length
ELT	extremely large telescope
E-ELT	European Extremely Large Telescope (project)
ESO	European Southern Observatory
FOV	field of view
FRD	focal ration degradation (in an optics fiber)
FTS	Fourier transform spectroscopy
fwhm	full width at half maximum
GMT	Giant Magellan Telescope (project)
GSMT	Giant Segmented Mirror Telescope (project)
GALEX	Galaxy Evolution Explorer (in ultraviolet)
GTC	Grand Telescopio Canarias
HST	Hubble Space Telescope
HET	Hobby Eberly Telescope
IFS	integral field spectrograph
IFU	integral field unit
IR	infrared
IS	image slicer
JELT	Japanese Extremely Large Telescope (project)
JWST	James Webb Space Telescope (project)
KPNO	Kitt Peak National Observatory

LADC	linear atmospheric dispersion compensator
LAMOST	Large Sky Area Multi-Object Fiber Spectroscopic Telescope
LBT	Large Binocular Telescope
LMT	liquid mirror telescope
LOOM	Laboratoire d'Optique de l'Observatoire de Marseille
LPMA	Laboratoire de Physique Moleculaire et Applications
LSST	Large Synoptic Survey Telescope (project)
MDM	multimode deformable mirror
NIAOT	Nanjing Institute of Astronomical Optics & Technology
NSO	National Solar Observatory
NTT	New Technology Telescope
OHP	Observatoire de Haute Provence
PE	paraboloid-ellipsoid
PH	paraboloid-hyperboloid
PP	paraboloid-paraboloid
PSF	point spread function
ptv	peak to valley
RC	Ritchey-Chrétien
rms	root mean square
SALT	South Africa Large Telescope
SAO	Special Astronomical Observatory (Russia)
SDSS	Sloan Digital Sky Survey
SOHO	Solar and Heliospheric Observatory
SS	spheroid-spheroid
TMT	Thirty Meter Telescope (project)
UV	ultraviolet
UVCS	Ultraviolet Coronal Spectrometer of SOHO
VCM	variable curvature mirror
VLT	Very Large Telescope
VLTI	Very Large Telescope Interferometer
VTD	variable thickness distribution
WS	Wolter-Schwarzschild
WHT	William Herschel Telescope
YAG (laser)	yttrium aluminum garnet laser crystal

Glossary

Abbe's number of a glass see reciprocal dispersive power.

Abbe sine condition or sine condition optical design condition which cancels spherical aberration and linear coma. In third-order theory of aberrations, a system which satisfies the sine condition is said to be aplanatic.

Abbe sphere locus of the intersection of input and output conjugate rays in a centered system satisfying the sine condition.

Acomatic grating diffraction grating which provides an aberration correction for coma.

Active compensator deformable mirror able to generate one or several flexure modes for correcting aberration modes. Such compensators are useful for the photosensitive recording of aberration corrected holographic gratings.

Active optics controlled process which is either non time-dependant or with low frequency bandpass. Used in optics for aspherization methods, replication methods, in-situ variable shape mirrors and telescope mirror alignments.

Active optics co-addition law law which determines an optical surface as resulting from co-addition of elastical flexure and spherical figuring.

Active submaster deformable substrate which is used as intermediate stage optics in the double replication process of obtaining aberration corrected gratings or, more generally, aspherized surfaces.

Adaptive optics high frequency bandpass control essentially concerned with the wavefront corrections caused by the atmospheric degradation.

Aerial extremely long focal length refracting telescopes built in the period 1640–1690 and made of a single lens objective.

Afocal system or afocal telescope null powered optical system which is mainly used as beam expander, beam compressor, or retro-reflective system.

Anamorphotic systems optical systems whose optical power – and imaging scale – differs in the two principal directions. Such systems generally include cylindrical components.

Anastigmatism property of an optical system which is free from third-order spherical aberration, coma, and astigmatism.

Anticlastic surface surface whose product of the two principal curvatures, generated by flexure, is negative. A product of opposite sign corresponds to a synclastic surface.

Aperture ratio or focal ratio or f-ratio For a surface alone: ratio f/D of its focal length over its clear aperture diameter. For a telescope: ratio of the efl over the light-gathered beam diameter. This ratio is usually denoted f/$\{f/D\}$, e.g. f/1 or f/5.

Aplanatism property of an optical system which is free from third-order spherical aberration and coma.

Aspherized reflective grating diffraction grating on a rigid substrate which has been aspherized via a double replication process by use of an active submaster.

Astatic levers counterweight and lever arrangements used as support systems for telescope mirrors.

Astigmatism length distance between the sagittal and tangential focii of an astigmatic beam of rays.

Bending or cambrure of a lens mean curvature of a lens.

Biconicoid surface optical surface whose conic sections differ in the x- and y-directions.

Biplate form or closed biplate form mirror substrate designed with two similar vase forms built-in together at their contour to form a closed cavity.

Blaze angle of a diffraction grating angle of the facets of the relief lines with respect to the local plane of the grating surface. In a given diffraction order (usually -1) blazed gratings – ruled or holographic – are much more efficient than sinusoidal line gratings.

Boundary conditions in elasticity free edge, (movable) simply supported edge, (movable) built-in edge, unmovable simply supported edge and unmovable built-in edge are the five basic boundaries at the edge of a solid plate.

Buckling elastic instability of a beam, plate, shell, or truss when overpassing a critical load.

Cambrure or bending of a lens mean curvature of a lens.

Camera obscura hole in a high-ceilinged room allowing to observe the sunspots without use of any lens. First mentioned by Al-Haytham (Alhazen), the imaging effect is provided by the Fraunhofer diffraction of the hole.

Cantilever beam or girder which is built-in at one end.

Characteristic function function derived by R.W. Hamilton to establish the analytic form of optics third-order aberrations.

Cassegrain focus focal surface of a reflecting telescope formed behind the primary mirror via a central hole in it.

Catadioptrics branch of optics involving optical systems made of both refractive and reflective surfaces.

Catoptrics branch of optics involving optical systems made of reflective surfaces or mirrors only.

Cat's eye system an afocal system where the light returns from were it comes. A moving cat's eye system translating on delay line carriage of a two-arm interferometer is generally a retro-reflective system equipped with a variable curvature at its focus for optical path compensation of the field of view.

Centered system optical system only formed of axisymmetric surfaces which are distributed on a common axis.

Chief ray of a beam see principal ray.

Clairaut doublet lens stigmatic and achromatic objective made of two cemented lenses of different glass. The French refer to this lens simply as a "Clairaut."

Clairaut-Mossotti doublet lens aplanatic and achromatic objective made of two cemented lenses of different glass.

Clear aperture (optical) aperture which physically determines the cross-section of an optical beam used by an optical surface. Usually specified by its diameter for a circular aperture.

Clebsch's polynomials infinite set of four-term series where each series is a radial component of the general non-axisymmetric solutions of Poisson's equation of thin plate theory when the holed or non-holed plate is axisymmetric.

Clebsch-Seidel modes superposable elastic modes which belong to a subclass of the optics triangular matrix.

Closed biplate form or biplate form mirror substrate designed with two similar vase forms built-in together at their contour to form a closed cavity.

Closed shell closed form made of two linked vase shells which can be bent by inner pressure or inner partial vacuum. When the two vase shells are identical, and linked to their flat outer bases, the arrangement is said to be a *flat closed shell*, otherwise it is said to be a *curved closed shell.*

Coma neutral point of a two-mirror telescope intersection point of the axes of the mirrors when the on-axis coma is nulled at the resulting focus.

Conicoid axisymmetric optical surface whose meridian section is a conic.

Conic constant or Schwarzschild constant constant which determines the asphericity of a conicoid surface.

Constringence of a glass see reciprocal dispersive power.

Coudé focus optical arrangement using mainly a set of folding plane mirrors – the coudé train – which allows the beams of a telescope to converge in an underground room – the coudé room.

Couder's law or flexure scaling law for mirrors under gravity law which scales the flexure of mirrors supported by horizontal point-pads as d^4/t^2, where d and t are the diameter and thickness of the mirror.

Ctd constant thickness distribution (opposite to variable thickness distribution – **vtd**).

Cycloid form axisymmetric geometry formed by a front plane surface and a variable thickness distribution which remains finite at the center and tends vertically towards zero at the edge – like a cycloid.

Deformed conicoid optical surface whose shape departs from a conicoid by a fifth or a higher-order term. Belongs to the spheroid class.

Descartes' ovoid or ovoid surface optical surface providing the stigmatism, i.e. correcting the spherical aberration at all orders. Introduced by Descartes for singlet lenses. The conicoids belong to a subclass of ovoids.

Dialytic system or dialytic telescope a system where the axial chromatism of a single objective lens is corrected by a smaller negative lens – or doublet lens – located near the middle of the system.

Dioptrics branch of optics involving optical systems made of refractive surfaces or lenses only.

Dispersive power of a glass relative variation of the refractive index of a glass with respect to the value of mean index minus unity over a given spectral range $[\delta n/(n_0 - 1)]$. The reciprocal of this number is the Abbe number ν, sometimes called V-number.

Double vase form or double vase shell single substrate for two distinct mirrors whose concentric surfaces form a continuous figure. The substrate is made of two built-in and concentric vase forms or vase shells.

Eccentricity dimensionless quantity which is related to the conic constant of conicoid.

Efl or effective focal length resulting focal length of an optical system for a given wavelength.

Effective focal length – often denoted efl resulting focal length of an optical system.

Eikonal function aberration function derived from Hamiltonian and reintroduced to optics by Bruns as related to the aberration wavefront function.

Elastic deformability ratio or flexibility ratio intrinsic ability of a material to provide an elongation or a flexure. This dimensionless ratio can be defined as the quotient of the usable limit tensile stress to the Young modulus $(\sigma_{\mathrm{T\,lim}}/E)$.

Elastic relaxation figuring or stress figuring active optics aspherization method which uses the elastic relaxation of the stresses applied during spherical figuring. Compared to the zonal figuring method which uses small tools, this method uses full-size aperture tools and thus produces extremely continuous surfaces, i.e. free from ripples.

Elastica or elasticae curves of the flexure of a thin circular rod or wire first introduced by Euler.

Enlarging monocular or spyglass early such systems were a chance arrangement of eyeglass lenses available on the 1600s market with positive objective lens and negative eye lens. This provided an amusing but poorly magnifying system from which Galilei invented the telescope.

Entrance pupil or input pupil pupil or aperture stop in which all axial and field beams enter through an optical system. The conjugate is the exit pupil, and also called output pupil. The entrance pupil of a two-mirror telescope can be either the primary or the secondary mirror.

Erect image image in which all directions are similar to the object. A classical different case is an inverted image.

Etendue invariant see optical etendue.

Ewing-Muir process stressing process of a material which allows one to extend the linearity range of the stress-strain relation.

Fermat's principle principle of least time which governs the light propagation and determines the geodesics of optical paths. It states that the light path – or ray trajectory – between two points is the path which corresponds to a minimal or a stationary path.

Field binodal astigmatism at the focal surface of a two-mirror telescope field distribution of the astigmatism showing two least confusion images when the on-axis coma is nulled. Caused by an angular error between the axes of the mirrors.

Finite element analysis computational modeling which, in elasticity, allows accurate determination of the deformations of a solid by basic use of the element stress-strain relations.

Fizeau test interferometric determination of an optical surface shape by use of a fringe pattern formed with respect to a reference surface.

Flexibility ratio or elastic deformability ratio intrinsic ability of a material to provide an elongation or a flexure. This dimensionless ratio can be defined as the quotient of the usable limit tensile stress to the Young modulus ($\sigma_{\mathrm{T\,lim}}/E$).

Flexural hysteresis stres-strain hysteresis loop which is shown by metallic alloys when the stressing reaches or slightly overpasses the elastic limit. Brittle materials, such as glass and vitrocerams, do not exhibit flexural hysteresis.

Flexural rigidity or rigidity of a plate quantity which involves the Young modulus, Poisson's ratio and third power of the thickness.

Flexure scaling law for mirrors under gravity or Couder's law law which scales the flexure of mirrors supported by horizontal point-pads as d^4/t^2, where d and t are the diameter and thickness of the mirror.

Focal ratio see aperture ratio.

Foucault test or knife-edge test wavefront analysis of the light distribution reflected by the mirror under test by use of a knife-edge partly occulting the convergent region of a beam.

Gaussian theory or paraxial theory of optics basic theory where any ray forms a small angle with respect to the system axis such that the sine and tangent functions are approximated by that angle itself.

Hamilton characteristic function first general form of the wavefront function.

Holosteric mirror mirror whose particular geometry is machined in a single piece.

Homocentric system optical system where the principal rays in the image space are converging towards a common center; so, the image focal surface is curved and normal to this center direction.

Homothetic transformation congruence of rays usual optical transformation where the conjugate of a paraxial ray remains a paraxial ray with the same homothetic height as that of the whole beam.

Huygens-Fresnel principle basic method for the explanation of diffraction effects of light. The principle states that any point of a wavefront give rise to secondary disturbances which are spherical wavelets and, at any later instant, the spacial light distribution may be regarded as the sum of these wavelets.

Hysteresis see flexural hysteresis.

Imager-spectrograph telescope post-focal instrument which exhibits both an imaging mode and a spectrographic mode.

In situ stressing deformable optical surface which is stressed during an astronomical observation.

Inverse square law (optics) law which governs the light energy propagation as the inverse square of the distance.

Inversion transformation congruence of rays optical transformation where a paraxial ray becomes a marginal ray. Systems realizing this transformation were first described by Mersenne for two confocal paraboloid mirrors used in the retroreflective form.

Inverted image image in which up and down and left and right are interchanged. A different case is an erect image. The human eye forms an inverted image on the retina.

Isoplanatic system optical system which is free from coma only.

Kerber's condition (1886) condition for minimizing the chromatic variation of spherical aberration – or spherochromatism. This expresses that the least confusion focus must be determined by a pupil ray height equal to $\sqrt{3}/2 = 0.866$. For the central color, this corresponds to the algebraic slope balance of the wavefront.

Kirchhoff's condition boundary condition at the free edge of a plate when the flexure is non-axisymmetric. This condition is set by nulling the net shearing force at the edge of the plate.

Large deformation theory elasticity theory of plates which takes into account the in-plane displacements of the neutral surface.

Large torsion theory elasticity theory elaborated by Saint-Venant which treats torsion cases where the cross-section of a non-circular beam does not remain plane.

Lagrange invariant invariant of the action of an optical system which is equivalent to a one-dimensional representation of the optical etendue, i.e. of the total energy transported by the optical system for a given spectral bandpass $\delta\lambda$.

Linear coma third-order field aberration which includes the odd sum of all image height terms.

Linear product law of a weakly conical shell law which relates the product of the thickness by the flexure of the shell as a linear function of the axial coordinate.

Littrow mounting of a diffraction grating geometrical arrangement where the diffraction angle for the central wavelength is equal and of opposite sign to the incidence angle.

Love-Kirchhoff hypotheses small deformation hypotheses which state that any point located on normal lines to the middle surface of a solid remains on normal lines to the deformed middle surface and that the thickness does not vary in flexure.

Marginal ray ray which passes at or near the edge of a pupil.

Mean curvature focal surface focal surface of least confusion imaging of an optical system. Merges into the Petzval surface once the spherical aberration, coma and astigmatism are nulled.

Meniscus form axisymmetric geometry resulting from the volume delimited by two axially separated surfaces of similar or quasi-similar curvature, or by two concentric surfaces.

Mersenne two-mirror telescopes afocal telescopes made of two confocal paraboloid mirrors.

Multimode deformable mirror meniscus form or vase-form mirror equipped with a set of radial arms on which are applied distributions of axial forces on each end. These configurations allow the superposition of the Clebsch-Seidel modes, a subclass of the triangle matrix optical modes.

Nasmyth focus focal surface of a reflective telescope formed on the platform of an altazimuth mount by a plane – Nasmyth – mirror operating 90° beam deviations about the telescope axis.

Navier's relations or stress-strain relations fundamental linear relations of elasticity which links the six stress components to the six strain components of the facets of a volume element of a solid.

Nebular f-ratio of a spectrograph camera f-ratio in the direction perpendicular to the dispersion.

Net shearing forces transversal forces in length unit which act in loaded plates. These forces, denoted V_x, V_y or V_r, V_θ, allow determination of the equilibrium reactions.

Neutral surface surface where the stress distribution through the thickness of the solid remains null during elastic bending. For small deformations the neutral surface is also the middle surface.

Non-centered system optical system including one or several non-axisymmetric surfaces where a general axis of symmetry does not exist.

Normal congruence a pencil of light can be seen as formed by a continuous set of propagation lines called a congruence. If a wavefront surface is orthogonal to all these lines, then the pencil is a normal congruence. If not so, the pencil is a skew congruence (e.g. electron optics).

Normal rectilinear congruence in photon optics and light propagation in isotropic media each curve of the normal congruence pencil is a straight line; the normal congruence is then said to be rectilinear.

Null-power zone ratio of a pseudo-plane-aspheric surface ratio of the null slope radius over its clear aperture radius.

Null test optical test requiring the use of a corrector made of one or more lenses or mirrors, or of a hologramatic correction – either used on or off axis – which provides the complete aberration compensation of the output wavefront so the resulting fringe deviation is zero.

Oblate ellipsoid flattened ellipsoid.

Optical etendue invariant (Jacquinot's) an invariant quantity when the light propagation occurs through a perfect optical system. For a given bandpass $\delta\lambda$, the etendue represents the total energy transported by an optical system.

Optical despaces the five basic displacements for the alignment of an axisymmetric optical piece which consist of piston (z translation), lateral centering (x and y translations), and tip-tilt (x and y rotations).

Optics triangular matrix in any centered system formed of axisymmetric optical surfaces belonging to the surface class with continuous tangent plane, any wavefront surface can be defined by a polynomial sum $\sum a_{nm} r^n \cos m\theta$ where $m \leq n$. Hence, the a_{nm} coefficients form a triangular matrix.

Orthoscopic eyepiece (Abbe's) an optical design which is distortion-free.

Ovoid surface or Descartes' ovoid optical surface providing the stigmatism, i.e. correcting the spherical aberration at all orders. Introduced by Descartes for singlet lenses. The conicoids belong to a subclass of ovoids.

Paraxial theory of optics see Gaussian optics.

Petzval condition optical design condition for obtaining a flat image field whilst the spherical aberration, coma and astigmatism are nulled.

Petzval theorem theorem which allows deriving the Petzval curvature of an optical system as a function of the curvatures of its surfaces and of the refractive indices separating them.

Plastic deformation permanent residual deformation of a material after a stress loading cycle whose level overpassed its limit elastic stress.

Point spread function or psf function representing the intensity distribution of the light in the image region of the Gaussian focal plane.

Poisson's equation of elasticity fourth derivative biharmonic equation of the thin plate theory.

Principal ray or chief ray of a beam a ray which starts from a given source point in the field, then passes through the vertex of an entrance pupil, and propagates through an optical system up to reaching an image focal plane.

Prime focus focal surface formed by the primary mirror of a telescope.

Prolate ellipsoid elongated ellipsoid.

Punctum proximum distance minimum distance at which the eye can accommodate. This distance is typically 15–18 cm for a young adult.

Pupil or stop aperture which delimits the common cross-section area of all the beams. Any optical system is defined by an input pupil or an output pupil.

Quarter-wave criterion (Rayleigh's) criterion which confers a centered system the ability to form diffraction limited images on its axis. The output wavefront must be included between two concentric surfaces separated of $\lambda/4$.

Retro-reflective system an afocal system where the light returns from were it comes. A moving cat's eye system translating on delay line of a two-arm interferometer is generally a retro-reflective system.

Reciprocal dispersive power of a glass this number defined by the Abbe number $\nu = (n_0 - 1)/\delta n$ expresses the intrinsic variation of the refractive index of a glass. This is the reciprocal number of the dispersive power. The Abbe number of a glass is sometimes called the constringence or V-number.

Ripple errors of a surface slope discontinuities which provide high spatial frequency errors of the wavefront and thus degrades the diffracted image quality. Usually caused by local retouches during the figuring of optical surfaces.

Rumsey telescope flat-field anastigmat three-reflection telescope.

Rupture stress or ultimate strength stress level at which a material ruptures.

Rupture tensile stress or ultimate tensile strength stress level at which a brittle material is brought to rupture by tension.

Sagittal ray fans meridian ray fans of an optical system which provide the tangential focii locus of the tangential focal surface.

Schwarzschild constant see conic constant

Saint-Venant's principle principle of equivalence of the elastic deformations of a solid in the application of equivalent boundary conditions applied over a small region. This principle is useful in active optics design for finding an optimal and practicable geometrical configuration and its associated load distribution.

Schwarzschild's eikonal function related to the aberration wavefront function and closely similar to Schwarzschild's perturbation function in the motion of planets, and that Schwarzschild called *Seidel's eikonal*.

Secondary spectrum residual chromatic error due to the inherent difficulty of perfectly achromatizing doublet lenses.

Seidel theory or third-order theory theory which allows determination of the amount of each primary aberration – or third-order aberration – of a complete optical system. The sine of any ray angle is approximated by the first two terms of its expansion.

Serrurier's truss mechanical truss of special concept which compensates for the decentering and tilt errors between the primary and secondary mirrors of large telescopes due to the tube flexure under gravity.

Shack-Hartmann test analytic determination of a wavefront shape from the two-dimensional lateral shifts of pencils autocollimated on subpupils of the surface to test. The wavefront is reconstructed from measurements of local slopes in the pupil.

Shear strain elastic relative displacement which occurs between two parallel plane facets of a solid when the facets remain parallel.

Shearing forces (radial) transversal forces in length unit, usually denoted Q_r, which act in concentric circular sections of bent axisymmetric plates.

Sine condition (Abbe's) see Abbe sine condition.

Sine refraction law (Snell's) sine law which determines the refractive angle of a ray when emerging in a new refractive medium.

Skew invariant property of a skew ray in a centered system which is used to establish Abbe's sine condition.

Speculum mirror mirror made of bell metal or CuSn25 alloy, i.e. bronze, which may include arsenic for a better optical polish.

Spherochromatism chromatic variation of spherical aberration.

Spheroid axisymmetric optical surface whose meridian section is expressed by an even polynomial series of the radius.

Spyglass see enlarging monocular.

Stacked triangles or whiffletrees assembly of pivoting triangles or beams arranged on successive levels for use as telescope mirror support systems.

Stigmatism property of an optical system which is – at least – free from third-order spherical aberration.

Stop or pupil aperture which delimits the common cross-section area of all the beams. Any optical system is defined by an input pupil or an output pupil.

Strehl intensity ratio intensity ratio which characterizes the light concentration of an optical system. If the ratio is at least 0.8 or greater, then the system is well corrected and so-called "diffraction limited."

Stress figuring or elastic relaxation figuring active optics aspherization method which uses the elastic relaxation of the stresses applied during spherical figuring. Compared to the zonal figuring method which uses small tools, this method uses full-size aperture tools and thus produces extremely continuous surfaces, i.e. free from ripples.

Stress lap polishing optical figuring alternative by controlled flexible tools – of smaller size than the surface clear aperture – which match the aspherical surface to minimize ripple errors.

Stress polishing last stage of stress figuring or elastic relaxation figuring.

Stress-strain relations or Navier's relations fundamental linear relations of elasticity which links the six stress components to the six strain components of the facets of a volume element of a solid.

Synclastic surface surface whose product of the two principal curvatures, generated by flexure, is positive. A product of opposite sign corresponds to an anticlastic surface.

Tangential ray fans ray fans laying in a perpendicular direction to a meridian plane of a system which provide the sagittal focii locus of the sagittal focal surface.

Telecentric system optical system where the principal rays in the image space are parallel to the system axis; so, the image focal surface is a plane.

Telephoto effect of a two-mirror telescope ratio which characterizes the effective focal length with respect to the telescope length. Allows classification of various designs as a function of the compactness.

Theory of shallow spherical shells elasticity theory which takes into account the significant curvature of the middle surface and the "in-plane" radial and tangential tensions appearing on it during flexure.

Theory of weakly conical shell elasticity theory elaborated for the aspherization of grazing incidence telescope mirrors.

Thick plate theory elasticity theory which takes into account all the shear components of the deformation of a plate.

Thin plate theory elasticity theory elaborated by Poisson and based on Love-Kirchhoff hypotheses.

Third-order theory of optics or Seidel theory theory which allows the determination of the amount of each primary aberration – or third-order aberration – of a complete optical system. The sine of any ray angle is approximated by the first two terms of its expansion.

Triangle matrix modes see optics triangular matrix.

Tulip form axisymmetric geometry formed by a front plane surface and a variable thickness distribution which tends towards infinity – like a stem – at the center and tends vertically towards zero at the edge.

Ultimate strength or rupture stress stress level at which a material is brought to rupture.

Ultimate tensile strength or rupture tensile stress stress level at which a brittle material is brought to rupture by tension.

Uni-constant theory early theory of elasticity where the Poisson's ratio of all materials were implicitly $\nu = 1/4$.

Vase form axisymmetric geometry formed by an inner plate or meniscus, of constant or quasi-constant thickness, built-in to a somewhat thicker outer cylinder.

Vase shell vase form whose inner meniscus is with significant curvature.

Variable thickness vase form vase-form mirror whose clear aperture is the surface of an inner built-in meniscus or plate of slightly variable thickness distribution.

V-number of a glass see reciprocal dispersive power.

Vtd variable thickness distribution (opposite to constant thickness distribution – ctd).

Whiffletrees or stacked triangles assembly of pivoting triangles or beams arranged on successive levels for use as telescope mirror support systems.

White pupil mounting spectrographic arrangement where the telescope pupil is transferred in a particular location. This mounting, originated by Baranne, is mainly used for high spectral resolution systems.

Wolter two-mirror systems grazing incidence systems for X-ray focusing. Type I of three is made of two convergent axisymmetric mirrors. Its compactness provides a basic design for all space telescopes.

Wolter-Schwarzschild (WS) telescope grazing incidence two-mirror system strictly satisfying Abbe's sine condition.

Yield strength stress level at which the deformation of a metal alloy exhibits a specified deviation from stress-strain linearity and thus reaches the plastic domain.

Zernike circle polynomials Zernike rms polynomials normalized representations of a wavefront or of an optical surface.

Author Index

Subject Index

About the Author

Gérard Lemaitre, born in France, a leading astronomer, received a degree in engineering from the École d'Arts et Métiers, Paris, in 1967, and focused on research in astronomical optics. In 1974, he earned the degree of Doctor es Sciences Physiques at the Université de Provence for his dissertation entitled *Astronomical Optics and Elasticity*.

His research concentrates on optical design and elasticity theory for improving the performance of telescopes and spectrographs by use of the minimum number of optical surfaces. These instruments mainly include aspherical surfaces which are difficult to obtain with accuracy. From Bernhard Schmidt's idea that an elastic deformation could provide an efficient method for generating asphercs, G. Lemaitre elaborated *Active Optics* and related theoretical conditions which are required for this method. The holder of eight patents in several countries, he developed this method to known aspheric cases and also to a large variety of other cases for new instrument designs.

Gérard Lemaitre has been an international member of many optical design committees, such as, the giant Schmidt telescope LAMOST of the Chinese Academy of Sciences, and more recently, the 42 m telescope project E-ELT of the European Southern Observatory. He was awarded the Grand Prix Andre Lallemand by the French Academy of Sciences.

MIX
Papier aus verantwortungsvollen Quellen
Paper from responsible sources
FSC® C105338

If you have any concerns about our products,
you can contact us on
ProductSafety@springernature.com

In case Publisher is established outside the EU,
the EU authorized representative is:
Springer Nature Customer Service Center GmbH
Europaplatz 3, 69115 Heidelberg, Germany

Printed by Libri Plureos GmbH
in Hamburg, Germany